Construction: Health and Safety

Wolters Kluwer (UK) Limited
145 London Road
Kingston upon Thames
Surrey KT2 6SR
Tel: 020 8247 1175
www.croner.co.uk

Published by
Wolters Kluwer (UK) Limited
145 London Road
Kingston upon Thames
Surrey KT2 6SR
Tel: 020 8247 1175
www.croner.co.uk

First published June 2008

British Library Cataloguing-in-Publication Data.

A catalogue record for this book is available from the British Library.

ISBN 978-1-85524-735-2

Printed by Hobbs the Printers, Southampton

Preface

by Keith Pickavance

It gives me great pleasure to write the preface for this excellent reference book on behalf of the CIOB — the associate publishers with Croner.

We all recognise that the safety and welfare of the workforce is crucial to the success of the construction industry. Indeed, poor performance in any part of the planning or implementation of risk prevention, by any member of the various consultant or construction teams, can have major consequences for the profitability of the project and also the image of industry. This book is designed to be used by all such team members.

In recent years, the increased health and safety accountability of companies and the severe penalties for management failure have been well publicised. The introduction of the Corporate Manslaughter and Corporate Homicide Act 2007 is a key milestone towards reinforcing the requirement for senior management and organisations to take account of their health and safety responsibilities.

Equally well publicised are the statistics that confirm the construction industry needs to continue to learn, absorb, communicate and deliver good practice in minimising death and injury risks. Recent accident statistics show that 32% of all UK worker deaths occurred in the construction industry and, in addition, 90,000 people suffered from an illness caused by or exacerbated by a construction related job.

Although the UK fatalities are among the lowest in Europe there have been no significant reductions in the last 5 years. We must not be complacent but strive for improved sharing of knowledge, skills and practice in worker welfare and safety within the industry.

Croner *Construction: Health and Safety* can greatly help us in achieving this goal, with its comprehensive index listing, up-to-date contents and user friendly layout of chapters and subject matter. Each section follows the same format of duties for employer, employee and responsible person, what has to be done in practice, training, lists of relevant legislation and further information.

The book includes chapters on issues that have current significance to the industry, ie the Building Regulations, the Construction (Design and Management) Regulations 2007, the Corporate Manslaughter and Corporate Homicide Act 2007 and Environmental Management, with a section on Site Waste Management Plans.

By increasing the knowledge within the construction industry in these areas, as well as providing clear reference support for all other health and safety topics, we should ensure the improved welfare of our construction colleagues and our clients.

Keith Pickavance FCIOB
President
The Chartered Institute of Building

Contents

Chapter 1

Chemical Hazards

COSHH for Construction

- Substances and preparations which can enter or affect the body with adverse health consequences are regarded as hazardous to health.
- To find out if a manufactured substance is hazardous, reference should be made to the material safety data sheet (MSDS) supplied by the manufacturer, and the hazard markings printed on the container in which the substance was supplied.
- There are some "hazardous substances" that occur in nature such as pathogens, micro-organisms and parasites which also fall within the scope of the regulations.
- A more stringent risk assessment is necessary where carcinogenic substances or substances that might cause occupational asthma are used.
- The COSHH charts are sample risk assessments for specific tasks and, among others, include:
  - cleaning brickwork using hydrochloric acid-based cleaners
  - cement when handled and transported
  - wet concrete
  - concrete dust
  - mixing and placing epoxy repair materials
  - petrol
  - sewage in small projects
  - hydralime

1.1 Both substances and preparations (ie mixtures such as mortar) which can enter or affect the body with adverse health consequences, are regarded as hazardous to health. When substances which are hazardous to health are used in the workplace, employers must comply with the Control of Substances Hazardous to Health Regulations 2002 (COSHH). It should be remembered that substances and preparations that are dangerous because of their risks to safety, for instance their flammability and explosive properties, are not covered by COSHH.

1.2 An assessment, relating the hazardous properties to the likelihood of harm to the user and others who might be affected, must be made. If the consequences to health are significant then the assessment must be recorded. The form, concentration and quantity are taken into account, together with the likely route of entry into the body.

1.3 When considering an example such as paint spraying, the route by which paint could affect the body may be by inhalation, absorption through the skin or other surfaces (such as the eye), ingestion or inoculation (through a cut). Paint splashes are not easy to breathe and may sit on the skin without penetrating. The route into the body is, therefore, closely related to the form of the substance.

Employers' Duties

1.4 Under the Control of Substances Hazardous to Health Regulations 2002 employers must:

- identify the hazardous substances present in the workplace
- assess the risks to health posed by the hazardous substances

- prevent employees being exposed to substances hazardous to health or, if prevention is not reasonably practicable, adequately control exposure
- provide personal protective equipment (eg face masks, respirators, protective clothing) only as a last resort and never as a replacement for other control measures which are required
- ensure, as far as reasonably practicable, the safe use, handling, storing and transporting of substances
- ensure control measures for dealing with hazardous substances are kept in efficient working order and good repair, and are properly used and maintained
- ensure local exhaust ventilation equipment is examined within the specific intervals set by COSHH, and records of examinations and tests carried out are kept for at least five years
- measure the concentration of hazardous substances in the air breathed in by workers where:
  - exposure limits might be exceeded
  - failure or deterioration of control measures could lead to serious risks to health
  - control measures may not operate correctly
- keep a record of any exposure monitoring for at least five years
- where appropriate, carry out health surveillance for their employees
- prepare plans and procedures in the case of an accident or emergency caused by a work activity
- provide employees with suitable and sufficient information, instruction, training and supervision.

Employees' Duties

1.5 Employees must:

- co-operate with their employers to enable their employers to comply with the Control of Substances Hazardous to Health Regulations 2002 (COSHH)
- make proper use of control measures, including personal protective equipment (PPE)
- return equipment after use to any storage place and report any defects found in the equipment
- attend medical examinations at the appointed time and give any information about their health as may be reasonable
- report any accident or incident which could have resulted in the release of a biological agent into the workplace and which could cause severe human disease.

In Practice

Substances Which Are Hazardous

1.6 The Control of Substances Hazardous to Health Regulations 2002 (COSHH) define a substance or preparation as hazardous to health:

- if it is listed in Part I of the approved supply list as dangerous for supply within the meaning of the Chemicals (Hazard Information and Packaging for Supply)

Regulations 2002 (CHIP), and for which an indication of danger specified for the substance is very toxic, toxic, harmful, corrosive or irritant

- if the Health and Safety Commission (now the HSE) has approved a workplace exposure limit (WEL)
- if it is a biological agent
- if it is a dust of any kind, except dust with a CHIP indication of danger or a WEL, if present at a concentration in air equal to or greater than:
  - 10 mg/m^3, as a time-weighted average over an 8-hour period, of inhalable dust
  - 4 mg/m^3, as a time-weighted average over an 8-hour period, of respirable dust.
- if, not being a substance falling into the categories above, it creates a risk to health because of its chemical or toxicological properties and the way it is used or is present at the workplace.

1.7 COSHH does not apply to asbestos, lead or substances or preparations that are dangerous because of their physical properties, flammability and explosive properties or extremes of temperature.

1.8 To find out if a manufactured substance is hazardous, reference should be made to the Material Safety Data Sheet (MSDS) supplied by the manufacturer, and the hazard markings printed on the container in which the substances was supplied. Other information is also available, eg from the HSE, on products which are not packaged, such as sawdust.

Biological Hazards

1.9 There are some "hazardous substances" that occur in nature such as pathogens, micro-organisms and parasites which also fall within the scope of the regulations. Examples include infections from contact with sewerage (eg mild cases of gastroenteritis and potentially fatal diseases, such as hepatitis), rat's urine (leptospirosis, or Weil's disease) and tick-borne diseases such as Lyme disease.

Carcinogens and Occupational Asthma

1.10 A more stringent risk assessment is necessary where substances are used that may cause occupational asthma and exposure must be reduced to as low a level as is reasonably practicable. For these substances the MSDS will have the risk phrase R42 (may cause sensitisation by inhalation) or R42/43 (may cause sensitisation by inhalation or by skin contact). Examples of substances that cause occupational asthma that may be encountered on the construction site include wood dust and disocyanates, which are used in some foams and paints.

1.11 Those substances accorded with either R45 (may cause cancer) or R49 (may cause cancer by inhalation) are categorised as carcinogenic under the terms of the COSHH Regulations. For these it is appropriate to use the COSHH assessment and, like occupational asthmagens, the exposure must be reduced to as low a level as is reasonably practicable.

1.12 COSHH also requires additional measures when carcinogens are used. Examples of carcinogens that may be encountered on construction sites include petrol, mineral oils, wood dust and used engine oil. However, additional information will be needed on the risks involved and the control measures.

What to do with the COSHH Assessment

1.13 The assessment is used by the employer to identify the risks and determine the controls required before the material is encountered. The assessment also provides a "method statement" on how the reduction of exposure to the substance to a safe level will be achieved. This makes it vital that the information is readily available to the person controlling the work.

1.14 Copies should be kept in a readily accessible file on site, in many cases with the health and safety plan which is provided in compliance with the Construction (Design and Management) Regulations 2007. This also enables easy inspection by the health and safety inspectors when they visit. The appropriate MSDSs should also be stored with the assessment.

1.15 When accidents involving hazardous substances occur, the necessary information (eg MSDS) should accompany the injured person to the hospital.

1.16 Information must also be available in a comprehensible form to the users and those affected by the substance. This could be achieved by toolbox talks.

Infection Risks

1.17 Biological hazards can be fatal. The need to assess significant biological risks is a specific requirement of COSHH. For the purposes of carrying out an assessment, infectious micro-organisms include bacteria, viruses, fungi and internal parasites all of which are hazardous to human health.

1.18 While some infections have long-term ill effects, many are mild and often do not require any medical intervention. The range of infections to be assessed applies only to those specifically relevant to the work activity. It does not apply to general infections such as influenza that are often acquired from colleagues, hence there is a separate risk assessment form (with guidance) for infections.

COSHH Charts

1.19 The COSHH charts are sample risk assessments for specific tasks. They cannot be reproduced and used without consideration of the actual circumstances of use.

1.20 It should be noted that other legislation and safety issues may have to be considered where the sample chart does not fit the specific risks on-site. For example, if large numbers of cement bags are to be moved by hand, a manual handling assessment will be needed, as required by the Manual Handling Operations Regulations 1992.

1.21 Information is also available from labels on product packaging, Material safety data sheets which are supplied direct from the producer, guidance leaflets published by the HSE, risk assessment reports published by the European Union, the previous experience of users, textbooks, professional institutions and the Internet.

1.22 The HSE has also made available *COSHH Essentials*, an online tool which provides advice on controlling the use of chemicals for a range of common tasks, eg mixing or drying. For most tasks, the website takes the user through a number of steps and asks for information about the exercise and chemicals.

COSHH Hierarchy of Controls

1.23 For each of the activities an attempt has been made to apply the COSHH hierarchy of controls.

1.24 The first priority is to eliminate exposure by not using the hazardous substance or by not carrying out the activity which generates the hazardous exposure. For example, removing road markings by burning exposes the operator of plant to a significant risk of inhaling fumes containing lead. This could be substituted for a process where road markings are removed by grinding. This can be done by a mobile device which produces a dust instead of a fume. The dust is handled by local exhaust ventilation while the driver sits inside a protected cab, removed from most of the risk of exposure.

1.25 If a product or process is re-designed then exposure may be eliminated or significantly reduced. If pre-cast concrete is supplied instead of *in-situ* concrete then site workers will not be exposed to cement. Exposure to hazardous substances can be reduced using controls such as wetting stone dust to reduce the inhalation of respirable silica.

1.26 The last option on the hierarchy of controls is the use of personal protective equipment (PPE) such as goggles and facemasks. PPE wears out quickly and generally "fails to danger". "Failing to danger" means that, eg when a glove fails the user may not be aware of its failure until after the substance has penetrated the glove.

Cleaning Brickwork Using Hydrochloric Acid-based Cleaners

1.27 As one of the most potentially hazardous materials used on site there should be careful control to ensure unauthorised persons do not have access to hydrochloric acid.

1.28 Staff should be instructed to control buckets of diluted material when not in use. Decanting should take place in the open air and barriers could be used to keep unprotected persons at a safe distance. If the work is carried out in cold weather staff are more likely to wear PPE. Alternative methods of cleaning brickwork should be sought if the work is to be carried out in confined spaces.

Risk Phrases

1.29 Irritating to the eyes, respiratory system and skin.

Note: This chemical attacks metals to produce potentially explosive hydrogen, and it reacts with oxidising agents to generate chlorine.

Hierarchy of Controls Considered

1.30 When selecting appropriate controls for operations involving potential exposure to brick cleaning acid, the COSHH Regulations require that the hierarchy of controls is fully considered, as detailed below.

- Regulation 7(2) Elimination: Where possible the brickwork will be protected, and mixers will be carefully sited to reduce spoilage.
- Regulation 7(3)(a) Design: Where possible use easy-to-clean bricks.
- Regulation 7(3)(b) Control: Minimise the need to spray by supplying suitable brushes, use only where there is effective air movement.
- Regulation 7(3)(c) PPE: Visors, disposable coveralls, gloves, face mask.

Cement when Handled and Transported

1.31 When stored in silos, the risk from cement is reduced until it is decanted for use, during which high winds can blow particles into people's faces causing eye injuries. Sweeping up cement dusts should be avoided and instead, a vacuum cleaner with a suitable filter should be used. Face masks should be worn when emptying dust-collection bags. If possible, disposable dust-collection bags which do not need to be emptied should be used. Bags should not be shaken to empty the residues.

1.32 Where bagged cement is handled, ensure all staff receive a toolbox talk on the precautions outlined in the HSE leaflet CIS26: *Cement*.

1.33 Regular monitoring is necessary for individuals who use significant quantities of cement, eg concreters and masons. In these instances an appointed person, able to recognise the symptoms of dermatitis, must check their skin condition regularly, in particular their hands. If the worker is found to be suffering from irritation they should be referred to an occupational health professional, as cement is a skin sensitiser.

1.34 Gloves should be chosen which are sufficiently durable and close fitting to prevent dry or wet cement from entering. Cotton glove liners are an effective way of controlling moisture inside gloves that are regularly worn for long periods. Hands should be washed with warm soapy water after use. Where work with cement involves wet concrete an additional assessment is required.

1.35 If large numbers of cement bags are to be moved by hand a manual handling assessment will be required.

Note: Cement additives must be assessed separately.

Risk Phrases

1.36 Contact with cement when wet may cause irritation, dermatitis or burns, and there is a risk of serious damage to the eyes.

Note: Damage increases with contact time. Skin may become itchy or sore, and may look red, scaly and cracked. Rhinitis can be linked to excessive inhalation of cement dust.

Hierarchy of Controls Considered

1.37 When selecting appropriate controls for operations involving potential exposure to cement dust the COSHH Regulations require that the hierarchy of controls is fully considered, as detailed below.

- Regulation 7(2) Elimination: Not normally applicable, unless off-site construction is possible.
- Regulation 7(3)(a) Design: Handling dry material is avoidable on contracts where ready-mixed mortars are viable.
- Regulation 7(3)(b) Control: Wetting down dusts or using mechanical ventilation attached to tool.
- Regulation 7(3)(c) PPE: Gloves. Goggles and face masks when windy.

Wet Concrete

1.38 Although wet concrete is not actually classified as a corrosive, the assessment has adopted this classification to draw attention to alkali burns which may result from prolonged skin contact. Injury from concrete burns is rarely preceded by pain.

1.39 Heavy exposure over a single day could result in serious alkali chemical burns which, if allowed to continue to worsen, may necessitate surgical intervention to remove the injured area. To minimise the risk to hands (usually most at risk), they must be washed with clean water if they are contaminated.

1.40 Wet concrete may easily become trapped inside PPE such as gloves, footwear and clothing and make contact with the skin. Contaminated clothing must, therefore, be immediately removed, washed and/or replaced. Footwear in particular should be regularly inspected to ensure protection is maintained.

1.41 For most operations, unless automated placing, vibration and finishing are employed, assessments may be needed for manual handling and vibration risk. Suitable protection may be needed to prevent accidental falls into wet material.

Risk Phrases

1.42 Irritating to the skin; may cause sensitisation by skin contact, risk of serious damage to the eyes.

Note: Damage increases with contact time.

Hierarchy of Controls Considered

1.43 When selecting appropriate controls for operations involving potential exposure to cement dust, the COSHH Regulations require that the hierarchy of controls is fully considered, as detailed below.

- Regulation 7(2) Elimination: Where possible, use off-site manufacture.
- Regulation 7(3)(a) Design: Move by pumping (where batch size makes this economic).
- Regulation 7(3)(b) Control: Avoid persons standing or kneeling in wet material. Move others away from areas where splashes may occur.
- Regulation 7(3)(c) PPE: Gloves. Waterproof clothing if the area may become contaminated. Safety glasses, if there is a risk of splashing, eg during vibration.

Concrete Dust

1.44 Concrete dust may be generated by many construction activities, including:

- hand sawing
- using petrol saws
- dry grinding using electric or air powered saws
- cutting and grinding operations with larger tools, such as floor saws and floor grinders.

1.45 Concrete and mortar contains 25–70% crystalline silica, mostly as quartz. If inhaled, very fine crystalline silica dust gets deep into the lungs and can lead to the development of silicosis. This results in the scarring of lung tissue and breathing difficulties.

1.46 Where high volumes of dust are generated, eg in the use of electric saws, dust suppression by two well established dust control techniques — wet dust suppression and local exhaust ventilation (LEV) — can be used to reduce exposure tenfold when used by trained operators. HSE leaflet *CIS54: Dust Control on Concrete Cutting Saws Used in the Construction Industry* provides further information.

1.47 Dust can be removed with a vacuum cleaner fitted with a suitable filter rather than with a sweeping brush. Airlines should not be used to move dust. Facemasks must be worn when emptying dust-collection bags. If possible, disposable dust collection bags, which do not need to be emptied, should be used. The dust suppression equipment must be inspected at suitable intervals as determined by the amount of use (to ensure that it is operating effectively). Wet systems that involve spraying water onto the rotating cutting disk to reduce dust emissions may be used on saws powered by combustion engines or compressed air. Wet dust suppression should not be used on saws that are electrically operated.

Risk Phrases

1.48 Breathing in fine dust containing crystalline silica can cause lung damage (silicosis). Silicosis is a slowly progressive, irreversible disease that usually takes some years to develop. Depending on the amount of dust inhaled, silicosis can cause breathing problems ranging in severity from mild through to severely disabling. In severe cases, silicosis leads to premature death.

1.49 Heavy and prolonged exposures to respirable crystalline silica under conditions that produce silicosis can also cause lung cancer.

Hierarchy of Controls Considered

1.50 When selecting appropriate controls for operations generating concrete dust by means other than petrol saws, the COSHH Regulations require that the hierarchy of controls is fully considered, as detailed below.

- Regulation 7(2) Elimination: Consider remote cutting or demolition using vehicle-mounted tooling.
- Regulation 7(3)(a) Design: Avoid cutting, scabbling by using hydraulic splitting, retarder paints.
- Regulation 7(3)(b) Control: Wetting down dusts or using mechanical ventilation attached to tools.
- Regulation 7(3)(c) PPE: Gloves, goggles and face masks (air-fed helmets are preferred). Waterproof clothing will be necessary if water sprays are employed.

Mixing and Placing Epoxy Repair Material

1.51 Epoxy materials have been developed to speed up the construction process and there are a wide variety now available. They normally involve the mixing of a base component and a hardener component.

1.52 The hazardous aspects of these materials are currently being researched so great care must be exercised in their use, in particular the use of PPE which is often the only reasonably practicable control.

1.53 The wide variety of epoxy resins available vary greatly in the health hazards they present. It is important that the MSDS is consulted and the hazards identified before use. Epoxy resins are known to cause skin sensitisation and photosensitisation, particularly for vulnerable persons, such as those with pre-existing skin problems. Hand, arm, face and throat skin sensitisation have been reported with epoxy systems. Some materials used in epoxy repair products are recognised as carcinogens and some may cause harm to the unborn child. Consequently, when possible, alternatives should be selected.

1.54 The assessment should also take account of the nature of the unmixed material as there may be some exposure to the individual components before mixing. If, during work, staff feel at all unwell they must be instructed to stop immediately and seek medical advice. Surface preparation is likely to require further assessment. In the event of a fire, all persons should be kept away from any products of combustion.

Risk Phrases

1.55 Irritating to skin and eyes, may cause sensitisation by skin contact, possible risk of impaired fertility, may cause harm to the unborn child, danger of very serious irreversible effects through inhalation, in contact with skin and if swallowed.

Hierarchy of Controls Considered

1.56 When selecting appropriate controls for operations involving exposure to materials that may harm the unborn child, the COSHH Regulations require that every reasonable effort must be made to avoid the use of this product and if use is unavoidable, that the hierarchy of controls is fully considered, as detailed below.

- Regulation 7(2) Elimination: Priority must be given to avoiding the use of this product either by mass repair (rather than local repair) or by complete redesign. Where possible, substitute the use of mechanical bonding techniques.

- Regulation 7(3)(a) Design: Where practicable, repairs should be effected before structures become enclosed to facilitate good ventilation, and practical ergonomic considerations, eg adequate access to the areas to be treated.
- Regulation 7(3)(b) Control: Experienced users, suitable quality brushes for application, powered mixing equipment, mixing area housekeeping.
- Regulation 7(3)(c) PPE: Full PPE must ensure that there is no possibility of skin contact or inhalation whatsoever: this must be determined on the basis of site conditions and discussion with the manufacturer/designer.

Expanding Foam Supplied in Aerosol Cans

1.57 Expanding foam requires skill to apply and is available with a disposable nozzle or a reusable metal applicator. (A solvent cleaner is available from the supplier and this requires separate assessment.)

1.58 Expanding foam may lead to skin sensitisation, particularly for vulnerable persons, such as those with pre-existing skin problems. Gloves must be selected which withstand the substance — the manufacturer should be able to offer advice on this. For bulk users it is worthwhile researching which gloves are best and which are preferred by the users.

1.59 Expanding foam is very difficult to remove from the skin and although eye contact is unlikely the material will drip if used overhead, and in such circumstances eye protection should be worn.

1.60 When carried in vehicles precautions must be taken to prevent inadvertent discharge. Cans should be carried upright.

Risk Phrases

1.61 Irritating to eyes, respiratory system and skin, may cause sensitisation by inhalation and skin contact. Although cases of dermatitis are rare the vapours may cause drowsiness, so if use is planned in confined spaces further precautions will be necessary.

Note: Expanding foam products are also extremely flammable. Although flammability is not covered by COSHH, steps should be taken to control the risks from this hazard.

Hierarchy of Controls Considered

1.62 When selecting appropriate controls for operations involving potential exposure to expanding foam the COSHH Regulations require that the hierarchy of controls is fully considered, as detailed below.

- Regulation 7(2) Elimination: Consider using conventional fixings or building in where possible.
- Regulation 7(3)(a) Design: None.
- Regulation 7(3)(b) Control: Experienced users.
- Regulation 7(3)(c) PPE: Gloves, safety glasses if working overhead, facemask if there are ventilation problems.

Frost-proofer Additive for Concrete

1.63 Obviously this product has seasonal use. Frost-proofer is best supplied in 5 litre cans. If mortar is supplied ready mixed the supplier should provide a safety data sheet for the mortar with the additive included. Adding direct to the mixer may prevent adequate mixing and is more likely to cause splashes.

Risk Phrases

1.64 Irritating to eyes and skin (may cause de-fatting of the skin).

Hierarchy of Controls Considered

1.65 When selecting appropriate controls for operations involving potential exposure to cement additives, the COSHH Regulations require that the hierarchy of controls is fully considered, as detailed below.

- Regulation 7(2) Elimination: Not reasonably practicable.
- Regulation 7(3)(a) Design: N/A.
- Regulation 7(3)(b) Control: Handling procedures.
- Regulation 7(3)(c) PPE: Safety glasses and gloves.

Fungicidal Solution Applied by Brush

1.66 Fungicidal solution is used to remove and kill active fungal growth on surfaces which are to be painted. It is a very thin liquid and is not easily held in the brush.

1.67 The liquid runs out of the brush easily, particularly when applying it above shoulder height so protection available for the hands, wrists and arms is particularly important. Safety goggles are essential.

1.68 Outdoor conditions rarely require the use of respiratory protection. For inside work good ventilation is essential. Care must also be taken to ensure that drink containers are not used to store this material.

Risk Phrases

1.69 Contact with acid liberates toxic gases (chlorine), irritating to skin and eyes.

Hierarchy of Controls Considered

1.70 When selecting appropriate controls for operations involving potential exposure to fungicidal solution, the COSHH Regulations require that the hierarchy of controls is fully considered, as detailed below.

- Regulation 7(2) Elimination: None.
- Regulation 7(3)(a) Design: None.
- Regulation 7(3)(b) Control: Not to be spray or roller applied, do not work with other persons below.
- Regulation 7(3)(c) PPE: Gloves, high specification eye protection, overalls, face mask, head protection if painting overhead.

Mixing and Applying Gypsum Plasters

1.71 A manual handling assessment is usually required for moving bags of plaster. If power sanding set plaster, then an additional COSHH assessment will be required. If machine mixing is undertaken, safety glasses should be worn to protect against splashes.

Risk Phrases

1.72 None, however it may irritate sensitive skin and respiratory tracts.

Hierarchy of Controls Considered

1.73 When selecting appropriate controls for operations involving potential exposure to plasters, the COSHH Regulations require that the hierarchy of controls is fully considered, as detailed below.

- Regulation 7(2) Elimination: Not applicable.
- Regulation 7(3)(a) Design: Dry lining may be possible where permitted by the designer.

- Regulation 7(3)(b) Control: Site the mixing equipment in a well-ventilated area. Training in handling procedures.
- Regulation 7(3)(c) PPE: Not normally required except for poorly-ventilated areas, gloves or barrier cream effective for sensitive skin.

Lime

1.74 Lime is a strong alkali and a caustic material. If it enters the eyes it is not only very painful but also corrosive and damaging, so every effort needs to be made to avoid raising dust. Sweeping up must not be done, instead a vacuum cleaner (with an appropriate and efficient filter) should be used and full precautions must be taken when the bag is disposed of. Staff employed to mix lime should be carefully supervised. It is easy to minimise the risk by carefully standing upwind of the mixer when loading, and siting mixing activities away from other workers.

Risk Phrases

1.75 Irritating to the skin, risk of serious damage to eyes, may cause irritation to the gastro-intestinal tract, prolonged and repeated skin contact may cause dermatitis.

Hierarchy of Controls Considered

1.76 When selecting appropriate controls for operations involving potential exposure to lime, the COSHH Regulations require that the hierarchy of controls is fully considered, as detailed below.

- Regulation 7(2) Elimination: Use ready-mixed material.
- Regulation 7(3)(a) Design: Not applicable.
- Regulation 7(3)(b) Control: Site any mixing equipment in a sheltered area. Training in handling procedures.
- Regulation 7(3)(c) PPE: Gloves, goggles and face masks, air-stream helmet preferred.

Fixing Timber with Mastic Adhesive

1.77 Solvent-free forms of mastics are now often specified because of the strong solvent odour given off by this product which can cause drowsiness. Some mastics may cause dermatitis.

Risk Phrases

1.78 Irritating (some products). Highly flammable (not relevant to COSHH, but steps must be take to control risks).

Note: Vapours may be irritating and cause drowsiness.

Hierarchy of Controls Considered

1.79 When selecting appropriate controls for operations involving potential exposure to mastic adhesive, the COSHH Regulations require that the hierarchy of controls is fully considered, as detailed below.

- Regulation 7(2) Elimination: Consider using conventional fixings or building in where possible.
- Regulation 7(3)(a) Design: None.
- Regulation 7(3)(b) Control: Experienced users.
- Regulation 7(3)(c) PPE: Gloves, safety glasses if working overhead, face mask if there are ventilation problems.

Petrol

1.80 The use of petrol on construction sites creates storage and transport problems as the lids, spouts and funnels often get lost or become unusable due to contamination.

1.81 Unlike most garages "hot works" are often carried out in close proximity to fuel. Some older handling techniques, such as siphoning by mouth, must be prohibited. Storage areas must be labelled with no smoking signs and constructed to allow pressure relief and protection from ignition and spread of fire (including specification of electrical fittings). Petrol and petrol engines must not be taken into confined spaces to prevent the build-up of petrol vapour (which is heavier than air) and the toxic exhaust fumes produced by petrol engines. The use of petrol as a fire-lighting agent must be prohibited, and it should be stored securely to prevent use by arsonists. Contaminated rags, etc must be stored in a sealed approved vessel.

Risk Phrases

1.82 May cause cancer, harmful, may cause lung damage if swallowed, irritating to the skin, vapours may cause drowsiness. Extremely flammable (not relevant to COSHH, but steps must be take to control risks).

Note: Petrol may be aspirated on swallowing or following regurgitation of stomach contents and it can cause severe or potentially fatal chemical pneumonitis.

Hierarchy of Controls Considered

1.83 When selecting appropriate controls for operations involving potential exposure to petrol, the COSHH Regulations require that the hierarchy of controls is fully considered, as detailed below.

- Regulation 7(2) Elimination: Consider use of alternative engines.
- Regulation 7(3)(a) Design: Supply into fuel tanks from explosion-proof containers.
- Regulation 7(3)(b) Control: No contaminated rags to be kept in clothing, refuelling to take place in well-ventilated areas away from ignition sources including unprotected electrical equipment. Minimise quantity in storage.
- Regulation 7(3)(c) PPE: Disposable, single use, dust-free, latex gloves for mechanics when draining down.

Sewage in Small Projects

1.84 Good standards of personal hygiene are essential prerequisites to health protection when working with even small quantities of sewage.

1.85 Those who suffer from skin problems should consult with their GP as to their suitability for work with sewage. The system of work may need to be rearranged if there is a risk of needle stick injury. To avoid infection from leptospirosis and hepatitis all skin injuries should be protected with waterproof dressings. In some circumstances tetanus and hepatitis inoculations should be considered.

1.86 Where high pressure or volumes of sewage are involved, sewage can become aerosolised into small airborne droplets.

1.87 Work in confined spaces where traces of sewage or sewer gas may collect will require careful attention to detail to ensure compliance with the Confined Spaces Regulations 1997.

1.88 Contaminated clothing should not be sent home for washing.

Risk Phrases

1.89 None allocated, however contact with sewage, and particularly sewage in aerosol form, is likely to cause sickness, diarrhoea and illness (this has the following non-specific symptoms — fever, breathlessness, dry cough, aching muscles and joints, jaundice, cramping stomach pains and infection of the eyes or skin).

1.90 In addition, exposure to sewage can lead to Weil's disease (a flu-like illness transmitted by rat urine), hepatitis, occupational asthma (produced by the inhalation of living or dead organisms), infection of skin or eyes and, rarely, allergic alveolitis (inflammation of the lung).

Hierarchy of Controls Considered

1.91 When selecting appropriate controls for operations involving potential exposure to sewage, the COSHH Regulations require that the hierarchy of controls is fully considered, as detailed below.

- Regulation 7(2) Elimination: Where possible remote methods of work should be used such as pressure washing.
- Regulation 7(3)(a) Design: Where possible divert the flow away from the work area.
- Regulation 7(3)(b) Control: Minimise all risks of falling into the sewage, and prevent the development of aerosol sprays.
- Regulation 7(3)(c) PPE: Waterproof footwear, clothing and gloves, respiratory protection required if there is exposure to aerosols and safety glasses if there is a risk of splashing.

Softwood Dust

1.92 The site use of softwood generates very little of the fine dust which may cause health problems. Sanders are the most likely to produce fine dust.

1.93 Where dust is generated in significant quantities it should be removed with a vacuum cleaner. If possible, disposable dust-collection bags, which do not need to be emptied, should be used.

1.94 The use of equipment, such as routers, is likely to expose workers to high levels of noise, for which an assessment will be required. In addition to hearing protection, safety glasses should be worn when power tools are used on wood.

1.95 Western red cedar and western red hemlock are not covered by the softwood assessments, but they may be associated with occupational asthma. Work on board materials may include exposure to hardwoods and resins. Where softwood is machined, then fixed local exhaust ventilation is required. This must be examined and a report prepared at no greater interval than once every 14 months.

1.96 Further information can be found in the HSE publications WIS23, WIS24, WIS25 and WIS26.

Risk Phrases

1.97 The following health problems are among the effects associated with exposure to wood dust.

- Skin disorders.
- Obstruction in the nose.
- Rhinitis.
- Asthma.
- A rare type of nasal cancer.

Hierarchy of Controls Considered

1.98 When selecting appropriate controls for operations involving potential exposure to softwood dust, the COSHH Regulations require that the hierarchy of controls is fully considered, as detailed below.

- Regulation 7(2) Elimination: Where possible work is carried out on machinery with fixed local exhaust ventilation.
- Regulation 7(3)(a) Design: Prefabricated components, eg roof trusses and door sets.
- Regulation 7(3)(b) Control: Using portable vacuum cleaners and hose attachments, or work could be undertaken in the open air.
- Regulation 7(3)(c) PPE: Facemasks may be necessary if sanding or router work is undertaken (see HSE leaflet WIS14).

White Spirit Used for Thinning Paint and Cleaning Brushes

1.99 White spirit should not be used in confined spaces without special precautions. As small quantities are normally required it is wise to obtain this product in small containers to ensure drinking bottles are not used.

1.100 A separate assessment should be carried out if used in connection with paint spraying.

1.101 To avoid the possibility of a chemical reaction, white spirit should be stored away from strong acids, alkalis and oxidising agents.

Risk Phrases

1.102 Toxic to aquatic organisms, harmful, may cause lung damage if swallowed, repeated exposure may cause skin dryness or cracking.

Note: Excessive inhalation of vapours may result in adverse effects such as irritation of the mucous membranes, coughing, headache, dizziness, fatigue and in extreme cases loss of consciousness.

Hierarchy of Controls Considered

1.103 When selecting appropriate controls for operations involving potential exposure to white spirit the COSHH Regulations require that the hierarchy of controls is fully considered, as detailed below.

- Regulation 7(2) Elimination: Where possible water-based paints will be used.
- Regulation 7(3)(a) Design: None.
- Regulation 7(3)(b) Control: No contaminated rags to be kept in clothing, where practical, single-use brushes will be considered.
- Regulation 7(3)(c) PPE: Gloves.

Wood Preserver Applied to Timber

1.104 This assessment is for a wood preserver supplied as low volatile organic compounds (voc) (ie containing a low level of voc).

1.105 Wood preserver should be applied before timber is fixed to ensure adequate ventilation and allowed to dry before handling. Care should be taken to avoid spray being generated when using radiator rollers to apply. If skin protection is not worn effectively then skin condition monitoring should be undertaken.

Risk Phrases

1.106 Irritating to skin and eyes, harmful, may cause lung damage if swallowed, repeated exposure may cause skin dryness or cracking, flammable.

Hierarchy of Controls Considered

1.107 When selecting appropriate controls for operations involving potential exposure to wood preserver, the COSHH Regulations require that the hierarchy of controls is fully considered, as detailed below.

- Regulation 7(2) Elimination: Not applicable.
- Regulation 7(3)(a) Design: Off-site treatment where practicable.
- Regulation 7(3)(b) Control: Do not spray, replace tin lids when not in use, ventilate.
- Regulation 7(3)(c) PPE: Gloves, safety glasses if there is a risk of splashes, coveralls if large quantities and in use.

Zinc-rich Paint Applied by Brush

1.108 Zinc-rich paint applied with a brush is generally used in small quantities.

1.109 The solvent cleaner is hazardous and will require a specific assessment, therefore single-use disposable brushes and masking tape to protect areas near to the painting are advocated.

1.110 As vapours are heavier than air, if sprayed the mist and vapours will not disperse easily and may cause irritation to the nose and respiratory tract. Vapours may cause headaches, dizziness and depression of the central nervous system.

Risk Phrases

1.111 Harmful by inhalation, irritating to eyes, respiratory system and skin, contact with water liberates highly flammable gases. Flammable (not relevant to COSHH, but steps must be taken to control risks).

Hierarchy of Controls Considered

1.112 When selecting appropriate controls for operations involving potential exposure to zinc-rich paint, the COSHH Regulations require that the hierarchy of controls is fully considered, as detailed below.

- Regulation 7(2) Elimination: Not applicable.
- Regulation 7(3)(a) Design: Off-site galvanising where practicable.
- Regulation 7(3)(b) Control: Replace lids when not in use, use disposable single-use brushes.
- Regulation 7(3)(c) PPE. Gloves, safety glasses if there is a risk of splashes, coveralls if large quantities are in use.

List of Relevant Legislation

- Construction (Design and Management) Regulations 2007
- Chemicals (Hazard Information and Packaging for Supply) Regulations 2002
- Control of Lead at Work Regulations 2002
- Control of Substances Hazardous to Health Regulations 2002
- Dangerous Substances and Explosive Atmospheres Regulations 2002
- Management of Health and Safety at Work Regulations 1999
- Manual Handling Operations Regulations 1992

- Personal Protective Equipment at Work Regulations 1992
- Health and Safety at Work, etc Act 1974

Further Information

Publications

HSE Publications

The following are available from *www.hsebooks.co.uk*.

- CIS26 *Cement*
- CIS54 *Dust Control on Concrete Cutting Saws Used in the Construction Industry*
- EH40 *Workplace Exposure Limits* (revised annually)
- MSA15 *Silica Dust and You*
- WIS14 *Selection of Respiratory Protective Equipment Suitable for Use with Wood Dust*
- WIS23 *LEV: General Principles of System Design*
- WIS24 *LEV: Dust Capture at Sawing Machines*
- WIS25 *LEV: Dust Capture at Fixed Belt Sanding Machines*
- WIS26 *LEV: Dust Capture at Fixed Drum and Disc Sanding Machines*

Organisations

- British Standards Institution (BSI)
  Web: *www.bsi-global.com*
  Founded in 1901, the BSI develops and provides supporting information on national and international standards in a wide range of areas.
- Building Research Establishment (BRE)
  Web: *www.bre.co.uk*
  The BRE provides consultancy, testing and commissioned research services covering all aspects of the built environment and associated industries.
- Chartered Institution of Building Services Engineers (CIBSE)
  Web: *www.cibse.org*
  The CIBSE is an international body which represents and provides services to the building services profession.
- Concrete Society
  Web: *www.concrete.org.uk*
  The society works through the co-operation of its members who come from all sectors of the industry to exchange information and experience and to enhance the performance, productivity and quality of concrete as a construction medium.
- Department for Environment, Food and Rural Affairs (Defra)
  Web: *www.defra.gov.uk*
  Defra is the main government department which deals with waste and other environmental issues. It consults on new regulations and provides guidance on legislation and best practice.
- Engineering Construction Industry Association (ECIA)
  Web: *www.ecia.co.uk*
  ECIA members build industrial assets such as power stations, oil refineries, chemical works, pharmaceutical plants, bridges and steel structures.

- Environment Agency
  Web: *www.environment-agency.gov.uk*
  The Environment Agency is the main environmental regulator in England and Wales and provides detailed information on legislative requirements, guidance for business and technical guidance on waste treatment and disposal.
- Federation of Building Specialist Contractors (FBSC)
  Web: *www.fbsc.org.uk*
  The FBSC seeks to bring together companies whose clients include organisations and businesses involved in the civil engineering, construction and house building markets.
- Glass and Glazing Federation (GGF)
  Web: *www.ggf.org.uk*
  The GGF is a trade association for all those who make, supply, or fit flat glass, such as windows, film or plastics.
- Health and Safety Executive (HSE)
  Web: *www.hse.gov.uk*
  The HSE is responsible for the regulation of almost all the risks to health and safety arising from work activity in the UK.
- Institution of Occupational Safety and Health (IOSH)
  Web: *www.iosh.co.uk*
  IOSH is Europe's leading body for health and safety professionals and provides guidance on health and safety issues.

Lead

- Employers are required to either prevent the exposure of employees, or others, to lead at work or, if prevention is not reasonably practicable, to adequately control the exposure.
- Employees who may be exposed to lead must be provided with appropriate information, instruction and training.
- Any equipment provided to control exposure to lead must be maintained in good working order, good repair and be kept clean. This includes local exhaust ventilation and personal protective equipment.
- In circumstances where employees are liable to be exposed to significant levels of lead, the concentration of lead in air should be measured.
- If employees are liable to be exposed to lead, the employer must provide suitable health surveillance.
- If employees are significantly exposed to lead, their blood-lead levels should be measured every three months. Employee blood-lead levels should not exceed specified lead action levels.
- If the lead in blood or urine concentrations reaches the appropriate suspension level, the employer must suspend the employee from the work that is liable to expose them to lead.
- Employers must ensure plans and procedures are in place to deal with accidents, incidents or emergencies related to the presence of lead.

1.113 Exposure to lead is a well-recognised health hazard and is one of the few areas where prohibitions still exist for certain specified workers, ie young people and women of reproductive capacity. Lead poisoning is most commonly caused through the inhalation of dust and fumes containing lead, although some absorption through the gut may also occur.

1.114 Lead has been widely used in the construction industry, eg in paints, solders, water pipes, roofing and flashing. It is no longer used in paints, with the exception of a few specialist paints, or in water pipes. The solder used for jointing copper pipes is now required to be lead free to comply with water regulations. Lead sheeting is still used in construction for roofing and flashing. Lead products are likely to be encountered in old premises.

1.115 The Control of Lead at Work Regulations 2002 require any health risks associated with exposure to lead at work to be assessed and appropriate measures implemented to control such risks. There are also provisions prohibiting certain types of lead work.

Employers' Duties

1.116 In order to comply with the Control of Lead at Work Regulations 2002, employers must do the following.

- Either prevent employees, or others, from exposure to lead at work or, if prevention is not reasonably practicable, adequately control the exposure.
- Prevent women of reproductive capacity and children from working with lead in certain prohibited processes.
- Where an employee is liable to be exposed to significant levels of lead, assess the risks to health arising from the exposure.

- Use the findings of the assessment to identify, and implement, the control measures needed to prevent or control exposure to lead.
- Record the results of the assessment if there are five or more employees.
- Maintain all equipment used to minimise exposure to lead.
- Ensure employees do not eat, drink or smoke in any place liable to be contaminated by lead.
- Provide every employee liable to be exposed to significant levels of lead with information, instruction and training.
- Measure the concentration of lead in the air if any of the employees are liable to be exposed to significant levels.
- Ensure every employee who is either liable to be exposed to significant levels of lead or has lead in blood or urine above specified concentrations is subject to medical surveillance.
- Put every employee who is liable to be exposed to significant levels of lead under regular biological monitoring.
- Develop plans and procedures to deal with accidents, incidents and emergencies related to the presence of lead.

Employees' Duties

1.117 Employees must:

- co-operate with their employers in order to enable compliance with the Control of Lead at Work Regulations 2002
- make proper use of control measures, including personal protective equipment
- return equipment after use to any storage place and report any defects found in the equipment
- not eat, drink or smoke in any place liable to be contaminated by lead
- attend medical examinations at the appointed time and provide any information about their health that may be reasonable
- provide samples of either blood or urine measurements for biological monitoring when required.

In Practice

Hazards of Lead

1.118 If the level of lead in the body gets too high, it can cause:

- headaches
- tiredness
- irritability
- constipation
- nausea
- stomach pains
- anaemia
- weight loss.

1.119 Continued high exposure can cause more serious symptoms, such as kidney damage or nerve and brain damage. The symptoms can have other causes and do not necessarily mean an individual is suffering from lead poisoning. Unborn children are at particular risk from exposure to lead, especially in the early weeks before the pregnancy becomes apparent.

1.120 Under the Chemicals (Hazard Information and Packaging for Supply) Regulations 2002, compounds of lead with a few exceptions, eg lead alkyls, are classified as the following.

- Toxic for reproduction Category 1, R61: May cause harm to the unborn child.
- Toxic for reproduction Category 3, R62: Possible risk of impaired fertility.
- R20/22: Harmful by inhalation and if swallowed.
- R33: Danger of cumulative effects.
- R50: Very toxic to aquatic organisms.
- R53: May cause long-term adverse effects in the aquatic environment.

1.121 Such compounds of lead should be labelled:

- "T" Toxic for reproduction
- "N" Dangerous for the environment
- R61, 20/22, 33, 50/53, 62
- S53, 43, 60, 61.

Significant Exposure to Lead

1.122 Employers must decide if any workers or any others who may be affected by the company's work processes are liable to be exposed to significant levels of lead.

1.123 Under the Control of Lead at Work Regulations 2002, an employee's exposure to lead will be significant if one of the following conditions is satisfied.

- Exposure exceeds half the workplace exposure limit for lead (0.15mg/m^3).
- There is a substantial risk of the employee taking lead into the body via the mouth.
- There is a risk of an employee's skin coming into contact with lead alkyls or any other substance containing lead in a form that can be absorbed through the skin, eg lead naphthenate.

1.124 Lead must be in a form in which it is likely to enter the body by:

- inhalation, eg lead or lead compounds of dust fume or vapour form
- ingestion, eg lead or lead compounds in the form of, or included in, dusts, powders, pastes or paints
- absorption through the skin, eg lead alkyls and lead naphthenate.

1.125 The Control of Lead at Work Regulations 2002 do not apply to materials or products containing lead which, because of their nature, cannot be inhaled, ingested or absorbed through the skin.

Prohibitions

1.126 People under the age of 18, or women who are medically and physically capable of becoming pregnant cannot be employed to work on any of the processes listed in schedule 1 to the Control of Lead at Work Regulations 2002. Schedule 1 covers certain activities carried out during:

- lead smelting and refining
- the manufacturing process of lead-acid battery.

Risk Assessment

1.127 The risk assessment should consider:

- the hazardous properties of the lead material or product used, including information on the health effects provided by the safety data sheet
- the amount of the lead material or product used
- the duration and frequency of work liable to result in exposure
- the nature of the work, and the likelihood that airborne lead will be produced or lead will be ingested or absorbed through the skin, including activities such as maintenance work with the potential to result in high levels of exposure
- how the lead could enter the body
- relevant workplace exposure limits, action levels and suspension levels
- the effect of existing control measures
- the results of relevant medical surveillance
- the results of any measurement of airborne lead concentrations.

1.128 The legal responsibility for the assessment rests with the employer, who must ensure that the person who carries out the assessment:

- has adequate knowledge, training and understanding of hazard and risk
- knows how the work activity uses or produces lead
- has the ability and the authority to collate all the necessary, relevant information
- has the knowledge, skills and experience to reach the right decisions about the risks and the precautions required.

1.129 In some situations it may be necessary to use specialist help. Employees using materials containing lead should be involved in the assessment. Safety representatives and safety committees should also be consulted.

Control Measures

1.130 Once the assessment has evaluated the risks of the work, the employer must identify the steps to be taken to provide adequate control of exposure. This will include consideration of:

- whether it is necessary to use lead, or if it can be substituted
- whether the amount of lead used can be reduced
- whether the number of people exposed and the duration of exposure can be reduced
- containment of the lead at source
- the use of engineering controls, such as local exhaust ventilation
- the precautions to be taken during the work process, including arrangements for:
  - handling
  - storage
  - transport of lead and lead waste
- the protective clothing and personal protective equipment to be provided
- changing and washing facilities required
- the extent to which measurement of lead in air concentrations should be carried out
- the need to place employees under medical surveillance
- procedures for dealing with accidents, incidents and emergencies involving lead.

1.131 All control measures, including local exhaust ventilation and personal protective equipment must be maintained in efficient working order, in good repair and in a clean condition.

1.132 Any engineering equipment used to control exposure must be thoroughly examined and tested, in the case of local exhaust ventilation, at least every 14 months and in other cases at suitable intervals.

1.133 Washing and changing facilities must be provided for employees in order to minimise the risk of ingesting or absorbing lead. The design of the washing facilities should reflect the type and level of exposure to lead. For significant exposure, this should include showers or baths.

1.134 Exposure must not exceed the workplace exposure limit for lead (0.15mg/m^3). If the workplace exposure limit is exceeded, the employer must identify the reasons for the limit being exceeded and take immediate steps to remedy the situation. Records of the tests must be kept for at least five years.

Information, Instruction and Training

1.135 When an employee is liable to be exposed to lead, the employer must provide the employee with suitable and sufficient information, instruction and training on:

- the risk presented by the lead containing materials, including the form of the lead and the risk to health it presents, including the risks to the employee's family from taking home contaminated clothing, etc
- any significant workplace exposure limit, action level and suspension levels
- the significant findings of the risk assessment, particularly the degree of risk
- the correct use and maintenance of control measures provided to safeguard the employee from exposure to lead including:
  - engineering controls
  - work methods
  - personal protective equipment
- the importance of good hygiene
- when the washing and changing facilities should be used
- the results and purpose of any air monitoring measurements
- the results and role of any medical surveillance, in a form that prevents the results being related to a specific individual
- the procedures for dealing with accidents, incidents or emergencies involving lead.

1.136 The employee should be given a copy of the HSE leaflet *Lead and You* and provided with access to any relevant safety data sheet.

Monitoring Lead in Air

1.137 If the risk assessment indicates that any employee is liable to receive significant exposure to lead, the employer must measure the concentration of lead in the air which employees are exposed to, to show compliance with the workplace exposure limit.

1.138 Measurements of lead in air concentrations should be made in the breathing zone by means of personal sampling equipment (usually worn on the individual's lapel or as close to the mouth as reasonable).

1.139 Static sampling alone, which is simply stationed somewhere in the workplace, is not enough, but may be useful in checking on the continuing efficiency of engineering control measures.

1.140 The measurement should be carried out at least every three months, unless there has been no material change in the work or the conditions of exposure since the previous monitoring and the lead in air concentration has not exceeded 0.10 mg/m^3 on the two previous tests.

1.141 A suitable record of the monitoring must be made and kept available for at least five years from the date of last entry.

1.142 If there is significant exposure to lead, the employees' exposure should be evaluated by both air sampling and by measuring the concentration of lead in their blood or their urine for work with lead alkyls.

Medical Surveillance

1.143 An employee should be placed under suitable medical surveillance by a doctor if:

- they are liable to be exposed to a significant level of lead
- the blood or urinary lead level is above:
  - for a woman capable of giving birth: 20 μg/dl lead in blood or 20 μgPb/g creatinine in urine
  - for any other employee 35μg/dl lead in blood or 40 μgPb/g creatinine
- a doctor recommends so.

1.144 Medical surveillance should be carried out before the employee starts work with lead, and at least every 12 months, or at shorter intervals if recommended by a doctor. The employer must ensure that adequate health records for those employees under medical surveillance are made and kept for at least 40 years.

Biological Monitoring

1.145 When employees are significantly exposed to lead, their blood-lead levels should be measured every three months. If a consistent blood-lead pattern is established, the interval can be extended, except for women of reproductive capacity and for young persons for whom the interval should not be extended to more than three months.

1.146 The method for monitoring the absorption of lead alkyls in the body is by measuring total urinary lead concentration, and the urinary creatinine concentration and correcting to allow for differences in the volume of urine produced.

Action Levels

1.147 If an employee's blood-lead level is above the appropriate action level (there is no action level for lead alkyls) the employer must investigate why it has been breached and review the range and effectiveness of control measures used. The employer must aim to reduce the employee's blood-lead below the action level and prevent it reaching the suspension level.

1.148 The lead in blood action levels are:

- women of reproductive capacity — 25 μg/dl
- young persons (aged under 18) — 40 μg/dl
- any other employee — 50 μg/dl.

Suspension Levels

1.149 If an employee's blood-lead concentration (or urinary-lead concentration where there is exposure to lead alkyls) exceeds the appropriate suspension level, the employee should be suspended from the work liable to expose them to lead.

1.150 The lead in blood suspension levels are:

- women of reproductive capacity — 30 μg/dl
- young persons (aged under 18) — 50 μg/dl
- any other employee — 60 μg/dl.

1.151 The urinary suspension levels (for lead alkyls) are:

- women of reproductive capacity — 25 μgPb/g creatinine
- young persons — 110 μgPb/g creatinine.

Emergency Procedures Relating to Lead

1.152 Employers are required to make arrangements to deal with events which might cause an employee to be exposed to lead to an extent well beyond that associated with normal day-to-day activity. Examples include a serious process fire, a serious spillage of molten lead, uncontrolled lead dust or fume, or any acute process failure liable to result in a release of lead dust, fume or vapour or a serious failure of local exhaust ventilation or other controls.

1.153 The arrangements should:

- reflect the scale of the release and the type and form of lead concerned
- alleviate the effects of the incident
- restore the situation to normal as soon as possible
- limit the risks to health of employees and anyone else likely to be affected by the incident.

Training

1.154 Regulation 5 of the Control of Lead at Work Regulations 2002 requires every employer to provide adequate information, instruction and training to employees who are liable to be exposed to lead. This training will ensure they are aware of the risks from lead and the precautions that should be observed. The regulations also require those responsible for specific duties to be able to carry out those duties effectively.

1.155 The objectives of the training include:

- ensuring adequate information, instruction and training is given to employees who are liable to be exposed to lead at work
- providing information on the risks and health hazards from lead
- familiarising employees who may be exposed to lead with the general control measures that are required to protect themselves and others
- familiarising employees with the specific control measures in relation to the tasks carried out
- ensuring that employees are aware of how to use the control measures required.
- providing information on air and biological monitoring and the role of medical surveillance
- ensuring that those who have responsibility to carry out specific tasks required by the Control of Lead at Work Regulations 2002, and the accompanying Approved Code of Practice, are competent to do so
- ensuring managers and supervisors have sufficient understanding to ensure processes are carried out within the legislative framework covering the use of lead at work.

Scope of Training

Employees

1.156 Training should ensure employees:

- know the hazards associated with exposure to lead
- understand the difference between hazard and risk
- know the routes by which lead can be absorbed into the body
- understand the importance of personal hygiene in reducing exposure
- know why smoking, eating and drinking are banned in the workplace
- understand the ways in which other control measures, such as local exhaust ventilation or personal protective equipment operate
- know how to carry out a simple examination of the protective equipment provided
- understand the procedures for, and importance of, reporting any defects or malfunction in protective equipment supplied
- know the purpose and function of the air monitoring and biological monitoring carried out
- understand the reasons for any medical surveillance programme
- understand the consultation arrangements within the organisation with regard to health and safety
- know the procedures to deal with emergencies related to the use of lead.

Managers and Supervisors

1.157 In addition to understanding employees' training, managers and supervisors should also understand:

- the requirements of the Control of Lead at Work Regulations 2002
- the provisions of the Approved Code of Practice to the Control of Lead at Work Regulations 2002
- the concept of risk assessment, as applied to the use of lead at work
- the workplace exposure limits for lead in air and the biological monitoring standards and their application
- the requirements for medical surveillance.

Those with Specific Duties Under the Regulations

1.158 In addition to other duties, certain specific duties are also placed on employers under the regulations. These duties include risk assessment, inspection and testing of control measures and workplace air monitoring. If carried out in-house, those carrying out these activities must be competent.

List of Relevant Legislation

- Chemicals (Hazard Information for Supply) Regulations 2002
- Control of Lead at Work Regulations 2002
- Data Protection Act 1998
- Health and Safety at Work, etc 1974

Further Information

Publications

HSE Publications

The following are available from *www.hsebooks.co.uk.*

- INDG305 *Lead and You: A Guide to Working Safely with Lead*
- L132 *Control of Lead at Work. Control of Lead at Work Regulations 2002. Approved Code of Practice and Guidance*

Other Publications

- *Old Lead Painted Surfaces: A Guide on Repainting and Removal for DIY and Professional Painters and Decorators*, 2005, British Coatings Federation

Storage of Hazardous Substances

- Employers are legally required to make arrangements for the safe storage of all substances, particularly substances hazardous to health, under the Control of Substances Hazardous to Health Regulations 2002.
- Several substances have specific legal requirements relating to their storage, including:
  - lead
  - explosive, flammable and oxidising substances
  - radioactive substances
  - oil
  - pesticides
  - substances that can cause environmental damage.
- The labelling of substances should identify any categories of danger, risk phrases and safety phrases, or carriage labelling. Any goods which are not readily identified must not be sent for storage.
- The risk from the storage of dangerous and hazardous substances is dependent on the level of danger the substance presents.
- The safe storage of dangerous and hazardous substances must take into account any particular hazards presented by the substance, eg the potential danger of gas cylinders.
- Any incompatibility between substances should be considered before storing different substances together.
- Employers should put in place procedures for dealing with emergencies related to the substances being stored.

1.159 There are a vast number of dangerous substances and products containing dangerous substances used in the workplace, varying in type, hazard, severity and the conditions under which they present a risk. The risk presented by substances is dependent on the substance itself and the quantity stored.

Employers' Duties

1.160

- The Health and Safety at Work, etc Act 1974 includes a general duty requiring arrangements for the storage of substances to ensure safety and the absence of risk to health.
- The Management of Health and Safety at Work Regulations 1999 create a duty on the employer to assess the risks from all of its operations, including the storage of hazardous and dangerous substances.
- The Control of Substances Hazardous to Health Regulations 2002 (COSHH) require the safe storage of substances hazardous to health where hazardous substances are present at the workplace.
- The COSHH Regulations also require the employer to carry out a risk assessment, including a consideration of how the substances are to be stored safely.
- The Control of Lead at Work Regulations 2002 require the safe storage of lead and lead compounds.

- The Dangerous Substances and Explosive Atmospheres Regulations 2002 require the safe storage of explosive, flammable and oxidising substances.
- The Ionising Radiations Regulations 1999 require the safe storage of radioactive substances.
- The Control of Pollution (Oil Storage) (England) Regulations 2001 set out requirements for the storage of oil in containers containing more than 200 litres. Similar requirements arise from the Water Environment (Oil Storage) (Scotland) Regulations 2006.
- The Planning (Hazardous Substances) Regulations 1992 apply to facilities where controlled hazardous substances are stored in quantities above a specified minimum.
- The Control of Major Accident Hazards Regulations 1999 apply to premises where certain dangerous substances are present above specified thresholds, and require operators to take measures to prevent major accidents. The steps include the storage of the dangerous substances.
- The Control of Pesticides Regulations 1986 require that sites where 200 kilograms or more of pesticides are stored must be under the control of a person holding a recognised certificate of competence.
- The Manufacture and Storage of Explosives Regulations 2005 require anyone who stores explosives to take appropriate measures to prevent fire or explosion and, with certain specified exceptions, to have a licence.
- Spillages or emissions of dangerous substances can cause environmental harm and such emissions may be subject to controls under the Environmental Protection Act 1990, and if water pollution occurs, under the Water Resources Act 1991 in England and Wales or the Control of Pollution Act 1974 in Scotland.
- Employers must provide adequate information, instruction and training to employees under the:
  - Management of Health and Safety at Work Regulations 1999
  - COSHH Regulations
  - Control of Lead at Work Regulations 2002 (CLAW)
  - Dangerous Substances and Explosive Atmospheres Regulations 2002 (DSEAR).
- Employers must examine and maintain stored substances in a safe condition, in particular examine containers for damage including leakage and deterioration of labelling. This duty comes from the Health and Safety at Work, etc Act 1974, DSEAR, CLAW and COSHH Regulations.
- Employers must put in place systems to deal with accidents, incidents and emergencies involving the stored materials under the Management of Health and Safety at Work Regulations 1999, COSHH Regulations, CLAW and DSEAR.

Employees' Duties

1.161 Employees must:

- co-operate with their employers to enable their employers to comply with the Control of Substances Hazardous to Health Regulations 2002 and other legislation governing the storage of hazardous materials
- ensure substances are stored safely in line with procedures given by their employers
- return substances to their storage places when work with them is completed.

In Practice

Legislation Affecting Storage

1.162 Before deciding on storage conditions, the employer should assess whether the substance is covered by legislation.

Legislation Dealing with the Storage of Hazardous and Dangerous Substances

Substance type	Legislation
Flammable	Dangerous Substances and Explosive Atmospheres Regulations 2002
Oxidising	Dangerous Substances and Explosive Atmospheres Regulations 2002
Explosive	Dangerous Substances and Explosive Atmospheres Regulations 2002
Explosives	Manufacture and Storage of Explosives Regulations 2005
Lead	Control of Lead at Work Regulations 2002
Radioactive substances	Ionising Radiations Regulations 1999
Pesticides	Control of Pesticides Regulations 1986
Hazardous waste	Waste Management Licensing Regulations 1994
Oil	Control of Pollution (Oil Storage) (England) Regulations 2001
Substances hazardous to health	Control of Substances Hazardous to Health Regulations 2002

Labelling and Storage

Chemicals (Hazard Information and Packaging for Supply) Regulations 2002

1.163 The Chemicals (Hazard Information and Packaging for Supply) Regulations 2002 (CHIP) require the classification of hazardous substances for use in the workplace into:

- categories of danger, eg toxic and corrosive
- risk phrases
- safety phrases.

1.164 This provides information that should be taken into account when deciding on the conditions required for storage.

1.165 Under CHIP, the supplier of chemicals for use at work must provide a safety data sheet (SDS) with each substance and the SDS must provide information on storage conditions.

1.166 Many substances arriving on site will be marked with carriage labelling required by the European Agreements concerning the International Carriage of Dangerous Goods by Road and Rail, which has been implemented in the UK by the Carriage of Dangerous Goods and Use of Transportable Pressure Equipment Regulations 2007. Any goods which are not readily identified must not be sent to store.

Principles of Safe Storage

1.167 Storage of hazardous substances must take into account:

- the hazard(s) presented by the substance
- the amount of hazardous substance
- the containers
- where the hazardous material will be stored
- procedures for emergencies.

Hazards

1.168 The precautions required to achieve a reasonable standard of control will vary but must take into account the properties of the substance to be stored. Different substances create very different risks because of their hazards. Storage conditions will require consideration of the physical and chemical properties, together with the health effects of the substances concerned.

Gases

1.169 Gas cylinders, because of their potential for danger, present specific problems.

1.170 Harm can be caused by:

- explosion of the cylinder, particularly in fires
- impacts causing violent release of compressed gas
- the release of the gas, which may:
  - cause fire if the gas is flammable
  - be asphyxiating if the release is into a confined space
  - be toxic, eg carbon monoxide
  - be corrosive, eg chlorine
- injury caused by badly stored cylinders falling.

1.171 Gas cylinders should be stored:

- in a dry, safe place on a flat surface in the open air, or, if this can not be done, in an adequately ventilated building or part of a building solely reserved for the purpose
- away from external heat sources
- away from sources of ignition and other flammable materials
- securely, to prevent cylinders falling or being knocked over
- away from vehicles such as fork-lift trucks.

1.172 Gas cylinders should also be clearly marked to show the contents and the hazards associated with their contents. The valve must be shut to stop contaminants entering. Gas cylinders should never be stored standing or lying in water.

Incompatibles

1.173 The possibility of interactions between different substances, especially those which are incompatible, creating hazards must be considered.

1.174 Examples of incompatibles include:

- acids which react with hypochlorites to generate chlorine gas
- acids which react with cyanides to generate hydrogen cyanide gas
- acids which react with alkalis to generate heat
- acids which react with sulphides to generate hydrogen sulphide
- nitric acid, which will react explosively with alcohol and other organic materials
- oxidising agents, which should not be stored with organic materials.

1.175 The SDS of a substance should provide information on substances it is incompatible with. Incompatible substances should not be stored together and it is good practice to segregate acids from other substances.

Oil

1.176 The most common form of water pollution is pollution by oil. It is an offence to pollute water in the environment with oil.

1.177 The requirements for the storage of oil in containers which carry more than 200 litres are set out in the Control of Pollution (Oil Storage) (England) Regulations 2001. These regulations, which do not apply to waste oil or oil stored in a building or wholly underground, require:

- secondary containment, a bund:
  - with a capacity of not less than 110% of the container's storage capacity
  - positioned to minimise any risk of damage by impact, eg by fork-lift trucks
  - with its base and walls impermeable to water and oil
  - without its base and walls penetrated by any valve, pipe or other opening used for draining the system
  - with the base or walls, where they are penetrated by any fill pipe, or draw off pipe, adequately sealed to prevent oil escaping from the system
- any valve, filter, sight gauge, vent pipe, etc to be situated within the secondary containment system
- fill pipes, not within the secondary containment system, to have a drip tray to catch any oil spilled when the container is being filled with oil.

1.178 The regulations also include requirements for the storage of oil in fixed tanks, both above ground and underground, which include particular requirements for ensuring that fill, draw-off and overflow pipes are not damaged, and that the tanks are fitted with automatic overfill protection devices.

Storing Large Quantities of Substances

1.179 The risk from the storage of dangerous substances is dependent on the amount of dangerous substance present. The following factors should be considered.

- Storage buildings and outdoor storage compounds for dangerous substances are subject to controls under building legislation. In England and Wales, Approved Document B under the Building Regulations 2000 sets standards for fire resistance and compartment size for industrial or storage buildings.
- Storage buildings and outdoor storage compounds for dangerous substances are subject to controls under planning legislation. The Planning (Hazardous Substances) Regulations 1992 apply to facilities where controlled hazardous substances are stored in quantities above a specified minimum.

- In premises where certain dangerous substances are present above specified thresholds, the Control of Major Accident Hazards Regulations 1999 (COMAH) require operators to take measures to prevent major accidents. The steps include the storage of the dangerous substances.
- If the COMAH Regulations apply, the risk assessment has to be very detailed, and cover storage facilities.

Emergencies

1.180 Procedures for dealing with emergencies related to the hazardous substances being stored need to be developed. Consideration needs to be given to the range of possible events, including:

- fire
- explosion
- releases, eg leakages or spillages.

1.181 The following factors must also be taken into account.

- The nature and quantities of the dangerous substances stored.
- The location and design of the storage facility.
- The people, both on and off the site, who may be affected.

1.182 Safe systems of work for dealing with spillages and leakages should be put in place and will depend on the nature of the substance involved. SDSs will give details of any specific action to be taken for dealing with spillages. These need to be available for all the substances stored on site.

Training

1.183 Employers must provide adequate information, instruction and training to employees on the:

- risks presented by the substances to be stored
- procedures for safe storage of hazardous substances
- procedures for fires involving stored hazardous substances
- procedures for spillages, leakages and other releases of hazardous substances
- procedures for other incidents involving the release of stored materials.

List of Relevant Legislation

- Carriage of Dangerous Goods and Use of Transportable Pressure Equipment Regulations 2007
- Manufacture and Storage of Explosives Regulations 2005
- Chemicals (Hazard Information and Packaging for Supply) Regulations 2002
- Control of Lead at Work Regulations 2002
- Control of Substances Hazardous to Health Regulations 2002
- Dangerous Substances and Explosive Atmospheres Regulations 2002
- Control of Pollution (Oil Storage) (England) Regulations 2001
- Control of Major Accident Hazards Regulations 1999

- Ionising Radiations Regulations 1999
- Management of Health and Safety at Work Regulations 1999
- Waste Management Licensing Regulations 1994
- Planning (Hazardous Substances) Regulations 1992
- Environmental Protection (Prescribed Processes and Substances) Regulations 1991
- Control of Pesticides Regulations 1986
- Water Resources Act 1991
- Environment Protection Act 1990
- Control of Pollution Act 1974 (Scotland)
- Health and Safety at Work, etc Act 1974

Further Information

Publications

HSE Publications

The following are available from *www.hsebooks.co.uk*.

- AIS16 *Guidance on Storing Pesticides for Farmers and Other Professional Users*
- CS3 *Storage and Use of Sodium Chlorate and Other Strong Oxidants*
- CS21 *Storage and Handling of Organic Peroxides*
- HSG51 (rev 1998) *Storage of Flammable Liquids in Containers*
- HSG71 (rev 1998) *Chemical Warehousing: The Storage of Packaged Dangerous Substances*
- HSG135 *Storage and Handling of Industrial Nitrocellulose*
- HSG176 *The Storage of Flammable Liquids in Tanks*
- INDG230(L) *Storage and Safe Handling of Ammonium Nitrate*
- L5 (rev 2005) *The Control of Substances Hazardous to Health Regulations 2002 (as amended). Approved Code of Practice and Guidance*
- L135 *Storage of Dangerous Substances. Dangerous Substances and Explosive Atmospheres Regulations 2002. Approved Code of Practice and Guidance*
- L139 *Manufacture and Storage of Explosives. Manufacture and Storage of Explosives Regulations 2005. Approved Code of Practice and Guidance*

Department of Communities and Local Government Publications

The following are available from *www.communities.gov.uk*.

- Approved Document B *Fire Safety: Volume 1: Dwellinghouses*
- Approved Document B *Fire Safety: Volume 2: Buildings other than Dwellinghouses*
- *Fire Safety Risk Assessment — Factories and Warehouses*

Other Publications

- *Guide to Good Practice for the Storage of Aerosols in Manufacturing/Wholesale Warehouses and Retail Stores*, British Aerosol Manufacturers Association, 1989
- *Storage and Handling of Drum and Intermediate Bulk Containers*, Pollution Prevention Guidance PPG26, Environment Agency 2004

Chapter 2

Contracts

Contracts

- It is helpful to spell out the specific obligations and responsibilities of the parties with regard to health and safety in each contract.
- In terms of function, each procurement route deals with the responsibilities of the parties in different ways; in terms of performance, each route apportions risk across the supply chain in a different way.
- Collateral warranties are a form of agreement used in the construction industry to ensure that a professional has a specific duty of care to parties other than the client.
- Approved Codes of Practice, which have a special legal status in relation to health and safety, can be incorporated into contracts.
- There are six recognised procurement routes for a client to choose from, with the addition of a seventh route called "partnering".

2.1 The purpose of a contract is to:

- set out the obligations and responsibilities of the parties involved
- outline the specific terms and conditions agreed by the parties in a clear and unambiguous manner.

2.2 The four essential elements in the creation of a contract are the:

- offer
- acceptance
- consideration (payment)
- intention to create a legal relationship.

2.3 While some schools of thought believe it is unnecessary to write health and safety into contracts (since the law always applies and cannot be modified by a contract and should be known to the contracting parties anyway), it can be helpful to spell out the specific obligations and responsibilities of the parties in each contract, avoiding the following dangers.

- Poorly drafted contracts may imply that one or both parties is required to do less than is actually required by statute.
- There is a danger of including outdated laws in contracts and omitting newer legislation.

In Practice

Contracts and the CDM Regulations

2.4 The Construction (Design and Management) Regulations 2007 (CDM Regulations), which came into force on 6 April 2007, apply to projects, not just contracts. There can be one contract or many contracts in a project.

2.5 The three main categories of construction work are:

- building
- civil engineering
- process engineering.

2.6 The three main types of construction within each category are:

- new build
- refurbishment
- maintenance.

2.7 Different series, or "families", of standard forms of contract have been developed for use in conjunction with each category and type of work.

Forms

2.8 A standard form can be used for the employment of:

- consultants
- contractors
- specialist subcontractors
- labour-only subcontractors
- various types of suppliers.

2.9 In their unamended state, the forms create an interlocking matrix which passes responsibilities and obligations across the supply chain in back-to-back fashion.

2.10 Although it is advisable to use an unamended matrix of standard forms appropriate to the category and type of work to be carried out, it is not compulsory to do so. The parties are free to use any combination of forms of contract they wish (or, as is sometimes the case in construction, no written contract at all). However, this does not mean that a contract does not exist. When the Construction (Design and Management) Regulations 2007 apply to a construction project, they create duties under the criminal code which must be fulfilled.

Procurement Routes

2.11 The duty holder is largely configured by the procurement route chosen by the client. The six recognised procurement routes are:

- traditional
- design and build
- design and manage
- management contracting
- construction management
- engineer, procure, construct (turnkey).

2.12 There is a seventh procurement route, "partnering", which may reasonably be applied to all of the above. This approach involves the various participants agreeing to work together in a spirit of "mutual trust and co-operation" and all parties are brought in at an earlier stage in the process. In theory, this means that those items which may have health and safety implications can be addressed sooner rather than later.

2.13 There has been an emergence of various partnering forms of contract, most notably the Association of Consultant Architects (ACA) publication *PPC 2000* and *Partnering Option X12*, the partnering option supplement to the New Engineering Contract (NEC) suite of publications.

2.14 The Joint Contracts Tribunal (JCT) has also published *Practice Note 4 — Partnering*, which contains a non-binding partnering charter designed for use with its standard forms of contract.

Choosing a Procurement Route

2.15 The procurement route is determined at concept stage by a rigorous risk assessment and precedes the choice of contract strategy.

2.16 Each route requires a complex matrix of contracts, subcontracts and supply agreements to link the various participants into the supply chains needed to ensure effective delivery of the project. There are many forms available, but one of the main forms in widespread use is *JCT 98 — Standard Form of Building Contract 1998 Edition* (available in many variations, including one for Scotland).

Traditional Route

2.17 The employer appoints a team of consultants to prepare a complete design for the project, which is then put out to competitive tender. Usually, the main contractor that submitted the lowest price is awarded the building contract and is then free to employ his or her own subcontractor/supplier network.

2.18 The contractor must then construct, complete and maintain the works to the time, cost and quality standards set out in the contract. The employer "owns" the start and finish dates and the main contractor "owns" the time in between.

2.19 Typical appointments under the Construction (Design and Management) Regulations 2007 are listed below.

- The employer is the client, and must carry out his or her duties under regulations 6, and 8–12.
- Each of the appointed consultants is a designer and must comply with regulation 13.
- The lead consultant, usually the architect, could be appointed by the client as the CDM co-ordinator for the duration of the design stage (regulations 14 and 15 are then directly relevant).
- The main contractor could be appointed by the client as the principal contractor and could also take over the CDM co-ordinator appointment for the construction stage of the project.
- Each of the main contractor's subcontractors is a contractor.

2.20 This arrangement is the default position assumed by the JCT, the main contract drafting body.

Design and Build

2.21 The employer prepares a statement, usually called the "employer's requirements", describing the building needs in terms of function and performance. This is put out to tender. The main contractor submitting the lowest bid is usually awarded the main contract, which includes:

- completing the design
- building the works.

2.22 In practice, many employers use consultants to prepare the employer's requirements, which sometimes include:

- the concept
- feasibility studies
- the outline design.

2.23 Substitutions of new obligations for old ones are mutually agreed, usually done to:

- ensure that the employer has a say in the design of the building

- obtain innovative design solutions from contractors
- expedite local authority approvals.

2.24 The configured appointments under this route are listed below.

- The employer is the client.
- The lead consultant is a designer and acts as the CDM co-ordinator until the time of the appointment of the design and build contractor.
- The design and build contractor is the principal contractor, designer and CDM co-ordinator from the time he or she is awarded the contract until completion and handover.
- Subcontractors are all contractors, and most will also be designers.
- Any novated consultants are designers, and the design and build contractor would benefit from making his or her own competence and resource enquiries, and make new appointments.

Design and Manage

2.25 This route is a variation of the design and build route. Most of the same conditions apply, however. The main contractor provides a design and manage service and he or she is unlikely to be doing any actual building work on site. This may lead to difficulties in achieving compliance with the Construction (Design and Management) Regulations 2007.

2.26 This route is used almost by default, and particularly for the procurement of maintenance. It tends to be used by client organisations, including:

- NHS Trusts
- local authorities
- housing associations.

2.27 Frequently, the responsible departments in the above organisations will issue a series of works or direct orders, which effectively mobilise a number of separate contractors on the same site to paint and decorate, install new mechanical and electrical services, etc. There is no main contractor; the client's organisation provides the design information and manages and co-ordinates the separate contractors.

2.28 Clients often use this route to save money, arguing that the effective work is done by the specialists and that a main contractor is therefore an unnecessary overhead. The clients expect the specialists to:

- provide their own attendances, eg small plant and equipment, scaffolding, etc
- undertake their own risk assessments
- make their own co-operation and co-ordination arrangements.

2.29 The danger for these clients lies in failing to recognise the part they are playing in procuring construction work that complies with the Construction (Design and Management) Regulations 2007. An important point to remember is that the CDM Regulations refer to projects, not contracts, and that there can be one or many contracts in a project.

2.30 Dispensing with the role of principal contractor on the grounds of saving costs is a danger of the design and manage route. The answer is to ensure that:

- contracts are not mistaken for projects
- a competent principal contractor is appointed on every project.

Management Contracting

2.31 The employer appoints consultants to prepare designs up to a certain point.

Following a tender competition based on partial designs, the employer enters into a direct contract with a management contractor. The employer then appoints a series of works or package contractors using competitive tendering to complete the design and construct the works contained in the packages.

2.32 The benefits of this procurement method include:

- time savings
- improved buildability
- open competition for the works packages
- reduced conflict
- ease of accommodating changes
- ease of achieving parallel working.

2.33 The disadvantages include:

- the essential requirement of producing a top-quality brief
- not being able to achieve a price until the last works package is let
- the necessity of having to put a good team together, otherwise the management contractor can default to a postbox operation between the client and the works contractors.

2.34 Appointments under the Construction (Design and Management) Regulations 2007 are listed below.

- The employer is the client and almost certainly a designer.
- The consultants are designers.
- One of the consultants is the CDM co-ordinator until the appointment of the management contractor.
- The management contractor is the principal contractor and a designer (he or she could also be the CDM co-ordinator).
- Works contractors are both contractors and designers.

2.35 There is a strong case for the appointment of an independent CDM co-ordinator to obtain an impartial overview of the designing process.

2.36 This route is best suited to:

- large, complex, heavily-serviced projects where time is of the essence
- projects where the client is experienced in the ways of the construction industry.

2.37 The management contractor will normally deploy his or her own resources on site to provide attendances for all of the works contractors (the cost of these resources is normally recovered through the contractor's fee).

2.38 There is potential for conflict arising from design responsibilities which are split between consultants and specialists. Where this happens, there is little incentive for the management contractor to oppose the works contractor's claims.

Construction Management

2.39 The employer enters into a series of direct appointments with consultants, designers and a construction manager who is paid a fee to co-ordinate the design activities of consultants and works with the consultants and the client to identify appropriate works or trades contracts packages.

2.40 Consultants may either produce a complete design or a partial design for later completion by a specialist contractor.

2.41 This is put out to competitive tender, and the employer then enters into a series of direct contracts with the successful trades contractors. The construction manager co-ordinates and manages the activities of the whole team.

2.42 Appointments falling naturally into place under the Construction (Design and Management) Regulations 2007 are listed below.

- The employer is the client and probably a designer.
- The construction manager is the CDM co-ordinator and principal contractor.
- Consultants are designers.
- Specialist contractors are contractors and designers.
- Any subcontractors may also be contractors and designers.

2.43 The construction manager acts as a consultant and does not undertake any work on site. This can lead to difficulties with the role of principal contractor, not so much from the point of view of competence, but in terms of adequacy of resources. Any of the consultants could be the CDM co-ordinator and there is a strong case for an independent CDM co-ordinator to be appointed in order to give an impartial view of the designing activity.

2.44 The route is best suited to large, complex projects and requires active and informed participation from the client. The claimed benefits are similar to management contracting, and so are the disadvantages, with the addition that it requires active participation from an informed client.

Engineer, Procure, Construct (Turnkey)

2.45 This route is used extensively in procuring process engineering, petrochemical and utilities related construction in the UK. It has a number of contract categories which reflect the range of differing applications that can be made (eg design, build, operate and maintain (DBOM), and design, build, operate and transfer (DBOT)).

2.46 In the UK, this route is often implemented using forms of contracts from the:

- Institution of Civil Engineers (ICE) (ie the engineering and construction contract (ECC))
- International Federation of Consulting Engineers (FIDIC)
- Institution of Chemical Engineers (IChemE).

2.47 The use of service contracts can be applied to most forms of construction, including:

- schools
- roads
- hospitals
- motorways
- other infrastructure projects.

2.48 In most arrangements, the natural configuration under the Construction (Design and Management) Regulations 2007 is similar to the design and build. These are listed below.

- The employer (the service purchaser) is the client, usually a designer and often best placed to be the CDM co-ordinator.
- The service enquiry, unless there is something exceptional about it, comprises design within the meaning given in regulation 2.
- The service-providing main contractor is usually best placed to be the principal contractor and the CDM co-ordinator, and will almost always be a designer.

- The main contractors and specialist contractors will be contractors and designers, and there will usually be a whole range of sub-subcontract and supply agreements below this level, where at least one, and often both, duties will be held.

2.49 Alternative arrangements can include one of the subcontractors undertaking the role of principal contractor.

2.50 Frequently, projects under this route are very large and complex. They involve hazardous processes and need to be brought on stream by section and in phases, or within a short timescale. Multiple ownership of the finished asset is not uncommon. For this reason, a combined client's agent/CDM co-ordinator role can be a sensible, compliant alternative, partly:

- because the identity of the clients may or may not be clear
- to obtain an independent overview of the design process.

2.51 Also, the complex nature of projects under this route may require a level of health and safety management expertise not available within the assembled supply chain and which can only be obtained from outside sources.

Appointing a CDM Co-ordinator

2.52 The CDM co-ordinator plays a key part in the management of health and safety issues at all stages of the construction process.

2.53 The CDM co-ordinator must be appointed as soon as preliminary design has finished in order to help the client fulfil his or her duties. There is no restriction on who is appointed as CDM co-ordinator; the only requirement is that the individual is competent to perform the functions of the CDM co-ordinator.

2.54 The appointment should be kept under review and may be terminated, changed or renewed as required so that there is always a CDM co-ordinator with the necessary competence to advise the client and the main contractor when required.

2.55 The two conditions regarding the services performed by the CDM co-ordinator are that they:

- must be provided
- are provided at the request of other duty holders (the so-called "optional services").

Optional Services

2.56 Certain sections of the CDM Regulations do not impose absolute obligations on the various duty holders to carry out certain functions. Therefore, if they are to be included in the contract arrangements, they must be explicitly stated, particularly where the duties of the CDM co-ordinator are concerned.

2.57 Contracts must clarify the exact responsibilities of the CDM co-ordinator. The following services will almost certainly need to be allocated to the CDM co-ordinator.

- The CDM co-ordinator is only required to ensure that pre-construction information is collated and provided so that contractors can plan and price for safety. The duty to actually compile the information is specified within the regulations as the client's responsibility.
- The client may require the CDM co-ordinator to advise on the competence and resources of designers and contractors (including principal contractors).
- The client is responsible for ensuring that a suitably developed health and safety plan has been prepared by the principal contractor before the beginning of construction work. The client should seek advice on the suitability of the plan from the CDM co-ordinator.

Forms for the Appointment of a CDM co-ordinator

2.58 Standard forms of appointment can be adapted for the purpose of appointing a CDM co-ordinator.

2.59 The document *Form of Appointment as CDM Co-ordinator* (PS/99), published by the Royal Institute of British Architects (RIBA), can be used to appoint suitably qualified construction professionals as CDM co-ordinators.

2.60 It is RIBA's intention that the PS/99 form should be used to appoint a CDM co-ordinator as distinct from the provision of architectural services under the following RIBA forms.

- *Standard Form of Appointment for an Architect* (SFA/99).
- *Conditions of Engagement for the Appointment of an Architect* (CE/99).

Appointing Designers

2.61 The designer has significant responsibilities under the CDM Regulations at all stages of the construction process. Like other parties involved, the duties of designers apply whatever the size or type of construction work.

2.62 There may be several people or design teams undertaking design work at any time for, or on behalf of:

- architects
- the principal contractor
- the engineer
- a design team within the client organisation.

2.63 They are all subject to the CDM Regulations as applicable to designers.

2.64 It is not necessary to interpret the regulations and outline obligations precisely when appointing designers — especially if the purpose is to impose a contractual obligation to comply with all requirements. A detailed breakdown may achieve the opposite purpose if the obligations under the CDM Regulations are determined to be broader than those set out in the contract.

2.65 A general obligation to comply with Acts of Parliament and regulations may be more appropriate, along with a list of certain specific duties stated to be "without prejudice to the generality of the above", or similar wording.

Appointing an Architect

2.66 The supplement issued with the *Standard Form of Appointment for an Architect* (SFA/99) sets out additional definitions, conditions and services relevant to compliance with the CDM Regulations. It also sets out notes on how the supplement is to be used and incorporated into a form of appointment.

2.67 A simplified document, *Conditions of Engagement for the Appointment of an Architect* (CE/99), takes account of the CDM Regulations solely in the context of the obligations of the architect as designer. The references to the regulations assume that a CDM co-ordinator is appointed and impose obligations on the client to comply with his or her duties under the regulations.

2.68 There is no specific contractual obligation for the architect to comply with the CDM Regulations in either SFA/99 or CE/99. Although this may be covered by the architect's general duty to exercise reasonable skill and care, the forms could be amended to accommodate clauses addressing this issue.

Tender Process and Appointing Contractors

2.69 The client's selection and appointment of the principal contractor is a key factor in ensuring that the construction phase of a project is managed successfully. Two health and safety issues to consider are the:

- adoption of pre-qualification procedures which take account of both health and safety and the competence of potential principal contractors
- preparation of pre-construction information.

2.70 The intention is to ensure that prospective principal contractors:

- are competent
- have made adequate provision for health and safety in their tenders
- have included health and safety in their project costings.

2.71 There is no standard procedure that must be adopted in order to take health and safety into account in the tendering stage. The HSE publication and Approved Code of Practice (ACOP) *Managing Health and Safety in Construction* suggests the following best practices.

- Tender documentation sent to principal contractors should be put together in such a way that it will be possible to determine how the principal contractor has planned and priced for health and safety.
- The tender documentation should include pre-construction information.
- The tender period should allow sufficient time for the prospective principal contractors to consider health and safety when preparing their tender submissions.

2.72 Prospective principal contractors can establish that health and safety issues have been adequately addressed by submitting the following with, or as part of, their tender submissions.

- Evidence of their competence, eg a record of experience on other projects.
- A health and safety policy that has already been adopted and which will be applied to all health and safety issues arising in relation to the project.
- Details of arrangements for dealing with matters highlighted in the pre-tender health and safety plan.
- Details of resources to be allocated to deal with identified health and safety risks.

Scope of the Pre-construction Information

2.73 The degree to which the pre-construction information is developed will largely be determined by the complexity of the project. In the most complex projects, eg chemical process plants, and in projects where the risks to health and safety are particularly severe, eg if the site is located next to a mainline railway, very detailed information may be needed.

2.74 The information should identify particular risks and state proposed management measures for control.

2.75 The prospective principal contractors could then be called upon to price the specific items and to suggest alternative approaches which may be beneficial to both health and safety and cost/time.

Tender Assessment

2.76 The client will need to evaluate the tenders submitted, possibly taking the advice of the CDM co-ordinator or other advisors. It may be necessary for clients to discuss the tenders with prospective principal contractors in order to ensure that health and safety has been fully covered.

2.77 Tenders that are significantly cheaper or have shorter completion times may need to be investigated further. The issue of health and safety should be discussed during post-tender interviews.

2.78 The format and extent of information required from the tenderers will depend on the nature of the project, but should be clearly outlined within the tender documents. In order to supply sufficient detail to demonstrate that they have considered the health and safety issues, tenderers may have to carry out more comprehensive enquiries and give deeper consideration to the methods by which they propose to execute the works than has previously been the case.

Standard Forms of Contract

2.79 The major standard forms of contract are issued by the organisations listed below.

- Joint Contract Tribunal: JCT issues the standard form of building contract series for use in building work (as opposed to civil works). There is a range of standard forms to cover the various situations which may arise.
- Institution of Civil Engineers (Conditions of Contract Joint Standing Committee (CCJSC)): ICE incorporated the CDM Regulations into the ICE series of conditions of contract by means of a general contractual obligation to comply with Acts of Parliament and regulations. A particular concern of the CCSJC — the body responsible for this series — was to avoid the possibility of the contractor being able to effectively pay itself for works that it would otherwise not be entitled to by claiming it was fulfilling a duty under the CDM Regulations (this has been achieved by issuing addenda modifying conditions of contract).
- Institution of Chemical Engineers: IChemE publishes three model forms of conditions of contract, referred to as the Red Book (for lump sum contracts), Green Book (for reimbursable contracts) and Yellow Book (for subcontracts). There is also a Purple Book, which is a guide to all of the above. The IChemE model, *Forms of Contract*, looks to cover the CDM Regulations by the existing requirements of the parties to the contract to comply with Acts of Parliament and regulations.

2.80 The standard contracts issued by these bodies already largely cover the CDM Regulations as they contain a general obligation on the part of the contractor to comply with Acts of Parliament and regulations.

Non-standard Documents

2.81 Those companies and organisations that have developed their own form of conditions of contract may desire to amend their standard forms. All such contracts should already contain a general clause requiring compliance with statutes and regulations.

2.82 However, additional clauses could also be included in the contract of those carrying out and managing the work.

Approved Codes of Practice and Contracts

2.83 ACOPs have a special legal status in relation to health and safety matters (guidance notes do not have the same legal status as ACOPs). In the case of a criminal prosecution for breach of health and safety law, if it is proved that an ACOP has not been followed, the party will be guilty of a breach unless it can be shown that the law was complied with in some other way.

2.84 ACOPs can be incorporated into contracts. This can be achieved by a contractual obligation to have due regard to them. In a civil action, as opposed to a criminal action, they may still be relevant even if they were not explicitly incorporated. A failure to comply with an ACOP is evidence of a breach of the relevant regulations unless it can be shown that they were complied with in some other way.

2.85 If a contract provides that failure to comply with the Construction (Design and Management) Regulations 2007 is a breach of contract, failure to follow an ACOP will be strong evidence of a breach of contract. The burden will then be on that party to show that compliance with the regulations had been achieved in some other way.

2.86 The courts may look to the ACOP (and perhaps to guidance issued by the Construction Industry Advisory Committee) in negligence actions in order to establish an objective standard. The parties will then be judged against that standard to determine whether their actions have been negligent.

ACOPs and Other Regulations

2.87 The Construction (Design and Management) Regulations 2007 are concerned with the management of construction projects. The detail for various aspects of the work to be carried out on construction sites will be governed by other, more specific, sets of regulations, including the:

- Work at Height Regulations 2005
- Control of Substances Hazardous to Health Regulations 2002
- Provision and Use of Work Equipment Regulations 1998.

2.88 ACOPs which accompany these regulations will be relevant to construction projects.

Collateral Warranties

2.89 Collateral warranties (or duty of care deeds) are a simple form of agreement commonly used in the construction industry to ensure that a professional has a specific duty of care to parties other than the client, ie the party with whom they are directly in contract and to whom they owe the contractual responsibilities.

2.90 Warranties are normally separate from an appointment and given in favour of:

- purchasers
- tenants
- individuals providing finance for given works.

2.91 Although they are for the benefit of these third parties, their format is normally agreed within, and described by, the documentation which forms the appointment with the client. This implies that warranties are normally sought from designers or principal contractors, but they may also be sought from specialist subcontractors, including:

- designers
- individuals installing mechanical and electrical services, eg heating, ventilation or lifts.

2.92 A simple form of warranty will express a duty of care to the non-client party, but most warranties also include provisions for:

- limitations of liability
- professional indemnity insurance
- other items.

Scheme for Construction Contracts

2.93 The Scheme for Construction Contracts (England and Wales) Regulations 1998 and Scheme for Construction Contracts (Scotland) Regulations 1998 both came into force in 1998. They implement the requirements of the Housing Grants, Construction and Regeneration Act 1996 (the Act).

2.94 The Act is not retrospective and only applies:

- to contracts entered into after 1 May 1998
- where the construction contract is in writing.

Main Provisions of the Housing Grants, Construction and Regeneration Act 1996

2.95 The main provisions of the Act are set out below.

- The Act gives every party to a construction contract the right to refer for adjudication any dispute arising under the contract.
- The Act confers the right to suspend work under certain circumstances.
- Construction contracts include:
  - carrying out construction operations as defined in the Act (which include all of the normal activities one might expect)
  - arranging for construction operations to be carried out by others
  - labour-only operations
  - consultancy agreements, eg design, cost consultancy or management.

2.96 The regulations include provisions dealing with:

- pay when paid
- pay if paid
- unfair contract terms
- interest on late payments
- optional provisions for advance payments
- optional provisions for payment for goods and materials off site
- alternative methods of valuing variations
- suspension of work rights
- adjudication.

2.97 The minimum provisions of the regulations will be implied in all construction contracts even if not expressly written into them. For that reason, all of the major contract drafting bodies are issuing their own versions of the scheme for construction contracts.

Adjudication

2.98 The following provisions have been introduced.

- Parties to a construction contract have a right to refer any dispute or difference to an adjudicator.
- The adjudicator acts as an expert and must act impartially in ascertaining the facts and the law, including any relevant terms and conditions of contract.
- The parties have seven days to agree the appointment of the adjudicator.
- The adjudicator has 28 days in which to give a decision (which he or she does need not to explain unless the parties request it).
- The timescale may be varied by agreement.
- The adjudication runs even if a party fails to comply with any reasonable request made by the adjudicator.

- The adjudicator's decision is binding on the parties until practical completion, when they can resort to arbitration or litigation if they wish.
- The adjudicator can issue instructions requiring action by either party.
- The adjudicator cannot be held liable for the consequences of his or her decisions unless he or she acts in bad faith.
- The adjudicator can decide his or her own payment terms and who pays.

Suspension of Work

2.99 There is no automatic right under English law to suspend work if the other party has not paid what it owes. Where a serious breach does occur, then the basic right is to determine the contract (although some contracts do provide an express right to suspend work in the event of non-payment).

2.100 The Act introduces the right of suspension in all construction contracts. The Act provides that where a sum due under a construction contract is not paid in full by the date for payment, and no effective notice to withhold payment has been given, the person to whom the sum is due has the right to suspend performance of his or her obligations under the contract.

2.101 The right may not be exercised without first giving the defaulting party at least seven days' notice of intention to suspend performance, stating the grounds on which it is intended to suspend performance. The right to suspend performance ceases when the party in default makes payment in full of the amount due.

Service of Notices

2.102 In the absence of any agreement between the parties, a notice or other document may be served by any effective means. It shall be treated as effectively served if it is addressed, pre-paid and delivered by post to:

- the addressee's last known principal residence
- in the case of a person carrying on a trade or profession, his or her last known principal business address
- in the event of a body corporate, the body's registered or principal office.

2.103 A notice, or any documents within the provisions of the Act, includes any requirements for a form of communication in writing.

Professional Services

2.104 Under the Act, the definition of construction contracts includes the professional work of:

- architects
- surveyors
- engineers
- other consultants.

2.105 Therefore, written agreements between clients and their professionals will fall within the Act and will have to contain the relevant provisions for adjudication, instalments or stage payments, set-off and the right to suspend work.

List of Relevant Legislation

- Construction (Design and Management) Regulations 2007
- Work at Height Regulations 2005
- Control of Substances Hazardous to Health Regulations 2002
- Provision and Use of Work Equipment Regulations 1998
- Scheme for Construction Contracts (England and Wales) Regulations 1998
- Scheme for Construction Contracts (Scotland) Regulations 1998
- Housing Grants, Construction and Regeneration Act 1996

Further Information

Publications

HSE Publications

The following is available from *www.hsebooks.co.uk*.

- L144 *Managing Health and Safety in Construction. Construction (Design and Management) Regulations 2007. Approved Code of Practice*

RIBA Publications

The following are available from *www.ribabookshops.com*.

- CE 99 *Conditions of Engagement for the Appointment of an Architect*, Royal Institute of British Architects
- JCT 98 *Standard Form of Building Contract 1998 Edition*, Joint Contracts Tribunal
- *Practice Note 4 (Series 2) — Partnering*, Joint Contracts Tribunal
- *Practice Note 6 (Series 2) — Main Contract Tendering*, Joint Contracts Tribunal
- PS 99 *Form of Appointment as Planning Supervisor*, Royal Institute of British Architects
- SFA 99 *Standard Form of Appointment for an Architect*, Royal Institute of British Architects

Thomas Telford Publications

The following are available from *www.thomastelford.com*.

- *Code of Practice for the Selection of Main Contractors*, Construction Industry Board
- *Code of Practice for the Selection of Subcontractors*, Construction Industry Board

Other Publications

- *Forms of Contract*, Institution of Chemical Engineers
- *NEC Partnering Option X12*, Institution of Civil Engineers
- PPC 2000 *ACA Standard Form of Contract for Project Partnering* (amended 2003), Association of Consultant Architects

Organisations

- Association of Consultant Architects (ACA)
  Web: *www.acarchitects.co.uk*
  ACA is the national professional body representing architects in private practice (consultant architects) throughout the UK.
- Construction Industry Advisory Committee (CONIAC)
  Web: *www.hse.gov.uk/aboutus/hsc/iacs/coniac/index.htm*
  CONIAC was reconstituted at the beginning of 2004, with a smaller membership, and a new approach, reflecting the desire of the Health and Safety Commission (HSC) to work in partnership with intermediary groups in the construction industry and to take the new HSC strategy forward.
- Construction Industry Training Board (CITB) and Construction Skills
  Web: *www.citb.org.uk*
  The CITB and Construction Skills provide assistance in all aspects of recruiting, training and qualifying the construction workforce.
- Health and Safety Executive (HSE)
  Web: *www.hse.gov.uk*
  The HSE is responsible for the regulation of almost all the risks to health and safety arising from work activity in the UK.
- Institution of Chemical Engineers (IChemE)
  Web: *www.icheme.org*
  The Institution of Chemical Engineers is the professional body for chemical and process engineers.
- Institution of Civil Engineers (ICE)
  Web: *www.ice.org.uk*
  The ICE is a charity that exists to promote and progress civil engineering.
- International Federation of Consulting Engineers (FIDIC)
  Web: *www.fidic.org*
  FIDIC's mission is to improve the business climate and promote the interests of consulting engineering firms, globally and locally, consistent with the responsibility to provide quality services for the benefit of society and the environment.
- Joint Contracts Tribunal (JCT)
  Web: *www.jctltd.co.uk*
  The JCT was established in 1931 and has since produced standard forms of contracts, guidance notes and other standard documentation for use in the construction industry.
- Royal Institute of British Architects (RIBA)
  Web: *www.riba.org*
  The RIBA is a member organisation promoting architecture and architects.

Chapter 3

Construction (Design and Management) Regulations 2007

Construction (Design and Management) Regulations 2007

- The Construction (Design and Management) Regulations 2007 (CDM) contain specific requirements to ensure the health and safety of those involved in construction work.
- All construction work falls under CDM, it is however a two-tier system and a project can be classed as notifiable or non-notifiable.
- The CDM Regulations set out roles and responsibilities for the various parties involved in construction activities: clients, designers, CDM co-ordinators, principal contractors and contractors.
- The client is any individual, partnership or organisation for whom a project is carried out.
- Designers are those who prepare designs (drawings, design details, specifications, etc) in relation to the structure.
- The CDM co-ordinator is an in-house or externally commissioned individual or organisation that has the knowledge and experience to make competent assessments of the design proposals and how they affect health and safety.
- The principal contractor is an experienced contractor who will have the responsibility for managing the construction works on site and has ultimate responsibility for any other contractors that are involved.
- The Health and Safety Executive enforces the duties and responsibilities of clients, designers and CDM co-ordinators.

3.1 There are many variables that make the construction industry one of the most difficult in terms of health and safety. Clients can range from a shopkeeper to a government department; sites can be a green field or a former chemical works; the type of work can be a small roof repair or the construction of a power station. Sites and workers on them change from day to day and often there are pressures imposed to complete the project to a tight deadline. In all of these circumstances, good management of construction projects is needed from conception through to execution and completion.

3.2 Construction work in all its forms involves some risk. Many of the deaths each year occur during maintenance and repair work. The Construction (Design and Management) Regulations 2007 apply to the widest definition of construction work. They are not restricted to projects for the construction of new buildings or structures of those that involve entering into contracts with construction companies but can apply, for example, to the "in-house" maintenance department.

3.3 The CDM Regulations place specific duties upon clients, designers and contractors to rethink their approach to health and safety. It should be taken into account and then co-ordinated and managed effectively throughout all stages of a construction project. This means from conception, design and planning through to the execution of works on site and subsequent maintenance and repair, and even to final demolition and removal.

In Practice

Application of CDM

3.4 The CDM Regulations apply following the initial design stage of the project. All construction work falls under CDM, it is however a two-tier system and a project can be classed as notifiable or non-notifiable. Until a project is confirmed the regulations have no formal effect, but when a decision is taken to build a particular structure, the designer must comply with the requirements of Regulation 11 and 18 if the project is notifiable.

3.5 The first question to ask is whether the work falls within the definition of construction work. This is defined as the carrying out of any building, civil engineering or engineering construction work and includes:

- construction
- alteration
- conversions
- fitting out
- commissioning and decommissioning
- renovations
- repairs
- maintenance (including high-pressure cleaning)
- redecoration
- demolition/dismantling
- site clearance prior to and at the end of construction work
- assembly/dismantling of prefabricated units
- installation, maintenance and removal of services associated with a structure.

3.6 A structure is defined as "any building, timber, masonry, metal or reinforced concrete structure, railway line or siding, tramway line, dock, harbour, inland navigation, tunnel, shaft, bridge, viaduct, waterworks, reservoir, pipe or pipe-line, cable, aqueduct, sewer, sewage works, gasholder, road, airfield, sea defence works, river works, drainage works, earthworks, lagoon, dam, wall, caisson, mast, tower, pylon, underground tank, earth retaining structure or structure designed to preserve or alter any natural feature, fixed plant and any structure similar to the foregoing; or any formwork, falsework, scaffold or other structure designed or used to provide support or means of access during construction work". Any reference to a structure includes a part of a structure.

3.7 Having confirmed that the work falls within the definition, it must then be established if the works are notifiable. The project is notifiable to the HSE if the construction work:

- is expected to last more than 30 working days
- involves work of shorter duration but is expected to involve more than 500 person days, eg 50 workers on site for 10 days or more.

3.8 If either of the above criteria is met, then the project is notifiable and increased duties apply to all parties.

3.9 The CDM Regulations refer to demolition or dismantling work. This is defined as the "removal of a structure or part of a structure or of any product or waste resulting from demolition or dismantling of a structure". The stripping of cladding, removal of roof tiles and other such operations are not deemed to come under the requirements of CDM (unless of course combined with other construction work).

3.10 Demolition work requires prior planning and carrying out in a safe manner. A written safe method of working is required prior to work commencing. The same criteria for notification apply to demolition work as other construction work.

Roles and Responsibilities for Those Involved in Construction Work

3.11 The Construction (Design and Management) Regulations 2007 set out roles and responsibilities for the following parties (duty holders) during construction activities:

- clients
- designers
- CDM co-ordinators
- principal contractors.

Clients

3.12 A client is defined as a person who, "in the course or furtherance of a business, seeks or accepts the services of another to carry out a project or carries out a project themselves".

Delegation

3.13 Clients can no longer delegate the duties imposed to an agent under the Construction (Design and Management) Regulations 2007. However, those appointed prior to 6 April 2007 can continue in this role for a period of up to five years or when the client terminates the appointment.

Lead Client

3.14 When several clients co-operate to commission a project, eg joint parking facilities for several businesses, they must agree on one lead client. Companies will often appoint one manager to represent them in dealing with a proposed project. The company will still be the client, not the named manager.

Clients Acting as Designers

3.15 The definition of design is broad and includes the specification of equipment or materials. Clients must therefore consider whether or not they are designers and subject to the designer's duties too.

Clients' Responsibilities

3.16 Clients have several responsibilities on notifiable projects, which are outlined below.

Appoint a Competent CDM Co-ordinator

3.17 The concept of the CDM co-ordinator was introduced in 2007 and therefore there are no organisations or individuals who can demonstrate a long history of acting in this capacity. There are, however, many individuals and organisations who can demonstrate their competence to be CDM co-ordinators by reference to their past experience in acting as planning supervisors within the context of the previous Construction (Design and Management) Regulations 1994.

3.18 Competent CDM co-ordinators will have experience and professional qualifications relevant to the proposed construction work. The CDM co-ordinator should be appointed following initial design. Although no precise definition of this has been given by the Health and Safety Executive, some common approaches include appointment of the CDM co-ordinator once:

- outline planning permission has been granted by the local authority
- budgetary provision has been agreed by the board of directors or finance committee
- a concept has received board approval and is to move forward to detailed design.

Provide the CDM Co-ordinator with Relevant Health and Safety Information

3.19 Relevant health and safety information may include drawings of existing structures/services or details of known areas of contaminated land affected by the proposed construction work. For existing premises that are in use by the client, it will include relevant details of the documented risk assessments that have been carried out.

Appoint Competent Designers with Sufficient Resources for Their Tasks

3.20 Clients may either decide to appoint the designers themselves, or seek the advice of the CDM co-ordinator.

Appoint a Competent Principal Contractor with Sufficient Resources

3.21 Clients may either decide to appoint the principal contractor themselves, or seek the advice of the planning supervisor.

Ensure that a Suitably Developed Construction Phase Health and Safety Plan is Prepared

3.22 The client does not prepare the health and safety plan (the CDM co-ordinator initiates the plan and the principal contractor develops it). However, the client can ensure that the plan precedes construction by refusing to allow the principal contractor construction access, or possession of the construction site, until a sufficiently detailed plan has been prepared and presented to the client.

3.23 It is now a requirement of the regulations to allow the principal contractor sufficient time to develop the plan – this information must be recorded on the notification to the HSE.

3.24 The client may ask for assistance from the CDM co-ordinator to assess whether the health and safety plan submitted by the principal contractor is suitably developed. This will be especially important in the event of a subsequent accident on site, as the client is open to civil proceedings for failing to ensure the existence of the health and safety plan.

Ensure that a Health and Safety File is Kept Available

3.25 On completion of the project, the client receives a health and safety file from the CDM co-ordinator, which contains "as-built" details and details of residual risks. The client then has a duty to keep this information available in the event of any future work on the structure.

3.26 There is no specific obligation to update the file, except for subsequent works which are applicable themselves under the Construction (Design and Management) Regulations 2007. However, it is recommended that the file is kept up to date, including minor maintenance alterations or small additional works, eg the provision of extra power points within an office. If the file is not kept up to date, the information contained within it may lead to a false sense of safety, which in turn may lead to accidents, eg removing a partition wall which contains additional electric wiring not recorded in the file.

3.27 Clients who frequently commission construction projects, eg large public companies or government departments, will receive many health and safety files. The storage, updating and retrieval of relevant data for future construction work, eg office refurbishments, will be a major task and will require a review of existing procedures to ensure that clients can fulfil their future duties.

In-house Appointments

3.28 If the client makes any internal appointments, then additional documentation will be necessary to demonstrate that the roles of designers, CDM co-ordinators, principal contractors or contractors have been properly assessed. The client should also be able to demonstrate independence between duty holder roles filled internally.

Designers

3.29 Designers are those who prepare or modify designs. Designs can include drawings, design details, specifications and bill of quantities (including specification of articles or substances) in relation to the structure.

3.30 Designers may therefore include:

- civil, structural, mechanical and electrical engineers and technicians
- architects and architectural technicians
- building services engineers
- building surveyors
- quantity surveyors
- project managers
- specifiers of materials or equipment (including clients).

3.31 The number and variety of designers employed on a project will depend upon the size and complexity of the works.

Competence

3.32 Clients may seek the advice of the CDM co-ordinator when assessing designers' competence. In order to demonstrate competence, it will be necessary for designers to respond to written enquiries. Designers should therefore prepare marketing information formulated specifically to demonstrate competence in a given discipline or area of work.

3.33 The designers will also be required to show that they have sufficient resources available to meet the requirements of the project. Technical and administrative back-up may be assessed, eg computer-aided design facilities and access to health and safety library information, to prove the overall capability of a design organisation.

Designers' Responsibilities

3.34 Designers' responsibilities apply to all designers, irrespective of the size, value or complexity of the work or the nature of the client.

Advise Clients of Their Duties

3.35 Designers are required to notify the client of the duties that the Construction

(Design and Management) Regulations 2007 place upon clients and the official guidance that is available on the regulations.

Consider Health and Safety in Designs

3.36 Designers are required to consider health and safety during the construction, future maintenance and final demolition of the structure. Consideration of future access requirements, eg to plant and equipment on a roof, is therefore essential. Designers are required to:

- identify the hazards that will occur during construction and maintenance
- eliminate risks where possible
- reduce the risks
- provide adequate information on the risks that cannot be eliminated.

3.37 Designers contribute key information to the pre-construction information and the health and safety file via the CDM co-ordinator. This can be done by discussion and by the designers providing the CDM co-ordinator with documentation summarising the designers' input.

CDM Co-ordinators

3.38 The CDM co-ordinator for a project can be:

- an individual
- an organisation
- a joint venture.

3.39 There is no single correct appointment procedure. Client organisations need to consider the nature of the proposed project and the skills, experience and resources required for taking on the role.

3.40 An individual can be the CDM co-ordinator provided that the person can show that he or she:

- has the time and availability to cope with the workload
- has the knowledge and experience to be able to make competent assessments of the design proposals and how they affect health and safety.

3.41 Many different types of organisation offer the services of CDM co-ordinator, including:

- engineers
- architects
- building surveyors
- quantity surveyors
- project managers
- civil engineering or building contractors.

3.42 These could be in-house or commissioned externally. Client organisations may decide to appoint in-house CDM co-ordinators for maintenance works and smaller contracts, such as conversions or refurbishment. In-house appointments may be particularly appropriate where the health and safety issues associated with an existing site are complicated and relatively high risk, such as chemical works or large manufacturing sites, and in-house specialists are knowledgeable of the ongoing processes.

3.43 The appointment of in-house CDM co-ordinators may offer advantages in saving administration time and costs, but raises questions over whether they have the competency and indeed the independency to perform this additional role and the time on top of existing duties.

3.44 Externally commissioned CDM co-ordinators should be able to offer specialist expertise and experience. This may be particularly necessary for larger works such as civil engineering projects, new building projects involving excavation, etc. Where the works are technically complex or high risk, the appropriate expertise and experience will be paramount.

Competence

3.45 When choosing a CDM co-ordinator, a client who does not commission construction work on a regular basis may be faced with a wide and confusing choice of organisations offering to act as CDM co-ordinator. The client should choose a competent organisation with adequate resources.

The CDM Co-ordinator's Responsibilities

3.46 The CDM co-ordinator is appointed directly by the client. The responsibilities of the CDM co-ordinator are as follows.

Notification of the Project to the Health and Safety Executive

3.47 Notification may be a two-stage process, comprising:

- stage one, at the start of the detailed design stage
- stage two, once a principal contractor has been appointed.

3.48 The Health and Safety Executive have produced a standard notification form F10(rev), which is available from Health and Safety Executive offices. It is also reproduced within the Approved Code of Practice (ACOP) *A Guide to Managing Health and Safety in Construction*. The form does not have to be used provided the required information as laid out in schedule 1 of the ACOP is sent to the HSE in writing.

Advise the Client on the Competence and Resources of Designers

3.49 Clients who do not regularly engage designers will probably rely heavily on the CDM co-ordinator for advice and reassurance regarding suitable designers for a particular project. However, clients who regularly use the same design team and therefore already trust their competence and resources may not require assistance.

Ensure that Designers Fulfil Their Duties

3.50 Designers are required to assess the risks contained within the proposed design. This includes risks both during construction and subsequent maintenance and demolition. It is not the CDM co-ordinator's role to carry out these risk assessments for designers, but to ensure that the designers carry out adequate risk assessments. The CDM co-ordinator must ensure that the results of these risk assessments are then incorporated into the design and the pre-construction information. CDM co-ordinators will therefore need to be satisfied that designers have:

- recognised significant risks
- eliminated foreseeable risks where reasonably practicable
- reduced risks where elimination is not practicable
- clearly identified significant remaining risks so that the CDM co-ordinator can ensure that these are included in the pre-tender information.

3.51 If the design continues after the appointment of a principal contractor or into the construction phase of a project, the CDM co-ordinator continues to be involved in checking designers' competence and provision for health and safety. Examples of this include design and build contracts, or a specialist mechanical engineering subcontractor designing and installing plant and equipment, eg heating and ventilation systems.

Ensure that Designers Co-operate with Each Other and with the CDM Co-ordinator

3.52 The CDM co-ordinator will probably attend design team meetings and may hold separate meetings with each design group in order to discuss progress and to be assured that the designers are fulfilling their duties. There is clearly an opportunity for conflict, eg the designer and CDM co-ordinator may disagree over the proposed means of minimising a hazard. Such a conflict could arise, for example, over whether to provide stairs or a ladder to give access to a flat roof with plant on it.

3.53 Ultimately, such a dispute must be resolved by a final decision from the client. However, multi-disciplinary design teams are now common and the integration of the CDM co-ordinator into such teams is another challenge for construction professionals to meet. Designers (or CDM co-ordinators) who do not adopt a constructive professional attitude and a co-operative team approach will quickly lose credibility and clients.

Ensure that Pre-construction Information is Prepared

3.54 The pre-construction information contains relevant health and safety information which tenderers will need in order to price for safety and plan for safe working. Once a principal contractor is appointed, they take over responsibility and develop the information into safe systems of work.

3.55 It is the CDM co-ordinator's responsibility to ensure that the pre-construction information is prepared and included in the tender documents. It is not the responsibility of the CDM co-ordinator to provide the information. However, they will need to collate the information from different sources. The CDM co-ordinator will require information from the client (eg site rules and access details) and designers (eg risk assessments and significant design hazards).

3.56 The contents of the pre-construction information are therefore the results of co-operation between the client, CDM co-ordinator and designers. Guidance on the information provided can be found within Appendix 2 of the ACOP.

Advise the Client on the Competence and Resources of Contractors

3.57 Clients who usually negotiate a contract with a preferred contractor and clients who already have standing lists of approved contractors may not require the help of the CDM co-ordinator in this respect. However, clients who do not regularly commission construction work may be unsure about suitable contractors and may rely heavily on the CDM co-ordinator for advice.

Advise the Client on the Construction Phase Health and Safety Plan and Welfare Facilities

3.58 Once a principal contractor has been appointed, it is the client's responsibility to ensure that a suitable construction phase health and safety plan is prepared by the principal contractor and is ready before construction begins. The CDM co-ordinator may be asked by the client to check the adequacy of the plan before the client authorises construction work to start. They may also be asked to check the adequacy of welfare facilities on site.

3.59 On fast-track projects, where the design is developed while the construction work continues, the whole plan will not be completed before construction work begins. However the initial phases of the plan, which affect the work that is taking place, must be sufficiently developed, ie the risks assessed and necessary control measures specified. For example, the construction phase health and safety plan for a new office building may only specify safe systems of work for operations involving excavations and building the foundations, but not for work on the outfitting of the upper floors, as the design detail is not yet available.

Ensure that a Health and Safety File is Prepared and Delivered to the Client

3.60 The principal contractor is responsible for the control and co-ordination of health and safety on site. The CDM co-ordinator still has an involvement, however, including the co-ordination and preparation of the health and safety file and ensuring that it is delivered to the client on completion of the works. The CDM co-ordinator is dependent on others for information for the file, eg "as-built" drawings, "as-used" materials and "as-installed" equipment details. It will therefore be necessary for the CDM co-ordinator to maintain a dialogue with the principal contractor and with designers who are:

- adjusting their specification, by making alterations to take account of unforeseen site circumstances, eg the relocation of a live service
- adding design detail, eg where the pre-contract information is in the form of a performance specification for mechanical services — eg lifts to be able to carry certain numbers of people — and where the actual design will be carried out by the lift engineering designers.

3.61 The CDM co-ordinator may wish to attend site progress meetings or to make specific arrangements to ensure that a flow of information for the health and safety file continues throughout the project. These arrangements should be made in a "continuing liaison" section within the agreed construction phase health and safety plan.

Principal Contractors

3.62 The principal contractor is an experienced and capable contractor. A contractor is anyone who conducts, manages or employs people to conduct construction work. This means that the principal contractor could be any of the following:

- a building or civil engineering contractor
- an in-house building/maintenance department
- a management contractor
- in-house or consultant project managers
- in-house or consultant facilities managers.

3.63 The most important factors in the appointment of the principal contractor are competence and resources, as appropriate for the proposed project. If, eg, a client wishes to use its own in-house project management team as principal contractor on an office refurbishment, the client must be certain and be able to demonstrate that the in-house project management team have both the competence (in terms of qualifications, experience and understanding) and the necessary resources (people, time, procedures and technical facilities) to be principal contractor.

3.64 Some clients use in-house organisations as principal contractor only for works up to a defined value, and external contractors above this value. However, there can only be one principal contractor on a project at any given time.

Competence

3.65 It should be relatively simple for experienced works contractors to demonstrate their competence as principal contractors. In-house departments may have more difficulty but the same approach should be used to demonstrate competence. Experience and a proven track record in the relevant type of construction are required, together with a sound management organisation.

Transitional Provisions

3.66 If a planning supervisor or principal contractor appointed by a client under the Construction (Design and Management) Regulations 1994 is appointed as the CDM co-ordinator under the Construction (Design and Management) Regulations 2007, then the client must have taken reasonable steps to ensure that the CDM co-ordinator and principal contractor are competent under the new regulations within 12 months of CDM 2007 being implemented, ie 7 April 2008.

3.67 Unless there is an express appointment by the client then existing planning supervisors and/or principal contractors appointed under the CDM Regulations shall be treated as though appointed as a competent CDM co-ordinator and/or the principal contractor respectively.

3.68 If such a person is treated as being appointed as CDM co-ordinator and/or principal contractor then they will need to have taken steps to ensure that they are competent within 12 months.

The Principal Contractor's Responsibilities

3.69 As employers, contractors have the prime responsibility for ensuring that safe systems of work are provided for their employees. The main duties for the principal contractor are outlined below.

Plan and Price for Safety at the Tender Stage

3.70 There is an implied duty for tenderers to take into account the information contained within the pre-construction information when putting together their tender bid. The purpose of the pre-construction information is to allow contractors to plan and price for health and safety in order to create the safe systems of work they will use on site.

3.71 Pricing specific safety-related bill of quantity items as "included" should be discouraged. For example, the bill of quantity may require specific prices for safety-related issues, eg the control and co-ordination of subcontractors' input to the construction phase plan. Clearly the client is interested in the importance (cost) that the principal contractor attaches to this activity. Pricing this as "included" may indicate to the client that the contractor is not giving adequate regard to safety issues.

Develop the Health and Safety Plan

3.72 The pre-construction information contains relevant information that will help the principal contractor decide the approach to the work. Once appointed, the principal contractor must use this information to develop proposed working methods to provide safe systems of work, as contained within method statements and detailed project management plans and arrangements. These will include details of the time allowed for the work, numbers of workers and types of equipment, safe access arrangements, supervision arrangements, emergency plans, etc.

Ensure that Subcontractors are Competent and Properly Co-ordinated

3.73 The principal contractor has responsibility for vetting the competence of those contractors it employs.

Provide Information from the Health and Safety Plan to Contractors

3.74 The principal contractor must provide relevant information from the plan to domestic contractors (ie those chosen directly by the principal contractor, unlike nominated contractors who are specified by the client or designers), to allow them to plan and price for safety, and to implement safe working procedures.

3.75 The principal contractor should work with other contractors to identify the hazards and assess the risks related to their work, including the risks they may create for others. Using this information and applying the general principles of prevention, the principal contractor, in discussion with the contractors involved, must plan, manage and co-ordinate the construction phase. This includes supervising and monitoring work to ensure that it is done safely and that it is safe for new activities to begin.

3.76 Even if the principal contractor has no direct contractual control over other contractors, the principal contractor can give reasonable direction to contractors in order to fulfil the duties of principal contractor.

Ensure that all Workers are Properly Trained and Attend an Induction

3.77 Site induction, training and information are vital to securing health and safety on site. The principal contractor has to ensure, so far as is reasonably practicable, that every worker has a suitable induction and any further information and training needed for the particular work. This does not mean that the principal contractor has to train everyone on the site — this will be the responsibility of individual contractors. The principal contractor must also ensure, so far as is reasonably practicable, that contractors are fulfilling their responsibilities to train their own employees. This could include requiring proof that contractors' employees who operate powered plant have the necessary training and certification, eg dumper drivers.

Enforce Site Rules and Site Access Arrangements

3.78 Principal contractors have to ensure that site rules are communicated, understood and enforced. Site rules must be written down, included within the construction phase health and safety plan and included in subcontract documentation. The rules may be included in site-specific induction training for all staff employed on a project.

3.79 Principal contractors are also responsible for controlling site access and limiting this to authorised persons only. On an existing industrial or commercial site, eg a factory extension or partial refurbishment of a multi-storey office, the control of site access will have to be agreed with the client.

Ensure that Notifications are Displayed On-site

3.80 It is the duty of the principal contractor to make sure that the notification of the project — setting out the names of the client, CDM co-ordinator and principal contractor, along with various project details — is displayed at the site for anyone to read.

Provide Information for the Health and Safety File

3.81 Irrespective of who actually prepares "as-built" drawings, the information to allow such drawings to be completed will come from the principal contractor. Similar information on materials used and types of mechanical or electrical plant should also be given by the principal contractor to the CDM co-ordinator for inclusion in the health and safety file.

3.82 It should also be noted that other contractors also have duties under the CDM Regulations. These are to:

- co-operate with the principal contractor
- provide the principal contractor with information which might affect health and safety
- comply with directions from the principal contractor

- comply with rules that are in the health and safety plan
- provide the principal contractor with information on RIDDOR (Reporting of Injuries, Diseases and Dangerous Occurrences Regulations 1995) accidents, dangerous occurrences and diseases
- ensure that employees and the self-employed are provided with the names of the CDM co-ordinator and principal contractor as well as the health and safety plan contents.

Enforcement of Construction Duties

3.83 The Health and Safety Executive (HSE) may ask to see the construction phase health and safety plan, and then inspect the site to ensure that the plan is being properly implemented.

3.84 The HSE enforces the duties and responsibilities of clients, designers and CDM co-ordinators. It is therefore possible for any of these organisations to be visited by the HSE following an accident, or simply to ensure that duties are being implemented. This may involve visits to design offices to inspect the preparation of drawings and tender documents, together with documents showing the risk assessments that designers have prepared for the project.

3.85 While the HSE provides advice through publications, meetings and to telephone callers, it must be stressed that inspectors are not required to perform risk assessments on behalf of any parties involved in construction.

3.86 There is no formal approval or certification of the construction phase health and safety plan or health and safety file by any statutory body.

Training

3.87

- The principal contractor, as an employer, has a clear and well-established duty to ensure its own employees are properly trained.
- Rules and access arrangements may be included in site-specific induction training for all staff employed on a project.

List of Relevant Legislation

- Construction (Design and Management) Regulations 2007
- Control of Asbestos Regulations 2006
- Work at Height Regulations 2005
- Building Regulations 2000
- Management of the Health and Safety at Work Regulations 1999
- Reporting of Injuries, Diseases and Dangerous Occurrences Regulations 1995
- Workplace (Health, Safety and Welfare) Regulations 1992
- Health and Safety at Work, etc Act 1974

Further Information

Publications

HSE Publications

The following are available from *www.hsebooks.co.uk*.

- HSG150 (3rd ed, 2006) *Health and Safety in Construction*
- INDG259 (rev 2003) *An Introduction to Health and Safety — Health and Safety in Small Businesses*
- INDG401 *The Work at Height Regulations 2005 (as amended): a Brief Guide*
- L21 (rev 2000) *Management of Health and Safety at Work. Management of Health and Safety at Work Regulations 1999. Approved Code of Practice and Guidance*
- L24 *Workplace (Health, Safety and Welfare) Regulations 1992 — Approved Code of Practice*
- L73 (rev 1999) *A Guide to the Reporting of Injuries, Diseases and Dangerous Occurrences Regulations 1995*
- L87 (rev 1996) *Safety Representatives and Safety Committees*
- L95 *A Guide to the Health and Safety (Consultation with Employees) Regulations 1996*
- L143 *Work with Materials Containing Asbestos. Control of Asbestos Regulations 2006: Approved Code of Practice and Guidance*
- L144 *Managing Health and Safety in Construction. Construction (Design and Management) Regulations 2007. Approved Code of Practice*

Organisations

- Association for Project Safety
  Web: *www.associationforprojectsafety.co.uk*
  The Association for Project Safety (APS) is a multi-disciplinary membership body for those who operate in the field of construction health and safety risk management.
- Construction Industry Training Board (CITB) and Construction Skills
  Web: *www.citb.org.uk*
  The CITB and Construction Skills provide assistance in all aspects of recruiting, training and qualifying the construction workforce.
- Health and Safety Executive (HSE)
  Web: *www.hse.gov.uk*
  The HSE is responsible for the regulation of almost all the risks to health and safety arising from work activity in the UK.

Designing for Safety

- The Construction (Design and Management) Regulations 2007 (CDM Regulations) impose legal requirements on designers to ensure the health and safety of individuals on site.
- Designers must form an approximate evaluation of risk in order to comply with the requirements under the CDM Regulations.
- Designers are required to pass on any relevant health and safety information to the CDM co-ordinator for incorporation into the construction phase plan and health and safety file.
- The best time to eliminate hazards from a construction project is during the initial design phases of feasibility and early concept design.
- Under the CDM Regulations, a designer must identify risks associated with construction work and, if possible, modify or alter the design in order to avoid the risk completely.

3.88 In the construction industry, the design element is frequently isolated from the actual fabrication process. Generally, architects do not:

- get involved with the actual construction
- tell the contractor how the construction should be carried out.

3.89 There is concern that such instructions could result in claims against the professional indemnity policy held by the practice. It is also thought that because contractors are the experts in construction methods/techniques, designers should not dictate construction methods. The concern is that this might adversely affect the contractor's freedom to use his or her expertise to the best advantage.

3.90 In some design areas, however, designers have been more involved in the fabrication process.

- Civil engineers, because of the nature of their design work, have had to consider how their projects are going to be built. Designing a complicated structure, eg a bridge, requires the engineer to decide, first of all, how the bridge is going to be constructed.
- Many other designers working in construction for specialist contractors, eg suppliers of window and roofing systems, have first-hand experience of the construction process.
- Architects working for design and build companies — which execute all stages of a project from conception to final completion — will be familiar with construction methods. It can be argued that an essential part of the practice of architecture is to understand how the designed work can be assembled on the construction site.

3.91 The CDM Regulations address an issue that is unique to the construction industry, ie that it is not under unified management control. This means that:

- the fabrication and assembly of construction work is carried out by organisations which are not responsible for the design
- the design itself may be produced by different design practices
- many major building contractors no longer employ large numbers of people, ie they employ key management staff but much of the construction work is carried out by subcontractors.

3.92 Prior to the introduction of the original CDM Regulations (in 1994), no single employer managed the construction and, consequently, no one held responsibility for the overall health and safety in what has proved to be a high-risk industry prone to accidents. The regulations attempt to address these problems by placing duties on designers to consider construction health and safety issues when designing construction work.

3.93 The duties of the designers are overseen by a CDM co-ordinator who co-ordinates health and safety considerations during the design phase of a project. The role of principal contractor was created to co-ordinate the health and safety of all individuals on a construction site, regardless of who employs them.

Designer's Duties

3.94 Under the CDM Regulations, designers of construction work must include adequate information about any aspect of the project which might affect the health and safety of anybody involved in the works. This will normally be achieved by passing information to the CDM co-ordinator for incorporation in the health and safety plan and the health and safety file.

3.95 Designers are also required to co-operate with the CDM co-ordinator and any other designer concerned with the project so that the overall objective of improving health and safety can be achieved.

3.96 Designers also have further duties under other legislation, including those parts of the Management of Health and Safety at Work (MHSWR) Regulations 1999 which require risk assessment. Compliance with regulation 11 of the CDM Regulations will usually be sufficient for designers to achieve compliance with regulations 3(1), (2) and (6) of the MHSWR Regulations, as they relate to the design of the structure.

In Practice

Overview

3.97 Anyone conducting an undertaking under the Health and Safety at Work, etc Act 1974 is required not to expose others to risks to their health and safety. For designers, this implies a duty to ensure that their designs do not expose individuals on site to risks.

3.98 However, the Construction (Design and Management) Regulations 2007 go even further and impose explicit legal requirements on designers for ensuring the health and safety of individuals on site.

Definition of Design

3.99 The definition of "design" under the CDM Regulations is wide-ranging and includes:

- drawing
- design details
- specification
- bill of quantities (including specification of articles or substances).

3.100 The definition of "construction work" and "structure" within the regulations is all-embracing, under which a designer may be:

- an architect
- an engineer, eg civil, electrical, services, structural
- a quantity surveyor
- a building surveyor
- any designer working for contracting firms or specialist suppliers, eg window and roofing systems
- a client, eg if the designer specifies materials.

3.101 Contractors may be designing all, or part of, proposed permanent and/or temporary works. All are designers and therefore subject to the regulations.

Design Considerations under the CDM Regulations

Health and Safety Status

3.102 Health and safety must be among the many design considerations, which means that the designer is not expected to deal with health and safety matters to the exclusion of all other considerations, including cost, appearance and performance.

3.103 Measures to improve health and safety should only be incorporated if they do not significantly increase costs or result in modifying performance or appearance in unacceptable ways. The need for implementing such measures must be balanced against the risk in cases where health and safety improvements will result in significant cost or modifications.

Adequate Regard

3.104 Designers must have adequate regard to health and safety matters under the CDM Regulations, eg identifying and documenting significant risks and dealing with them.

Future Maintenance and Cleaning at Any Time

3.105 Under the CDM Regulations, the designer is required to consider how:

- foreseeable risks may be avoided and combatted at source
- protection can be given to all persons, rather than just the individual, at any time.

3.106 Health and safety considerations must be given not only to the initial construction but to construction work that might take place on the structure at any future date. Therefore, the designer is required to consider hazards that might occur:

- during the maintenance of the building, eg access to plant rooms
- when any refurbishment or remodelling takes place, eg when insulation materials may be exposed
- at an eventual date, when the structure is demolished or the component has reached the end of its useful life.

3.107 The above requirements are very onerous. The method for dealing with work which will be carried out after the initial structure has been completed is set out in the regulations and depends upon the proper completion of a health and safety file.

3.108 For any future construction work where the CDM Regulations apply — particularly demolition — the client is required to appoint a CDM co-ordinator and a principal contractor. The duties of the client will apply to that later work, ie information relevant to health and safety must be passed along to the CDM co-ordinator and, subsequently, to the principal contractor who has been appointed for the new work. This includes the:

- health and safety file as held by the client (if the structure has been created since the introduction of the CDM Regulations)
- designer's information in the health and safety file (this information may also come under new scrutiny and its validity tested as it is used in the new plans).

Designing for Future Cleaning and Maintenance Work

3.109 The Construction (Design and Management) Regulations 2007 define cleaning and maintenance work as:

- cleaning any window or any transparent or translucent wall, ceiling or roof in or on a structure
- maintaining the permanent fixtures and fittings of a structure.

3.110 The designer is required under the regulations to avoid foreseeable risks in cleaning and maintenance work by:

- considering how the windows, roof lights and ceiling lights in a building are to be cleaned
- making adequate provision for the health and safety of individuals who will be involved in conducting the cleaning and/or maintenance work.

3.111 This requirement applies to both the inside and outside of any glazing. The performance of window cleaning itself is not subject to the CDM Regulations but the client and the employer of the window cleaners will be subject to the Management of Health and Safety at Work Regulations 1999.

3.112 A designer might be at risk from claims for personal injury if adequate regard has not been given as to how cleaning could be done. Therefore, the designer should agree with the client on which procedures should be adopted in order to provide safe access to the windows.

3.113 Under the CDM Regulations, some specialised cleaning is included in the definition of "construction work", ie cleaning which involves the use of high pressure water, grit blasting or any substance classified as toxic or corrosive under the Carriage of Dangerous Goods and Use of Transportable Pressure Equipment Regulations 2004.

Minimising Risk (Reasonably Practicable)

3.114 The designer, under the CDM Regulations, must minimise risks:

- where it is reasonable to expect the designer to address the risks at the time the design is prepared
- to the extent that it is otherwise reasonably practicable to do so.

3.115 The use of the term "reasonable" is qualified by the designer's knowledge and what is meant by "reasonably practicable".

- *Designer's knowledge.* Designers need to provide information based upon what he or she may be reasonably expected to know at the time the design is prepared. For example, technical information about the hazardous properties of certain substances may, as yet, be undocumented, as was the case with asbestos 50 years ago. It may not be reasonable to expect designers to know of as yet unanticipated or unsubstantiated scientific developments. However, it may be reasonable to expect designers to know about information which is currently available, eg in their trade press.

- *Reasonably practicable.* In deciding what is "reasonably practicable", a reasoned judgment has to be made, taking into consideration the extent of the risks to health and safety posed by a design hazard relative to the time, effort and resources that would need to be expended if the hazard were to be designed out. Obviously, the greater the risk, the more resources could reasonably be expected to be provided in dealing with the risk.

3.116 The designer of construction work is not in the same position as the employer, who can directly observe work activities taking place and make a risk evaluation. Designers of construction work have to use their imagination to anticipate the type of workplace the contractor will create to carry out the assembly of the design.

Workplace Regulations

3.117 The Workplace (Health, Safety and Welfare) Regulations 1992 (Workplace Regulations) apply to the employer in control of a workplace and refer to:

- the need for maintenance
- effective and suitable ventilation
- temperatures during working hours
- lighting
- cleanliness
- facilities for disabled people
- the provision of satisfactory space for the work to be carried out.

3.118 The employer will not be able to comply with these regulations unless the design of the workplace takes account of the requirements to the extent that they can be satisfied by the actual structure of the building. Normal design considerations for the utility of the building (together with the need to satisfy the requirements of the Building Regulations 2000 in new work) should enable the regulations to be satisfied, but designers should be familiar with the requirements in any case.

3.119 Although construction sites are exempt from the provisions of the Workplace Regulations, the Construction (Design and Management) Regulations 2007 provide for health and welfare to a similar standard.

Principles of Prevention and Protection

3.120 When an evaluation of risk has been considered, the principles of prevention and protection should be applied, as outlined in the Approved Code of Practice *Managing Health and Safety in Construction.*

3.121 Briefly, the principles include:

- avoiding risks
- evaluating the risks which cannot be avoided
- combating the risks at source
- adapting the work to the individual
- adapting to technical progress
- replacing the dangerous with the non-dangerous or the less dangerous
- developing a coherent overall prevention policy which covers technology, organisation of work, working conditions, social relationships and the influence of factors relating to the working environment
- giving collective protective measures priority over individual protective measures
- giving appropriate instructions to employees.

Avoiding Risk

3.122 A designer is required by the CDM Regulations to identify risks associated with the construction work and, if possible, by modifying or altering the design, to avoid the risk completely. For example, if an area of contaminated ground has been found on a site which is to be developed, it may be possible to locate the new buildings in such a position that the contaminated ground does not have to be disturbed. The area of contaminated ground could be sealed, covered with top soil and planted to form a green area on the site. The risk to human health that the contaminated land could pose during construction work is therefore avoided.

Evaluating the Risks Which Cannot be Avoided

3.123 Where risks to the construction workforce cannot be avoided an evaluation of the residual risk must be made and this must be communicated to the principal contractor/contractor in the pre-tender information. Where significant risks remain when they have done what they can, designers should provide information with the design to ensure that the CDM co-ordinator, other designers and contractors are aware of these risks and can take account of them.

Combating the Risks at Source

3.124 If it is not possible to avoid the risk then risks should be combated at source. If there will be a need to lift an awkward shaped structural member into position by crane, then the designer should incorporate attachment points in the design so that the item can be lifted simply into position.

Adapting the Work to the Individual

3.125 This applies especially as regards the design of workplaces, the choice of work equipment and the choice of working and production methods, with a view, in particular, to alleviating monotonous work and work at a predetermined work-rate and to reducing their effect on the health of the individuals concerned.

Adapting to Technical Progress

3.126 There are many techniques available for controlling risks. Designers should be aware of the technology available in terms of work equipment, etc. Controls introduced by designers should be co-ordinated for best effect, which is one of the main functions of the CDM co-ordinator and the pre-tender information.

Replacing the Dangerous with the Non-dangerous

3.127 This requires designers to make provision in their design to overcome the risk rather than leaving it to the contractor. For example, it may be safer to specify a water-borne paint rather than a solvent-based one and thereby avoid the exposure of people on site to high levels of solvents in the paint.

Developing a Coherent Policy

3.128 Designers also need to take account of other relevant health and safety requirements when carrying out design work. Where the structure will be used as a workplace, (ie factories, offices, schools, hospitals) they need to take account of the provisions of the Workplace (Health, Safety and Welfare) Regulations 1992. This means taking account of risks directly related to the proposed use of the structure, including associated private roadways and pedestrian routes, and risks arising from the need to clean and maintain the permanent fixtures and fittings. For example, hospitals will need to be designed in a way which will accommodate the safe lifting and movement of patients as well as food preparation.

Giving Priority to Protective Measures

3.129 Any measures to control the risk should give priority to measures which would protect all people in a workplace rather than the individual person. Hence when, for example, a risk of falling is considered serious, it would be preferable to provide a protective rail at the edge of a roof, ie protecting everybody who came onto the roof, rather than rely on a safety harness with an attachment point.

Giving Appropriate Information to Employees

3.130 Where remaining hazards will present a risk on-site, the designer should pass on information to the CDM co-ordinator, so that information may be passed onto the contractors. This will allow them to develop safe systems of work which take account of the risks.

Managing Health and Safety in the Design Process

3.131 Design is often considered in the following terms.

- Feasibility and early concept.
- Elemental design and detail layouts.
- Detail design and specification.

3.132 The best time to eliminate hazards from a project is during the initial design phases of feasibility and early concept design. This applies to:

- a new building
- an engineering project
- the refurbishment of an existing structure
- the maintenance of an existing structure.

3.133 Although demolition may be regarded as the last consideration, it is frequently the first consideration when moving onto a new construction site.

3.134 As the project develops, the opportunities to eliminate hazards will decrease. While it may be possible to reduce the risk posed by hazards during detailed design, it is rarely possible to do more than specify control measures as the end of the pre-contract period approaches.

3.135 It is important to apply the principles of risk assessment at the earliest stages of the project. This can be very time consuming, with the added complications of finding ways:

- to apply the principles so that hazards can be addressed collectively
- for all designers to contribute to the process.

3.136 One method of achieving this is to include health and safety as part of the design review process (carried out periodically) from the earliest stages through to the completion of detailed design and specification.

Design Review

3.137 The following procedures should take place at each review.

- Identify the hazards within the design. It is useful if each hazard, when identified, is given a reference number. Newly identified hazards should then be added to the end of the list.
- Make an assessment of the risk associated with the hazard.

- Decide on action to be taken. If possible, actions required in response to hazards should be taken immediately, although this may not always be possible. Alternatively, the hazard can be dealt with at a later review stage when either more information is available, at which point action on the designer's part can be specified, or the hazard is left unresolved for contractors to deal with.
- Identify who needs to be informed about the hazard. For example, the CDM co-ordinator will need to be informed of the hazard as it may need to be incorporated into the construction phase health and safety plan and the health and safety file. Other designers may need to know about the hazard as it may affect their work.

3.138 Any unresolved hazards should be reconsidered at each design review. A more comprehensive description of risk will be required when the final design is completed and the information is prepared for the pre-construction information.

Managing Health and Safety in the Design Process — Feasibility and Early Scheme Design

3.139 The early design reviews at feasibility and early sketch planning stages should concentrate on hazards that might impact on the positioning of the building on-site and the overall form of the building.

3.140 This is the stage in the process when the hazards associated with the site can be taken into account and be avoided or their associated risks reduced.

3.141 Issues that should be examined include:

- locations/nature of retained structures and surrounding structures
- locations/nature of existing services/systems
- ground conditions.

Information from Clients

3.142 Clients have a duty under the CDM Regulations to make any information they have about the building(s) or site available to the CDM co-ordinator. The CDM co-ordinator should make this information, as well as the result of any surveys which may have been commissioned, available to the designer. Where the available information is inadequate or insufficient, the client should be requested to commission appropriate surveys or investigations.

Surrounding Property

3.143 Hazards can arise from property surrounding the construction site. The design should highlight this information, where it is known, and take it into account. For example, the CDM co-ordinator should be informed if there are premises near the site which are sensitive to vibration or noise, eg a court building, and are likely to make a request for the noise to be stopped. Stopping planned operations at any time will cause disruption which could result in unanticipated hazards.

Alteration and Refurbishment

3.144 The site could be an existing building which can raise different issues affecting health and safety, including:

- access and segregation of the construction activities from existing activities on-site
- means of escape in the case of emergency
- waste disposal
- the interruption of services, eg electricity and water

- phasing of the work operations
- temporary stability
- hazardous operations, eg hot work.

3.145 All of the above will need to take account of:

- the employer's existing staff
- members of the public who may be on the premises
- construction workers.

3.146 The health and safety policies of resident management will also need to be considered. For example, if the building under design is to be located on a refinery site, the oil company will have very stringent health and safety site rules that will affect contractors' workers while they are on-site. Designers should ask the project's CDM co-ordinator for advice on how the resident company's health and safety policy may affect the design.

Site Arrangements

3.147 Consideration should be given to:

- problems contractors may have in gaining access to the site
- providing for the siting of cabins
- providing for the storage of materials
- providing welfare facilities.

3.148 The positioning of buildings on the site can facilitate safe access and enable vehicles to leave the site without reversing (thereby reducing the possibility of accidents).

Future Cleaning

3.149 Cleaning the windows of any building often creates difficulties. However, consideration should be given to buildings which are unconventionally shaped because conventional means of access are not possible. In these cases, the designer should find alternatives to ensure access.

3.150 It should also be remembered that windows and atria need to be cleaned both inside and outside, and design solutions must address this, eg by providing gantries, attachment points, parking for mobile towers, etc.

Managing Health and Safety in the Design Process — Elemental Design and Detail Layouts

3.151 Once the overall form and layout of the buildings has reached a degree of finality, consideration will need to be given to the erection of the primary structure and the form of spaces within the building.

3.152 A requirement of good design is "buildability". Buildability means that the workers can operate in a safe working environment to carry out the assembly process.

3.153 Many hazards might be eliminated by prefabrication, where hazardous operations can be carried out under controlled conditions, eg a dedicated manufacturing site where workers are not exposed to climate extremes. If it is not possible to prefabricate off-site, it is better to assemble at ground level or on a firm upper floor rather than working from less stable (and therefore more hazardous) platforms, eg scaffolding and tops of ladders.

3.154 Provision in the design for the installation of floors, stairs, perimeter walls and parapets to follow closely behind the erection of the primary structure will reduce the need for temporary access and edge protection measures. Consideration must also be given to temporary instability during construction and any temporary works required to facilitate construction. Any special features of the design which affect these temporary measures, eg load bearing capacity, must be noted and drawn to the attention of the CDM co-ordinator.

Future Maintenance

3.155 Proper provision should be made for safe maintenance. Consideration should be given to:

- ensuring future access, ie having the space and arrangements in place in order to carry out maintenance on plant, equipment and systems
- replacing worn out items, eg the ability to change the light bulbs in an auditorium 20m high with a sloping floor and fixed seating
- highlighting any special requirements of the design that require regular inspections. The CDM co-ordinator will include this information in the health and safety file.

Managing Health and Safety in the Design Process — Detail Design and Specification

3.156 A number of more detailed factors which will have an impact on health and safety need to be considered once the outline scheme has been agreed. Attention should be given to the following questions before the detailed design process begins.

- Is it possible to avoid heavy building blocks which require manual handling?
- Can arrangements be made for elements to be fixed from secure floors rather than external scaffolding?
- Can wet trades, particularly those involving cement (which may give rise to dermatitis), be designed out?
- Can designing and specifying processes that produce a lot of dust and dirt be avoided, eg cutting and chasing newly formed surfaces?
- Can specifying materials that may adversely affect health be avoided?

3.157 Manufacturers have a statutory duty with regard to health and safety to provide relevant information about their products, and, in the case of dangerous substances, safety data sheets must be provided by chemical suppliers.

3.158 Processes which produce dust and dirt on-site may not be immediately apparent to the designer. Research published by the HSE, *Dust and Noise in the Construction Process*, can provide some guidance on ways to avoid and control the hazards and risks associated with dust and noise on construction sites. Hazard checklists are a useful tool for designers carrying out health and safety reviews.

Provision of Information

3.159 Designers are required to pass on any relevant health and safety information, usually to the CDM co-ordinator for incorporation into the construction phase health and safety plan and the health and safety file.

3.160 It is important for the designer to keep a record of any significant hazards identified at any time throughout the design process. Every designer working on writing specifications or the drawing board should be familiar with the need to eliminate or reduce risks in construction.

3.161 Many minor hazards can be eliminated through careful thought as the design takes place. Nevertheless, a record should be kept of:

- hazards which cannot be eliminated
- action to be taken regarding the hazard
- people who need to be informed about the hazard.

3.162 Recording this information serves the following purposes.

- Recording hazards will alert other designers to hazards which have been identified and may have an impact on the way they carry out their own design work.
- The CDM co-ordinator will need information on hazards so that he or she can decide if it needs to be incorporated into the pre-construction information or health and safety file.
- There may be a cost implication in dealing with the hazard which the quantity surveyor will need to allow for when preparing the bill of quantities.
- A record of carrying out health and safety reviews on the project should be maintained for the designer's protection in the event of an accident occurring on the building site. Contemporary records showing that a risk analysis programme was conducted at the time the design was prepared should give a great deal of protection from criminal prosecution by the HSE and from potential civil action by an injured party.

Designer's Risk Evaluations

3.163 The pre-construction information, which is supplied with tender documents for potential contractors, is collated and prepared by the CDM co-ordinator. The CDM co-ordinator might set out the format for any information required, eg hazard identification forms.

3.164 A more comprehensive description of the hazards will need to be documented upon final completion of the design. This will help the CDM co-ordinator to establish priorities for the pre-construction information.

Drawings

3.165 The designer can make a large contribution to health and safety by the way information is conveyed to people working on the site.

3.166 People working on-site will not take a bill of quantities, specification or the health and safety plan to the workplace. However, people do work from drawings so it would be beneficial if there were a section on the drawings set aside for health and safety notes providing details about any particular hazards associated with the design information that workers should know about.

3.167 This type of information is likely to be more comprehensive, specific and detailed than the information that will generally be included in the pre-construction information.

Designer Liaison with the CDM Co-ordinator

3.168 The CDM co-ordinator will need to be satisfied that the designers:

- are carrying out a risk analysis of those hazards identified on the construction site
- are discussing design issues with regard to health and safety with other designers
- will provide appropriate information for the health and safety plan and health and safety file.

3.169 The CDM co-ordinator may not necessarily be an expert in a designer's area of speciality and will depend on the designer to identify hazards associated with that design. However, the CDM co-ordinator will be aware of the general legislation affecting construction safety and be familiar with the causes of accidents on the building site. The CDM co-ordinator may:

- ask a designer to carry out a risk assessment for particular hazards which the designer may have overlooked
- wish to discuss the results of any risk analysis which the designer may have carried out.

3.170 If agreement cannot be reached on how to resolve a particular hazard, the CDM co-ordinator may refer the matter to the client or its representative, who may then give instructions to the designer.

3.171 Design work does not end when the contract is placed with the principal contractor. There may be continuing development of the design or changes in design because the:

- tender price may be too high and the specification will need to be changed in order to allow for cheaper materials
- client's requirements may have changed
- conditions on-site may not be as predicted and live services may need to be relocated.

3.172 It is necessary to take health and safety into consideration whenever there is a change in design or specification. The CDM co-ordinator will need to be notified of any risks to health and safety which might require a revision to the health and safety plan or file. These will need to be allowed for when the contractor prepares estimates for any variation.

Responsibilities of Contractors

3.173 The contractor is responsible for the way in which work is executed on-site. A designer may illustrate how it was assumed the work might be carried out to overcome a hazard, but the contractor does not have to do it that way. The principal contractor is entirely responsible for managing health and safety on the site.

3.174 In some cases, the designer will need to specify the way in which work is to be undertaken, eg sequence of steel erection, temporary supports, etc. In order to appreciate the risks involved, the designer must have considered at least one safe method for the construction, maintenance or removal of his or her design. However, unless specified, the choice of method is entirely up to the competent contractor who will have expertise based on experience in such matters.

3.175 Under the Health and Safety at Work, etc Act 1974 and the Management of Health and Safety at Work Regulations 1999, individuals who observe a dangerous situation on-site must report the incident to a responsible person. Designers with specialist knowledge may be expected to recognise hazardous situations where their own work is involved.

List of Relevant Legislation

- Construction (Design and Management) Regulations 2007
- Control of Asbestos Regulations 2006
- Work at Height Regulations 2005
- Carriage of Dangerous Goods and Use of Transportable Pressure Equipment Regulations 2004

- Building Regulations 2000
- Management of Health and Safety at Work Regulations 1999
- Workplace (Health, Safety and Welfare) Regulations 1992
- Health and Safety at Work, etc Act 1974

Further Information

Publications

HSE Publications

The following are available from *www.hsebooks.co.uk*.

- CRR73 *Dust and Noise in the Construction Process*
- INDG401 *The Work at Height Regulations 2005 (as amended): a Brief Guide*
- L144 *Managing Health and Safety in Construction. Construction (Design and Management) Regulations 2007. Approved Code of Practice*
- L21 (rev 2000) *Management of Health and Safety at Work. Management of Health and Safety at Work Regulations 1999. Approved Code of Practice and Guidance*
- L24 *Workplace (Health, Safety and Welfare) Regulations 1992 — Approved Code of Practice*
- L143 *Work with Materials Containing Asbestos. Control of Asbestos Regulations 2006: Approved Code of Practice and Guidance*

Organisations

- Association for Project Safety
  Web: *www.associationforprojectsafety.co.uk*
  The Association for Project Safety (APS) is a multi-disciplinary membership body for those who operate in the field of construction health and safety risk management.
- Association of Consultant Architects (ACA)
  Web: *www.acarchitects.co.uk*
  ACA is the national professional body representing architects in private practice (consultant architects) throughout the UK.
- Chartered Institute of Architectural Technologists (CIAT)
  Web: *www.ciat.org.uk*
  The CIAT represents over 8,000 professionals working and studying in the field of Architectural Technology. CIAT is internationally recognised as the qualifying body for Chartered Architectural Technologists and Architectural Technicians.
- Health and Safety Executive (HSE)
  Web: *www.hse.gov.uk*
  The HSE and the Health and Safety Commission are responsible for the regulation of almost all the risks to health and safety arising from work activity in the UK.
- Royal Institute of British Architects (RIBA)
  Web: *www.riba.org*
  The RIBA is a member organisation promoting architecture and architects.

Health and Safety Files

- The purpose of a health and safety file is to create a permanent record of information relevant to health and safety, which will be used after construction is completed.
- The health and safety file must be given to the client by the CDM co-ordinator at practical completion.
- All new structures created since the full implementation of the Construction (Design and Management) Regulations 1994 should have health and safety files.
- The health and safety file has to be given to the new owner when a client sells a property.
- The scope, structure and format for the file should be agreed between the client and the CDM co-ordinator at the start of the project.
- The CDM co-ordinator is responsible for the management and preparation of the health and safety file.

3.176 Under the Construction (Design and Management) Regulations 2007 (CDM Regulations), the health and safety file is defined as a permanent record of those aspects of a construction project which might affect the health and safety of any person:

- carrying out future construction, maintenance, refurbishment or demolition work
- occupying the building and who may be affected by those carrying out work on it.

3.177 Therefore, the file should record issues relating to hazards inherent in the maintenance or ultimate demolition of the building. It also provides an opportunity for recording:

- normal maintenance procedures
- the location of incoming mains services
- general services commissioning data.

3.178 The regulations are flexible enough to cope with relatively minor maintenance projects as well as multimillion-pound civil engineering projects. For this reason, the regulations are not prescriptive about the format and content of health and safety files as these will vary greatly between projects. The file can be combined with the Building Regulations log book as long as this does not result in health and safety information being misplaced.

In Practice

Duties of Parties Involved (Client, CDM Co-ordinator and Principal Contractor)

3.179

- Under the CDM Regulations the client is required to ensure that the health and safety file is available for inspection.
- The CDM co-ordinator is responsible for ensuring, among other duties, that the health and safety file is prepared for each structure comprised within a construction project.

- The principal contractor must make certain that the CDM co-ordinator is supplied with any relevant information for inclusion in the health and safety file.

Client

3.180 Under the CDM Regulations, the client must ensure that the health and safety file is available for inspection. Therefore, the client should have the file in an accessible format so that relevant details can be passed to anyone who has cause to need the information contained in it, including:

- future CDM co-ordinators, designers and principal contractors who would need the file when any new work which is subject to the CDM Regulations is undertaken
- those responsible for maintaining the building.

3.181 Where a client disposes of a property, the file has to be handed over to the new owner. The original client must also ensure that the new keeper of the file is aware of the contents of the file and how the information should be used.

CDM Co-ordinator

3.182 The CDM co-ordinator is obliged, under the CDM Regulations, to ensure that:

- the scope, structure and format for the file is agreed with the client
- the file contains the information identified by designers as affecting health and safety in any future maintenance and demolition.

3.183 The CDM co-ordinator is also required to make certain that the file contains any other information — not necessarily resulting from the design — which could affect the health and safety of any person working in or on the structure. This means providing information on aspects of the structure which have health and safety implications for future maintenance programmes, including:

- cleaning facilities
- specific access arrangements.

Updating the File During the Construction Phase

3.184 During the construction phase (ie when the project is actually being built), the CDM co-ordinator must revise the health and safety file as necessary based upon new design and design amendment information supplied by designers and the principal contractor. This ensures that the file accurately reflects the project as built, and not necessarily as designed.

3.185 A contract administrator can withhold the certificate of practical completion until all information required to complete the file has been supplied to the CDM co-ordinator.

Handover at Completion

3.186 The health and safety file must be given to the client by the CDM co-ordinator at practical completion, ie when the principal contractor has finished all of the work, thereby enabling the client to take possession.

3.187 The Health and Safety Executive (HSE) publication and Approved Code of Practice (ACOP) *Managing Health and Safety in Construction* notes that it is possible that the file may not be ready at this time, especially if there has been a rush to complete the job on time. The guidance suggests that the file is handed over as soon as is practical.

Multi-structure Projects

3.188 The scope, structure and format for the file should be agreed between the client and the CDM co-ordinator at the start of the project.

3.189 There can be a separate file for each structure, one for the entire project or site or one for a group of related structures. It is essential though that people can find the information they need easily and that any differences between similar structures are clearly shown.

Principal Contractor

3.190 The Construction (Design and Management) Regulations 2007 place an obligation on the principal contractor to ensure that the CDM co-ordinator is supplied with any relevant information for inclusion in the health and safety file. This applies to:

- amendments to the original design that occur on-site, eg a live service is moved to overcome a particular difficulty
- new design work that is carried out by the principal contractor's subcontractors, eg specialist contractors who design and install lifts and air-conditioning systems (the details of which are not included in the original design).

Existing Structures

3.191 While all new structures created since the full implementation of the CDM Regulations in 1994 have a health and safety file, buildings or other structures constructed before the regulations came into force will almost certainly not have a health and safety file.

3.192 Where work needs to be undertaken on these existing structures — work which is subject to the CDM Regulations — the HSE only expects a health and safety file to be created in relation to the new work actually being carried out. It is not essential to create a health and safety file that applies to the whole structure, although many clients may see the advantages of providing a file for the entire building.

Contents of the Health and Safety File

3.193 While the format of the health and safety file will vary greatly between different types of project (eg the file for a processing plant is likely to be far more extensive than that for an office building), the sections set out below indicate the basic elements that need to be included in any file.

Section 1 — Drawings

3.194 A brief description of the work carried out.

Section 2 — Residual Hazards

3.195 Information on any residual hazards which remain and how they have been dealt with. For example, surveys or other information concerning asbestos, contaminated land, water bearing strata, buried services, any working at height that may be required, etc.

Section 3 — Key Structural Principles

3.196 The health and safety file should contain details of the design concepts behind various elements of the structure. For example, bracing, sources of substantial stored energy — including pre- or post-tensioned members — and safe working loads for floors and roofs, particularly where these may preclude placing scaffolding or heavy machinery there.

3.197 A useful checklist might include:

- structural frame/load-bearing walls
- cladding/infill
- curtain walling and window systems
- floor structures
- roof structure/covering
- mechanical services' design concept, eg whether natural ventilation or air conditioning is used
- electrical services' design concept, eg whether all electric lighting is on one circuit or on a floor-by-floor basis.

3.198 The file should identify any specific sequence that was used in the erection of the building and which might need to be reversed during alterations or demolition. This is particularly relevant to prefabricated buildings or structural elements (eg portal frames) that are inherently unstable in isolation during erection and have specific temporary propping requirements.

3.199 Detailing the sequence in which external cladding panels were assembled could also be included in the file, which would indicate how they might be safely disassembled.

Section 4 — Hazardous Materials Used

3.200 Potentially hazardous materials that may be identified on the drawings or as a separate piece of documentation include:

- flammable finishes
- various types of insulating materials
- lead paint
- pesticides
- special coatings which should not be burnt off.

Section 5 — Information Regarding Installed Plant

3.201 Any special arrangements for lifting, order or other special instructions for dismantling installed plant and equipment. This should also include maintenance requirements and if lock off procedures will be required.

Section 6 — Health and Safety Information Required for Maintaining the Structure

3.202 The file must set out the various elements within the building that are provided for maintenance and cleaning purposes and which have health and safety implications for those using them, including:

- facilities for roof access
- gantries
- window-cleaning cradles
- remote window opening gear
- permanent fixings for fastening ladders.

3.203 The file must outline the health and safety issues with regard to the overall structure and its finishes. This might include procedures for decorating the outside of the structure, clearing gutters or renewing air-conditioning unit filters. These procedures should include details about the required frequency of cleaning and the types of cleaning materials to be used and those to be avoided.

Section 7 — Significant Services

3.204 Included in the file should be information on the nature, location and markings of significant services, including underground cables, gas supply equipment, fire-fighting services, etc.

3.205 A useful checklist includes:

- mains distribution, eg the location, size and termination of the gas main, water main and telecommunications
- emergency backup facilities, eg standby generators
- security alarms
- fire-fighting systems, eg sprinkler systems, drencher systems and fire shutters.

Section 8 — Information and As-built Drawings

3.206 Information and as-built drawings should be included of the structure, its plant and equipment (for example, the means of safe access to and from service voids, fire doors and compartmentalisation etc).

3.207 The drawings should be the final, "as-built" version (ie as amended from the originals through the construction process). These will represent the final structure as it actually exists and not just as it was conceived. The drawings should:

- indicate the position of incoming services and distribution (any or all of which may be concealed)
- indicate the location and details of various building materials used
- identify the various types of insulation material, flammable finishes, etc that may represent hazards if they are disturbed
- make cross-references to the information on hazards where appropriate.

The Health and Safety File Should NOT Contain:

3.208

- the pre-construction information, or construction phase plan
- construction phase risk assessments, written systems of work and COSHH assessments
- details about the normal operation of the completed structure, eg operation and maintenance manuals
- construction phase accident statistics
- details of all the contractors and designers involved in the project (though it may be useful to include details of the principal contractor and CDM co-ordinator)
- contractual documents
- information about structures, or parts of structures, that have been demolished, unless there are any implications for remaining or future structures, for example voids
- information contained in other documents, but relevant cross-references can be included.

Managing the Preparation of the Health and Safety File

3.209 Under the Construction (Design and Management) Regulations 2007, the CDM co-ordinator is responsible for ensuring that the health and safety file contains all of the necessary information when it is handed to the client upon completion of the project.

3.210 Preparation of the file should begin at the same time as the pre-construction information is being prepared. If the file is left until the end of the project, it will result in a great deal of duplication of effort to trace the same information sources and will delay the completion of the file unnecessarily.

3.211 The file involves compiling information from a variety of sources, including:

- the client, who can provide existing information, eg drawings and location of services, as well as information on how he/she would like the final maintenance procedures to be arranged
- the designers, eg architects, structural engineers and quantity surveyors
- the principal contractor and subcontractors
- statutory/private undertakers for utilities, eg gas, electricity, water and telecommunications.

3.212 Certain information may be readily available, including:

- designers' drawings
- operation and maintenance manuals from specialist equipment suppliers.

3.213 The CDM co-ordinator may need to be proactive in order to obtain relevant information by petitioning the various parties to supply the necessary details to complete the contents outlined above.

3.214 Compilation information for the health and safety file and health and safety plan should occur simultaneously. As with the health and safety plan, the two stages in the development of the file are the pre-construction preparation of the file and the construction phase development of the file.

Pre-construction Preparation of the File

3.215 The starting point of the health and safety file is the pre-construction information. Two sources of information for the file are the drawings and the layouts.

3.216 Designers are obliged under the CDM Regulations to undertake risk evaluations for the health and safety implications of their designs in order to design out risks or identify any residual risks that could not be eliminated at design stage. These will highlight any remaining hazards for inclusion within the pre-construction information.

3.217 It would be helpful for designers to distinguish the risk evaluations that have implications for future maintenance and demolition so that they may be readily accessed by the CDM co-ordinator for the purposes of compiling the health and safety file. This may even be conducted as a separate exercise by designers.

3.218 With regard to designers, the management responsibilities of the CDM co-ordinator will vary greatly between projects, as follows.

- If the design is being carried out by a team within a particular organisation, designers can liaise directly with one another.
- The responsibilities of the CDM co-ordinator will be more onerous on projects where the work of several individual designers not directly in contact with each other needs to be co-ordinated.

3.219 In either case, the CDM co-ordinator will need to:

- review the interaction of various elements of the design for their health and safety implications
- extract those elements which will affect future maintenance and demolition for inclusion in the health and safety file.

Updating the File During Construction

3.220 The CDM co-ordinator will need to liaise with the principal contractor and subcontractors during the construction phase of a project in order to assess any design variations or new design elements for possible inclusion in the health and safety file.

3.221 The compilation of the health and safety file should be properly managed in order to prevent it from becoming an unstructured dumping ground for miscellaneous data. Such management requires:

- organising regular meetings between the CDM co-ordinator and the principal contractor to review design variations
- checking that all variations to the work content are recorded, even when there is no financial effect
- verifying that all variations are assessed in health and safety terms, risk assessments are carried out and that risks to health and safety for future maintenance or demolition are recorded in the file.

Function of the File After Construction is Complete

3.222 The CDM Regulations require the client to keep the health and safety file once it has been delivered by the CDM co-ordinator at the conclusion of a project. The client must make the file available to the CDM co-ordinator and designers on subsequent projects and to anyone else who has need of it.

3.223 There is no legal requirement under the CDM Regulations for the file to be updated other than when a new project that is subject to the regulations is initiated. In this case, the file provides basic data for a new health and safety plan and a new health and safety file, which reflects issues raised during the new project.

3.224 The process of passing on the information contained within the file should take place whether or not the regulations apply. For example, a maintenance task which involves only a few individuals for a short period of time does not require the appointment of a CDM co-ordinator and principal contractor. Nevertheless, information could be vital to the safety of those individuals.

3.225 When work has been completed, the file should record any new circumstances that arise. The file will essentially remain unchanged provided the same methods and materials adopted previously are used again. However, changes that affect the ways in which safe systems of work are set up must be recorded.

Multi-occupied Buildings

3.226 In multi-occupied buildings there could, after a period of time, be several files for the various sections where work has been carried out. A decision may be taken to draw all of the files together as a single point of reference. It may even be decided to carry out retrospective audits on parts of the building which do not have a health and safety file.

3.227 The retrospective audit could include:

- an examination of earlier specifications and drawings
- a detailed site inspection
- minor investigations or opening up works.

3.228 This would provide a suitable baseline for future activity as there would be a single file for adaptation by whoever carries out the work. There is no obligation for such files to be prepared retrospectively, although some clients may consider this to be a good investment.

Storing the File

3.229 The information for the health and safety file can be recorded in a variety of ways, including:

- in paper format, with all of the necessary information bound into a single or series of folders, properly cross-indexed
- on computer, ensuring that all information will be retrievable in future.

3.230 The safe keeping of the file should be treated with as much care as other important legal documents. In multi-occupancy situations, eg where a housing association owns a block of flats, the owner should keep and maintain the file but ensure that individual flat occupiers are supplied with health and safety information concerning their home.

List of Relevant Legislation

- Construction (Design and Management) Regulations 2007
- Control of Substances Hazardous to Health Regulations 2002

Further Information

Publications

HSE Publications

The following is available from *www.hsebooks.co.uk*.

- L144 *Managing Health and Safety in Construction. Construction (Design and Management) Regulations 2007. Approved Code of Practice*

Organisations

- Health and Safety Executive (HSE)
  Web: *www.hse.gov.uk*
  The HSE is responsible for the regulation of almost all the risks to health and safety arising from work activity in the UK.

Pre-construction Information and Construction Phase Plans

- The Construction (Design and Management) Regulations 2007 emphasise that the control of health and safety on a construction site starts at the planning, design and specification stages of the construction phase.
- Planning for the construction phase comprises a collection of information which evolves throughout the construction project with contributions from various parties. There are two very distinct stages of its evolution:
  - the pre-construction information
  - the construction phase plan.
- Both the pre-construction information and the construction phase plan represent the communication of essential information between duty holders.
- The CDM co-ordinator usually collates the pre-construction information as part of the CDM co-ordinator function.
- The provision of client information to the pre-construction information must occur early in the supply chain to allow greater opportunity for the design phase to address and reduce reported risks.
- The construction phase health and safety plan is a site-based document which any health and safety inspector would request to see if he or she came onto the site.
- Workplace risk assessments and method statements are key elements in the development of the construction phase health and safety plan.

3.231 Clients and designers can make significant contributions to the recognition of hazards and the control of risks in construction projects by developing pre-construction information at the earliest opportunity.

3.232 The pre-construction information is a key component in the recognition of hazards, control of risks and the provision of information to those who will be involved in the construction phase.

In Practice

3.233 Planning for the construction phase involves compiling information to identify hazards during construction. The two distinct stages are the pre-construction information and the construction phase health and safety plan.

3.234 The pre-construction information contains relevant information (eg background conditions and hazards associated with the project) to allow tenderers to plan for safety and cost the project accurately.

3.235 The construction phase health and safety plan uses the information from the pre-construction information and is developed further to produce safe systems of work through:

- detailed procedures
- method statements
- working instructions relevant to the actual construction activities.

Health and Safety Plans — Duties of Parties Involved (CDM Co-ordinator, Client, Designer(s), Principal Contractor)

3.236

- Under the Construction (Design and Management) Regulations 2007 (CDM Regulations), the CDM co-ordinator is responsible for ensuring that the pre-construction information is collated, complete and goes out to designers and tendering contractors.
- It is the responsibility of both the client and CDM co-ordinator to ensure that pre-construction information is made available to tenderers.
- Under the CDM Regulations, the client has a duty to provide information about the site.
- There is a basic requirement under the CDM Regulations for designers to avoid foreseeable risks which, if unavoidable, should be dealt with at source, so far as is reasonably practicable.
- This construction phase health and safety plan is drafted, developed and owned by the principal contractor, who initially compiles it in response to the information provided in the pre-construction information.
- The client must be satisfied that the construction phase health and safety plan is suitable before the construction phase begins. The CDM co-ordinator is to advise the client on the adequacy of the plan.

Pre-construction Information

3.237 The purpose of the pre-construction information is to provide relevant information about the existing health and safety hazards associated with a construction project. Designers and tendering contractors use the information to assimilate and subsequently provide adequate resources to manage the issues effectively (which is then reflected in the tender price submitted).

3.238 The information is intended to be dispatched together with the tender documentation to allow:

- sufficient time for its contents to be appreciated
- solutions to be formulated prior to the return of a tender submission.

3.239 The focus of the information is on issues over and above those a competent contractor would normally expect to deal with on a construction site. Two key providers of information for this document are the client and the designer(s).

CDM Co-ordinator

3.240 The role of the CDM co-ordinator is to ensure that the health and safety risk management process is fully owned and effectively discharged by all those duty holders appointed under the CDM Regulations. Therefore, it is usually the CDM co-ordinator who collates the information received from the client and the relevant design teams, advises on any further requirements and abstracts the relevant items into the pre-construction information.

Client's Input

3.241 Clients cannot leave it to contractors to discover hazards. Clients must provide sufficient information to facilitate the design/planning stage undertaken by the designer(s) and those preparing the construction information. Therefore, significant risks during the construction phase (including any demolition) can be anticipated, avoided and properly controlled.

3.242 Client information may include:

- the project duration and commencement date
- the amount of time allowed between award of contract and the start of the construction phase
- information about the previous legacy of the site, eg industrial usage materials, contamination
- adjacent activities, eg additional traffic generation, pedestrians, vehicles, neighbouring sports stadia, nearby factory estates, retail parks and schools
- ongoing activities, eg occupancy of the site and overlapping projects
- restrictions, eg access, logistical restraints such as height and weight, available space for lay-down areas and/or compounds, parking restrictions and one-way streets
- existing information, eg surveys (including asbestos, lead, condition, geotechnical and structural integrity), health and safety files (residual issues), contamination, statutory undertakers, existing drawings
- services, eg underground and overhead
- traffic routes and designated routes to the site.

3.243 This information is based on reasonable enquiries made by the client through his or her own efforts or via his or her advisory team. Where relevant information is not available, strong recommendations must be made to the client by the CDM co-ordinator.

Designer(s)' Input

3.244 The designer(s)' health and safety risk assessment strategy provides information that feeds into the pre-construction information and/or the health and safety file (after which the designer will have eliminated or reduced some of the hazards associated with the project based on an "as far as is reasonable" approach).

3.245 However, some residual issues need to be communicated and managed by others further down the supply chain. This vital information about significant hazards needs to go into both of the documents mentioned above.

3.246 The Approved Code of Practice (ACOP) *Managing Health and Safety in Construction* notes that such hazards are not always those that may result in the greatest risks, but which are not likely to be obvious to a competent contractor or other designers, unusual or likely to be difficult to manage effectively.

3.247 For the purpose of the pre-construction information, the design information (in addition to the information supplied by the client) is linked to the concept of "constructability/buildability". Therefore, designers must be aware of the construction methods and procedures associated with the asset or facility being designed. For example:

- sequential erection requirements to account for temporary instability, eg maximum wind speeds, erection sequences and bracing
- weights of components and identification of eccentricity of centres of gravity
- hazardous materials, eg flammable, toxic and carcinogenic material
- envisaged erection methodologies, eg how the design will be executed on-site
- construction methodologies, including extensive use of vibrating equipment and manual handling.

3.248 The designer is required to give consideration not only to safety issues (eg work at height and deep excavations) but also to occupational health issues which greatly reduce the quality of life of those affected (eg hand-arm vibration syndrome and allergic dermatitis).

Contents of the Pre-construction Information

3.249 The ACOP *Managing Health and Safety in Construction* provides guidance on the format and detail of the pre-tender stage health and safety plan, with the level of detail proportionate to the health and safety risk complexity of the project.

3.250 The pre-construction information should incorporate the following sections.

- Section 1 — Project Description
- Section 2 — Client's Considerations and Management Requirements
- Section 3 — Environmental Restrictions and Existing On-site Risks
- Section 4 — Significant Design and Construction Hazards
- Section 5 — Health and Safety File

Section 1 — Project Description

3.251 This section provides background information and identifies the various parties involved in the project. It is not always obvious who functions as the client, but guidance is given in the ACOP mentioned above on factors to be taken into account, including who:

- is at the head of the procurement chain
- arranges the design work
- engages the contractors.

3.252 Additional effort is also needed to identify all those who fulfil designer duties, including sub-designers and work package contractors who are involved in both the design and construction phases.

3.253 This section should include:

- a description of the project and programme details
- the minimum time allowed between appointment of the principal contractor and the instruction to commence work on-site
- details about the client, designers, CDM co-ordinator and other consultants
- the extent and location of existing records and plans
- whether or not the structure will be used as a workplace, in which case the finished design will need to take into account the provisions of the Workplace (Health, Safety and Welfare) Regulations 1992.

Section 2 — Client's Considerations and Management Requirements

3.254 The revised ACOP mentioned above introduced the concept of safety goals to be identified by the client as a means of measuring health and safety performance during construction. As such, the approach seeks to monitor and measure the performance of the principal contractor during this phase.

3.255 Other considerations include the following.

- Permits and authorisation requirements have particular relevance where construction work is to be undertaken within occupied premises.
- Welfare provision for the site.
- Emergency procedures should be in place.

- It is important to gain a perspective on what activities are taking place around or adjacent to the site that might have an impact on health and safety management.
- Co-ordination and liaison arrangements between parties remain prerequisites for effective communication.
- Site rules carry additional relevance where work is being undertaken on an occupied site, with liaison initiatives required in order to promote communication and co-ordination. (These should also address the phased handover or partial completion situation, where site boundaries change and have implications for security arrangements.)

3.256 This section should cover:

- structure and organisation
- welfare arrangements
- safety goals for the project and arrangements for monitoring/reviewing progress
- permits and authorisation requirements
- emergency procedures
- site rules and other restrictions on contractors, suppliers and others, eg access arrangements to those parts of the site which continue to be used by the client
- activities on or adjacent to the site during the works
- arrangements for communication between the parties
- security arrangements
- any areas designated as "confined spaces"
- smoking and parking restrictions.

Section 3 — Environmental Restrictions and Existing On-site Risks

3.257 This section contains input from the client, either from archive information or as a result of further surveys undertaken to provide the salient information. The law requires clients to proactively provide essential information upfront so that the principal contractor and his or her team can develop safe and suitable systems of work.

3.258 Boundaries change during the life of a project to accommodate sectional and partial handover, as well as the phasing of the works. Again, this constitutes vital information to enable the principal contractor to secure the site.

3.259 This section should cover the following hazards.

- Safety hazards, including:
  - boundaries and access, including temporary access
  - any restrictions on deliveries or waste collection or storage
  - adjacent land uses
  - existing storage of hazardous materials
  - the location of existing services particularly those that are concealed ground conditions, underground structures or water courses where this might affect the safe use of plant
  - information about existing structures, including stability, structural form, fragile or hazardous materials, anchorage points for fall arrest systems (particularly where demolition is involved)
  - previous structural modifications, including weakening or strengthening of the structure (particularly where demolition is involved)
  - fire damage, ground shrinkage, movement or poor maintenance which may have adversely affected the structure
  - any difficulties relating to plant and equipment in the premises
  - health and safety information contained in earlier design, construction or "as-built" drawings

- Health hazards, including:
  - asbestos
  - existing storage of hazardous material
  - contaminated land
  - existing structures
  - health and safety risks arising from the client's activities.

Section 4 — Significant Design and Construction Hazards

3.260 This section involves input from design team members and provides the link between permanent and temporary design. The design assumptions, where relevant, help in the appreciation of strategies and ensure that issues designed out are not designed back in.

3.261 The provision of information is again focused on significant risks, ie over and above what a competent contractor would ordinarily expect to deal with. However, the word "significant" is difficult to interpret so it is easier to consider significant and principal risks.

3.262 Design includes specification writing and, therefore, the designer needs to have access to the associated material hazard data sheets (which by law must be available from the supplier) when specifying materials so that residual hazard information can again be relayed to those who need it.

3.263 This section should cover:

- design assumptions and control measures
- arrangements for the co-ordination of ongoing design work and handling changes
- information on significant risks identified during design (health and safety risks)
- materials requiring particular attention.

Section 5 — Health and Safety File

3.264 The health and safety file is the document that will eventually be handed over to the client for the future health and safety management of the facility or asset being designed and constructed.

3.265 It therefore becomes a client document and, as such, it is prudent for the client to have an input into the format and content of this document, guided by the CDM co-ordinator and/or other professional advisors.

3.266 This section is helpful in providing guidance for the provision of relevant information. There are resource implications in the provision of such information, which should be provided upfront so that tendering contractors can take resource costs into account.

3.267 The pre-construction information should make it clear that the principal contractor needs to address the site specific health and safety issues identified in the pre-construction information in their construction phase plan.

Construction Phase Health and Safety Plan

3.268 The construction phase health and safety plan is drafted, developed and owned by the principal contractor, who initially compiles it in response to the information provided in the pre-construction information.

3.269 The plan is a site-based document which, like the pre-construction information, represents a project-specific approach and addresses issues related to the site. Specifically, the plan:

- outlines the principal contractor's management arrangements for the effective health and safety control of the construction phase
- continues to be developed in parallel with the programme of works.

3.270 Primary control for ensuring the principal contractor does not start the construction phase before an adequate construction phase plan has been produced rests with the client. This process can be facilitated by the CDM co-ordinator, who must be available to give advice if it is required by the client. The client must allow sufficient time between appointment of the principal contractor and the commencement of works on-site to prepare other arrangements such as welfare.

3.271 The client must:

- monitor the process prior to the principal contractor beginning work (since legal duties cannot be transferred, it is only the client who can approve its suitability)
- be satisfied that all of the early phases of a project are adequately controlled through the outline of corresponding arrangements within the construction phase plan.

3.272 Failure by the principal contractor to demonstrate satisfactory arrangements as outlined in the construction phase health and safety plan will result in a delayed start due to not receiving the relevant approval to start from the client.

3.273 As a tendering contractor, the principal contractor will be aware of the content of the pre-construction information. However, it is not until appointment (after tender scrutiny) that the detail required would be developed into the construction phase health and safety plan.

3.274 The construction phase health and safety plan:

- is an articulation of the principal contractor's statutory duties to adequately resource those health and safety issues identified in the pre-construction information
- must be developed in conjunction with the discharge of duties in compliance with the requirements of the Construction (Design and Management) Regulations 2007 and other construction-related legislation.

3.275 Information for the construction phase health and safety plan will be collected and collated by the principal contractor. Sources of information include:

- the pre-construction information
- consultants to the principal contractor
- contractors, eg subcontractors
- material suppliers
- temporary works designers.

3.276 Once approval to start has been received from the client, it is the principal contractor's duty to continue to function competently and to develop the construction phase health and safety plan throughout the construction phase. Such development continues up to and including all works associated with project completion.

3.277 Repair and replacement work (as distinct from items on the snagging list) undertaken within the maintenance period (defects correction period) by virtue of contractual arrangements will require a new and separate construction phase health and safety plan.

Format of the Construction Phase Health and Safety Plan

3.278 The ACOP *Managing Health and Safety in Construction* provides guidance on the format and detail of the construction phase health and safety plan, with the level of detail being proportionate to the health and safety risk complexity of the project.

Section 1 — Project Description

3.279 This section describes background information already contained in the relevant first part of the pre-construction information. The information is extended to include details about the principal contractor and other appointed designers, contractors, etc.

3.280 This section should include:

- a description of the project and programme details
- details about the client, designers, CDM co-ordinator, principal contractor and other consultants
- the extent and location of existing records and plans.

Section 2 — Communication and Management of the Work

3.281 This section should cover the following areas.

- Management structure and responsibilities. Individual responsibility, particularly for health and safety management, needs to be identified within the management structure on-site. This responsibility must embrace management from the top down to include all subcontractors and relevant health and safety communication links.
- Health and safety goals for the project and arrangements for monitoring and reviewing health and safety performance. The principal contractor needs to focus on those health and safety goals already outlined by the client and embrace them further into his or her own systems for proactive monitoring and reviewing of effectiveness on-site.
- Arrangements (in conjunction with the Management of Health and Safety at Work Regulations 1999 and the Reporting of Injuries, Diseases and Dangerous Occurrences Regulations 1995) must be made for:
  - ensuring regular liaison between parties on-site, eg induction talks, briefings, etc
  - consulting with the workforce (communication with the workforce ensures that information is transmitted/received in the most efficient manner)
  - exchanging design information between the client, designers, CDM co-ordinator and contractors on-site
  - handling design changes during the project
  - selecting and controlling contractors
  - exchanging health and safety information between contractors
  - site security
  - site induction and on-site training
  - putting welfare and first-aid facilities in place (facilities for personal hygiene must be on-site from the beginning of any construction project)
  - reporting and investigating accidents and incidents, including near misses
  - producing and approving risk assessments and method statements (these form a key element in the development of the construction phase health and safety plan and need to be presented and accepted before the associated work begins).
- Site rules. Principal contractors will enforce their own site rules. Where work is undertaken within an ongoing facility, the client's site rules may also have to be embraced.

- Fire and emergency procedures. These procedures need to account for overlapping projects and changes brought about by partial and sectional handovers, as well as shared sites. The question of whose procedures take precedence in an occupied site when there is an emergency needs to be resolved before work starts. For example, escape routes and assembly areas should be identified and made known to individuals on-site.

Section 3 — Arrangements for Controlling Significant Site Risks

3.282 All work activities will be the subject of work-based risk assessments and method statements, which continue to contribute to the construction phase health and safety plan. It is essential that modern technological developments are employed as far as is reasonably practicable to minimise occupational health risks (eg craneage for the movement of heavy and/or awkward loads).

3.283 This section should cover safety and health risks, as outlined below.

- Safety risks involve:
  - delivery and removal of materials (including waste) and work equipment taking account of any risks to the public, for example during access to or egress from the site
  - dealing with services – water, electricity and gas, including overhead powerlines and temporary electrical installations
  - accommodating adjacent land use
  - stability of structures whilst carrying out construction work, including temporary structures and existing unstable structures
  - preventing falls
  - work with or near fragile materials
  - control of lifting operations
  - maintenance of plant and equipment
  - work on excavations and work where there are poor ground conditions
  - work on wells, underground earthworks and tunnels
  - work on or near water where there is a risk of drowning
  - traffic routes and segregation of vehicles and pedestrians
  - storage of materials (particularly hazardous materials) and work equipment.
- Health risks involve:
  - removing asbestos
  - dealing with contaminated land
  - manual handling
  - using hazardous substances
  - reducing noise and vibration
  - working with ionising radiation
  - exposure to UV radiation from the sun.

Section 4 — Health and Safety File

3.284 The principal contractor must collate and store relevant information arising out of his or her own processes, as well as those of his or her contractors, designers, etc. It is preferable to document the information into an agreed format, but this information will eventually be passed to the CDM co-ordinator who, by virtue of his or her duties, will amend and add information before ensuring the health and safety file is handed over to the client.

3.285 This section should cover:

- layout and format
- arrangements for collecting and gathering information
- storage of information.

Designers

3.286 Much design work is ongoing during the construction phase and undertaken by both permanent and temporary designers. All designers must embrace an effective health and safety design risk strategy and proactively provide relevant information to the principal contractor controlling the work. Similarly, information must also be relayed to the CDM co-ordinator.

Contractors and Self-employed

3.287 The ACOP *Managing Health and Safety in Construction* states that contractors must comply with the construction phase health and safety plan in order to ensure, so far as is reasonably practicable, the health and safety of their employees and the public.

3.288 It also states that the principal contractor must supply sections in good time as relevant to the contractor's work and requires the principal contractor to allow the contractor time to prepare properly for the work they have to undertake.

List of Relevant Legislation

- Construction (Design and Management) Regulations 2007
- Control of Substances Hazardous to Health Regulations 2002
- Management of Health and Safety at Work Regulations 1999
- Reporting of Injuries, Diseases and Dangerous Occurrences Regulations 1995

Further Information

Publications

HSE Publications

The following is available from *www.hsebooks.co.uk*.

- L144 *Managing Health and Safety in Construction. Construction (Design and Management) Regulations 2007. Approved Code of Practice*

Organisations

- Environment Agency
  Web: *www.environment-agency.gov.uk*
  The Environment Agency is the main environmental regulator in England and Wales and provides detailed information on legislative requirements, guidance for business and technical guidance on waste treatment and disposal.
- Health and Safety Executive (HSE)
  Web: *www.hse.gov.uk*
  The HSE is responsible for the regulation of almost all the risks to health and safety arising from work activity in the UK.

Chapter 4

Employees, Contractors and Visitors

Consultation With Employees

- All employers are required to consult with employees on health and safety arrangements.
- In a unionised workplace, employers must consult with the union safety representative under the Safety Representatives and Safety Committees Regulations 1977.
- In a non-unionised workplace, employers must either consult with employees directly or through a representative of employee safety under the Health and Safety (Consultation with Employees) Regulations 1996.
- Consultation is not the same as providing information. It involves:
  - inviting feedback
  - listening to the views of employees
  - taking viewpoints into account before making decisions.
- Employers should consult on:
  - any new measures that might affect health and safety
  - the appointment of competent persons
  - any information that must be provided under health and safety law
  - health and safety training
  - the health and safety implications of new technologies.
- Employers must provide representatives with relevant information, facilities and reasonable assistance to fulfil their duties.
- Representatives are allowed paid time off to fulfil their duties.
- Union safety representatives will usually have their training arranged by the relevant union.
- Employers may have to arrange training for representatives of employee safety in non-unionised workplaces.

4.1 Consultation with employees on arrangements for health and safety is a legal requirement in both unionised and non-unionised workplaces. Effective consultation can make a significant contribution to a positive health and safety culture within the workplace.

4.2 A proactive approach will assist in meeting legal requirements, and has the added potential for reducing accidents and increasing efficiency.

4.3 However, there is now greater emphasis on *involving* employees. According to the Health and Safety Executive (HSE) involvement means “active participation where the workforce and their representatives participate in the key elements of health and safety management such as setting targets and reviewing performance”.

Employers’ Duties

4.4

- Under the Health and Safety at Work, etc Act 1974, the employer is required to consult any safety representative appointed by a recognised independent trade union. The Act also places a duty on the employer to establish a safety committee, if requested to do so by a safety representative.
- Under the Safety Representatives and Safety Committees Regulations 1977, as amended by the Management of Health and Safety at Work Regulations 1999,

employers must consult any recognised trade union safety representative on health and safety arrangements. In particular, employers should consult safety representatives on:
  - the introduction of any measure that will affect the health and safety of employees
  - arrangements for appointing persons to assist the employer in complying with legislation, and to assist in emergency procedures
  - the provision of information as required under health and safety legislation
  - any planning and organisation of training required to be provided to employees under health and safety legislation
  - the health and safety consequences of introducing new technologies into the employees' workplace.
- In non-unionised workplaces, the Health and Safety (Consultation with Employees) Regulations 1996 require employers to consult with employees directly, or through nominated representatives of employee safety, on arrangements for health and safety. This includes:
  - supplying any information representatives might need to participate in consultation
  - providing appropriate and reasonable training to representatives
  - giving paid time off for representatives to perform their functions and attend training.

Employees' Duties

4.5 There are no specific legal duties placed on employees acting as safety representatives under the Safety Representatives and Safety Committees Regulations 1977 or representatives of employee safety under the Health and Safety (Consultation with Employees) Regulations 1996.

4.6 Such representatives do not incur criminal or civil liability for any act or omission whilst performing their representative duties.

4.7 However, like all employees, representatives have a duty to co-operate with their employer in relation to health and safety.

- Under the Health and Safety at Work, etc Act 1974, employees must take reasonable care of their own health and safety and that of others.
- Under the Management of Health and Safety at Work Regulations 1999, all employees, including safety representatives, must:
  - use machinery, equipment, dangerous substances, transport equipment, means of production or safety devices provided by their employer in accordance with any training given
  - inform their employer of any work situation they reasonably consider to represent a serious and immediate danger to health and safety.
- Under the Safety Representatives and Safety Committees Regulations 1977 trade unions are required to inform the employer of the groups of employees represented by the safety representative.

In Practice

Difference Between Unionised and Non-unionised Workplaces

4.8 If the workforce is represented by a recognised trade union, then the union may appoint a safety representative from among a group or groups of workers. The elected person's main function is to represent employees in consultation with employers on health and safety.

4.9 If there is not a recognised union representing employees, then the employer is still legally required to consult employees on health and safety. This can be carried out with each employee directly, or via one or more nominated representatives of employee safety.

4.10 It is important to clarify the terminology used. A representative of employee safety is someone elected under the Health and Safety (Consultation with Employees) Regulations 1996. A safety representative is someone appointed by a trade union under the Safety Representatives and Safety Committees Regulations 1977.

4.11 A significant difference between the two pieces of legislation relates to training. Where a safety representative is appointed by a trade union, the union will usually arrange whatever health and safety training is necessary. The responsibility for meeting the training needs of a representative of employee safety rests with the employer.

The Difference Between Informing and Consulting

4.12 An employer is already required by law to provide employees with information on:

- the health and safety risks identified by risk assessments
- the protective and preventive measures designed to ensure their health and safety at work
- emergency procedures for the organisation
- who is the competent person for health and safety
- any risks highlighted by another employer in the same workplace arising from that employer's business.

4.13 The crucial aspect of consultation is that employers are not just required to pass on information, but also to:

- actively invite feedback from employees
- listen to the views of employees
- take these views into account before any decision is taken.

4.14 Time must be built in to allow proper consultation to take place before a decision is made on any issue that might affect the health and safety of employees.

Union Safety Representatives

4.15 Union safety representatives must be members of the recognised union for their workplace. They must have been employed by the employer or have been in similar employment for at least two years.

4.16 Union safety representatives must also be able to keep up with any legislative and technological developments applicable to their workplace. It is vital that they understand hazards and control measures associated with the work activities of the employees they represent. This is of paramount importance where the industry is potentially high-risk (eg demolition, steel erection, etc).

Functions of a Union Safety Representative

4.17 Union safety representatives have a number of functions. They can:

- consult with employers on health and safety matters
- investigate and make appropriate representations on hazards, dangerous occurrences and accidents in the workplace they represent
- raise issues relating to general health and safety matters
- carry out three-monthly (or if necessary more frequent) inspections of the workplace
- consult with enforcing inspectors and be made party to any relevant enforcement-related information
- attend safety committee meetings.

What Should Representatives Be Consulted On?

4.18 Employers must consult on:

- the introduction of any new measures which may affect the health and safety of employees
- the appointment of competent people to assist the employer in complying with health and safety laws and with implementing emergency procedures
- any information that must be given to employees under health and safety law
- the planning and organisation of any health and safety issues
- the health and safety consequences of introducing new technologies into the workplace.

Information Employers Must Make Available

4.19 Safety representatives, upon request, are entitled to see and receive copies of any relevant health and safety documents relating to their workplace, including plans of proposed changes. Such documents may include:

- accident statistics
- accident reports
- cases of ill health
- reports of dangerous occurrences
- monitoring procedures.

4.20 Information which breaches national security, is involved in legal proceedings or breaks personal confidentiality, etc need not be made available.

Assistance Employers Must Provide

4.21 Employers must provide reasonable facilities and assistance for the safety representatives to perform their duties. This could include providing office space, secretarial support or time away from their normal work activities.

Time Off

4.22 Safety representatives are legally entitled to paid time off (during working hours) for performing their duties and attending any training associated with those duties. If an employer refused time off with pay, the representative could take the complaint to an employment tribunal. If the complaint were proven, the employer would have to pay compensation.

Safety Committees

4.23 The employer must establish a safety committee if it is requested by two or more trade union safety representatives. The committee must be set up within three months of the request. Details of membership and work areas covered must be posted around the workplace. The committee should be independent of both management and safety representatives.

Membership of Safety Committees

4.24 There are no definite rules on committee membership, although it should be compact and representative of the whole workforce (both employees and management). The number of management members should not exceed the number of safety representatives, but should include line managers and supervisors.

4.25 Specialists, eg works managers, personnel managers, etc may be invited to join the committee or may attend individual meetings as necessary.

4.26 Attendance at meetings should be considered part of the members' normal work. Meetings should be held as often as necessary, planned well in advance and made known to the workforce. The meetings should be minuted, with copies given to each member and distributed around the workplace.

Functions of Safety Committees

4.27 Safety committees should have the following objectives and functions.

- Study accident statistics and trends to develop reports for management.
- Examine safety audit reports.
- Analyse enforcement reports and establish a link with the enforcing officers.
- Consider safety representative reports.
- Develop, introduce and monitor safety rules and safe systems of work.
- Appraise safety training.
- Monitor the effectiveness of health and safety communication in the workplace.

4.28 Managers should ensure the committee has full and proper authority and should consider any points raised.

Non-unionised Consultation with Employees

4.29 In non-unionised workplaces, employers may consult with their employees either directly or through elected representatives of employee safety. In very small workplaces, direct consultation with employees may be more beneficial.

4.30 Self-employed people and non-employees, eg agency staff, volunteers, etc, do not have to be consulted. However, including them in consultation is good practice, particularly if they are working long-term at the employer's workplace. Trainees on work experience are considered to be employees and must therefore be consulted.

4.31 Consultation methods will vary. Examples include:

- health and safety committees
- newsletters

- notice boards
- employee surveys.

Functions of a Representative of Employee Safety

4.32 A representative of employee safety is entitled to:

- raise with the employer any general matters affecting the health and safety of the employees he or she represents
- highlight to the employer any potential hazards and dangerous occurrences at the workplace that affect the employees that he or she represents
- represent the interests of employees in any consultations at the workplace or with the enforcing authorities.

4.33 Representatives of employee safety should be consulted on the same matters as union safety representatives.

4.34 The employer must inform employees of the names of representatives and when consultation with these representatives stops, eg if the representative leaves the company.

Information Employers Must Make Available

4.35 Employers must provide any information necessary for employees or their representatives to effectively participate in consultation on health and safety. This information includes:

- likely hazards and risks associated with the work
- risks that might arise from proposed changes to the work
- implemented and proposed control measures, and safe working practices.

4.36 In addition, representatives of employee safety should be provided with information on reportable accidents in relation to the group of employees they represent.

4.37 Employers do not have to provide information if:

- there are legal proceedings involved
- it would breach national security
- personal confidentiality would be broken
- it would harm the employer's undertaking.

Training, Time Off and Facilities

4.38 In order to fulfil their duties and responsibilities, representatives of employee safety must be:

- appropriately trained
- allowed paid time off from their regular work for training, canvassing in elections
- provided with any reasonable facilities
- paid for any reasonable expenses, eg for travelling, training, etc.

Electing Representatives of Employee Safety

4.39 The representatives of employee safety must be elected from within the group of employees they represent. There is no guidance on the length of office for representatives of employee safety. Where there is a rapidly changing workforce, a representative may need to be elected more frequently in order to continue to remain the choice of those workers.

4.40 There is no limit on the number of representatives elected. The number selected should take into account the:

- size of the workforce
- different activities and/or shift work undertaken
- nature of the work.

4.41 All of the above points should be clarified and made known before an election is held.

Employee Involvement

4.42 In the modern workplace, to ensure the successful organisation of health and safety, the co-operation of employees and communication with them is fundamental. HSG65: *Successful Health and Safety Management* states that "participation by employees supports risk control by encouraging their ownership of health and safety policies. Pooling knowledge and experience through participation, commitment and involvement means that health and safety really becomes everybody's business."

4.43 The Health and Safety Commission (HSC) has stated that "an organisation's greatest asset is its workforce. Employees are often best able to spot issues and bring about real improvements. We need to expand the base of employee involvement in health and safety management to cover the whole workforce."

4.44 In 2004 the *HSC Statement on Worker Involvement and Consultation* was published. The purpose of this statement is to make the case for worker involvement and consultation and to highlight what a collaborative approach between partners can achieve. To be effective, the following principles should be applied.

- Value the contribution employees can make.
- Actively seek the view of employees.
- Involve employees as equal partners.
- Involve employees in all areas of health and safety management.
- Nurture, support and sustain involvement.
- Be prepared to adapt and change current practices and procedures.

Training

4.45 Where a safety representative is appointed by a trade union, the particular union will usually arrange whatever health and safety training is necessary for the individual.

4.46 The responsibility for identifying and meeting training needs for the representative of employee safety rests with the employer.

Health and Safety Training for Representatives of Employee Safety

4.47 The training needs for representatives of employee safety will vary widely from one organisation to another, and probably from one individual to another. Training needs should reflect the particular individual's previous knowledge and experience, and the extent of their role.

4.48 It is obviously important that any representative undertakes the same health and safety training that is offered to those that he or she represents. Additionally, a representative is likely to require more detailed training on:

- the legal requirements relating to health and safety at work and, in particular, any specific aspects that relate to the group of workers he or she represents
- the type and extent of hazards within the workplace, and the controls that are put in place to minimise the risks
- the employer's health and safety policy, and the arrangements in place to fulfil the policy commitments.

4.49 In most cases representatives of employee safety will need training in the skills to enable them to fulfil their function as two-way communicators. Additionally, the representative may be involved with any or all of the following, each of which may generate specific training needs:

- promoting workplace health and safety initiatives
- helping to develop the organisation's health and safety policy
- taking part in accident and incident investigations
- participating in safety audits
- helping in plant, equipment and premises inspections
- preparing safe working procedures
- membership of safety committees
- improving communication about health and safety
- evaluating the understanding of health and safety issues amongst colleagues
- identifying training needs
- maintaining competence through further training and self-development.

Appropriate Training Methods for Representatives of Employee Safety

4.50 Once training needs have been established there are a number of options available, including:

- in-company courses that can be delivered at convenient times and locations by a specialist training provider
- training by other employees within the organisation
- courses offered by trade associations, educational establishments, etc
- courses offered by trade unions for safety representatives, where the union is agreeable (although it is important to realise that the role of a union safety representative may be more extensive than that of a non-union representative and therefore the course may not be appropriate)
- distance learning.

4.51 Any training that is modular, or well-spaced over a period of time, is often likely to be more productive than an initial, very intensive block of training.

4.52 The employer is obliged to meet both the costs of training and any reasonable travel/subsistence costs. The employer is also obliged to provide reasonable time off with pay, during working hours, for relevant training.

List of Relevant Legislation

- Management of Health and Safety at Work Regulations 1999
- Health and Safety (Consultation with Employees) Regulations 1996
- Safety Representatives and Safety Committees Regulations 1977
- Health and Safety at Work, etc Act 1974

Further Information

Publications

The following are available from *www.hsebooks.co.uk*.

- HSG217 *Involving Employees in Health and Safety. Forming Partnerships in the Chemical Industry*
- INDG232(L) *Consulting Employees on Health and Safety: A Guide to the Law*
- L87 (rev 1996) *Safety Representatives and Safety Committees*
- L95 *A Guide to the Health and Safety (Consultation with Employees) Regulations 1996*

Contractors

- Contractors provide a service ranging from the performance of relatively simple tasks, such as maintenance of office machinery, to more hazardous activities, eg window cleaning, construction work or work with electrical systems.
- When employing contractors, both the client (usually the employer) and the contractor have legal responsibilities under the Health and Safety at Work, etc Act 1974. Each party has a shared duty of care to safeguard, as far as is reasonably practicable, the health, safety and welfare of employees and others who may be affected by the work activities (eg other persons on-site, the general public).
- Consideration should be given to the careful selection of competent contractors through a variety of methods such as recommendation, trade association links, questionnaires, etc.
- Where contractors are used regularly, employers may wish to keep a list of approved contractors who satisfy the chosen criteria.
- Prior to contractors starting work, a risk assessment should be carried out to determine the risks involved. The level of detail required in the assessment will depend on the level of expertise an employer has in relation to the contracting work being undertaken.
- From the outset, employers should ensure that accountabilities are clearly defined, so that all relevant parties agree and understand what they are responsible for.
- Employers should agree what amenities are to be available to the contractor's staff, eg catering, washing facilities, first aid, storage, etc.
- Contractors should be notified of any hazards and risks that are specific to the workplace.
- Employers might wish to ask contractors to prepare a method statement that could form the basis of a formal written safe system of work.
- The employer and the contractor must discuss any safety precautions that are necessary to the performance of the contract work.
- The two parties must agree on the method statement, the safe system of work to be followed, any permits-to-work that might be necessary and reporting procedures for accidents. Any other relevant information should also be exchanged to ensure the health and safety of all involved.
- Regular monitoring and review is an essential aspect of managing contract work, to make sure problems are dealt with swiftly and work has been carried out satisfactorily.

4.53 Contractors can be engaged to perform many activities. This could include relatively simple tasks, such as building maintenance and maintenance of office machinery. It also includes more hazardous work, eg entering into confined spaces, window cleaning or maintenance of electrical systems.

4.54 The use of contracting by organisations has greatly increased in recent years, allowing businesses to concentrate on their core activities. Accidents — and subsequent criminal prosecutions — have led to a plethora of case law, as well as huge sums having to be paid out in fines and compensation. Consequently, there is a need to manage the risks of contracting.

Employers' Duties

4.55 Employers are required, as far as is reasonably practicable, to:

- ensure the health, safety and welfare of their employees (including protecting them from contractors' activities)
- provide and maintain systems of work that are, as far as is reasonably practicable, safe and without risks to health
- provide adequate instruction and training (this duty has been extended to non-employees, ie contractors, by case law)
- carry out their activities in such a way as not to harm the health and safety of those not in their employment
- ensure the health and safety of persons other than their employees, including contractors.

4.56 The Management of Health and Safety at Work Regulations 1999 require employers to:

- carry out an assessment of the risks to the health and safety of their own employees and persons not in their employment, ie contractors, that might arise as a result of their (the employer's) work activities
- ensure co-operation with other employers, who have employees working in the same premises, on health and safety arrangements
- supply any necessary health and safety information to the employers of any visiting employees
- inform any such employees of any qualifications or skills necessary for them to carry out their work safely
- inform contractors of any health surveillance arrangements
- inform any employment agency of any qualifications or skills necessary for that employment agency's staff to carry out their work safely, and of any specific health and safety features of the work to be carried out
- be reasonably satisfied that any contractor assigned to construction work is competent to carry out the job.

4.57 The Construction (Design and Management) Regulations 2007 require that no persons — which will include employers — shall arrange for a contractor to carry out construction work unless they are reasonably satisfied that they are competent to carry out the work.

Contractors' Duties

4.58 Contractors, as employers in their own right, have health and safety statutory duties that are identical to those of an employer.

Self-employed Contractors' Duties

4.59 The Health and Safety at Work, etc Act 1974 requires the self-employed to carry out their activities, as far as is reasonably practicable, in such a way as not to harm the health and safety of those not in their employment.

4.60 The Management of Health and Safety at Work Regulations 1999 require the self-employed to:

- carry out an assessment of the risks to the health and safety of persons not in their employment that might arise as a result of their — the self-employed person's — work activities
- ensure co-operation with employers, who have employees working on the same premises, on health and safety arrangements
- supply any necessary health and safety information to the employers of any visiting employees
- inform any visiting employees of any skills necessary for them to carry out their work safely
- inform visiting employees of any health surveillance arrangements
- inform any employment agency of any qualifications or skills necessary for that employment agency's staff to carry out their work safely, and of any specific health and safety features of the work to be carried out.

Employees' Duties

4.61 Employees must:

- take reasonable care of their own health and safety and that of other people who may be affected by their work
- use machinery, equipment, dangerous substances, transport equipment, means of production or safety devices in accordance with any instruction and training given by the employer
- inform their employer of any danger to health and safety posed by a work activity
- inform their employer of any shortcomings in the employer's protection arrangements.

In Practice

Setting the Organisation's Policy for Contract Work

4.62 The organisation's policy and arrangements for contract work can take the form of a policy statement. The policy should include elements such as:

- a clear definition of roles within the organisation, including personnel responsible for contract work
- other arrangements, details of which will include:
  - procedures for assessing contractors
  - safety rules and procedures
  - equipment requirements
  - co-ordination of each contract
  - reporting of problems or safety issues
  - arrangements for meeting requirements under the Construction (Design and Management) Regulations 2007 (CDM 2007).

Determining What Activities Are to be Contracted Out

4.63 This will be determined as much by business considerations as by health and safety ones, eg the need to control operating costs and the availability of resources.

4.64 The main health and safety advantage of contracting activities out will be in the area of maintenance and operation of specialist plant and equipment. The use of contractors can also be advantageous for activities where the employer has little expertise.

4.65 Construction work, including major refurbishments or extensions, will almost always be carried out by contractors and the CDM Regulations will apply. These regulations set out roles and responsibilities for the various parties involved in the construction activities: namely the client, the designer(s), the CDM co-ordinator, as well as the principal contractor and (sub)contractors.

4.66 Some activities are contracted out for economic reasons, even though the employer has the expertise to carry out these activities themselves. These include:

- maintenance of fixed plant and equipment
- staff catering
- office cleaning
- window cleaning
- grounds maintenance
- routine building maintenance.

Disadvantages in Using Contractors

4.67 The main disadvantages involved with the use of contractors include the unfamiliarity of the contractor with the employer's:

- business
- management systems
- procedures
- work processes
- premises
- plant and equipment.

4.68 Activities such as these are within the control of the employer and hence a much greater degree of supervision will be required. Factors such as increased costs associated with extra supervision, together with the factors outlined above, need to be balanced against the costs of carrying out such activities in-house. This will determine whether a contractor should be used.

Risk Assessments of Contract Work

4.69 Regardless of the type of contract, the employer should carry out some form of risk assessment. The extent of the risk assessment will depend on three factors, namely whether:

- the employer directs the contractor in a similar way to an employee
- the contractor will be expected to control the work
- existing work is being transferred to contractors.

4.70 It should be noted that many smaller contractors, and the self-employed, will not have much experience in carrying out risk assessments. This is despite the fact that there is a legal requirement to carry them out. In these cases, the employer's assessments might be adapted by the contractor for their own use.

Where the Employer Directs the Contractor in a Similar Way to an Employee

4.71 Since the contractor is being treated almost as an employee, the employer will need to carry out the full risk assessment. If there is already an assessment in place, then it will still need to be reviewed.

Where the Contractor Will be Expected to Control the Work

4.72 In this case, the contractor has specialist knowledge, skills and equipment and the employer will expect the contractor to control the work. However, the employer will still need to carry out a simple risk assessment considering:

- the scope of the work
- the required tasks
- the hazards of the work as understood by the employer, eg will the contractor be exposed to hazardous substances?
- how the work might interface with the employer's existing safety management systems, procedures, work premises and plant
- the nature of the work, eg will it be necessary to enter into a confined space?
- whether or not the employer's work activities will affect the health and safety of the contractor's employees
- if the performance of the contracting work will impact on the health and safety of the employer's staff.

4.73 The findings of the employer's risk assessment will be forwarded to the contractor, possibly as part of the tendering process.

Where Existing Work is Being Transferred to Contractors

4.74 This type of contract could be the most difficult to manage. If the employer is already carrying out the work, it implies that he or she has a knowledge of the work. The employer's degree of control over the contract will subsequently be greater.

4.75 In these cases, the existing risk assessment should be reviewed.

Reviewing Existing Risk Assessments

4.76 The review should take the following factors into account and consider whether:

- any additional control measures needed over risks relating to the work, as a result of the transfer of work
- those who might be put additionally at risk by the contracting work
- any implications of the contracting work for the employer's remaining activities
- situations caused by the transfer of work that might lead to serious and imminent danger.

4.77 These assessments should form part of the documentation sent to prospective contractors. A pro forma risk assessment can be used and, if necessary, adapted for these assessments.

Selecting Competent Contractors

4.78 There are several considerations when selecting contractors, one of which is economic. For all construction work, there is a duty on the client to appoint competent contractors and so health and safety has to be a part of any employer's considerations, but it need not dictate the choice exclusively. For example, a contractor may have a "gold plated" health and safety arrangement and a good safety record. These arrangements may, however, be excessive for the purposes of the contract.

4.79 The key is to select a contractor with satisfactory health and safety arrangements, while being competitive and satisfactory in other respects. It is the whole package that has to be considered.

4.80 Contractors can be sourced via:

- personal recommendations
- trade organisations
- local authorities
- agencies and telephone directories
- questionnaires.

Personal Recommendations

4.81 One of the best ways of starting the process will be to obtain advice from others who have employed contractors. They may recommend a contractor who has worked, or is still working, for them. They may also eliminate some contractors altogether. You should not rely on personal recommendations alone, however. Some employers can get "cosy" with contractors and ignore, or not be aware of, their limitations.

Trade Organisations

4.82 Many trade organisations will have lists of recommended contractors, some of which they might have already put through the selection process.

Local Authorities

4.83 Some local authorities have a list of recommended contractors who have gone through a common selection process. Again, this should not be the only criterion against which a contractor should be selected; such contractors may not have maintained their standards and the extensive use of subcontractors may have led to a dilution of satisfactory arrangements.

Agencies and Telephone Directories

4.84 For relatively simple work, it might be necessary only to consult a telephone directory, or a similar publication. Such cases will include a one-off repair of a window, servicing a gas boiler or short-term office work.

4.85 Alternatively, employment agencies can be a valuable — although sometimes an expensive — source of contractors. In these cases, it might be necessary only to ensure that the contractor is a member of a trade association.

Questionnaires

4.86 If personal recommendations, trade organisations, local authorities, agencies and telephone directories prove unhelpful, prospective contractors will need to be vetted. The principal method of doing this is to send them a questionnaire that, on completion, can be evaluated.

4.87 Questionnaire responses will yield a "like-for-like" comparison between prospective contractors, in order for them to be measured against each other. This "score" can then be used as part of the wider process of selecting a satisfactory contractor.

4.88 Even if a prospective contractor emerges from a personal or trade association contact, questionnaires can be a sensible method of confirming the choice.

Provision of Information to Contractors

4.89 Many of the answers to a questionnaire sent to a contractor will depend upon the information provided to them by the employer. Before the questionnaire is sent to prospective contractors, therefore, the details and scope of the work required need to be formulated. In particular, any risk assessments must be included with this information. The employer should also supply information on any special control measures that it might require.

4.90 Under CDM 2007, there is a duty on the client to provide the contractor with suitable pre-construction information. This information must include all of the information in the client's possession (or which is reasonably obtainable), including:

- any information about, or affecting, the site or the construction work
- any information concerning the proposed use of the structure as a workplace
- the minimum amount of time before the construction phase which will be allowed to the contractors for planning and preparation for the construction work
- any information in the existing health and safety file.

Components of the Questionnaire

4.91 Any questionnaire used as part of the contractor recruitment process should include requests for the following detailed information:

- the company details, including size and number of employees
- any experience in the type of work to be carried out, including safety arrangements for this type of work
- personnel who would be involved in the project — in particular, the name and qualifications of the safety advisor/officer
- copies of risk assessments and method statements relative to the project
- details of health and safety training for managers, employees and any subcontractors who might be involved in the project
- outline of selection and control processes for any subcontractors
- details of procedures for accident and incident reporting
- information on any enforcement action(s) taken against the contractor by the Health and Safety Executive (HSE)/local authority
- summary of any accidents/incidents reported to the enforcing authority
- the contractor's arrangements for dealing with emergencies.

4.92 Care should be taken to request only the information that is strictly relevant to the project. For example, a request for the health and safety policy of a contractor might result in a huge document containing a mass of generic safety information with little relevance to the particular project.

4.93 Similarly, questions such as, "How many accidents have you had in the last three years?" might result in answers of "none". All this might prove, however, is that the contractor in question has no accident-reporting procedures.

Evaluating the Questionnaire

4.94 An evaluation form can be used to complement the questionnaire. Use of it will ensure a consistent approach to the contractor selection process.

4.95 The following are examples of questions that could be included as part of an evaluation form.

- Does the contractor have the necessary technical competence and experience for this particular project?
- Do the contractor's safety policy statement and arrangements for safety meet the requirements of the contract?
- Are the safety arrangements suitable and sufficient for the work to be carried out?
- Do the risk assessments adequately address the risks posed by the work?
- Have the personnel involved in the project received suitable health and safety training?
- Is a satisfactory system for selecting subcontractors (if applicable) in place?
- Are the accident-reporting and investigation procedures satisfactory, and will they meet the requirements of the project?
- Are the number of reportable accidents/incidents commensurate with the size and activities of the organisation? A simple scoring scheme that discriminates against the contractor with the most reportable incidents could be misleading.
- Does the contractor have satisfactory arrangements in place for dealing with emergencies?

4.96 The scoring of the evaluation can be simple and subjective, ie yes/no. Alternatively, scoring could be quantitative, by allocating marks to each response.

Other Forms of Evaluation

4.97 It must be emphasised that the questionnaire and evaluation process is only one part of the contractor selection process. For example, it might also be necessary to check all references and to confirm membership of a trade organisation.

4.98 Where the contractors will be carrying out construction work, this selection process could also be used for the selection of contractors under the Construction (Design and Management) Regulations 2007 and will help to demonstrate the competence of the contractors and compliance with these regulations.

4.99 It might be necessary, for some projects, to ask the prospective contractor(s) for safety method statements, an integral component of developing safe systems of work. This is common for one-off tasks, particularly in the construction industry, but can be used for almost any work where more detailed information about a particular task is required. Safety method statement evaluation will also be needed.

Use of Approved Contractor Lists

4.100 Once an organisation has identified a satisfactory contractor, it makes sense to use the same contractor for subsequent tasks. Many organisations, particularly larger ones, will add the names of such contractors to an approved list. This has obvious advantages in that the selection process does not have to be repeated.

4.101 The use of approved contractor lists is widespread. Indeed, many trade organisations, and other employers' organisations, keep such lists. These are available to their members.

4.102 These lists need to be regularly reviewed if they are to be successful. However, if a contractor's name is removed from a list after a review, do not automatically assume that this is because of the poor performance of the contractor. It could be that the employer managed the contract badly.

Managing the Contract Arrangements

4.103 The risk assessment process will determine how much control employers should have over the contractor. There are three main levels of control, although not all projects fit easily into one of these levels. The main levels of control are where:

- the contractor is integrated into the employer's organisation
- the employer exerts considerable control over the work
- the contractor exerts considerable control over the work.

Contractor is Integrated into the Employer's Organisation

4.104 Examples of where the employer almost exclusively controls the work performed include temporary office workers and maintenance workers permanently on-site. In these cases, it would be sensible to fully include them within the employer's own health and safety arrangements, if possible.

4.105 For example, the employer will be responsible for the provision of personal protective equipment or will at least ensure that the contractor obtains satisfactory equipment — and uses this equipment.

4.106 The employer should specify what training is required. This training can either be delivered by the employer or undertaken by the contractor prior to work commencing.

Employer Exerts Considerable Control Over the Work

4.107 In this situation, the contractor will have health and safety arrangements in place that are already approved by the employer. However, the employer will maintain considerable control over the work and, therefore, the safety of the work.

4.108 A contractor supplying catering services on-site, including food preparation, is an example of such a contractor. In this situation, employers must ensure two things. They must ensure that the contractor's own procedures for the maintenance of cooking equipment, and the contractor's own Control of Substances Hazardous to Health Regulations 2002 (COSHH) procedures, are both satisfactory and being implemented.

4.109 Similarly, in the event of a window cleaner arriving at an employer's premises in order to clean the company's windows, the employer has the same responsibilities. In addition, the employer must ensure that the premises are safe, in order for the window cleaning to take place. For example, safety hooks, used by the window cleaner to connect a safety harness when cleaning the outside of second storey windows, must be provided.

Contractor Exerts Considerable Control Over the Work

4.110 These are the types of contract where the employer has procured contractors for specialised work of which the employer has little knowledge or experience. Subsequently, the control over the work largely rests with the contractor.

4.111 Examples of such a situation are specialist activities, including the maintenance of high voltage electrical equipment, the cleaning of tanks that involve entry into confined spaces and the removal of asbestos.

4.112 Also included in this category are principal contractors appointed for the purposes of the CDM Regulations. These regulations, however, require co-operation and co-ordination, therefore making all persons involved in the construction work responsible for health and safety. It is not the sole responsibility of the (principal) contractor.

4.113 In these situations, the main consideration for the employer is the correct selection of a competent contractor. The employer will still have to carry out initial risk assessments and will also have to monitor the safety situation.

Supplying Contractors with a Safety Rule Book

4.114 One strategy for mitigating the risks involved in contracting out work is to issue contractors with a booklet providing guidance on safety issues. Such a booklet would include details of the employer's:

- security arrangements
- fire and emergency procedures
- health and safety rules
- general site facilities
- code of behaviour
- confidentiality policy.

4.115 The issue of a booklet of this sort should be accompanied by a training session. It might include a simple test to confirm understanding of the arrangements.

Monitoring and Reviewing the Contract Arrangements

4.116 This is an essential tool in managing contract work, regardless of the extent of control the employer has over the work. Monitoring should be in two parts:

- active monitoring
- reactive monitoring.

Active Monitoring

4.117 Active monitoring ensures that the requirements of the contract are being met. This can be broken down into two separate components:

- a desktop exercise
- workplace inspections.

Desktop Exercise

4.118 A desktop exercise measures the performance of the contract against the original specification. It can be part of a general performance review of the contract, not just a health and safety review. The main indicators against which the performance can be measured are:

- the production of risk assessments and subsequent method statements required as the contract progresses
- the quality and compliance with method statements
- specification of materials used, ie less hazardous substances
- accident/incident and near miss statistics.

Workplace Inspections

4.119 Workplace inspections should be carried out, preferably with a representative from the contractor. The inspections can be formal or informal. In the case of specialist work where the employer has little experience or control, however, these may be of somewhat limited value.

4.120 A workplace inspection considers the following points.

- Is the required work being carried out correctly?
- Are agreed work practices being used, ie using safe systems of work or permits to work?
- Are the agreed site rules being adhered to?
- Are there changes in personnel that might impact on training or information requirements?
- Are there any unforeseen issues that require changes in work patterns or in the actual function being provided by the contractor?

Reactive Monitoring

4.121 This type of monitoring examines the health and safety performance of the contractor in terms of accident/incidents that may have occurred. The effectiveness of reactive monitoring will depend on there being adequate — and agreed — methods of accident reporting for the project.

4.122 The accident-reporting procedure should include events that might result in reports being necessary under the Reporting of Incidents, Diseases and Dangerous Occurrences Regulations 1995, and also accidents that result in other injuries and near misses.

4.123 Statistics derived from this procedure can be a useful indicator of safety performance. Over a period of time, they can give a measure of improvement — or otherwise. On some major projects, these figures can be compared with targets set at the commencement of the contract.

Training

4.124 Training for contract work, in most instances, will cover the procedures for managing the contracting process.

4.125 The vetting of contractors, using a questionnaire and an evaluation form, will generally be carried out by a safety advisor/officer. This person should have the necessary skills to carry out this work without additional training. The safety advisor/officer will also monitor the contract.

4.126 Managers and supervisors (including the safety personnel) with responsibility for managing contract work will need to know all details of the project and its agreed safety measures. This can generally be achieved through a briefing exercise.

4.127 All staff coming into contact with contractors and their contract activities should have some basic training as part of their induction. Matters that should be the subject of this training include details of:

- any contract work in place
- any hazards arising from the work that might involve them
- control measures for the risks that might arise from these hazards
- individuals' responsibilities in relation to contract work
- how to report concerns regarding such work
- emergency arrangements in addition to the normal workplace arrangements.

List of Relevant Legislation

- Construction (Design and Management) Regulations 2007
- Management of Health and Safety at Work Regulations 1999
- Reporting of Incidents, Diseases and Dangerous Occurrences Regulations 1995
- Health and Safety at Work, etc Act 1974

Further Information

Publications

HSE Publications

The following are available from *www.hsebooks.co.uk*.

- HSG159 *Managing Contractors: A Guide for Employers*
- INDG268 *Working Together: Guidance on Health and Safety for Contractors and Suppliers*
- INDG368 *The Use of Contractors: A Joint Responsibility*
- L144 *Managing Health and Safety in Construction. Construction (Design and Management) Regulations 2007: Approved Code of Practice and Guidance*

Organisations

- Asbestos Removal Contractors Association (ARCA)
  Web: *www.arca.org.uk*
  The ARCA aims to promote and maintain the safe working standards required for the handling and removal of asbestos and other hazardous materials.
- Confederation of Roofing Contractors (CORC)
  Web: *www.corc.co.uk*
  The CORC is a membership-based roofing organisation. Members undergo a vetting procedure.
- Council for Registered Gas Installers (CORGI)
  Web: *www.trustcorgi.com*
  CORGI is the national watchdog for gas safety in the UK, charged by the HSE to maintain a register of competent gas installers. It aims to improve standards in the gas industry and help consumers to find and use safe trades people.

- Lifting Equipment Engineers Association (LEEA)
  Web: *www.leea.co.uk*
  The LEEA represents, supports and promotes organisations engaged in the design, manufacture, testing, examination, supply, hire, service and repair of lifting equipment. LEEA is also involved in the training of those performing the functions.
- National Inspection Council for Electrical Installation Contracting (NICEIC)
  Web: *www.niceic.org.uk*
  The NICEIC maintains a list of approved contractors.
- Painting and Decorating Association (PDA)
  Web: *www.paintingdecoratingassociation.co.uk*
  The PDA provides information on technical issues, training, employment and contractual issues.

Lone Workers

- The HSE defines lone workers as those who work by themselves without close or direct supervision.
- There is no general legal prohibition on lone working, but it is prohibited in some work situations and activities.
- Certain regulations specify requirements for supervision, assistance or accompanied working if work is hazardous or involves certain categories of workers.
- Lone working can bring additional risks, such as the inability to summon help in an emergency, so a risk assessment must be conducted, taking into account the hazards of the specific job and employees involved. Certain employees may be at particular risk when working alone, such as new and expectant mothers.
- Employers must develop procedures to control risks and protect employees from the dangers of lone working.
- Employers should ensure:
  - lone workers have full knowledge of the hazards and risks they are exposed to
  - lone workers know what to do if something goes wrong
  - someone else knows the whereabouts of the lone workers and what they are doing.
- The level of supervision for lone workers should be determined by a risk assessment to ensure it is consistent with the possible risks and that there is a system for maintaining contact.
- Arrangements should be in place to protect or assist lone workers in the event of fire, accident or illness or an incident of violence.
- Lone workers should receive appropriate training to ensure they are competent and able to deal with foreseeable problems.
- Lone workers should receive training in the use of any necessary tools and equipment. The tools or equipment should be safe and correct for their intended use, bearing in mind any increased risk to lone workers.

4.128 The Health and Safety Executive (HSE) defines lone workers as those who work by themselves without close or direct supervision.

4.129 Public sector union UNISON defines lone workers as those "whose activities involve a large percentage of their working time operating in situations without the benefit of interaction with other workers or without supervision".

4.130 This definition clearly covers those workers who carry out their work in isolation from others for significant periods, such as:

- people working at different times from others, eg:
  - cleaners
  - caretakers
  - security staff
- people who work with other people, such as members of the public, but without contact with other workers or supervision
- people who travel alone for significant periods or in circumstances that may give rise to significant risks, such as:
  - driving long distances
  - travelling at night
  - travelling in dangerous conditions.

4.131 Employers must ensure that lone workers are at no more risk than other workers. Therefore, the risk factors which relate specifically to lone workers must be assessed by employers and appropriate action taken to reduce the risks.

Employers' Duties

4.132 There is no general legal prohibition on lone working, but it is prohibited in some work situations and activities.

4.133 Certain regulations set out specific requirements for supervision, assistance or accompanied working if:

- hazardous work is carried out
- particular categories of workers, such as young or inexperienced workers are carrying out activities that require direct supervision.

4.134 These activities include:

- hazardous electrical work
- some manual handling activities
- erection of scaffolding
- use of unsupported temporary access equipment
- demolition on construction sites
- diving operations
- young people doing hazardous woodworking
- work with certain chemicals
- work in confined spaces.

4.135 The Provision and Use of Work Equipment Regulations 1998 (PUWER) and the Lifting Operations and Lifting Equipment Regulations 1998 (LOLER) may apply to some activities where lone working, or lone working by certain categories of workers, may be hazardous and therefore prohibited, eg:

- lone use of some temporary access equipment
- unsupervised use of wood lathes by young or inexperienced workers.

General Duties

4.136

- Under the Health and Safety at Work, etc Act 1974, employers must ensure:
  - so far as is reasonably practicable, the health, safety and welfare at work of all employees
  - a safe system of work for all workers, including lone workers
  - that any lone working activities they control do not endanger others not in their employment.
- The Management of Health and Safety at Work Regulations 1999 require employers and the self-employed to:
  - carry out suitable and sufficient risk assessments of all hazardous activities and put in place measures to prevent and control the risks to both physical and mental health
  - identify vulnerable groups who may be at particular risk of harm, and record these in the risk assessment findings
  - consult employees and their representatives about health, safety and welfare matters, including issues affecting lone workers

  - co-operate with other employers where a lone worker is working at another employer's workplace.
- The Workplace (Health, Safety and Welfare) Regulations 1992 require employers, amongst other things, to provide a safe workplace and adequate welfare facilities whenever employees are working. Domestic premises are excluded.
- The Health and Safety (First-aid) Regulations 1981 require employers to provide adequate and appropriate equipment, facilities and personnel to administer first aid and to make these available to all employees, including lone workers, at all times people are at work.
- Employers and the self-employed must:
  - assess risks of fire and inform all employees, including lone workers and contractors and visitors on-site, of fire precautions, first-aid arrangements and emergency procedures
  - co-operate with other employers in shared workplaces or managed premises where lone workers are working
  - Fire risk assessments must cover fire risks to all workers, including lone workers. Lone workers such as night cleaners should be included in all the arrangements made to inform, instruct and train workers in fire precautions and emergency evacuation procedures.
- The Personal Protective Equipment at Work Regulations 1992 and the Provision and Use of Work Equipment Regulations 1998 apply to equipment provided for the lone worker's protection.
- Employers must ensure that any equipment and communications systems designed to protect the lone worker or to provide effective supervision, eg mobile phones, personal alarms or closed-circuit television, are:
  - correctly installed and used
  - maintained in good working order
  - regularly tested to ensure they are working properly.
- Employers must establish safe systems for the purposes of:
  - responding to requests for help from lone workers using equipment
  - supervising lone workers' activities using communications equipment provided.

4.137 The systems must be operated correctly and properly staffed by the designated person at all times that lone work is taking place, to ensure a safe system of work as required by the Health and Safety at Work, etc Act 1974.

Employees' Duties

4.138 Under the Health and Safety at Work, etc Act 1974 and the Management of Health and Safety at Work Regulations 1999, employees have a duty to:

- take reasonable care of their own health and safety and that of other people who may be affected by their activities at work
- co-operate with their employer to enable the employer to comply with health and safety duties.

In Practice

Preliminary Audit of Lone Working

4.139 The purpose of the preliminary audit is to:

- determine when and where lone working is taking place
- determine who is working alone
- determine what work lone workers are doing
- identify where lone working risk assessments are needed.

4.140 Strict definitions of what constitutes lone working do not necessarily serve a useful purpose. How far away must other people be, and what intervals must lapse between contacts, before special arrangements for lone working are required?

4.141 Lone working can be the defining characteristic of a job, or simply contingent circumstances in which workers may occasionally find themselves. For instance, in every workplace there will always be somebody who is the first to arrive, and somebody who is the last to leave. Likewise, an employee may have to go to a storeroom to collect items unaccompanied. In all such circumstances, the individual is unarguably a lone worker in the strict definition of the term. Therefore, when considering this subject, it is important to distinguish between the chance or random occurrence of finding oneself on one's own, and work which is specifically intended to be carried out in isolation, and may last for some time.

4.142 Lone working is often not recognised. If lone workers are out of sight, they may be out of mind when other risk assessments are conducted. They may be forgotten or slip through the safety net. Host employers may fail to include them in the risk assessments or they may be included in generic risk assessments. The impact of lone working may be overlooked.

4.143 Employers should:

- identify all lone working situations involving employees
- identify everyone who works alone for any significant period of time, including vulnerable groups such as young and inexperienced workers, remembering that they may not be working alone the whole time
- identify any work activities where regulations prohibit lone working.

4.144 This can involve using several approaches as outlined below.

Workplace Inspections

4.145 Workplace inspections should be conducted at all relevant times, in order to:

- observe situations and locations in which lone working is taking place
- observe the type of work activities, equipment, materials and work environments involved
- check workplace health, safety and welfare, including access arrangements and access to welfare provisions and first aid, outside normal hours
- check who is on-site, including visitors, contractors and members of the public, at times when employees are working.

Mapping Techniques

4.146 Mapping techniques can be used to produce a visual record of when and where people are working alone, and which workers are involved.

4.147 These techniques can be used in conjunction with other activities, but examples include workplace, working time and occupational mapping.

Workplace Mapping

4.148 Workplace mapping is used to identify and record locations and mobile work patterns where lone working is taking place. This involves sketching a diagram of the workplaces involved and marking it to show where lone working is taking place.

Working Time Mapping

4.149 Working time mapping is used to identify when lone working occurs in a 24-hour period, a 7-day week and over the year, especially in the case of seasonal or fluctuating demand, or where bank holiday working is required. This involves marking the relevant times on a chart or calendar.

Occupational Mapping

4.150 Occupational mapping is used to identify and record any occupational groups involved in lone working activities, eg drivers, community health workers and security guards. This involves charting the occupational groups involved in lone working activities and/or marking the occupational groups involved on the workplace map.

Consultation with Employees and/or their Representatives

4.151 Consultation with employees and/or their representatives is important in order to:

- gather information about any problems employees have experienced or witnessed regarding lone working
- seek employees' views about possible solutions to any problems related to lone working.

Generic Risk Assessments

4.152 It might be useful to refer to a generic risk assessment of work activities or locations to identify:

- activities involving lone working, including movements between sites
- locations where lone working is required, including work outdoors or in the community
- activities involving possible risks of violence where employees may be working alone from time to time
- activities where lone working is either prohibited or allowed only by experienced workers or workers with special training or qualifications
- activities involving increased risks of harm where hazardous work is carried out by one person rather than two, or in isolation from others, or unsupervised.

Risk Assessments for Individuals and Personal Safety Plans

4.153 It might also be useful to refer to individual risk assessments or personal safety plans for employees who are particularly vulnerable to risk if working alone because of individual factors such as complications of pregnancy, disability or health problems.

4.154 Other sources of information include time sheets, overtime records and documentation relating to travel and transport arrangements.

4.155 Findings can also be recorded using a form or checklist.

4.156 Any lone working activities identified as prohibited activities should be stopped immediately. The identified lone working activities which remain must then be evaluated to determine what action is required to prevent and control any risks posed, taking account of the effectiveness of any control measures already in place. This involves conducting a full lone working risk assessment.

Conducting a Full Lone Working Risk Assessment

4.157 A risk assessment must be suitable and sufficient, meaning it must look at the lone working situation and anticipate all reasonably foreseeable factors. It is important to ensure the person carrying out the lone working risk assessment is competent to do so.

4.158 A full risk assessment should take into account:

- work activities
- workload
- job design
- working environment
- timing of work
- working conditions of lone workers
- relevant individual factors.

4.159 It should also consider:

- the combined effects of any risks, such as working alone and at night in direct contact with members of the public
- the effectiveness of any control measures already in place.

4.160 A full risk assessment of lone working involves applying the HSE's *5 Steps to Risk Assessment* as follows.

- Identify the hazards of the job or activity, and of lone working in relation to it.
- Identify the people at risk, including people who may work alone for part of the day, vulnerable groups and people other than employees who may be at risk if a lone worker is unable to carry out their tasks alone without putting either themselves or others at risk.
- Evaluate the risks and decide whether existing control measures are adequate or whether further action is needed, and consult employees and their representatives.
- Record the findings and implement the measures.
- Monitor the situation and review the assessment on a regular basis and check whether there are changes. Revise if necessary.

Identify the Hazards of Lone Working

4.161 Lone working may be both:

- a risk factor — something which increases the likelihood of harm occurring because it makes the worker more vulnerable or susceptible to harm
- a hazard — something with the potential for harm, eg carrying out a task single-handed that requires two people to carry it out safely.

4.162 Risks to lone workers include:

- increased risks of accidents, injury or ill health, such as musculoskeletal injuries or stress-related ill health
- increased vulnerability in the event of:
  - violence, eg involving service users or members of the public
  - injury, sudden illness, fire or other emergency due to lack of assistance
  - ill health, eg stress-related illness and complications arising from lack of assistance in a medical emergency.

4.163 The risks of lone working can be associated with:

- what people are doing (their work activities)
- organisational working practices (staffing arrangements, mobility, contractual arrangements)
- where people are working (the workplace or premises involved)
- when people are working (shift patterns or working hours, including overtime).

4.164 Further risks arise where lone workers are required to carry out work in addition to their normal duties on their own, such as covering for an absent colleague. This may mean:

- workload intensifies
- working hours increase
- the lone worker cannot leave the workstation safely
- existing control measures cannot be implemented.

4.165 Sources of information about the hazards of lone working include employer's records and relevant guidance, such as:

- accident, incident and sickness records relating to lone working
- staff records relating to:
  - expectant, new and nursing mothers
  - young and inexperienced workers
  - workers with disabilities or health problems that increase their vulnerability
- complaints and information from staff involved in lone working, including feedback from:
  - performance appraisal
  - exit interviews
  - anonymous staff surveys
  - staff meetings
  - team meetings
  - group discussions
- guidance produced by the HSE, employers' associations, unions and organisations such as the Suzy Lamplugh Trust
- information or guidance produced by local community safety officers/police.

4.166 The following factors should be taken into consideration in identifying the hazards associated with lone working.

Work Activities

4.167 Work activities include all the tasks and responsibilities carried out by the employee in the course of their employment, and the working methods used, including:

- travelling between sites
- driving
- going into and out of workplaces
- communicating with others, ie face-to-face, in writing, by telephone or electronically.

4.168 The following features of the lone working activities should be considered as part of the risk assessment.

Tasks

4.169 Assessors should ensure that any tasks undertaken by the lone worker can be done safely and without risks to health.

4.170 Tasks comprise:

- specific actions
- postures
- movements
- behaviours
- information processing
- decision making.

4.171 Hazardous tasks for lone workers include:

- working at height or in confined spaces
- hazardous manual handling
- handling certain hazardous substances, eg fumigants
- use of some temporary access equipment
- use of hazardous work equipment, such as lathes or mills
- diving operations
- travelling to and from remote locations
- conducting home visits or working in the community
- dealing with the public, especially difficult, distressed or disturbed people
- dealing with complaints or conflict.

4.172 Assessors should take into account the control measures in place, but also recognise that when working alone, the assistance normally available to workers from co-workers or supervisors cannot be summoned immediately.

4.173 The lone worker may feel obliged to undertake activities or tasks to which they are unsuited or know to be risky, in the absence of assistance or support.

Roles and Responsibilities

4.174 Specific jobs and tasks may involve responsibilities which make the task more hazardous for lone workers. Responsibilities cover the duties the employee has towards others.

4.175 Responsibilities which can increase risks to lone workers include:

- responsibilities for people, property or premises, such as:
  - protecting people, cash or valuables
  - caring for children or for older, frail, sick or disabled people
  - dealing with emergencies or life-threatening situations
  - safety-critical work
  - taking decisions that significantly affect people's lives or livelihoods
- enforcing rules and regulations, ie:
  - taking children into care
  - taking people into custody
  - guarding prisoners or property
  - controlling access to premises, services, benefits or places of entertainment
  - controlling crowds and public behaviour
  - controlling access to drugs and alcohol
  - serving court orders or injunctions
  - recovering debts.

4.176 Assessors should take account of the lone worker's role and responsibilities, and their ability to fulfil these without risks to their own or other people's health and safety.

Hazardous Workloads

4.177 Assessors should ensure that workloads are taken into account when assessing risks to lone workers. The workload is the sum of work demands made upon the lone worker from all sources in the course of their work. It includes physical, mental and emotional demands.

4.178 Aspects of workloads that can be particularly hazardous to lone workers include:

- excessive, intensive, unpredictable or unmanageable demands
- lack of control or autonomy over the workload or how demands are met
- timing and duration of demands which result in extended periods of lone working without adequate support, supervision or constructive social interaction
- arduous physical work, such as climbing or strenuous manual handling which increases demands for physical effort and balance
- stressful work, such as dealing with difficult problems, situations or people, which increases demands for mental effort and judgment, eg when dealing with the public.

Job Design

4.179 Job design covers the way the different demands are organised and prioritised.

4.180 Poor job design can mean risks to lone workers if they are required to carry out potentially hazardous tasks alone or for long periods without adequate support, supervision or constructive social interaction, or if lone working increases risks of work intensification, stress, repetitive strain or other injury.

Hazardous Working Environments

4.181 The working environment covers both the physical and psychosocial work environments which can impact on the health, safety and welfare of lone workers.

4.182 The working environment should be safe for all workers, but lone workers may be especially at risk compared with other workers. Working alone may mean they cannot rotate tasks with others or vary their working environments to limit exposure to environmental hazards.

4.183 Assessors should evaluate all lone working environments at all times that work is carried out, and consider:

- the impact of lone working on the duration, frequency and timing of exposures to physical, biological, chemical, psychosocial and environmental hazards
- the relevant control measures in place for lone workers.

4.184 No worker should be placed at significant risk from their work environment, and lone workers should be at no greater risk than any other worker.

4.185 If the evaluation shows it is not possible to work alone safely in the work environment concerned the lone working should be stopped. If that is not possible, then appropriate control measures should be implemented to minimise the risk.

Hazardous Working Time Arrangements

4.186 Working times are the times when work takes place. These can affect risks to all workers, but may pose particular risks to lone workers.

4.187 Assessors should consider the times when employees are working alone, for the following reasons.

- The work activities of the lone worker may take place after other people have left the workplace, leaving lone workers isolated and unsupervised, eg cleaners, office workers working outside normal hours, people with responsibility for securing premises.
- The fact that an employee or contractor works at different times from other people may mean they are forgotten or excluded from risk assessments.
- Working outside normal working hours, especially during hours of darkness or when there are few people on the streets, can make lone workers more vulnerable to assault, or fear of assault, when travelling or when entering and leaving the workplace.
- Weather conditions can vary significantly at different times of the year, and can affect the risks associated with the work activities of lone workers, such as agricultural, construction and transport workers, and other people who work outdoors, such as workers in large chemical processing plants or utilities.
- Access to transport facilities can vary at different times of the day, week and during holiday periods.
- There may be no access to welfare facilities outside normal working hours.
- Heating and ventilation systems may be shut off when buildings are unoccupied, and windows may be alarmed or locked.
- Emergency procedures and first-aid provisions may not be operative outside normal working hours.
- Security arrangements may not operate at certain times.
- Assistance, back-up and remote supervision may be limited to certain times of the day or week, and holiday periods may be excluded.

Hazardous Working Conditions

4.188 Working conditions are the terms and conditions of employment and the organisational working practices governing the employment relationship. They include:

- shift patterns and working hours
- contractual arrangements, and terms and conditions of employment
- staffing levels and skill mix
- management and supervisory arrangements (including management of health, safety and welfare at work)
- information and communications systems.

Shift Patterns and Working Hours

4.189 Shift patterns and working hours can determine the working times of lone workers and their relationship to workplace facilities and other people's activities — both in the workplace and the community.

4.190 The duration and patterns of shifts, rest and meal breaks, the length of the working week and any overtime or unsociable hours working can affect the risks to lone workers, eg starting and finishing times may place lone workers at greater risk.

Contractual Arrangements, and Terms and Conditions of Employment

4.191 Contractual arrangements, and terms and conditions of employment govern the relationship between employer and employee and any:

- contractor
- supplier
- premises manager
- employment agency.

4.192 Contractual arrangements, and terms and conditions of employment can affect lone working in the following ways.

- They govern working hours, leave arrangements, overtime and any facilities provided for transportation, health care and vocational training, as well as pay and pensions, employees' rights and responsibilities.
- They govern the times when other people are working on their own site, how employees work on third-party sites, and the facilities available to them, including first aid and welfare facilities.
- They may affect the timing of deliveries and arrival and departure of other workers on-site, including security and cleaning staff.
- If the evaluation shows that these governing factors are associated with risks to lone workers in particular, after taking into account any control measures in place, steps should be taken to make adjustments or changes where necessary to avoid and control the risks.

Staffing Levels and Skill Mix

4.193 Staffing levels and skill mix can have implications for lone working and can determine how and when employees work alone, as well as what they do and the degree of supervision provided.

4.194 Employers must ensure they provide appropriate levels of competent supervision for workers undertaking hazardous activities where the control measures required include direct supervision or accompanied working. If the worker concerned is young or inexperienced, their vulnerability should be considered.

Management and Supervisory Arrangements and Communications

4.195 Management and supervisory arrangements include the management of:

- people
- premises
- resources
- equipment
- health, safety and welfare at work.

4.196 The matters to be considered include:

- line management and supervisory arrangements
- maintenance regimes
- responsibilities for arrangements with contractors and service providers, eg premises management, security and cleaning
- safety systems and occupational health arrangements
- safety representatives and safety committee arrangements
- workplace inspections and accident reporting procedures
- policy statements and relevant procedures
- risk assessment, monitoring, audits and reviews
- information and communications systems, including lone workers' access to any company intranet or safety information and consultative procedures.

4.197 Assessors should consider whether there are any arrangements which fail to address the needs of lone workers, or specifically exclude them. If so, steps must be taken to ensure lone workers are included within the scope and operation of all relevant arrangements.

Identifying People at Risk

4.198 Those at risk may include:

- lone workers in general
- employees particularly at risk when lone working because of individual factors
- workers classed as vulnerable and who are therefore at greater risk when lone working
- non-employees who may be at risk of the activities of a lone worker.

Lone Workers in General

4.199 Lone workers may be:

- working alone at a fixed establishment
- working alone away from a fixed base
- mobile, covering a number of workplaces, including third-party sites
- doing outreach activities, eg in community-based work
- travelling alone between sites or locations.

4.200 Lone workers may be:

- alone for part or all of their time at work, eg cinema projectionists, security guards on patrol, night cleaners and postal workers
- alone at particular times, eg interviewers conducting one-to-one interviews or treatments in a private room
- alone when travelling or moving between sites, eg maintenance engineers covering third-party sites, delivery drivers and travelling sales representatives
- working alone in unpopulated or remote workplaces, outdoors or empty premises, which may be controlled by their employer or by a third party, eg night cleaners, security guards, agricultural workers and telecommunications engineers
- working alone from home for part or all of the time, eg homeworkers and teleworkers
- working alone from temporary locations or public facilities, eg travelling sales representatives
- working alone when travelling abroad or attending meetings and conferences on behalf of their employer, but not temporarily based with a host employer.

4.201 The risks to lone workers vary depending on the work activities, the hazards and the work situation.

Individual Factors

4.202 Each individual is different in terms of their physical and mental capabilities. Individual factors should be assessed in consultation with individual lone workers.

4.203 If the assessment shows that the employee's individual circumstances, health, condition, disability or capabilities mean they would be at increased risk when working alone, then, after taking account of any control measures in place, lone working should be avoided.

4.204 If the risks can be effectively controlled, steps should be taken to adjust the work or working conditions or provide the individual with additional support, back-up and protection.

Examples of Lone Workers in Different Categories

Type of Work	Typical Type of Lone Worker
One person in a fixed establishment	Small workshop staff, petrol station staff, kiosk attendants, shop workers and homeworkers
One person working separately from others in a fixed establishment	Factory workers, warehouse workers, research and training workers, leisure centre and fairground staff
One person working outside normal hours in a fixed establishment	Cleaners, security staff, special production staff, maintenance and repair staff
One mobile worker working away from a fixed base	Construction workers, plant installation workers, maintenance and cleaning workers, lift repair staff, painting and decorating staff, vehicle recovery workers, agricultural and forestry workers, rent collectors, postal staff, social workers, home helps, district nurses, pest control workers, drivers, engineers, architects, estate agents, sales representatives, professional staff and journalists

Vulnerable Groups

4.205 Certain groups of workers likely to be at particular risk are classed as vulnerable. Vulnerable groups include:

- women
- new and expectant mothers
- disabled workers
- people with health problems
- young and inexperienced workers.

4.206 Lone working may pose particular risks to these groups.

Women

4.207 Generally speaking, women are particularly vulnerable to risks of violence, especially when working alone outside normal working hours.

4.208 Women working alone may be seen as soft targets, both by strangers and perpetrators of domestic violence. Young women may be at particular risk of sexual harassment, stalking, abduction or rape. Although these risks are not confined to lone workers, women are more vulnerable when working alone.

New and Expectant Mothers

4.209 Employers are required to carry out both a generic risk assessment for hazards to reproductive health and a specific risk assessment for the individual expectant, new or nursing mother.

4.210 Employers should ensure lone working is included in this risk assessment. When assessing the risks of lone working for new and expectant mothers, assessors should consider:

- whether lone working will involve manual handling, work at height or using temporary access equipment, as lone working may increase the risks of unassisted manual handling that would not be hazardous but for pregnancy or recent childbirth
- the presence of any known or foreseeable stressors, as lone working is also a potential stressor and may increase these risks
- the woman's condition and any medical advice
- working hours and shift patterns
- whether the woman is working alone, pregnant or a new mother, these factors may make the woman more susceptible to stress, accidents and ill health or more vulnerable in the event of a fire, injury, medical emergency or other unforeseen event
- whether the woman will be able to summon help immediately if necessary, and whether that help will be forthcoming at all work times.

People with Health Problems or Disabilities

4.211 Generic risk assessment should already take into account those workers with health problems or disabled workers. However, if individuals are suffering from health problems or disabilities, lone working may place them at greater risk. Advice should be sought and the worker consulted.

4.212 For example, lone working activities that would be perfectly safe for others may pose particular risks to an employee who needs supervision or assistance in the event of an emergency.

4.213 To assess medical suitability, the assessor should first consider the following questions.

- Does the job impose any extra demands on the lone worker's physical or mental stamina?
- Does the lone worker suffer from any illness or other condition that might increase the risks of the job?

4.214 The employee, and medical advisor as appropriate, should be consulted to identify any medical issues which might increase the risks or make it unsuitable for them to work alone.

4.215 Health records and medical reports should be handled in strict confidence in accordance with statutory requirements and data protection law.

4.216 Employees with sensory disabilities may need to be advised not to work alone unless the risks of difficulties arising from an emergency can be avoided or controlled.

4.217 A personal safety plan or special communications provisions may help to ensure they can use a means of escape safely and promptly, but it will be important to ensure that other people can assist them if necessary. However, employers must:

- avoid blanket restrictions on where, when and whether the disabled employee can work
- take practical steps to control the risks or make reasonable adjustments to staffing levels, work activities, equipment, communications systems or locations.

Young and Inexperienced Workers

4.218 Young and inexperienced workers are generally more at risk than older, more experienced workers. These risks may be increased when working alone or without competent supervision. Young and inexperienced workers may be at greater risk because:

- of their inexperience of the job or their lack of familiarity with the working environment

- of their vulnerability to acts of violence (they are popular targets for violence and may not have enough experience to handle conflict in a way that reduces the risks of escalation or assault)
- they may lack knowledge and understanding of safe systems of work and the reasons for using them
- they may be unfamiliar with hazardous or complex work equipment, the safety procedure or personal protective equipment (PPE) or how to use them correctly
- they may not know how to get help, or when.

4.219 Employers are required to assess the risks to inexperienced or young workers and to ensure they are provided with appropriate levels of supervision, training and support.

4.220 Assessors should take account of any control measures in place and any legal provisions prohibiting lone or unsupervised working by young or inexperienced workers.

Non-employees at Risk from Lone Workers

4.221 Non-employees may be at increased risk from employees working alone and this should be taken into account during the risk assessment, eg:

- if a home carer is unable to carry out a lifting operation safely involving a client with mobility problems, those at risk may be both the lone worker and the client
- if the lone worker is an overworked childcare worker, the people at risk if a problem occurs at work are the children in care
- if the lone worker is an overtired coach driver, the people at risk are the coach passengers and others on the road, as well as the driver.

Examples of Hazards and Risks of Lone Working

Hazard	Risk to Lone Worker
Unassisted manual handling	Musculoskeletal injury
Handling cash or valuables, including valuable portable work equipment	Increased risks of being targeted by violent people because of working alone, possibly resulting in physical or mental injury, potential fatality, stress-related ill health
Driving long distances without adequate rest breaks/change of driver	Increased risks of road traffic accidents/falling asleep at the wheel
Poor working relationships/poor organisational culture	Increased risks of lone workers being bullied or harassed in the absence of witnesses, possibly resulting in stress-related illness or assault

Evaluate the Risks of Lone Working

4.222 Risks should be evaluated in relation to their source — the hazard. The process involves:

- identifying the risks
- identifying the people at risk
- evaluating the likely severity and extent of the risk.

Control Measures

4.223 Control measures should address risks at the source wherever practicable. The first task is to eliminate lone working in any circumstances in which it is prohibited.

4.224 In general, the measures should:

- address risks at the source
- reduce risks to a minimum if they cannot be removed completely
- prevent lone workers being exposed to foreseeable risks, or minimise exposure if the risks cannot be avoided entirely
- enable them to deal with any problems, eg threats of violence, should they be encountered.

4.225 Finally, employers should consider measures, such as first aid, recovery measures and counselling, for dealing with lone workers who have experienced violence or stress at work. In this way, an overall preventive and protective strategy can be developed.

4.226 A number of precautions can be taken to reduce the risks of working alone.

4.227 Lone working should be avoided if there is a risk of violence which cannot be adequately controlled. Depending on the situation, two-person working may be required, or additional measures may be needed to provide the necessary support when the lone worker is involved in complex decision-making, problem-solving or conflict resolution.

Example of Control Measures for Lone Working

Issue	Examples of Possible Control Measures
Ensuring effective management and supervision	• active supervision of all workers, remotely or indirectly as necessary • effective risk management arrangements • monitoring workers' activities and whereabouts as appropriate when away from base • appropriate levels of support and assistance and prompt response • site visits involving two people to locations, homes or people where services are to be delivered • incident reporting systems and systems for reporting incidents away from base • co-operating with host employers, employment agencies and contractors • monitoring incident records and information provided by community safety officers regarding risks of violence • including homeworkers and lone mobile workers in display screen equipment (DSE), manual handling, fire and COSHH assessments • periodic site visits and workplace inspections • consultation with lone workers and their representatives • ensuring lone workers are included in staff meetings and communications, wherever they are based

Issue	Examples of Possible Control Measures
Ensuring a safe system of work	• extra precautions for lone workers carrying out hazardous activities, or specific controls on when, where and how those activities take place if working alone • preventive measures to avoid manual handling injuries, such as equipment at outreach sites to enable lifting operations to be carried out by the lone worker • effective communications systems for lone workers
Providing information, instruction, and training for workers and managers	• develop a lone working policy • permission for lone workers to remove themselves from situations they feel unsafe in due to threats of violence • raising awareness of the risks to lone workers of stress-related illness and violence at work and what can be done to prevent them • training in use of protective equipment or communication systems designed for lone workers' protection • information about fire instructions and emergency procedures for lone workers away from base • explicit instructions to all workers about prohibited lone working activities
Ensuring workplace health, safety and welfare	• avoiding or minimising lone working in unoccupied premises • first aid and welfare provisions for all workers, including lone workers, accessible at all times • including lone workers in fire drills, emergency procedures and fire risk assessments • adequate security and monitoring arrangements for lone workers, including protective barriers and communications systems (with responders) if appropriate
Ensuring safety for remote working (off-site, between sites, in the community)	• effective supervision of lone workers through use of PPE, surveillance and communications systems, checks on lone workers' safe arrival/return, etc if necessary • effective personal safety plans and back-up systems for lone workers with named responders at all necessary times • checks on-sites, clients, locations and early warning systems to inform managers and lone workers of any known trouble spots and risks of violence, eg violent clients
Reducing risks of lone working time arrangements	• avoiding the need for lone working at certain times • adjusting start and finish times when working alone • providing extra support, back-up and supervision for lone workers • relocating the work to avoid it being done in isolation • limiting or redistributing overtime and emergency duties, including on-call arrangements • avoiding the need for lone working after hours in empty buildings • avoiding lone working at hazardous times, eg late at night in places with public access or poor security • adjusting working times and/or staffing levels or skill mix • improving lighting and security at times of darkness and in remote areas • ensuring safe access and exit at all times, including hours of darkness

Supervision and Communication

4.228 Although by definition lone workers cannot be subject to direct supervision, the employer still needs to provide a system of *effective* supervision. This involves:

- knowing the risks involved
- knowing the competence and capabilities of the lone worker
- having an effective means of communication with the lone worker.

4.229 Employees who lack the necessary training, experience and qualifications to carry out their work unsupervised should not be permitted to work alone.

4.230 Passive systems of communication when the lone worker has to make contact if facing a threat, eg violent assault, will be ineffective and may escalate the threat level or even provoke an assault.

4.231 For lone workers, supervision may be:

- remote — at a distance, or by electronic means such as electronic surveillance or e-contact
- discreet — where a competent co-worker or supervisor is close at hand, contactable in an emergency or request for back-up, but not in direct line of sight or in the immediate vicinity.

4.232 Effective monitoring arrangements for lone workers include:

- periodic visits and observations
- regular contact via telephone or two-way radio and, in the case of a teleworker, also by e-mail
- automatic warning devices which activate if periodic signals are not received from the lone worker
- manual or automatically operated devices which raise the alarm in an emergency
- checks to ensure lone workers have returned to either their base or home on completion of their work activity.

Implementing Control Measures

4.233 When implementing control measures for lone workers, employers should tackle the most serious risks first.

4.234 When planning action and consulting workers, representatives and managers about the measures being introduced, the reasons for the measures being taken should be clearly explained.

Recording the Findings

4.235 A written record of the main findings of the risk assessment for lone working should be kept where it can be readily accessed and reviewed by:

- employees
- representatives
- supervisors
- managers.

4.236 The findings should be communicated to all concerned, including:

- lone workers
- safety representatives
- supervisors
- host employers.

4.237 Electronic records, accessible via the Internet, may be helpful for lone workers working away from base.

Monitoring and Review of Assessment

4.238 Lone working should be closely monitored to ensure control measures are correctly implemented and effective in preventing and controlling risks. This can be done in several ways, by:

- using staff surveys, meeting with lone workers and holding team meetings to gather information
- monitoring accident and incident reports, tie sheets, travel records and sickness absence records to establish whether there are adverse effects on lone workers and whether measures concerning working time arrangements and travel are being taken
- consulting safety representatives and supervisors.

4.239 If there are changes in work activities, locations, staffing levels or personnel, or if an adverse incident is reported, the assessment should be reviewed and revised. Assessments should be reviewed periodically, or if:

- there is reason to think they are no longer valid
- there are changes in the work, workers or workplace that could affect the risks.

4.240 Where necessary, assessments should be revised.

Incident and Accident Reporting for Lone Workers

4.241 Lone workers must be encouraged to report all adverse incidents and accidents to their employer.

4.242 Incidents and accidents which happen on third-party sites must be reported to the host employer. The responsibility for investigating the incident lies primarily with the lone worker's employer, but should also be undertaken by the host employer too (possibly jointly) since the host employer's health and safety arrangements may be relevant.

4.243 Some violent incidents fall within the reporting requirements of the Reporting of Injuries, Diseases and Dangerous Occurrences Regulations 1995.

4.244 Incident reporting arrangements should be kept as simple as possible to facilitate use by lone workers.

4.245 Since both violence and stress are closely associated with lone working, employers should make lone workers aware that reporting such an incident will not result in disapproval or criticism. It should also be made clear that reports should not be confined to physical acts of violence involving injury.

4.246 In the case of violence involving bullying or other forms of abuse and harassment or intimidation, formal and informal reporting routes must be provided and confidentiality assured where possible.

4.247 Employers should consult the employee about any proposed action and, if possible, allow them to influence the action taken by the employer. The primary role of the employer is to ensure the employee's protection and to ensure there is no victimisation because of the report or action taken.

4.248 It is recommended that employers give special consideration to procedures for reporting incidents which occur when working away from base to ensure all such incidents are reported to the employer as well as third parties.

Training

4.249 Employers must provide training and instructions to staff and managers on preventing and avoiding risks, including risks that apply when working alone or managing lone workers.

4.250 Control measures include, and must be accompanied by, appropriate arrangements for the provision of information, instruction, training and supervision. The measures should include:

- provision of understandable information to employees
- suitable instruction, training and supervision for all relevant employees, including managers, supervisors and team leaders.

Information

4.251 Employers must ensure lone workers have access to all relevant information about their work and any risks attached to it before undertaking a job alone.

4.252 They may need additional information about the risks of doing their job when alone, but they should not have less information, or less frequent information and updates, than other workers even if they are working off-site.

4.253 The lone worker must be informed of all the factors affecting their personal safety when beginning work for a host employer or away from base.

4.254 The host employer is legally obliged to inform the lone worker's employer of any risks and the control measures to be followed. This information must be passed on to the lone worker by their employer. This also applies to employment agencies and contractors.

Training

4.255 Training should include a safe system of work for lone workers, safe use of work equipment when working alone and correct use of any precautionary measures that the employer has provided to ensure lone workers' safety and personal security, including:

- use of any PPE or alarms provided
- emergency procedures
- reporting procedures to be used when working away from base for mobile and travelling workers
- measures to reduce work-related stress.

4.256 If there is a risk of violence or if the work involves dealing with members of the public, the training should include:

- use of measures provided by the employer to prevent and control the risks of violence
- techniques for dealing with difficult, distressed or disturbed people; defusing or resolving conflict, and how to handle complaints.

4.257 Lone workers should also be given training on how to avoid panic reactions. Training should also ensure the lone worker is aware of the available control measures and is capable of dealing with aggression or seeking help where necessary. It should also set clear limits on what is expected of a lone worker, so they are aware of what they can and cannot do on their own. It should stress that their personal safety comes first.

Training for All Employees on Lone Working

4.258 Employers should provide general training and information on the risks of working alone to all workers, including at induction, since there may be times when they find themselves inadvertently working alone, eg in the unexpected absence of a colleague.

4.259 This general training should cover:

- what can and cannot be done when working alone
- how to deal with unusual situations
- when to stop work and seek advice or assistance from a supervisor or emergency service
- how and when to use any communications equipment provided
- procedures for checking back to base
- what to do in the event of an emergency or medical emergency
- where to locate first-aid assistance and fire instructions.

Managers' Training

4.260 Specific training should be given to managers and others responsible for the health and safety of lone workers. This should cover:

- how to provide support to lone workers to reduce risks of stress or social isolation
- how to use remote supervision systems and equipment
- the importance of maintaining effective early warning systems and implementing any protocols or procedures for preventing and monitoring risks of violence in the community
- how to provide effective back-up, eg in the event of an alarm call or request for help, how the system will operate and who is responsible at different times if a rota system is in place for designated responders
- how to report incidents involving the lone workers for whom they have responsibility, including incidents that occur away from base.

Risk Assessors' Training

4.261 A competent person must carry out any risk assessment. The assessor's abilities should include:

- an understanding of the relevant regulations
- a knowledge of the work activities, environment and risk factors to be assessed
- an awareness of any relevant individual factors and limitations
- an ability to recognise particular risks
- an ability to recommend reasonably practicable solutions
- a judgment of what constitutes an acceptable residual risk.

4.262 Assessors should also be trained in particular risk assessment methods for lone working, covering:

- specific training about risks of lone working
- the way these risks may interact with individual factors or with risks arising from other hazards.

List of Relevant Legislation

- Management of Health and Safety at Work Regulations 1999
- Provision and Use of Work Equipment Regulations 1998
- Lifting Operations and Lifting Equipment Regulations 1998
- Workplace (Health, Safety and Welfare) Regulations 1992
- Personal Protective Equipment at Work Regulations 1992
- Health and Safety (First-aid) Regulations 1981
- Health and Safety at Work, etc Act 1974

Further Information

Publications

HSE Publications

The following are available from *www.hsebooks.co.uk*.

- INDG73(L) (rev 2005) *Working Alone in Safety: Controlling the Risks of Solitary Work*
- INDG163(L) *5 Steps to Risk Assessment*
- L21 (rev 2000) *Management of Health and Safety at Work. Management of Health and Safety at Work Regulations 1999. Approved Code of Practice and Guidance*

Suzy Lamplugh Trust Publications

The following are available from *www.suzylamplugh.org*.

- *Loneworking 2008: Special Report*, D Collins, C Hamer and L Smail
- *Personal Safety at Work: Lone Working* (G19)
- *First Steps to Personal Safety at Work Video and Poster* (V11)
- *Personal Safety at Work: Planning for Safety & Conflict Management* (V22)

Other Publications

- *Barefoot Research: A Workers' Manual for Organising on Work Security*, Margaret Keith, James Brophy, Peter Kirby and Ellen Rosskam. International Labour Organisation In-Focus Programme on Socio-Economic Security publication, 2002
- *Working Alone: A Health and Safety Guide on Lone Working for Safety Representatives.* UNISON brochure, London, 2002, stock number: 1750

Organisations

- GMB
  Web: *www.gmb.org.uk*
  The GMB is a general union of 700,000 people. Forty per cent of the union's members are women.

- Suzy Lamplugh Trust
  Web: *www.suzylamplugh.org*
  The Suzy Lamplugh Trust, a registered charity, is considered a leading authority on personal safety.
- UNISON
  Web: *www.unison.org.uk*
  UNISON is Britain's biggest trade union with over 1.3 million members working in the public services, for private contractors providing public services and the essential utilities. These include frontline staff and managers working full or part time in local authorities, the NHS, the police service, leges and schools, the electricity, gas and water industries, transport and the voluntary sector.

Migrant Workers

- Many migrant workers do not have a fluent grasp of English and may be particularly vulnerable to failings in health and safety practices.
- The employment of migrant workers is subject to the same health and safety legal standards as the employment of any other worker in the UK.
- For UK companies employing migrant workers, the health and safety risks can be considerable due to the problems that arise from language, cultural and skills issues.
- When carrying out a risk assessment for migrant workers, consider any prior work experience and the extent to which it is relevant in relation to health and safety practices in the work activity being undertaken.
- The most useful control measure for minimising the risks for migrant workers on-site is close supervision by a competent person speaking both the migrant worker's language and English.
- If significant numbers of foreign workers are expected, employers should consider preparing material in the appropriate language.
- Employers should try to ensure that they employ *bona fide* migrant workers and not those who may have an illegal status.
- Each individual's competency should match as closely as possible the required level of competency for the work activities to be undertaken.
- Migrant workers have the same rights and responsibilities as UK workers under UK legislation.
- Employers have a duty to provide all migrant workers with appropriate, clear and understandable safety training to enable them to do their jobs without risk of illness or injury whether or not they speak English.

4.263 There are, broadly speaking, three categories of migrant workers currently employed in the UK.

- The first, most numerous and lawful, are those from European Union (EU) Member States who have the right to work in the UK under EU freedom of movement rules. (Though special rules currently apply to workers from Romania and Bulgaria.)
- The second group comprises lawful migrants from countries outside the EU. Complex immigration rules apply to this group of workers.
- The third group consists of illegal immigrants. The fact that they are in the UK secretly means that their numbers cannot be accurately assessed.

4.264 Many of these workers do not have a fluent grasp of English and may be particularly vulnerable to failings in health and safety practices.

4.265 The Health and Safety Executive (HSE) is aware of these problems and has issued detailed advice and guidance on the proper management of migrant workers' health and safety. It recognises that factors such as poor language skills and unfamiliarity with the workplace can magnify the effects of existing health and safety problems, and it advises that migrant workers with better English should be asked to interpret for their less fluent colleagues. In addition, the HSE recommends the use of internationally recognised signs, videos and audio materials to communicate health and safety messages.

Employers' Duties

4.266 The employment of migrant workers is subject to the same health and safety legal standards as the employment of any other worker in the UK. Migrant workers have the same rights and responsibilities as UK workers under UK legislation.

4.267 Under the Health and Safety at Work, etc Act 1974, employers have a duty to inform migrant workers about:

- risks to their health and safety from current or proposed working practices
- things or changes that may harm or affect employees' health and safety
- how to do their jobs safely
- what is done to protect their health and safety
- how to get first-aid treatment
- what to do in an emergency.

4.268 Employers must also provide:

- sufficient training to migrant workers to enable them to do their jobs safely
- protection for migrant workers when necessary (eg clothing, shoes or boots, eye and ear protection, gloves, masks)
- health checks to migrant workers if there is a danger of ill health because of their work
- regular health checks if migrants work nights and a health check prior to starting night work.

Employees' Duties

4.269 In general, employees have a duty to:

- take care of their own health and safety and that of other people who may be affected by what they do (or do not do)
- co-operate with others on health and safety and not interfere with, or misuse, anything provided for health, safety or welfare purposes
- inform their employer of any shortcomings in the employer's protection arrangements.

In Practice

4.270 Migrant workers come to the UK from all parts of the world, particularly Asia, the Middle East, and Eastern Europe. The number of migrant workers coming into the UK from the EU has increased since EU enlargement in May 2004.

4.271 While many migrants come from developing countries to work in the construction industry in the UK, standards of health, safety and workers' rights are far less established than in the UK. Many are vulnerable, dependent on their jobs to stay in the country, and are therefore obliged to accept lower standards of pay and conditions.

4.272 For UK companies employing migrant workers, the health and safety risks can be considerable due to the problems that arise from language, skills and cultural issues. The key is to identify those who may be most at risk, to identify what the risks could be and then to control those risks accordingly.

Vulnerability of Migrant Workers

4.273 All workers, not just migrants, can be exposed to some level of risk on-site. However, factors that may make migrant workers more vulnerable to risk include:

- working relatively short periods of time in the UK
- limited knowledge of the UK's health and safety system
- different experiences of health and safety regimes in countries of origin
- motivations in coming to the UK (eg high earnings in a short period of time)
- lack of ability to communicate effectively with other workers and with supervisors, particularly in relation to an understanding of risk
- access to limited health and safety training and, where proficiency in English is limited, difficulties in understanding what is being offered
- the failure of employers to check their work abilities and their language skills
- where workers are supplied by recruitment agencies, labour providers or are self-employed, their employment relationships are unclear and there is uncertainty concerning who has responsibility for health and safety
- lack of knowledge about health and safety rights, how to raise them and the channels through which they can be represented.

Risk Assessment

4.274 When carrying out a risk assessment, thought will have to be given to the approach to be taken. For example, is it necessary to review all risk assessments for all activities or is it possible to carry out a more generic risk assessment to cover migrant workers working throughout the organisation? It may be necessary to determine how many migrant workers there are in the organisation, where they are working and what they are doing (eg there could be a particular influx of migrant workers during the organisation's peak periods).

4.275 It should be determined if migrant workers are at greater risk than the working population generally and, if so, why. In addition to the vulnerability issues, other factors to consider will include the following.

- What the employment status of the migrant worker is (eg employed, self-employed, agency).
- Language and literacy skills with regard to their ability to communicate and to understand information (written and oral) in relation to the work activities and general work environment health and safety.
- The level of information, instruction, training and supervision required in relation to the work.
- Prior work experience and the extent to which it is relevant in relation to health and safety practices in the work activity being undertaken.
- Perception of the risk and extent to which this may differ due to experience of another country's health and safety workplace culture.
- Interrelationships between migrant workers and UK workers and the potential impact this could have on health and safety.

4.276 The assessment may also highlight issues within the organisation that could enhance the risk factor, including the following.

- Is health and safety information provided in languages other than English?
- Are induction and ongoing training provided and how are these tailored to make sure migrant workers understand them?
- Is personal protective equipment provided and are steps taken to ensure that migrant workers use it appropriately?
- Are migrant workers informed of their rights in relation to health and safety?
- Are there systems in place to ensure that migrant workers are able to report accidents or raise concerns?
- Are responsibilities for health and safety clearly defined when temporary agency and/or casual migrant labour are engaged?

Language Barriers and Understanding

4.277 By far the biggest risk is the language barrier. The failure of workers to understand even the most basic of health and safety requirements will inevitably be compromised by their inability to understand, read and, to a lesser extent, speak English.

4.278 Most migrant workers who cannot understand oral instructions will probably not understand written ones, which are necessary for their health and safety. For example, the use of machinery often requires an understanding of English — a common control measure for the safe use of electrical appliances is the "next inspection date" or "not to be used after" date label. This is normally attached to the equipment to indicate the expiration of an inspection and/or test period. A worker without a grasp of the language will have no understanding of the significance of such a warning.

4.279 Where safety is dependent on safety signs, compliance with the Health and Safety (Safety Signs and Signals) Regulations 1996 will assist, particularly in respect of those signs with symbols and pictograms whose message or warnings are internationally known and more easily understood. Many such signs, however, are accompanied by text in English to complete the message. Employers have a responsibility under the regulations to train all staff in the meaning of such signs.

4.280 Much new work equipment is to be supplied with instructions in many languages. Despite this, there will be an increased need to ensure understanding of the instructions and in this regard the use of a competent interpreter will be vital.

Control Measures

4.281 The principal control measure in respect of migrant workers will be key training that accounts for their lack of understanding of the English language.

4.282 The most useful method, though not always the most cost-effective, is close supervision by a competent person speaking both the migrant worker's language and English. Where migrant workers work in groups on low-risk activities, this will probably be the most effective control measure.

Translations

4.283 Induction and job-specific training involving lectures or videos are unlikely to be effective with foreign workers who have limited command of English. Simultaneous translation by use of an interpreter is possible, but may not work well in practice. If significant numbers of foreign workers are expected, employers should consider preparing material in the appropriate language and amending the delivery method, eg by using bilingual employees.

4.284 It may also be necessary to have health- and safety-related documents, such as risk assessments and operating procedures, translated into the appropriate languages. This may also apply to procedures such as permit-to-work systems.

Managing Migrant Workers' Health and Safety

4.285 The approach to managing migrant workers' health and safety should be based on the basic principles detailed in HSG65: *Successful Health and Safety Management*, ie:

- developing an effective health and safety policy to set a clear direction for the organisation to follow
- implementing an effective management structure and ensuring suitable and sufficient arrangements are in place for delivering the policy
- putting in place a planned and systematic approach to implementing the health and safety policy through an effective health and safety management system with the aim of minimising risks
- measuring performance against agreed standards to reveal when and where improvement is needed
- regularly auditing and reviewing performance.

4.286 It may be necessary to amend the organisation's health and safety policy to ensure that, where large numbers of migrant workers are employed, the procedures adopted in controlling any risks particular to this group are recorded and made known to all relevant parties.

4.287 It may also be necessary to make arrangements to ensure that recruitment procedures reflect the need to consider migrant workers' health and safety. Risk assessments may have to be reviewed and monitored to ensure their effectiveness.

Reducing Risk from Employing Migrant Workers

4.288 It is always preferable to eliminate or to reduce risk at source and much can be achieved by ensuring that those to be appointed are suitable for the work activities to be undertaken. It is best to eliminate or to reduce risk at the "input stage" to the organisation, and so employers should try to ensure that they employ *bona fide* migrant workers and not those who may have an illegal status (eg those without a work permit, illegal immigrants).

Competency

4.289 Clearly, each individual's competency should match as closely as possible the required competency for the work activities to be undertaken. Reasonable attempts should be made to determine whether or not the individual has the necessary skills, qualifications and experience required, including the necessary language skills.

4.290 Even where competency levels are adequate, migrant workers, like all employees, should receive an appropriate level of information, instruction and training.

Rights of Migrant Workers

4.291 The employment of migrant workers is subject to the same health and safety standards as the employment of any other worker in the UK. It follows that such workers have the same rights and responsibilities as UK workers under UK legislation.

4.292 In respect of workers' rights, the HSE, in conjunction with the TUC, has issued a series of publications entitled *Your Health and Safety: A Guide for Workers*. This guide is available in a number of eastern European as well as many Asian and other languages, and includes:

- the rights of workers to complain about unsafe working conditions to both their employers and the HSE without the risk of losing their jobs
- the rights of workers to join trade unions and to become safety representatives
- details of working time and rest breaks.

4.293 There is also a statutory requirement to post the health and safety law poster: *What You Should Know*, although the posters are currently available only in English and some Asian languages.

Training

4.294 Employers have a duty under the Health and Safety at Work, etc Act 1974 to provide all their employees — including migrant workers — with appropriate, clear and understandable safety training to enable them to do their jobs without risk of illness or injury, whether or not they speak English.

4.295 The level of information, instruction, training and supervision provided should be appropriate to the work being carried out.

4.296 Induction and job-specific training involving lectures or videos are unlikely to be effective with migrant workers who have a limited command of English. Simultaneous translation by use of an interpreter is possible, but may not work well in practice. If significant numbers of migrant workers are employed it might be necessary to prepare material in the appropriate language and to amend the delivery method of any training given, eg by the use of bilingual employees.

4.297 It may also be necessary to have health- and safety-related documents, such as risk assessments and operating procedures, translated into the appropriate languages. This may also apply to procedures such as permit-to-work systems.

Language Training

4.298 It is impractical in most cases to attempt to teach migrant workers the English language, although training in the workplace to achieve a basic understanding of English is recommended.

4.299 The Government has launched a suite of workplace-based English language qualifications for migrant workers and employers. The English for Speakers of Other Languages (ESOL) for Work qualifications are designed to make it easier for employers and migrant workers to attain the functional English language skills they need.

List of Relevant Legislation

- Construction (Design and Management) Regulations 2007
- Management of Health and Safety at Work Regulations 1999
- Employers' Liability (Compulsory Insurance) Regulations 1998
- Provision and Use of Work Equipment Regulations 1998
- Health and Safety (Safety Signs and Signals) Regulations 1996
- Personal Protective Equipment at Work Regulations 1992
- Workplace (Health, Safety and Welfare) Regulations 1992
- Health and Safety Information for Employees Regulations 1989
- Health and Safety (First-Aid) Regulations 1981
- Health and Safety at Work, etc Act 1974

Further Information

Publications

HSE Publications

The following are available from *www.hsebooks.co.uk*.

- HSE27E (rev1) *Your Health, Your Safety — A Guide for Workers* (also available in Bengali/Punjabi/Gujarati/Urdu/Hindi)
- HSG65 (rev 1997) *Successful Health and Safety Management*
- RR502 *Migrant Workers in England and Wales — An Assessment of Migrant Worker Health and Safety Risks*

Organisations

- Health and Safety Executive (HSE)
  Web: *www.hse.gov.uk*
  The HSE is responsible for the regulation of almost all the risks to health and safety arising from work activity in the UK.
- Trades Union Congress (TUC)
  Web: *www.tuc.org.uk*
  The TUC describes itself as "the voice of Britain at work". With 71 affiliated unions representing nearly 7 million working people from all walks of life, it campaigns for a fair deal at work and for social justice at home and abroad.

Temporary Workers

- Temporary workers employed by an organisation on a fixed-term contract are considered to be employees.
- If an employment agency is used to provide temporary workers, those workers are considered to be employees of the agency rather than the "host" employer.
- Temporary workers may also be self-employed.
- Workers that are only temporary are likely to be unfamiliar with the premises, safety rules and procedures, which means they are at greater risk.
- Employers are legally required to ensure that workers are not exposed to risks to their health and safety, so far as is reasonably practicable. This includes temporary workers.
- Employers must conduct a risk assessment to determine the risks associated with temporary workers and the necessary control measures to minimise those risks.
- Employers are specifically required to provide all temporary workers with the information they need to carry out their work safely.
- Where the temporary worker is not an employee (ie they are supplied by an agency), the host employer is responsible for:
  - conducting display screen equipment (DSE) assessments
  - providing information and training on fire safety procedures
  - assessing and minimising the risks in relation to hazardous substances and any other workplace activities that are to be undertaken.
- The employing agency is responsible for:
  - the provision of first-aid facilities
  - providing personal protective equipment
  - eyesight tests and the provision of glasses for DSE use
  - informing the enforcing authority of any reportable accidents or incidents
  - providing any necessary health surveillance.

4.300 Employers may use temporary workers in one of two ways, for example:

- taking on new staff employed directly on temporary contracts
- using agencies hiring out their workers on a temporary basis.

4.301 Temporary workers in the first category should be regarded as employees and therefore be treated like any other member of staff under health and safety legislation. However, temporary workers provided by an agency are not considered employees of the host employer.

4.302 There are particular risks associated with the use of temporary workers as a result of their unfamiliarity with the workplace. Employers are required to assess and minimise these risks.

Employers' Duties

4.303

- Employers have a general duty to ensure, so far as is reasonably practicable, the health, safety and welfare at work of all employees and non-employees under the Health and Safety at Work, etc Act 1974. This includes any temporary workers on the premises, whether they are considered to be employees or not.

- Under the Management of Health and Safety at Work Regulations 1999, employers should:
  - consider temporary workers within the company risk assessments
  - set up suitable and sufficient arrangements to protect the health and safety of people at work, including temporary workers
  - provide information to temporary workers on risks and control measures
  - give details of procedures, site rules, safe systems of work, etc
  - provide details of any health surveillance required
  - give information on what qualifications and skills the temporary worker must have to undertake the proposed work safely.
- If a temporary worker has an accident while working for a host employer, the host employer should notify the worker's employer as soon as possible, under the Reporting of Injuries, Diseases and Dangerous Occurrences Regulations 1995.
- Under the Personal Protective Equipment at Work Regulations 1992, the duty to provide personal protective equipment (PPE) to employees does not extend to temporary workers employed by an agency. The employment agency is responsible for ensuring any necessary PPE is provided.
- It is the host employer's responsibility to conduct display screen equipment assessments under the Health and Safety (Display Screen Equipment) Regulations 1992, and to implement any necessary control measures to ensure the safe use of DSE.
- Under the Health and Safety Information for Employees Regulations 1989, the host employer is responsible for informing temporary workers or self-employed contractors:
  - of the risks involved with the work and any control measures
  - that they must provide PPE (if the host employer decides to provide the PPE, then that employer is responsible for ensuring it is suitable and maintained in effective working order)
  - of the exact nature, purpose and location (if known) of the work
  - of specific site details
  - about any equipment provision
  - of the emergency action procedures
  - of any (technical) language expectations.
- Under the Employers' Liability (Compulsory Insurance) Act 1969, organisations should provide cover for temporary staff who work on their premises.

Employees' Duties

4.304

- Under the Health and Safety at Work, etc Act 1974, employees have a duty to take reasonable care of their own health and safety and that of other people who may be affected by their activities while at work.
- Employees, irrespective of permanent or temporary status, also have a duty to co-operate with their employer to enable the employer to meet their health and safety responsibilities.

In Practice

4.305 Temporary workers pose an important safety management issue due to their unfamiliarity with the workplace and company rules. Temporary working can give rise to many potential risks as:

- the type of work may require irregular or unusual working hours, which the worker may not be familiar with
- the workplace may be different to any the worker has previously experienced, requiring periods of adaptation
- the working environment may include health hazards that are unfamiliar to the temporary worker.

Risk Assessment of Temporary Workers

4.306 Employers must carry out a risk assessment of the risks posed by the work activities of temporary workers. These assessments should include the risks to temporary workers on-site who may be affected by the employers' work activities.

4.307 The starting point is to identify all possible activities and operations undertaken by temporary workers. Consulting previous temporary workers and other employees can help with the assessment. It will also be useful to consider the organisation's general risk assessments when evaluating the risk associated with temporary staff. It should always be remembered when evaluating the risks and considering the appropriate control measures that temporary workers may be unfamiliar with workplace hazards and are therefore at greater risk.

Information Employers Must Provide

4.308 Employers must give comprehensible information to the employees of others who are working on their premises, including temporary workers, on

- any relevant risks
- the control measures in place to safeguard their health and safety
- emergency procedures
- the qualifications or skills they need to carry out their job safely
- any health surveillance required.

4.309 It may be useful to draw up a written specification for each job carried out by a temporary worker, which contains all the necessary information. This can be provided before the employer chooses a worker for the job, or the self-employed person before he or she agrees to take on the work.

Information Required by the Host Employer

4.310 The employer of the temporary worker (eg an agency) should provide the host employer with information on:

- the health of the worker, eg whether he or she has a history of back problems or sensitisation to a particular chemical
- the training the worker has received
- whether the worker is experienced in the type of work they will be doing for the host employer.

4.311 It would be useful for employment agencies to have a written record containing this information, which should be provided to the host employer before the worker starts the job.

4.312 Temporary workers should also be provided with a copy of the site safety rules. Their supervisor or line manager should outline these orally before the work starts.

Accident Reporting

4.313 Employers and those in control of premises are required to report certain accidents to the Health and Safety Executive (HSE) or the local authority. If a temporary worker is injured while working on the premises of another employer, the accident must be reported by the worker's employer in most cases (eg the employment agency), or the workers themselves if they are self-employed. Temporary workers should also be informed of accident-reporting procedures.

First Aid

4.314 Employers must provide their employees with adequate and appropriate first aid. In the case of temporary employees, the worker's employer would be responsible for first-aid arrangements, rather than the host employer. Self-employed people must ensure adequate first-aid provision for themselves. However, for practical reasons the host employer may agree that temporary workers can use the first-aid facilities and personnel provided for their own employees.

Display Screen Equipment

4.315 Employers are required to assess the use of display screen equipment (DSE). They must assess all the workstations they provide, whether these workstations are to be used by their own employees or by temporary workers. The temporary worker's employer also has duties to:

- provide the employee with an eyesight test
- pay for glasses (where required) for use with DSE
- ensure that the employee has the appropriate training on DSE.

Personal Protective Equipment

4.316 Employers must provide employees with suitable personal protective equipment (PPE). In the case of temporary workers, the worker's employer is responsible for providing PPE. Self-employed workers must supply their own PPE.

4.317 The host employer must give the employer of temporary workers information on health and safety risks and arrangements, including what PPE they will need.

Work Equipment

4.318 Careful checks must be made of all work equipment to be used by temporary workers. Employers should ensure that full instruction and training is available on all equipment prior to being used, and that proper visual safety inspections are performed before use.

Fire

4.319 The fire safety arrangements required for individual premises, including training, should be identified in the fire risk assessment. As temporary workers are unfamiliar with the premises and procedures, they should be provided with fire safety training when they first arrive on-site. Such training should include:

- the action they should take on discovery of a fire
- how to raise the alarm
- procedures for evacuating the building safely
- the location of fire-fighting equipment and how to use it
- where fire escape routes are located.

Hazardous Substances

4.320 The Control of Substances Hazardous to Health Regulations 2002 (COSHH) cover the safety of chemicals and preparations used at work. In relation to temporary workers, the responsibilities are divided. The host employer is responsible for:

- carrying out a COSHH assessment
- setting up and implementing control measures
- maintaining engineering controls
- monitoring
- information, instruction and training.

4.321 The employer of the worker is responsible for providing and recording health surveillance.

Electrical Safety

4.322 Staff should be informed that electrical equipment is periodically tested and should carry a dated test label. Only equipment that bears an "in date" label must be used. Employers should examine any item of equipment the temporary worker will use prior to use. Any item that appears unsafe must not be used. No personal item of electrical equipment should be used in the organisation without permission.

Health and Well-being

4.323 Temporary workers should be notified if they can access the organisation's occupational health unit, and that self-referral can be made if the worker believes him or herself to be suffering from:

- a condition which is related to their current work
- a pre-existing condition which may affect their health and safety whilst working.

4.324 Temporary staff must also be reminded that such consultations are confidential.

Gangmasters Licensing

4.325 The Gangmasters Licensing Authority (GLA) was set up to curb the exploitation of workers in the agriculture, horticulture, shellfish gathering and associated processing and packaging industries. It is now illegal to supply workers to the agriculture and food processing and packaging sectors without a GLA licence. In general terms, a person is acting as a gangmaster if they:

- supply workers to another person to do work to which the Act applies (including subcontracting)
- use a worker to do work to which the Act applies in connection with services provided by him to another person
- use a worker to do certain types of work (in particular gathering shellfish).

List of Relevant Legislation

- Management of Health and Safety at Work Regulations 1999
- Reporting of Injuries, Diseases and Dangerous Occurrences Regulations 1995
- Health and Safety Information for Employees Regulations 1989
- Health and Safety at Work, etc Act 1974

Further Information

Publications

HSE Publications

The following are available from *www.hsebooks.co.uk*.

- HSG65 (rev 1997) *Successful Health and Safety Management*
- L21 (rev 2000) *Management of Health and Safety at Work. Management of Health and Safety at Work Regulations 1999. Approved Code of Practice and Guidance*

Visitors

- Visitors can be at greater risk than employees because visitors are not familiar with the hazards of the workplace.
- An employer must carry out risk assessments that take into account the risks to visitors and to others caused by having visitors in the workplace.
- Good communication is necessary to advise visitors of any particular hazards in the workplace and of emergency procedures.
- An employer should also consider any particular requirements or vulnerabilities of visitors.
- Prior arrangements should be agreed if the visitor is bringing any particular hazard with them, such as a delivery of chemicals to site.
- Records should be kept of who is visiting the site.
- Visitors should be supervised, monitored or accompanied as necessary and appropriate, particularly in restricted or hazardous areas.
- Employers should keep a record of any accidents.

4.326 Employers have a duty of care to ensure the health, safety and welfare of visitors to the workplace. This comes from the Health and Safety at Work, etc Act 1974 and from various other acts and regulations, most notably the Occupiers' Liability Acts (which impose duties on employers as occupiers of buildings) and from regulations such as the Management of Health and Safety at Work Regulations 1999.

4.327 Workplace hazards can put visitors even more at risk than employees, since visitors are likely to be unfamiliar with the particular hazards and activities of the workplace.

Employers' Duties

4.328

- Employers have a general duty to ensure, so far as is reasonably practicable, the health, safety and welfare at work of all employees under the Health and Safety at Work, etc Act 1974.
- Employers have a duty to ensure the health and safety of their visitors as well as their employees.
- Similarly, occupiers of buildings, eg employers, have duties to protect visitors.
- Employers must implement arrangements to identify and control risks to visitors, not just risks to employees.
- Employers must report accidents involving visitors, not just accidents involving employees.
- Employers who operate particularly high-risk premises, such as chemical installations, have other specific duties to visitors, eg as contained in the Control of Major Accident Hazards Regulations 1999.
- The employer (or the occupier of the premises) has a common law duty to take reasonable precautions for the safety of legitimate visitors (Occupiers' Liability Act 1957) and of illegitimate visitors, such as trespassers, (Occupiers' Liability Act 1984) to the premises.

Employees' Duties

4.329

- Employees have a general duty to take reasonable care of their own health and safety and that of other people who may be affected by their work under the Health and Safety at Work, etc Act 1974.
- This means that they need to comply with their employer's instructions about visitors, eg supervising them and making sure they sign the visitors' book.

In Practice

4.330 How an employer keeps visitors safe from harm will depend on the particular hazards in the employer's workplace and on any particular needs or characteristics of the visitor.

Who are Visitors?

4.331 Visitors are all non-employees who are visiting the employer, such as:

- salespeople
- customers
- delivery drivers
- interviewees
- contractors
- guests and friends of employees
- emergency services
- trespassers
- members of the public
- school children (eg day trips, etc).

Risk Assessment

4.332 In addition to risk assessments for its employees, an employer must assess the risks to:

- visitors from its work activities
- employees created by the presence of visitors.

4.333 Existing risk assessments for employees should help assess the risks to visitors from work activities. The unfamiliarity of the visitors to the workplace should be taken into account. If children are present, the level of risk increases.

Review of Risk Assessments

4.334 Risk assessments should be reviewed at least once a year. A review should also be undertaken if circumstances in the workplace change or if there is any reason to suspect that the assessment is no longer valid, eg if an accident, near-accident or some ill health occurs.

Information and Communication

Communication to Visitors of the Hazards of the Workplace

4.335 Visitors must be informed before, or on, arrival of the risks to which they may be exposed while on-site and of the control measures and emergency arrangements in place, including the location of assembly points. The Code of Practice which accompanies the Management of Health and Safety at Work Regulations 1999 asserts that this information must be comprehensible.

Ascertaining Any Needs of Visitors

4.336 An employer should ascertain before the visitor arrives whether any special arrangements need to be put in place, eg:

- for the safe handling of any chemicals that the visitor may be bringing on-site
- to accommodate any disability of the visitor.

Personal Protective Equipment

4.337 Employers should provide personal protective equipment to visitors where the risk assessment has shown this to be necessary, eg head protection for visitors to construction sites. Employers should bear in mind that visitors will come in all shapes and sizes and that therefore protective equipment should be available in a full range of sizes. If children might visit, eg on a school trip, child sizes will also be needed.

Supervision and Accompaniment of Visitors

4.338 Adequate supervision should be maintained whilst the visitor is on-site, particularly if the visitor will be handling or transporting hazardous articles or substances. Supervision should be adequate to prevent the visitor from straying into hazardous areas and becoming exposed to danger. Entry to hazardous areas should be permitted only where necessary and by a prior arrangement agreed to by the appropriate person who is responsible for health and safety.

Workplaces Open to the Public

4.339 In shops, leisure centres and other workplaces open to the public it will usually be impractical to keep track of all visitors on-site. Adequate notices should indicate emergency procedures and escape routes and staff training should include regular evacuation practices simulating the involvement of the general public. Hazardous areas should be inaccessible to visitors where supervision is impractical.

Shared Buildings

4.340 Those employers who are one of a number of occupiers of a building are usually responsible for the safety of visitors only in those areas that they occupy.

Some Special Categories of Visitor

Disabled Visitors

4.341 Employers should ascertain in advance any particular requirements of a disabled visitor so that appropriate arrangements can be made. Employers should bear in mind that a disabled person may take longer to evacuate premises in an emergency. Generally speaking the needs and requirements of disabled visitors mirror the needs and requirements of disabled workers.

Contractors

4.342 Contractors bring particular and very specific risks into the workplace depending on the job that they are contracted to do. Arrangements that employers should put into place should take into account issues such as:

- procedures for assessing contractors
- safety rules and procedures
- equipment
- co-ordination of each contract
- reporting of problems or safety issues
- arrangements for meeting requirements under the Construction (Design and Management) Regulations 2007.

Trespassers

4.343 Employers who are "occupiers" of buildings, as defined in the Occupiers' Liability Act 1957, owe duties of care to trespassers under the Occupiers' Liability Act 1984, no matter how little the employer might think the intruder deserves the employer's care.

4.344 Occupiers, eg many employers, will be liable to an injured trespasser if three conditions apply.

- The occupier knew of the risk or had reasonable grounds to believe that it existed.
- The occupier knew or had reasonable grounds to believe that trespassing may occur.
- The risk is one that the occupier may reasonably be expected to offer some protection against.

4.345 Occupiers are not expected to make their premises burglar-proof. They should, however, take steps to ensure that there are no hidden or unexpected dangers on-site. Steps to deter entry, eg warning signs and the like, will often be sufficient to protect the occupier from liability.

Accident Reporting

4.346 Employers are under a duty to report significant accidents. The Reporting of Injuries, Diseases and Dangerous Occurrences Regulations 1995 cover visitors and require the employer to report:

- fatalities
- injuries to visitors that require hospital treatment
- dangerous occurrences
- certain accidents arising from work which require hospital treatment.

Training

4.347 Many staff receive visitors and therefore all staff should be trained in the care of visitors.

4.348 Where the visitors' book or register is kept by reception or security, further specific training of reception or security staff will be needed.

4.349 Employers must not only make sure that their employees attend training but also that the employees have understood what they have been taught.

4.350 Employers should consider refresher training from time to time.

List of Relevant Legislation

- Control of Major Accident Hazards Regulations 1999
- Management of Health and Safety at Work Regulations 1999
- Reporting of Injuries, Diseases and Dangerous Occurrences Regulations 1995
- Occupiers' Liability Act 1984
- Health and Safety at Work, etc Act 1974
- Occupiers' Liability Act 1957

Further Information

Publications

HSE Publications

The following are available from *www.hsebooks.co.uk*.

- HSG76 *Health and Safety in Retail and Wholesale Warehouses*
- HSG154 *Managing Crowds Safely*
- INDG246(L) *Prepared for Emergency — Chemical Industry*
- L73 (rev 1999) *A Guide to the Reporting of Injuries, Diseases and Dangerous Occurrences Regulations 1995*

Young Workers

- A "young person" is someone between the minimum school leaving age (usually 16 years) and 18 years.
- The Management of Health and Safety at Work Regulations 1999 require employers to identify, assess and address risks to young persons before the young person starts work.
- Employers must protect young people from any risks associated with their:
  - inexperience
  - lack of awareness of risks or danger
  - physical and/or psychological immaturity.
- Young people may not be employed where the work involves harmful exposure to specified chemical, biological and physical agents, or where the work is beyond their physical and psychological ability.
- The Control of Lead at Work Regulations 2002 also prohibit young people from certain lead processes.
- These prohibitions on employment do not apply where:
  - the work is part of the young person's training
  - the task is carried out under the supervision of a competent person
  - any risks have been reduced to the lowest level reasonably practicable.
- The Working Time Regulations 1998 place restrictions on the hours worked by young people. Night work is generally banned for young workers, with some exceptions.
- A "child" is defined as someone who is below the minimum school leaving age, ie usually under 16 years of age. There are heavy restrictions on the employment of children.
- If children are employed, the employer must provide a copy of the risk assessment to the child's parents or guardian.

4.351 The Management of Health and Safety at Work Regulations 1999 define a young person as someone who is under the age of 18. In practice, a young person is accepted as someone who is between the minimum school leaving age (usually 16 years, although they can be 15 years, depending on their birth date) and 18 years.

4.352 A child is defined as someone below the minimum school leaving age. The employment of children is covered under very specific and separate legislation.

4.353 In order to ensure that young people can work without any unacceptable risks to their health and safety, employers must be aware of:

- any workers under the age of 18 years, particularly children
- the young person's inexperience, lack of awareness or perception of danger and their physical and/or psychological immaturity
- any work activities or aspects of the workplace that may pose a particular risk to young persons.

Employers' Duties

4.354 Under the Health and Safety at Work, etc Act 1974, employers have a general duty to ensure, so far as is reasonably practicable, the health and safety at work of all employees, and any other people who may be adversely affected by the employer's work activities or workplace.

4.355 Under the Management of Health and Safety at Work Regulations 1999, employers are required to:
- carry out a risk assessment of all the risks to their employees, including any particular or additional risks to young people
- take into account the capability of employees to carry out intended tasks before they are assigned to them, and to provide necessary information, instruction and training to perform these tasks safely (this may be of particular relevance for tasks assigned to young people).

- Two exemptions from the above requirements exist in the Management of Health and Safety at Work Regulations 1999. These are where work is occasional or short-term in:
  - family undertakings, provided the work is not considered harmful, damaging or dangerous to young people
  - domestic service within a private household.
- Employers must provide sufficient and understandable information, instruction, supervision and training to allow young people to work safely.
- The Control of Lead at Work Regulations 2002 and the Ionising Radiations Regulations 1999 require employers to:
  - prohibit young people from undertaking defined activities, eg certain lead processes
  - restrict certain activities, eg those involving exposure to hazardous substances.
- There are also some prohibitions, restrictions and conditions in relation to young people working within certain industries, eg agriculture, ship building, mines and quarries, and in the carriage of dangerous goods or with certain dangerous equipment/machinery, eg power presses and various woodworking machinery.
- The Working Time Regulations 1998 (as amended) impose restrictions on the hours of work engaged in by young persons. They:
  - may not work undertake work at night, with some exceptions
  - should have a daily rest period of at least 12 consecutive hours in each 24-hour period
  - should have a weekly rest period of at least 48 hours in each 7-day period
  - are entitled to a rest break of 30 minutes if their working time is more than 4.5 hours.

Employees' Duties

4.356

- Employees, including young persons, have a duty to take reasonable care of their own health and safety and that of other people who may be affected by their work under the Health and Safety at Work, etc Act 1974.
- Employees have a duty to co-operate with the employer's health and safety arrangements under the Management of Health and Safety at Work Regulations 1999.
- Employees must act in accordance with any training, instructions and information provided to them by their employers under the Health and Safety at Work, etc Act 1974 and the Management of Health and Safety at Work Regulations 1999.

In Practice

4.357 Employers must understand the factors that make young people particularly vulnerable and take these into account when conducting risk assessments.

Protection of Young People

4.358 Young people may be vulnerable to harm associated with their work because of their inexperience, lack of perception of danger, and physical and/or psychological immaturity (see the table below).

Young Workers: Risk Factors and Control Measures

Physical/ Emotional Factor	Work Factors	Control Measures
Inexperience and lack of perception of danger	• Moving and handling • Any work task, but especially where the dangers are not obvious, eg noise • Work with chemical, biological or physical agents • Work with equipment and machinery • Influence (good and bad) of colleagues • Young person's attitude and aptitude	• Risk assessments • Clear, simple information and instructions for: – each work task – the hazards present and control measures required • Training to achieve and assess minimum levels of competency • Supervision by a competent person
Physical immaturity (eg risk of musculoskeletal damage if the muscles and bones are not fully developed)	• Prolonged strenuous, physical work • Repetitive work • Work requiring forceful movements • Prolonged, awkward postures • Work rates imposed by machinery • Shift work, especially rotating day/night shifts	• Risk assessment • Clear, simple information and instructions for each work task, the hazards present and control measures required • Regular and sufficient work/rest breaks — reduce exposure times to these hazards • Workplace layout • Regular task rotation • Assessment of young person's physical fitness • Remove imposed work rates for young people until they are able to reach the required standard • Remove need to work shifts, where possible, and limit unsociable hours

Physical/ Emotional Factor	Work Factors	Control Measures
Psychological immaturity	• Work involving pre-determined productivity rates (peer pressure to meet targets) • Exposure to: – aggression and violence – bullying and harassment • Being required to make decisions outside of their ability • Shift work, especially where it disrupts their social life and normal sleep patterns	• Clear, simple information and instructions for each work task, the hazards present and control measures required • Risk assessments including assessment of work-related stress • Remove imposed work rates for young people until they are able to reach the required standards • Supervision • Remove need to work shifts, where possible, and limit unsociable hours • Counselling

Risk Assessments for Young People

4.359 The developing and maturing body undergoes physiological, hormonal and psychological changes. The risk assessment process must reflect this, ie be regularly reviewed and, if necessary, revised in order to pick up the effects of these changes as they occur.

4.360 There are also other important points to bear in mind.

- Young people may never have been in a work environment before and may not be familiar with appropriate or safe behaviour.
- They may be easily influenced or pressured into unsafe work practices by older colleagues.
- Young people, particularly children, may be curious and act unpredictably without any regard to instructions, etc that have been given.
- Information, instructions and training provided for older workers might not be comprehensible to young people and may need to be adapted so that they are understood.

Initial Assessment

4.361 The purpose of an initial assessment is to identify:

- the presence of any young persons (these will usually be employees but may be trainees, work placement students or visitors)
- work activities and/or areas in the workplace that may pose a risk of harm to young people and, therefore, warrant a full risk assessment.

4.362 These activities, and any actions taken as a result of them, should be recorded.

Conducting a Full Risk Assessment for Young People

4.363 The full risk assessment will evaluate the work activities identified as potentially harmful in the initial assessment and determine which of these present a significant risk to young people.

4.364 These risk assessments should take into account the:

- inexperience, lack of perception of danger and immaturity of young people
- fitting out and layout of their workstation and workplace
- nature, degree and duration of any exposure to physical, biological or chemical agents
- form, range, use and handling of work equipment
- organisation of work processes and activities
- health and safety training provided, or intended to be provided, by the employer.

4.365 Certain other information can also be used to help decide whether a significant risk exists, including examining:

- the presence of any particular hazards
- compliance with prescribed exposure levels, eg lead and ionising radiation
- accident and sickness records
- complaints and information from staff.

4.366 Those carrying out the risk assessment must be competent to do so.

Specified Hazards

4.367 While young persons are generally at no greater risk than other employees in the workplace, there are some defined hazards that employers must take into account. The Health and Safety Executive (HSE) guidance book, HSG165 *Young People at Work — A Guide for Employers*, contains examples of physical, biological and chemical agents, as well as various work situations, which employers must consider as part of any risk assessment for young persons.

Physical Hazards

4.368 Physical agents can cause long-term damage (eg occupational deafness through exposure to noise, or genetic damage through exposure to ionising radiation) and this may have more significant consequences for young people.

4.369 Physical agents identified in the HSE guidance include:

- temperature extremes
- noise
- vibration (hand/arm and whole body)
- radiation (ionising and non-ionising)
- work in increased atmospheric pressures, eg diving.

4.370 There is no evidence that young people are at greater risk from physical agents than their adult colleagues. However, some exceptions are:

- whole-body vibration
- ionising radiation
- work in high-pressure atmospheres.

Exposure to Vibration

4.371 Exposure to vibration — especially low-frequency whole-body vibration, such as occurs when driving tractors, etc — may cause spinal damage in young people as their bones and supporting muscles are not fully developed. Exposure to this type of vibration should be avoided or, at least, controlled by restricting the exposure time and/or level. Young people with non-occupational Raynaud's disease should not carry out tasks involving exposure to hand/arm vibration.

Exposure to Ionising Radiation

4.372 Exposure to ionising radiation is covered by the Ionising Radiations Regulations 1999, which both define specific exposure limits for young people and lay down strict control measures. The associated risks of cancer and genetic damage appear to be greater for young people.

4.373 One point of possible future significance is the effect of exposure to non-ionising radiation from the use of mobile phones. While there is currently no evidence of significant risk, it is still too early for any long-term effects to have been established or researched, and future health issues could arise. Employers have to be aware of the possibility of past exposures.

Exposure to High-pressure Atmospheres

4.374 Exposure to high-pressure atmospheres may create a greater long-term risk to young people as the air bubbles formed in the body during decompression illness can damage bones that are not yet fully developed. Industry best practice recommends that young people should not be exposed to high-pressure atmospheres. There are no age restrictions on diving, although divers have to be certified as medically fit and appropriately qualified.

Exposure to Physical Agents Unrelated to Work

4.375 Exposure to non-work-related physical agents (eg exposure to loud noises from pop concerts and clubs, etc) might cause possible adverse health effects to young people. Some employers require all new employees (regardless of their age) to undergo an initial health assessment (eg to determine current hearing levels) at the start of their employment.

Exposure to Other Physical Agents

4.376 Young people may be at greater risk of musculoskeletal damage from work involving strenuous and/or awkward moving and handling activities, prolonged or poor postures, etc while their muscles and bones are still developing and maturing. Such activities should be avoided, or at least controlled, by restricting exposure time, introducing task rotation, ensuring adequate recovery periods and rest breaks, etc.

Biological Agents

4.377 Biological agents are, basically, infections and include bacteria, viruses and other micro-organisms. They are assigned a hazard group classification from 1 to 4 (with 1 being the least hazardous and 4 the most) depending on the level of risk they pose to individuals and society.

4.378 The occupations most likely to involve exposure to biological agents are health care, animal care/treatment, laboratories, farming or other land management related work, etc. Controls, as covered in detail by the Control of Substances Hazardous to Health Regulations 2002 (COSHH), implemented for the workforce as a whole, will generally provide acceptable protection to young people. Control measures might include:

- avoiding exposure to the biological agent
- containment of the work
- providing personal protective equipment (PPE) and clothing
- high standards of personal hygiene (with provision of the necessary facilities, eg wash hand basins, showers)
- vaccines.

4.379 The potential exposure to biological agents in hazard groups 3 and 4 should be included in the risk assessment for young people. In addition, guidance from the Advisory Committee on Dangerous Pathogens (ACDP) recommends that young people do not handle animals infected with biological agents assigned to hazard group 4. Hazard group 4 biological agents are those that cause severe human disease, pose a serious risk to employees, are likely to spread to the community and that have no effective prophylaxis, or treatment, available.

Chemical Agents

4.380 Many chemical agents are capable of having adverse health effects on young people, although young people are not generally considered at greater risk than are other employees. The risk of such chemicals is determined by the chemical agent involved, the level of exposure and any particular circumstances at individual workplaces. Employers should consider the following chemicals specifically in their risk assessments for young people:

- chemicals classified by the Chemicals (Hazard Information and Packaging for Supply) Regulations 2002 (CHIP)
- lead and lead compounds
- asbestos.

Chemicals Classified by the Regulations

4.381 Employers should consider chemicals classified by the CHIP Regulations as:

- "very toxic", "toxic", "corrosive" or "explosive" (consideration should also be given to substances that may cause allergic reactions such as asthma and dermatitis)
- "harmful" with one or more of the following risk phrases:
  - R39 — danger of very serious irreversible effects
  - R40 — limited evidence of a carcinogenic effect
  - R42 — may cause sensitisation by inhalation
  - R43 — may cause sensitisation by skin contact
  - R45 — may cause cancer
  - R46 — may cause heritable genetic damage
  - R48 — danger of serious damage to health by prolonged exposure
  - R60 — may impair fertility
  - R61 — may cause harm to an unborn child
- "irritant" with one or more of the following risk phrases:
  - R12 — extremely flammable
  - R42 — may cause sensitisation by inhalation
  - R43 — may cause sensitisation by skin contact

4.382 Carcinogens, and certain related work processes, should also be carefully considered.

4.383 In many cases, there are legal limits placed on exposures to chemicals. Again, the Control of Substances Hazardous to Health Regulations 2002 (COSHH), and the accompanying codes of practice, provide details on assessing and controlling risks associated with exposure to chemicals. Common control measures include:

- avoiding exposure
- engineering controls, eg local exhaust ventilation
- good personal hygiene
- personal protective equipment (PPE).

4.384 Inexperience, immaturity and a lack of perception of danger may prevent young people from recognising or accepting "invisible" long-term dangers — ie health effects that may take many years to develop — and the need to follow strict controls.

Other Work Activities

4.385 In addition to exposure to physical, biological and chemical agents, HSE guidance also lists other work activities, equipment and situations that employers should consider in their risk assessments for young people. These are:

- flammable liquids and gases
- gas cylinders
- work with vats, tanks, reservoirs or carboys containing chemical agents
- work involving a risk of structural collapse
- work involving high voltage electrical hazards
- manufacture and handling of devices, fireworks or other objects containing explosives
- work with fierce, or poisonous, animals
- animal slaughtering on an industrial scale.

Review of Assessment

4.386 A review, and possibly a revision, of the assessment will also be required if there is:

- any relevant health and safety issue that arises during the young person's employment
- reason to believe that the previous assessment is no longer valid (eg following a change in work activities and/or equipment)
- an injury or accident.

Control Measures

4.387 Each young person will experience different effects as a result of their work due to personal differences, ie experience, fitness, etc. Employers should take these capabilities into account when assigning tasks to young people. As young people are generally at no greater risk than other employees at work, the control measures put into place to protect the workforce as a whole should also be sufficient to protect them. However, where the employment of young people is subject to specified legal controls, eg prohibition from certain lead work, these must be complied with. The risk assessment may also identify situations where additional and individual measures, over and above those required for the general workforce, may be necessary to protect young people.

Fitness to Work

4.388 Factory inspectors may issue a notice instructing employers to move a young person to more suitable work or to end their employment within seven days if the health of the young person is at risk. Employers must comply with the notice unless an appointed factory doctor, or an employment medical advisor, has examined the young person and certified them fit to continue in their work.

Working Hours

4.389 The Working Time Regulations 1998 place certain conditions and restrictions on the hours that young people can work, in particular in relation to the hours of night work and rest periods for young persons.

Night Work

4.390 Young workers are generally banned from working at night, ie between 10pm and 6am or between 11pm and 7am. There are some exceptions for certain kinds of employment and in certain circumstances, eg if it is necessary to either maintain continuity of service or production, or respond to a surge in demand.

4.391 If this is the case, certain rules must be followed:

- there should be no adult available to perform the task instead
- the training needs of the young worker must not be adversely affected
- the young worker must be allowed compensatory rest of a period of time equivalent to the night work
- there must be adequate supervision of the young worker.

Daily Rest Periods

4.392 Young people are entitled to a daily rest period of at least 12 consecutive hours in each 24-hour period during which they are at work. This may be interrupted in situations where the work activities are split up over the day, or are of short duration.

Weekly Rest Periods

4.393 Young people are entitled to a weekly rest period of at least 48 hours in each seven-day period during which they are at work. This weekly rest period may be interrupted where the work activities are split up over the day, or are of short duration, or where there are technical or organisational reasons. It may not be reduced to less than 36 consecutive hours.

Rest Breaks

4.394 Young people are entitled to a rest break if they work for more than 4.5 hours. The rest break should be of at least 30 minutes duration, consecutive if possible, and may be taken away from their workstations. Where the young person works for more than one employer, the hours worked for each different employer are aggregated.

Children

4.395 The Children and Young Persons Acts 1933 and 1963 prohibit the employment of children in the following circumstances:

- where the child is under 14 years of age
- where the work is any work apart from light work, ie work which could harm the health, safety, development or education of children
- before the close of school hours on any day on which children are required to attend school
- for more than two hours on any day on which they are required to attend school
- before 7am or after 7pm on any day
- for more than two hours on any Sunday
- for more than eight hours (or five hours, if the children are under 15 years) in any day in which they are not required to attend school and which is not a Sunday
- for more than 35 hours (or 25 hours, if the children are under 15 years old) in any week in which they are not required to attend school
- for more than four hours in any day without a rest break of one hour
- at any time in a year when the children are not required to attend school unless they have had at least two consecutive weeks without employment.

Training

4.396 Training, including the provision of relevant information and instructions, is important in ensuring young persons are aware of:

- their own duties and responsibilities
- what is going on and why
- changes, revisions, etc to working arrangements.

4.397 The training should cover the following areas:

- what risks have been identified and why they are significant
- what control measures are needed and how they will be implemented
- what is expected from the young person.

List of Relevant Legislation

- Control of Lead at Work Regulations 2002
- Control of Substances Hazardous to Health Regulations 2002
- Ionising Radiations Regulations 1999
- Management of Health and Safety at Work Regulations 1999
- Working Time Regulations 1998
- Health and Safety at Work, etc Act 1974

Further Information

Publications

HSE Publications

The following are available from *www.hsebooks.co.uk*.

- EH40 *Workplace Exposure Limits* (revised annually)
- HSG165 (rev 2000) *Young People at Work: A Guide for Employers*
- L5 (rev 2005) *The Control of Substances Hazardous to Health Regulations 2002 (as amended). Approved Code of Practice and Guidance*
- L21 (rev 2000) *Management of Health and Safety at Work. Management of Health and Safety at Work Regulations 1999. Approved Code of Practice and Guidance*

Chapter 5

Environmental Management

Contaminated Land

- The Environment Agency estimates that there may be around 300,000 hectares of land affected by some sort of industrial or natural contamination in the UK.
- The contaminated land regime enables direct regulatory action to ensure that land is identified and remediated to an appropriate standard.
- Under the contaminated land regime, contaminated land is defined with the aim that action is targeted at those pieces of land that present the greatest risk to human health and/or the environment.
- Land can be contaminated by a number of different sources.
- The Environment Agency or the Scottish Environment Protection Agency deals with the most hazardous sites and local authorities with the remainder.
- A pollutant linkage must exist for the land to be classified as contaminated under the regime.
- It is the role of local authorities to determine whether land falls under the contaminated land regime.
- Once a contaminated site is identified, the polluter is responsible for remediation. If no polluter can be found, liability for remediation passes to the owner or occupier of the land.
- Remediation is either carried out voluntarily or through a remediation notice.
- The contaminated land regime does not apply to the cleaning up of brownfield sites for redevelopment, which is dealt with under the planning system.

5.1 The Environment Agency (the Agency) estimates that there may be around 300,000 hectares of land affected by some sort of industrial or natural contamination in the UK.

5.2 The purpose of the contaminated land regime is to identify land that, because of the presence of contaminants, presents in its existing use a significant threat to human health, property or the natural environment. The regime has specific criteria for classifying land as contaminated — the clean-up of other derelict and brownfield sites for redevelopment is dealt with under the planning system.

5.3 The provisions for the contaminated land regime are contained in Part IIA of the Environmental Protection Act (EPA) 1990 (which was inserted by the Environment Act 1995) and the:

- Contaminated Land (England) Regulations 2006
- Contaminated Land (Scotland) Regulations 2005
- Contaminated Land (Wales) Regulations 2001.

Employers' Duties

5.4 The contaminated land regime imposes duties and powers on local authorities, the Environment Agency and the Scottish Environment Protection Agency (SEPA) to identify seriously contaminated sites and to ensure that they are cleaned up by those responsible for the pollution.

5.5 If a remediation notice is served by a local authority or one of the environment agencies, it must be complied with by the employer. Failure to comply with a remediation notice is an offence.

5.6 Employers who have knowingly permitted land to become contaminated in the past may become liable for clean-up measures for sites which are designated under the new regime.

In Practice

The Contaminated Land Regime

5.7 The traditional approach to dealing with contaminated land in the UK has been to leave it alone until a situation arises that will require some sort of legal intervention. Such situations arise:

- when the site is to be redeveloped
- where the substances in, on or under land give rise to a situation that is prejudicial to health, or a nuisance
- where there is a breach of one of the pollution control regimes.

5.8 In each of these cases, legal intervention is relevant but not specific to contaminated land.

5.9 These approaches have been complemented by a proactive mechanism for the identification and remediation of land where contamination is causing unacceptable risks to human health and/or the wider environment on the basis of the current use and circumstances of the land.

5.10 The contaminated land regime enables direct regulatory action to ensure that the land is identified and remediated to the "suitable for use" standard.

"Suitable for Use" Standard

5.11 The UK Government favours the "suitable for use" approach to dealing with contaminated land, considering that it provides the best means of reconciling the various environmental, social and economic needs in relation to contaminated land.

5.12 The alternative approach of "multifunctionality" is not used because of the prohibitive costs of remediating land to a standard suitable for any use.

Soil Guideline Values

5.13 Soil guideline values (SGVs) have been issued to help determine whether the concentration of a contaminant poses a significant risk to human health or the environment. They represent intervention values, which indicate to an assessor that soil concentrations above this level could pose an unacceptable risk to the health of site users and that further investigation and/or remediation is required. SGVs have been derived using the Contaminated Land Exposure Assessment (CLEA) model according to three typical land uses:

- residential (with and without vegetable growing)
- allotments
- commercial/industrial.

5.14 Exceeding a SGV indicates the need for either further investigation and/or remediation under the formal requirements of Part IIA of the EPA 1990.

5.15 SGVs have superseded the values developed by ICRCL (Interdepartmental Committee on the Redevelopment of Contaminated Land), until recently the basis for contaminated land assessment. The Department for Environment, Food and Rural Affairs (Defra) has announced that no further SGV or toxicological reports will be published for the time being. The SGV task force has identified several issues which need to be addressed, and Defra plans to develop a new package of policy and practical options. It has issued a new contaminated land advice note CLAN 6/06 *Soil Guideline Values: The Way Forward*, which outlines the latest thinking on the use of guideline values in assessing land contamination.

The EU Soil Framework Directive

5.16 In July 2007, Defra and the devolved administrations issued a *consultation paper* on a European Commission proposal for a European Union (EU) Soil Framework Directive. The proposed directive aims to provide a framework for EU Member States to identify threats to soil quality and resources, and requires measures to overcome identified issues. The proposal arises from the EU Soil Thematic Strategy adopted in 2006.

5.17 Under the proposed directive, Member States would be required to:

- address the impacts of policies likely to exacerbate or to reduce soil degradation processes
- ensure that land users whose actions might hamper, significantly, soil functions take precautions to prevent or to minimise their impacts
- take appropriate measures to limit soil sealing or to mitigate the effects through the use of appropriate construction techniques
- identify risk areas with regard to soil erosion, loss of organic matter, compaction, salinisation and landslides
- set risk-reduction targets and to draw up a programme of cost-effective measures for reaching these targets
- take appropriate action to prevent soil contamination
- establish an inventory of contaminated sites
- require sellers of land on which a potentially soil polluting activity has taken place to supply a soil status report
- remediate all sites over a specified time period
- set up mechanisms to fund remediation.

Definition of Contaminated Land

5.18 Contaminated land is defined in the EPA 1990 as "land in such condition by reason of substances in, on or under the land, that (a) significant harm is being caused or there is a significant possibility of such harm being caused; or (b) significant pollution of controlled waters is being or is likely to be caused".

5.19 This definition aims to ensure that action is targeted at those pieces of land that present the greatest risk.

Sources of Contaminated Land

5.20 Land may become contaminated through:

- leakage
- accidental spillage
- uncontrolled waste disposal

- release of hazardous substances during poorly managed demolition or redevelopment
- deposition of airborne substances
- inappropriate application of substances to land
- flooding events.

5.21 The contamination may be severe or relatively minor, and can remain undetected even during redevelopment. All contaminated land, however, has the potential to:

- present risks to the health and safety of those living and working in the vicinity
- present risks to the health and safety of anyone working on the site
- present risks to the integrity of any proposed or existing development
- give rise to detrimental environmental effects, such as air and water pollution, poor plant growth, contamination of foodstuffs, adverse impact on habitats and its associated fauna and flora
- impact adversely on the value of the land
- give rise to the blighting of land and property in the vicinity
- require protracted and expensive remediation work.

5.22 Typical examples of potentially contaminating uses of land include:

- smelters, foundries, steelworks and metal processing and finishing works
- coal and mineral mining and processing
- heavy engineering and engineering works
- gasworks, coal carbonisation plant and power stations
- oil refineries, petroleum storage and distribution-sites
- the manufacture and use of asbestos, cement, lime and gypsum
- the rubber industry
- munitions and explosives production, testing and storage
- the manufacture of organic and inorganic chemicals
- glass making and ceramics manufacture
- the textile industry, including tanning and dyestuffs
- paper and pulp manufacture
- timber treatment
- food processing, abattoirs and animal waste processing
- transport depots for air, road, rail and sea
- military and defence-related land
- landfill, incineration of waste, sewage treatment and scrap yards.

5.23 On 4 August 2006, new regulations came into force in England extending the contaminated land regime to land contaminated by radioactive substances.

Special Sites

5.24 More hazardous contaminated land sites are classified as "special sites". Essentially, such sites have particular characteristics which mean that they should more appropriately be dealt with by the Agency or SEPA. Such sites include those contaminated by waste acid tars, where the processing of crude petroleum or the manufacture or processing of explosives has occurred, eg land within a nuclear site.

5.25 In the case of special sites, remediation, determining liability and the apportionment of costs is effected by the relevant environment agency. For all other sites, the local authority has responsibility for these duties.

Designation of Contaminated Land

5.26 The main features of the contaminated land regime are:

- inspection by local authorities of their area to identify contamination, verify pollutant linkages, determine that land is contaminated and decide whether it should be designated a "special site"
- establishment of responsibility for the remediation of identified contaminated land
- determination and enforcement, where necessary, of the remediation required
- apportionment of liability for the cost of remediation
- public accessibility to information concerning contaminated land.

5.27 Local authorities are the key players in giving effect to the regime. The Agency or SEPA assists local authorities in their role but they are also responsible for special sites.

Pollutant Linkage

5.28 One of the key elements in identifying whether land falls under the contaminated land regime is the identification of a "pollutant linkage". A pollutant linkage is the linkage between a contaminant and a receptor by means of a pathway.

5.29 All three elements (contaminant, environmental pathway, receptor) must be present before a local authority can consider whether or not the land falls under the contaminated land regime. Receptors are:

- human beings
- ecological systems or living organisms that form part of an ecological system
- crops, produce, livestock, owned or domesticated animals and wild animals which are the subject of shooting or fishing rights
- buildings.

5.30 The pollutant linkage must be considered for each contaminant or combination of contaminants. In respect of controlled waters, the local authority must identify a significant pollutant linkage where controlled waters form the receptor.

Determining Whether Land Falls Under the Contaminated Land Regime

5.31 Local authorities are required to prepare a written strategy setting out their approach to dealing with contaminated land. They must inspect their area from time to time for the purpose of identifying contaminated land. The contaminated land strategy will enable local authorities to identify sites within their area where contamination is likely to be found and where a detailed inspection of the site is needed.

5.32 This detailed inspection may include the collation and assessment of documentary information, including that from other parties. This will largely be a desk activity and may involve the review of current and archived materials. Many local authorities employ geographic information systems (GIS) to map data.

5.33 There may also be a visual inspection of the land and limited sampling of surface deposits or a more thorough and intrusive investigation. Where access to a site is required for a visual or intrusive investigation, the local authority has wide-ranging powers to enter and to carry out its investigation. Local authorities must be satisfied that there is a reasonable possibility of a pollutant linkage.

5.34 The decision-making process is essentially a risk assessment involving:

- the identification and characterisation of pollutant linkages
- an estimation of the nature and magnitude of the risks and effects
- an evaluation of the acceptability of those risks.

5.35 A determination that land is contaminated cannot be made where the situation can more appropriately be dealt with under another piece of legislation, eg pollution prevention and control.

Establishment of Responsibility for Remediation

5.36 Once a contaminated site is discovered, the local authority will serve a notice on the appropriate person responsible for the pollution — normally the person who knowingly caused or permitted the presence of the substance causing the contamination, known as the Class A appropriate person. If no polluter can be found, liability for remediation passes to the owner or occupier of the land — the Class B appropriate person.

5.37 Where there is more than one Class A appropriate person, the local authority must decide how to share out the costs of remediation. This is usually based on the relative contribution to the overall contamination by each appropriate person.

5.38 The local authority may have to pay for the remediation of "orphan sites" where the appropriate person cannot be found or is unable to pay.

Determination and Enforcement of Remediation

5.39 The enforcing authority has the power to serve a remediation notice on the appropriate person where contaminated land has been identified. The Government's preferred approach, however, is to secure remediation through agreement with interested persons, rather than formally through the service of a remediation notice.

5.40 Where voluntary remediation occurs, the appropriate person must prepare and publish a remediation statement. This sets out the:

- things that are being, have been or are expected to be done by way of remediation
- name and address of the person who is doing, has done or is expected to do such things
- relevant periods of time over which the activities will be done.

5.41 A copy of the remediation statement is kept on the public register. Where a remediation statement has not been followed, the enforcing authority must consider whether the service of a remediation notice is required.

5.42 Whichever approach is adopted, the local authority (or environment agency in the case of a special site) will consult with those with an interest in the land concerning the particular remediation actions that would achieve the necessary degree of remediation. Such consultation will:

- identify options for remediation
- identify appropriate timescales over which remediation might take place
- attempt to resolve as many disagreements as possible prior to the service of a remediation notice or a situation of agreed voluntary remediation.

5.43 Remediation packages must be considered for each pollutant linkage and must represent the best practicable techniques of remediation.

Urgent Remediation Action

5.44 The need for urgent remediation action may occur at any time during an investigation and will be applicable where the enforcing authority considers there is an imminent danger of serious harm or serious pollution of controlled waters. Under such circumstances, a remediation notice is served on an urgent basis — there is no need for prior consultation and the usual three-month interval between notification and service of the notice does not apply.

Completion of Remediation

5.45 The enforcing authority will keep the progress with remediation under review, whether voluntary or following the service of a remediation notice. The authority will assess whether the remediation has been carried out as specified in the remediation notice or in the remediation statement.

5.46 The person carrying out the remediation notifies the enforcing authority that remediation has been effected through a statement of claimed remediation, ie a claim regarding what has been done. This statement of claimed remediation must be entered on the public register by the enforcing authority. There is no formal signing-off procedure by the enforcing authority.

Recovery of Costs

5.47 Where an enforcing authority has carried out remediation, it is entitled to recover any reasonable costs incurred in doing so, provided it was not itself the appropriate person. The enforcing authority can decide to recover the costs immediately, or can postpone cost recovery and impose a charge on the land that was subject to the remediation.

Contaminated Land Registers

5.48 Enforcing authorities are required to maintain a publicly accessible register of information pertaining to statutorily contaminated land. Information in the register includes:

- the designation of special sites
- prescribed details in relation to remediation notices
- remediation statements
- notifications of claimed remediation
- convictions.

Water Pollution from Contaminated Land

5.49 Only "significant" water pollution triggers the contaminated land regime. This avoids land being identified as contaminated land on the basis of small amounts of matter entering controlled waters.

Management of Contaminated Land

5.50 The *Model Procedures for the Management of Land Contamination* (CLR 11) have been developed to provide the technical framework for applying a risk management process when dealing with land affected by contamination. The process involves identifying, making decisions on and taking appropriate action to deal with land contamination in a way that is consistent with government policies and legislation within the UK.

5.51 The model procedures consist of three parts — the procedures, supporting information and the information map. These provide a hierarchy of information in which part 1 sets out the framework of the risk management process, part 2 provides further technical detail to support the process and part 3 contains sources of further information and guidance. These procedures are intended to assist all of those involved in dealing with land contamination, including landowners, developers, professional advisors, regulatory bodies and financial providers.

Redevelopment of Contaminated Land

5.52 Land which does not meet the definition of contamination under the regulations (for instance, some so-called "brownfield sites") may still need to be cleaned up prior to redevelopment. This is dealt with under the planning system. Annex 2 to Planning Policy Statement 23: *Planning and Pollution Control* gives guidance on using the planning and development process to address wider aspects of land clean up.

5.53 In general, where there is a proposal for certain prescribed forms of development, an application for planning permission must be made. Since contaminated land may present a threat to public health and safety, the natural environment, the built environment and economic activities through its impact on users of land, it is a material consideration in decisions concerning planning applications.

5.54 Local planning authorities must therefore be assured that, in granting planning permission, the extent of any contamination has been assessed and that the proposed development incorporates appropriate remediation or is required to through the use of conditions attached to the approval.

5.55 The onus will be on the applicant to assess the suitability of the land for the proposed development and to provide such evidence that the land is, or will be, suitable for use. Information furnished to the local planning authority should be sufficient to enable the local planning authority to decide whether to grant or to refuse planning permission.

List of Relevant Legislation

- Contaminated Land (England) Regulations 2006
- Contaminated Land (Wales) Regulations 2006
- Radioactive Contaminated Land (Modification of Enactments) (England) Regulations 2006
- Contaminated Land (Scotland) Regulations 2005
- Water Act 2003
- Environment Act 1995
- Environmental Protection Act 1990
- Town and Country Planning Act 1990

Further Information

Publications

Defra Publications

The following are available from *www.defra.gov.uk*.

- Defra Circular 01/2006: *Contaminated Land*
- CLAN 6/06 *Soil Guideline Values: The Way Forward*
- CLR11: *Model Procedures for the Management of Land Contamination*

Communities and Local Government Publications

The following are available from *www.communities.gov.uk*.

- PPS10 *Planning for Sustainable Waste Management*
- PPS23 *Planning and Pollution Control*
- PPS23 *Planning and Pollution Control — Annex 1: Pollution Control, Air and Water Quality*
- PPS23 *Planning and Pollution Control — Annex 2: Development on Land Affected by Contamination*

Organisations

- CIRIA
  Web: *www.ciria.org.uk*
  CIRIA provides information and publications on contaminated land as well as sustainable land use and construction.
- Contaminated Land: Applications in Real Environments
  Web: *www.claire.co.uk*
  Contaminated Land: Applications in Real Environments is a public-private partnership which aims to help develop cost-effective methods of investigating and remediating contaminated land in a sustainable way.
- Department for Environment, Food and Rural Affairs (Defra)
  Web: *www.defra.gov.uk*
  Defra is the main government department which deals with waste and other environmental issues. It consults on new regulations and provides guidance on legislation and best practice.
- Environment Agency
  Web: *www.environment-agency.gov.uk*
  The Environment Agency is the main environmental regulator in England and Wales and provides detailed information on legislative requirements, guidance for business and technical guidance on waste treatment and disposal.
- Network for Industrially Contaminated Land in Europe
  Web: *www.nie.org*
  The Network for Industrially Contaminated Land in Europe is an industry-led European forum for the exchange and dissemination of scientific and technical knowledge and ideas relating to all aspects of industrially contaminated land. Its key objective is to assist in the management of soil and groundwater problems.
- Scottish Environment Protection Agency (SEPA)
  Web: *www.sepa.org.uk*
  SEPA is the main environmental regulator for Scotland. It provides detailed information on legislative requirements, guidance for business and technical guidance on waste treatment and disposal.

Discharge to Sewer

- Any trade or business that produces liquid waste matter in the course of carrying out its trade or business is producing a trade effluent.
- The sewerage service is provided by the water and sewerage undertakers (ie water companies) in England and Wales. In Scotland the sewerage service is provided by Scottish Water and in Northern Ireland by Northern Ireland Water Ltd.
- Trade effluent discharged into a sewer must have the consent of the sewerage undertaker.
- Special category effluents must be referred to the Environment Agency or the Scottish Environment Protection Agency before determination of applications for consent to discharge to the sewerage system.
- The trade effluent consent will include a number of terms and conditions that must be complied with when discharging to sewer.
- Trade effluent discharges to sewer are subject to a charge for its reception, conveyance, treatment and disposal of residual sludge.
- There are a number of potential offences relating to the discharge of sewage effluent.

5.56 Any trade or business that produces liquid waste matter in the course of carrying out its trade or business is producing a trade effluent. Effluents may be discharged to the sewerage system or, in some circumstances, directly to a watercourse, groundwater or the sea (controlled waters). Discharge to sewer and discharge to water are covered by different legislative regimes.

5.57 Trade effluent discharges are regulated by sewerage undertakers (ie water service companies) to ensure the protection of people, the environment and the sewer system. Regulation of discharges to sewers is via a consent system. It is an offence to discharge into a sewer without consent or to breach the conditions of a discharge consent.

5.58 Legislation controlling trade effluent discharges to sewers was consolidated in England and Wales in the Water Industry Act 1991. In Scotland the relevant legislation is part of the Sewerage (Scotland) Act 1968 and in Northern Ireland the Water and Sewerage Services (Northern Ireland) Order 1973. The requirements of these sets of legislation are very similar.

5.59 The Urban Waste Water Treatment Directive 91/271/EEC requires Member States to have adequate trade effluent control. The directive was transposed by the Urban Waste Water Regulations 1994 (there are separate regulations for England and Wales, Scotland and Northern Ireland), which require sewerage undertakers to have trade effluent schemes and to modify discharge consents to ensure that treatment works meet the requirements of the directive. These regulations reinforce powers already available under previous legislation.

5.60 If the process discharging the effluent is authorised under the pollution prevention and control (PPC) regime, discharges to sewer will be considered under the terms of the PPC authorisation.

Employers' Duties

5.61 Employers who discharge to sewer from trade and business premises require a trade effluent consent or agreement from the sewerage undertaker to do so.

Employees' Duties

5.62 Employees should comply with the conditions of their employer's trade effluent consent. They should not discharge substances into a sewer without approval.

In Practice

The Sewer System

5.63 Sewerage is the network of pipes and associated equipment used for the collection and transport of sewage. Sewerage includes all sewers (which normally serve more than one building) and drains (excluding drains used for the drainage of one building or buildings within one curtilage) which are used for the drainage of buildings and yards that are part of buildings.

5.64 Sewerage services include the treatment and disposal of sewage and any other services that are required to be provided by a sewerage undertaker for carrying out its functions. The sewerage service is provided by the water and sewerage undertakers in England and Wales. In Scotland the sewerage service is provided by Scottish Water and in Northern Ireland by Northern Ireland Water Ltd.

5.65 A public sewer is one that is vested in the sewerage undertaker (the privatised water company for the area). A private sewer is one that is not vested in the sewerage undertaker, but serves a number of buildings. Such a sewer is the responsibility of the owners and occupiers of those buildings that drain into the private sewer.

Trade Effluent

5.66 Trade effluent is any liquid, either with or without suspended particles, which is wholly or partly produced in the course of any trade or industry carried out at a trade premises. Examples are liquid process wastes, cooling water, rinse water and washing down waters. It does not include domestic sewage from toilets and washbasins or separated uncontaminated rainwater from site drainage.

5.67 Trade effluent discharges must be controlled to:

- protect the health and safety of personnel working in the sewer system
- protect the fabric of the sewerage system

- ensure that the waste water treatment works operate satisfactorily (normally biological processes are used and these can be inhibited by certain chemicals)
- ensure that the sewerage system and treatment works is not hydraulically overloaded
- prevent pollution of watercourses via storm overflow systems
- ensure that the requirements of the discharge consent from the treatment works are met, as removal rates of chemicals vary
- ensure that treatment and recycling of residual solids (sewage sludge) from the waste water treatment works is not compromised.

Special Category Trade Effluent

5.68 In England and Wales, if the effluent is considered to be special category trade effluent, the water company could refuse to accept it. If it does consider acceptance, the water company must refer to the Environment Agency (the Agency) within two months to discuss if and how, in terms of any restrictions, the discharge might be accepted.

5.69 Special category effluents contain prescribed substances above background concentrations. The prescribed substances are:

- mercury and its compounds
- cadmium and its compounds
- gamma-hexachlorocyclohexane
- DDT
- pentachlorophenol and its compounds
- hexachlorobenzene
- hexachlorobutadiene
- aldrin
- dieldrin
- endrin
- carbon tetrachloride
- polychlorinated biphenyls
- dichlorvos
- 1,2-dichloroethane
- trichlorobenzene
- atrazine
- simazine
- tributyltin compounds
- triphenyltin compounds
- trifluralin
- fenitrothion
- azinphos-methyl
- malathion
- endosulfan
- trichloroethylene (above 30kg per year)
- perchloroethylene (above 30kg per year).

5.70 Alternatively they are discharges from the following prescribed processes:

- production of chlorinated organic chemicals
- manufacture of paper pulp
- manufacture of asbestos cement

- manufacture of asbestos paper or board
- processes involving the use of more than 100kg of raw asbestos in any 12-month period.

5.71 The water company is obliged to keep the applicant apprised of progress following referral to the Agency.

5.72 In Scotland, Scottish Water would discuss potential receipt of such effluents with the Scottish Environment Protection Agency (SEPA) prior to agreeing acceptance, but there is no legal requirement. In Northern Ireland it would be necessary to discuss the discharge with the Environment and Heritage Service.

Trade Effluent Consents

5.73 No discharge can be made to the sewerage system without the authority of the sewerage undertaker. Sewerage undertakers have a duty to accept trade effluent as far as they are able to in respect of their facilities.

5.74 When the sewerage undertaker does not permit a discharge to sewer, the effluent must be treated as waste and subject to the Environmental Protection (Duty of Care) Regulations. Where the effluent has hazardous properties, it must be treated as hazardous waste.

Application to Discharge Trade Effluent to a Sewer

5.75 If an application to discharge trade effluent to a sewer is being considered for a new premises or activity, the sewerage undertaker should be contacted as early as possible to ensure that the discharge is possible and to establish what limitations there might be.

5.76 The prospective discharger must then apply for a consent to discharge to the sewer in the form of a trade effluent notice to the sewerage undertaker. This would require the following information as a minimum:

- the nature or composition of the trade effluent
- the maximum quantity of the trade effluent which it is proposed to discharge on any one day
- the highest rate at which it is proposed to discharge the trade effluent.

5.77 If the sewerage undertaker is happy with the application it would then issue a trade effluent consent. It is an offence to contravene the conditions of a trade effluent discharge consent and sewerage undertakers have the power to enter premises to take compliance samples.

Consent Conditions

5.78 The consent will include the following discharge consent conditions.

- A description of the sewer or sewers into which the trade effluent may be discharged.
- The nature or composition of the trade effluent that may be discharged, which will include prohibition of some substances and limits on others. Typical conditions are:
  - pH between 6 and 10
  - sulphide shall not exceed 1mg/litre as S
  - cyanide shall not exceed 1mg/litre as CN
  - sulphate shall not exceed 1mg/litre as SO_4
  - separable oil and grease shall not exceed 100mg/litre
  - metals: zinc, lead, copper, chromium, nickel, antimony, arsenic, silver, beryllium, tin, selenium and vanadium shall not exceed individually or in total 10mg/litre
  - the temperature shall not exceed 43.3°C

  - maximum biochemical oxygen demand, chemical oxygen demand and concentration of suspended solids.
  - Petroleum spirit, calcium carbide and carbon disulphide are not allowed to be discharged due to risk of fire or explosion. Nothing is allowed to be discharged that may interfere with flows in the sewer or pumping stations.
- The maximum quantity of trade effluent that may be discharged on any one day, either generally or into a particular sewer.
- The highest rate at which trade effluent may be discharged.

5.79 The volume and composition of trade effluent allowed to be discharged depends on the capacity of the waste water treatment works and sewerage system, and the impact of other discharges.

5.80 The consent might also include other conditions, such as the:

- period or periods of the day during which the trade effluent may be discharged from the trade premises into the sewer
- exclusion from the trade effluent of all condensing water
- elimination or diminution of any specified constituent of the trade effluent before it enters the sewer
- temperature of the trade effluent at the time when it is discharged into the sewer and its acidity or alkalinity at that time
- payment by the occupier of the trade premises to the undertaker of charges for the reception of the trade effluent into the sewer and for the disposal of the effluent
- provision and maintenance of an inspection chamber or manhole to enable a person readily to take samples, at any time, of what is passing into the sewer from the trade premises
- provision, testing and maintenance of such meters, as may be required, to measure the volume and rate of discharge of any trade effluent being discharged from the trade premises into the sewer
- keeping of records of the volume, rate of discharge, nature and composition of any trade effluent being discharged and, in particular, the keeping of records of readings of meters and other recording apparatus provided in compliance with any other condition attached to the consent
- making of returns and giving of other information to the sewerage undertaker concerning the volume, rate of discharge, nature and composition of any trade effluent discharged from the trade premises into the sewer.

Change of Ownership

5.81 In England and Wales a new owner or occupier wishing to continue a trade effluent discharge must inform the sewerage undertaker in order that a new consent may be issued. Consents are not directly transferable.

5.82 In Scotland and Northern Ireland the consent relates to the premises/discharge and is transferable.

Variation of Conditions

5.83 The sewerage undertaker has the power to issue a direction to the occupier or owner of a trade premises giving new consent conditions. This is usually done if the consent conditions of the waste water treatment works discharge changes, necessitating further controls on what enters the sewerage system.

5.84 Such variations can be made only after a two-year interval, unless the discharger agrees in writing. Similarly the occupier of the trade premises must give notice to the sewerage undertaker if the effluent is likely to change as a result of process changes or production levels.

Appeals Against Consent Conditions

5.85 A discharger has a right of appeal against conditions imposed in the original direction or subsequent variations. There is also a right of appeal against refusal to accept special category effluent.

5.86 In England and Wales the appeal is to the Water Services Regulation Authority (Ofwat) and in Scotland to Scottish ministers. Ofwat or Scottish ministers may impose revised conditions on appeal.

Public Register

5.87 Sewerage undertakers are required to maintain a register of information relating to trade effluent consents and agreements. These may be inspected by members of the public at all reasonable times. A small charge may be payable.

Charging Scheme

5.88 Trade effluent discharges to sewer are subject to a charge for their reception, conveyance, treatment and disposal of residual sludge. The charge is calculated using a formula which takes account of the volume and strength of the effluent, and the level of treatment provided.

5.89 These may also include an element reflecting discharge costs which may be imposed on the water service company.

5.90 In Scotland, the Water Services, etc (Scotland) Act 2005 requires Scottish Water to set up a separate business to provide retail services to non-domestic customers. From April 2008 competition for this business will be allowed.

Radioactivity

5.91 Approval for discharge of low levels of radioactivity into the sewer must be obtained from the Agency or SEPA.

Offences Relating to Sewage Effluent Discharge

5.92 There are a number of potential offences relating to the discharge of sewage effluent. It is an offence to:

- cause, or knowingly permit, poisonous, noxious or polluting matter, or any solid waste, to enter controlled waters, or for trade or sewage effluent to be discharged into controlled waters
- put or pass into any drain or sewer connecting with a public sewer any matter likely to damage the sewer or interfere with the free flow of its contents, or to prejudice the treatment and disposal of the sewer contents
- put into a public sewer:
  - any chemical refuse or waste stream
  - any liquid of a temperature higher than 43°C
  - any petroleum spirit (including crude petroleum, oil made from petroleum or from coal, shale, peat or other bituminous substances)
  - calcium carbide.

Pollution Prevention and Control

5.93 Controls on discharges of trade effluent from Part A prescribed processes under the pollution prevention and control (PPC) regime are included within the site permit, which is issued by the Agency, SEPA or the Northern Ireland Environment and Heritage Service.

5.94 Discharges from processes authorised under PPC are subject to treatment by best available techniques but the Agency or SEPA are able to take into account the effect of waste water treatment in setting emission limits to sewer. The environmental control then takes place through the waste water treatment works discharge consent. Such discharges are, in practice, controlled by the water company.

List of Relevant Legislation

- Pollution Prevention and Control Regulations 2000
- Urban Waste Water Treatment (England and Wales) Regulations 1994
- Urban Waste Water Treatment (Scotland) Regulations 1994
- Trade Effluents (Prescribed Processes and Substances) Regulations 1992
- Water Services etc (Scotland) Act 2005
- Water Act 2003
- Water Environment and Water Services (Scotland) Act 2003
- Water Industry (Scotland) Act 2002
- Water Industry Act 1991
- Water Resources Act 1991
- Control of Pollution Act 1974
- Sewerage (Scotland) Act 1968
- Water and Sewerage Services (Northern Ireland) Order 1973

Further Information

Publications

Croner Publications

The following is available from *www.croner.co.uk*.

- Waste Management

Environment Agency Publications

The following is available from *www.environment-agency.gov.uk*.

- *Treatment and Disposal of Sewage Where no Foul Sewer is Available*: PPG4

Organisations

- Environment Agency
  Web: *www.environment-agency.gov.uk*
  The Environment Agency is the main environmental regulator in England and Wales and provides detailed information on legislative requirements, guidance for business and technical guidance on waste treatment and disposal.
- Environment and Heritage Service
  Web: *www.ehsni.gov.uk*
  The Environment and Heritage Service is the largest agency within the Department of the Environment in Northern Ireland. It takes the lead in advising on, and implementing, environmental policy in Northern Ireland.
- Northern Ireland Authority for Utility Regulation
  Web: *http://ofreg.nics.gov.uk*
  The Northern Ireland Authority for Utility Regulation, known as the Utility Regulator, has amongst its responsibilities the regulation of Northern Ireland Water Ltd. It oversees most of the functions of the Directors General of Electricity Supply and Gas for Northern Ireland, and the water and sewerage functions.
- Northern Ireland Water Ltd
  Web: *www.niwater.com*
  In April 2007, the NI Water Service became Northern Ireland Water Ltd, a government-owned company. NI Water provides the water and sewerage services in Northern Ireland.
- Ofwat
  Web: *www.ofwat.gov.uk*
  Ofwat is the economic regulator for the water and sewerage industry in England and Wales.
- Scottish Environment Protection Agency (SEPA)
  Web: *www.sepa.org.uk*
  SEPA is the main environmental regulator for Scotland. It provides detailed information on legislative requirements, guidance for business and technical guidance on waste treatment and disposal.
- Water Industry Commission for Scotland
  Web: *www.watercommissioner.co.uk*
  The Water Industry Commission for Scotland is responsible for regulating the activities of Scottish Water. It has powers to determine charges including those for trade effluent and to licence providers of retail services to non-domestic customers.
- Water UK
  Web: *www.water.org.uk*
  Water UK is the trade association for the water industry. Its members are the water and waste water service suppliers of England, Scotland, Wales and Northern Ireland.

Environmental Management

- Sound environmental management, based on tools such as environmental management systems or environmental audits, can bring about a number of business benefits — including reduced environmental risk and liability — and significant cost savings.
- Environmental issues of importance to the construction industry include air pollution, contaminated land, use of greenfield and brownfield sites, noise and waste streams.
- Although ISO 14001 is voluntary, it requires organisations who sign up to it to publish an environmental policy, identify significant environmental impacts, and set about controlling them to improve environmental performance.
- In 1993, an EU regulation allowing voluntary participation by companies in the industrial sector in a Community Eco-Management and Audit Scheme (EMAS) was adopted by the European Commission.
- An environmental audit is the systematic examination of how any business operation affects the environment.
- Environmental impact assessments (EIAs) require systematic analysis and presentation of information in a form which provides a focus for public scrutiny of a project.
- For those projects subject to EIA, an environmental impact statement will have to be prepared.

5.95 The growing magnitude of global environmental concerns has driven environmental issues steadily up the political agenda. All companies should consider the full impact of their activities on the environment, particularly construction industry companies and clients.

5.96 The Environment Agency for England and Wales regards the construction industry as a significant potential source of environmental pollution and a potential major contributor to water and pollution incidents. The Government has pushed the environment higher up the construction industry's agenda by publishing guidance and running campaigns to improve environmental performance and management of waste.

5.97 The management of the environmental impacts of a company's operations can bring about a number of business benefits, including guaranteeing compliance with legislation, cost savings, and improved corporate image.

Employers' Duties

5.98 Although ISO 14001 and the Eco-Management and Audit Scheme (EMAS) are both voluntary standards, employers who decide to meet their specifications have to commit themselves to comply with legislation and to provide sufficient resources to improve performance continually.

Employees' Duties

5.99 Environmental protection involves contributions from everyone. In the case of construction projects, this is even more important. Employees often spot opportunities for improvements in work practices and for reducing waste, and these should be shared with a manager where identified.

In Practice

Environmental Regulation

5.100 During the last decade, the amount of environmental regulation in the UK has increased since the introduction of the Environmental Protection Act 1990 and the Environment Act 1995.

5.101 The European Union (EU) has also introduced tighter controls with the adoption in recent years of new directives on integrated pollution prevention and control (IPPC), the landfilling of waste, and the requirement for an environmental impact assessment (EIA) to be carried out prior to some types of development.

Environmental Management Systems and Audits

5.102 A number of tools are available to ensure compliance with environmental legislation and to identify and manage the environmental aspects and impacts of operations. For example, a company may choose to implement an environmental management system (EMS) or carry out an environmental audit of its activities.

5.103 Sound environmental management, based on tools such as EMS or EIA can bring about a number of business benefits including reduced environmental risk and liability and significant cost savings. EMS and EIA can also assist in achieving sustainable development, a concept that has become a hot topic within the construction industry.

Benefits of Environmental Management

5.104 Sound environmental management can offer opportunities for:

- identifying environmental risks
- targeting controls and improvements
- reducing the risk of breaking laws
- saving costs through energy and waste reduction
- meeting a higher standard of environmental requirements from clients
- improving shareholder value and corporate image

Saving Costs through Energy and Waste Reduction

5.105 The most common benefits of environmental management come from a review of the use and efficiency of resources and energy, which may lead to the consideration of the cost effectiveness of using alternative energy sources.

5.106 In addition, waste minimisation can result in savings on the cost of disposal and can lead to a reduction in the amount of raw materials used.

A Higher Standard of Environmental Requirements from Clients

5.107 The extent and variety of customer needs and expectations is constantly growing. Many clients now prefer to use suppliers and subcontractors that can demonstrate environmental credentials to avoid tarnishing their reputation or risking non-compliance with legislation. In an industry where client requirement is the main focus of a project, suppliers cannot afford to risk losing a customer through the inability to meet environmental best practice demands.

Improved Shareholder Value and Corporate Image

5.108 There is a growing awareness within the construction industry that companies which actively manage environmental impacts are lower risk than those which do not. This can inspire confidence and therefore have a positive effect on share prices. Demonstration of environmental credentials and a good environmental image can boost profits and enhance business opportunities, since reduced environmental performance can improve profitability.

5.109 Financial institutions — such as banks and insurance companies — may also be more willing to deal with a company if they can be assured that their environmental liabilities have been minimised by good environmental practice. For example, the Association of British Insurers has published guidelines that, if adopted, will allow companies who effectively manage their environmental risks to pay lower insurance premiums.

Environmental Responsibility within a Project

5.110 Environmental control ultimately reverts to cost control. It is a major undertaking to improve the environmental efficiency of a company within a set budget. As a result, unless a client makes environmental issues a priority within their business objectives and project brief, it is not usual practice for a construction company to take environmental issues further than the minimum necessary to meet legislative requirements. The architect will include environmental issues within a design if it has been specified as part of the project brief, and the cost planner will account for this within the budget.

Environmental Manager

5.111 It is useful within a project to have one person with an identifiable environmental portfolio who has direct access to the client or client's representative so that decisions can be made when necessary. This person must not be seen as the only source of environmental expertise, however, but as the co-ordinator and facilitator.

5.112 This role may go to the project manager, the quality manager, or an appointed environmental manager, who must communicate effectively with all members of the workforce. In this way, change in effective environmental practice may be brought about through education and persuasion rather than by client dictate — this should be saved as a necessary final sanction.

Employee Responsibility

5.113 Environmental protection involves contributions from everyone. In the case of construction projects this is even more important. People working on-site can often spot opportunities for improvements in work practices and for reducing waste. This responsibility should be emphasised at all briefing meetings and be re-stated in training courses.

Management Philosophy

5.114 It is a major responsibility to change the heart of a company's operations, involving an examination of production efficiency, the acceptance of new products and services, ensuring the optimum use of raw materials and by-products and reducing the costs of achieving statutory environmental standards.

5.115 As with all changes in a company philosophy, the introduction of environmental issues will not be accepted unless management is fully committed — a token gesture is never enough.

5.116 Environmental control requires continual assessment, and permanent prevention of a pollution problem is much cheaper than repetitive cure. Maintenance should aim to design out faults and not just repair them. If a problem is seen to recur from project to project, records should show the fault so that it may be effectively dealt with before tackling the next project. All companies — client and industry — must stay ahead of the game. Competition is fierce and a company will need to monitor the market conditions not only at home, but also abroad, including the prevailing and expected environmental controls and customer awareness.

5.117 A company must give itself sufficient time to phase-in improvements, aiming to introduce the most effective measures first and then continue with the fine tuning to achieve overall improvement over the long term.

Environmental Issues

5.118 Environmental issues of importance to the construction industry are as follows.

- Primary environmental issues:
  - air pollution, eg dust nuisance from demolition
  - energy use, eg emissions to air, climate change, resource depletion
  - water pollution, eg surface water run-off containing silt, spillage of hazardous materials or leaks from fuel tanks
  - noise pollution, eg noise from construction machinery
  - contaminated land, eg migration of contaminated land off-site on tyres of vehicles, disposal of contaminated material
  - ozone depletion, eg ozone-depleting substances such as CFCs and HCFCs, refrigerants, air conditioning
  - species diversity, eg harm to endangered species, use of certain construction materials
  - sensitivity of site, eg rural heritage, sites of special scientific interest (SSSIs), areas of outstanding natural beauty (AONB), national parks.
- Secondary environmental issues:
  - architectural quality, eg compatibility with local environment
  - landscape, eg change of land use
  - planning, eg use of greenfield and brownfield sites, transport implications of sites, town planning
  - local wildlife, eg damage to local ecosystems
  - materials use, eg specification, recycling, reuse, demolition

  - materials supply chain, eg raw material extraction/depletion, manufacturing processes, transport
  - flexibility of fabric and plant, eg change of use of buildings
  - water conservation.
- Health issues:
  - internal and external air and water quality
  - lighting
  - internal layout
  - noise
  - comfort.
- Key operational issues, eg management of:
  - energy systems
  - waste streams (materials and emissions)
  - the site
  - transport policy for the company and the site
  - structures within companies which are related to environmental outputs.

Environmental Management Systems

5.119 The aim of an environmental management system (EMS) is to provide a risk-based management tool which:

- identifies significant environmental impacts and risks
- targets areas for control
- provides a mechanism for continual improvement.

5.120 In 1993, the European Commission adopted the first pan-European standard on environmental management, the EU's Eco-Management and Audit Scheme (EMAS).

ISO 14001

5.121 The International Standards Organisation (ISO) published ISO 14001: Environmental Management Systems – Specification with Guidance which has superseded BS 7750. More recently, ISO has developed various supplementary environmental management standards in what is generally referred to as the ISO 14000 series.

5.122 ISO 14001 is the only certifiable standard in the ISO 14000 series. There is no legal requirement for organisations to participate in the scheme. ISO 14001 requires an organisation to publish an environmental policy, identify significant environmental impacts, and set about controlling them to improve environmental performance. It also requires compliance with applicable legislation and regulations, and continual improvement in an organisation's overall environmental performance. Environmental performance targets are determined by each organisation individually.

Key Stages in ISO 14001

5.123 The key stages in ISO 14001 are as follows.

- An environmental policy should be established.
- Planning of the environmental management system will involve:
  - identification of environmental aspects and evaluation of associated environmental impacts
  - establishments of relevant legal and regulatory requirements
  - development and maintenance of environmental objectives and targets
  - establishment and maintenance of an environmental programme to achieve the organisation's objectives and targets.

- Implementation and operation of the system will involve:
  - establishment of an organised management structure
  - training of the workforce and the establishment of effective methods of communication within the workforce
  - documentation of the environmental management system
  - operational control to ensure that the environmental management system is operating as it should be
  - emergency preparedness and response.
- Checking and corrective action should be undertaken, such as:
  - monitoring and measurement of operational activities, including record keeping
  - environmental management systems audit procedures.
- Management review of the environmental management system should be undertaken to determine its suitability, adequacy and effectiveness.

5.124 A review of ISO 14001 was completed late in 2004. The standard was revised against a policy of "no substantive change", but it does contain subtle changes to improve its accessibility and, in certain areas, to continue the gradual convergency in vocabulary and structure with ISO 9001: 2000, the international quality management system standard. The limited changes include:

- relocation of, and re-emphasis on, the scope of the system
- suppression of reference to an environmental programme
- emphasis on competence rather than qualification.

EMAS

5.125 The EU's Eco-Management and Audit Scheme (EMAS) was introduced by an EU regulation in 1993 and updated five years later to dovetail EMAS with ISO 14001. EMAS now requires ISO 14001 as a foundation and, in simple terms, can be viewed as an enhanced specification for an environmental management system (EMS).

5.126 EMAS is a voluntary scheme with the following requirements.

- An environmental policy must be adopted which meets the requirements.
- An environmental review of the site must be conducted.
- An environmental programme geared toward continuous improvement must be set up.
- Environmental audits must be carried out which meet the requirements.
- Objectives must be set in the light of the findings of the audit, and the programme revised to include the achievement of the objectives.
- A site specific environmental statement must be drawn up.
- The policy, programme, management system, review, audit procedures and statement must demonstrate that they possess the necessary qualifications, training and experience.
- The validated statement must be sent to the competent body, who would then make the information available to the public, as appropriate.

Third-party Certification and Verification of ISO 14001 and EMAS

5.127 Formal approval of an EMS, whether through third party ISO 14001 certification or EMAS verification, provides an organisation with environmental credibility which, in turn, can lead to a competitive advantage applying for registration. The organisation should provide all relevant information to the certification/verification body and back-up support during site visits, data analysis, staff interviews and report writing.

5.128 Where there are problems — such as infringement of regulations — certification or verification can be withheld until the issues have been resolved.

Environmental Auditing

5.129 An environmental audit is the systematic examination of how any business operation affects the environment. It can include all emissions to air, land and water, legal constraints, the effects of the neighbouring community, landscape and ecology, and considers products as well as processes.

5.130 It is a total strategic approach to an organisation's activities, from design through to recycling. It does not stop at compliance with legislation, nor is it a public relations exercise, although if it is done well it can improve a company's image.

5.131 The main purpose of an audit is to provide information on environmental performance against predetermined standards, helping to ensure those standards are met and enabling the company to stay ahead of legislation. A full audit will give a detailed analysis which can then be used to decide what should be done to improve performance and monitor and manage types and rates of change.

ISO 19011

5.132 Since October 2002 there has been one standard, ISO 19011, that provides guidance on carrying out quality systems and EMS audits. In addition, there is a professional association and listing of environmental auditors through the Environmental Auditors Registration Association (EARA), managed by the Institute of Environmental Management and Assessment.

Environmental Impact Assessments

5.133 The rationale behind most environmental impact assessment (EIA) methods has been to raise the profile of environmental concerns in the planning and operating phases of a project, and to provide methods for quantifying environmental impacts.

5.134 EU Member States are required to provide legislation for EIA, making certain that a full EIA is carried out to prevent or reduce environmental damage at the earliest stage. It ensures that the likely events of new development are as well understood as possible and potential impacts are mitigated at the design stage, before the project actually commences. Information about the environmental effects of a project will be collected both by the developer and from other sources and taken into account by the competent authority before granting an authorisation for the development.

5.135 EIA requires systematic analysis and presentation of information in a form which provides a focus for public scrutiny of a project.

5.136 The system of EIA has been implemented in the UK largely via the land use planning system. The Town and Country Planning (Environmental Impact Assessment) (England and Wales) Regulations 1999 divide the types of projects affected by the requirement of EIA into two lists:

- schedule 1, listing those projects where an EIA is mandatory
- schedule 2, listing those that may require an EIA if they are likely to give rise to significant effects on the environment.

5.137 Significance in this case is determined by the local planning authority (LPA) or the competent authority (CA) covering the area in question.

5.138 The first practical step for those seeking development planning permission is to discover if their particular project is subject to the regulations concerning EIA, and which authority will grant permission. Projects for which an EIA is considered mandatory include:

- nuclear power stations or reactors
- installations for permanent storage or final disposal of radioactive waste
- thermal power stations where heat output is more than 300MW
- an integrated chemical installation
- gasification and liquefaction of coal or bituminous shale
- crude oil refineries
- extraction, processing and transformation of asbestos
- trading ports
- inland waterways capable of handling vessels over 1350 tonnes
- aerodromes (where runways are longer than 2100m)
- special roads (ie motorways)
- long-distance railway lines.

5.139 As there is a requirement for the LPA to exercise judgment and discretion in areas where the regulations indicate that an impact assessment is not mandatory, but may still be necessary, many developers seek a preliminary opinion from the LPA before filing a full planning application.

5.140 Projects for which an LPA must decide if an EIA is required include:

- extractive, eg extraction of coal, lignite, peat, petroleum or natural gas, ores, shale and minerals, etc
- energy production, eg thermal power station, hydroelectric schemes, wind farms, transmission by overhead power cables, surface storage of fuels, production of electricity, hot water and/or steam, etc
- metal processing activities/installations
- chemical industry activities/installations
- food industry activities/installations
- textile, leather, paper and wood industry activities/installations
- rubber industry activities/installations
- infrastructure projects activities/installations.

Environmental Impact Statement

5.141 For those projects subject to EIA, an environmental impact statement (EIS) will have to be prepared. In short, the EIS should cover:

- the environmental impacts of the proposed activities/action
- adverse environmental impacts that cannot be avoided if the development/project proceeds
- any alternatives to the proposed activities/action
- any relationship between the long-term productivity associated with the project and the local short-term uses of the local environment
- any irretrievable commitments of resources, or their irreversible use in the proposed activities/action.

5.142 Once the EIS has been received and considered, the LPA will, as part of the decision-making process, consult with the Secretary of State, the Environment Agency, the Countryside Commission, English Heritage and English Nature. Applicants should also note that details of the planning application will be included in the local planning register, the EIS itself should be made publicly available, and notice of the application should also appear on the development site and in the local newspaper.

Building Research Establishment's Environmental Assessment Method

5.143 The Building Research Establishment's Environmental Assessment Method (BREEAM) is a voluntary, self-financing scheme for the assessment of the environmental performance of buildings, especially regarding their energy use.

5.144 BREEAM grades compliant buildings with a "pass", "good", or "excellent" rating, depending on the degree of compliance with specified criteria.

5.145 The scheme has become all the more relevant since the Government announced its intention to provide tax breaks for new homes with good environmental performance, whereas there could be penalties for new homes which fail to meet "sustainability" targets. A new, energy efficient building, for example, can use about 10% of the energy of a traditional building.

5.146 The scheme awards certificates to individual buildings that have been designed to maximise environmental performance. The performance of each building assessed under the scheme is recorded against a set of defined environmental criteria. The assessment and resulting certificate can form part of the developer or occupier's overall environmental policy statement and management system.

5.147 A key advantage of the scheme is that different models exist for different building types and these are continually being updated to reflect changes in technology, research and regulation. The scheme also enables businesses to demonstrate the application of environmental intentions and objectives to others.

5.148 The assessment covers a wide range of issues affecting:

- the global environment (ozone depletion, global warming, resource depletion)
- the neighbourhood (legionnaires' disease from cooling towers, local wind effects, reuse of existing site)
- the internal environment (legionnaires' disease from water supplies, lighting, indoor air quality, hazardous materials).

5.149 It gives credits for each aspect of the design where specific targets are met.

5.150 Assessments are carried out by independent certifiers approved by the British Research Establishment. The assessment certificate for the final design forms part of the building documentation for consideration by potential purchasers or users.

5.151 The BRE also has an award scheme to encourage developers to build homes which have reduced environmental impacts.

List of Relevant Legislation

- Town and Country Planning (Environmental Impact Assessment) (England and Wales) Regulations 1999
- Environment Act 1995
- Environmental Protection Act 1990

Further Information

Publications

- ISO 14001: 2004 *Environmental Management Systems — Requirements with Guidance for Use*
- ISO 19011: 2002 *Guidelines for Quality and Environmental Management Systems Auditing*

Organisations

- Building Research Establishment (BRE)
  Web: *www.bre.co.uk*
  The BRE provides consultancy, testing and commissioned research services covering all aspects of the built environment and associated industries.
- Department for Environment, Food and Rural Affairs (Defra)
  Web: *www.defra.gov.uk*
  Defra is the main government department which deals with waste and other environmental issues. It consults on new regulations and provides guidance on legislation and best practice.
- Environment Agency
  Web: *www.environment-agency.gov.uk*
  The Environment Agency is the main environmental regulator in England and Wales and provides detailed information on legislative requirements, guidance for business and technical guidance on waste treatment and disposal.
- Institute of Environmental Management and Assessment (IEMA)
  Web: *www.iema.net*
  IEMA is a not-for-profit organisation established to promote professional and best practice standards in environmental management, auditing and assessment.

Environmental Noise

- Environmental noise (also known as ambient noise) is not usually so loud that it causes hearing impairment. However, it can be stressful, result in loss of sleep and cause considerable annoyance if it interferes with quality of life.
- Noise emitted from premises can be a statutory nuisance, causing personal discomfort or interference with the enjoyment of one's property.
- Local authorities can set up noise abatement zones to protect against gradual increases in environmental noise, and improve noise levels where possible.
- The new Environmental Noise Directive will introduce noise mapping and noise action plans to major towns and cities over the next few years.
- Companies subject to integrated pollution prevention and control (IPPC) need to control noise and vibration as part of their IPPC permit conditions.
- Planning authorities will consider the impact of noise when making planning decisions.
- Control measures for environmental noise will normally involve one or more of the following:
  - remove the noise source
  - modify/maintain the noise source
  - modify the path between source and receiver.
- The production of a noise management plan may be a useful exercise for any company with an environmental noise problem.

5.152 Unlike occupational noise with established links between sound levels and hearing damage, the effects of environmental noise relate more to personal discomfort or interference with the enjoyment of one's property.

5.153 The main provisions for controlling environmental noise exist under the statutory nuisance provisions of the Environmental Protection Act 1990, the Control of Pollution Act 1974 and the Noise and Statutory Nuisance Act 1993. The Noise Act 1996, the Anti-Social Behaviour Act 2003, and the Clean Neighbourhoods and Environment Act 2005 also include measures to control local noise nuisance.

Employers' Duties

5.154 Employers should:

- reduce noise from machines, plant or equipment at source, as far as is reasonably practicable
- comply with any requirements set out in abatement notices, if issued.

Employees' Duties

5.155 Employees should ensure that environmental noise is kept to a minimum.

5.156 If an abatement notice is issued, employees must comply with its requirements.

In Practice

Environmental Noise

5.157 Unlike occupational noise with established links between sound levels and hearing damage, the effects of environmental noise are largely subjective.

5.158 Environmental noise is not usually so loud that it causes hearing impairment. However, it can be stressful, result in loss of sleep and cause considerable annoyance if it interferes with quality of life. It can also be harmful to animals.

5.159 Environmental noise problems commonly arise from a variety of sources including:

- excavation work
- demolition
- compression and concrete mixing
- transportation activities, eg lorry movements.

Statutory Nuisance and Noise

5.160 Noise emitted from premises or sites can also be a statutory nuisance. Statutory nuisance is an extremely wide area of law and covers a number of areas, all matters which may be prejudicial to health or a nuisance. The essence of this is personal discomfort or interference with the enjoyment of one's property.

5.161 Prime responsibility for enforcement of statutory nuisance lies with local authorities, and the prime method they deal with this is through noise abatement zones (NAZs).

Noise Abatement Notices

Business Implications of Noise Abatement Zones

5.162 The large majority of businesses and process operators are unlikely to be affected by NAZs, however there are around 60 established around the UK. In order to maintain a NAZ, local authorities have to continue to monitor and enforce the regulations. Where new buildings are proposed in potentially noisy areas, councils tend to use planning powers and guidance such as *(PPG) 24: Planning and Noise.*

5.163 Where specific sites cause nuisance, action is usually taken under the Environmental Protection Act or, if the premises are controlled under the integrated pollution prevention and control (IPPC) regime through the authorisation and permitting process.

5.164 If the local authority is still taking an active interest in maintaining its NAZ, it may be worth discussing the implications with local noise control officers. The Control of Noise (Measurement and Registers) Regulations 1976 and the Noise Levels (Measurement and Registers) (Scotland) Regulations 1982 give details of methods of noise measurement and local authority responsibility for maintenance of the NAZ register.

Appeals

5.165 An appeal may be lodged if the requirements of a NAZ are felt to be too unrealistic or onerous. Appeals may be made against:

- entry on the noise level register
- the authority's refusal to allow exceedence of a registered level
- any proposed fixing of noise levels from a new building.

5.166 Appeals are determined by the Secretary of State under the Control of Noise (Appeal) Regulations 1975. Under s.70 there is also a right of appeal to the magistrates' court for three months from the date on which a noise reduction notice is served. The Control of Pollution Act 1974 (COPA) does not list grounds for objection, but appeals have been successful on the following grounds.

- Confirmation of the order would involve unnecessary expenditure by the council.
- The council may not have sufficiently trained staff to administer the order.
- Benefits obtained from the order are doubtful.
- No complaints have been received in relation to the premises scheduled for control.
- The type of premises controlled by the order causes negligible noise nuisance.
- Noise can be controlled more effectively under other legislation.
- Order is an unfair burden on small businesses.
- Order could affect property values.
- Order could prevent subsequent changes of land use which would otherwise be allowed under planning and noise guidance.

5.167 It is a defence to prove that the best practicable means were used to prevent or counteract the effects of the noise.

Forthcoming Requirements for Noise Action Zones

5.168 Although NAZs are now relatively rare, new and much larger noise action zones are likely to be created under new EU legislation.

5.169 The EU Commission Fifth Action Programme on the Environment proposed a number of measures aimed at reducing people's exposure to night-time noise, including:

- exposure to more than 65dB(A) should be phased out and at no time should 85dB(A) be exceeded
- those exposed to levels of between 55–65dB(A), and those currently exposed to less than 55dB(A), should not suffer any increase.

5.170 The Commission also proposed a number of measures to achieve these noise levels, including:

- noise abatement programmes
- standardisation of noise measurement and rating
- directives aimed at further reducing noise from transportation
- noise control measures related to infrastructure and planning, including airports, industrial areas, major roads and railways.

Environmental Noise Directive

5.171 The most immediate effect of EU interest in noise is Directive 2002/49/EC, adopted in 2002, and known as the Environmental Noise Directive (END). While acknowledging that noise is essentially a local issue, the EC identified a number of areas where an EU-wide approach is needed.

5.172 The aim of the END is to define a common approach across the EU with the intention of avoiding, preventing or reducing — on a prioritised basis — the harmful effects, including annoyance, caused by exposure to environmental noise. This will involve:

- informing the public about environmental noise and its effects
- preparing strategic noise maps for:
  - large urban areas (referred to as "agglomerations" in the END)
  - major roads
  - major railways
  - major airports as defined in the END.
- preparing action plans based on the results of the noise mapping exercise to manage and reduce environmental noise where necessary, and preserve environmental noise quality where it is good.

5.173 The noise mapping and action planning process is to be taken forward on a five-yearly rolling programme. The first round of mapping and action planning applies to the largest of the agglomerations (including the industries and ports within them), the busiest major roads and railways and all major airports.

5.174 Noise maps are currently being produced around the UK, with action plans required to be produced later in 2008, although the UK is behind schedule in this respect.

5.175 During the second round (2012–2013) and in subsequent rounds, all agglomerations, major roads, major railways and major airports as defined by the END will be mapped and then action plans will be developed for them.

5.176 The END seeks to limit or reduce individual exposure to environmental noise, in particular in urban areas, recreational zones, or other quiet areas, and in noise-sensitive buildings such as schools and hospitals. END defines environmental noise as "unwanted or harmful outdoor sound created by human activities, including noise emitted by means of transport, road traffic, rail traffic, air traffic, and from sites of industrial activity…" to which humans are exposed in the domestic environment (eg in and near the home, in public parks, in schools and hospitals, etc). It does not cover noise from domestic activities, noise created by neighbours, noise at workplaces, noise inside means of transport or due to military activities in military areas.

What Will the END Require?

5.177 END will introduce a number of new anti-noise measures.

- Noise maps, using common indicators and assessment methods, will be created to cover all large agglomerations (defined as areas containing more than 250,000 inhabitants), major roads (defined as those with over 6 million vehicle movements per year), major railways (over 60,000 rail movements per year) and civil airports (over 50,000 aircraft movements per year). Since June 2005, and from then every five years, Member States must advise the Commission of all major roads, major railways and major airports and agglomerations.
  Note: The UK is behind schedule in this process.
- Technical noise assessment indicators known as Lden and Lnight are to be used for benchmark measurement in EU noise policy, in new legislation on noise mapping, acoustical planning or noise zoning and in any revision to existing legislation. Lden is the day-evening-night level in decibels and is an indicator for annoyance caused by noise. Lnight is the "overall night-time noise indicator". A reduction in Lnight would reduce sleep disturbance. Annex II to the directive sets out assessment methods for both indicators. The limits are based on the World Health Organization's *Guidelines for Community Noise*.

- Over the longer term, noise maps for agglomerations between 100,000 and 250,000 inhabitants, roads with more than 3 million vehicle movements per year and railways with more than 30,000 movements per year must be prepared and approved by June 2012.
- The noise maps should include data on noise levels and identify the number of people affected by excessive noise levels as defined in the END. Noise maps are to be reviewed every five years.
- Action plans are to be prepared for the mapped areas (by July 2008 for large agglomerations or 2013 for agglomerations over 100,000), to include measures for protecting the relatively quiet areas and for reducing noise in noisier areas. Action plans are to be reviewed and revised should there be a major development which affects the noise situation, and in any case, at least every five years.
- The noise maps are to be published, and the public must be consulted during the preparation of action plans.
- Member States are responsible for forwarding noise maps and action plans to the Commission, who will set up a data bank of information on noise maps and publish a summary report every five years.

Implications for Business of the END

5.178 Businesses located in an area defined as a major agglomeration should analyse the local noise map, when it is published. Possible effects could include:

- impact on property prices in areas perceived as "noisy"
- local public interest in reducing ambient noise levels, particularly around noise-sensitive locations
- concern from employees about potential health impacts from consistently high levels of ambient noise
- political interest in the possible introduction of noise action plans, and the potential impacts on business and community.

5.179 There may be increased pressure to introduce quiet equipment and processes. However, the major sources of ambient noise tend to be transport-related, especially road vehicles, and heavy duty vehicles in particular. There may be incentives or requirements to adopt environmentally enhanced vehicles (EEVs) — a new classification being developed by the EU — which have very low noise characteristics.

5.180 Businesses should contact their local authority's noise officers to find out:

- when the local noise map will be published
- what plans they have to consult on the results
- whether an END noise action plan is planned for the area.

Burglar Alarms

5.181 If premises are fitted with audible intruder alarms, they may be subject to local authority control, and they must comply with the relevant standards, including cut-out mechanisms.

5.182 For example, if the local authority has designated an area an "alarm notification area", under the 2005 Clean Neighbourhoods and Environment Act those with burglar alarms in England and Wales are required to register keyholders with the police.

Noise and IPPC

5.183 Integrated pollution prevention and control (IPPC) covers a wider range of environmental impacts than integrated pollution control (IPC). This includes noise and vibration.

5.184 Best available techniques (BAT) should be used to prevent or reduce noise, a condition that should appear in all IPPC permits. Specific conditions on noise reduction measures may also be included.

5.185 Local authorities already control noise from industrial installations and in order to avoid operators being subject to two separate control regimes, local authorities will not be able to take statutory nuisance action in respect of noise from an IPPC installation. However, they will be able to enforce the standard permit condition by issuing appropriate warnings and prosecuting if required.

5.186 The new arrangements under IPPC aim to produce a more proactive approach to noise abatement issues, rather than nuisances being addressed after they have arisen.

5.187 Noise at IPPC installations not covered by the IPPC regime (eg a barking dog or a burglar alarm) will still be regulated as a statutory nuisance by local authorities.

Planning and Noise

5.188 It has been well established that noise is a planning issue, a fact recognised by the publication of planning policy and guidance note *(PPG) 24: Planning and Noise.*

5.189 This guidance outlines that, wherever practicable, noise-sensitive developments should be separated from major sources of noise, such as road, rail and air transport and certain types of industrial development. Similarly, new developments involving noisy activities should, if possible, be sited away from noise-sensitive land uses.

5.190 To assist planning authorities, a system of noise exposure categories has been developed to assist in the consideration of proposed noise-sensitive developments likely to be affected by transport-related noise. The categories are as follows.

- Category A: noise unlikely to be a determining factor.
- Category B: noise should be considered and conditions may be attached.
- Category C: consent not normally granted due to excessive noise.
- Category D: consent normally refused.

5.191 While the focus of PPG24 is on controlling proposed noise-sensitive developments from existing noise, it does recognise that problems can arise from new noisy developments. The guidance contains examples of planning conditions and other advice for planning authorities that may help to minimise the impact of noisy developments.

Environmental Assessment

5.192 Certain types of development may be subject to an environmental assessment. The developments that are subject to environmental assessment tend to be the larger, more contentious types, such as certain industrial installations, with the potential to cause harm to the environment.

5.193 Noise will be considered when an environmental assessment is carried out and should be included in the environmental statement submitted by the developer to the planning authority.

Noise Emissions by Outdoor Equipment

5.194 The Noise Emission in the Environment by Equipment for Use Outdoors Regulations 2001 control noise emissions by equipment used outdoors, eg cranes, dumpers, lift trucks, etc. The regulations also deal with labelling requirements for noise emissions from outdoor equipment.

Control of Environmental Noise

5.195 Environmental noise can be very difficult to control effectively but there are a number of steps that can be taken to reduce or prevent it.

5.196 To establish whether there is a potential environmental noise problem that needs controlling, noise emissions should be monitored near the boundary of the site during different operating conditions and at different times of the day.

5.197 Establishing a good relationship with neighbours is beneficial to all parties. Advising neighbours in advance of a potentially noisy period can reduce the chance of a complaint and make the business seem more considerate.

5.198 Night-time noise should be avoided, or at least kept to a minimum, with as much as activity as possible taking place during normal work hours.

Noise Control Measures

5.199 Control measures for environmental noise will normally involve one or more of the following:

- remove the noise source
- modify/maintain the noise source
- modify the path of noise from the source to the receiver.

Remove the Noise Source

5.200 Clearly, the most effective means of reducing or preventing an environmental noise problem is to remove the noise source. There are a number of ways this can be achieved.

5.201 Replacing a noisy item of plant or equipment with a quieter item may prove successful. For example, replacing an air-cooled heat exchange unit with a water-cooled heat exchange unit can greatly reduce the noise levels, as can replacing diesel engine plant on a construction site with petrol engine or even electric motor plant (concrete mixers, lift trucks, etc).

5.202 Relocating noisy plant and equipment can also be a very effective control measure but care must be taken to ensure that the problem is relocated at the same time. Taking the example of the heat exchange unit, it might be possible to relocate to an area where the noise it produces does not cause any disturbance, eg on top of a roof, at a different part of the building, etc.

5.203 Increasing the distance of the source from the receiver will reduce the noise levels measured at the reception point. However, the distances necessary before this technique will be successful might by unrealistic, particularly for lower frequency noise.

Modify/Maintain the Noise Source

5.204 Modifying/maintaining the noise source involves treating the noise source in such a way that the amount of noise produced by it is reduced. One obvious example of this is the fitting of silencers to air intakes and exhausts. Vibrating lightweight panels can be dampened by the application of suitable materials, and drive chains can be replaced by drive belts.

5.205 Bins, chutes, etc can all be lined with energy-absorbing material, a treatment that has been successfully used for bottle banks. Flexible pipe and duct couplings can help to minimise the amount of noise that is transmitted through a structure. Examples of this include:

- flexible pipe couplings
- isolating pumps and other noisy and vibrating components of communal central heating systems
- flexible duct connectors isolating the ventilation fan from the rest of the structure.

5.206 Maintenance of existing plant and equipment can have a very important role to play in minimising noise. Worn gears and bearings, loose or ill-fitting parts and panels, stretched drive belts, inadequate lubrication, unbalanced rotating parts, etc can all contribute to excessive noise levels. Improved maintenance can result in a noticeable reduction in noise levels.

5.207 Active noise attenuation is a relatively new control measure that relies on the fact that sound travels in waves, with areas of compression and rarefaction. If an identical sound wave is produced (same intensity and frequency) but 180° out of phase with the original, then the area of compression of the new sound wave will coincide with the area of rarefaction of the original. They should then cancel each other out, resulting in silence.

5.208 Active noise attenuation systems are commercially available, particularly for air handling systems, but they are still very expensive and perhaps not as flexible as traditional control methods. However, active noise attenuation has great potential and it is likely that its role in noise control will become much greater in years to come.

Modify the Path Between Source and Receiver

5.209 There are three main ways that the sound path between the source and the receiver can be modified:

- use of enclosures
- use of barriers
- room treatments.

Enclosures

5.210 Complete or partial enclosures are widely used to control noise but they rarely achieve their theoretical levels of attenuation and, as a consequence of this, frequently fail to satisfactorily resolve the noise problem. Enclosures can be custom-built or purchased as modular units.

5.211 Enclosure manufacturers should be able to provide attenuation data allowing more effective selection. The attenuation provided by custom-built enclosures should be able to be calculated.

5.212 Enclosures can be less effective where an airflow is required for ventilation, cooling, a combustion process, etc. Any opening will substantially reduce the attenuation provided by the enclosure, although this can be mitigated to some extent with the use of acoustic louvres.

Barriers

5.213 Barriers can be effective in some circumstances and they have been extensively used to control road traffic noise. To be effective, barriers should be as long (or wide) and as tall as possible. They will also be more effective if they are located very close to the source of the noise or to the receiver.

5.214 Generally, the greater the mass of the barrier, the more effective it will be, particularly at lower frequencies, although it is more important that the barrier is complete, ie there are no holes or gaps in it.

5.215 Although purpose-built barriers are the most common types of barrier seen, particularly at the side of busy roads and railway lines, temporary barriers using readily available material can also be very effective. For example, the use of building materials or temporary structures as barriers on construction sites can be a very effective way of minimising the noise produced by the activities carried out there.

Room Treatments

5.216 Room treatments can be used to control noise, but this type of control measure will normally be used to improve the noise environment within a given area. For example, adequate sound insulation is required for new noise-sensitive properties. It may also be useful to reduce the noise escaping from an existing noisy enclosed area or protecting an existing noise-sensitive area.

5.217 Sound insulation is normally improved by the addition of mass to the structure and/or the use of a cavity wall construction technique where panels are isolated from each other by means of an air-filled cavity. Both techniques can be applied retrospectively but the theoretical improvement in sound insulation is rarely achieved.

5.218 An extremely high standard of materials and construction is required for sound insulation works to be effective. Sound insulation tends to be more effective at high rather than low frequencies.

5.219 The acoustic environment within a room can be improved by increasing the absorption provided by the walls, floor and ceiling. Absorption is very dependent on frequency and improved sound absorption will tend to be more effective at higher frequencies than lower frequencies.

Noise Management Plans

5.220 Noise management plans will need to be produced by any organisation making an integrated pollution prevention and control application. However, the production of a noise management plan may be a useful exercise for any company with an environmental noise problem.

5.221 The noise management plan identifies and describes:

- all of the significant noise and/or vibration sources
- noise-sensitive locations that may be affected by the noise
- the type of monitoring carried out
- results of the monitoring and the control measures in place or planned.

5.222 The noise management plan should also include a noise purchasing policy. A noise purchasing policy will consider the level of noise produced by any plant or equipment intended for purchase and may already exist for the benefit of reducing noise exposure of workers. If so, it could be extended to include minimising the environmental noise emissions.

5.223 It is always easier to prevent environmental noise problems than trying to eliminate them after they occur.

List of Relevant Legislation

- *Directive 2002/49/EC* relating to the assessment and management of environmental noise
- Environmental Noise (Identification of Noise Sources) (England) Regulations 2007
- Environmental Noise (Identification of Noise Sources) (Wales) Regulations 2007
- Environmental Noise (England) Regulations 2006
- Environmental Noise (Scotland) Regulations 2006
- Environmental Noise (Wales) Regulations 2006
- Noise Emission in the Environment by Equipment for Use Outdoors Regulations 2001
- Building Regulations 2000
- Control of Noise (Measurement and Registers) Regulations 1976
- Control of Noise (Appeals) Regulations 1975
- Pollution Prevention and Control Act 1999
- Noise and Statutory Nuisance Act 1993
- Environmental Protection Act 1990
- Building Act 1984
- Control of Pollution Act 1974

Further Information

Publications

British Standards Publications

The following are available from *www.bsi-global.com.*

- BS 4142: 1997 *Method for Rating Industrial Noise Affecting Mixed Residential and Industrial Areas*
- BS 5228–1: 1997 *Noise and Vibrations Control on Construction and Open Sites*
- BS 7445–1: 2003 *Description and Measurement of Environmental Noise. Guide to Quantities and Procedures*
- BS 7445–2: 1991 *Description and Measurement of Environmental Noise. Guide to the Acquisition of Data Pertinent to Land Use*
- BS 7445–3: 1991 *Description and Measurement of Environmental Noise. Guide to Application to Noise Limits*
- BS 61252: 1997 *Electroacoustics. Specifications for Personal Sound Exposure Meters*

Environment Agency Publications

The following are available from *www.environment-agency.gov.uk.*

- IPPC H3 *Horizontal Guidance for Noise: Part 1 — Regulating and Permitting,* Environment Agency

- IPPC H3 *Horizontal Guidance for Noise: Part 2 — Noise Assessment and Control*, Environment Agency

Communities and Local Government Publications

The following is available from *www.communities.gov.uk*.

- Planning Policy Guidance Note 24: *Planning and Noise*

CIRIA Publications

The following are available from *www.ciriabooks.com*.

- C650: *Environmental Good Practice on Site*
- PR070: *How Much Noise Do You Make? A Guide to Assessing and Managing Noise On Site*
- R120: *A Guide to Reducing the Exposure of Construction Workers to Noise*
- TN138M: *Planning to Reduce Noise Exposure in Construction*

Other Publications

- *Good Practice Guide for Strategic Noise Mapping and the Production of Associated Data on Noise Exposure*, European Commission Working Group Assessment of Exposure to Noise (WG-AEN)
- *Guidelines for Community Noise*, World Health Organization

Organisations

- Association of Noise Consultants
  Web: *www.association-of-noise-consultants.co.uk*
  The Association of Noise Consultants consists of independent consultant firms offering professional services in the fields of noise and vibration.
- CIRIA
  Web: *www.ciria.org.uk*
  CIRIA provides information and publications on contaminated land as well as sustainable land use and construction.
- Department for Environment, Food and Rural Affairs (Defra)
  Web: *www.defra.gov.uk*
  Defra is the main government department which deals with waste and other environmental issues. It consults on new regulations and provides guidance on legislation and best practice.
- Environmental Protection UK
  Web: *www.environmental-protection.org.uk*
  Environmental Protection UK (formerly known as the National Society for Clean Air and Environmental Protection) is a charity that brings together organisations across the public, private and voluntary sectors to promote the understanding of and solving of environmental problems.
- Institute of Acoustics
  Web: *www.ioa.org.uk*
  The institute administers and sets the standards for its Diploma in Acoustics and Noise Control and for a one-week course leading to a certificate of competence in workplace noise assessment. Various classes of membership are available.
- National Physical Laboratory
  Web: *www.npl.co.uk*
  The National Physical Laboratory is the UK's national standards laboratory, an internationally respected and independent centre for research and development, and knowledge transfer in measurement and materials science.

Managing Waste

- Anyone who imports, produces, carries, keeps, treats or disposes of industrial or commercial waste has a duty of care to take reasonable steps to ensure their waste does not cause harm to the environment, even after it has been passed on to a carrier or contractor.
- Waste should not be consigned for disposal until all opportunities for reuse, recovery and recycling have been considered.
- In order to comply with the duty of care, the following measures should be undertaken.
  - Determine whether the duty applies.
  - Decide if a separate description of the waste is necessary.
  - Complete a transfer note which will be sent with the waste.
  - If the waste is being stored on-site, ensure that this is done safely.
  - Ensure that the waste is transferred to an authorised person.
- Waste is generally defined as any substance or object that the holder discards, or intends to, or is required to discard.
- The European Waste Catalogue (EWC) is an EU-wide classification system for waste. It also defines which wastes should be deemed hazardous.
- All duty holders should be alert for signs of a breach of the duty by others in the chain. Action should be taken if appropriate.
- The Environment Agency and the Scottish Environment Protection Agency will pursue breaches of the duty and prosecute if necessary.
- There is a compulsory registration scheme for waste carriers which is closely allied to the duty of care regime.
- Construction projects over £300,000 are required to have a Site Waste Management Plan in place which provides a structure for waste delivery and disposal at all stages during a construction project.

5.224 The duty of care regime, which aims to reduce the threat to the environment and human health from the improper management and unsafe recovery or disposal of waste, is a key aspect of the Environmental Protection Act 1990 as far as waste producers are concerned. The Act places a duty of care on all persons who import, produce, carry, keep, treat or dispose of "controlled waste" or who, as brokers, have control of such waste. In practice, the duty applies to all industries and businesses in the UK.

Employers' Duties

5.225 Under s.34 of the Environmental Protection Act 1990, anyone who imports, produces, carries, keeps, treats or disposes of controlled waste (ie household, industrial and commercial waste) has a duty to take all reasonable and applicable measures:

- to prevent anybody else, whom they may or may not know, from illegally treating, keeping, depositing or otherwise disposing of the waste
- to prevent the escape of waste
- on transfer of the waste to ensure that transfer only occurs to an authorised person or for authorised transport

- to ensure that transfer of the waste is accompanied by a written description of the waste where appropriate and to complete a transfer note sufficiently to enable others to avoid contravention of the duty of care
- to ensure that all employees and contractors are instructed on how each type of waste should be handled and disposed of.

5.226 Construction projects over £300,000 are required to operate a Site Waste Management Plan.

5.227 In addition, employers must ensure they:

- classify waste in accordance with the European Waste Catalogue and waste acceptance criteria
- comply with the waste producer's duty of care
- have a pollution prevention and control permit, a waste management licence or are exempt from licensing if they operate waste on-site storage, treatment and disposal facilities
- establish company procedures for segregation and storage of different waste streams
- comply with general health and safety legislation as applicable to waste, eg Control of Substances Hazardous to Health Regulations 2002 (COSHH), personal protective equipment (PPE) and manual handling
- select appropriate waste management options for each waste stream
- review services offered by contractors and engage a reputable contractor who offers an appropriate package, seeking to move away from landfill where possible
- provide contractor(s) with information required under landfill legislation, eg test results and details of pre-treatment.

Employees' Duties

5.228 Employees are required to:

- follow company waste procedures
- comply with relevant aspects of the duty of care for waste, eg ensuring waste does not escape from containers or vehicles
- report breaches of procedure, eg where hazardous waste is incorrectly placed in a skip for general waste.

In Practice

Compliance with the Duty of Care

5.229 Compliance with the duty of care for waste follows a step-by-step process.

Determine Whether the Duty Applies

5.230 The duty of care applies only to controlled waste (ie household, industrial and commercial waste). Explosive wastes and most radioactive wastes are therefore excluded.

5.231 The types of person to whom the duty applies are anyone who imports, produces, carries, keeps, treats or disposes of controlled waste.

5.232 On a construction site, where it may be difficult to decide who is the waste producer, the waste producer is the person doing the work that gives rise to the waste, not the person who issues instructions or establishes contracts that give rise to waste. If a main contractor makes arrangements for a haulier to remove a subcontractor's waste, the main contractor is acting as a broker and all three parties are therefore subject to the duty.

Note: Since April 2008 Site Waste Management Plans are required for all construction projects over £300,000.

Describe the Waste

5.233 The description of the waste may be part of the transfer note if fairly simple, or it may be a separate document. Drums, containers, skips, etc should also be labelled with a suitable description. The description of waste must include the relevant six-digit code from the European Waste Catalogue.

5.234 The length and complexity of the description will depend on the nature of the waste. For mixed waste with no special handling or disposal requirements (eg much commercial waste), a statement of the type of premises or business of origin may suffice. If the waste consists of only a few substances, or a single substance or material, the name of the substance(s) (either the chemical name or the common name) should be given.

5.235 For most industrial wastes and some commercial wastes, the process producing the waste should be described. Such a description should include:

- details of materials used and/or processed
- equipment used
- changes which produced the waste
- relevant information from the suppliers of the material and equipment, eg chemical hazard data sheets often contain waste disposal guidelines.

5.236 In addition to the process description, some industrial waste streams may warrant a chemical and physical analysis. This is necessary for mixed wastes originating from different processes or activities, and wastes where the process has altered the properties or composition of the raw materials. Analysis is also advisable where the exact source of the waste is unknown.

5.237 A description should always contain information on any special problems associated with the waste, eg:

- any special containers which are required
- whether the waste is incompatible with other substances/wastes
- whether the waste can be safely crushed and transferred between vehicles
- whether the waste can be safely incinerated and, if so, the conditions which apply, eg minimum temperature and combustion time
- whether the waste can be safely landfilled with other wastes
- any hazardous substances included in the waste
- whether the waste is likely to change its physical state during storage or transport
- problems previously encountered with the waste
- any recommendations given by the Environment Agency or the Scottish Environment Protection Agency on handling or disposal.

5.238 Each party to a transfer must keep a copy of the waste description and retain it for at least two years.

Waste Definitions

Directive Waste

5.239 Based on the definition of waste in the EU Waste Framework Directive 1975 (known as "directive waste"), the Environment Act 1995 defines waste as any substance or object that the holder discards, or intends to, or is required to discard, and which is listed in the following 16 categories:

- residues from production or consumption
- off-specification products
- products that have passed their "use by" date
- materials that have been spilled, lost, etc and any other materials or items that have been contaminated by the spillage
- contaminated materials which have been used in cleaning up a spillage, etc (eg residues from cleaning operations, containers or packaging)
- reject batteries, exhausted catalysts and other unusable parts
- chemicals and other substances that no longer perform satisfactorily, eg contaminated solvents and acids
- slags, still bottoms and other industrial process residues
- scrubber sludges, spent filters, baghouse dusts and other residues from pollution abatement processes
- residues from machining or finishing, eg lathe turnings and mill scales
- residues from the extraction and processing of raw materials, eg mining residues and oil field slops
- materials that have been adulterated, eg oil contaminated with polychlorinated biphenyls
- substances, materials and products that have been banned
- products that the holder no longer requires (including agricultural, household, office, commercial and shop waste)
- contaminated substances, materials and products that result from the remediation of contaminated land
- any materials, substances or products which are not contained in the above categories.

5.240 Directive waste does not include:

- gaseous emissions (this exemption does not apply to gaseous emissions from waste disposal and recovery operations, such as landfill gas)
- radioactive waste
- waste from mining and quarrying operations (this exemption does not apply to non-mineral waste from buildings at mines or quarries)
- animal carcasses, manure and other natural, non-dangerous materials used in farming (this exemption does not apply to non-natural agricultural waste such as discarded pesticide containers)
- waste waters (eg effluents, etc which are subject to the discharge consent or trade effluent consent systems)
- decommissioned explosives.

Controlled Waste

5.241 All provisions which apply to controlled waste now apply only to controlled waste which is *also* directive waste. Controlled waste is divided into a number of categories such as household, industrial or hazardous as follows.

Household Waste

5.242 Wastes to be treated as household waste include:

- waste from any land associated with domestic property, including a caravan or residential home
- waste from a private garage which either:
  - has a floor area of 25m^2 or less
  - is used wholly or mainly to accommodate a private motor vehicle
- waste from private storage premises used wholly or mainly for the storage of domestic articles
- waste from a moored vessel which is used only for living accommodation
- waste from a place of worship, etc which is exempted from business rates
- waste from premises occupied by a charity and used for charitable purposes
- waste from a campsite
- waste from a prison or other penal institution
- waste from a hall or other premises used wholly or mainly for public meetings
- waste arising from street cleaning by the local authority.

5.243 Waste from construction or demolition on domestic property, and septic tank sludge, are classified as household waste only for the purpose of exemption from the duty of care (ie people producing construction waste, etc from their own houses are not subject to the duty of care). However, if a builder is used then the waste produced is subject to the duty of care.

Industrial Waste

5.244 Wastes to be treated as industrial waste include:

- waste from premises used for maintaining vehicles, vessels or aircraft (except waste from small private garages, etc)
- waste from a laboratory
- waste from a workshop or similar premises (not including factories or premises at which the principal activities are computer operation or the copying of documents by photographic or lithographic means)
- waste from dredging operations
- waste arising from tunnelling or any other excavation
- clinical waste other than:
  - clinical waste from a domestic property, caravan or residential home, or from a moored vessel which is only used as living accommodation
  - waste collected by the local authority under the litter and street cleaning provisions of the Environmental Protection Act 1990
- waste removed from land on which it has previously been deposited and any soil with which such waste has been in contact, other than waste collected under the litter and street cleaning provisions of the Environmental Protection Act 1990
- leachate from a deposit of waste
- poisonous or noxious waste arising from any of the following processes undertaken on business or trade premises:
  - mixing or selling paints
  - sign writing
  - laundering or dry cleaning
  - developing or printing photographs
  - selling petrol, diesel fuel, paraffin, kerosene, heating oil or similar substances
  - selling pesticides, herbicides or fungicides
- waste oil, waste solvent or scrap metal, other than household waste
- waste arising from the cleaning of roads for which the Secretary of State is responsible
- waste imported into Great Britain

- waste from works of construction or demolition (except where classified as household waste for the purposes of the duty of care)
- agricultural waste.

5.245 The following are not industrial waste:

- sewage or sludge that is treated, kept or disposed of at a sewage treatment works
- sludge or septic tank sludge used in accordance with the Sludge (Use in Agriculture) Regulations 1989
- sewage kept or treated in a privy, cesspool or septic tank.

5.246 Sewage, sludge or septic tank sludge is industrial waste if it does not fall within the above categories and is treated, kept or disposed of on land or by means of mobile plant.

Agricultural Waste

5.247 Agricultural waste is waste from premises used for agriculture (within the meaning of the Agriculture Act 1947) and that is classified as industrial waste. Agriculture includes:

- horticulture
- fruit growing
- seed growing
- dairy farming and livestock breeding and keeping
- the use of land as grazing land, meadow land, osier land, market gardens and nursery grounds
- the use of land for woodlands, where that use is ancillary to the farming of land for other agricultural purposes.

5.248 Agricultural wastes include, eg waste silage wrap, waste pesticides and their containers, scrap machinery, waste oils and waste veterinary medicines from farms.

Commercial Waste

5.249 Wastes to be treated as commercial waste include:

- waste from an office or showroom
- waste from a hotel
- waste from the business part of a building which is used for both domestic and trade purposes
- waste from a private garage with a floor area exceeding $25m^2$ or which is not used wholly or mainly to accommodate a private motor vehicle
- waste from premises occupied by a club, society or association
- waste from premises occupied by a:
  - court
  - government department or local authority
  - body corporate or an individual appointed to discharge any public functions
  - body incorporated by a Royal Charter unless classified by legislation as household or industrial waste
- waste from a tent pitched on land other than a campsite
- waste from a market or fair
- waste from open land collected by the local authority.

Hazardous Waste

5.250 Hazardous waste is the term used for waste which presents a hazard to human health or the environment because it contains dangerous substances. It is the term used in England, Wales and Northern Ireland — "special waste" is the term used in Scotland (although it has the same meaning).

5.251 A waste is deemed hazardous if it is described as such in the European Waste Catalogue. Some types of waste are hazardous outright (listed as "absolute entries" in the European Waste Catalogue), while others depend on the amount of dangerous substances above threshold concentrations "mirror entries".

Non-hazardous Waste

5.252 For the purposes of the Landfill (England and Wales) Regulations 2002 and the Landfill (Scotland) Regulations 2003, non-hazardous waste is waste which is neither hazardous nor inert. It includes most biodegradable municipal waste.

Inert Waste

5.253 Under the Landfill (England and Wales) Regulations 2002 and the Landfill (Scotland) Regulations 2003, a waste is inert if:

- it does not undergo any significant physical, chemical or biological transformations
- when in contact with other matter, it does not react in any way which would cause environmental pollution or harm to human health (eg by dissolving, burning or biodegrading)
- it produces only an insignificant quantity of leachate, which would not cause pollution or endanger the quality of surface or groundwater (much construction and demolition waste falls into this category).

Municipal Waste

5.254 Municipal waste is waste collected by or on behalf of a local authority. This includes waste from households and waste which, because of its nature or composition, is similar to waste from households.

5.255 The view of the Department for Environment, Food and Rural Affairs is that municipal waste applies only to waste collected by a waste collection authority or through a waste collection authority's contractors.

European Waste Catalogue

5.256 The European Commission has established an EU-wide classification system for waste based on the European Waste Catalogue in order to ensure that legislation is implemented in a consistent way. England, Northern Ireland and Wales have transposed the European Waste Catalogue into their own laws through the List of Wastes Regulations (LoWR) for each region (EWC codes are also known as LoW codes).

5.257 The European Waste Catalogue assigns a six-digit code to each listed waste type, and hazardous wastes are identified with an asterisk (*).

5.258 Waste producers and managers must classify their wastes in accordance with the European Waste Catalogue to meet the requirements of the following regulatory regimes.

- The six-digit code must be entered on the duty of care waste transfer note and hazardous waste consignment note.
- Hazardous waste, eg landfill sites must be classified as sites for hazardous, non-hazardous or inert waste.
- Classification of hazardous waste when it is being imported or exported.

Complete the Transfer Note

5.259 As well as the written description, a transfer note must be transferred with the waste and the transfer note must be signed by both parties to the transaction. It is reasonable for a single transfer note to cover multiple consignments of waste transferred within a one-year period, provided that the description and all other details on the note are the same for all the consignments and the parties involved in the transfer do not change. Copies of the written description and the transfer note must be kept by both parties for two years from the date of the transfer. A copy of the transfer note must be made available to the Environment Agencies or waste collection authorities if they ask to see it. For hazardous waste, the consignment note takes the place of the transfer note. The following information must be included on the transfer note:

- identity of the waste, including the six-digit code from the European Waste Catalogue
- quantity of waste
- type of container, if applicable
- time and place of transfer
- name and address of the transferor and transferee
- whether the transferor is the producer or importer of the waste
- which (if any) authorised transport purpose applies
- the categories into which the transferor and transferee fall, eg waste producer, registered carrier, etc
- for a waste management licence holder, the licence number and name of the licensing authority
- for a registered carrier, the registration number
- the reasons for any exemption from registration or licensing
- the name and address of any broker involved in the waste transfer.

Keep the Waste Safely

5.260 Whether waste is being stored on the producer's premises, at a transfer station or at a disposal site, it must not be allowed to escape. Carriers are responsible for ensuring that waste does not escape during transportation.

5.261 Containers should therefore be of good construction, in good condition and covered or otherwise secured. They should also be accurately labelled.

5.262 Waste producers must ensure that waste is properly managed prior to collection by the disposal contractor.

- Storage areas for hazardous waste will require a waste management licence if more than 23,000 litres of liquid waste or 50m3 of solid waste are stored (80m3 in a secure container). Other stores may need to be registered as exempt from licensing.
- Duty of care documentation must be completed, including an accurate description of the waste. Waste must not be allowed to escape.
- Hazardous waste must be consigned in compliance with the hazardous waste regulatory regime.
- Waste should not be overlooked when carrying out health and safety assessments, eg for compliance with the Control of Substances Hazardous to Health Regulations 2002.

5.263 When packaging waste, the handling and transport that it will undergo in the future should be considered: the security of the packaging remains the producer's responsibility once the carrier has removed the waste. If an intermediate holder treats or repacks the waste the responsibility is transferred to that person. Segregation of different waste types may be necessary in order to prevent the mixing of incompatible wastes. The waste holder — whether the waste producer or a licensed waste manager — should ensure that employees and contractors are aware of the location and use of each segregated waste container. A suggested format for the transfer note is illustrated below.

Section A — Description of Waste

1. Please describe the waste being transferred: *Six-digit code:*
2. How is the waste contained?
   Loose ☐ *Sacks* ☐ *Skip* ☐ *Drum* ☐ *Other* ☐ → *please describe:*
3. What is the quantity of waste (number of sacks, weight etc):

Section B — Current holder of the waste (Transferor)

1. Full Name (BLOCK CAPITALS):
2. Name and address of Company:
3. Which of the following are you? (Please tick one or more boxes)

producer of the waste ☐	*holder of waste disposal or waste management licence* ☐	→	*Licence number:* *Issued by:*
importer of the waste ☐	*exempt from requirement to have a waste disposal or waste management licence* ☐	→	*Give reason:*
Waste Collection Authority ☐	*registered waste carrier* ☐	→	*Registration number:* *Issued by:*
Waste Disposal Authority (Scotland only) ☐	*exempt from requirement to register* ☐	→	*Give reason:*

Section C — Person collecting the waste (Transferee)

1. Full Name (BLOCK CAPITALS):
2. Name and address of Company:
3. Which of the following are you? (Please tick one or more boxes)

Waste Collection Authority ☐	*authorised for transport purposes* ☐	→	*Specify which of those purposes:*
	holder of waste disposal or waste management licence ☐	→	*Licence number:* *Issued by:*
Waste Disposal Authority (Scotland only) ☐	*exempt from requirement to have a waste management licence* ☐	→	*Give reason:*
	registered waste carrier ☐	→	*Registration number:* *Issued by:*
	exempt from requirement to register ☐	→	*Give reason:*

Section D

1. Address of place of transfer/collection point:
2. Date of transfer: 3. Time(s) of transfer (for multiple consignments, give 'between' dates):
4. Name and address of broker who arranged this waste transfer (if applicable):

	Transferor	**Transferee**
5.	Signed:	Signed:
	Full name (BLOCK CAPITALS)	Full name: (BLOCK CAPITALS)
	Representing:	Representing:

Reproduced from *Waste Management: the Duty of Care — a Code of Practice* with the permission of the Controller of Her Majesty's Stationery Office

Decide on the Best Option for Waste Disposal

5.264 Waste producers are responsible for choosing the most appropriate waste management option for each of their waste streams. In doing this they should take into account both environmental and economic considerations. Ideally, as much waste as possible should be recycled, reused and recovered, with disposal as the option of last resort.

5.265 Waste management options should be reviewed regularly to take account of new developments in the market. For example, landfill costs are rising significantly, particularly for hazardous wastes, making it more economic to treat and recover them.

Landfill

5.266 Some types of waste have been banned from landfill sites. These include:

- liquid wastes from hazardous and non-hazardous waste landfill sites
- waste which is explosive, highly flammable, oxidising or corrosive, and chemicals whose effects on humans or the environment are unknown
- infectious clinical waste
- tyres (including shredded tyres)
- waste which has been mixed or diluted purely to meet the waste acceptance criteria.

5.267 In July 2004 the co-disposal of hazardous waste with non-hazardous waste was banned. This means that hazardous waste may only be landfilled in a dedicated "hazardous" site (of which there are only around 10–15 in the UK).

Waste Acceptance Criteria

5.268 Waste may not be landfilled unless it meets certain waste acceptance criteria and procedures. Details of these criteria and procedures can be found in the Environment Agency's guidance document, Guidance for Waste Destined for Disposal in Landfill.

5.269 Waste producers, with the help of their contractors, are responsible for carrying out the basic characterisation of the waste. This will generally involve testing.

Basic Characterisation

5.270 Before waste is accepted for landfill, the following information must be provided:

- source and origin of the waste
- process producing the waste
- treatment applied
- composition, including an assessment of the waste against the limit values for leaching and organic composition
- appearance, including smell, colour, consistency and physical form
- six-digit code from the European Waste Catalogue
- properties that render the waste hazardous (if applicable)
- evidence demonstrating that the waste is not banned from landfill
- landfill class at which the waste may be accepted
- likely behaviour of the waste in landfill, and precautions which need to be taken
- whether the waste can be recovered or recycled.

5.271 Where the waste is regularly generated in the same process, information is required regarding the range and variability of the waste's properties and the key variables to be tested during compliance testing.

Testing Requirements

5.272 In general, waste must be tested prior to acceptance at a landfill although specified inert wastes and non-hazardous municipal wastes do not require testing.

5.273 Where waste is regularly generated in the same process, it is not necessary to test each batch but compliance testing must be carried out. This must be done at least once a year (more often if the characterisation indicates that this is necessary).

5.274 The Landfill (England and Wales) Regulations 2002 lay down detailed requirements for the sampling and testing of waste as part of its characterisation. The testing must be carried out in accordance with the specified European standards.

5.275 The characterisation will determine which type of landfill must be used, and the type and extent of pre-treatment required. Pre-treatment is required even if a waste already meets the waste acceptance criteria, as long as it provides some environmental benefit.

Municipal Waste Diversion Targets

5.276 Local authorities have been set demanding targets for the diversion of biodegradable municipal waste from landfill. They will have to find alternative options to landfill, eg:

- increased recycling
- mechanical/biological treatment
- composting
- energy recovery.

5.277 This will affect producers of commercial waste who rely on a local authority collection service. It is likely that the local authorities will raise their charges, encouraging commercial waste producers to switch to private sector contractors.

Incineration and Co-incineration

5.278 Incineration is an appropriate disposal option for combustible wastes which cannot be recycled. It may be the only available option for some hazardous wastes which are banned from landfill, eg highly flammable wastes. Waste producers may consider:

- incineration with energy recovery, for the type of waste which is collected by local authorities
- co-incineration in a cement kiln, lime kiln or power station: this is currently used to dispose of suitable flammable solvent wastes, non-recyclable waste oils and tyres, and is likely to be used for a wider range of wastes in the future
- high temperature merchant incineration for particularly difficult wastes such as polychlorinated biphenyls (there are currently only two such incinerators in operation in the UK — Cleanaway's at Ellesmere Port and Shanks' at Fawley)
- in-house incineration for smaller quantities of combustible, non-hazardous waste.

Liquid Wastes

5.279 Liquid wastes are banned from landfill. In the past, small quantities have sometimes been disposed of to sewer. Although not strictly illegal, this has never been encouraged by the water companies. Any discharge must comply with the conditions of the company's trade effluent discharge consent.

5.280 It is prohibited to dilute wastes to meet waste acceptance criteria or render the wastes non-hazardous.

5.281 Treatment, either at an in-house facility or commercial treatment works, is the preferred option for liquid wastes. Landfill operators increasingly offer treatment as part of their package. The treated effluent can then be discharged to sewer in compliance with the trade effluent discharge consent.

On-site Disposal Facilities

5.282 Companies with their own landfill, incineration or treatment plant must ensure that it has the relevant pollution prevention and control permit or waste management licence. Smaller-scale recovery activities, such as land spreading, may qualify for an exemption from licensing.

5.283 In-house landfills must be operated by a fit and proper person. There are three elements of a fit and proper person:

- absence of relevant convictions
- technical competence, ie the person in charge of the facility must have the relevant Certificate of Technical Competence from the Waste Management Industry Training and Advisory Board
- financial provision: funds or credit must be available to meet the costs associated with the site throughout its lifetime, including closure and aftercare costs.

5.284 In-house treatment plants covered by a pollution prevention and control permit do not require a fit and proper person.

5.285 The waste acceptance criteria apply to in-house landfills, but in certain circumstances, where the waste is very uniform, some of the criteria may be relaxed.

Transfer to the Right Person

5.286 Under the duty of care regime, transfer of waste must be to an authorised person. The transferee should ask to see proof of authorisation before transferring the waste, particularly for the first time. Authorised persons are:

- waste collection authorities
- registered carriers of controlled waste or carriers who are exempt from the registration requirements
- holders of waste management licences
- persons exempted from waste management licensing
- Scottish waste disposal authorities.

Waste Collection Authority

5.287 If waste is to be collected by the local authority (ie the waste collection authority in England and Wales or the waste disposal authority in Scotland), it must be suitable for them to collect, eg not include any unexpected hazardous substances.

Registered Carrier

5.288 Before a contract is drawn up with a carrier, it is important to check with the Environment Agency or the Scottish Environment Protection Agency that the carrier is registered or exempt from registration. The Agencies maintain public registers of registered waste carriers and exempt carriers. Carriers can be asked to produce their registration certificate or official copy certificate (note that photocopies are not valid) in order that the expiry date can be checked.

5.289 Certain carriers are exempt from the requirement to register. These are:

- charities and voluntary organisations
- waste collection authorities collecting waste themselves (authorities' contractors are not exempt)
- ship operators, where the waste is to be disposed of under licence at sea

- persons authorised under the Animal By-products Regulations 2003 to hold or deal with animal waste; or holders of a knacker's yard licence.

5.290 If a carrier claims to be exempt, it is advisable to ask for details of the type of exemption and evidence that the exemption is valid for the waste in question.

Waste Management Contractor

5.291 The waste producer should ensure that the contractor has a licence or permit and, if so, examine it to check that it is valid for the type and amount of waste under consideration. Some contractors are exempt from licensing, eg those who only bale waste paper or handle small quantities of scrap metal.

5.292 If a contractor claims exemption, the producer should check that the exemption is valid and applicable to their particular waste load. If in doubt about the exemption of a particular activity, the Environment Agency or the Scottish Environment Protection Agency will give advice.

5.293 If the carrier has a contract with either the producer or the waste management contractor, he or she may also have a duty to check the contractor's licence or exemption. However, if the only contract is between the producer and the waste management contractor, responsibility for checking the licence lies with the producer alone.

5.294 It is not necessary to check the licence for every load of waste. While annual checks are advisable, repeat checks are also advisable where:

- the nature, source or destination of the waste has changed
- several different carriers come to a site and there is a risk of confusing different waste loads
- the carrier has changed
- the licence conditions of the destination have changed.

Authorised Transport Purposes

5.295 Waste does not have to be transferred to an authorised person if it is being transferred for authorised transport purposes. Essentially, these are imports and exports of waste, or movements between two sites under the same ownership. Separate legislation applies to imports and exports of waste.

Checking Up

5.296 All duty holders should be alert for signs of a breach of the duty by others in the chain. Indications of a breach include:

- wrongly or inadequately described waste arriving at a contractor's site
- waste being delivered or removed without proper secure packing
- damaged or insecure containers
- an incomplete or inadequate transfer note
- an unsupported claim of exemption from registration or licensing
- failure of the waste to arrive at the agreed destination
- observing a carrier fly-tipping someone else's waste
- carriers returning to the producer's site without sufficient time having elapsed for them to have reached the licensed waste management site.

5.297 If a breach of the duty is suspected, the following action can be taken:

- check the facts to ensure that a breach has actually occurred
- cancel further dealings with the offender

- if cancellation is impossible, eg because it would result in a breach of contract, then do whatever is practicable to prevent further offences:
  - if the waste was wrongly described, analyse further consignments
  - if it was badly packed, inspect subsequent loads
  - if a load does not arrive at its destination, ensure that subsequent loads are monitored and arrive safely.

5.298 New waste contracts could provide for termination if a breach of the duty occurs and is not rectified.

5.299 There is no legal requirement for waste producers to carry out a detailed audit of the contractor's site, although this is often advisable, especially where hazardous waste is involved.

Responsibilities of Each Waste Holder

Responsibilities of Different Waste Holders

Waste Producer	Waste Carrier	Broker	Waste Manager
Describe waste fully and accurately.	Ensure adequacy of packaging while waste is in the waste carrier's control.	Be well informed about nature of waste.	Check description, eg by sampling.
Store it safely on-site.	Ensure waste does not escape.	Ensure correct and adequate description is transferred.	Check documentation.
Select an appropriate treatment/disposal method.	Repack waste if necessary.	Ensure site licence is valid.	Follow up previous misconduct.
Ensure waste falls within the terms of the contractor's waste management licence.	Make a quick visual inspection to check accuracy of description.	Ensure carrier is registered.	—
Pack waste securely.	Re-describe waste which is treated or repacked (if necessary).	Check that documentation has been completed.	—
Check the carrier's registration.	Check waste manager's licence if the waste producer does not have a contract with the waste manager.	Act on causes for suspicion.	—
Make reasonable checks on the carrier/ waste manager.	—	—	—
Report offences to the Environment Agency or the Scottish Environment Protection Agency.	—	—	—

5.300 It should be borne in mind that although it is not possible for a waste producer to abdicate all responsibility for the waste once it is transferred to a carrier or disposal contractor, a breach of the duty of care is not an offence of strict liability. Strict liability imposes liability even where there is no negligence.

5.301 If the producer has taken all reasonable measures to ensure that the duty is complied with, he or she is not guilty of an offence even if the waste ends up being illegally tipped (although the producer may not find it easy to satisfy the investigating officer that all reasonable steps have been taken in such a case).

5.302 The problem lies in ascertaining exactly what the reasonable measures are in each circumstance. Circumstances to be taken into account include:

- the nature of the waste
- whether the waste is hazardous and, if so, the type of hazard
- how the waste will be dealt with
- what the holder might reasonably be expected to know or foresee.

5.303 The Government's guidance, *Waste Management: The Duty of Care: A Code of Practice*, outlines the responsibilities of different kinds of waste holder and these are summarised in the previous table.

Enforcement of the Duty of Care

5.304 Enforcement of the duty of care should primarily be achieved by waste holders checking on each other. However, the Environment Agencies are expected to pursue breaches of the duty and prosecute if necessary. They will seek access to waste holders' records if there is a suspicion that waste has not been transferred to an authorised person, or there has been illegal treating, keeping, depositing or disposing of the waste.

5.305 The Environment Agency may require the production of either the written description or the transfer note by serving a notice on any person who is under a duty to keep that document. The notice must specify:

- the document required
- the Agency office to which it must be sent
- the period within which the document must be supplied (not less than seven days from the notice being served).

5.306 Refusal to answer a notice, keeping inaccurate or incomplete records and furnishing false copies of documents are all offences under the Environmental Protection Act 1990.

5.307 By obtaining copies of all the transfer notes along a waste chain, the Environment Agency should be able to trace the point at which an offence was committed and hopefully identify the offender. Implementation of the duty is expected to help the Agencies to detect related offences, particularly fly-tipping and other unlicensed waste management activities.

Role of Local Authorities

5.308 Local authorities are themselves subject to the duty of care by virtue of being waste producers, carriers or brokers. Waste collection authorities are deemed to be the producers, as well as the carriers, of the household waste which they collect.

5.309 Waste collection authorities and waste disposal authorities that award contracts for waste collection, transport and disposal are brokers and must ensure that the transfers which they arrange are properly documented. They may assist waste producers by providing transfer note/description forms for the producers to complete and sign.

5.310 Waste collection authorities have the power to serve notices requiring anyone in the waste chain to produce their copy of the transfer note. This gives them more power to track down fly-tippers.

Waste Carriers

5.311 A compulsory registration scheme exists for waste carriers, with the aim of curtailing fly-tipping. The registration is closely allied to the duty of care regime.

5.312 It is an offence for any person who is not a registered carrier to transport controlled waste to or from any place in Great Britain, in the course of any business, with a view to profit. (It is a defence to prove that waste was carried in an emergency of which the Environment Agency was notified, or that the carrier was acting under instructions from an employer.) Transport includes transport by road, rail, air, sea or inland waterway, but not the moving of waste by pipelines or direct transport into or out of the UK.

5.313 Certain carriers are exempt from registration, including:

- waste producers carrying their own waste, except where it is building or demolition waste
- holders of a knacker's yard licence or licence under the Animal By-products Regulations 2003 (or Scottish or Welsh equivalents)
- charities and voluntary organisations
- waste collection authorities
- ferry operators in relation to carriage of waste by vehicles on the ferry.

5.314 The Agency will be able to refuse registration or subsequently revoke registration if the carrier or a relevant person (eg another partner in a partnership) has been convicted of a prescribed offence (as listed in schedule 1 of the Controlled Waste (Registration of Carriers and Seizure of Vehicles) Regulations 1991) and in the opinion of the Agency it is undesirable for the carrier to continue to be authorised to transport controlled waste.

Seizure of Vehicles

5.315 The purpose of seizing vehicles is to aid in the detection and prosecution of those involved in fly-tipping or other illegal disposal of waste. It is not concerned with the punitive confiscation of vehicles.

5.316 The Environment Agency has the power to seize vehicles when it has reason to believe that controlled waste has been treated, deposited or disposed of illegally, that a vehicle was used in the commission of the offence and that proceedings for the offence have not been brought.

Waste Brokers

5.317 Waste brokers are:

- traders or dealers who buy and sell scrap metal or other recoverables, or arrange such deals for money
- waste management contractors or carriers who may make disposal arrangements for waste they cannot accept at their own sites
- brokers or environmental consultants who arrange for the disposal of waste on behalf of their clients.

5.318 Brokers must apply for registration with the Environment Agency or the Scottish Environment Protection Agency. Brokers who hold a waste management licence will not be required to register and it is possible to register jointly as a carrier and broker. In practice, it will mainly be those in the third category above who will be required to register as brokers.

5.319 The broker registration scheme is very similar to the registration scheme for carriers and the same exemptions apply. As with carriers, a simplified scheme applies to waste collection authorities, waste disposal authorities and charities or voluntary organisations.

Site Waste Management Plans

5.320 A Site Waste Management Plan (SWMP) provides a structure for waste delivery and disposal at all stages during a construction project.

Training

5.321 Employees who deal directly with waste management contractors should be aware of their duty of care requirements, such as the need to fill in a transfer note or consignment note. They can be encouraged to look out for breaches of the duty of care by carriers and contractors, eg a badly maintained vehicle from which waste could escape.

5.322 The Chartered Institution of Wastes Management runs a training course which leads to employees gaining a Waste Awareness Certificate. The course aims to provide an introduction to waste management and to provide the knowledge, skills and understanding to improve the management of waste.

5.323 Where there is an on-site landfill, the person with operational control (the fit and proper person) of the site must obtain a Certificate of Technical Competence from the Waste Management Industry Training and Advisory Board.

5.324 Managers may require specialised training in order to assess whether waste is hazardous, interpret analytical data and determine treatment requirements. (Alternatively, a contractor or consultant could take on these responsibilities.)

5.325 Those responsible for a Site Waste Management Plan (SWMP) will need appropriate training for them to fully understand their responsibilities. It may also be necessary for them to develop a SWMP training programme to ensure that everyone on-site is aware of the importance of asking for and recording the correct paperwork, receipts, destinations for materials, etc.

List of Relevant Legislation

- Hazardous Waste (England and Wales) Regulations 2005
- Hazardous Waste (Wales) Regulations 2005
- Landfill Allowances Scheme (Scotland) Regulations 2005
- Landfill Allowances and Trading (England) Regulations 2004
- Landfill Allowances Scheme (Wales) Regulations 2004
- Landfill (Scotland) Regulations 2003
- Waste Incineration (Scotland) Regulation 2003
- Landfill (England and Wales) Regulations 2002
- Waste Incineration (England and Wales) Regulations 2002
- Waste Management Licensing Regulations 1994
- Controlled Waste (Registration of Carriers and Seizure of Vehicles) Regulations 1991
- Environmental Protection (Duty of Care) Regulations 1991
- Waste and Emissions Trading Act 2003
- Environmental Protection Act 1990
- Control of Pollution (Amendment) Act 1989

Further Information

Publications

Croner Publications

The following is available from *www.croner.co.uk*.

- *Waste Management*

Defra Publications

The following are available from *www.defra.gov.uk*.

- *Waste Management: The Duty of Care: A Code of Practice*
- *Waste Strategy for England 2007*

Environment Agency Publications

The following are available from *www.environment-agency.gov.uk*.

- *European Waste Catalogue* (Consolidated text version), Office for Official Publications of the European Communities *http://europa.eu.int/eur-lex/en/consleg/pdf/2000/en_2000D0532_do_001.pdf*
- *Guidance for Waste Destined for Disposal in Landfill: Interpretation of the Waste Acceptance Requirements of the Landfill (England and Wales) Regulations 2002 (as amended)*

SEPA Publications

The following is available from *www.sepa.org.uk.*

- *National Waste Strategy for Scotland*

Other Publications

- *Wise About Waste: National Waste Strategy for Wales*, National Assembly for Wales
- *Towards Resource Management: The Northern Ireland Waste Management Strategy 2006–2020*, Environment and Heritage Service, Northern Ireland

Organisations

- Chartered Institution of Wastes Management
  Web: *www.ciwm.org.uk*
  The Chartered Institution of Wastes Management is the professional body representing waste professionals.
- Department for Environment, Food and Rural Affairs (Defra)
  Web: *www.defra.gov.uk*
  Defra is the main government department which deals with waste and other environmental issues. It consults on new regulations and provides guidance on legislation and best practice.
- Environment Agency
  Web: *www.environment-agency.gov.uk*
  The Environment Agency is the main environmental regulator in England and Wales and provides detailed information on legislative requirements, guidance for business and technical guidance on waste treatment and disposal.
- Environmental Services Association
  Web: *www.esauk.org*
  The Environmental Services Association is the trade association for the waste management sector in the UK.
- Envirowise
  Web: *www.envirowise.gov.uk*
  Envirowise provides free advice and assistance on waste reduction, clean production and general environmental management.
- Scottish Environment Protection Agency (SEPA)
  Web: *www.sepa.org.uk*
  SEPA is the main environmental regulator for Scotland. It provides detailed information on legislative requirements, guidance for business and technical guidance on waste treatment and disposal.
- Waste Management Industry Training and Advisory Board
  Web: *www.wamitab.org.uk*
  The Waste Management Industry Training and Advisory Board provides training and qualifications for the waste industry.

Site Waste Management Plans

- The UK's construction output is the second largest in the EU. Each year, 400 million tonnes of solid materials are used in the UK construction industry, but only two-thirds is added to the building stock. The rest is sent directly to landfill.
- Any client who intends to carry out a project on a construction site with an estimated cost greater than £300,000 must prepare a Site Waste Management Plan.
- SWMPs aim to reduce the amount of waste produced on construction sites by setting out how building materials, and resulting waste, is to be managed during the project.
- While SWMPs are legally required for those construction projects in excess of £300,000, smaller projects can also reduce both resource use and cost by having one.
- The SWMP needs to be written at the construction design stage, but it is a requirement of the SWMP Regulations to maintain it during the entire project.
- The client, or principal contractor, is responsible for updating the plan with the site's day-to-day activity.
- The principal contractor must ensure that the SWMP is kept at the site office or, if there is no site office, at the site.
- The Environment Agency advises that a step-by-step methodology for preparing and maintaining a SWMP is used.
- Those responsible for SWMPs will need appropriate training for them to fully understand their responsibilities.
- Contractors and employees must ensure that they ask for and record the correct paperwork, receipts, destinations for materials, etc when disposing of on-site materials.
- By the end of the project, the SWMP should provide an accurate record of how effectively the materials on-site were managed and how well targets for waste management were met.

5.326 A SWMP is simply a plan that details the amount and type of waste that will be produced on a construction site and how it will be reused, recycled or disposed of. The plan is then updated during the construction process to record how the waste is managed and to confirm the disposal of any materials that cannot be reused or recycled at a legitimate site.

Employers' Duties

5.327 Under the Site Waste Management Plans Regulations 2008, any client who intends to carry out a project on a construction site with an estimated cost greater than £300,000 (excluding VAT) must prepare a site waste management plan (SWMP) conforming to the regulations before construction work begins.

5.328 The SWMP needs to be written at the construction design stage, but it is a requirement of the SWMP Regulations to maintain it during the entire project. Therefore, the client (or principal contractor) is also responsible for updating the plan with the site's day-to-day activity.

Employees' Duties

5.329 Employees must ensure that they ask for and record the correct paperwork, receipts, destinations for materials, etc when disposing of on-site materials.

In Practice

What are Site Waste Management Plans

5.330 The UK's construction output is the second largest in the EU. Each year, 400 million tonnes of solid materials are used in the UK construction industry, but only two-thirds is added to the building stock. The rest is sent directly to landfill.

5.331 Site waste management plans (SWMPs) aim to reduce the amount of waste produced on construction sites and to prevent fly-tipping. They do this by setting out how building materials, and resulting waste, is to be managed during the project.

5.332 A SWMP is simply a plan that details the amount and type of waste that will be produced on a construction site and how it will be reused, recycled or disposed of. The plan is then updated during the construction process to record how the waste is managed and to confirm the disposal of any materials that cannot be reused or recycled at a legitimate site.

Which Projects Need an SWMP

5.333 SWMPs are required for all those construction projects with estimated costs in excess of £300,000 (excluding VAT). The cost of the project is the price in the accepted tender, or if there is no tender, the cost of labour, plant and materials, overheads and profit.

5.334 While SWMPs are legally required for those construction projects in excess of £300,000, smaller projects can also reduce both resource use and cost by having one.

5.335 Examples of those projects that will require a SWMP to be in place include:

- new build
- maintenance
- alteration or installation/removal of services such as sewerage, water, etc.

Key Elements of the SWMP Process

5.336

- One individual, usually the principal contractor, will be responsible for writing and implementing the SWMP.
- There are two levels of SWMP — standard and detailed:
  - the standard SWMP applies to projects costing less than £500,000
  - the detailed SWMP applies to projects costing more than £500,000 and will require more detailed reporting.

- Local authorities and the Environment Agency will enforce SWMPs and they will impose penalties for failure to make, keep or produce a SWMP.

5.337 SWMPs are intended to change the construction industry's attitude to waste by raising the profile of waste planning.

5.338 The SWMP will:

- identify the different types of waste that will be produced by the project and note any changes in the design and materials specification that seek to minimise this waste
- consider how to reuse, recycle or recover the different wastes produced by the project
- require the construction company to demonstrate that it is complying with the duty of care regime
- record the quantities of waste produced.

Requirements of an SWMP

5.339 The SWMP needs to be written at the construction design stage, but it is a requirement of the SWMP Regulations to maintain it during the entire project. Therefore, the client, or principal contractor, (as defined in the Construction (Design and Management) Regulations 2007) is also responsible for updating the plan with the site's day-to-day activity.

5.340 A suitable SWMP must identify the:

- client
- principal contractor
- person who drafted the SWMP.

5.341 It must describe the construction work proposed, including the:

- location of the site
- estimated cost of the project.

5.342 It must record any decision taken before the SWMP was drafted on the nature of the project, its design, construction method or materials employed in order to minimise the quantity of waste produced on-site.

5.343 The SWMP must:

- describe each waste type expected to be produced in the course of the project
- estimate the quantity of each different waste type expected to be produced
- identify the waste management action proposed for each different waste type, including reusing, recycling, recovery and disposal.

5.344 The principal contractor and the client must declare that all reasonable steps will be taken to ensure:

- all waste from the site is dealt with in accordance with the waste duty of care
- materials will be handled efficiently and waste managed appropriately.

5.345 Breaches of the regulations will be an offence, punishable on summary conviction by a fine not exceeding £50,000, or on conviction on indictment by a fine.

Updating SWMPs

Projects with Estimated Costs of Less than £500,000

5.346 If a project has an estimated cost of £500,000 or less, whenever waste is removed from the site, the principal contractor must update the SWMP by recording the:

- identity of the person removing the waste
- types of waste removed
- site that the waste is being taken to.

5.347 In addition, within three months of the work being completed, the principal contractor must add to the plan:

- a confirmation that the plan has been monitored on a regular basis to ensure that work is progressing according to the plan and that the plan was updated in accordance with the regulations
- an explanation of any deviation from the plan.

Projects with Estimated Costs of More than £500,000

5.348 If the project has an estimated cost of greater than £500,000, the principal contractor must update the SWMP by recording on the plan whenever waste is removed:

- the identify of the person removing the waste
- the waste carrier registration number of the carrier
- a copy of, or reference to, the written description of the waste (as required by s.34 of the Environmental Protection Act 1990)
- the site the waste is being taken to and whether the operator of that site holds a permit under the Environmental Permitting (England and Wales) Regulations 2007 or is registered as a waste operation exempt from the need for such a permit.

5.349 The principal contractor must, as often as necessary, and no less than every six months, do the following to ensure that the plan accurately reflects the progress of the project.

- Review the plan.
- Record the types and quantities of waste produced.
- Record the types and quantities of waste that have been:
  - reused (and whether this was on or off-site)
  - recycled (and whether this was on or off-site)
  - sent for another form of recovery (and whether this was on or off-site)
  - sent to landfill
  - otherwise disposed of.
- Update the plan to reflect the progress of the project.

5.350 Within three months of the work being completed, the principal contractor must add to the plan:

- confirmation that the plan has been monitored on a regular basis to ensure that work is progressing according to the plan and that the plan was updated in accordance with the regulations
- a comparison of the estimated quantities of each waste type against the actual quantities of each waste type
- an explanation of any deviation from the plan
- an estimate of the cost savings that have been achieved by completing and implementing the plan.

Storing the SWMP

5.351 The principal contractor must ensure that the SWMP is kept at the site office or, if there is no site office, at the site.

5.352 The principal contractor must make sure that every contractor knows where it is kept, and must make it available to any contractor carrying out work described in the plan.

SWMP Methodology

5.353 A SWMP provides a structure for waste delivery and disposal at all stages during a construction project. The Environment Agency advises that a step-by-step methodology for preparing and maintaining a SWMP is used.

Appoint a Responsible Person

5.354 Appoint someone to take overall responsibility for the site's SWMP.

5.355 Any number of individuals can be involved in the delivery of the plan, but just one person is required to be in charge and responsible for updating the plan throughout the life of the project. That person needs to be clear of their responsibilities and also have enough authority to ensure that everyone co-operates.

Identify Types and Quantities of Waste that will be Produced

5.356 Identify the types (such as "controlled waste", "hazardous waste" and "directive waste"), and quantities of waste that will be produced during the project.

5.357 This requires thinking through every stage of the project and working out in advance what materials will be used. It also requires estimating how much waste will be able to be reused, recycled or disposed of. This should include the waste hierarchy (eliminate, reduce, reuse, recycle, dispose) and any special arrangements for hazardous wastes produced.

Identify Waste Management Options

5.358 Work out the best options available for recycling and disposal, such as incineration or landfill, and ensure that:

- all waste is stored and disposed of responsibly
- a record is kept of all waste disposed of or transferred through a system of signed waste transfer notes.

Identify Where and How to Dispose of Waste

5.359 Ensure that how and where waste is disposed of is known. When using contractors for waste disposal it is necessary to ensure they comply with all legal responsibilities.

Ensure On-site Materials and Waste Handling Is Well Organised

5.360 Careful planning of the materials needed for the project is required, but this also brings benefits, eg pre-ordering materials to specification at the design stage could create time savings later on. Avoid over-ordering and consider using recycled or previously-used materials. State all SWMP targets on the data sheet.

Communicate the Plan and Carry Out Training

5.361 Ensure that everyone on-site knows about the SWMP. Hold meetings with staff and contractors to explain why the SWMP is important.

5.362 It may also be necessary to develop a training programme to ensure that everyone is aware of the importance of asking for and recording the correct paperwork, receipts, destinations for materials, etc.

Measure Waste

5.363 Once the plan is being used:

- measure how well it is working by assessing how much and what type of waste is being produced
- think about setting up measurements to compare with future projects, eg:
  - volume (number of skips)
  - value (cost of disposal)
  - weight (weighbridge tickets).
- record costs and measurements on the data sheet
- update the SWMP as necessary.

Monitor and Review Success of SWMP

5.364 Make sure everything runs to plan, but be prepared to make changes. Learn lessons for future projects.

5.365 By the end of the project, the SWMP should provide an accurate record of how effectively the materials on-site were managed and how well targets for waste management were met.

Training

5.366 Those responsible for a site waste management plan (SWMP) will need appropriate training for them to fully understand their responsibilities. It may also be necessary for them to develop a SWMP training programme to ensure that everyone on-site is aware of the importance of asking for and recording the correct paperwork, receipts, destinations for materials, etc.

5.367 Specific SWMP training courses, such as that offered from the South East Centre for the Built Environment (SECBE), are available. The training focuses on improving the competence of the SWMP practitioner, particularly through guiding the drafting of site-based SWMP.

List of Relevant Legislation

- Site Waste Management Plans Regulations 2008
- Construction (Design and Management) Regulations 2007
- Environmental Permitting (England and Wales) Regulations 2007
- Clean Neighbourhoods and Environment Act 2005
- Environmental Protection Act 1990

Further Information

Publications

CIRIA Publications

The following is available from *www.ciria.org*.

- RP863: *Environmental Good Practice: Training Resource*

Constructing Excellence Publications

The following is available from *www.constructingexcellence.org.uk*.

- *Site Waste Management Plans Code of Practice and Guidance*, Department of Trade and Industry, 2004

NetRegs Publications

The following is available from *www.netregs.gov.uk*.

- *Site Waste — It's Criminal. A Simple Guide to Site Waste Management Plans*, NetRegs

Organisations

- Constructing Excellence
  Web: *www.constructingexcellence.org.uk*
  Constructing Excellence aims to deliver improved industry performance resulting in a demonstrably better built environment. It acts as a bridge between industry, clients, government and the research community, delivering government programmes, membership programmes and other commercial contracts.
- Department for Environment, Food and Rural Affairs (Defra)
  Web: *www.defra.gov.uk*
  Defra is the main government department which deals with waste and other environmental issues. It consults on new regulations and provides guidance on legislation and best practice.
- NetRegs
  Web: *www.netregs.gov.uk*
  NetRegs provides free environmental guidance for small businesses in the UK, helping them to comply with environmental legislation and understand how to protect the environment.
- South East Centre for the Built Environment (SECBE)
  Web: *www.secbe.org.uk*
  The SECBE is a consortium of business leaders that exists to inform policy and drive business-to-business learning and networking. It takes regional strategies and industry issues and develops action plans to improve business performance throughout the sector.
- Waste and Resources Action Programme (WRAP)
  Web: *www.wrap.org.uk*
  The Waste and Resources Action Programme aims to accelerate resources efficiency by creating efficient markets for recycled materials and products, while removing barriers to waste minimisation, reuse and recovery.

Sustainability in Construction

- Sustainable development involves balancing and integrating the economic, social and environmental considerations for any policy or decision.
- Sustainability in construction is driven primarily by stakeholder and business interests, although there are specific legislative requirements.
- Any construction project that wants to achieve sustainable development will need to have a good understanding of current and emerging risks.
- The construction industry annually produces three times the amount of waste generated by all UK households combined.
- Site Waste Management Plans are a statutory requirement for construction projects above a specified value.
- With the construction, occupation and maintenance of buildings being responsible for around half the UK's emissions of CO_2 (by far the most important greenhouse gas) it is critical that buildings are designed and constructed in a way which will minimise resultant emissions.
- Although the initial onus is on housing, the Government's long-term intention is to provide the framework to support a move towards making all new developments carbon neutral.
- The construction industry has set itself tough goals to improve its track record in relation to health and safety.
- The Construction Skills Certification Scheme provides a means of registering and ensuring continuing competence for an individual construction worker.
- There are significant opportunities for soil management improvement if consideration is given earlier in the construction process.
- Construction projects need to embrace concepts of sustainability at the design stage, eg what is being built, how it is being built, with which products and methods.

5.368 The goal of sustainable development is to enable everyone to satisfy their basic needs and enjoy a better quality of life without compromising the quality of life of future generations. It involves balancing and integrating the economic, social and environmental considerations for any policy or decision. Sustainable development represents the process of attaining sustainability.

Employers' Duties

5.369 Sustainability in construction is driven primarily by stakeholder and business interests, although there are specific legislative requirements including:

- putting in place Site Waste Management Plans (SWMPs) for construction projects above a specified value
- ensuring that all construction waste is disposed of appropriately
- ensuring, so far as is reasonably practicable, the health, safety and welfare at work of all employees under the Health and Safety at Work, etc Act 1974.

Employees' Duties

5.370 Employees must:

- know, understand and comply with the organisation's sustainability policy
- follow procedures and other instructions that apply to their work.

In Practice

Achieving Sustainable Development

5.371 Any construction project that wants to achieve sustainable development will need to have a good understanding of current and emerging risks. For construction projects, issues that will need to be considered include:

- employee relations
- health and safety
- community development
- protection of the public
- marketplace practices
- supply chain practices
- fiscal responsibility
- accountability.

5.372 An important element in achieving sustainable development is to promote a built environment that:

- minimises adverse impacts on the environment, during construction and in use, while enhancing the natural surroundings
- maximises the positive contribution to business activity through the entire life of the building
- helps to encourage productivity through being flexible for future use, building cost-efficiently and improving people's working environment
- takes fully into account the impact of construction on the surrounding environment by seeking to maintain biodiversity within the location and avoiding any unnecessary pollution
- makes use of modern methods of construction, wherever possible, to improve building efficiency and to minimise environmental effects of construction sites.

5.373 *Sustainable Building in Practice: An Essential Guide* was published in February 2008, aiming to provide hands-on, practical advice to developers in the area of new-build residential homes. It contains several contributions from UK Green Building Council (UK-GBC) members and includes best practice suggestions in areas such as construction materials, energy and water. It also summarises the Code for Sustainable Homes and considers zero carbon for the mass market. Case studies include:

- the Oxley Woods Development (Taylor Wimpey's award-winning development)
- Broughton Square (a large-scale sustainable development)

- the Kingspan Lighthouse, the prototype which has achieved post-completion Level 6 of the Code.

Waste Reduction

5.374 The construction sector generates 109 million tonnes of waste per year, of which at least 13 million tonnes are unused new materials, ie materials delivered to the site, unused and then sent away for disposal. Demolition materials and soil amounts to 91 million tonnes in England and Wales annually.

5.375 Over 90% of non-energy minerals extracted in the UK are supplied as construction materials, and the industry annually produces three times the amount of waste generated by all UK households combined. Pollution has major sources in the construction process: waste, materials, noise, vehicle emissions, and contaminant release into the atmosphere, ground and water.

5.376 The European Union framework directive on waste continues to cause concern within the construction industry, including the aggregates and construction products sectors. In achieving sustainable waste management the Government is working closely with organisations such as the Environment Agency, Waste and Resources Action Programme and local authorities. It has identified the need to reduce the amount of illegal activity, or waste crime, that is currently a feature of this sector.

5.377 Construction and demolition waste accounts for 16% of all fly-tipping incidents reported to Flycapture, the national fly tipping database, and 30% of all incidents are dealt with by the Environment Agency (the Agency).

5.378 The Agency seeks to demonstrate good practice in its new construction projects. This is supported by its sustainable procurement practices and by incorporation of sustainable procurement practices and sustainable construction principles into its specifications.

Site Waste Management Plans

5.379 Site Waste Management Plans (SWMPs) are a statutory requirement for construction projects above a specified value. Originally developed as a Department of Trade and Industry (DTI) Voluntary Code of Practice, SWMPs provide a structure for systematic waste management at all stages of a project's delivery. These plans became a legal requirement for all construction projects over £300,000 in April 2008 under the Site Waste Management Plans Regulations 2008.

5.380 The plans focus largely on on-site operations and primarily identify:

- an individual responsible for resource management
- the types of waste that will be generated
- resource management options for these wastes
- the use of appropriate and licensed waste management contractors
- a plan for monitoring and reporting on resource use and the quantity of waste.

Demolition Protocol

5.381 The Institution of Civil Engineers Demolition Protocol was launched in 2003. The protocol is a resource-efficiency model that shows how the production of demolition material can be linked to its subsequent specification and procurement as a high value material in new builds. In addition, the protocol shows how resource efficiency can be driven through the planning process.

Review of Waste Strategy 2000

5.382 Waste Strategy 2000 established the Government's vision of sustainable waste management in England until 2020. This sets out the rationale for more sustainable management of the huge quantities of waste produced each year with the focus on:

- reducing the impact of waste on climate change
- conserving limited natural resources
- reducing the risks to health and the environment from potentially harmful substances within waste.

5.383 The past five years have seen a growing awareness of the economic and environmental benefits that can be delivered through this more sustainable approach to managing waste. A further substantial shift in direction is required, in particular with a stronger emphasis on:

- managing waste as part of a wider resource economy
- reducing the growth in the amount of waste produced
- the reduction, reuse and recycling of non-municipal waste (including from the construction sector)
- stimulating investment in waste management infrastructure.

Landfill Tax

5.384 Landfill tax contributes to the development of an environmentally sustainable economy by tackling the UK's over-reliance on landfill across all sectors and encouraging more sustainable waste management options, including recycling and reuse.

5.385 Landfill tax applies to all waste disposed of by way of landfill by businesses and local authorities at licensed landfill sites on or after 1 October 1996, unless the waste is specifically exempt and is paid by the landfill site operator.

Sustainability and the Olympics

5.386 The Olympic Delivery Authority (ODA) announced in January 2008 that it was beating its target to ensure sustainable principles are at the heart of cleaning and clearing the Olympic Park in London.

5.387 A year after the ODA Sustainable Development Strategy set out how the Park, venues and infrastructure for the 2012 Games would be created using the most sustainable methods possible, the ODA said that it was ahead of its target of reclaiming 90% of demolition material for recycling or reuse.

5.388 The aim was to make the project an exemplar which set new standards for sustainable development. In the process of clearing the land and demolishing over 200 buildings, the ODA has been reclaiming as much material as possible to re-use in the design of the venues and Park. Soil has been cleaned to help create the right ground levels, rubble used to create roads and wood chipped to add to soil or used to create log walls. A large number of bricks, paving slabs, tiles, lampposts and bollards have also been saved to be re-used. In addition, three buildings have already been reclaimed in their entirety.

Reducing Carbon Dioxide Emissions

5.389 The construction, occupation and maintenance of buildings account for around 50% of UK emissions of carbon dioxide (CO_2) therefore contributing to climate change, the depletion of non-renewable resources and adding to pollution.

5.390 The urgent need to reduce the emissions of CO_2, which are largely responsible for causing climate change, is a major factor behind the Government's drive to raise standards of construction.

5.391 With the construction, occupation and maintenance of buildings being responsible for around half the UK's emissions of CO_2 (by far the most important greenhouse gas) it is critical that buildings are designed and constructed in a way which will minimise resultant emissions. This not only applies to the creation of new buildings, but also to the refurbishment and redevelopment of existing ones.

Climate Change Levy

5.392 The Climate Change Levy applies to energy used in the non-domestic sector (industry, commerce, public sector). The aim of the levy is to encourage these sectors to improve energy efficiency and reduce emissions of greenhouse gases. The levy is administered by HM Revenue & Customs.

Energy Review 2006

5.393 The report of the *Energy Review* sets the framework for delivering the Government's two key energy objectives: secure, affordable energy supplies and cutting carbon emissions.

5.394 In addition to allowing new nuclear power stations to be built, the review gives a major push to both renewable energy sources and energy efficiency, including a major focus on buildings and the way they are used. The last of these is obviously very pertinent to the *Review of Sustainable Construction*. The review also sets out a series of measures — a mixture of regulation, guidance, encouragement, and demonstration — that are already in hand to move significantly towards the Government's carbon neutral targets.

Code for Sustainable Homes

5.395 In July 2007, the Government consulted on proposals for making it mandatory for new homes to be rated against the Code for Sustainable Homes. This would mean that, once introduced, all homes would either:

- have to be assessed against the Code and given a certificate indicating the rating they had achieved
- not be assessed and would be deemed to have achieved a zero rating against the Code.

5.396 In all instances, as a result of the mandatory rating policy, the purchasers of new homes would be given clear information about the sustainability of their home and house builders would have a clear and consistent basis on which to compare and market their products.

5.397 In November 2007, the DCLG published a *summary of responses* to the July consultation, setting out the Government's final policy and confirming its intention to proceed with the implementation of mandatory rating against the Code for all new homes in April 2008.

5.398 The Government also intends to introduce a document that makes it clear where a new home has not been assessed against the Code and a zero-star certificate for non-assessed properties. Documentation relating to the Code ratings are to be included in House Information Packs (HIPs), but it will not be mandatory for all Code assessor organisations (or self-employed individuals) to be able to provide Code and EPC (Energy performance Certificate) services as a single package. It will however "allow the market to drive services in this direction" if customers require it.

5.399 In future, the Government will update the Code at appropriate points to reflect changes to building and other regulations, including those changes proposed for 2010.

Carbon Neutral Developments

5.400 The Government's long-term intention is to provide the framework to support a move towards making all new developments carbon neutral. While the onus will initially be on housing, this will also include emissions from business and transport connected with new developments.

5.401 For example, a report from the Green Commercial Buildings Task Group, published in December 2007, suggests that a challenging yet achievable timeframe for achieving zero carbon new non-domestic buildings, along the lines set for housing, is needed. With a trajectory in place similar to that adopted for the Code for Sustainable Homes, it argues that a deadline of 2020 could be adopted. A key recommendation of the report is the construction of a national database on energy use in non-domestic buildings to improve on the existing incomplete and inconsistent data.

The EU and Sustainable Construction

5.402 In recent years, more and more of the UK's environmental protection laws have derived from European Union Directives. The EU is developing a number of what it calls *Thematic Strategies* within its Sixth Environment Action Programme (2002–12), which cover some of the areas of most interest to construction, including waste, water and soil.

5.403 The European Commission has also named sustainable construction among six important markets in which it wants to unlock the potential for innovative goods and services. According to the Commission, these markets have high economic and societal value with an annual turnover of more than €120 billion and providing nearly two million jobs in the EU. With the new *Lead Markets Initiative for Europe* (LMI) these figures may increase to over €300 billion and over three million jobs by 2020.

5.404 In order to give European companies the chance to profit from fairer and better chances of entering new fast-growing world-wide markets with a competitive advantage as lead producers, the Commission intends to improve legislation, encourage public procurement and develop interoperable standards.

5.405 With regard to sustainable construction, the Commission noted that buildings account for the largest share of the total EU final energy consumption (42%) and produce about 35% of all greenhouse emissions, hence it views developing sustainable solutions as crucial. Orientation towards innovative solutions and cutting administrative burdens, particularly for small and medium-sized businesses, are some of the proposed measures.

Building Regulations

5.406 There are several provisions in the Government's climate change programme of direct relevance to the construction industry. Key among these is the uptake of the Building Regulations which, together with the 2002 and 2005 revisions, will result in a 40% saving of carbon emissions from homes built from 2006 compared with pre-April 2006 new-build.

Soil Management

5.407 The consideration of sustainable soil management and use is embedded within the agricultural sector.

5.408 Within the urban environment the construction sector has seen the largest impact on the sustainable use of soils, but soils are often only a last minute consideration in landscaping a development. However, evidence suggests that there are significant opportunities for improvement if consideration is given earlier in the construction process.

5.409 Soil degradation occurs as a result of a range of construction practices and these have additional impacts on water quality, aquifer recharger and flood risk. Evidence has shown that soil erosion from construction sites is a major component of urban diffuse water pollution incidents handled by the Agency. Soil compaction from the use of heavy machinery reduces the infiltration capacity of the soil leading to excess run-off, and this may be beyond the capacity of existing storm drains and can lead to local flooding. Compacted soil may also compromise Sustainable Drainage Systems and provides a poor growing medium for plants leading to loss of landscaping schemes and higher maintenance costs.

5.410 The impacts of development are addressed in a large part through the planning system.

Sustainable Design

5.411 Construction projects need to embrace concepts of sustainability at the design stage. This involves not just considering what is being built, but how it is being built, with which products and methods, and which functions the project will perform or facilitate once completed.

5.412 The Commission for Architecture and the Built Environment (CABE) is the government-funded agency promoting better designed buildings to improve long-term sustainability and to provide the UK with a better designed built environment. CABE indicates that architecture's contribution to sustainability comes in three main areas.

1. Good architectural design leading to better buildings helps generate social, economic and environmental value over the long term, helping to sustain and improve communities by encouraging community spirit, healthier lifestyles and lower crime.
2. Good architectural design can reduce CO_2 emissions through specification of materials and manufacturing methods — both reducing waste, the design of buildings to maximise the use of the sun or wind in lighting, heating, cooling and power generation, and the specification of high standards of thermal insulation. The imaginative conversion of existing buildings to other uses can also be seen as promoting sustainability through the use of existing structures leading to a reduction in the consumption of building materials.
3. Strong architectural input to the planning of new local developments can promote higher-density mixed-use development (residential and business on one site or mixed-use buildings) where retaining green spaces, designing-in integrated public transport thereby discouraging car use and commuting.

Building Contracts and Sustainability

5.413 An industry consultation was launched in February 2008 by the Joint Contracts Tribunal (JCT) to ask for comments regarding the suggestion that building contracts should include stronger sustainability performance provisions. The month-long consultation followed successful discussions in December 2007 between the JCT and senior executives from organisations across the industry, as well as government departments and agencies.

5.414 The question at issue was whether additional contractual provisions and guidance would be effective in improving the industry's sustainability. JCT, pointing out that over 70% of all building contracts were under a JCT form, stated that there was an opportunity to gain consensus from, and perhaps provide leadership to, the industry to help improve its record on sustainability.

5.415 The construction industry generally appears to accept that something has to be done about sustainability and, as it is the largest industry in the economy, it should take responsibility for introducing changes. Other issues brought forward by the consultation included:

- whether contractual provisions could be effective
- what remedies could be sought for failure to meet sustainability requirements
- how contracts and guidance could assist and encourage professional advisers to give priority to sustainable design features.

Sustainable Development Action Plan

5.416 In January 2008, the Department for Communities and Local Government (DCLG) prepared a revised Sustainable Development Action Plan (SDAP). It builds on the SDAP produced in 2006 by its predecessor department, the Office of the Deputy Prime Minister.

5.417 The new report explains what sustainable development means to the DCLG, a department with a leading role in promoting the social dimensions of sustainable development. It also sets out the actions that the DCLG will take in 2008 to promote sustainable development and to integrate its principles throughout its policies, operations, procurement and people.

5.418 It sets out priorities in the areas of:

- sustainable consumption and production
- climate change and energy
- natural resource protection and environmental enhancement
- sustainable communities.

List of Relevant Legislation

- Directive 2006/12/EC on waste
- Building (Amendment) (No. 2) Regulations 2004
- Building (Amendment) Regulations 2002
- Building Regulations 2000

Further Information

Publications

BERR Publications

The following is available from *www.berr.gov.uk*.

- Review of Sustainable Construction 2006

Department for Communities and Local Government Publications

The following is available from *www.communities.gov.uk*.

- *Sustainable Development Action Plan 2007–08*

Other Publications

- *Sustainable Building in Practice: An Essential Guide*, NewzEye, 2008

Organisations

- Commission for Architecture and the Built Environment (CABE)
  Web: *www.cabe.org.uk*
  The CABE promotes high-quality design and architecture to raise standards of the built environment.
- Department for Environment, Food and Rural Affairs (Defra)
  Web: *www.defra.gov.uk*
  Defra is the main government department which deals with waste and other environmental issues. It consults on new regulations and provides guidance on legislation and best practice.
- Environment Agency
  Web: *www.environment-agency.gov.uk*
  The Environment Agency is the main environmental regulator in England and Wales and provides detailed information on legislative requirements, guidance for business and technical guidance on waste treatment and disposal.
- HM Revenue & Customs (HMRC)
  Web: *www.hmrc.gov.uk*
  HMRC is the department responsible for the business of the former Inland Revenue and HM Customs and Excise. The HMRC is responsible for lecting the bulk of tax revenue as well as paying tax credits and child benefit.
- Institution of Civil Engineers (ICE)
  Web: *www.ice.org.uk*
  The ICE is a charity that exists to promote and progress civil engineering.

Zero Carbon Housing

- The UK emitted more than 550 million tonnes of carbon dioxide ($MtCO_2$) in 2005. Energy use in buildings accounted for nearly half these emissions, and more than a quarter came from the energy used to heat, light and run homes.
- The Government believes a total of three million new homes are needed by 2020, two million of these by 2016.
- This is why it has introduced the Code for Sustainable Homes. It is intended to play a key role in helping to build a future housing stock which both meets these needs and protects the environment.
- The first zero carbon houses are now being built in a number of environmentally sustainable housing developments. In July 2007, the Government invited bids for at least five new zero carbon eco-towns, each to provide 5,000 to 20,000 new homes by 2016.
- The Government also confirmed that, from 2016, all new homes should be zero carbon, with milestones that by 2010 they should emit 25% less carbon, and, by 2013, 44% less carbon.
- Sustainable housing is increasingly seen as being vital to other areas of Government policy.
- Zero carbon means that, over a year, the net carbon emissions from all energy use in the home would be professionally assessed as zero.

5.419 Currently, the energy used to heat, light and run homes accounts for 27% of all of the UK's carbon emissions — around 40 million tonnes. If the UK is to meet the challenge of reducing greenhouse gas emissions by 60% by 2050, action in both new and existing housing is vital. By that date, as much as one-third of the then existing house stock will have been built since 2007. What is built in the next few years will therefore have a huge impact on the country's response to climate change.

5.420 In order to tackle this, the Government is proposing that all new homes be zero carbon by 2016 and has increased its support of this through measures such as the Code for Sustainable Homes and the publication of *The Callcutt Review*, commissioned to look at the capability of the housebuilding industry to deliver the type of homes needed for future generations. As a result of *The Callcutt Review*, the Government aims to set up a new body to monitor and co-ordinate work on delivering the target, and has asked the UK Green Building Council to lead the establishment of the work programme and priorities for the new body once it is created.

5.421 Energy efficient and insulated buildings, which draw their energy from zero or low carbon technologies and therefore produce no net carbon emissions from all energy use over the course of a year, will help reduce carbon emissions as well as lowering fuel bills for households.

5.422 Reports show that if the country does not increase house building above previous plans, the percentage of 30–34 year old couples able to afford to buy will worsen significantly in the long term, falling from over half today to around 35% in 2026. The impact of increased house building on the environment is becoming clearer however and the Government has to balance the need for more homes, often in already crowded areas, with its stated determination to cut greenhouse gas emissions as part of the fight against climate change.

5.423 An important part of the answer is seen as building houses which minimise the use of energy and reduce harmful emissions both in their construction phase and during use.

In Practice

The Code for Sustainable Homes

5.424 Improved energy efficiency is a key area of focus in the bid to provide zero carbon homes, but sustainability also implies action to tackle areas such as water use, waste generation and the use of polluting construction materials. To encourage the construction industry towards the required new ways of working, the Government produced the Code for Sustainable Homes (the Code) in together with various technical guides. A requirement for mandatory rating against the code was implemented on 1 May 2008.

5.425 The Code resulted from co-operation between the Government, the Building Research Establishment (BRE) and Construction Industry Research and Information Association (CIRIA) and involved close consultation with a Senior Steering Group consisting of Government, industry and NGO (non-governmental organisation) representatives.

5.426 Minimum standards for compliance with the Code have been set above the requirements of the Building and Approved Inspectors (Amendment) Regulations 2006. This is intended to signal the future direction of building regulations in relation to carbon emissions from, and energy use in, homes, providing greater regulatory certainty for the homebuilding industry.

5.427 The Code complements the system of Energy Performance Certificates introduced in June 2007 as part of the UK's compliance with the EU Energy Performance of Buildings Directive (2002/91). This Directive requires that all new homes (and in due course other homes, when they are sold or leased) should have an Energy Performance Certificate providing key information about the energy efficiency/carbon performance of the home. To avoid the need for duplication, energy assessment under the Code uses the same calculation methodology.

5.428 Demonstration buildings have already been constructed on the BRE (Building Research Establishment) Innovation Park and some new developments are already applying the Code.

Non-domestic Buildings

5.429 The Green Commercial Buildings Task Group commissioned the UK Green Building Council (UK-GBC) to investigate the costs and benefits of raising the energy performance standards in new non-domestic buildings above those currently set out in the Building Regulations.

5.430 The report prepared by UK-GBC, which was published in December 2007, investigates the opportunities for achieving zero carbon in such buildings. Following on from the targets set out in the Code to achieve radical emissions reductions in new homes, the report aims to add to the understanding of whether similar targets in the commercial sector can be set and achieved, and in what timescale.

5.431 What makes the report unusual is that several of those who participated are competitors who nevertheless shared data and collaborated over the project. One of their key findings is that it is possible to reduce carbon emissions from energy use down to zero in the majority of new non-domestic buildings, as long as on-site, near-site and off-site renewable solutions are employed.

5.432 There is a cost associated with building to zero carbon, the report concedes. While this varies widely with both the form and the use of the building, preliminary modeling suggests that the premium could range from over 30% down to as low as 5–10% of current baseline costs.

5.433 A challenging yet achievable timeframe for achieving zero carbon new non-domestic buildings along the lines set for housing is needed, the UK-GBC concludes. With a trajectory in place similar to that adopted for the Code, it argues that a deadline of 2020 could be adopted. A key recommendation of the report is the construction of a national database on energy use in non-domestic buildings to improve on the existing incomplete and inconsistent data.

The Code in Theory

5.434 The Code for Sustainable Homes was developed using the Building Research Establishment's (BRE) EcoHomes System, which was itself intended to reduce the impact of affordable housing projects, in particular within the social housing sector.

5.435 The Code builds upon EcoHomes in a number of ways, for example it:

- introduces minimum standards for energy and water efficiency at every level, therefore requiring high levels of sustainability performance in these areas for achievement of a high Code rating
- uses a simpler system of awarding points, with more complex weightings removed
- includes new areas of sustainability design, such as lifetime homes and the inclusion of composting facilities.

5.436 The Code is intended to measure the whole home and takes account of the following design categories, each with its own degree of flexibility.

Design Category	Level of Standards
Energy/CO_2	Minimum standards at each level of the Code
Pollution	No minimum standards
Water	Minimum standards at each level of the Code
Health and well-being	No minimum standards
Materials	Minimum standard at entry level
Management	No minimum standards
Surface water run-off	Minimum standard at entry level
Ecology	No minimum standards
Waste	Minimum standard at entry level

The Code in Practice

5.437 A star rating system is used, extending from one star (entry level, but still above the level of the Building Regulations) to six (the maximum), depending on the extent to which the building has achieved the required code standards. In order to achieve a particular code level, and the associated sustainability rating, a home must integrate minimum standards, and additional points for other design features must be attained.

Minimum Standards for Star Ratings

Code Level	Energy Standard*	Points Awarded	Water Standard — Litres per Person/ Day	Points Awarded	Other Points Required
1 star	10	1.2	120	1.5	33.3
2 stars	18	3.5	120	1.5	43
3 stars	25	5.8	105	4.5	46.7
4 stars	44	9.4	105	4.5	54.1
5 stars	100**	16.4	80	7.5	60.1
6 stars	Zero carbon***	17.6	80	7.5	64.9

* This figure is the percentage by which the home is better than Building Regulations: Approved Document L (2006) — *Conservation of Fuel and Power.*

** This figure relates to zero emissions in relation to Building Regulations issues (that is, from heating, hot water, ventilation and lighting).

*** This means a completely zero carbon home (that is, zero net emissions of carbon dioxide (CO_2) from all energy use in the home).

Assessment Procedures

5.438 Assessment procedures will be similar to BRE's EcoHomes System, which depends on a network of specifically trained and accredited independent assessors. BRE will retrain and accredit assessors for the new code. Existing EcoHomes Assessors who have qualified in the past will inherit UKAS "grandfather rights" to operate as code assessors, subject to attendance at a training update session.

5.439 The assessors will conduct initial design stage checks, recommend a sustainability rating and issue an interim code certificate based on design drawings, specifications and commitments. A post-completion check will verify the rating before a final code certificate of compliance is issued. A design stage assessment will only need to be carried out on each home type within any development — not every single home. Similarly, post-completion checks will be carried out on a sample basis.

5.440 The Department for Communities and Local Government (DCLG) suggest that the levels can be considered as follows.

- One Star: above regulatory standards and a similar standard to BRE's EcoHomes "pass" level and the Energy Saving Trust's (EST's) Good Practice Standard for energy efficiency.
- Two Stars: a similar standard to BRE's EcoHomes "good" level.
- Three Stars: a broadly similar standard to BRE's EcoHomes "very good" level and the EST's Best Practice Standard for energy efficiency.
- Four Stars: broadly set at current exemplary performance.
- Five Stars: based on exemplary performance with high standards of energy and water efficiency.
- Six Stars: the "aspirational" standard based on zero carbon emissions for the dwelling and high performance across all environmental categories.

How Home Builders Will Benefit

5.441 The Government suggests that the Code will provide the following.

- Mark of quality: increasing media attention and public concern over environmental issues, notably climate change, has given rise to a growing appetite among consumers for more sustainable products and services. The Code for Sustainable Homes can be used by home builders to demonstrate the sustainability performance of their homes, and to differentiate themselves from their competitors.
- Regulatory certainty: the levels of performance for energy efficiency indicate the future direction of building regulations, bringing greater regulatory certainty for home builders, and acting as a guide to support effective business and investment planning.
- Flexibility: the Code is based on performance, which means it sets levels for sustainability performance against each element but does not prescribe how to achieve each level. Home builders can innovate to find cost-effective solutions to meet and exceed minimum requirements.

Practical Examples

5.442 The DCLG has issued details of how some of the levels will actually be calculated.

Code Level 1

5.443 To be graded Level 1, a home will have to be 10% more energy efficient than one built to the standards of the Building and Approved Inspectors (Amendment) Regulations 2006. This could be achieved by:

- improving the thermal efficiency of the walls, windows, and roof (by using more insulation or better glass)
- reducing air permeability, that is by improving the control of the fresh air into a home, and the stale air out of a home — a certain amount of air ventilation is needed in a home for health reasons
- installing a high-efficiency condensing boiler
- carefully designing the fabric of the home to reduce thermal bridging (thermal bridging allows heat to escape easily between the inner walls and the outer walls of a home).

5.444 The home will have to be designed to use no more than about 120 litres of water per person per day. This could be achieved by installing a number of items such as:

- a 6/4 litre dual flush WC
- flow-reducing/aerating taps throughout
- a shower using 6–9 litres per minute (an average electric shower uses about 6–7 litres per minute)
- a 18-litre maximum volume dishwasher
- a 60-litre maximum volume washing machine.

5.445 Other minimum requirements are required for:

- surface water management — this may mean the provision of soakaways and areas of porous paving
- materials — this means a minimum number of materials meeting at least a ‘D’ grade in the Building Research Establishment’s Green Guide (the scale goes from A+ to E)
- waste management — this means having a site waste management plan in place during the home’s construction, and adequate space for waste storage during its use.

5.446 However, this still does not complete the list of requirements. To be awarded Level 1, the home must amass a further 33.3 points. The builder/developer must therefore consider:

- providing accessible drying space (so that tumble dryers need not be used)
- providing more energy efficient lighting (taking into account the needs of disabled people with visual impairments)
- providing cycle storage
- providing a room that can be easily set up as a home office
- reducing the amount of water than runs off the site into the storm drains
- using environmentally-friendly materials
- providing recycling capacity either inside or outside the home.

Code Level 6

5.447 At the other extreme, a home meeting the exemplary Level 6 status will have to be completely zero carbon with zero net emissions of carbon dioxide from all energy use in the home.

5.448 This could be achieved by:

- improving the thermal efficiency of the walls, windows, and roof as far as is practically possible (eg by using more insulation or better glass)
- reducing air permeability to the minimum consistent with health requirements (a certain amount of air ventilation is needed in a home for health reasons)
- installing a high-efficiency condensing boiler, or being on a district heating system
- carefully designing the fabric of the home to reduce thermal bridging (thermal bridging allows heat to escape easily between the inner walls and the outer walls of a home)
- using low and zero carbon technologies such as solar thermal panels, biomass boilers, wind turbines, and combined heat and power systems. For example, it would mean that energy taken from the national grid would have to be replaced by low or zero carbon generated energy, so that over a year the net emissions were zero.

5.449 Furthermore, the home will have to be designed to use no more than 80 litres of water per person per day. This could be achieved by fitting such items as:

- a 6/4 dual flush WC
- flow-reducing/aerating taps throughout
- a shower using 6–9 litres per minute
- a smaller, shaped bath that is still long enough to lie down in, but requires less water to fill it to a level consistent with personal comfort
- a 18-litre maximum volume dishwasher
- a 60-litre maximum volume washing machine.

5.450 To achieve the standard would also mean that approximately 30% of the water requirement of the home was provided from non-potable sources such as rainwater harvesting systems or grey water recycling systems.

5.451 Other minimum requirements are required for:

- surface water management
- materials
- waste management.

5.452 Having done some or all of the above will leave the builder/developer still needing a further 64.9 points to get to Level 6, leaving them with much still to do to obtain the remaining points. In fact, they will need to address 90% of everything in the Code to achieve Level 6, including:

- energy-efficient appliances and lighting
- supplying accessible water butts

- reducing surface water run-off as much as possible
- using highly environmentally friendly materials
- minimising construction waste
- maximum, accessible provision for recycling
- improved daylighting, sound insulation and security
- building to the Lifetime Homes standard
- assessing and minimising the ecological impact of the construction of the home.

Putting Theory into Practice

5.453 The first home in the country to be rated zero carbon is at Hockney Green, Testway Housing's innovative housing development in Andover. Internal and external eco-features of the development include:

- eco-paint
- recycled furniture
- roof-top solar panels
- wind cowls.

5.454 Three of the houses will be monitored for their carbon consumption and electricity output and one will be fitted with both solar panels and a wind turbine to create electricity. This is the house that hopes to gain the first six-star Code level rating. A second house has been fitted with the electricity-making solar panels and should be virtually zero carbon — broadly the standard for a five-star rating in the Code.

5.455 All homes on the site have a range of energy-saving features to reduce the amount of energy they consume. The homes are designed to require as little heat input as possible. They all have a sun space the height of the building, similar to a conservatory, with a wall of glazing separating it from the outside.

5.456 Elsewhere, five demonstration buildings have been constructed on the BRE Innovation Park, and the "One Gallions" consortium is working on a zero carbon development at Gallions Park, in the Royal Docks in East London. Some 200 homes will be built, all designed as net zero carbon and built to the highest environmental standards, including renewable energy sources.

5.457 In July 2007, local councils and developers were invited by the Government to bid to host at least five new zero carbon eco-towns, each providing between 5,000 and 20,000 new homes, by 2016. A Government-led design competition is intended to boost the architectural standards of the eco-towns.

5.458 In December 2007, the Department for Communities and Local Government (DCLG) announced that Barratt Developments plc had been selected by English Partnerships, the Government's National Regeneration Agency, as the preferred developer to create a new community at the site of the former Hanham Hall Hospital near Bristol. This was the first site identified under the Carbon Challenge, being run by English Partnerships as part of the Government's commitment to tackle climate change, and homes on the site will meet Level 6 of the Code for Sustainable Homes. An on-site biomass CHP (combined heat and power) plant will deliver energy to all of the homes.

5.459 The Housing and Planning Minister also unveiled the six shortlisted bidders for the second Carbon Challenge site — South Bank Phase 1 in Peterborough — and two new sites in the North of England. Brodsworth Colliery in Doncaster and Bickershaw Colliery in Wigan are both in former coalfield communities, devastated by pit closures in the 1980s. The shortlist for the Peterborough project includes the Gladedale Group, Barratt Homes Ltd and a consortium of Galliford Try and Cross Key Homes.

2016 Commitment

5.460 The Government has welcomed industry support for the zero carbon homes 2016 target and the proposed changes to Building Regulations in 2010 and 2013 by formally launching the "2016 Commitment". This requires the signatories to work in partnership together, with the aim that:

- by 2016, zero carbon homes are a reality at the scale required to meet housing targets
- over the interim period, new homes will meet the increasingly higher environmental standards as set out in building regulations
- the issues that need to be tackled to achieve the 2016 objectives will be resolved.

5.461 While not in any sense legally binding on the organisations concerned, the Commitment includes a shared endeavour to:

- undertake and commission the necessary research, using sound science, to find practical solutions that deliver homes that are affordable, durable, desirable and healthy places to live
- consult on and contribute to the key policy and regulatory changes and how they will be implemented, and ensure compliance with the agreed national standards
- build a proper skills base in the relevant sectors and industries
- support efforts to communicate the importance of higher environmental standards to the wider industry and consumers, and raise awareness of the benefits
- be constructive and co-operative in approaches to delivering higher environmental standards in partnership, including comparing and sharing best practice.

5.462 An updated list of signatories was published in February 2008 and includes well over 100 organisations, ranging from the Confederation of British Industry (CBI), the Home Builders Federation and the Construction Products Association to Arup, Taylor Wimpey plc and NG Bailey Ltd. Universities and local authorities also feature on the list.

Guidance

5.463 The DCLG published full technical guidance on how to comply with the Code in March 2007. The guidance includes, in Part One, explanations of the:

- procedure for the assessment of houses and credit verification
- scoring and weighting system.

5.464 Part Two includes:

- details of the new credit definitions and their evidence base
- specification of methods of credit verification
- calculation algorithms, checklists and other tools which are part of the assessment method.

List of Relevant Legislation

- Energy Performance of Buildings (Certificates and Inspections) (England and Wales) Regulations 2007
- Building and Approved Inspectors (Amendment) Regulations 2006
- Building (Amendment) (No. 2) Regulations 2004
- Building (Amendment) Regulations 2002
- Building Regulations 2000

- *Directive 2002/91/EC* on the energy performance of buildings

Further Information

Publications

Department for Communities and Local Government Publications

The following is available from *www.communities.gov.uk*.

- *Code for Sustainable Homes — Technical Guide*, 2007

Planning Portal Publications

The following are available from *www.planningportal.gov.uk*.

- Approved Document L1A: *Conservation of Fuel and Power (New Dwellings)*, 2006 edition
- Approved Document L1B: *Conservation of Fuel and Power (Existing Dwellings)*, 2006 edition
- Approved Document L2A: *Conservation of Fuel and Power (New Buildings other than Dwellings)*, 2006 edition
- Approved Document L2B: *Conservation of Fuel and Power (Existing Buildings other than Dwellings)*, 2006 edition

Organisations

- Building Research Establishment (BRE)
  Web: *www.bre.co.uk*
  The BRE provides consultancy, testing and commissioned research services covering all aspects of the built environment and associated industries.
- CIRIA
  Web: *www.ciria.org.uk*
  CIRIA provides information and publications on contaminated land as well as sustainable land use and construction.
- Department of Communities and Local Government (DCLG)
  Web: *www.communities.gov.uk*
  The Department of Communities and Local Government is the Government department responsible for policy on housing, planning, devolution, regional and local government and has a remit to promote community cohesion and equality. This department was formerly known as the Office of the Deputy Prime Minister.
- Energy Saving Trust (EST)
  Web: *www.est.org.uk*
  The EST provides advice on saving energy in homes and businesses.

Chapter 6

Fire Safety

Explosives: Application and Risk Management

- The majority of applications of explosives used in the construction industry relate to detonation.
- The local geology, wind direction, proximity of other buildings, etc all determine the nature, quantity and position of the explosive device and the control measures required.
- The initial stage in carrying out a risk assessment is to identify the hazards associated with the explosives and their handling, storage and use in the workplace.
- When evaluating the risks from explosives, the risk assessment should cover the life of the explosive from arriving on-site, storage and transport through to its preparation for use.
- Typical control measures employed in managing the risk from explosives concern their storage and transport so that they are free form sources of ignition, whether that comes from temperature, friction and impact, electrical impulse, or chemical sources.
- Employers are under a duty to ensure that suitable separation distances exist between the explosives store and any inhabited buildings, etc.
- Before any work commences on-site, a method statement based on pre-construction surveys, the blast design and associated risk assessment should be prepared.
- The storage of explosives will need to be either registered or licensed, unless only a small quantity of explosives is involved.
- Along with other transport requirements, the Classification and Labelling of Explosives Regulations 1983 require explosives to be classified before they are conveyed, kept, sold or otherwise supplied or imported.
- One of the conditions for obtaining an explosives certificate is that the applicant will "take all reasonable precautions to prevent access to the explosives by unauthorised persons and to prevent loss of the explosives".
- The disposal of explosives should be risk assessed and conducted in the same way as any other explosive operation.
- Employers should ensure that suitable and sufficient information, instruction and training is provided to all employees on the storage, handling and use of explosives.

6.1 Whether faced with an unexploded bomb during excavation or preparing a blasting charge during demolition, the use of explosives must be managed with the utmost care. Fatal explosive-related incidences are increasingly rare, and less than 2% of all notifiable dangerous occurrences, across all industries, are caused by the unintentional ignition of explosives.

6.2 Examples of explosives include blasting explosives, detonators, ammunition, propellants and fireworks. An "explosive article" is an article containing one or more explosive substances. An "explosive substance" is a solid or liquid substance, or a mixture of solid or liquid substances, which is capable, by chemical reaction, in itself of producing gas:

- at such a temperature and pressure and at such a speed as could cause damage to surroundings
- which is designed to produce an effect by heat, light, sound, gas or smoke, or a combination of these, as a result of non-detonative self-sustaining exothermic chemical reactions.

6.3 Various regulations apply to the manufacture, supply, use and transport of explosive materials. The Dangerous Substances and Explosive Atmospheres Regulations 2002 apply to all hazards arising from both the manufacture and storage of explosives and from the other dangerous substances present on-site (eg including substances not being used or those in storage waiting to be used).

Employers' Duties

6.4 Under the Health and Safety at Work, etc Act 1974, employers have a general duty to ensure, so far as is reasonably practicable, the health, safety and welfare at work of all employees and other people on the premises.

6.5 Under the Manufacture and Storage of Explosives Regulations 2005, employers are required to:

- introduce appropriate control measures to prevent fires and explosions
- take appropriate actions to limit the extent of any fires and explosions
- protect people from the effects of fire or explosion (including provision of means of escape and fire-fighting equipment)
- make suitable arrangements for the storage of explosives
- ensure that suitable separation distances exist between the explosives store and any inhabited buildings
- make suitable arrangements for the disposal and decontamination of explosive-contaminated items
- prohibit unauthorised access to the premises where explosives are stored
- obtain a suitable licence for the storage and manufacture of explosives.

6.6 Under the Control of Explosives Regulations 1991, employers are required:

- to be competent to acquire and to keep explosives and be licensed appropriately
- to hold an explosives certificate if required
- not to transfer explosives without an explosives certificate
- to maintain current records of all people involved in the handling, use, etc of explosives in the workplace
- to maintain current records of the type and quantities of explosives in the workplace, a description of the explosives and where they are kept
- to ensure that the site and premises where explosives are used, kept and stored are secure
- to ensure drivers of vehicles carrying explosives are suitable and sufficiently trained.

6.7 The Manufacture and Storage of Explosives Regulations 2005 do not apply to explosives in transportation (unless they are kept in one place for longer than 24 hours), but they do apply to the on-site transport of explosives, including transport on public roads between buildings on the same site. The regulations do not apply to explosives that are in use.

Employees' Duties

6.8 Employees have a duty to:

- co-operate with employers to enable the employer to comply with legal requirements
- make proper use of control measures, including personal protective equipment (PPE)
- report any shortcomings in working practices to supervisors/managers.

In Practice

Application

6.9 The table below shows some typical examples of the application of explosives within the construction industry which fall under the British Standard *Code of Practice for the Safe Use of Explosives in the Construction Industry* (BS 5607).

Application of Explosives in the Construction Industry

Area of Construction	Application of Explosive	Further Guidance
Demolition	Use of explosive charges to weaken a structure to the extent that it collapses under its own weight	BS 6187: 2000 *Code of Practice for Demolition*
Tunnelling/ shaft sinking	Use of explosive charge to break up rock face	BS 6164: 2001 *Code of Practice for Safety in Tunnelling in the Construction Industry*
Demolition/ construction	Use of cutting charges to cut through metallic and non-metallic structures	Manufacturers' guides
Land excavation	Use of explosive charges to blast and shape surface rock for pipe and road laying, constructing trenches, building foundations, landscaping, etc	BS 6031: 1981 *Code of Practice for Earthworks*
Fabrication	Joining of dissimilar metals	Manufacturers' guides
Signalling	Use of explosive as a coloured flare to signal for help and to indicate location	Manufacturers' guides
Fabrication	Caps/cartridges for fixing cartridge operated fixing tools (the licensing rules for stored cartridges are slightly different to that for bulk explosives, see BS 4078)	BS 4078: 1989 *Powder Actuated Fixing Systems: Part 1: Code of Practice for Safe Use*

6.10 Most, but not all, of the applications above are achieved by detonation. During a detonation, the chemical decomposition of the high explosives (HE) occurs so violently that it produces a shock wave. The main charge — sometimes referred to as the secondary explosive — may take the physical form of a powder, gelatine, soft plastic, slurry or an emulsion.

6.11 Ammonium nitrate and fuel oil mixes, RDX (cyclotrimethylenetrinitramine) and nitroglycerine are typical base compounds of explosive charges.

6.12 For safety reasons, the main charge is formulated so that it requires a substantial shock to initiate a detonation. This ignition shock or impulse is provided by the detonator or a priming charge that contains less HE, but which is more sensitive (a primary explosive) to ignition. The detonator (also an explosive item) may be initiated or fired by means of a flame, spark, electric current or shock tube.

Planning

6.13 There are very few explosives tasks which fall directly into the scenarios analysed in textbooks. The local geology, wind direction, proximity of other buildings, etc all determine the nature, quantity and position of the explosive device and the control measure required.

6.14 A shock wave may travel many kilometres if the rock strata is aligned in certain operations to the blast site. A detailed survey of the construction site and the local area will provide explosives engineers with this essential information.

6.15 In addition, the planning phase will involve informing local services such as water, gas and electricity utility suppliers, as well as the public transport authorities and police. The noise and vibration of repeated shot firing may constitute a nuisance, and local residents may need to be kept informed.

Risk Assessment of Explosives in the Workplace

Deployment

6.16 The useful deployment of an explosive can occur in milliseconds, but the safety considerations for the storage, transport and physical preparation of the explosive device play a significant part in the risk assessment.

6.17 The risk assessment of explosives in the workplace should take into account normal work activities and any non-routine or out-of-the-ordinary work, eg maintenance or equipment repairs. Maintenance or repair work can often involve a higher degree of risk of fire or explosion. Normal work activities should take place only when a risk assessment of the current working status has been made.

Identifying the Hazards

6.18 The initial stage in carrying out a risk assessment is to identify the hazards associated with the explosives and their handling, storage and use in the workplace. Information to help identify the hazards is available in the relevant regulations, from the manufacturer/supplier of the product and a number of other sources. All relevant sources of information should be examined to help with the risk assessment process.

6.19 The methods adopted for initiating explosives are frequently flame, spark, electric current or shock tube. For those who physically handle explosives, the techniques to avoid premature ignition must be embedded by robust training. The following are examples of sources of ignition which should be considered hazards on a construction site.

- Temperature: exposure of the explosive (detonation cord, detonator or main charge) to naked flames; contact with hot surfaces.
- Friction and impact: nipping and trapping of an explosive during its preparation; impact of falling objects; impact of projected objects; impact by vehicles; drops, vibrations and blast impulses.
- Electrical impulse: discharge of static electricity from operator or machinery; lightning strike; stray electrical currents from buried cables; electromagnetic impulse.
- Chemical: incompatibility between certain substances; contamination; water ingress.

6.20 The sensitivity of an explosive to any of the above stimuli may vary because of its temperature, the presence of any contamination or its chemical properties altered by the passage of time.

6.21 Protecting the electrical ignition system from exposure to unwanted electrical or electromagnetic impulse is generally more important than protecting the secondary explosive itself. In electrical storms, the direct lightning strike or its discharge via surface objects (pipes or rails) or through the ground itself can be a major hazard.

6.22 Radio-operated equipment emitting high-frequency electromagnetic impulses can provide enough energy to initiate the detonator. *The Assessment of Inadvertent Initiation of Bridge Wire Electro-explosive Devices by Radio-frequency Radiation* (BS 6657: 2002) provides more information about this hazard.

Evaluating the Risks from Explosives

6.23 This part of the risk assessment should cover the life of the explosive from arriving on-site, storage and transport through to its preparation for use. The risk can range from minor injuries to catastrophic damage incurring fatalities over an area of many square miles.

6.24 In considering storage and transport, the risk factors should consider:

- the possibility and likelihood of fire or explosion
- the quantities of explosive materials (and other dangerous substances) stored/used on the premises
- the structures and property that could be affected
- the number of people on-site
- the potential severity of the damage to people and property
- any existing safety or control measures
- the competency of people in the workplace
- any accident and emergency procedures in place
- information available to employees and other people on-site.

6.25 When evaluating the risks associated with explosive tasks, the assessment must consider a host of potential effects including, but not limited to, the:

- projection of debris or fall area of a collapsing structure
- effect of blast and overpressure
- effect of dust or gases released
- potential ignition of flammable substances
- consequences of misfire
- effect on observers.

6.26 Under the Reporting of Injuries, Diseases and Dangerous Occurrences Regulations 1995, flying debris falling outside of a designated exclusion zone must be reported to the Health and Safety Executive (HSE). Metal fragments from either explosive demolition or cutting charges can potentially travel long distances at extremely high velocities.

6.27 Blast waves and vibration, as well as being a potential nuisance, can initiate more major ground disturbances, such as flow slides. Such events are more likely to occur in loose-to-medium dense saturated soils, or in loose, uniform fine sands.

6.28 During underwater blasting, the ability of water to transfer shock must not be underestimated. A compression wave can cause severe injuries to divers or swimmers in the area of the blast. In addition, the blast may affect surface vessels in the vicinity.

6.29 A less obvious hazard might be the distraction caused from a large demolition project. Drivers on roads adjacent to the blasting site may lose concentration and cause a road traffic accident.

6.30 Misfires can pose a very real hazard and, to some extent, can be prevented by careful and competent preparation. Misfires should be approached with a patient and logical methodology as defined in BS 5607: 1998. All misfires which are not immediately rectifiable after the first attempt must be reported to the HSE.

Control Measures

6.31 Due to the sensitivity of explosives and the potentially devastating consequences of premature ignition, trained operators only should have access to explosives.

6.32 Control measures concern the storage and transport of explosives so that they are free from sources of ignition (as listed above), and may include:

- no smoking in certain areas
- a permit-to-work for any procedure which could potentially start a fire
- ensuring a suitable distance between explosive stores and site traffic or active areas of the site
- utilisation of packaging design to reduce static and shock when transporting or storing explosives
- limitation on the types of hand tools used to those made of wood, soft non-ferrous material and anti-static plastic
- ventilation of areas where explosives are being stored to prevent build up of a flammable atmosphere
- suspension of explosives activities during thunderstorms
- restrictions on the use of radios or mobile phones
- use of non-electrical or high-energy initiation systems in areas where there is a high risk of discharge from stray electrical currents
- information and communication methods to keep those on-site aware of restrictions and hazards
- restricted access to competent operators
- methods to minimise the time that the ignition system and secondary explosive are in contact.

6.33 To mitigate the effect of an explosive event, the aim is to ensure that personnel and equipment are protected to as low a level as is reasonably practicable. Typical measures include:

- restricting the amount of explosives to a level appropriate to the work that is being undertaken
- keeping the number of personnel working within the blast zone of the explosive store/preparation area to a minimum

- using protective clothing and screens during preparation
- implementing emergency procedures.

6.34 The experience of an explosive contractor should ensure that the positioning, timing and choice of explosive devices minimises the requirement for additional control measures. However, even the most experienced operator will appreciate the limitations of their predictions and what they can achieve without additional control measures.

6.35 Additional control measures include:

- an exclusion zone
- deflection or absorption of blast by use of objects such as sandbags, netting wire and steel mesh, rubber belting, tyres and water bags
- inspection of surrounding structures prior to lifting exclusion zones
- wait times to allow for ground to settle (particularly in tunnelling)
- procedures to check atmosphere is fit for respiration before re-entry to blasting site
- use of water-filled bags placed to absorb dust
- preparation with appropriate fire-containment techniques
- road closure during time of blasting
- crowd control.

6.36 An exclusion zone must be designed around the area where the explosive device is to be deployed. All persons, except for possibly the shot-firer, must be outside the exclusion zone at the time of the blast. The zone is designed to ensure there is a buffer area between those outside of the exclusion zone and the maximum anticipated fall of debris or hazardous overpressure.

6.37 In extreme cases, such as the demolition of a high-rise structure, the exclusion zone may require the evacuation of a neighbourhood, a major communication plan, additional policing and crowd control. The HSE's information sheet CIS45: *Establishing Exclusion Zones when Using Explosives in Demolition* contains further guidance on this topic.

6.38 In preparing a suitable safety plan, the experience of the explosive engineer in predicting the blast and throw of debris cannot be over-emphasised. Verification and greater confidence in blasting predictions can be gained by conducting smaller isolated test blasts or by using advanced techniques such as computer modelling.

6.39 An exclusion zone is established by physical barriers or signage and enforced appropriately by perimeter guards or "sentinels". Visual and audible signalling will keep workers and the public informed as to when blasting is about to take place and when the exclusion zone can be lifted (the "all clear").

6.40 Human error must also be considered and various procedural arrangements can provide an extra check of safety-critical aspects of explosives preparation.

Separation Distances

6.41 Employers have a duty to ensure that suitable separation distances exist between the explosives store and any inhabited buildings. In the Manufacture and Storage of Explosives Regulations 2005, separation distances are required for different types of explosives and different storage arrangements. The distances apply between the explosive store and inhabited buildings. There are also distance requirements for public traffic routes and public places. Exceptions to these requirements include:

- stores holding very small quantities of explosive (100g or less)
- stores holding 30kg of shooters powders or less and/or up to 100g of primers for small arms ammunition, subject to certain conditions

- stores holding up to 200 detonators and 5kg of water-based explosive and detonating cord, or 5kg of water-based explosive or detonating cord, subject to certain conditions.

Method Statements

6.42 Adherence to well-considered procedures or method statements is of utmost importance in ensuring the safety of all. Before any work commences on-site, a method statement based on pre-construction surveys, the blast design and associated risk assessment should be prepared.

6.43 By way of example, the following questions should be answered in a typical demolition method statement.

1. How will the structure collapse?
2. What type of charge, and how much, will be used?
3. How will the charges be placed and delayed?
4. What protection measures will be deployed?
5. What is the size and location of the exclusion zones?
6. What are the evacuation procedures?
7. How will potential crowds and personnel be controlled and kept informed?
8. Do any roads need to be closed?
9. What are the transport arrangements for explosives?
10. What is the sequence of events?
11. What will happen if there is a misfire?
12. What happens if the desired effect is not achieved as planned or what could go wrong (contingency plans)?
13. How will the procedure be monitored and managed?

Handling Redundant/Unexploded Ordnance

6.44 It has been estimated that up to 10% of the bombs that fell on the UK during World War II failed to function. The discovery of unexploded ordnance (UXO) can cause a major disruption on a construction site.

6.45 In built-up areas, particularly in heavily bombed cities, a paper-based survey and risk assessment should be conducted before proceeding with any ground works. This risk assessment may determine whether or not a more localised survey should be instigated, eg where piles are to be driven.

6.46 On contaminated land, such as defence properties (or near such locations), a land survey will be required. Experience has taught developers of such areas to be ready for surprises. Fortunately, there have been no fatalities recorded in relation to UXOs for over 60 years.

6.47 On discovering a suspected bomb, contact with the police should be made immediately. The police will normally make contact with the bomb disposal unit and advise on the severity of the situation and any immediate exclusion zone that should be set up. Especially in built-up areas, the bomb disposal squad are likely to have remedied the situation in a matter of days.

Control of Unauthorised Access

6.48 The Manufacture and Storage of Explosives Regulations 2005 place prohibitions on entering an explosives building or area without permission from the licensee or their representative. The regulations also require anyone who has entered without permission to leave when requested to do so and permits the occupier to remove them. It is recommended that the police are called if the person refuses to leave.

6.49 The occupier should only remove unauthorised people themselves in situations where they consider that there is an imminent threat to the safety of the explosives. Only reasonable means may be used and these will depend on the severity and imminence of the threat to the explosives.

6.50 The operator should normally make suitable arrangements to mark the boundaries of the site. The decision on whether to erect boundary markers and warning signs will take into account the circumstances of the explosives stored, especially their location. Where further guidance is required, this may be sought from the police explosives liaison officer.

Storage

6.51 The storage of explosives will need to be either registered or licensed, unless only a small quantity of explosives is involved. The Manufacture and Storage of Explosives Regulations 2005 detail the quantities of explosives that can be kept without a licence or registration. The quantity depends on the amount and type of explosives.

6.52 Sources of ignition should not be taken within 25m of explosive stores, and detonators should be stored separately from explosives.

Transport

6.53 The transport of explosives from the place of purchase to the site must comply with the Transport of Dangerous Goods by Road Regulations 1996 (certain regulations were revoked by the Carriage of Dangerous Goods and Use of Transportable Pressure Equipment Regulations 2004).

6.54 The regulations aim to ensure the following.

- Vehicles are suitably furnished and designed to minimise the risk of premature ignition.
- Steps are made to reduce the risk of accident to the vehicle.
- Steps are made to ensure the security of the vehicle on its route.
- In the event of an accident the explosive load is clearly identifiable.

6.55 Once on-site, the transport of explosives must be in a vehicle intended for this purpose and under the control of an authorised person. The vehicle must be marked appropriately and many of the control measures associated with explosives storage will be deployed. For example, the movement of explosives within the site should be in the sealed packaging specifically designed by the manufacturer to minimise exposure to sources of ignition and contamination.

Classification and Authorisation for Transport

6.56 The Classification and Labelling of Explosives Regulations 1983 require explosives to be classified before they are:

- conveyed
- kept
- sold or otherwise supplied
- imported.

Transfer Control of Explosives

6.57 A recipient competent authority document should accompany explosives at all times when supplied and transported. These documents are issued through the Explosives Policy Division of the HSE.

- Recipient competent authority documents are issued to the person/company in possession of explosives following a transfer of the explosives.
- Recipient competent authority documents are usually valid for three years.
- Consignors of explosives should ensure that consignees have a valid recipient competent authority document.
- The consignee should send the valid recipient competent authority or certified true copy to the supplier when ordering explosives, and the document should accompany the explosives during transit.
- Employers transporting their own explosives should ensure that the document accompanies the explosives during transit.

Security of Explosives

6.58 Explosives invite interest from terrorist organisations as well as anti-social or criminal groups. Those working with explosives have a duty to ensure the security of explosives so they do not fall into the hands of those who could do harm.

6.59 Under the Control of Explosives Regulations 1991, an explosives certificate is required for the purchase or storage of explosives. Some explosives cannot be bought or used by persons who have committed certain offences. Theft or loss of explosives should be reported immediately to the police.

6.60 One of the conditions for obtaining an explosives certificate is that the applicant will "take all reasonable precautions to prevent access to the explosives by unauthorised persons and to prevent loss of the explosives". An explosives certificate can be revoked by the police if the licensee fails to demonstrate adequate control.

6.61 The Control of Explosives Regulations 1991 cover the security of explosives and include the following provisions.

- An explosives certificate is required for the purchase or storage of explosives.
- Restrictions apply to the transfer of explosives.
- Certain explosives (or substances which could be used as explosives) cannot be acquired, kept, handled or stored by anyone who has committed certain offences or been sentenced to certain terms of imprisonment.
- Restrictions apply to the types and quantities of explosives for private use and not for resale.
- Occupiers of licensed explosives stores should appoint a competent person to be responsible for explosives security.
- Comprehensive, accurate and up-to-date records should be kept of certain explosives.
- The loss or theft of explosives should be reported immediately to the police.
- Plastic explosives should be marked with a chemical to enhance detection.
- Unmarked explosives should not be purchased, transferred or imported.

Disposal

6.62 The disposal of explosives is an important topic, especially when considering the cost of transport, limits on storage and degradation of explosives over time. Four main methods are deployed for the disposal or destruction of explosives, namely:

- burning
- detonation
- dissolution or dilution
- chemical destruction.

6.63 Explosive burning can produce toxic fumes or create additional fire hazards, and suitable measures to protect staff will need to form part of the risk assessment.

6.64 The disposal of explosives should be risk assessed and conducted in the same way as any other explosive operation. While guideline operating instructions are normally supplied with the explosive item, these do not remove the need for the user to carry out their own risk assessment. Operators must be competent and authorised for the operation and be aware of the requirement to maintain a log of what disposal activities are undertaken.

Operating Provisions for Disposal Facilities for Explosive Waste

6.65

- The disposal or destruction of explosive waste should be carried out by competent staff.
- Burning-site operators should be provided with suitable protective clothing, without pockets and preferably made with heavy-duty fire-retardant wool.
- All tools used should be constructed from materials that do not present a spark or friction hazard.
- Vehicles used to convey waste should be specifically designed for the purpose.
- Logbooks detailing the daily activities of the disposal site should be kept.

Operating Instructions for Disposal Facilities for Explosive Waste

6.66 Detailed operating instructions should be displayed at the explosive waste disposal site. Employers should design operating instructions specific to each site. Instructions and information should be included on:

- explosive limits
- limits on the number of people present
- dress requirements
- safety precautions
- communication systems
- instructions and procedures to be followed in the event of an accident or emergency
- medical arrangements
- orders on prohibited articles
- authorised means of ignition
- records to be kept.

Duties Under the Construction (Design and Management) Regulations 2007

6.67 Under the Construction (Design and Management) Regulations 2007, Part 4 (paragraph 30) outlines the general requirements for the client, employer and employee, as follows.

1. So far as is reasonably practicable, explosives shall be stored, transported and used safely and securely.
2. Without prejudice to paragraph (1), an explosive charge shall be used or fired only if suitable and sufficient steps have been taken to ensure that no person is exposed to risk of injury from the explosion or from projected or flying material caused thereby.

6.68 The table below outlines the typical duties which may be required when using explosives on a construction site.

Duties Relating to the Use of Explosives

Duty Holder	Examples of Specific Duties Relating to the Use of Explosives
Clients	Ensure and check that the CDM co-ordinator and the principal contractor have suitable knowledge and experience in the use of explosives. Ensure that project funding and resource does not put undue pressure on the timeliness of planning and implementation of explosive operations. Hasty planning and pressure to meet an unrealistic deadline can dramatically increase the risk of a major incident. Provide pre-construction information to designers and contractors. This will require consideration, among other things, of the adjacent land usage, local ground conditions and information about existing structures.
CDM co-ordinators	Advise and assist the client with his/her legal duties. Ensure appropriate surveys of the local areas have been undertaken for demolition, earthworks, tunnelling, etc. Identify, collect and pass on this information to those affected. Ensure appropriate notifications are given to the HSE, local authorities, services, police, airports, etc.
Designers (assume explosive engineering)	Where possible, eliminate hazards and reduce risks during design, eg the design should be engineered (ie through the position and sequencing of blasts, use of test blasts, etc) to minimise fly of debris, rather than relying solely on blast protection. Provide any information needed for the health and safety file (blast modelling, residual risks, etc). Minimise the quantity of explosive (eg an efficient design) by techniques such as structural pre-weakening. Prepare an outline method statement in co-operation with all parties.

Duty Holder	Examples of Specific Duties Relating to the Use of Explosives
Principal contractors	Notify the HSE as well as other agencies, eg local authorities, services, police and airports, of planned explosive operations. Prepare, develop and implement a written plan and site rules. Ensure rules are effectively communicated by use of signs, toolbox talks, induction training, evacuation drills, etc. Ensure site rules are kept and that all workers on-site are properly informed. Assess the contractor's risk assessment and their method for implementing high-risk operations. Check the competence of all subcontractors through vetting of documentation, certification, method statements, accident history, etc. Secure the site (employees and subcontractors) with due regard to the Control of Explosives Regulations 1991. Ensure the storage of explosives adheres to the Manufacture and Storage of Explosives Regulations 2005 and the Control of Explosive Regulations 1991.
Contractors (assumed final users of explosives)	Ensure all work is conducted in accordance with the agreed method statement. Check competence and, if necessary, train all appointees and workers. Provide information about health and safety to workers. Co-operate with the principal contractor in planning and managing work, including reasonable directions and site rules. Discuss and agree with the principal contractor the risk assessment and detailed method statement. Inform of any problems or changes to any earlier plan. Inform the principal contractor of reportable accidents, diseases and dangerous occurrences.
Workers/ everyone	Check their own competence. Co-operate with others and co-ordinate work to ensure the health and safety of construction workers and others who may be affected by the work. Obey site rules and signage and do not endanger themselves or others.

Competence

6.69 The competence of a company, a contractor, a CDM co-ordinator or a designer involved with explosive tasks is assessed on a variety of criteria, including:

- policies, management and organisation for health and safety
- access to information, expert advice and training
- individual qualification and experience
- ability to monitor, audit, report and review
- workforce involvement/consultation
- subcontractor procedures
- risk control and development of safe systems of work
- co-operations.

6.70 As well as assessing evidence demonstrating the standard of the individual or company against the above criteria, work experience and track record should also be assessed.

Training

6.71 Annex D of the *Code of Practice for the Safe Use of Explosives in the Construction Industry* (BS 5607) lists the topics of knowledge which must be addressed during the training of a "qualified" explosives engineer. This training includes hazard assessment and safety management, legislation relating to the purchase, transport, storage and disposal of explosives, etc.

6.72 A record of all training given should be kept and maintained during employment and for three years afterwards.

6.73 Employers should ensure that suitable and sufficient information, instruction and training is provided to all employees on the storage, handling and use of explosives. Training should include:

- the names of explosives in the workplace
- associated hazards
- instruction on safety data sheets (and how to access them) and other relevant information
- current relevant legislation
- results of relevant risk assessments
- the control measures to be in place, including the use of PPE if necessary
- safe working practices
- procedures to be followed on the safe storage, handling and use of relevant explosives
- the use of permit-to-work systems if applicable
- personal protective clothing
- in-house rules, eg no smoking, no fooling around, the avoidance of activities that could cause sparks generated by friction, etc
- accident, emergency and evacuation procedures
- communications systems
- reporting procedures.

Driver Training

6.74 The Carriage of Dangerous Goods by Road (Driver Training) Regulations 1996 provides information on driver training requirements for the transport of explosives. Employers should ensure that drivers of vehicles carrying explosives are suitably and sufficiently trained.

List of Relevant Legislation

- Construction (Design and Management) Regulations 2007
- Manufacture and Storage of Explosives Regulations (Northern Ireland) 2006
- Manufacture and Storage of Explosives Regulations 2005
- Dangerous Substances and Explosive Atmospheres Regulations 2002
- Control of Major Accident Hazards Regulations 1999

- Quarries Explosives Regulations 1999
- Carriage of Dangerous Goods by Road Regulations 1996
- Carriage of Dangerous Goods by Road (Driver Training) Regulations 1996
- Transport of Dangerous Goods by Road Regulations 1996
- Reporting of Injuries, Diseases and Dangerous Occurrences Regulations 1995
- Control of Explosives Regulations 1991
- Classification and Labelling of Explosives Regulations 1983
- Control of Pollution Act 1974
- Health and safety at Work, etc Act 1974

Further Information

Publications

HSE Publications

The following are available from *www.hsebooks.co.uk*.

- CIS45 *Establishing Exclusion Zones when Using Explosives in Demolition*
- INDG335 *Is it Explosive? Dangers of Explosives in Metal Recycling*
- INDG370 *Fire and Explosion — How Safe is Your Workplace? A Short Guide to the Dangerous Substances and Explosive Atmospheres Regulations*
- L10 *A Guide to the Control of Explosives Regulations 1991: Guidance on Regulations*
- L138 *Dangerous Substances and Explosive Atmospheres: Approved Code of Practice and Guidance. Dangerous Substances and Explosive Atmospheres Regulations 2002*

HSE Explosives information sheets are available at *www.hse.gov.uk/explosives*.

British Standards Publications

The following are available from *www.bsi-global.com*.

- BS 5607: 1998 *Code of Practice for the Safe Use of Explosives in the Construction Industry*
- BS 6187: 2000 *Code of Practice for Demolition*
- BS 6657: 2002 *Assessment of Inadvertent Initiation of Bridge Wire Electro-explosive Devices by Radio-frequency Radiation*

Other Publications

- *Guidance for the Safe Management of the Disposal of Explosives*, Confederation of British Industry

Fire Emergency Procedures and Means of Escape

- To safeguard the safety of employees in the event of fire, the Regulatory Reform (Fire Safety) Order 2005 requires the establishment of procedures for serious and imminent danger and the provision and maintenance of suitable means of escape.
- Under the provisions of the Regulatory Reform (Fire Safety) Order 2005, all organisations must appoint a "responsible person" who must ensure that the organisation complies with all aspects of the Regulatory Reform (Fire Safety) Order 2005.
- The organisation's arrangements for fire safety and the provision of means of escape are determined by risk assessment, as required under the Regulatory Reform (Fire Safety) Order 2005.
- The Building Regulations 2000 cover the means of escape and recommend that the building should be designed and constructed so there are appropriate provisions for the early warning of fire and appropriate means of escape in case of fire.
- Employers must ensure staff are aware of the fire evacuation procedures, including location of fire exits and assembly points.
- Fire exits and emergency routes must be adequately provided with signs and, where necessary, adequate emergency lighting.
- It is the duty of the employer to ensure that there are sufficient numbers of escape routes to handle all the employees and any contractors or visitors who may be on the site.
- Each employee should be aware of the company's policies on action to take in the event of fire and the recommended evacuation policies.
- Assembly points outside the building should be indicated clearly. These points should have been selected in consultation with the fire authority and routes to them signposted with appropriate notices.

6.75 There are two main reasons for evacuating premises, ie in response to a fire alarm and in response to a bomb threat. There are some less common reasons, eg storm damage, flooding or gas leaks (or fire drills), but the main reasons are a real fire or a threatened bomb. There should not be a single set of evacuation procedures to cover both types of evacuations, because there are different risks. However, there could be a single emergency plan which encompasses both sets of evacuation procedures.

6.76 One of the fundamental requirements of fire safety is to ensure that persons within a building can evacuate safely in the event of an emergency. To ensure this, each workplace should have emergency procedures which all employees must be made aware of. These procedures should include:

- the action to be taken by staff in the event of a fire
- the evacuation procedure
- arrangements for calling the fire brigade.

6.77 For small workplaces this could take the form of a simple fire action notice posted where staff can readily read it and become familiar with it. For larger workplaces or places with special fire risks (eg hotels, hospitals), a more detailed plan will be required which will take account of the risk assessment previously carried out.

6.78 The ability of the occupants of a building to evacuate in the case of fire is a fundamental aspect of fire safety. In the case of a fire, or indeed any other emergency, people should be able to turn away from the hazard and escape to the open air or other place of safety. To enable this, certain provisions need to be made and maintained during the time that the building is occupied.

Employers' Duties

6.79 The following duties are imposed on employers.

- Under the Health and Safety at Work, etc Act 1974, the Management of Health and Safety at Work Regulations 1999 and the Regulatory Reform (Fire Safety) Order 2005, employers have a general duty of care to provide a safe working environment in relation to fire safety.
- Under the Regulatory Reform (Fire Safety) Order 2005, employers must:
  - carry out a fire risk assessment, identifying possible dangers and risks
  - consider who may be especially at risk
  - get rid of or reduce the risk from fire as far as is reasonably possible and provide general fire precautions (including fire detection and fire alarms, if appropriate) to deal with any possible risk left
  - take other measures to make sure there is protection if flammable or explosive materials are used or stored
  - create a plan to deal with any emergency and, in most cases, keep a record of the findings
  - review the findings when necessary.

6.80 Employers must also provide:

- exit routes and exits that lead as directly as possible to a place of safety
- satisfactory evacuation procedures
- satisfactory means of escape
- emergency exit doors that open in the direction of escape (sliding or revolving doors are *not* acceptable for fire exits)
- emergency doors which are easily and readily openable without the use of a key in the event of an emergency
- emergency routes and exits provided with signs and where necessary adequate emergency lighting, to provide for the failure of normal lighting.

* The Building Regulations 2000 cover the means of escape and require that:
  - buildings should be designed and constructed so there are appropriate provisions for the early warning of fire and appropriate means of escape in case of fire
  - there be a means of escape from the building to a place of safety outside the building. This designated place should be capable of being safely and effectively used at all material times.
* Under the Management of Health and Safety at Work Regulations 1999 employers must:
  - prepare and implement procedures to be followed in the event of serious and imminent danger to persons at work in their undertaking
  - ensure that any necessary contacts with external emergency services are arranged, particularly in regard to first aid, emergency medical care and rescue work
  - provide employees with information on emergency procedures and the means of escape.

Employees' Duties

6.81 The following duties are imposed on employees.

- Under the Health and Safety at Work, etc Act 1974, employees have a duty to take reasonable care of their own health and safety and that of other people who may be affected by their activities at work.
- Employees also have a duty to co-operate with their employer to enable the employer to comply with health and safety duties.
- Under the Management of Health and Safety at Work Regulations 1999, employees are required to:
  - use any equipment or safety devices in accordance with any training and instructions provided by the employer
  - inform the employer of any shortcomings in the employer's health and safety arrangements.

Duties of the Responsible Person

6.82 The responsible person must ensure that general fire precautions are taken to ensure, so far as is reasonably practicable, the safety of employees. In relation to relevant persons who are not employees, the responsible person must take such general fire precautions "as may reasonably be required in the circumstances of the case" to ensure that the premises are safe.

6.83 The responsible person also has duties to:

- carry out a suitable and sufficient (fire safety) assessment of the risks to which persons are exposed
- ensure that appropriate arrangements for the effective planning, organisation, control, monitoring and review of the preventive and protective measures are in place
- ensure that where a dangerous substance is present in or on the premises, risks from that dangerous substance are either eliminated or reduced
- ensure that premises are equipped with appropriate fire-fighting equipment and with fire detectors and alarms (any non-automatic fire-fighting equipment provided must be easily accessible, simple to use and indicated by signs)
- ensure that routes to emergency exits from premises, and the exits themselves, are kept clear at all times and emergency routes and exits lead as directly as possible to a place of safety
- ensure that procedures to follow in the event of serious and imminent danger are established
- ensure that any facilities, equipment or devices provided under the Regulatory Reform (Fire Safety) Order 2005 are maintained in an efficient state, in working order and good repair
- appoint one or more competent persons to assist in undertaking the preventive and protective measures
- ensure that employees are provided with comprehensible and relevant information
- ensure that the employers of any other employees working on the premises are provided with comprehensible and relevant information on the risks
- ensure that employees are provided with adequate fire safety training at the time when they are first employed and on their being exposed to new or increased risks

- co-operate with any other responsible person in cases where two or more responsible persons share or have duties in respect of the premises
- ensure that every employee is aware that while at work they must take reasonable care for the safety of themselves and of other relevant persons who may be affected by their acts or omissions at work.

In Practice

6.84 The danger which may threaten people if fire breaks out depends on many different factors. Consequently it is not possible to construct a model procedure for action in the event of fire suitable for all premises. However, the following plan of action shows the points that should normally be covered. This plan can be adapted accordingly to suit different premises or risks. Typically a small office, shop or other premises may require only a very simple plan, while a larger premises may require a more in-depth plan.

(1) **Raising the alarm**. All employees should be familiar with the procedures for operating the alarm. The alarm may be raised automatically via a fire detector system or manually by operation of a manual call point or by other means. The emergency plan should cover how the alarm is raised and the subsequent actions. It should be ensured that delays in response are minimised.

(2) **Calling the fire service**. The duty of informing the fire service immediately an alarm is sounded must be specifically allocated to a designated person or persons. Facilities must be made available at all times when people are on the premises.

(3) **Stopping of machinery, isolation of power supplies**. These tasks should be carried out by previously designated people to ensure the safety of all those concerned.

(4) **Evacuating premises**. Everyone must be able to escape from danger. Personnel who do not have specific designated duties should start to leave the building as soon as the alarm sounds, unless instructions have been given to the contrary (eg as may be the case if phased evacuation is employed). Everyone should leave in a calm, orderly manner, by the most direct route avoiding the use of lifts. Their egress should not be delayed by stopping to collect belongings. Depending on the circumstances of the building, fire marshals may have been nominated to ensure each area is evacuated.

(5) **Assembly points**. An assembly point should be pre-determined and everyone made aware of its location. These points should be in a safe place (an enclosed courtyard forming part of the premises is not classed as a safe place), preferably under cover.

(6) **Roll call**. One person in each department of the building should have the duty of maintaining a roll so that a quick check can be made. The fire service should be informed on arrival if anyone is not accounted for.
Note: In circumstances where it is not possible to clarify the whereabouts of staff during the working day (eg if staff are constantly exiting and returning to the premises), a roll-call system may be ineffective.

(7) **Fire-fighting**. Sufficient numbers of persons throughout the building(s) should be trained in the use of fire extinguishers. For small premises, this may mean that all employees are trained. Where possible, a member of staff designated for fire-fighting purposes should attack the fire with a suitable extinguisher. It must be stressed that fire-fighting should only be carried out secondary to personal safety. If there is any doubt about the ability to extinguish the fire safely, it should not be tackled but left for the fire service to deal with.

(8) **Responsibilities in the event of fire**. In workplaces employing large numbers of employees, it may be appropriate to nominate certain employees to carry out specific tasks in the event of fire. For example, these tasks might include:
- acting as floor marshals, ensuring that the floor is completely evacuated during a fire evacuation and reporting this fact to a previously established control point
- ensuring that security of the building is maintained
- ensuring that disabled people receive any assistance required.

(9) **Information and training**. For an efficient fire routine it is essential that every person has received adequate instructions and fully understands them. Instruction must leave no room for doubt as to the action to be taken. It should be as brief as practicable and expressed clearly in simple language.

Phased Evacuations

6.85 It is possible to plan for a phased evacuation from a building in the event of fire, to avoid a large number of panic-stricken people blocking the key escape routes. A phased plan means training people to take actions that are not instinctive. Phased evacuations cannot be achieved via the use of siren-only alarms and there are always occasions when the arranged plans have to be disregarded, eg a major fire following an explosion.

Fire Safety Signs and Emergency Lighting

6.86 Under the Regulatory Reform (Fire Safety) Order 2005, emergency routes and exits must be provided with signs and where necessary adequate emergency lighting. The following list indicates some measures that should form part of the emergency plan in respect of signs, lighting and other mechanisms that assist in fire evacuations.

- The employer should designate a person to supervise the efficiency of the evacuation, (eg the effectiveness of the control measures and who is responsible for building management actions, such as stopping the lifts or any escalators).
- Emergency lighting should be provided in line with the requirements of BS 5266: Part 1: 1999, *Code of Practice for Emergency Lighting*.
- Safety signs should be clear and unambiguous, in accordance with the Health and Safety (Safety Signs and Signals) Regulations 1996 and BS 5499–4:2000 *Safety Signs, Including Fire Safety Signs. Code of Practice for Escape Route Signing*. The signs should be suitable for use by those who:
  - have poor vision
  - suffer from dyslexia
  - do not have English as their first language.
- It is recommended that the fire safety directions are in picture form for ease of comprehension.
- Both emergency lighting and fire safety signs require the continuous supply of uninterrupted power, preferably via fire-resistant cables.
- Smoke control requires effective ventilation measures, as per BS 5588–9:1999 *Fire Precautions in the Design, Construction and Use of Buildings. Code of Practice for Ventilation and Air Conditioning Ductwork*.

Soot Particles Covering Signs

6.87 It is common for fire safety signs and emergency lights to be clouded by smoke or soot particles during a fire. Some companies are experimenting with the use of powered way lighting in smoke-filled areas, as an alternative to traditional emergency lighting. For example, stair rails can be lit by battery-powered electroluminescent systems to guide people towards exits or as directional route signs. Such luminous systems fall under the remit of BS 5266–2:1998 *Code of Practice for Electrical Low-mounted Way Guidance Systems for Emergency Use*.

Escape Routes and Assembly Points

6.88 A means of escape can be defined as: the structural means whereby a safe route is provided for people to travel from any location in a building or structure, to a place of safety without the need of outside assistance.

6.89 Ideally, persons should be able to turn their backs on a fire and walk away from it, rather than towards it, to reach safety. Escape in one direction, such as would occur in a dead-end corridor, is only acceptable in specific circumstances. Thus, except in certain circumstances, there should be alternative escape routes, defined as: "escape routes sufficiently separated by either direction and space, or by fire resisting construction, to ensure that one is still available should the other be affected by fire" (BS 4422).

6.90 This means that people should be able to make their own way out of the building and not have to be rescued by the fire brigade, albeit in some cases special provision will still need to be made for people with special needs.

6.91 Part B1 of the Building Regulations 2000, together with the accompanying Approved Document B, *Fire Safety* (2000 edition), covers means of warning and escape. It advises that the building should be designed and constructed so there are:

- appropriate provisions for the early warning of fire
- appropriate means of escape in case of fire.

6.92 The means of escape should be from the building to a place of safety outside the building. This designated place should be available to be used safely and effectively at all times. Escape routes should be planned in consultation with the local fire authority and with reference to the workplace fire risk assessment.

Escape Route and Approved Document B

6.93 The important factors highlighted by Approved Document B include the following.

- Each escape route should be protected and enclosed by fire and smoke-resistant materials.
- The route should be lit by suitable emergency lighting.
- The entrances and exits should have suitable signs.
- There must be suitable measures in place to restrict the spread of smoke in the escape route.
- There should be ventilators to remove smoke from the stairwells.
- No escape route should run close to a hazardous area, eg a chemicals store.
- Any changes in the location or use of escape routes must be notified to the fire authority.
- The fire authority must be notified of any new or proposed escape routes before they are put into effect.

6.94 Ideally there will always be at least two separate escape routes from each room, compartment or storey in a building — if possible these should be diagonally opposite to each other. A simple calculation to determine whether escape routes could be considered alternatives to each other is by applying the "45 degree" rule. Stand at any point in a room and draw a line from that position to each of the routes from which escape is possible. If the angle within the lines drawn is 45 degrees or more, then they can be considered alternative routes. Less than 45 degrees indicates a reasonable risk that the same fire or emergency could affect both exits.

Final Exits

6.95 Ultimately, all escape routes lead to a final exit from the premises. Common requirements in respect of final exits are as follows.

- The exits should be obvious and/or signposted.
- The exits must open easily without use of a key; panic bars should be fitted to locked exits that are likely to be used by a significant number (eg 50 or more) of staff or members of the public.
- Revolving doors are normally required to have conventional exit doors sited adjacent to them, unless they fold flat.
- In modern codes of practice, wicket doors, goods delivery shutters and window exits are not normally acceptable as final exits and are generally regarded as unsuitable for members of the public under any circumstances.
- On escape through a final exit, it must be possible to disperse from the vicinity of the building without re-entering it.
- There is an obligation on the employer or premises owner/occupier to keep the means of escape free at all times, eg not allowing employees to prop fire doors open.

Assembly Points

6.96 Fire escape routes should lead to a well-designated assembly point. As there are several important details to note concerning assembly points, the main advice includes the following.

- Assembly points outside the building should be indicated clearly. These points should have been selected in consultation with the fire authority and routes to them will be signposted with appropriate notices.
- For larger sites, a well-disciplined procedure should be in place to handle hundreds of people, both employees and visitors, moving from various exits to a single assembly point.
- The designation of the assembly points is important, because disabled people should not be disadvantaged by being expected to assemble at points that are too far from the evacuated premises. On the other hand, disabled people need to be far away enough from the premises so they are not put at risk.
- It is a good idea to have designated assembly areas in sheltered facilities, in case of a forced evacuation in poor weather.
- Employees and other persons leaving the premises must be advised which assembly areas they are to use or not use. In this respect, a voice evacuation or public address system is invaluable.

6.97 The relevant standard is BS 5588–6. 1997 *Fire Precautions in the Design, Construction and Use of Buildings. Code of Practice for Places of Assembly.*

Disabled Persons and Fire Evacuation

6.98 Emergency procedures, such as in the event of a fire or security alert, should also consider the safety of those who have physical or sensory impairment. It is worth remembering that employers without disabled employees may occasionally have staff with temporary incapacities (eg a worker with a broken leg, who uses crutches), and may receive disabled visitors from time to time.

6.99 For those with a disability, consideration must be given to a number of factors including:

- the disability of the person or persons likely to be at risk
- unfamiliarity with the premises and/or the evacuation procedures
- position of the person in the building (does evacuation take longer?)
- inability to recognise alarms/evacuate the building without assistance
- characteristics of the building that may affect evacuation.

6.100 Emergency evacuation of a disabled person can usually be achieved by devising simple procedures, but specialist equipment may be needed in some cases. For example, it may be appropriate to install flashing lights that are linked to the alarm system, if people who have severe hearing impediments are employed. Risk assessment and practice drills will help to identify if any special equipment is needed and whether the emergency procedures are effective.

6.101 Evacuation lifts are specially designed for use by disabled persons and will reduce dependence upon physical help and will ensure the rapid evacuation of the person in question. The design of such lifts is described in BS 5588 Part 8. Fire-fighting lifts may also be used as a means of evacuation. The purchase and fitting of evacuation lifts may be cost prohibitive and not practicable in older premises.

6.102 Consideration should be given to the use of other methods of ensuring the safety of the disabled person. This should include the provision of refuges, particularly for those in a wheelchair where they can wait in relative safety until the evacuation lift is available:

- to await assistance
- to wait until escape stairs can be used safely.

6.103 Refuges should be provided for each protected stairway at each storey level and should have at least 30 minutes fire resistance. In larger premises refuges may be fitted with two-way communication systems. All refuges should be clearly identified and kept free from obstruction.

6.104 There is a common misunderstanding that the refuge is a place to leave disabled people to wait for the fire service for their escape. BS 5588 Part 8 actually states that "refuges are relatively safe waiting areas for short periods. They are not areas where disabled people should be left indefinitely until rescued by the fire brigade, or until the fire is extinguished. (This should not be confused with the use of refuges in progressive horizontal evacuation, eg in hospitals or care homes)".

6.105 For those with mobility, a simple "buddy system" may suffice. This system has a nominated person(s) who will assist the disabled person in the event of an evacuation. This should preferably be a close work colleague.

Personal Emergency Evacuation Plans for Disabled Employees and Regular Visitors

6.106 In cases where there are disabled employees or regular visitors to the premises, they require an evacuation plan. The evacuation plan must be tailored to their individual needs and is likely to give detailed information on their movements during an escape. It is also possible that there will be some building adaptation to facilitate their escape and to reduce the need for personal assistance.

Standard Plans for Disabled Occasional Visitors

6.107 A standard plan should be devised and used where there are visitors or casual users of the building who may be present infrequently or on only one occasion. This plan should consider the:

- disabled person's movements within the building
- operational procedures within the building
- types of escape that can be made available from different parts of the premises
- building systems, such as the fire alarm system
- existing emergency plan for the premises.

Training

6.108

- The Regulatory Reform (Fire Safety) Order 2005 requires that those employed by the business on the premises are trained in the action to be taken in the event of fire and that appropriate records should be kept.
- It is recommended that fire drills, eg drills that involve evacuation of the business, should occur on a regular basis, at least twice a year.
- In many companies there are fire wardens who are trained to use fire extinguishers correctly (at least until all the people in that zone have been evacuated).

Notifying Staff, Visitors and Contractors of Emergency Procedures

6.109 The following information should be displayed prominently for the benefit of staff, visitors and contractors.

- Different types of alarms and their meanings.
- Required responses to the alarms.
- Location and direction for escape routes.
- Location of alarm points.
- Location of fire-fighting appliances and their use.
- Restrictions on the use of lifts.
- Location of assembly points.
- A reminder to leave behind all belongings if the fire alarm should sound.
- A reminder to take all personal belongings, if the bomb threat alarm should sound.

6.110 Staff members should be given:

- basic training on the location and use of escape routes
- basic training on evacuation procedures
- a copy of the written fire policy of the business.

List of Relevant Legislation

- Regulatory Reform (Fire Safety) Order 2005
- Building Regulations 2000
- Management of Health and Safety at Work Regulations 1999
- Health and Safety (Safety Signs and Signals) Regulations 1996
- Health and Safety at Work, etc Act 1974

Further Information

Publications

Department for Communities and Local Government Publications

The following are available from *www.communities.gov.uk*.

- Approved Document B *Fire Safety: Volume 1: Dwellinghouses*, Department for Communities and Local Government
- Approved Document B *Fire Safety: Volume 2: Buildings Other Than Dwellinghouses*, Department for Communities and Local Government
- *A Short Guide to Making Your Premises Safe From Fire*

British Standards Publications

The following are available from *www.bsi-global.com*.

- BS 5266–1: 2005 *Emergency Lighting. Code of Practice for the Emergency Lighting of Premises Other than Cinemas and Certain Other Specified Premises Used for Entertainment*
- BS 5266–2: 1998 *Emergency Lighting. Code of Practice for Electrical Low-mounted Way Guidance Systems for Emergency Use*
- BS 5499–4: 2000 *Safety Signs, Including Fire Safety Signs. Code of Practice for Escape Route Signing*
- BS 5588–0: 1996 *Fire Precautions in the Design, Construction and Use of Buildings. Guide to Fire Safety Codes of Practice for Particular Premises/Applications*
- BS 5588–1: 1990 *Fire Precautions in the Design, Construction and Use of Buildings. Code of Practice for Residential Buildings*
- BS 5588–6: 1991 *Fire Precautions in the Design, Construction and Use of Buildings. Code of Practice for Assembly*
- BS 5588–8: 1999 *Fire Precautions in the Design, Construction and Use of Buildings. Code of Practice for Means of Escape for Disabled People*
- BS 5588–9: 1999 *Fire Precautions in the Design, Construction and Use of Buildings. Code of Practice for Ventilation and Air Conditioning Ductwork*
- BS 5588–10: 1991 *Fire Precautions in the Design, Construction and Use of Buildings. Code of Practice for Shopping Complexes*
- BS 5588–11: 1997 *Fire Precautions in the Design, Construction and Use of Buildings. Code of Practice for Shops, Offices, Industrial, Storage and Other Similar Buildings*
- BS 5588–12: 2004 *Fire Precautions in the Design, Construction and Use of Buildings. Managing Fire Safety*

Organisations

- Emergency Planning Society
  Web: *www.the-eps.org*
  The Emergency Planning Society is a member-led organisation for professionals dealing with emergency planning and crisis and disaster management.

- Fire Brigades Union (FBU)
  Web: *www.fbu.org.uk*
  The FBU is a union for the main trades and industries local authority fire brigades.
- Fire Protection Association (FPA)
  Web: *www.thefpa.co.uk*
  The FPA is the UK's national fire safety organisation. It provides a range of fire safety audit and fire risk assessment services.
- Fire Service College
  Web: *www.fireservicelege.ac.uk*
  The Fire Service College provides facilities for both practical and theoretical fire-fighting, fire safety and accident emergency training.

Fire Risk Assessment

- The main legislation in the UK requiring employers to undertake fire risk assessments is the Regulatory Reform (Fire Safety) Order 2005.
- The process of carrying out a fire risk assessment is made up of several stages. These are as follows.
  - Establish the objectives of the assessment.
  - Look for the hazards.
  - Decide who might be harmed.
  - Evaluate the risks.
  - Record the findings.
  - Review the assessment.
- Employers are required to conduct fire risk assessments at work, in order to determine the appropriate level of fire safety measures necessary to control fire risks.
- Employers should use competent employees for advice and assistance on fire safety issues in preference to external sources.
- If the person responsible for the fire risk assessment is not the employer, but the *landlord*, it is still the employer who is responsible for the workplace risk assessment of their (rented) part of the building. However, it is the landlord's duties to carry out risk assessments for communal parts of the building.
- Basic fire training should be provided for all staff.
- The fire risk assessment will be more effective if conducted as a separate exercise to health and safety risk assessments.

6.111 Employers are required to conduct fire risk assessments at work, in order to determine the appropriate level of fire safety measures. Such risk assessment is a legal obligation under the Regulatory Reform (Fire Safety) Order 2005.

Employers' Duties

6.112 Under the Management of Health and Safety at Work Regulations 1999, employers must:

- undertake suitable and sufficient assessments of the risks to employees and non-employees
- make appropriate arrangements for the planning, organisation, control, monitoring and review of preventive and protective measures necessary for the safety of employees in the event of fire while they are at work
- appoint one or more competent persons to assist in undertaking required measures (a competent person is defined in the regulations as "one who has sufficient training and experience or knowledge and other qualities to enable him properly to assist in undertaking the appropriate measures")
- provide understandable and relevant information to employees on:
  - measures identified by the risk assessment as necessary for the health and safety of employees in the event of fire
  - the imminent danger procedures
  - the fire-fighting measures required under the Regulatory Reform (Fire Safety) Order 2005
  - the identity of those persons nominated to implement these measures
  - any risks notified by any other employer sharing the workplace

- arrange any necessary contacts with external services, ie regarding first aid, emergency medical care and rescue work
- co-operate and co-ordinate measures with other employers sharing the same workplace
- ensure that employers of any employees from outside undertakings are provided with comprehensible information on risks to the health and safety of their employees and the measures taken to comply with the regulations
- ensure employees are provided with adequate safety training which:
  - is repeated periodically
  - is adapted to take account of changed risks
  - takes place during working hours.

6.113 Under the Regulatory Reform (Fire Safety) Order 2005, employers must:

- carry out a fire risk assessment identifying possible dangers and risks
- consider who may be especially at risk
- get rid of or reduce the risk from fire as far as is reasonably possible and provide general fire precautions (including fire detection and fire alarms, if appropriate) to deal with any possible risk left
- take other measures to make sure there is protection if flammable or explosive materials are used or stored
- create a plan to deal with any emergency and, in most cases, keep a record of the findings
- review the findings when necessary.

Landlord Versus Employer Duties

6.114 If the person responsible for the premises' fire risk assessment is not the employer, but the *landlord*, there are several differences to note.

6.115 Most commonly, it is still the employer responsible for the workplace risk assessment of their (rented) part of the building. However, it is the landlord's duty to carry out risk assessments for the communal parts of the building, eg reception areas, stairwells, delivery points and so on. It is the landlord's duty to ensure that means of escape are suitable, that the fire detection and suppression systems are maintained and that adequate fire compartmentation has occurred, eg between tenants.

Employees' Duties

6.116 Under the Health and Safety at Work, etc Act 1974 employees have a general duty to:

- take reasonable care of their own health and safety and that of other people who may be affected by their work
- co-operate with the employer as much as is necessary to ensure that any duties imposed on the employer are performed. This includes:
  - each employee being aware of and co-operating with the company's policies on action to take in the event of fire and the recommended evacuation policies
  - employees proceeding to the external assembly points in an orderly fashion, as directed by the fire wardens
  - employees being prepared to offer assistance to a fire warden, if required, eg helping the evacuation of a disabled worker.

6.117 Under the Management of Health and Safety at Work Regulations 1999, employees have a duty to inform their manager or fire warden of any new or unidentified hazards.

Duties of the Responsible Person

6.118 The responsible person must ensure that general fire precautions are taken to ensure, so far as is reasonably practicable, the safety of employees. In relation to relevant persons who are not employees, the responsible person must take such general fire precautions "as may reasonably be required in the circumstances of the case" to ensure that the premises are safe.

6.119 The responsible person also has duties to:

- carry out a suitable and sufficient (fire safety) assessment of the risks to which persons are exposed
- ensure that appropriate arrangements for the effective planning, organisation, control, monitoring and review of the preventive and protective measures are in place
- ensure that where a dangerous substance is present in or on the premises, risks from that dangerous substance are either eliminated or reduced
- ensure that premises are equipped with appropriate fire-fighting equipment and with fire detectors and alarms (any non-automatic fire-fighting equipment provided must be easily accessible, simple to use and indicated by signs)
- ensure that routes to emergency exits from premises, and the exits themselves, are kept clear at all times and emergency routes and exits lead as directly as possible to a place of safety
- ensure that procedures to follow in the event of serious and imminent danger are established
- ensure that any facilities, equipment or devices provided under the Regulatory Reform (Fire Safety) Order 2005 are maintained in an efficient state, in working order and good repair
- appoint one or more competent persons to assist in undertaking the preventive and protective measures
- ensure that employees are provided with comprehensible and relevant information
- ensure that the employers of any other employees working on the premises are provided with comprehensible and relevant information on the risks
- ensure that employees are provided with adequate fire safety training at the time when they are first employed and on their being exposed to new or increased risks
- co-operate with any other responsible person in cases where two or more responsible persons share or have duties in respect of the premises
- ensure that every employee is aware that while at work they must take reasonable care for the safety of themselves and of other relevant persons who may be affected by their acts or omissions at work.

In Practice

6.120 When looking to undertake a risk assessment, it is essential that the process established is systematic and thorough and follows a logical pattern through a number of stages.

6.121 *Fire Safety: An Employer's Guide* and *Regulatory Reform (Fire Safety) Order 2005: A Short Guide to Making Your Premises Comply* (and any of the Department for Communities and Local Government fire safety publications) details how a fire risk assessment may be performed. Broadly, with the same approach as that used in general health and safety legislation, it can be carried out either as part of a more general risk assessment or as a separate exercise.

6.122 The suggested five steps are as follows.

- Step 1: identify fire hazards.
- Step 2: identify the location of persons at significant risk.
- Step 3: evaluate the risks.
- Step 4: record the findings and action plan.
- Step 5: review and revise the risk assessment.

6.123 It is also useful before starting the risk assessment to establish and define the objectives of the fire risk assessment.

Establishing Objectives

6.124 The overriding purpose of the fire risk assessment will usually be to:

- comply with statutory requirements
- meet insurance requirements
- meet business continuity and financial requirements
- meet environmental protection requirements.

6.125 Before starting the fire risk assessment, it is wise to clearly establish the objectives. They should be written down so a check can be made later to ensure all of them have been achieved. The objectives will refer to the specific items and areas that require assessment, and to the actions and outcomes necessary to complete the task satisfactorily.

6.126 The fire risk assessment objectives will need to include the following.

- Identifying fire hazards and persons that may be at risk from such hazards.
- Examining the provision of emergency escape facilities and routes.
- Examining the provision of fire-fighting equipment and where installed, detection and alarm systems.
- Assessing the results of the stages by:
  - evaluating the risks
  - identifying those requiring action to eliminate or reduce them
  - putting into action an appropriate plan to eliminate or reduce risks
  - correcting deficiencies in escape facilities and providing any necessary fire-fighting equipment.

6.127 These are the minimum objectives for an assessment designed to satisfy the requirements of the Regulatory Reform (Fire Safety) Order 2005. However, the objectives could easily be extended to cover:

- protection of property
- providing for business continuity
- life safety of persons other than employees
- limiting the environmental impact of fire and fire protection measures.

Identifying Hazards

6.128 To cause a fire or for it to spread, three elements must be brought together: fuel (something to burn), air (to support combustion) and source of ignition (heat).

6.129 In most workplaces, air is always present, so employers must, therefore, identify sources of fuel and ignition and prevent them coming together. In most cases, the process of identifying fire hazards should be simple and straightforward. It will essentially involve a careful examination of the entire workplace, including cupboards, other similar enclosed areas and void spaces, to identify:

- what could cause a fire
- what would enable it to develop and spread.

6.130 A fire hazard is something that has the potential to cause a fire. A hazard could be either an explosive or flammable material or a situation in which it is possible that a fire could occur, eg an industrial process which:

- produces flammable dusts
- involves high temperatures
- involves a naked flame.

6.131 A list of fire hazards is essential for accurate assessment of risks. Fire hazards can be present in equipment, substances or working practices which:

- act as an ignition source
- are combustible
- provide a source of heat.

6.132 The main aim is not only to evaluate how and where a fire could occur, but also to consider how quickly a fire could develop into a serious incident.

Recording Hazards

6.133 To enable accurate assessment of risks, a clear identification of hazards is vital. When listing identified hazards, it will be useful to describe the nature of the harm that may be caused by them. This will help to focus on the issues that are of importance and to identify clearly the aspect of the hazard which needs to be assessed.

6.134 All hazards identified should be recorded, even if the assessor feels they are very trivial or irrelevant. If a hazard is present, it should be identified and listed. Decisions on the relevance of any particular hazard come later in the process.

Toxic Smoke and Gases

6.135 One area that should be included in the fire risk assessment is the production of toxic smoke or gases in a fire situation. Depending upon the nature of fuel in the workplace, a wide variety of toxic gases could arise in a fire. Carbon monoxide and soot particles are almost certainties. Some fires could cause such poisonous gases as:

- nitrogen dioxide
- sulphur dioxide
- sulphur trioxide.

6.136 The assessor should consider the potential for toxic gas production and determine how these gases would spread within the premises. This risk could be analysed via computer modelling and comparing the premises' ventilation with recommended standards, eg BS 5588: *Fire Precautions in the Design, Construction and Use of Buildings*: Part 4: 1998, *Code of Practice for Smoke Control Using Pressure Differentials*, and Part 9: 1999, *Code of Practice for Ventilation and Air Conditioning Ductwork*.

Identifying Persons at Significant Risk

6.137 Once hazards have been identified, it is important to establish who is at risk. Also, as required under the Management of Health and Safety at Work Regulations 1999, it is important to work out who is especially at risk.

6.138 In regard to fire risks, as well as employees, persons at risk will include non-employees such as:

- employees of other employers, eg contractors, persons sharing premises, etc
- invited and uninvited visitors, eg customers, sales people, delivery drivers, trespassers, etc
- members of the public, eg in shops, etc
- others, eg passers-by, neighbours, drivers on the road, etc.

6.139 It will also be necessary to look at special reasons for such persons being at risk from a fire on the premises. There may be reasons which warrant special provisions to manage the risks posed. Such reasons will include:

- number of people — large groups can pose evacuation problems
- distribution — uneven spread between floors in a multi-storey building or individuals or small groups of people working in isolated parts of a building
- familiarity with the building — people in an unfamiliar environment, eg concert halls
- state of wakefulness — sleeping accommodation, such as hotels and hostels, pose special risks as sleepers are particularly vulnerable to fire and smoke as their ability to react is inhibited and they can be overcome by smoke before they are even aware of a fire
- responsiveness — this may be slow for those who are:
  - sleeping
  - unfamiliar with fire-warning systems
  - involved in tasks requiring deep concentration
- mobility — persons with disabilities of all kinds, also including:
  - age, both young and old
  - state of health, eg heart disease, asthma
  - special circumstances, eg pregnant women.

Note: Neither of the lists is exhaustive and care should be taken to identify any other groups or circumstances where persons could be at risk from fire.

Evaluating the Risks

6.140 The next stage is to analyse and evaluate the fire risks. A fire risk is:

- the likelihood a fire will occur as a result of a fire hazard
- the extent and severity of the damage (harm potential) which may be caused.

6.141 It will be necessary to evaluate the likelihood of a fire occurring and the harm potential of each hazard identified. From this, it will be possible to determine what, if any, measures are needed for the control of the hazard and the ensuing risk.

6.142 The main factors affecting severity are the potential for development of a fire and the number of people who could be harmed. Another factor affecting both likelihood and severity is the effectiveness of the existing control measures in place.

6.143 If the risk is found to be negligible or acceptable, no further action is required, although a simple written statement may be necessary for compliance with legislation. If the risk is found to be unacceptable, control measures will be required.

Likelihood

6.144 The main factor affecting likelihood is the potential for ignition. Ignition can be caused by a number of sources including:

- smoking and smoking materials
- cooking appliances
- area heating appliances
- faulty electrical equipment or incorrect installation of electrical equipment, eg if allowed to deteriorate or incorrectly installed
- work processes, particularly those involving the use and/or handling of flammable or combustible materials and those where heat is a feature of the process, such as hot work.

6.145 If the factors are found to be present, they should be listed as a hazard and a description provided as to why they are considered to be a hazard.

6.146 The assessor will consider various factors for each workplace, including:

- type of building materials used in each workplace
- type and location of machinery in the workplace
- machinery use, ie nature of the work
- usual air quality and temperatures
- traffic flows, ie movements of the workers
- worst case scenarios, eg what happens if fire doors are left open or if the detection system fails?

6.147 Other factors should be considered in conjunction with the potential for ignition, including the:

- number of times a situation could occur
- position of the hazard
- environmental conditions
- competence of people involved
- condition of equipment
- quantities of materials involved.

6.148 Assessors use these facts and their knowledge to show the:

- places where a fire is most likely to start
- best places for fire control measures
- types of fire control measures.

Harm Potential

6.149 To work out the harm potential of a hazard, the assessor will need to make a judgment of the possible outcomes of the hazard. The harm potential of a fire hazard will depend on the potential:

- for development of a fire originating from the hazard
- consequences in terms of life risk and property loss.

6.150 The potential for development of a fire will be affected by issues such as:

- building construction, combustible materials, lack of compartmentation, etc
- contents, combustible and flammable materials
- delay in detection and warning of the fire
- existing control measures.

6.151 When looking at life risk, the persons likely to be at risk will have been listed before the risk evaluation stage. However, this stage of working out the potential for development of a fire may well identify others who might be at risk. It might also identify others who are at a greater risk than previously thought.

6.152 In terms of life risk, the assessor will need to assess the risks to the safety of employees in the event of fire. To do this, the existing escape facilities must be considered. Points to look out for include:

- adequacy and location of signs
- the condition of escape facilities and routes, eg distance of travel, width, height above ground, stairways, etc
- adequacy of illumination, including from emergency systems
- fire doors, eg can they be opened immediately and in the direction of travel?
- presence of obstruction in escape routes and exits, including trip hazards
- existence of adequate procedures and signals.

6.153 Consideration also needs to be given to those with disabilities, including those with impaired mobility, vision or hearing and those with learning difficulties.

6.154 Key factors affecting the severity of harm potential which must be considered carefully include:

- the number of people who might be affected in one incident
- the presence of individuals especially at risk because of disabilities or medical conditions.

6.155 Once the likelihood and harm potential of a fire resulting from a hazard have been studied, the assessor should know:

- areas in which the risk of fire is low and general fire precautions are sufficient, eg offices
- areas in which there are processes that might give rise to a fire, eg a kitchen area
- areas in which there are few hazardous processes, but in which a fire could increase quickly, eg there might be strong through currents of fresh air or the presence of stored flammable liquids or waste paper
- areas in which there are both processes that could give rise to a fire and fuel or oxygen is present — these are high-risk areas and could be present in factories or workshops.

Implementing Control Measures

6.156 Once the risk level of identified hazards has been established, it will be necessary to decide if any further control measures are needed to eliminate or reduce the risks.

6.157 The main aim when implementing control measures is to:

- reduce the likelihood of ignition
- minimise any potential fuel in the workplace
- optimise fire escape possibilities.

6.158 Some control measures may be straightforward. For example, adequate control measures might simply mean:

- keeping combustible materials separate from ignition sources
- providing suitable fire extinguishers for use in an emergency
- encouraging tidiness in the workplace to keep escape routes free from obstructions.

6.159 Deciding on control measures should follow a systematic approach. Risks should be prioritised into high, medium and low categories. The assessor needs to decide *what* needs to be done and *when* to do it; timescales should be given for achieving any new control measure.

6.160 Details of all existing fire measures should be included in the fire risk assessment. These will provide the starting point for deciding whether or not any further control measures are necessary.

Qualitative and Quantitative Techniques for Fire Risk Assessment

6.161 Qualitative and quantitative techniques of risk analysis and evaluation exist.

- **Qualitative techniques** — these are based on judgments using generalised data on risk and these techniques express risk in descriptive terms, such as high risk or trivial risk.
- **Quantitative techniques** — these use probability and severity calculations (based on data obtained from tests, statistical evidence and experience) and express risk in mathematical probability terms. These techniques are used in circumstances where there are particularly high or complex risk situations, eg in the chemical, nuclear and transport industries.

6.162 Use of the qualitative techniques satisfies most employers' duties under the Regulatory Reform (Fire Safety) Order 2005 and the Management of Health and Safety at Work Regulations 1999. There are a number of qualitative techniques available, which use simple scales for the measure of likelihood and severity. The technique chosen should be the one which is most appropriate for the workplace and which the assessors understand.

Recording the Assessment Findings

6.163 Record keeping is a very important aspect of the risk assessment process. Employers must keep records in order to comply with the Regulatory Reform (Fire Safety) Order 2005 and the Management of Health and Safety at Work Regulations 1999. They also need full records in case of legal proceedings.

6.164 Every employer should:

- make and keep full records irrespective of the number of employees
- ensure a copy of all the relevant documentation is kept in a safe place well away from the building or buildings to which it relates, as these documents may be the best means of defence in the event of a serious fire.

6.165 The record of the risk assessment does not need to be complicated. It should be a simple working tool to record risk factors and decisions made on the arrangements necessary for control. The aim is to provide:

- improved control of fire risks in the workplace
- a means of reminding the appropriate people of the actions that should be taken
- the means of confirming that these actions have been taken
- a demonstration that adequate control remains.

Details Included in the Record

6.166 Under the Management of Health and Safety at Work Regulations 1999, employers should list *significant findings*. These are:

- significant hazards identified
- existing control measures in place and the extent to which they control the risk
- the persons who may be affected by these significant risks, including those especially at risk.

6.167 With this in mind, fire risk assessment records should contain the following information:

- location of the area being assessed (site, building, floor, department, etc)
- date of the assessment and name of the assessor
- hazards identified (a brief description)
- potential for harm (details of the likelihood and severity of a fire arising from the hazard)
- persons/property exposed to a risk (this should be listed separately for each identified hazard)
- the risk analysis and evaluation — can be determined with a simple first ranking exercise followed, if necessary, by a more detailed analysis and evaluation if the risk is shown to be other than negligible or acceptable. Details of any existing control measures should be recorded. Adequacy of provisions relating to means of escape must be included
- risk control measures (details of any additional or changed control measures considered necessary to reduce the risk)
- final situation (confirmation that any required measures have been implemented and that a revised assessment has shown that the risks are acceptable).

Reviewing the Assessment

6.168 The final stage in the risk assessment process is to continuously review the fire risk assessment, to ensure fire risks are adequately managed on an ongoing basis.

6.169 Monitoring activities after the initial risk assessment is an important part of the overall fire risk assessment. They may take the form of routine inspections, tests and maintenance of all the procedures and equipment provided to eliminate or reduce risk. In this way, employers can ensure that the measures implemented are the correct way to reduce risk.

6.170 To review fire risk assessments effectively, it will be necessary to:

- monitor the measures intended to eliminate or reduce the risk
- regularly test, check and maintain fire protection equipment
- monitor all procedures connected with fire safety
- regularly hold general fire safety inspections/audits.

6.171 Under the Management of Health and Safety at Work Regulations 1999 and the Regulatory Reform (Fire Safety) Order 2005, employers should undertake risk assessment reviews:

- if the number of employees increases, decreases or changes significantly
- if there is some reason to suppose the original assessment is out of date
- whenever there is a change in the operation of the business that could affect the safety of workers in the event of a fire, including changes in:
  - the premises
  - the processes
  - plant and equipment installed and used
  - storage arrangements.

Training

6.172 Basic training will be required for all staff. Employers need to ensure their employees:

- understand fire risks
- know what they should do in the event of a fire
- know how to control fire hazards inherent in their jobs.

6.173 Under the Management of Health and Safety at Work Regulations 1999, training should be provided when:

- an employee is first recruited
- there are new or increased risks as a result of:
  - job change or new responsibilities
  - new or changed work equipment
  - new technology
  - new or changed systems of work.

6.174 Each employee should be aware of the fire safety policy and of any fire safety control measures that are required. If the employer implements a policy of general fire safety training, it is the duty of each employee to attend such training.

List of Relevant Legislation

- Regulatory Reform (Fire Safety) Order 2005
- Building Regulations 2000
- Management of Health and Safety at Work Regulations 1999
- Health and Safety at Work, etc Act 1974

Further Information

Publications

HSE Publications

The following is available from *www.hsebooks.co.uk*.

- INDG163(L) *5 Steps to Risk Assessment*

Department for Communities and Local Government Publications

The following are available from *www.communities.gov.uk*.

- *A Short Guide to Making Your Premises Safe from Fire*
- *Fire Safety Risk Assessment — Means of Escape for Disabled People*
- Approved Document B *Fire Safety: Volume 1: Dwellinghouses*, Department for Communities and Local Government

- Approved Document B *Fire Safety: Volume 2: Buildings Other Than Dwellinghouses*, Department for Communities and Local Government

British Standards

The following are available from *www.bsi-global.com.*

- BS EN 12101–6: 2005 *Smoke and Heat Control Systems. Specification for Pressure Differential Systems. Kits*
- BS 5588–9: 1999 *Fire Precautions in the Design, Construction and Use of Buildings. Code of Practice for Ventilation and Air Conditioning Ductwork*

Other Publications

- *Fire Safety — An Employer's Guide*

Organisations

- Fire Protection Association (FPA)
  Web: *www.thefpa.co.uk*
  The FPA is the UK's national fire safety organisation. It provides a range of fire safety audit and fire risk assessment services.
- Fire Service College
  Web: *www.fireservicelege.ac.uk*
  The Fire Service College provides facilities for both practical and theoretical fire-fighting, fire safety and accident emergency training.
- Institution of Fire Engineers (IFE)
  Web: *www.ife.org.uk*
  The IFE is the international qualifying organisation for fire engineering and fire safety professionals. It was founded in 1918 to promote, encourage and improve the science and practice of fire extinction, fire prevention and fire engineering.

Chapter 7

High Risk Activities and Hazards

Access and Egress

- The Health and Safety at Work, etc Act 1974 requires safe access and egress to every workplace.
- The principal control against the most likely cause of injury at places of access and egress (slips and trips) is the provision and maintenance of suitable surfaces.
- Attention must be paid to dust spreading across the entrance to a site which can obscure sight lines, road markings and signs for pedestrians attempting to cross junctions.
- Speed limits and parking arrangements should be included in the site rules. In particular, certain times where traffic movement and deliveries are prohibited should be explained.
- All warning signs should be reflective and designed to meet highway specifications.
- It may be necessary to erect guard rails to stop pedestrians crossing on corners where visibility is reduced.
- It may be necessary to create a window of time for crane operations, particularly where they over-sail the site entrance/exit area and stop all access and egress to the site.
- The area around the entrance should be separated from parking and storage areas.
- All possible efforts should be made to avoid unloading in the site entrance.
- The road layout, signing and marking should be as similar to the public road network as possible and be such that all road users can reasonably understand what is required of them.
- The risk assessment for access and egress should take the form of a traffic management plan which recognises the various hazards created by physical and behavioural features.

7.1 The Health and Safety at Work, etc Act 1974 requires safe access and egress to every workplace.

Employers' Duties

7.2 The Construction (Design and Management) Regulations 2007 require:

- each construction site to be organised so that pedestrians and vehicles "can move safely and without risks to health".
- suitable signs to be displayed where necessary for reasons of health and safety.

7.3 Gates must not obscure the pedestrian's view of traffic and if vehicles are positioned where they could trap pedestrians there must be effective means of warning. This reflects the generally stated requirement for a gap of at least 600mm around the rear swing of a slewing machine.

7.4 The Management of Health and Safety Regulations 1999 require that employers who share a workplace must have effective procedures for communication and co-operation to ensure safety is managed in those areas where their individual and joint actions affect the safety of either party. Pre-contract meetings and reasonable accommodations go a long way to ensuring that there is a harmonious long-term relationship.

Employees' Duties

7.5

- Under the Health and Safety at Work, etc Act 1974 employees have a duty to:
  - take reasonable care of their own health and safety and that of other people who may be affected by their activities at work
  - co-operate with their employer to enable the employer to comply with health and safety duties.
- Under the Management of Health and Safety at Work Regulations 1999 employees are required to:
  - use any machinery, equipment, dangerous substances, transport, safety devices and means of production in accordance with any training and instructions provided by the employer
  - inform the employer of any serious and imminent dangers to health and safety
  - inform the employer of any shortcomings in the employer's health and safety arrangements.

In Practice

Slips and Trips

7.6 The most likely injury at places of access and egress to the site are those associated with slips and trips. For site staff footwear must be regarded as an essential control measure. All persons under the control of the principal contractor must have footwear in good condition. Very few designs of sole actually fulfil the requirement for non-slip on construction site surfaces, as determined by the industry expert SATRA. The principal control is the provision and maintenance of suitable surfaces. For an entrance into an existing building undergoing refurbishment, barrier matting can remove contamination. In wet conditions staff must either wipe their feet or the mat will need to be large enough to take approximately seven or eight paces to traverse. Every effort should be made to prevent movement of the matting so careful selection is essential.

7.7 Trenches covered with steel road plates could be painted with a sanded primer paint, or bituminised paint, to provide grip. Chequerplate generally offers very little improvement. Where lightweight panels are suitable then proprietary textured covers should be secured. Walkways made of plywood rapidly become slippery, particularly with contamination from some construction materials such as lime and clay. Self adhesive non-slip tapes could be used, provided they are maintained and do not loosen. Care must be taken to provide a fillet to the edges of plates as a 10mm variation in profile is regarded as a significant trip hazard — and margins of this size are relevant in the event of any litigation.

7.8 The New Roads and Street Works Act 1991 suggests that 1.5m clear width should be arranged around any barrier. Even though the pedestrian count may be low this width has to be considered if double pushchairs might use the footway. An absolute minimum of 1m can be permitted in exceptions. Below this an alternative route has to be provided — but it must be protected against traffic.

7.9 If eye protection is mandatory on the site then thought should be given to manage the risk of sunlight or reflected glare. Wearers of sunglasses can experience temporary blind spots.

Dust

7.10 Dust spreading across the entrance to a site can obscure sight lines, road markings and signs for pedestrians attempting to cross junctions. If it rains the settled dust can then become very slippery. If a road has not been surfaced then commercially available binders can be spray applied to reduce the generation of dust and to prevent rutting. Work on scaffolds near to site entrances which create a mess will need particular attention, such as wetting, provided the run-off is ducted away. Debris netting or sheeting might be useful.

Maintenance

7.11 Adequate spreading equipment and a supply of grit/sand, etc should be readily available on-site during winter in anticipation of snow and icy conditions. Daily cleaning, particularly during any bulk excavation, is the minimum, either by hand or preferably by powered brush. The brush can be attached to a forklift, although that is likely to be too busy on most sites. Some brushes can be fitted to skid steers. Specialists are available who can provide a regular on-site service with road sweepers which remove the material. Where there are high levels of traffic movement a wheel-wash should be used to remove tyre contamination before vehicles leave the site. Newer equipment allows recycling of the wash water, minimising the waste water disposal problem.

Site Rules

7.12 Speed limits and parking arrangements should be included in the site rules. An acceptable speed limit for unmade roads is 5 mph and 10 mph for surfaced roads. During induction all personnel must have the traffic management plan appropriately communicated to them. In particular, it should be ensured that the types of vehicle movements they will be involved in are explained in detail. This may include certain times where traffic movement and deliveries are prohibited.

Warning Signs

7.13 All warning signs should be reflective and designed to meet highway specifications. Advice should be sought from the highways authority. If a sign is needed in position for an extended period then, rather than weigh it down with bags of granular material, it should be properly mounted on posts as this increases the visibility.

7.14 A schematic site plan should be available at the site entrance to assist navigation. The site speed limit should be clearly identified and adequate signposting should be provided to assist delivery vehicles unfamiliar with the site layout.

Barriers

7.15 It may be necessary to erect guard rails to stop pedestrians crossing on corners where visibility is reduced. Guardrails should be set back a minimum of 500mm from the kerb. Temporary guardrails may need to be concreted in.

Driver Training

7.16 Trained mobile plant drivers should have, for example, CPCS (construction plant certification scheme) cards issued by the CITB to prove competence. Key control is an essential strategy to prohibit unauthorised vehicle use. In addition to recognising national competence systems, all drivers should be assessed for skill and knowledge by their own employer to ensure continuing competency.

Craneage

7.17 On some occasions it will be necessary to create a window of time for crane operations, particularly where they over-sail the site entrance/exit area. This may mean Sunday or after hours working, particularly if any loads have to be unloaded in the highway. Such work will require diversions approved by the highways authority.

Site Hoardings

7.18 Great care must be taken when installing and painting hoardings in order to avoid personal injury claims from members of the public. In addition, significant damage to parked cars and pedestrians has occurred when hoardings are blown over by strong winds. To avoid interested members of the public from clogging up site entrances, specific viewing panels can be made available provided they:

- do not compromise site security
- prevent vandals throwing items onto the site
- prevent sparks and other hazardous emissions from exiting the site.

Waste and Parking

7.19 Forklift skips, clearly marked with the type of waste they are to accommodate, should be sited close to the work area but not around the site entrance. The area around the entrance should be separated from parking and storage areas. Bin areas soon generate their own trail of polypropylene banding and rubbish lying around creating trip hazards. Visible rubbish may also increase the chances of arson or vandalism.

Disability and Temporary Walkways

7.20 Where road crossings are wide, it is good practice to provide central refuges to allow the roadway to be crossed safely in two or more movements. No installed slope should have a greater incline than 1 in 12, and where handrails are provided these should be in a contrasting colour to other features to assist the visually impaired.

Events

7.21 If there are to be demonstrations, carnivals, etc which might make for traffic problems and public safety issues then the police should be consulted.

Client's Security and Safety Requirements

7.22 For reasons of national security, there are special security arrangements at premises involved in research, the military, etc and early negotiations with both suppliers and the client are essential. Similar considerations may apply to COMAH or chemical sites.

Unloading at the Entrance

7.23 All possible efforts should be made — in terms of order of build, component design, pallet size and wagon size — to avoid unloading in the site entrance.

7.24 Fatal accidents may occur if, for example, a load falls from the wagon, or a forklift reverses into uncontrolled pedestrians. Delivery plans which are sent at tender and again at order stages could include the following details.

- The load description — quantity, size, weight, lifting points and methods.
- Driver instructions — where and who to report to in addition to the site address and any prohibited times of arrival.
- Names of responsible persons.
- The sequence of unloading.
- Site specific hazards such as overhead obstacles, cables, etc.
- Load labelling.
- Site rules regarding PPE such as high-visibility clothing, limits on use of mobile phones, prohibitions on reversing or conditions for reversing such as the use of a banksman, requirements for drivers to stay in their vehicles, accident reporting.
- Fall prevention where open sides of vehicles have to be accessed by personnel.
- Requirements regarding unloading from vehicles near traffic and avoiding the need for vehicles moving around once they have removed ropes and load restraints.
- Rules for turning away or isolating loads that have shifted during transit.
- Prohibition on uncoupling in the entrance.
- Prohibited types of wagon for goods to arrive on.
- Arrangements for the arrival of left-hand drive vehicles and signs in the relevant language.

Lighting and Visibility

7.25 Where there is opportunity to design the entrance, either as enabling works or as part of the final design, early installation of lighting and layout will pay dividends.

7.26 Vehicle lighting systems, reversing alarms and on-board CCTV must be maintained.

7.27 Where the entrance is restricted, convex mirrors can assist early detection of other traffic and pedestrians. Such mirrors should be highlighted.

Temporary Bridges

7.28 Temporary bridges will need significant lead-in times to assist planning. Installation requires large make-up and unload areas. Services and water courses often have to be diverted so early planning with specialist suppliers is essential.

Railways

7.29 The Rail Safety and Standards Board issue Railway Group Standards which are applicable wherever there are likely conflicts between track and its environs. The highways authority and police authority for the area will also need to be consulted as the setting up of traffic control near to a railway requires special permission.

Clients, Staff and Visitors

7.30 The client bears the major responsibility for controlling their staff and visitors, but the contractor is able to provide information, because the contractor understands the feasibility of separation, footways signage and other controls.

Traffic Layout and Traffic Flow

7.31 In principle the road layout, signing and marking should be as similar to the public road network as possible and be such that all road users can reasonably understand what is required of them. Roads should be designed to accommodate the largest vehicles that may have to use them in respect of pavement construction, width, radius, gradient, clearance and visibility.

7.32 If at all possible adequate turning facilities should be provided, which may be in the form of a hammerhead but preferably a turning circle.

7.33 Two-way industrial roads need a minimum width of 7.3m. Where there are curves they should allow HGVs to pass without the need for riding the kerb, otherwise local widening should be considered. If possible the maximum longitudinal gradient should be no greater than 1 in 12 and the minimum channel gradient should be 1 in 125. A crossfall or camber of about 1 in 40 is sufficient for drainage, otherwise surface water will tend to pond. In winter these ponds could then freeze. Standing water can also obscure road markings.

7.34 Where congestion is anticipated traffic lights should be installed. If at all possible a separate entrance and exit are preferred as these can be incorporated in creating a one way system. For very large sites, if a roundabout is to be built at a later stage then temporary roads can incorporate a roundabout layout using barriers.

7.35 Care should be taken, especially at the site entrance, to avoid creating blind spots. Mirrors can be helpful as a secondary measure. Pedestrians should be prevented, perhaps by barriers, from simply stepping out into the traffic flow.

Obstructions

7.36 If there are any low level obstructions, such as cables and pipe bridges, they need to be effectively signed and illuminated. The use of goalposts is an approved method for warning of overhead cables.

7.37 Drivers expect to have the same clearance they are used to on the highway where 5.03m is the normal minimum. A container on a flatbed may well be 4.2m above the ground. Since vehicles may bump up and down, the actual clearance available as measured on-site should be 75mm more than the published clearance. Clearance information, like the delivery plans, should form part of the health and safety plan and reminders should be issued to all contractors, particularly those such as demolition contractors who may need to bring heavy plant on-site. The principal contractor's own transport department also needs the information. Signs will be needed if there are temporary changes which might reduce the clearance such as ramps. Some fire tenders need significant clearance — the principal contractor should discuss these needs with the fire brigade.

Site Speed Limits

7.38 A site may appear to restrict the acceptable speed by its physical characteristics. A speed limit prominently displayed has some advantages — it can remind drivers of the site rules and it can be used to support disciplinary measures. Those who repeatedly transgress could lose the privilege of parking on-site. If necessary, the sign could be supplemented by traffic calming devices such as width restrictors, speed bumps or rumble strips. Fork trucks may have difficulty negotiating these obstructions when carrying a load, so they need intelligent siting.

Risk Assessment for Access and Egress

7.39 The risk assessment should take the form of a traffic management plan which recognises the various hazards created by physical and behavioural features. This should take into account:

- weight or space limitations
- the age and vulnerability profile of road and footway users
- the build programme including provisions for enabling works to avoid disruption to traffic management due to late appointments for service connections, etc
- the commissioning or handover programme
- the time of day of arrival of workers and materials movements
- the visibility and ease of entering the site
- the available space on-site for offloading and storage
- how congestion will be avoided
- relevant site rules
- drivers' qualifications restrictions
- special arrangements for the disabled and for those who may have poor or no knowledge of English or are unable to read
- security arrangements
- how roadways and footpaths will be maintained and cleaned, particularly during disruption such as excavations and installation of scaffolds.

7.40 The plan should state how reversing is to be managed, what methods will be used to direct traffic, where loading and unloading is to take place, and any restrictions to accommodate the client or neighbours. On housing estates, phased occupation should take into account alternative routes for homeowners on foot, cycle and vehicle. This may require liaison with public transport providers, particularly for the school run.

7.41 When the site is closed for long periods, such as weekends, a close down procedure needs to be drawn up and followed.

List of Relevant Legislation

- Construction (Design and Management) Regulations 2007
- Management of Health and Safety Regulations 1999
- Workplace (Health and Safety at Work) Regulations 1992
- New Roads and Street Works Act 1991

Further Information

Publications

HSE Publications

The following are available from *www.hsebooks.co.uk*.

- HSG136 *Workplace Transport Safety: An Employers Guide*

- HSG151 *Protecting the Public — Your Next Move*
- INDG199 (rev 1) *Workplace Transport Safety: An Overview*
- L64 *Safety Signs and Signals: Guidance on Regulations 1996*

British Standards Publications

The following are available from *www.bsi-global.com*.

- BS 5489: *Road Lighting*
- BS 5400: *Steel, Concrete and Composite Bridges*
- BS 7958: 2005 *Closed Circuit Television (CCTV). Management and Operation Code of Practice*

The Stationery Office Publications

The following are available from *www.tso.co.uk*.

- *Traffic Signs Manual*
- *The Safety of Loads on Vehicles: Code of Practice*

Other Publications

- *Designing for Deliveries*, Freight Transport Association 1999
- C652 *Safer Surfaces to Walk On — Reducing the Risk of Slipping*, CIRIA
- *Code of Practice for the Safe Application and Operation of Lorry Loaders*, (rev) 9, Association of Lorry Loader Manufacturers and Importers

Organisations

- Health and Safety Executive (HSE)
  Web: *www.hse.gov.uk*
  The HSE is responsible for the regulation of almost all the risks to health and safety arising from work activity in the UK.
- Rail Safety and Standards Board
  Web: *www.rssb.co.uk*
  A not-for-profit company operating as a centre of excellence for all matters relating to railway safety.
- SATRA Footwear Testing
  Web: *www.satra.co.uk*
  SATRA has a comprehensive range of physical and chemical tests available in the world for footwear and footwear materials. It can test all types of materials and their properties used in footwear, as well as components and grindery. SATRA's experts will advise which tests are required to determine if a material is suitable for a specific shoe style.

Confined Spaces

- A confined space is defined as any place, including any chamber, tank, silo, pit, trench, pipe, sewer, flue or other similar space, which, by virtue of its enclosed nature, gives rise to a risk.
- Confined spaces should not be thought of as simply small, enclosed areas. A confined space could be enormous, eg a freight container. Alternatively, a normal workroom could become a confined space, if there is little ventilation and hazardous substances are in use.
- The Confined Spaces Regulations 1997 require employers to ensure that no person enters a confined space for work purposes, unless it is not reasonably practicable to perform the work externally.
- If entry to a confined space is unavoidable, employers should carry out a risk assessment of the work activity before anyone enters the confined space.
- The risk assessment should consider the general condition of the confined space.
- The confined space should be isolated to prevent ingress (entrance) of substances, eg liquids, gases and steam into the confined space.
- It may be necessary to limit the time period that a person can work in a confined space.
- The degree of supervision should be based on the risk present.
- Flammable materials generally should not be stored in confined spaces, except in special circumstances.
- Adequate and suitable lighting, including emergency lighting, should be provided.
- If a confined space is not safe to work in without the need for personal protective equipment (PPE) or respiratory protective equipment (RPE), such equipment should be designed to protect against the risks identified.
- An adequate communication system will be needed for those carrying out the work and in order to summon help in the case of an emergency.
- No entry to, or work in, a confined space should occur without adequate emergency arrangements being put into place.
- All those involved in working in confined spaces must be adequately trained in the safe system of work and any permit-to-work systems.

7.42 Entry into confined spaces is extremely hazardous. An average of 15 people die each year in confined spaces as a result of such things as lack of oxygen, poisonous gases, fumes or vapours, fire and explosions, and excessive heat.

7.43 One of the more tragic consequences of accidents in confined spaces is the number of people who die effecting rescue of their colleagues, falling victim to the very conditions they are attempting to rescue their colleagues from.

Employers' Duties

7.44 The Confined Spaces Regulations 1997 define confined spaces as any place — including any chamber, tank, silo, pit, trench, pipe, sewer, flue or other similar space — which, by virtue of its enclosed nature, gives rise to a reasonably foreseeable specified risk.

7.45 The regulations require employers to:

- prohibit entry and work in confined spaces, unless it is not reasonably practicable for the work to be carried out from outside of the space, or if the work cannot be avoided
- ensure that, if entry into a confined space is unavoidable, then any entry or work in a confined space shall, as far as is reasonably practicable, be in accordance with a safe system of work
- ensure that the safe system of work controls the "relevant specified risks" (ie injury from fire or explosion; loss of consciousness arising from excessive heat, gas, fume or vapour, or lack of oxygen; drowning arising from an increase in the level of a liquid, or asphyxiation, as a result of a free-flowing solid, or entrapment preventing access to a respirable environment)
- ensure that, in the event of an emergency, suitable and sufficient arrangements are in place for the rescue of persons working in a confined space (these arrangements must include measures to control the risks to those putting the rescue operation into place and the means to effect resuscitation, where necessary).

7.46 The Dangerous Substances and Explosive Atmospheres Regulations 2002 (DSEAR) require employers to carry out risk assessments and to introduce appropriate control measures, in respect of any dangerous substances present in the workplace that may lead to the formation of a flammable or explosive atmosphere.

7.47 Employers have a general duty to ensure, so far as is reasonably practicable, the health, safety and welfare at work of all employees under the Health and Safety at Work, etc Act 1974.

7.48 The Management of Health and Safety at Work Regulations 1999 require employers to:

- make an assessment of the risks to health and safety of their employees while at work
- make a similar assessment of the risks to health and safety of persons not in their employment that may arise as a result of their (ie the employer's) work activities.

Employees' Duties

7.49

- Employees have a duty to take reasonable care of their own health and safety and that of other people who may be affected by their work under the Health and Safety at Work, etc Act 1974.
- The Management of Health and Safety at Work Regulations 1999 require employees to:
  - use machinery, equipment, dangerous substances, transport equipment, means of production or safety devices in accordance with any instruction and training given by the employer
  - inform their employer of any danger to health and safety posed by a work activity
  - inform their employer of any shortcomings in the employer's protection arrangements.

In Practice

Identifying Relevant Activities

7.50 In order to identify relevant activities, it is first necessary to determine what constitutes a confined space.

7.51 A confined space is an area that is mostly enclosed, although some confined spaces are not entirely enclosed (eg trenches, which are open at the top). The enclosed area is a confined space if there is a risk of serious injury due to a substance or condition causing, eg a fire or explosion, loss of consciousness or asphyxiation.

7.52 Examples of confined spaces include:

- sewers
- tanks
- silos, such as those used in agriculture
- trenches
- inspection pits
- engine rooms
- tunnels
- freight containers.

7.53 Where any doubt exists, a definition from the Confined Spaces Regulations 1997 should be used. These define a confined space as any place, which, by virtue of its enclosed nature, gives rise to a reasonably foreseeable, specified risk. Employers must identify foreseeable specified risks.

7.54 A common misconception is that confined spaces are only small, enclosed areas. However, size is not a part of the definition of a confined space. A normal workroom can become a confined space if there is little ventilation and hazardous substances are being used. A confined space might be enormous — such as a large freight container — but people working in it might be at risk, because of a lack of oxygen.

Determine if Entry Can be Avoided

7.55 Having identified work areas that could be confined spaces, the first consideration must be whether entry to the work area can be avoided.

Does the Work Have to be Carried Out?

7.56 Is the work really necessary? In most cases, even if for economic reasons only, work should not be considered unless it is absolutely necessary. One strategy to minimise the need for entry might be to extend maintenance frequencies, reducing risks by limiting the occasions needed for entry.

Can the Work be Carried Out Without Entry?

7.57 If the work has to be carried out, then consideration must be given to carrying out the work without entry into the confined space, eg:

- carrying out inspections of ducts via remote-controlled cameras

- installing closed-circuit television (CCTV) to monitor conditions within a confined space without having to enter
- installing instruments to monitor atmosphere, etc from outside of the confined space
- opening up the space so that it is no longer enclosed
- "designing out" the need for entry by changing work methods
- clearing blockages by the use of remote-controlled flail devices (devices with rotating arms used to clear obstructions by smashing, battering or cutting through them), vibrators or air purgers
- cleaning tanks by using water jets and tools, with long handles, from outside of the contained space.

Risk Assessments of Work in Confined Spaces

7.58 If it is not reasonably practicable to avoid entry into a confined space, then the employer must carry out a risk assessment of any work to be performed before anyone enters the confined space. The assessment should take into account:

- any substances that might be present in the confined space
- the possibility of oxygen deficiency and the risks from water
- asphyxiants and free-flowing solids, etc.

Risks to be Taken Into Account

7.59 Risks to be taken into account include the following.

- If flammable substances are present, or the air is oxygen enriched for some reason, a fire or explosion could result.
- There may be fumes, gases or vapours present that pose a risk to workers if they are inhaled, eg:
  - substances that the worker is using
  - substances that are kept in the space, eg a tank containing chemical residues
  - substances from natural sources or contaminated land
  - substances from other, less obvious sources, such as from equipment that is being used outside of the confined space (ie as fumes from a generator running nearby the entrance to the space).
- There might be insufficient oxygen for the workers to breathe, causing unconsciousness and possible death. Lack of oxygen could be due to several factors, including:
  - the air being displaced by heavier gases
  - lack of fresh air entering the confined space and the oxygen being used up by the person breathing in the confined space
  - natural processes using up the oxygen, such as biodegradation of organic matter in silos and sewers or rusting of steel in tanks
  - carrying out activities that use up oxygen in a confined space, such as welding.
- Liquids might get into confined spaces and cause drowning, eg raised water levels in a sewer.
- Free-flowing solids may either cause asphyxiation or trap the worker so that they are unable to leave the confined space. An example of this is grain flowing into a silo while workers are present in the vessel.
- Heat in a confined space could cause adverse effects to workers inside the space. A confined space might be hot for a number of reasons, including:
  - the type of work being carried out, eg welding, or if it is a furnace that is still hot when the workers enter
  - the work itself might cause workers to overheat, eg asbestos-stripping while wearing protective clothing and using breathing apparatus.

Assessing the Risks

7.60 The assessment should look at the confined space and consider:

- the presence of residues, such as chemicals, sludge, scale or rust
- the possibility of a lack of oxygen in the space, either initially or increasingly as work is carried out in the space
- the presence of too much oxygen in the space, which would lead to a risk of fire or explosion
- what has been kept in the space, regardless of how recently or how briefly it was stored there (because of the nature of confined spaces, fumes, etc from residues can be present for a long time after the material has been removed)
- how the design of the confined space might be such that dangerous fumes, etc, might accumulate in one area, eg if there is a part of the space that is below the level of the rest of the space
- the air in the confined space might be contaminated by outside sources, such as from nearby equipment or methane from landfill sites nearby
- the hazards associated with the work activity itself, such as use of welding equipment, or use of substances for cleaning.

7.61 A standard risk assessment format can be used for this purpose.

7.62 Where dangerous substances are present, the particular requirements of the DSEAR apply. These regulations lay down the requirements for minimising the risks of explosive atmospheres.

Control Measures

Organisation of Work

7.63 Work in confined spaces should be carried out only by competent persons with the experience and training for that particular type of work. The employer should check that anyone who is to work in a confined space is competent to do so, especially if they will not be under direct supervision.

Supervising Work in a Confined Space

7.64 In most cases, it will be necessary to have a supervisor nearby while the work in a confined space is being carried out. The supervisor should be made responsible for:

- ensuring any permits to work are followed properly
- carrying out any atmosphere testing, or ensuring that it is carried out by a competent person
- enforcing the designated safety procedures
- checking that all safety precautions are in effect
- carrying out any necessary briefings and/or instructions for all employees prior to work commencing.

Communications Within a Confined Space

7.65 There should be a communication system for the people inside the confined space: both between themselves and with people outside. There should also be a way of summoning help in the event of an emergency.

7.66 There are a number of options for a communications system, including mobile phones, two-way radios or, in some cases where the use of such equipment is not practicable, pre-arranged signals using tugs on a rope.

7.67 Some communications systems will need to be specially designed for use in the confined space, eg a communications system that is intrinsically safe for use in flammable atmospheres. Consideration of the means for those outside of the confined space to summon the emergency services quickly will also be necessary. This is especially important if the confined space is in an isolated area.

Restricting Time Spent in a Confined Space

7.68 In some cases, the amount of time any one worker spends working in a confined space needs to be restricted. An example of circumstances where this is necessary is asbestos stripping where the work, in combination with protective clothing, breathing apparatus and a very small space, can cause overheating and fatigue.

Testing the Atmosphere

7.69 If there is any uncertainty over whether the confined space contains hazardous substances or lacks oxygen, it will be necessary to test the atmosphere. Regular, or continuous, monitoring might be needed while work is being carried out in a confined space if there is a possibility that the atmosphere may deteriorate during this time.

7.70 Testing equipment must be suitably calibrated and well maintained. Wherever possible, the tests should be carried out without entering the confined space, eg by using instruments with long handles.

7.71 The tests should be carried out only by people who:

- are competent to do so
- understand what constitutes a safe atmosphere to work in
- have received instruction and training regarding the risks
- are able to interpret the results
- have the qualities and authority to allow them to take appropriate action as a result of the testing.

Purging the Confined Space

7.72 Flammable or toxic gases might have to be purged by flushing the confined space with inert gas (suitable for flammable or toxic substances) or with air (suitable for toxic substances only). Consideration should be given to the risks posed to people outside of the confined space as a result of any exposure to the gases being purged.

7.73 Once the confined space has been purged, the atmosphere should be tested, before work commences, to check the purge has been successful.

Use of Respiratory Protective Equipment

7.74 Where residues are removed from a confined space, appropriate precautions must be taken, such as respiratory protective equipment and powered ventilation. Ventilation will be required for confined spaces where they are so enclosed that there is insufficient circulation of fresh air. Normal, fresh air only (without added oxygen) should be used and it is necessary to check that there is no contamination at the source, eg traffic fumes.

7.75 In many cases, the confined space must be isolated in such a way as to prevent gases, liquids, etc from entering the space.

Appropriate Equipment

7.76 All equipment used in confined spaces must be suitable for the work involved. Where there might be flammable or explosive atmospheres, specially designed equipment only should be used in order to reduce the risk of igniting gases, vapours or fumes. Examples include non-sparking tools and specially protected lighting and, in some cases, intrinsically safe electrical equipment.

7.77 There are detailed classification and control measures necessary, including the provision of special equipment and protective systems, for places where explosive atmospheres may be present. These will not be discussed in detail here.

7.78 When work in confined spaces is carried out, it should be isolated from sources of power, with the exception of essential facilities, such as lighting, communication, etc. Petrol-fuelled internal combustion engines must never be used in confined spaces.

7.79 Where other types of internal combustion engines are used, ventilation will be required and continuous monitoring should be carried out. The exhaust fumes should be vented a good distance from the confined space. Sources of fresh air used for ventilation in the confined space should be protected from any source of contaminant.

7.80 Wherever reasonably practicable, all portable equipment with engines must be refuelled outside of the confined space.

7.81 The use of gas cylinders is not recommended. However, if there is no safer option available, extensive precautions are needed.

Fire Prevention

7.82 In addition to the measures previously described, there are a number of more general fire prevention measures that need to be taken into account with confined spaces.

- Flammable substances should not be stored in a confined space, unless the confined space is specifically designated for that purpose, eg a petrol tanker.
- If flammable substances have to be stored in a confined space, eg in some tunnelling work, the amount stored should be minimised and kept in fire-resistant containers.
- If there is a risk of flammable substances being present in a confined space, measures should be taken to prevent the risk of ignition from static electricity.
- Smoking should never be allowed in confined spaces.

Personal Protective Equipment

7.83 Wherever reasonably practicable, work in confined spaces should be made safe without the use of personal protective equipment (PPE). If it is necessary to use PPE, it should be suitable for the work being carried out and be designed to protect against the risks identified.

7.84 All PPE should be well maintained and anyone using this equipment should be trained in its proper use. Examples of PPE that might be needed for work in confined spaces include respiratory protective equipment (RPE), safety harnesses and lines.

7.85 Particular care needs to be taken on the choice of RPE. Where oxygen deficiency is evident, fresh-air breathing apparatus will be necessary.

7.86 Self-contained breathing apparatus is bulky and can be impractical for entry into some confined spaces. Such apparatus might introduce fresh hazards, such as entanglement and trip hazards, and impede escape in an emergency. In such cases, airline-breathing apparatus may be preferable.

7.87 PPE should meet relevant product standards under the Personal Protective Equipment Regulations 2002.

7.88 The provision of PPE must comply with the requirements of the Personal Protective Equipment at Work Regulations 1992.

The Workplace

7.89 There must be a safe means of entry to and exit from a confined space. A sign should be put up next to all points of access stating that authorised people only may enter the confined space.

7.90 Where practicable, the access must allow people to enter and exit the space quickly. It should be free from obstruction whenever anyone is in the confined space. The entrance must be large enough — the guidance to the Confined Spaces Regulations 1997 recommends a minimum diameter of 575mm, in order to allow fast access and exit to someone wearing breathing apparatus. If the size of the access is smaller than 575mm, eg in tank containers, special precautions will be needed.

7.91 Adequate lighting must be provided. The lighting might need to be specially designed, eg for use in flammable atmospheres or damp/wet conditions.

Developing Permit-to-Work Systems

7.92 If there is a reasonably foreseeable risk of serious injury from working in a confined space, a permit to work will probably be needed. A permit to work is a written system that formalises a safe system of work.

7.93 The permit to work is designed to check that all eventualities have been considered when planning and organising work in confined spaces and is an important means of minimising any risks involved. The permit to work will involve following the steps shown below.

- List items that need to be checked, as per the permit-to-work form, before work can proceed.
- Prevent work if conditions fail to meet the accepted standard.
- Limit times of work before repeat testing is due to be undertaken (original tests will need to be repeated, in case the atmosphere deteriorates after work is under way).
- Prohibit other types of work in adjacent areas when a permit is in operation, eg use of hazardous substances near to the ventilation intake for a confined space.
- Specify precautions to be taken, eg:
  - use of RPF
  - the display of signs such as "no unauthorised entry"
  - the availability of rescue equipment, etc
  - tests that need to carried out, such as testing for flammable or toxic gases or for the presence of oxygen
  - any necessary isolations, such as power supplies.
- Display the permit to work at the work site and work control area to ensure that all employees are made aware of the permit operational requirements.
- Cancel the permit to work when work in the confined space is stopped.
- Ensure that the permit to work applies to all staff on the site, ie employees and contractors.

Emergency Rescue Procedures

7.94 Emergency arrangements for those working in a confined space should include the following.

- Someone should be stationed outside of the confined space to keep watch and to communicate with those inside. This person may also summon help and take charge of the rescue arrangements.
- Rescue equipment, such as lifelines, equipment to lift unconscious people, first-aid equipment and extra sets of breathing apparatus, should be provided.
- Appropriate people trained in resuscitation procedures must be available.
- Artificial respiration equipment, for use only by specially trained people, must be provided.
- There must be a means of summoning help in an emergency, eg by mobile phone, using a rope signal, radio or alarm.
- Equipment must be provided that protects rescuers against the conditions causing the emergency, eg breathing apparatus for confined spaces deficient in oxygen.
- Fire extinguishers and a sprinkler system may be needed. These must, however, be suitable for use in a confined space.
- Designated rescuers must be trained in the action and procedures to follow in an emergency, including how to use emergency equipment and how to administer first aid.

Training

All Those Who Will Work in Confined Spaces

7.95 Those who work in confined spaces must undergo specific training that includes:

- an outline of the requirements of the Confined Spaces Regulations 1997
- the need to avoid entry into a confined space unless authorised to do so
- details of the work to be carried out
- the hazards and risks associated with the work
- the safety precautions and procedures
- the safe system of work, including any permits to work
- information on emergencies, including how they arise and the importance of following the emergency procedures
- specific instruction to never attempt a rescue, unless trained and authorised to do so
- personal responsibilities of all those involved in entry into confined spaces.

7.96 Training should be provided during induction, and before entry into confined spaces is authorised. Refresher training will be necessary at regular intervals. This should be part of regular health and safety refresher training.

Those Supervising and Authorising Work in Confined Spaces

7.97 In addition to the training needing to be provided to those working in confined spaces, those supervising and authorising work in confined spaces will need specific training in:

- the hazards, risks and control measures necessary for work in confined spaces
- their particular supervisory responsibilities
- the use of instrumentation to test for hazardous environments.

Those Carrying Out Risk Assessments and Devising Safe Systems of Work

7.98 These persons should, in addition to general training in working and/or supervising work in a confined space, have specific training in risk assessment.

7.99 Very often, this person will be a health and safety practitioner. However, because of the technical nature of the risks involved, such a practitioner might need the assistance of a suitably qualified engineer.

External Courses

7.100 In addition to in-house training, many health and safety training providers, and colleges of further education, provide confined spaces training, both for those supervising such work and for those who simply need an understanding of how to work in such situations.

7.101 For those supervising such work, one-day courses, with examinations, are available. General training in working in confined spaces is usually provided within more general courses, particularly those associated with construction.

List of Relevant Legislation

- Dangerous Substances and Explosive Atmospheres Regulations 2002
- Personal Protective Equipment Regulations 2002
- Management of Health and Safety at Work Regulations 1999
- Confined Spaces Regulations 1997
- Personal Protective Equipment at Work Regulations 1992
- Health and Safety at Work, etc Act 1974

Further Information

Publications

HSE Publications

The following are available from *www.hsebooks.co.uk*.

- AIS26 *Managing Confined Spaces on Farms*
- HSG53 *The Selection, Use and Maintenance of Respiratory Protective Equipment — A Practical Guide*
- HSG251 *Fumigation — Health and Safety Guidance for Employers and Technicians Carrying Out Fumigation Operations*
- INDG258 *Safe Work in Confined Spaces* (revision 7)
- L101 *Safe Work in Confined Spaces — Confined Spaces Regulations 1997*

Demolition

- Demolition work is the province of specialist contractors and is rarely within the competency of normal building contractors.
- Site managers should hold a five-day safety qualification in site management, and all staff working on a demolition-site should ideally hold a CSCS card (Construction Skills Certification Scheme).
- All sites should have security fences and controlled gates with appropriate signs. In some areas there may be the need for night security watchman or CCTV.
- If excessive dust is likely neighbours should be consulted so that suitable times for this work can be worked out.
- All premises occupied after 2002 should have an asbestos management plan which might be useful in indicating the possibility of asbestos-containing materials. However, a type 3 survey is essential to avoid unplanned discovery and potential release of fibres.
- All temporary works should be inspected by the engineer prior to loading.
- Constant reinforcement by formal means such as training and constant spot checks and supervisory tours are essential to emphasise the importance of safe working.
- Only qualified persons should operate loading plant and consideration should be given to providing a gantry from which lorries can be sheeted and loads secured.
- Visitors should not be permitted onto an operational demolition-site unless they are accompanied and a specific risk assessment is carried out with designated routes to pre-planned viewing positions.
- A suitable form of audible emergency alarm or personal alarm system must be installed which can override site noise levels and other conflicting distractions.
- The scope of the decommissioning report should be related to the scope of the demolition project to ensure that the whole of the site is covered by the report.
- Any scaffolding requirement should be installed by a scaffold engineer and be designed with demolition in mind.
- Guard rails or easily handled covers of sufficient strength must be installed to all holes where persons may be at risk of falls.

7.102 Despite years of experience, demolition managers, supervisors and workers are often faced with new combinations of problems which require variations to custom and practice. As legislation and technology change, working methods have also had to be adapted and revised.

Employers' Duties

7.103

- The Construction (Design and Management) Regulations 2007 state that the demolition and dismantling of structures must be planned and carried out in such a way as to prevent danger or, where this is not practicable, to reduce danger to as low a level as reasonably practicable. Arrangements for the carrying out of demolition work must be recorded in writing before the demolition or dismantling work begins.

- The Control of Substances Hazardous to Health Regulations 2002 require employers to ensure exposure to any substance, either naturally occurring or man made, is prevented so far as is reasonably practicable if there is a possibility of adverse health effects. Control should be by methods which avoid human contact wherever possible. Where this is not possible fail safe methods should be selected. If it is not reasonable to use these methods then control by wearing PPE is permitted provided the employer takes steps to remove reliance on such measures. Whatever type of PPE is selected must be monitored and kept in an effective condition. If a risk of exposure remains employers should carry out suitable health surveillance.
- The Lifting Operations and Lifting Equipment Regulations 1998 require any operation which involves lifting by mechanical means to be properly planned and risk assessed. All equipment is to be under a scheme of examination. The amount of detail in planning and the frequency of inspection should be linked to the likelihood and consequences of failure. Only competent persons may be involved in lifting operations.
- The Control of Major Accidents Hazards Regulations 1999 cover sites which exceed stated threshold quantities of substances which are deemed to be extremely dangerous to life and the environment. The operators of such sites are required to take all measures necessary to prevent major accidents and limit their consequences to people and the environment. Measures include the preparation of emergency plans within a major accident prevention policy (MAPP).
- The Road Traffic (Temporary Restrictions) Act 1991 legislates for temporary closure of the highway to facilitate safer working methods.
- The Control of Asbestos Regulations 2006 ensure employers use reasonably practicable measures to prevent the inhalation of asbestos fibres. Anyone who works on asbestos insulation board, asbestos coatings or asbestos lagging must be licensed by the HSE to do that work. There are a few minor exemptions.
- The Confined Spaces Regulations 1997 require employers to carry out risk assessments and implement suitable control measures for any work on or in spaces where specified risks such as asphyxiation may be experienced from liquids, gases, fumes, free-flowing solids or other materials. The primary requirement of the regulations is to avoid the need for entry. If entry is unavoidable then adequate precautions should be in place wherever possible to remove the likelihood of ingress of hazardous substances, eg by purging, isolating and providing temporary supports.
- The Control of Noise at Work Regulations 2005 require employers to reduce the exposure to noise where noise generated by work is likely to cause injury. Normally precautions include providing hearing protection, but they should also include providing information and seeking to find quieter ways of working, or removing workers from the noisy areas. In some cases workers will have to participate in a programme of audiometric testing.
- The Work at Height Regulations 2005 require assessment to identify suitable equipment and have in place controls to make sure it is used properly. Where possible work at height should be avoided or if not an existing workplace might be suitable to work from. If there is a risk of falling from a work platform then collective methods, such as handrails, should be used to prevent falls. If edge protection is not practicable then the consequences of a fall should be minimised. Safety netting or bean bags would fulfil this requirement. If these solutions are not possible then methods which protect the user might be acceptable such as harnesses which restrain the user from getting to a position from which they could fall. Alternatively, as a last resort a fall-arrest harness may be worn to minimise the consequences of a fall. It is not permitted to provide no precautions at all, where a fall could cause injury.

- In all cases the work should be properly planned and appropriately supervised. Those working at height should be competent to do so or if being trained, they should be supervised by a competent person.
- Working at height safely also includes avoiding the danger of falling through fragile surfaces or falling into holes by not working near them, covering them with secure and robust covers. Alternatively it might be acceptable to prevent approach to them or reduce the degree of injury which would result from falling through.
- The Control of Vibration at Work Regulations 2005 set limits to daily exposure to vibration to prevent nerve and blood circulation problems which can result in long-term disability. Employees at risk might need health surveillance. Employers must provide information, select methods and choose tooling which reduce exposure.
- The Electricity at Work Regulations 1989 require maintenance of all electrical equipment to anticipate breakdown of insulation and security of all connections. The regulations do not permit work on or near live services unless very detailed precautions are drawn up and recorded by a competent person. It is normal in demolition work to make dead all such services as part of the enabling works. Power requirements for the work are then designed as a temporary supply, carried in specially protected cables routed away from interference or carried within ducts.
- The Manual Handling Operations Regulations 1992 prohibit manual handling which might cause injury unless an assessment has been made and suitable control measures have been put in place. In all cases reasonable alternatives to handling should be employed. Actual assessments may have to be generic where a wide variety of short-term tasks have to be performed.

Employees' Duties

7.104

- Employees have a duty to take reasonable care of their own health and safety and that of other people who may be affected by their work under the Health and Safety at Work, etc Act 1974.
- The Management of Health and Safety at Work Regulations 1999 require employees to:
  - use any machinery, equipment, transport, safety devices and means of production in accordance with any training and instructions provided by the employer
  - inform the employer of any serious and imminent dangers to health and safety
  - inform the employer of any shortcomings in the employer's health and safety arrangements.

In Practice

Survey and Planning

7.105 A surveyor attending a site for tendering purposes may be in great personal danger. As much information as possible should be made available. Lone working is rarely acceptable. It is useful to seek the assistance of a manager associated with the previous use of the property or its design. Any drawings will assist the surveyor, although their accuracy will need to be confirmed as the survey proceeds.

Competency and Qualifications

7.106 Skills such as the ability to read and interpret drawings, a nature which is naturally inquisitive, and years of construction experience make surveyors competent.

7.107 Research into methods of securing structural elements, panels, chimneys and stressed skin structures pays off.

7.108 A competent person must appreciate the loadings arising from backfill, the consequences of non-vertical alignment or other forms of poor construction, eg missing ties or bonding in masonry. They should be adept at spotting defects from ingress of water or moisture, the consequences of drying out or impact, corrosion and deterioration from insects, rot or chemical attack, or anything which might now or in the future influence safe and systematic controlled demolition.

7.109 All staff engaged in the work on-site should hold either a CSCS (Construction Scheme Certification Scheme) card, have demolition-site operative, labourer or experienced worker, mattockman or topman qualifications, or have plant operating qualifications which are valid and match the plant they are required to use. Site managers should hold a five-day safety qualification in site management.

Site Security

7.110 All sites should have security fences and controlled gates with appropriate signs. In some areas there may be the need for night security watchman or CCTV. A daily inspection for gaps in site fencing is essential. All plant should be immobilised and any fuel supply protected from interference. All access equipment should be isolated eg ladders removed at night or secured against climbing.

Site Induction

7.111 The site induction should be site specific and relevant. Whatever method is chosen to convey initial warnings and hazard appreciation it should be a memorable and meaningful experience, and ought to include a tour of the work area.

Flooding

7.112 If any of the work could affect the local water table the correct siting for flood defences such as sandbags and the effect of localised pumping on other structures must be considered. Water movement, particularly from any excavations, can rapidly reduce the effectiveness of temporary support and pre-weakened foundations.

Method Statement

7.113 A method statement should be as short as possible. It must relate to drawings and be able to be understood by all who will be involved. Generic method statements are of little use to demolition work.

Services

7.114 All existing services should be made dead and proved dead.

7.115 However, if any services which cross the site have to be maintained every effort should be made to organise a permanent or temporary diversion. Service drawings and piping and instrumentation diagrams should not be expected to be accurate, and will need verification. If services cannot be diverted then physical measures must be installed to prevent contact such as ducting and installing goal posts and devices to limit plant from making contact or entering the zone of danger where the service might arc or receive damage.

7.116 When services have to be stripped out then prior to work commencing management must obtain suitable certificates of isolation, purging and/or disconnection from statutory undertakings or other service owners.

Pipework and Vessels

7.117 The procedures to be followed for isolation and purging of pipework and vessels prior to removal must follow the guidance laid down in the Health and Safety Executive (HSE) publication HSG253.

7.118 Any attempt to isolate without a category 1 positive isolation should be strongly resisted.

7.119 Fluids should be drained and removed off-site by specialist contractors provided suitably controlled venting is in place. Effective filtering of vented fumes must be considered.

Control of Traffic Movements

7.120 It is worthwhile checking that the highways authority and police have no objections to the scheme to be adopted for managing the movement of heavy plant.

7.121 If at all possible the design of the site should be orientated to avoid reversing and preserve visibility splays. All staff should keep to designated and barriered footways. High-visibility (hi-vis) clothing is a minimum requirement. Adequate lighting may be needed (designed to avoid glare) and plans/signs should show how emergency services access the site.

Dust

7.122 Although the provision of suitable respiratory protective equipment, goggles or safety glasses is an easy solution, their effectiveness is limited by the skill and diligence of the wearer. If it is possible to damp down dusty work this should be adopted. Some methods of demolition use water as a power source. Dust levels might also by reduced by restricting speed limits, particularly on haul roads. Vehicle washing facilities can be useful. If the roadways are sufficiently resistant regular vehicle mounted road brushing can be effective. If excessive dust is likely neighbours should be consulted so that suitable times for this work can be worked out.

Air Overpressure

7.123 When large quantities of material impact the ground, the air pressure wave generated can affect the stability of other structures. A specialist can predict the value of air pressures and advise on the maximum accepted threshold. Air pressure waves may only damage windows but this can of course result in injury.

Noise

7.124 Noise on-site is very difficult to reduce without changing work methods. It might be possible to mitigate the effect of noise to neighbours by installing acoustic screens.

Fire Prevention

7.125 The presence of any oxidising agents such as bottled oxygen, weedkiller, etc should be established and these items removed or protected from fire. Effective means of warning must be available and all staff must have suitable training in how to fight fires. Portable fire points should be made available, with all staff trained in usage. Where they are sited will depend on the possibility of damage, but if hot working is taking place they must be close to the work.

Hazardous Substances

7.126 Information should be sought on the type of materials previously processed or handled at the premises. The by products of such activities and any ingress from neighbouring sites must be discovered. A desktop study might also throw up historical information pre-dating the previous occupancy.

Biological Substances

7.127 Some plant can give rise to a risk of contracting legionnaires' disease, and where rats might have been present the risk of contracting Weil's disease must be considered. Staff should be issued with HSE cards and given a short toolbox talk. If biological wastes such as bird refuse are encountered specialist removal is required.

Asbestos

7.128 All premises occupied after 2002 should have an asbestos management plan which might be useful in indicating the possibility of ACMs (asbestos-containing materials). However a type 3 survey is essential to avoid unplanned discovery and potential release of fibres. Specialists will, in almost all cases, be required to carry out any removal work, which should be part of the enabling works completed before structural demolition begins.

Other Hazards

7.129 Process fluids may be subject to decomposition or evaporation. Stagnant fluids may give rise to harmful vapours or gases. An occupational hygienist should give advice following safe sampling.

7.130 Anthrax may be retained in horse hair plaster.

Structural Stability

7.131 Calculations by a qualified demolition engineer are necessary following detailed site investigation. The investigation will probably need to be intrusive to find the interior of concrete or timber structures suspected of hidden internal decay. The engineer should then specify the planned sequence of operations and provide details of any temporary support needed.

7.132 All temporary works should be inspected by the engineer prior to loading. A system for the transfer of demolition materials to minimise surcharges to the structure will be needed. Careful control needs to be exercised to understand the whereabouts of each person on-site. Care must be taken in the controlled release of stored energy such as counterweights. As the demolition proceeds the engineer will need to continue a re-assessment of load paths.

Human Factors

7.133 Some demolition work is monotonous and the methods for safe working can be tedious. For such work, constant reinforcement by formal means such as training, constant spot checks and supervisory tours are essential to emphasise the importance of safe working and adhering to the method statement. A white-helmeted safety officer should be an unannounced, but frequent visitor. Toolbox talks may be a useful method of reinforcing the need to wear personal protective equipment (PPE), tie ladders or comply with other straightforward safety messages. Receipt for copies of site-specific directives or instructions should be accounted for by signature and date.

Arisings

7.134 All arisings should be moved to a safe storage area as soon as possible so that safe routes around the site can be maintained.

Hygiene

7.135 The standard of personal hygiene and the type of welfare unit necessary might need to be determined by an occupational hygienist, depending on the presence of hazardous substances.

Loading

7.136 If possible a designated area should be cleared and kept clear from early on in the project. Only qualified persons should operate loading plant and considerations should be given to provide a gantry from which lorries can be sheeted and loads secured.

Visitors

7.137 Visitors should not be permitted onto an operational demolition-site unless they are accompanied and a specific risk assessment is carried out with designated routes to pre-planned viewing positions. If a more detailed visit is necessary then work should cease and a survey undertaken by an experienced person just before the visit. Visitors must wear all necessary PPE.

Vibration

7.138 All staff operating any form of vibrating equipment, including sitting or standing on such equipment, should regularly complete a health questionnaire regarding vibration symptoms.

7.139 Legislation requires employers to assess the level of risk and this will involve reference to manufacturer's data. In addition, staff must be trained to recognise symptoms and use any control methods such as sharing work with colleagues, avoiding smoking and keeping their hands warm. Employers must ensure that low vibration tools and equipment are always selected and working methods are adopted which minimise the risks from vibration.

Slips and Trips

7.140 There should be a cleaning/tidying regime to ensure all designated pedestrian access routes are kept in good condition. It may be wiser to move staff around the site in easily identifiable vehicles. Good lighting will be needed if winter or shift work is anticipated. All staff should wear good quality safety footwear.

Emergency Planning

7.141 A number of clear signs should be displayed indicating the location of the assembly point. Liaison meetings with the local emergency services should be arranged, preferably on-site, to enable them to appreciate the scale of the operations being undertaken. A suitable form of audible alarm or personal alarm system must be installed which can override site noise levels and other conflicting distractions. A first-aid point should be established and an effective method put in place for summoning the first aider(s).

Ejected Material

7.142 The establishment of exclusion zones where personnel are not permitted to venture is one of the surest methods of preventing injury from material ejected from the demolition process. Protected staff will be permitted within a designated safe working spaces. In addition, work could be programmed to avoid concentrations of people, vehicles, etc. BS 6187 gives details of how to plan exclusion zones.

Dismantling Machinery

7.143 An engineer with knowledge of the particular plant and possession of operation and maintenance manuals will need to specify the sequence, method of dismantling as well as any access requirements. The engineer will need to determine any scotching necessary to act as additional precautions against uncovenanted motion, and schedule all the isolations that are required.

Dismantling Process Plant

7.144 A decommissioning report forms the essential foundation for developing site-specific safe systems of work for work on process plant. The scope of the report should be related to the scope of the demolition project to ensure that the whole of the site is covered by the report. Where decommissioning is not complete an action plan must be provided by the client detailing who is responsible. There should be an information flow diagram to ensure accurate and timely communications. The extent of cannibalising existing plant to use components for elsewhere must be established as this may have involved hasty alterations to the structure and services to facilitate removal. Special care will be needed in handling gases, fumes residues and by-products which may lurk in dark corners, deadlegs in pipelines and hidden sumps.

Scaffolding

7.145 Any scaffolding requirement should be installed by a scaffold engineer and be designed with demolition in mind. The normal weekly inspection should be supplemented by interim inspections. Edge protection may only be removed for the passage of materials, and only if alternative fall protection is put in place.

Rubbish Chutes

7.146 Rubbish chutes greatly minimise fly of material and dust generation, however they must be secured against movement and be of adequate size. The apertures must be positioned to avoid staff falling into them. Exit points are often best shrouded to minimise dust. During skip movement a clear system for closing the chute should be adopted.

Gas Cutting

7.147 Users of gas cutting equipment must be trained in choosing the correct type of cut (eg sit cut, dog leg cut, v cut, hinge cut, key cut) and careful handling of equipment. Correct lighting up and leak detection procedures are vital, as is maintenance of safety devices, hoses and regulators. All bottles, whether in use or not should be secured against falling and sited where damage and interference is not possible.

Arches

7.148 Where a range of interconnected arches or a series of barrel vaulting is to be demolished it is often necessary to install some form of temporary support. In any instance the structure to be demolished cannot be classed as a working platform. Remote machine methods of demolition are preferred using suitable long reach machines.

Manual Handling

7.149 Much salvageable material such as radiators, doors, windows fixtures and fittings cannot be machine handled so staff will need to be trained in the correct manual handling technique. Adequate numbers of staff who work well as a team are essential. The use of pallets, genie type lifts and pallet trucks is recommended.

Drainage and Pits

7.150 During industrial processes spillages occur leading to drains and pits being contaminated. Such contamination must be identified and the fluids (or gases) should be extracted safely.

Radioactivity

7.151 If radioactive components or wastes are likely to be encountered this work should be carried out by specially trained persons before other work proceeds.

Contaminated Land

7.152 Where there is reason to believe that the ground is contaminated with hazardous substances specialist advice is needed. A BOHS (British Occupational Hygiene Society) qualified occupational hygienist will be able to assess the risks and advise on suitable controls.

Holes in Floors

7.153 Guardrails or easily handled covers of sufficient strength must be installed to all holes where persons may be at risk of falls. All covers should be secured in place and labelled to avoid mistaken removal.

Hot Plant

7.154 It is rarely acceptable to dismantle hot plant. Normally all plant should be allowed to cool and fail safe methods adopted to prevent re-heating. On no account should stubborn residues be warmed up to ease removal. Differential melting across the material may trap pressurised vapours and result in fires and explosion.

Fragile Roofs

7.155 Fragile roofs should be regarded as no go areas. All work on removing them should be by remote means or carried out carefully from below.

List of Relevant Legislation

- Construction (Design and Management) Regulations 2007
- Control of Asbestos Regulations 2006
- Control of Noise at Work Regulations 2005
- Control of Vibration at Work Regulations 2005
- Work at Height Regulations 2005
- Control of Substances Hazardous to Health Regulations 2002
- Control of Major Accident Hazards Regulations 1999
- Management of Health and Safety at Work Regulations 1999
- Lifting Operations and Lifting Equipment Regulations 1998
- Confined Spaces Regulations 1997
- Manual Handling Operations Regulations 1992
- Electricity at Work Regulations 1989
- Road Traffic (Temporary Restrictions) Act 1991
- Health and Safety at Work, etc Act 1974

Further Information

Publications

British Standard Publications

The following are available from *www.bsi-global.com*.

- BS 6187: 2000 *Code of Practice for Demolition*
- BS 5607: 1998 *Code of Practice for the Safe Use of Explosives in the Construction Industry*

HSE Publications

The following are available from *www.hsebooks.co.uk*.

- HSG47 (rev 2000) *Avoiding Danger from Underground Services*
- HSG141 *Electrical Safety on Construction Sites*
- HSG150 (3rd ed, 2006) *Health and Safety in Construction*

- HSG151 *Protecting the Public — Your Next Move*
- HSG168 *Fire Safety in Construction Work*
- HSG253 *The Safe Isolation of Plant and Equipment*

Other Publications

- *Anthrax and Historic Plaster: Managing Minor Risks in Historic Building Refurbishment*, English Heritage
- *Control of Dust from Construction and Demolition Activities*, 2003, BRE Publications
- *Decommissioning, Mothballing and Revamping*, 1997, IChemE
- *Do's and Don'ts for Demolition*, National Federation of Demolition Contractors

Organisations

- British Occupational Hygiene Society (BOHS)
  Web: *www.bohs.org*
  The BOHS seeks to promote and protect occupational and environmental health and hygiene and offers training in occupational hygiene and related subjects.
- Health and Safety Executive (HSE)
  Web: *www.hse.gov.uk*
  The HSE is responsible for the regulation of almost all the risks to health and safety arising from work activity in the UK.
- The Highways Agency
  Web: *www.highways.gov.uk*
  The Highways Agency is responsible for the construction, maintenance and improvement of the strategic road network (motorways and trunk roads) in England. All other roads in England are the responsibility of local authorities.
- National Federation of Demolition Contractors
  Web: *www.demolition-nfdc.com*
  The National Federation of Demolition Contractors aims to establish new and safe procedures for demolition, dismantling and associated works and to represent the industry on government matters affecting the industry.

Electricity

- Employers and the self-employed must control the risks associated with electricity under the Electricity at Work Regulations 1989.
- All electrical systems and electrical equipment must be maintained so as not to give rise to danger.
- All equipment and systems must be suitable for its purpose, protected against adverse environments and adequately insulated or otherwise protected.
- Electrical systems and equipment must be protected from giving rise to danger in the event of faults. This includes adequate means of earthing or similar protection and the provision of over current protection.
- Adequate means must be provided for cutting off supply and isolation.
- Adequate means must be provided for working safely on systems that have been made dead, including the provision of adequate space and lighting.
- No work should be carried out on live equipment, except in exceptional circumstances.
- No person should work on electrical equipment unless competent to do so.

7.156 This topic covers the following.

- Procedures for Working on Electrical Systems and Equipment
- Live Working
- Reducing the Risks Posed by Electrical Equipment
- Maintenance and Testing of Portable Electrical Appliances and In Service
- Examination and Testing
- Maintenance and Testing of Fixed Electrical Installations

Employers' Duties

7.157

- Employers are required to provide and maintain plant and systems that are, as far as is reasonably practicable, safe and without risks to health under the Health and Safety at Work, etc Act 1974.
- The Electricity at Work Regulations 1989 stipulate the following.
  - All electrical systems, including electrical equipment connected to them, must be used and maintained in such a way to prevent a risk of injury, including any work being carried out on systems.
  - All equipment provided for the protection of employees working on electrical systems must be suitable, maintained and used properly.
  - Electrical equipment should not be installed if its strength and capability may be exceeded.
  - Electrical equipment used in hazardous environments should be adequate for the environment.
  - All conductors should be adequately insulated or placed in a position which will not pose a risk of injury.
  - Adequate measures, such as earthing of equipment or other protective devices, must be taken to ensure systems that develop faults will not give rise to a risk of injury.

- Adequate means must be provided to protect against the risk of injury from excess currents, eg adequate fuses or circuit breakers.
- Provision must be made for the isolation of electrical equipment so work can be safely carried out on it. This must include precautions to prevent inadvertent reconnection of the supply through locking off and, where necessary, earthing devices.
- No work should be carried out on live conductors, except under certain conditions.
- Adequate working space, lighting and access should be provided for work on electrical systems.
- All persons working on electrical systems where knowledge and/or experience is needed to prevent risk of injury should be competent or adequately supervised for the work.

- Under the Building (Amendment) (No.3) Regulations 2004, employers must ensure that all domestic electrical wiring installation work complies with *Approved Document P*.

Employees' Duties

7.158

- Employees must take reasonable care of their own health and safety and that of others who may be affected by their activities at work under the Health and Safety at Work, etc Act 1974.
- Employees also have a duty to co-operate with the employer's health and safety arrangements under the Management of Health and Safety at Work Regulations 1999.
- Under the Electricity at Work Regulations 1989 employees have a duty to co-operate with the employer's health and safety arrangements. In addition, employees have a general duty to comply with the regulations in so far as they apply to their own activities.

In Practice

7.159 The use of electricity at work is widespread and when used and managed safely, is an essential tool of industry. Poorly maintained, sub-standard or damaged equipment can, however, result in:

- electrocution through direct or indirect contact with live conductors
- arcing, caused by poor working practices or overloading of systems and equipment
- fire, principally from damaged, overloaded or worn conductors and/or equipment.

7.160 All three risks can result in death or major injury, and also cause extensive property damage.

7.161 Key legislation relating to the area includes the Electricity at Work Regulations 1989. Many of the requirements of the regulations relate to the standard of electrical installations and their use and maintenance. So, designers and installers, as well as employers and the self-employed, need to comply.

Procedures for Working on Electrical Systems and Equipment

7.162 Work on electrical systems and equipment can vary from changing a 13 amp plug on a portable appliance or replacing a cartridge fuse in a switchboard, to installing or maintaining high voltage switchgear. The procedures and control measures needed for each will be very different.

7.163 Procedures need to be developed to ensure suitably trained staff carry out all such work safely. The level of competence required for wiring installations will be different from those required for changing a lightbulb.

7.164 A risk assessment process should be used to determine the risks from various electrical work. An inventory of all relevant tasks should be made and appropriate control measures determined.

7.165 Low-risk activities, eg changing fuses, lightbulbs or plugs could be carried out by competent maintenance personnel, but work which requires technical knowledge and experience, eg maintenance of high voltage switchgear, will require a qualified electrician. Control measures that include safe systems of work and on occasion, a permit to work, will be required for such work.

7.166 A recommended procedure for work on electrical systems follows.

1. Determine and record all work activities carried out on electrical equipment and systems.
2. Carry out a risk assessment on these activities, taking the following factors into account.
   - Is the work to be carried out on normally live equipment? For example, changing or maintaining an isolator.
   - Is there any possibility of coming into contact with live conductors? For example, when changing a fuse in a distribution cabinet with live terminals.
   - Have the environmental factors been considered? For example, work on electrical systems in damp, wet environments can increase the risks.
   - Has access to the work been considered? Consideration should be given to subsidiary risks, eg using equipment to access certain areas.
   - Is there a necessity for live working?
3. Determine the control measures necessary for each level of risk. For high-risk activities a safe system of work will be needed that may incorporate a permit to work. Only suitably qualified and competent persons can carry out such work.

7.167 The level of risk for each task should be determined through the risk assessment process.

Safe System of Work

7.168 Guidelines for a safe system of work for high-risk activity follow.

1. Identify the scope of the task.
2. Determine the level of competence required and allocate the task or components of the task accordingly.
3. Determine what equipment will be required, eg live line, detectors, barriers or insulated tools.
4. Determine any access arrangements.
5. Isolate apparatus to be worked on and lock off any isolators that affect isolation.
6. Ensure any adjacent equipment is also isolated if access to it is possible during the work.
7. Test conductors/apparatus to be worked on to ensure that they are dead, using a live line detector or similar device.

8. On completion of the work, check it has been carried out correctly and that all tools, etc have been removed.
9. Return equipment and test.

7.169 On some occasions it may be necessary to carry out work on high voltage equipment, such as 11kv circuit breakers. This is highly specialised work, usually carried out by the electrical supply authorities. Such work should always incorporate a permit to work and the conductors to be worked on would be earthed, normally at the point of supply.

Live Working

7.170 On some occasions, continuity of supply is considered such a high priority that maintenance on live equipment may be necessary. This is an extremely high-risk activity and should be avoided if possible. It is recommended that the provisions of the Electricity at Work Regulations 1989 be strictly followed if this course of action is considered.

7.171 The regulations carry a general prohibition on live working. However, such work will be permitted if all three following conditions are met.

1. It is unreasonable in all circumstances for the system to be dead.
2. It is reasonable in all circumstances for the work to be carried out live.
3. Suitable precautions are taken to prevent injury.

7.172 A special risk assessment is carried out to determine if all three conditions are met before work is allowed to proceed.

7.173 For example, the testing of systems and equipment would meet these criteria. Use of a multimeter to detect voltages and read circuit currents can usually only be carried out on a live system, during which the operator is exposed to live equipment. The use of properly designed test equipment by competent persons, incorporating current limiting devices and properly shrouded test prods with insulated handles, is an essential control measure for such work.

Reducing the Risks Posed by Electrical Equipment

7.174 Careful selection of electrical equipment and its supply system can significantly reduce the risk of harm from electricity.

Reduced Voltage

7.175 One of the principal methods of reducing risk is to reduce the voltage used for the appliance. The harmful component of electricity is the current generated in the circuit. Since the current is proportional to the voltage provided, providing other factors remain constant, then reducing the voltage will reduce the current, and hence, the risk.

7.176 110v systems with the 240/110v transformer secondary winding centre tapped to earth effectively reduces the normal 240v supply to 55v, a much safer voltage. These are commonly used in construction and some industrial situations.

Double Insulation

7.177 The principle is to separate the user from live conductors by two sets of insulation. Such equipment is safer to use because of this double protection and, as a consequence, these devices are not normally earthed. These devices are known as Class II devices, while those that are earthed are known as Class I.

7.178 A double insulated Class II will carry the double insulation symbol (a square inside another square).

Correct Selection of Fuses and/or Circuit Breakers

7.179 The insulation of a fuse or circuit breaker with too high a value means that the circuit that it is protecting will not be disconnected until the higher value of current is reached, increasing the risk of harm. Reducing the value of the device will reduce the current at which the circuit will be protected. However, there is sometimes a balance to be found, because if the device is set too low for the equipment being used, then it may operate at or below operating current and cause nuisance trips.

7.180 One of the principal factors in the protection of circuits is the speed of the protective device. A wire fuse is the least effective and a cartridge fuse is quicker. A circuit breaker is superior to a fuse and is more easily reset without having to be replaced. Fuses and circuit breakers are designed to protect the circuit, not prevent or mitigate the effects of electric shock to humans.

Residual Current Circuit Breaker

7.181 Residual current circuit breakers are superior to fuses and circuit breakers because they can mitigate against the effects of electric shock by disconnecting the circuit before significant harm occurs.

7.182 They operate extremely quickly, at very low levels of earth current. They are not a substitute for fuses or circuit breakers, because they are not always reliable. They should be used in addition to fuses and circuit breakers. They should always be subject to a testing regime.

Maintenance and Testing of Portable Electrical Appliances and In Service Electrical Equipment

7.183 A portable electrical appliance is any equipment with a plug and lead that is easily moveable or could be moved. Portable appliances are extremely varied, from the obvious, eg power tools used on construction sites and on assembly lines in factories to the less obvious, eg photocopiers, printers and personal computers (PCs). Other office equipment and cleaning tools, eg such as vacuum cleaners and more mundane but significant risk items, eg kettles and toasters, are also portable electrical appliances.

7.184 This section includes all equipment connected to fixed wiring of the installation. It includes such items as water heaters and radiators and production work equipment, eg lathes and devices that are normally hard wired.

7.185 The varied nature of portable electrical appliances means they pose different risks and require a range of control measures.

7.186 A typical procedure to control the risks from portable appliances follows.

1. Identify all portable appliances that need to be maintained and tested. An inventory of such equipment should be made. Many organisations have inventories as part of an asset recording procedures, which may be used. Alternatively, an inventory of electrical equipment could be used.
2. Carry out an assessment of the risk posed by each type of equipment, taking into account:
   - type of equipment, ie Class I or Class II
   - frequency of use and movement
   - the environment in which equipment is used
   - if equipment is hand-held or stationary.

3. Categorise into high, medium or low-risk, eg a Class I appliance frequently used in the hands in a harsh environment would be high-risk and a PC that is rarely, if ever, moved would be low-risk.
4. Determine if the appliance needs to be tested and examined or examined only, taking into account that the tests that can be carried out on Class II appliances are very limited.
5. Determine the frequency of examination/testing. There are no statutory requirements relating to frequency of testing, but there is guidance available.

7.187 *Table 1 — Equipment and Testing* shows the testing required for different types of equipment.

7.188 Leads to all electrical equipment, regardless of the level of risk of the appliance, should be examined at least yearly. Extension leads and leads to Class I devices should be treated as Class I.

Examination and Testing

User Checks

7.189 The examination can be carried out by the user and involves a simple visual check. Things to look for are:

- frayed or damaged leads
- placement of leads, ensuring they are secure in plug and apparatus
- physical damage, ie dents or cracks
- any signs of overheating
- testing history, ensuring equipment is in date for formal visual inspection and test.

Formal Visual Examination

7.190 This is a more formal examination of the equipment than a user check. Only a competent person should carry it out.

7.191 Repeat user checks and:

- check cables and their runs, including leads from socket outlets
- check there is not excessive use of adapters and/or extension leads
- ensure equipment is suitable for its use
- ensure equipment is being used in an environment it is designed for
- ensure leads are of the correct rating for equipment
- ensure plugs are correctly wired and fitted with the correct fuse for the equipment.

Testing

7.192 A purpose-made portable appliance tester should be used. Many models are available, from very simple devices to more sophisticated ones which include data recording and print facilities. The advantage of recording individual apparatus details is that a comparison of results can be made over time to detect deterioration.

7.193 Any competent person can normally carry out testing using such devices, but some formal training is recommended.

7.194 Tests that should be carried out, normally on Class I equipment only, are earth continuity and insulation resistance. More complex electrical apparatus may need specialist testing recommended by the manufacturer. In some of these cases, more specialist equipment may be necessary and must be used by qualified electrical engineers.

7.195 A record should be kept of all inspections and tests.

Table 1 — Equipment and Testing

Equipment	Class	Inspections and Test	Frequency of Testing/ Examination
Hand-held, eg drills and soldering irons	Class I and II	User checks	Before use
	Class I	Formal visual inspection	Every 4 months
		Combined inspection and test	Every year
	Class II	Formal visual inspection	Every 6 months
		Combined inspection and test	None
Portable or movable	Class I and II	User checks	Weekly
	Class I	Formal visual inspection	Every year
		Combined inspection and test	Every year
	Class II	Formal visual inspection	Every year
		Combined inspection and test	None
Stationary equipment not intended to be moved	Class I and II	User checks	None
	Class I	Formal visual inspection	Every 2 years
		Combined inspection and test	Every 2 years
	Class II	Formal visual inspection	Every 2 years
		Combined inspection and test	None
Fixed equipment which remains in place	Class I and II	User checks	None
	Class I	Formal visual inspection	Every 2 years
		Combined inspection and test	Every 4 years
	Class II	Formal visual inspection	Every 4 years
		Combined inspection and test	None
Information technology equipment	Class I and II	User checks	None
	Class I	Formal visual inspection	Every 2 years
		Combined inspection and test	Every 4 years
	Class II	Formal visual inspection	Every 2 years
		Combined inspection and test	None

Maintenance and Testing of Fixed Electrical Installations

7.196 A fixed electrical installation includes the main supply input to the premises, the associated switchgear and distribution boards, the fixed wiring and cables, and all socket outlets and switches.

7.197 The risk of harm from fixed installations is somewhat less than that posed by portable equipment, as it is not usually moved or disturbed. However, such installations can be hazardous if not properly maintained.

7.198 A typical procedure to control the risks from fixed electrical installations will include the following.

1. Carry out inventory of all fixed electrical equipment.
2. Carry out routine checks on the installation as recommended in guidance to the Institute of Electrical Engineers (IEE) Wiring Regulations (17th edition).
3. Carry out inspection and testing also as recommended in guidance to the IEE Wiring Regulations

Routine Checks, Inspection and Testing

7.199 There are no statutory requirements for the frequency of inspection, testing or methodology, but the IEE has produced the following recommendations for routine checks, inspection and testing.

Installation	Routine Check	Maximum Period Between Inspections and Testing
Domestic	None	Change of occupancy or 10 years
Commercial	1 year	Change of occupancy or 5 years
Industrial	1 year	3 years
Offices, shops and laboratories	1 year	5 years
Cinemas and leisure complexes	1 year	3 years
Public houses	1 year	5 years
Caravans	1 year	3 years
Swimming pools	4 months	1 year
Construction-sites	3 months	3 months

Routine Checks

7.200 A suitably trained and competent person without formal qualifications can carry out the examination. The check, using a list derived from the equipment inventory, should include:

- checking for visual signs of damage
- checking for any signs of overheating
- ensuring cables are secure in switchgear
- ensuring all switchgear and isolators are secure
- ensuring all signage is in place
- checking that access to all equipment is readily available, in particular to emergency isolators
- ensuring all equipment is clear of flammable materials
- checking for signs of interference
- checking the operation of switchgear
- a test of any residual current devices (RCDs) by operating the test button.

Periodic Inspection and Testing

7.201 Testing should be carried out in accordance with Guidance Note 3 to the IEE Wiring Regulations. The test will include an invasive inspection and test, and may include a supply interruption.

7.202 On completion of the test, a periodic inspection report will be produced, usually recommending actions necessary for bringing the installation up to standard. Items marked "requires urgent attention" may put those using the installation at risk.

7.203 A suitably trained and competent person, usually a qualified and experienced electrician, should carry out this exercise. There is a course available for such work.

Selection of Electrical Equipment

7.204 Correct selection of electrical equipment and systems is paramount in controlling the risks from electricity.

7.205 Points to consider before the purchase of equipment include the following.

- Is the apparatus suitable for the task?
- Can 110v equipment be used for the task?
- Are double insulated devices available?
- Is the equipment CE marked?
- Does the existing supply system have the capacity to safely operate the equipment?

7.206 A word of caution about CE marking. Manufacturers have a duty to construct electrical equipment in accordance with certain European, British or combined standards and the CE mark is an indication that the equipment meets the relevant standard. The Health and Safety Executive (HSE) has stated that the CE mark is only a claim by the manufacturer that the device complies with the appropriate standard. The CE mark should not be considered a mark of quality. The purchase of electrical apparatus cannot be separated from the purchase of work equipment generally.

Training

7.207 Most employees will require some level of training for working with electricity. It may mean an instruction not to carry out certain work or that some personnel will need qualifications.

All Staff

7.208 General risks and instructions on the use of electricity should be given at induction and/or general health and safety training sessions. Items that should be covered include:

- risks associated with electricity — a short video can be useful
- how to carry out general user checks as required by portable appliance maintenance procedures
- emphasis on what work can and cannot be carried out on electrical systems and apparatus, which would generally include a complete ban
- what to report regarding found and suspected faults and how to do so.

Staff Authorised to Carry Out Minor Work on Electrical Equipment

7.209 This training session is for those not formally qualified as electrical technicians. It is only intended for those carrying out work on equipment that does not need to be isolated before work. Such tasks will have been determined through the risk assessment procedure. Training should focus on:

- general principles of electricity
- precautions
- particular knowledge necessary for the tasks to be carried out
- what work can and cannot be carried out.

Inspection and Testing of Portable Appliances and Other In Service Electrical Equipment

7.210 The inspection and testing of portable appliances and other in service electrical equipment needs specific training, but may not require an electrical technician. A competent person selected to carry out portable appliance testing should attend a suitable course, eg the City & Guilds 2377. This is a one-day practical course, with an hour-long examination. A certificate is awarded upon completion.

Electricians

7.211 For electricians required to do inspection, testing and certification of electrical installations, City & Guilds 2391 course is suitable. This can be a three- or four-day course depending on knowledge and experience. There is an examination and a certificate is awarded.

Staff to Carry Out Work on High-risk Activities

7.212 All such staff must be qualified and trained electrical technicians. City & Guilds qualifications will be necessary, with a particular emphasis that all qualifications are updated to include the latest edition of the IEE Regulations.

Managers

7.213 Managers responsible for electrical work should be familiar with the general principles and understand the legal and practical requirements.

Contractor's Employees Carrying Out Electrical Installations in Domestic Premises

7.214 Contractor's employees who carry out electrical installations in domestic premises must be trained in accordance with the conditions of the contractor's registration with a competent persons scheme. This registration gives the contractor the responsibility of ensuring that their electricians enable the contractor to meet the requirements of self-certification.

List of Relevant Legislation

- Building (Amendment) (No.3) Regulations 2004
- Management of Health and Safety at Work Regulations 1999

- Provision and Use of Work Equipment Regulations 1998
- Electricity at Work Regulations 1989

Further Information

Publications

HSE Publications

The following are available from *www.hsebooks.co.uk*.

- HSG85 (rev 2003) *Electricity at Work: Safe Working Practices*
- HSG107 *Maintaining Portable and Transportable Electrical Equipment*
- HSG230 *Selection, Use, Care and Maintenance of Electrical Switchgear*
- HSR25 *Memorandum of Guidance on the Electricity at Work Regulations 1989*
- INDG231(L) *Electrical Safety and You*
- INDG236 *Maintaining Portable Electrical Equipment in Offices and Other Low Risk Environments*
- INDG372 *Electrical Switchgear and Safety: A Concise Guide for Users*

British Standards

The following is available from *www.bsi-global.com*.

- BS 7671: 2008 *Requirements for Electrical Installations: IEE Wiring Regulations.* 17th Edition

Other Publications

- *IEE Guidance Note 3 to the IEE Wiring Regulations: Inspection and Testing*

Organisations

- British Standards Institution (BSI)
  Web: *www.bsi-global.com*
  Founded in 1901, the BSI develops and provides supporting information on national and international standards in a wide range of areas.
- City & Guilds
  Web: *www.city-and-guilds.co.uk*
  Established in 1878, City & Guilds is the leading vocational awarding body in the UK and its qualifications assess skills that are of practical value in the workplace.
- Institution of Engineering and Technology (IET)
  Web: *www.theiet.org*
  The IET is a not-for-profit organisation and professional society for the engineering and technology community worldwide.

Electricity — Underground Cables and Overhead Lines

- The installation of underground cables and overhead lines is a specialist activity and should only be carried out by specifically trained personnel.
- Cables should be laid at a depth appropriate to the situation and future use of the site.
- Each high voltage cable should be run in a separate duct and separated from all other cables.
- Cables should be laid below the finished site level and covered, unless otherwise specifically indicated on the drawings which must exist for every installation.
- Separating cables to certain distances represents good practice.
- Every year, thousands of underground cables are accidentally disturbed. It is essential that, wherever possible, all underground cables in the vicinity of digging operations should be made dead.
- If supply cables are to be run between buildings or other structures and above areas where vehicles may operate, the cables should be at least 5.2m above ground level.
- Contact (or near contact) with overhead lines can be fatal whether they are carrying high voltage (eg 400,000V) or low voltage (eg 240V).
- When working with overhead lines, the pre-planning of safe working practices is essential.

7.215 The installation of underground cables and overhead lines and work near such installations, will often involve several work groups. Some of these workers may not be engaged in electrical work and may consequently have little awareness of the substantial dangers associated with their activities. Contractors may also be involved and their unfamiliarity with the site increases the potential for danger to arise. In these cases, it is crucial that close control is exercised by a competent person.

7.216 There are several methods of working and requirements to consider that should help ensure that work with underground cables and overhead lines is as safe as possible.

Employers' Duties

7.217 The Electricity at Work Regulations 1989 (EAWR) apply to all those persons working on or near electricity. Under these regulations:

- all electrical systems, including electrical equipment connected to them, must be used and maintained in such a way as to prevent a risk of injury. This equally applies during any work being carried out on systems
- all equipment provided for the protection of employees working on electrical systems must be suitable, maintained and used properly
- no electrical equipment should be installed if its strength and capability may be exceeded
- electrical equipment used in hazardous environments should be adequate for the environment

- all conductors should be adequately insulated or placed in a position which will not pose a risk of injury
- adequate measures, such as earthing of equipment or other protective devices, must be taken to ensure systems that develop faults will not give rise to a risk of injury
- adequate means must be provided to protect against the risk of injury from excess currents, eg adequate fuses or circuit breakers
- provision must be made for the isolation of electrical equipment so work can be safely carried out on it. This must include precautions to prevent inadvertent reconnection of the supply through locking off and, where necessary, earthing devices
- no work should be carried out on live conductors, except under certain conditions
- adequate working space, lighting and access should be provided for work on electrical systems
- all persons working on electrical systems where knowledge and/or experience is needed to prevent risk of injury should be competent or adequately supervised for the work.

Employees' Duties

7.218

- Under the Health and Safety at Work, etc Act 1974 employees must take reasonable care of their own health and safety and that of others who may be affected by their activities at work.
- Under the Management of Health and Safety at Work Regulations 1999 employees have a duty to co-operate with their employer's health and safety arrangements.
- Under the Electricity at Work Regulations 1989 employees have a duty to co-operate with the employer's health and safety arrangements. In addition, employees have a general duty to comply with the regulations in so far as they apply to their own activities.

In Practice

Underground Cables

7.219 The installation of underground cables is a specialist activity and should only be carried out by specifically trained personnel. However, the following guidance represents good safety practice and may be used as a general guide to assess the quality of such work.

Depth and Protection

7.220 Cables should be laid at a depth appropriate to the situation and future use of the site. Crossovers should be eliminated unless absolutely necessary and no electrical and mechanical damage should be sustained during the laying process.

7.221 At road crossings and entries to buildings, cables should be drawn through glazed earthenware ducts or similar components.

7.222 The laying of the ducts should be supervised to ensure that they are in the correct position and are suitable in every respect for the cable installation. The total cross-sectional area of all cables installed in a duct should not exceed about 35% of the cross-sectional area of the duct.

Ducts

7.223 Each high voltage cable should be run in a separate duct and separated from all other cables. After the installation of cables, the ducts should be sealed with a hessian plug combined with a proprietary sealing compound. Spare ducts should also be sealed by plugging with the approved material.

Cable Depth

7.224 Cables should be laid below the finished site level and covered, unless otherwise specifically indicated on the drawings which must exist for every installation. The following depths are recommended.

High voltage cables	1000mm
Extra low and low voltage cables in open ground	500mm
Extra low and low voltage cables under carriageways (to account for extra weight and vibration)	750mm

Cable Separation

7.225 The following separation between cables represents good practice and should be observed where practicable. A contractor installing the cables should submit details of any proposed deviations for approval prior to the start of the installation process.

Between high voltage cables and high voltage cables	50mm
Between extra low and low voltage cables and extra low and low voltage cables	50mm
Between high voltage cables and extra low and low voltage cables	300mm
Between medium voltage cables and PO and Telecom Cables	150mm
Between high voltage cables and PO and Telecom cables	350mm
Between all cables and gas and water mains	200mm
Between low voltage cables and computer cables	500mm
Between high voltage cables and computer cables	1000mm

Cable Identification and Marking

7.226 Unless otherwise indicated on the design plans, all cables laid direct in the ground should have interlocking cover tiles to BS 2484 covering them. Tiles should be of such a width that they overlap the cable or cables on each side by at least 50mm and should bear the words “DANGER — ELECTRICITY”. Additionally, a polythene tape coloured yellow and carrying a black legend bearing the words “CAUTION ELECTRIC CABLE BELOW” should be laid over the cables(s) at a depth of approximately 250mm below ground level.

Cable Bends

7.227 The radii of bends should not be less than those shown in IEE Guidance Note No. 1 *Selection and Erection* or those recommended by the cable manufacturer, whichever is greater.

Cable Markers

7.228 Weatherproof marker posts should be erected at approximately 60m intervals on straight runs and elsewhere where the cables change direction, to indicate the route and depth of the cables. Durable identification tags should be fitted:

- to all cables at intervals of approximately 30m along the route length
- at either end of the ducts
- at every cable gland.

Cable Trenches

7.229 Cable trenches may be excavated by persons other than those who will lay the cables. The excavation contractor should therefore allow for trimming the bottoms of trenches to make them level and the laying of a 75mm thick layer of sand before laying the cables. After doing so, a further 75mm thick layer of sand should be placed in the trenches. Protective cover tiles should then be laid in the trench in line over the cables.

7.230 The backfilling of the trenches may then be carried out to 250mm below the finished level and the polythene warning tape should then be installed; backfilling may then be completed.

Cable Joints

7.231 Cables should preferably be laid in one length from terminal point to terminal point and no through joints should be allowed unless specifically authorised in writing by a competent electrical engineer. Exceptionally, if a joint is allowed, it should be made by a competent cable jointer since specific standards relating to both mechanical and electrical strength must be achieved.

Working Temperatures

7.232 Cables should not be handled at temperatures below 0°C and should not be handled until the temperature has been above this level for at least 24 hours.

Digging Near Underground Cables

The Accident Potential

7.233 Every year, thousands of underground cables are accidentally disturbed, often with serious injuries being inflicted. Buried cables are often in such a position within the fixed electrical distribution system that very high levels of energy are present. If a metal digging tool accidentally causes a short circuit, large amounts of energy may therefore be released. This can result in the vaporisation of a portion of the cable, its insulation and part of the offending digging tool. The vapour will be expelled from the hole with considerable force (called explosive arcing), an effect somewhere between the action of a very large blowtorch and a substantial explosion.

7.234 People engaged in digging are often not electrocuted but are usually very severely burned on the front of their bodies and are often blown backwards for several yards. The severity of these injuries can prove fatal.

Precautions Required

7.235 It is essential that, wherever possible, all underground cables in the vicinity of digging operations should be made dead. Where this is not possible, if the excavation will reach a depth of approximately 300mm then the route of all underground cables in the area to be excavated should be clearly ascertained. The identification process may first be carried out by referring to the approved site drawings or cable diagrams in order to locate the approximate position of the cables. Such diagrams should indicate both the line and depth of the cable.

7.236 Each company should have a nominated person who will hold the drawings of all underground and overhead cable runs. All persons who will supervise activities or contracts involving work in the vicinity of such cables should know the identity of this person. The name of the nominated person should also be displayed on switchgear enclosures. It should be part of the safe system of work to contact the drawings holder before work begins and this task should be written into the excavation permit used.

7.237 The complete excavation process should be carefully controlled by a competent person and should be carried out under an excavation permit. This is a permit to work specially designed to ensure that all of the correct safety procedures are implemented. The permit should form part of the company's safety policy document. It is often appropriate to employ an excavation permit whenever any excavation is expected to extend below 300mm, irrespective of who is carrying out the task or for what purpose.

7.238 If the company operates on other premises or in public areas, advice should always be sought before excavation begins and the sources of information may include the following, depending on the site location and area:

- the site owner
- the site developer
- the local authority
- the local electricity board
- the electricity generating authority
- the highway authority
- the street lighting authority
- British Rail
- London Transport or London Underground.

7.239 Clearly, other service providers should also be contacted to prevent damage to their services. Damage can cost many millions of pounds to repair and create hazards such as explosions. These services and buried cables must not be expected to be positioned exactly as indicated on drawings and diagrams. They will usually "snake" within the trench and may also be displaced by ground movements or by previous work. Therefore, they may no longer be beneath the marker tapes.

7.240 Once the desired location has been identified, a cable detector with appropriate sensitivity should be used to refine the location accuracy. The approximate route should then be marked. Some cable detectors/locators may have difficulty identifying the presence of a cable that is only carrying a small current or perhaps no current, even though the cable is live. The next safety procedure, therefore, should be to carefully hand dig trial holes near the cable in order to locate it, remembering that it may be higher in the ground than expected. Finally, the cable route should be carefully marked before other site work begins.

Note: It should always be remembered that the location of a cable is only known for certain when it is actually found. Even when a service is exposed and appears to be the right size, colour and position it may not be the one expected.

7.241 The persons using cable locators should be specially trained in their use and limitations and those required to hand dig also require special instructions regarding safety procedures. Once exposed, cables should not be stood on or used as hand or foot holds. They should be protected from movement and damage by being temporarily covered (eg by wooden planks) and supported by temporary slings or other supports where necessary.

7.242 More people die from cable explosions in the summer than the winter. The reason is simply that the best protection from the heat of the explosion is a thick cotton or woollen garment. In hot weather those in the vicinity of possible cable explosions must wear white cotton clothing which covers their arms, chests and legs. Clothing made from synthetic materials is likely to melt and increase the area of burnt flesh, as well as complicating the healing process.

Overhead Lines

7.243 The erection of overhead lines is a specialist activity and should only be carried out by specifically trained personnel. However, the following guidance represents good safety practice when low voltage (less than 1000V) electrical cables are installed between buildings or across areas such as construction sites. The installation of high voltage is beyond the scope of this text.

Positioning Overhead Lines

7.244 If supply cables are to be run between buildings or other structures and above areas where vehicles may operate, the cables should be at least 5.2m above ground level. The cable should be attached to a suitable catenary wire capable of safely supporting its weight. If any doubt exists regarding the actual height of a cable for a particular location, the advice of the local supply authority should be sought. For example, on farms, the use of harvesters and elevators must be anticipated.

7.245 If the level of an access is changed at some future time, the cables must be reinstalled at the required height as part of the same project. This task should precede any alterations to the level of the ground.

7.246 Where the cable enters a building, it should approach the structure in an upwards direction in order to allow water to drain from the cable. This reduces the possibility of moisture entering the building.

7.247 Manufacturers should be consulted regarding the choice of suitable cables, taking into account working conditions and anticipated ambient temperatures. Some thermoplastic insulation will crack in sub-zero temperatures, especially if subjected to high winds.

Work Near Overhead Cables

7.248 Overhead lines are often not insulated and contact (or near contact) with them can be fatal whether they are carrying high voltage (eg 400,000V) or low voltage (eg 240V). Contact (or near contact) with a metal object such as a scaffold pole, ladder, crane jib or excavator arm may cause sufficient electrical energy to discharge to inflict severe or fatal burns or electric shock to any persons in the immediate vicinity. HSE figures show one in three such accidents is fatal.

7.249 Work near overhead cables is particularly hazardous for the following reasons.

- When looking into the air, no scale exists and cables may appear higher or lower than they really are.

- If the ground beneath the cables is uneven, site plant may bounce and lurch therefore adding significantly to its effective height.
- Electrical energy can jump gaps, the distance depending on the voltage present and the humidity of the surrounding air.

Work Practice

7.250 The pre-planning of safe working practices is essential. The first task should be to contact owners of overhead lines and seek their advice or recommendations for work in their vicinity. The owner will often be the electricity supply authority.

7.251 If the lines have not been made dead or the supply diverted, then barriers should be erected in their proximity to prevent contact. In all cases, unless a representative of the supply authority offers alternative advice, no part of a vehicle, plant or equipment should be allowed to come within:

- 15m of lines suspended from steel towers
- 9m of lines suspended from wooden poles.

7.252 The first essential step is to make a detailed inspection of the site in order to assess the risk and carry out a hazard analysis.

7.253 If early discussions with the owner reveal that the lines cannot be diverted or made dead, it may be possible to arrange that a certain part of the work, such as the passage of tall plant or work nearest to the line, can be carried out at times when the power is removed by the owner for other reasons. This will require the closest co-operation and should be controlled by a formal permit system.

7.254 If the lines are to remain live at all times, the following precautions are required in the following situations.

1. No passage of plant under the line(s) or work near the line(s) shall be allowed. Substantial barriers must be built to prevent this occurring.
2. Plant will be allowed to pass under the line(s). Defined passageways within the barriers must be constructed.
3. Work will be carried out beneath the line(s). Further precautions are required in this case, such as solid overhead barriers, warning notices and the careful selection of suitable plant and equipment, in addition to specific workforce training.

No Passage of Plant or Work Under Lines

7.255 Substantial barriers running parallel to the line should be erected at ground level. The distance between the barrier and the line will vary depending on the line voltage but should never be less then 6m when measured horizontally to the vertical plane of the line as shown in the diagram below.

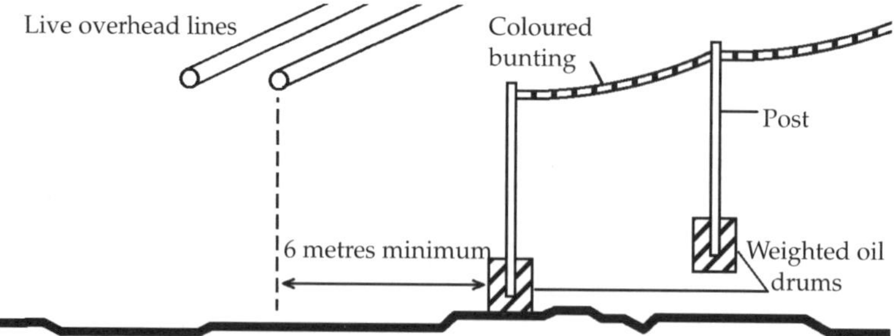

7.256 Where the plant or equipment has high extensions, such as a crane jib or excavator arm, a high level indication of the barrier distance will be required. This should take the form of coloured plastic tape/flags or bunting. If possible, the horizontal distance between the barrier and the lines when a crane is to be used should be the jib length plus 6m, in which case the bunting becomes necessary.

7.257 The barrier should be constructed using one of the following:

- a substantial non-conducting post and rail fence
- large steel drums (eg 40 gallon oil drums) filled with rubble and with posts inserted
- an earth bank not less than 1m high and with posts inserted.

7.258 Additional timber balks to act as wheel stops will enhance safety. The barrier arrangement chosen should be of a prominent colour. Yellow/black or red/white striped materials are often used.

7.259 The area enclosed should be a "no-go" area for all purposes and to all persons.

Plant Allowed to Pass Under Lines

7.260 Where the passage of plant under lines is required, the passage areas should:

- be kept to the very minimum in number
- be as small as possible
- cross the lines at right angles.

7.261 Structures resembling goalposts should be erected at either side of the lines. The "goalposts" should be of a non-conducting material and the path across the lines should be fenced at either side between the posts. The height of the "crossbar" will depend on the overhead line voltage; the advice of the supply authority should be sought with regard to this dimension.

7.262 The "goalposts" will act as gates for the passage of plant. They should be:

- of a substantial construction
- marked in the same distinctive colours as the general parallel barrier
- constructed from a non-conducting material.

7.263 Notices should be fixed at both ends of the passageway informing those using the crossing of the height of the barrier. Notices giving instructions to lower crane jibs, etc should be fixed about 30m before the barriers. Both the crossing and the notices should be lit if work is to continue into the hours of darkness.

7.264 The surface of the passageway should be level and firm to prevent plant bouncing or tilting.

Work to be Conducted Beneath the Line

7.265 The maximum safe working height should be ascertained from the supply authority. Suitable plant and equipment (eg ladders, scaffold poles, etc) should be chosen so that the maximum working height cannot be exceeded. If specially designed plant is not available, this may require existing plant to be fitted with restraining devices in order to limit the range of movement. Safety devices should be designed so that they "fail to safety".

7.266 A horizontal barrier in the form of a grid of insulating material should be constructed beneath the lines at the safe clearance distance in order to form a semblance of a roof above the working area.

Note: It is vitally important that all barrier structures are erected under a safe system of work since this task will be the one that is conducted at the closest point to the lines. It is preferable that the work is carried out with the lines made dead.

7.267 All work underneath overhead lines should be strictly supervised by a competent person whose only duties are to ensure safety at all times.

7.268 It should be noted that the HSE has stated that the use of proximity warning devices and insulated guards for jibs, etc will not be considered as an adequate means of complying with the Electricity at Work Regulations 1989.

List of Relevant Legislation

- Construction (Design and Management) Regulations 2007
- Electricity (Safety, Quality and Continuity) Regulations 2002
- Provision and Use of Work Equipment Regulations 1998
- Electromagnetic Compatibility Regulations 1992
- Management of Health and Safety at Work Regulations 1999
- Electricity at Work Regulations 1989

Further Information

Publications

The following are available from *www.hsebooks.co.uk*.

- HSG47 (rev 2000) *Avoiding Danger from Underground Services*
- HSG85 (rev 2003) *Electricity at Work: Safe Working Practices*
- HSG141 *Electrical Safety on Construction Sites*
- HSR25 *Memorandum of Guidance on the Electricity at Work Regulations 1989*
- GS6 *Avoidance of Danger from Overhead Electrical Lines*

Excavation

- Excavation work is one of the most unpredictable construction areas to work in, and for this reason there are a variety of hazards associated with it.
- Staining, smells or material protruding from the trench sides might be the first indication of land contamination.
- Wheeled excavators can spin their wheels clean if there is a safe area to do this where projected material cannot injure anyone.
- If support structures are used they must be in good condition and proven by calculation to be able to withstand all imposed forces, during installation, use and removal.
- Radar surveys can detect voids, and trial holes can be dug to confirm the extent of the void.
- The location of all underground services should be determined prior to any excavation work being carried out.
- A competent person should be appointed to inspect excavations. Excavations must be inspected at the beginning of each shift and after any event likely to affect stability.
- Site conditions will dictate if warning signs are needed for dangers such as the risk of contacting overhead power lines.
- Exposure to vibration must be minimised.
- A qualified first aider should be readily available to anyone involved in significant excavation, particularly where there is a risk of contact with buried services.
- No one should be involved in any excavation work unless they have received training.

7.269 Excavation work is one of the most unpredictable construction areas to work in. The principal reasons for this are local variations in water table and the presence of underground services, eg electricity cables or gas supplies.

Employers' Duties

7.270

- The Construction (Design and Management) Regulations 2007 require that steps are taken to:
  - prevent danger to any person by providing supports to ensure no part of the excavation collapses
  - ensure no person or material can fall into the excavation
  - prevent any part of the excavation being overloaded (ie with work equipment)
  - ensure that all work equipment and materials used are inspected by a competent person at the start of each shift; after any event that may have affected its strength and stability; and after any material has fallen unintentionally or become dislodged. (The person who has inspected the work equipment and materials must be satisfied that the work can be carried out safely. If they are not, they must notify the person they were inspecting for and where they have done so, work cannot be carried out until the matter has been "successfully remedied".)
- The Lifting Operations and Lifting Equipment Regulations 1998 (LOLER) require employers to ensure:
  - that all equipment used in lifting operations is fully tested and inspected

- that loads do not become displaced or that equipment becomes overturned by thorough planning
- that employees working in mechanical handling are suitably qualified.

- The New Roads and Streetworks Act 1991 require all persons carrying out or supervising excavation in the highway to be qualified. Training requirements vary according to the role, but includes avoidance of underground services and setting out appropriate road signage.
- Under the Control of Vibration at Work Regulations 2005 employers must instruct and protect employees from harmful vibration, including whole-body vibration. The regulations set limits to an individual's daily exposure to vibration. This is in order to prevent nerve and blood circulation problems which can result in long-term disability. Employees at risk might need health surveillance.
- Employers must also provide information, and select methods and tooling which will reduce exposure to vibration. Measurement of vibration is not mandatory, but knowledgeable interpretation of published data is essential in order to provide written exposure assessments. (For example, the use of a 225mm angle grinder for two hours would be credited as high risk.)
- Under the Manual Handling Operations Regulations 1992, manual handling which might cause injury is prohibited unless an assessment has been made and suitable control measures are in place. In all cases, reasonable alternatives to handling should be employed. Actual assessments may have to be generic where there are a variety of short-term tasks. The legislation does not state a maximum safe load as there are too many variables to consider, eg the strength of the individual, size of the load, and working conditions.
- The Work at Height Regulations 2005 require assessments to identify suitable equipment and have in place controls to make sure it is used properly. If there is a risk of falling from height, then collective methods (eg handrails) should be used in preference to other methods to either prevent falls or minimise the consequences of a fall. Handrails may be removed for the passage of materials provided they are replaced as soon as possible afterwards.
- The Control of Substances Hazardous to Health Regulations 2002 state that exposure to any substance, either naturally occurring or man-made must be prevented so far as is reasonably practicable if there is a possibility of adverse health effects. Control should be maintained by methods which avoid human contact wherever possible. Where this is not possible fail-safe methods should be selected. If it is not reasonable to use these methods then the use of personal protective equipment (PPE) is permitted as a control measure provided the employer takes steps to remove reliance on such measures.
- Control measures must be monitored and kept in an effective condition. If a risk of exposure remains employers should carry out suitable health surveillance.
- The Confined Spaces Regulations 1997 require risk assessment and suitable control measures to be taken for any work on, or in, spaces where risks such as asphyxiation may be experienced from liquids, gases, fumes, free-flowing solids or other materials.
- Under the Control of Noise at Work Regulations 2005, where noise generated by work is likely to cause injury, employers must reduce the exposure. For example, if the noise experienced by the worker averages out as over 80 decibels (dB) across an 8-hour day this is likely to cause a degree of permanent hearing loss. Typical precautions include providing hearing protection, providing information and seeking to find quieter ways of working, or removing workers from the noisy areas. In some cases, the precautions will involve carrying out a programme of audiometric testing of persons at risk.

Employees' Duties

7.271

- Employees have a duty to take reasonable care of their own health and safety and that of other people who may be affected by their work under the Health and Safety at Work, etc Act 1974.
- The Management of Health and Safety at Work Regulations 1999 require employees to:
  - use any machinery, equipment, transport, safety devices and means of production in accordance with any training and instructions provided by the employer
  - inform the employer of any serious and imminent dangers to health and safety
  - inform the employer of any shortcomings in the employer's health and safety arrangements.

In Practice

Risks Involved in Excavation

7.272 Some of the main risks to consider related to excavation work include:

- contaminated land
- disturbance of underground services.

Contaminated Land

7.273 Pre-contract documentation should identify a risk of encountering contamination, but records are often sketchy. Staining, smells or material protruding from the trench sides might be the first indication of land contamination. Work should stop and employees should evacuate and wash their hands before eating.

7.274 An occupational hygienist will be needed to take samples for analysis. An industry standard has been developed for ensuring representative sampling. After analysis, the hygienist will provide advice on a safe method of working.

Wheel Washing

7.275 Many excavations will not have room for a washing plant. In this case regular road cleaning will be needed.

7.276 Wheeled excavators can spin their wheels clean if there is a safe area to do this where projected material cannot injure anyone. Shovels can be used if the item of plant is isolated.

7.277 Every effort should be made to prevent the need for unprotected workers to clear mud from the highway or footway.

Support and Existing Structures

7.278 Support is determined by the type and depth of material requiring support, the activity of the water table and the amount of available space. If support structures are used they must be in good condition and proven by calculation to be able to withstand all imposed forces during installation, use and removal. Support systems are available for hire, and the larger suppliers are able to provide suitable designs with supporting calculations.

7.279 Hazards associated with installation include the risk of falling into the excavation and the risk of contact with moving materials and equipment. Despite installing suitable support or battering back or creating steps in the sidewalls (faster installation speed) the support needs at least daily monitoring.

7.280 Movement of the trench can be avoided by:

- re-routing trenches away from previously disturbed ground or other structures
- installing supports before excavating
- installing support in cohesive soils as soon as possible after excavating
- digging true to line and level
- carefully maintaining fill behind supports
- intelligent pumping to lower the local water table.

7.281 The use of timbering for support in normal excavation work has almost ceased as it has to be more carefully inspected and is difficult to install from a place of safety.

7.282 Work carried out close to boundary walls, buildings, poles and other structures might cause their movement or spontaneous collapse. Specialist advice might be appropriate if these structures are closer than twice the final maximum depth of the excavation.

7.283 Where slate rock is encountered it may prove difficult to "cut" a clean trenchline without serious undermining taking place. Diamond sawing might be an effective precaution.

Burning

7.284 Vegetation often gets disrupted during stripping of the topsoil. If there are ways to avoid removing hedges these should be explored. The dangers of burning scrub, etc are particularly high during the summer. Not only can fire spread, but employees may be tempted to use accelerants such as tyres or fuel to get the fire started resulting in fumes, and the risk of burns. If the opportunity exists to move the material to wasteland where it can rot, this is likely to be safer and cheaper. Larger material could be chipped by a mobile chipper operated by trained employees.

Voids

7.285 Although voids seem to occur almost anywhere, they are encountered mostly in built-up areas. Radar surveys can pick them up, and trial holes dug to confirm the extent of the void. Many old wells are poorly capped, and redundant culverts uncharted. A structural engineer may be able to authorise the filling of such voids with pumped and perhaps foam concrete, but the engineer must be sure that the inlets and outlets to the void have been CCTV surveyed and sealed first.

7.286 Since most supports rely on homogenous material, support where there are voids will need forethought to enable materials to be available before full excavation.

Dewatering and Waste Water

7.287 Poorly handled waste water causes pollution, creates slippery conditions, and if allowed to pond can create unsafe lagoons where children might be at risk. Arrangements should be made to obtain the necessary discharge consents. Secure metal settlement tanks are a safer alternative.

Underground Services

7.288 No excavation work can ever rely on the statement, "we know where the services are — we put them in". All excavation work should be controlled by a written scheme. A permit to dig is normally the most satisfactory system.

7.289 HSG47 is a vital guide for all those who control excavation work. This guide contains sample texts and information on protecting most types of underground services. The broad principles are as follows.

- Study desktop information and plans from statutory undertakings on likely service routes.
- Examine on-site evidence such as manhole covers, street lights and changes in hard landscape which might indicate buried services.
- Using a cable-detecting tool to carry out a survey for possible underground services.
- Mark on the ground any suspected services detected (do not use anything which could penetrate a service).
- Wherever possible plan the excavation route to avoid services.
- Carefully excavate sufficient small trial pits, using a safe digging technique to find the actual location and suggested extent of services.
- Backfill or guard the trial pits as necessary.
- Mark the run of these services outside the projected excavation using spray paint or any other effective method which will survive traffic movement, etc (not using anything likely to penetrate a service).
- Draw up a permit to dig.
- Carry out the excavation, recording for future reference the actual location of existing services.

Rock

7.290 Where hard rock is encountered machine-mounted breakers and drills should be used as much as possible. The use of explosives in the UK for excavation is rare.

Backfilling

7.291 Backfilling by tipping has been associated with many fatalities. Despite the requirement for stop blocks to be provided to prevent machines tumbling into excavations this requirement is often ignored. Even the use of side-tipping dumpers requires great skill and planning to ensure that the trench sidewalls can support the surcharge.

7.292 Side-tip dumpers become less resistant to tipping over when the skip is raised. Special side discharge trailers equipped with conveyor belts are useful on long trench runs. Most sites can easily utilise the excavator bucket to backfill without too much complication. This reduces the level of traffic movement and can permit some preliminary compaction.

7.293 Compaction in layers is normally best achieved by remote-controlled machines.

Trenching Machines

7.294 Trenching machines need almost uninterrupted runs in suitable soils in order to be effective. When these conditions are present they reduce many hazards by passing spoil through a conveyor directly into haulage trucks. Precautions to prevent contact with underground and overhead services need to be extremely thorough. The trencher should be supplemented by conventional methods near to services. All staff must be trained and an effective method used for dealing with overspill. This type of machinery is very noisy and road closure is normally essential. A high level of whole-body vibration is common.

Inspection

7.295 A competent person should be appointed to inspect excavations. Excavations must be inspected at the beginning of each shift and after any event likely to affect stability. In addition, one of these inspections must result in a written report if the excavation is left open. This report is required within seven days and following that at no more than seven-day intervals. A significant improvement to the statutory minimum is achieved by training all staff to carry out continuous monitoring of trench support.

Bridges

7.296 Where excavations have to be crossed by pedestrians, alternative routes are the preferred choice. If this is not possible properly secured, non-slip bridges must be secured with full edge protection.

Noise

7.297 High noise levels are commonplace with all excavation work. There are high noise levels particularly when piling equipment is in use, and hearing protection will have to be matched to the A and C weighted noise levels. A competent person should carry out measurement using a type 2 meter. Many people assume that they will not be able to hear instructions if they protect their hearing, however this is not true. All shouted warnings have to exceed the ambient noise level at the time. Hearing protection approximately reduces both shouting and the background noise level. A very effective system for giving warnings is for machine operators to install a PA system for the driver. This amplifies the driver's voice so that people can be moved out of the way easily when they are entering a danger zone.

Safety Signs

7.298 Apart from signs installed to comply with the streetworks code, it is useful to display generic warning signs. Normally these remind parents to keep children away and state which forms of PPE are mandatory. Site conditions will dictate if warning signs are needed for other dangers such as the risk of contacting overhead power lines. If the works are in excess of 1.5m deep it is normal to display "danger deep excavation".

7.299 All machines should have signs affixed to them in accordance with the manufacturer's specification eg, "Warning Crush Zone".

Visibility Around Excavators

7.300 Health and Safety Executive (HSE) inspectors will require drivers to be able to see employees at all times all around the machine. This is tested by someone moving all the way round the machine at a distance of 1m. CCTV or extra mirrors are often needed to achieve this.

Vibration

7.301 All employees, particularly those who make regular use of hand-held vibrating equipment should be made aware of the risk of injury. Vibration can cause permanent disability. The use of hand tools which vibrate should be avoided where possible. If the work is essential then hand tools can be replaced by machine-mounted tools.

7.302 If using hand tools employees will need to know how long they are permitted to work with a particular tool (this is regulated, and advice must be made available by suppliers). Users of vibrating tools should keep their hands warm, avoid smoking and share the work with other employees. Owners of tools need to maintain the anti-vibration components installed in the tool.

Access into the Excavation

7.303 Ladders are traditionally difficult to secure into excavations. Often the backfill slope is used even though this is unstable and normally has reduced support. Depending on the depth of access required and space limitations a tower or man-riding basket are far superior to a ladder. If a ladder has to be used there must be a protected stepping-off point with suitable edge protection. The ladder must rise about five rungs above ground level to provide an easy handhold. As a minimum, clean granular material or hard standings should be located around the top and bottom of the ladder to reduce mud being transferred onto the rungs.

Other Excavation Hazards

Working Space

7.304 Every method statement for excavation involving more than two staff members can and should be accompanied by a simple schematic layout drawing to aid in sequencing. If consideration is not given to the space required for plant, pipe bedding, spoil and materials then double handling is inevitable, and the risk of injury from moving plant is substantially increased.

Spoil Heap

7.305 Boulders and loose material which could become dislodged should be briefly compacted by the digging machine. Surface water may also need to be managed to control the effects of run off.

Surface Water

7.306 Where streams and ditches have to be diverted it is important to consider the disruption to natural drainage, as this can lead to movement of the ground. French drains might be needed. Sandbags can be used to divert flows.

Bentonite

7.307 If a lagoon is formed for the storage of bentonite, warning signs must be displayed and a very secure fence capable of being leaned on but not capable of being climbed on should be concreted into position. The same precautions apply if groundwater is in filter beds prior to returning to a watercourse.

Welfare Facilities

7.308 The following welfare considerations should be taken into account when planning excavation work.

- Welfare facilities are likely to become contaminated due to the muddy or dusty nature of the work.
- If workers are staying on-site showers should be available. Where there is contamination in the ground advice should be sought from an occupational hygienist on additional facilities such as decontamination units.
- All staff should be made aware of the risk of Weil's disease and employers could hand out the HSE's INDG84.
- Except where the work is of very short duration (two days or less) workers should not be expected to travel to obtain hot food. An eye wash station is usefully kept in a clean place adjacent to the working area as staff may have grout or fine dust entering their eyes. If the eating room is some way from the work a temporary foul weather shelter or shade could house the eye wash.
- Portable toilets without mains connection should be the exception, used only where there are good reasons, such as the short nature of the work or an absence of drainage. Hot water is essential for both hand washing and drinks.
- If employees are required to work in the rain they should be provided with a drying facility, and somewhere to store and change their clothes.

Air Tooling

7.309 Air tooling should be maintained to ensure that tooling is sharp and vibration dampers are in sound condition. Blunt tools increase vibration exposure and are inefficient. Compressors, unless they are of the Hydrovane, screw type should have a current pressure systems inspection certificate. Anti-whip cables can be used to prevent whiplash if hose unions are at risk of becoming dislodged.

Emergency Procedures

7.310 A qualified first aider (someone who has been on a four-day course) should be readily available to anyone involved in significant excavation, particularly where there is a risk of contact with buried services. A suitable dry area will be needed for the first-aid kit. The first aider might need special training if there are unusual hazards nearby such as pipelines on chemical works.

7.311 The risk of drowning is normally small or inapplicable but sudden ground movement due to flooding can block normal escape and rescue routes. The possibility of flooding can be monitored by a competent person who appreciates the local effects of tidal water movement, disruption of the water table and ground water run off. Weather can also have an immediate or delayed effect depending on local hydrology.

7.312 For most excavation work there will be employees present above ground who should be aware of what to do in the event of an injury. There should be reliable means at hand for contacting the emergency services. For remote locations the air ambulance could be needed.

7.313 Ground that has been contaminated with flammable materials such as hydrocarbons may ignite. Although a fire extinguisher might appear under-sized for such an event, an extinguisher could be vital in preventing a small fire developing near such work. A foam extinguisher ought to be sited close to fuel stores.

Handling Pipes

7.314 Pipe hooks attached to a set of chain slings are adequate for steel pipes, otherwise strops are commonly used. Excavation work is very demanding and equipment and accessories should be checked far more frequently than in normal service. Coiled pipes are best handled on specialist coil trailers.

Confined Spaces

7.315 All supervisors of excavation work should have a foundation of knowledge on how to identify a confined space. If there is a risk of an excavation becoming a confined space, work must cease until a site-specific method statement has been drawn up.

Personal Protective Equipment

7.316 Basic personal protective equipment (PPE) for excavation work includes:

- a helmet
- hearing and eye protection (some are incorporated into helmets)
- high-visibility (hi-vis) clothing
- wet weather clothing
- safety footwear
- gloves.

7.317 Other PPE which might be required might include:

- knee pads
- gauntlets
- respiratory protective equipment (such as disposable face masks)
- harnesses.

Training

7.318 No one should be involved in any excavation work unless they have received training. All staff must have a site-specific briefing to appreciate special hazards peculiar to the site, such as where to park and cross the road, etc.

7.319 Other training could include information on:

- confined spaces
- the banksman/signaller (formerly called a slinger)
- the plant operator
- first aid
- fire
- fusion equipment
- abrasive wheels

- new roads and streetworks
- excavation support.

7.320 Some training in occupational health hazards will also be needed, for example noise, vibration, COSHH and manual handling.

List of Relevant Legislation

- Construction (Design and Management) Regulations 2007
- Control of Vibration at Work Regulations 2005
- Control of Noise at Work Regulations 2005
- Work at Height Regulations 2005
- Control of Substances Hazardous to Health Regulations 2002
- Lifting Operations and Lifting Equipment Regulations 1998
- Confined Spaces Regulations 1997
- Hedgerows Regulations 1997
- Manual Handling Operations Regulations 1992
- New Roads and Streetworks Act 1991

Further Information

Publications

HSE Publications

The following are available from *www.hsebooks.co.uk*.

- CIS8 *Safety in Excavations*
- GS6 *Avoidance of Danger from Overhead Electrical Lines*
- HSG47 (rev 2000) *Avoiding Danger from Underground Services*
- HSG185 *Health and Safety in Excavations: Be Safe and Shore*
- INDG84(L) *Leptospirosis: Are You at Risk?*

CIRIA Publications

The following are available from *www.ciria.org.uk*.

- R097: *Trenching Practice*
- R113: *Control of Groundwater for Temporary Works*
- TNO95M: *Proprietary Trench Support Systems*

Organisations

- Health and Safety Executive (HSE)
  Web: *www.hse.gov.uk*
  The HSE is responsible for the regulation of almost all the risks to health and safety arising from work activity in the UK.

Hot Work

- Hot work is any process which generates flames, sparks or heat, such as:
  - arc welding
  - brazing
  - gouging
  - flame cutting.
- Employers must identify any processes that may involve hot work and the related risks from:
  - fumes
  - gases
  - radiation
  - spatter
  - hot components
  - fire
  - explosions
  - electric shock
  - electric burns.
- Employers must ensure the workplace environment is suitable for the task.
- If necessary, employers must provide employees with suitable personal protective equipment (PPE) and ensure they are trained and competent to use it correctly.
- For work with gases, employers must ensure that:
  - equipment and installations meet recognised standards
  - safe cylinder handling techniques are used
  - the potential for the formation of flammable and explosive atmospheres is controlled
  - employees have appropriate training, instruction and supervision to follow safe practices.

7.321 Hot work is any activity that generates flames, sparks or heat. It includes arc welding, brazing, cutting and grinding.

7.322 Employers must identify any processes that may involve hot work and the related risks (eg fumes, fire, electric shock, etc). They must also ensure that the workplace environment is suitable for any tasks involving hot work.

Employers' Duties

7.323

- Employers have a general duty to ensure, so far as is reasonably practicable, the health, safety and welfare at work of all employees under the Health and Safety at Work, etc Act 1974.
- The Management of Health and Safety at Work Regulations 1999 require risk assessments to be carried out to identify risks to the health and safety of employees and for procedures to be in place to protect employees from serious and imminent danger.
- The Control of Substances Hazardous to Health Regulations 2002 place a specific duty on-site management to assess the risks to health created by an activity in which a hazardous substance is used, produced or encountered.

- The Provision and Use of Work Equipment Regulations 1998 require employers to ensure that all work equipment provided for use in welding and hot cutting operations is suitable for use.
- The Dangerous Substances and Explosive Atmospheres Regulations 2002 require employers to assess the risks of fire and explosives associated with (potentially) hazardous atmospheres.
- The Regulatory Reform (Fire Safety) Order 2005 requires the employer (through their responsible person) to carry out a fire risk assessment that should identify all potential ignition sources, including hot work.

Employees' Duties

7.324

- Employees have a duty to take reasonable care of their own health and safety and that of other people who may be affected by their work under the Health and Safety at Work, etc Act 1974.
- The Management of Health and Safety at Work Regulations 1999, require employees to:
  - use any machinery, equipment, transport, safety devices and means of production in accordance with any training and instructions provided by the employer
  - inform the employer of any serious and imminent dangers to health and safety
  - inform the employer of any shortcomings in the employer's health and safety arrangements.

In Practice

Definition of Hot Work

7.325 Hot work is any process which generates flames, sparks or heat, eg:

- arc welding
- brazing
- gouging
- flame cutting.

7.326 Hot work processes include all types of:

- welding
- cutting
- grinding
- sawing.

7.327 Although welding and hot cutting operations take place in a wide range of workplace locations, each time welding or hot cutting takes place, it exposes employees and others nearby to risks to their health and safety.

Main Hazards and Control Measures

7.328 The main health and safety hazards, and relevant control measures, include the following.

- Fumes — work in a well ventilated area.
- Gases — choose a less toxic gas if possible, work in a well ventilated area, eg with local exhaust ventilation (LEV) and provide employees with appropriate respiratory protective equipment (RPE) if necessary.
- Radiation — use protection such as personal protective equipment (PPE).
- Fire and explosion — ensure a safe working environment, ie keep flammable substances well away, keep appropriate fire-fighting equipment nearby, remain vigilant and work in a well ventilated area.
- Electric shock and burns — ensure that:
  - work equipment conforms to the appropriate standards
  - only qualified staff install the equipment and check that it is safe for use
  - electrode holders and connectors are insulated
  - employees are fully trained and check the equipment before use for damage or worn components.

Risk Assessment

7.329 Before carrying out any hot work, an appropriate risk assessment must be carried out to see what action should be taken to ensure the control of exposure to workers from any hazardous conditions.

7.330 Factors to consider include:

- the type of equipment to be used
- what it is to be used for
- where it is to be used (eg indoors or a confined space)
- when it is to be used
- the materials involved.

Welding and Cutting

7.331 Employees carrying out the welding or cutting task should ensure that the work environment is suitable for the task. The following guidelines should be considered.

- The working position should be dry, secure and free from any dangerous obstructions.
- Personnel in the immediate area must be protected from the work activities, including arc flash.
- Ample ventilation of the work area must be provided at all times.
- If necessary, consideration should be given for the provision of an air mover.
- Any confined spaces where work is being carried out should be well ventilated while the work is in progress.
- To reduce the risk of fire, sparks should be confined to the close proximity of the work area by the provision of fireproof shields, etc.
- All combustible materials must be removed from the work area if possible.
- Fire extinguishers must be provided adjacent to the working area and where necessary fire watchmen should be provided.
- No personnel must be allowed to work below elevated welding or burning working areas and no cylinders should be placed below any such areas.

- Approved respirators must be used when flame cleaning, paint burning or welding on painted or galvanised surfaces for extended time periods and must always be used when working on such surfaces in confined spaces.
- Finished hot work must carry warning notices and be guarded until it has cooled.
- The equipment user must be provided with adequate eye protection for the welding process and the chipping of slag.

Personal Protective Equipment

7.332 Suitable personal protective clothing, which is grease and oil free, must be provided and the equipment user must ensure that it is worn when carrying out welding or burning operations. The recommended minimum requirements for personal protective clothing are:

- flame-retardant gauntlets
- flame-resistant or leather aprons
- rigger safety boots or gaiters worn over safety boots
- welding safety helmets/anti-glare screens
- flame-retardant overalls, which should have button-up collars, no open pockets or turn-ups and sleeves which should be kept buttoned.

Arc Welding

7.333 There are a number of forms of arc welding to which the general guidance applies, including:

- manual metal arc
- flux cored arc
- metal inert gas
- tungsten inert gas.

Arc Welding Equipment Checks

7.334 Equipment checks which apply to all types of arc welding equipment include the following.

- The arc welding equipment and cables should be inspected and electrically tested by a competent person at regular intervals.
- Welding leads and return leads must be:
  - insulated
  - of robust construction
  - sized to carry the maximum welding current safely.
- The welding current return leads must have a cross-sectional area no less than the welding current supply leads.
- Welding lead connections must be suitably insulated and should not allow inadvertent access to live conductors when parts of the connector are separated.
- If practicable, the welding cables must be bound together up to a point compatible with the work in progress.
- Cables must be examined for damaged insulation by the equipment user immediately before use and any defects found repaired immediately.
- Cable joints must be properly constructed, ensuring that live metal is shrouded, even when disconnected.
- The cable from the power source to the welding unit must be kept as short as possible and the cable ends must be suitably restrained.

- The electrode holder must hold electrodes firmly in any working position and be provided with overall insulation so there is no bare metal that can be inadvertently touched.
- Electrode holders should be suitably shaped to avoid damage to the cable insulation where the cable enters the holder and where it is subject to bending when the holder is in use.
- Electrode holders must be provided with a means of disconnection from the welding supply lead close to the holder. It is recommended that a plug and socket be used for this purpose, with the male plug being on the cable attached to the electrode holder.
- Earth clamps and welding return lead clamps should be purpose-built and connected to the work with a bolt for strip conductors and a proper cable plug or a "G" clamp for stranded cables. It is not permissible to attach bare strands of welding cable to the work by means of a bolt.
- The welding supply unit must be supplied from a circuit protected by a sensitive fault detector having a maximum sensitivity of 30 milliamps (mA).
- If practicable, the secondary windings of the welding transformer should not be earthed at the welding unit. One side of the secondary windings should be connected to the work piece by its welding lead and the other side by the welding return lead.
- The case of the welding transformer should be connected to the system earth either through a dedicated earth core and armourings/screen or by direct bonding using a bolted connection.
- If conditions can cause damage to the cable, the earth core within the supply cable should be monitored, using a monitoring unit.
- Earthing of the work piece should be used to provide protection against internal insulation failure of the welding transformer, by keeping the work piece at or near earth potential until the protective device operates to cut off the mains supply.
- The work piece earthing conductor should be robust enough to withstand possible mechanical damage and should be connected to the work piece and a suitable earth terminal by bolted lugs or screw clamps.
- In all cases, an earth connection must be applied at the work piece and connected to a suitable earth point adjacent to the work piece. The welding return lead must be clamped as close to the point of welding as is practicable.

Equipment Examination

7.335 In addition to daily user checks, the equipment should be examined regularly throughout its service life by a competent person to ensure it remains safe and in good condition.

7.336 The safety features, which form part of the mains supply to welding equipment, in particular bonding, earthing and circuit protection devices, should be checked and tested regularly by a competent person.

Arc Welding Work Practices

7.337 The following work practices are applicable to all forms of arc welding.

- Welders should remove personal items of jewellery, in particular rings, bracelets and metallic watch straps before starting work.
- When welding outdoors, the equipment should have an appropriate level of waterproofing, an ingress protection rating of IP X4 is preferable.
- Welding in heavy rain should be avoided by erecting a fire-resistant cover over the work piece and the welder.

- The equipment user must ensure that the welding return lead is connected to the work piece and not to adjacent steel unless the type of work being carried out does not allow direct connection on to the work piece. Under these circumstances, the welding return lead should be connected as close as possible to the work piece and every effort made to ensure good earth continuity between the work piece and the welding earth lead.
- When welding on-sites where there is a system earth ring main, the welding earth should be connected to this or to structures that have been bonded to the system earth.
- When the electrical supply for the welder is taken from the site, a separate welding earth should be taken between the work piece and the earthing on the transformer case.
- Provision should be made for storing the electrode holder on the job. An insulated container or an insulated hook should be provided to prevent the electrode being inadvertently short circuited. It should not be normal practice to lay them on or in face screens or gloves.
- The equipment user should be provided with adequate eye protection for the welding process to prevent injury from "eye flash". This condition is caused by ultraviolet light from the welding arc. It is a temporary but painful sensation which may take a day or two to disappear. Persistent and prolonged exposure without protection can cause more serious eye damage.
- Protective clothing must be worn and adjusted so that no part of the skin is exposed to radiation from welding operations.

Arc Welding Additional Measures

7.338 Where practicable, the following additional measures should also be observed.

- Non-reflecting screens (eg matt green) must be placed around the welding area to shield other personnel in the vicinity from arc flash.
- Reflected glare should be reduced by the use of non-reflecting surfaces for walls. Consideration should be given to reflections from aluminium plate and white clothing which can all contribute to glare.
- Notices must be displayed warning of the possibility of arc flash — welders must warn personnel in the immediate vicinity before striking the arc.
- The user, paying particular attention to electrode holders, cables, plugs, sockets, clamps and earthing, should carry out daily checks of the equipment.
- Any defects found during the inspection or during use should be repaired immediately.
- Damaged or worn parts should be replaced. Repairs should only be considered temporary measures.

Gas Welding and Burning

7.339 There is a general requirement for all equipment used in gas welding and burning to be gas tight to prevent leakage to the atmosphere. These requirements are specified in BS EN 2909–0: 1992 *Specification for Gas Tightness of Equipment for Gas Welding and Allied Processes*.

Gas Welding and Burning Equipment Checks

7.340 General checks which should be made on equipment to ensure that it is in good condition and suitable for the work task include the following.

- Gas cylinders should be seamless or welded steel and all new cylinders must comply with the design standards, approval and certification requirements in the Carriage of Dangerous Goods (Classification, Packaging and Labelling) and Use of Transportable Pressure Receptacles Regulations 1996 or the Carriage of Dangerous Goods and Use of Transportable Pressure Equipment Regulations 2004.
- No cylinders should be allowed on-site unless they clearly identify the gas contained, the cylinder inspection date, etc. Cylinders must be stored upright at all times and only be transported in correctly designed trolleys.
- It is important to ensure the pressure regulator in use can handle the maximum supply pressure. For supply pressures up to 300 bar, regulators should comply with BS EN ISO 2503: 1998 *Gas Welding Equipment.* Pressure regulators for gas cylinders used in welding, cutting and allied processes up to 300 bar.
  (Note: Many existing regulators were made to standards which have now been withdrawn such as BS 5741, BS 7650 and BS EN 585. These regulators may not be suitable for pressures above 200 bar. If in doubt, check with the supplier or manufacturer.)
- Only regulators which are designed and specified for use with a particular gas should be used with that gas.
- Any regulators which show "creep", an inability to control the outlet pressure under no flow conditions, must be replaced immediately.
- Regulators which are to be used for oxygen must be free from oil and grease.
- Pressure gauges should comply with BS EN 562:2003 *Gas Welding Equipment. Pressure Gauges Used in Welding, Cutting and Allied Processes.* This standard incorporates safety features including safe venting if the bourdon tube fails and releases gas.
- All hoses must comply with BS EN 559:2003 *Gas Welding Equipment. Rubber Hoses for Welding, Cutting and Allied Processes.*
- Working hose lengths should not be excessive and should be kept to 5m for welding and a maximum of 20m for cutting where practicable.
- Hose joints should be properly designed couplers with hoses being crimped. Wire must not be used for securing hoses.
- Hose repairs are allowed, but only by removal of the damaged section and use of approved joints. No hose can contain more than two joints in its length.
- All hoses should be fitted with flashback arrestors, ie flame arresters incorporating pressure or temperature actuated cut-off valves.
- The welding set operator must be familiar with the stopping devices associated with the equipment.

7.341 For further guidance, refer to the Health and Safety Executive (HSE) publication *The Safe Use of Compressed Gases in Welding, Flame Cutting and Allied Processes* (HSG139).

7.342 Prior to starting work, checks should be carried out to determine that:

- all the correct items of equipment are available for the gases being used
- all necessary safety devices are fitted
- the equipment is undamaged.

7.343 Care should be taken when laying out gas hoses to avoid areas likely to damage them, ie hot pipes, places where burning or welding sparks may fall, etc.

7.344 Care should be taken when inert gas welding is being used, as inert gases can kill by suffocation as effectively as flammable gases.

7.345 Effective maintenance of gas welding and cutting equipment is essential to ensure safe operation. In addition to the pre-use checks, all the equipment should be regularly examined for the following defects.

- Leaks at any connection. Leak testing is also advisable on any occasion when connections are made, when gas leaks are suspected and after flashbacks and other incidents.
- Cuts, cracking and abrasion damage to hoses.
- Malfunction of non-return valves.
- Internal leakage in pressure regulators, possibly resulting in rising pressure in the outlet side when the outlet valve is closed.
- Damage to bull nose connections of pressure regulators.
- Incorrect operation of pressure gauges.
- Build-up of deposits of combustion products in flame arrestors, resulting in low gas flow rates.
- Damage to or malfunction of any components in the system.

7.346 The frequency of the examinations depends on the frequency of use of the equipment and the conditions it is used in. The aggressive nature of the working environment should be taken into account.

Permit-to-work Systems

7.347 In all cases, with the exception of welding or cutting in the workshop, welding operations should be subject to a permit-to-work system. This is a formal written system used to control high-risk activities.

Confined Spaces

7.348 Entry to confined spaces to carry out welding or hot cutting operations should always be subject to a permit to work.

7.349 If work is carried out in confined spaces the person in charge of the supervising work must ensure that:

- there is adequate ventilation
- the electrode holder is fully insulated to prevent accidental arcing onto metal surfaces
- low voltage protection is provided
- gas cylinders are not taken into any confined space
- unlit welding torches are not taken into a confined space
- torches are lit outside and then passed into the work area
- welding or burning equipment does not remain in a confined space once the work has been stopped
- oxygen is not used for ventilating a confined space or for clearing fumes.

Hot Work on Plant Which Contains Flammable Materials

7.350 In all cases where work is required to be carried out on equipment which contains flammable materials and where it is practicable, the flammable material must be removed and the equipment made safe.

7.351 The Dangerous Substances and Explosive Atmospheres Regulations 2002 place a duty on the employer to prevent the formation of flammable and explosive atmospheres where it is reasonably practicable to do so. If it is not reasonably practicable to do so, then the employer must assess the risks of fire and explosion arising from the (potential) flammable/explosive atmosphere and reduce these risks to as low a level as is reasonably practicable. Hence, hot work should not be carried out in areas where flammable/explosive atmospheres are likely to exist.

7.352 On no account should welding or hot cutting operations be allowed on:

- any equipment which contains a flammable mixture of gas or vapour and oxygen
- any equipment which contains oxygen rich atmospheres combined with hydrocarbons or other combustible materials
- any equipment which contains compressed air combined with hydrocarbons or other combustible materials
- any equipment which contains material liable to stress corrosion cracking, unless the equipment is subsequently stress relieved
- any pressure vessel in service which could require stress relieving or any vessel designed to high stress codes such as the American Society Mechanical Engineering Section, Division 2
- any flare system which cannot be isolated during the welding or cutting operation
- any equipment which contains substances which may undergo reaction or decomposition leading to either a dangerous increase in pressure, explosion or attack of the containing metal.

Note: Under certain conditions of concentration, temperature and pressure, acetylene, ethylene and other unsaturated hydrocarbons may decompose explosively when subject to heat.

Fumes and Gases

7.353 Welding fumes and gases are a varying mixture of airborne gases and fine particles, which if inhaled or swallowed, may be a health risk. The degree of risk will depend on the:

- composition of the fume or gas
- concentration of the fume or gas
- duration of exposure.

7.354 The main health effects are:

- irritation of the respiratory tract
- metal fume fever
- systemic poisoning
- long-term chronic effects.

7.355 A risk assessment is required to determine the risk to employees from exposure by inhalation of substances hazardous to health. In determining the level of risk and the control measures required, the Control of Substances Hazardous to Health Regulations 2002 have introduced specific workplace exposure limits.

7.356 Substances present in welding fumes may have a workplace exposure limit (WEL) indicated in EH40: *Workplace Exposure Limits*.

7.357 If welding fumes and gases cannot be eliminated, control measures need to be adopted in the following order of priority:

- choice of welding process
- improvement in working practices
- local exhaust ventilation and/or general ventilation
- use of respiratory protection equipment.

Contractors

7.358 The employer should ensure that any contractor employed to carry out welding and hot cutting operations is competent to carry out the work.

7.359 Welding and burning equipment brought to a site by contractors must be examined and must comply with minimum standards. All electrical equipment must be examined as far as is reasonably practical to ensure that:

- it is of the correct operating voltage
- it is complete in itself
- cables and plugs are intact and safe
- the required electrical testing has been carried out
- all guards are secure.

7.360 All mechanical equipment must be examined as far as is reasonably practical to ensure that it is safe for use at the location where it is used and it has, where required, correct flame traps, shut off devices, all guards are present, hoses are safe, etc.

Training

7.361 Employers must ensure that any employee or contractor involved in hot work is competent and has received appropriate information and training.

7.362 The Provision and Use of Work Equipment Regulations 1998 require anyone using work equipment to receive adequate training in its use, which should cover:

- the methods that might be adopted when using the equipment
- any risks use entails
- the precautions to be taken.

7.363 It is also necessary for supervisors to have training relating to the work equipment.

7.364 The Personal Protective Equipment at Work Regulations 1992 require that if employers deem it a necessary control measure to provide employees with personal protective equipment, that the employees are fully trained on its safe and correct usage.

List of Relevant Legislation

- Regulatory Reform (Fire Safety) Order 2005
- Carriage of Dangerous Goods and Use of Transportable Pressure Equipment Regulations 2004
- Dangerous Substances and Explosive Atmospheres Regulations 2002
- Control of Substances Hazardous to Health 2002
- Management of Health and Safety at Work Regulations 1999
- Pressure Equipment Regulations 1999
- Provision and Use of Work Equipment Regulations 1998

- Confined Spaces Regulations 1997
- Pipelines Safety Regulations 1996
- Carriage of Dangerous Goods (Classification, Packaging and Labelling) and Use of Transportable Pressure Receptacles Regulations 1996
- Reporting of Injuries, Disease and Dangerous Occurrences Regulations 1995
- Workplace (Health, Safety and Welfare) Regulations 1992
- Personal Protective Equipment at Work Regulation 1992
- Health and Safety at Work, etc Act 1974

Further Information

Publications

HSE Publications

The following are available from *www.hsebooks.co.uk*.

- EH40 *Workplace Exposure Limits* (revised annually)
- EH54 *Assessment of Exposure to Fume From Welding and Allied Processes*
- EH55 *The Control of Exposure to Fume From Welding, Brazing and Similar Processes*
- HSG53 *The Selection, Use and Maintenance of Respiratory Protective Equipment — A Practical Guide*
- HSG107 *Maintaining Portable and Transportable Electrical Equipment*
- HSG118 *Electrical Safety in Arc Welding*
- HSG139 *The Safe Use of Compressed Gases in Welding, Cutters and Allied Processes*
- HSG193 (rev 2003) *COSHH Essentials: Easy Steps to Control Chemicals*
- HSG204 *Health and Safety in Arc Welding*

British Standards

The following are available from *www.bsi-global.com*.

- BS EN 559: 2003 *Gas Welding Equipment — Rubber Hoses for Welding, Cutting and Allied Processes*
- BS EN 562: 2003 *Gas Welding Equipment — Pressure Gauges Used in Welding, Cutting and Allied Processes*
- BS EN ISO 2503: 1998 *Gas Welding Equipment — Pressure Regulators for Gas Cylinders Used in Welding, Cutting and Allied Processes up to 300 Bar*
- BS EN 14717: 2005 *Welding and Allied Processes. Environmental Checklist*
- BS EN ISO 15609–1: 2004 *Specification and Qualification of Welding Procedures for Metallic Materials. Welding Procedure Specification. Arc Welding*
- BS 29090: 1992 *Specification for Gas Tightness of Equipment for Gas Welding and Allied Processes*

Manual Handling

- Employers are required to reduce the risks from manual handling operations under the Manual Handling Operations Regulations 1992.
- Employers must identify all manual handling activities that may pose a risk of injury.
- A manual handling activity that represents a significant risk should be avoided, automated or mechanised, wherever possible.
- If a manual handling activity cannot be avoided, a risk assessment should be carried out.
- The risk assessment should take into account the task, the load, the environment in which the activity takes place and the individual's capabilities.
- Through risk control measures, employers should reduce the risk of injury to as low as reasonably practicable.
- Employees must be provided with the necessary information, instruction and training.
- All manual handling activities, controls and risk assessments must be monitored and reviewed.

7.365 This topic covers the following.

- Preliminary Manual Handling Assessment
- Conducting a Full Manual Handling Risk Assessment
- Control Measures

Employers' Duties

7.366

- Employers have a general duty to ensure, so far as is reasonably practicable, the health and safety at work of all employees under the Health and Safety at Work, etc Act 1974.
- Under the Manual Handling Operations Regulations 1992, so far as is reasonably practicable, the employer must *avoid* the need for hazardous manual handling operations.
- Where a hazardous manual handling operation cannot be avoided, a thorough assessment must be undertaken. Assessments should:
  - be suitable and sufficient, ie they must look at the complete handling operation and have anticipated all reasonably foreseeable factors
  - be carried out by a competent person whose abilities should include:
    (i) an understanding of the regulations
    (ii) a knowledge of the handling operations that are to be assessed
    (iii) an awareness of human (individual) capabilities and limitations
    (iv) an ability to recognise risks
    (v) an ability to recommend reasonably practicable solutions
    (vi) a judgment of what constitutes an acceptable residual risk
  - be kept up-to-date and revised when a significant change occurs, or in the light of experience
  - be recorded (at least the significant findings) in a retrievable medium except when:
    (i) the assessment is simple, obvious and easily repeatable
    (ii) the risks can be shown to be insignificant

(iii) the handling operation is low-risk and short-lived, and the time taken to compile a record can be shown to be disproportionate
  - take into account the tasks, the loads, the working environment, individual capability and other factors.
- Following the assessment, measures must be introduced to reduce the risk of injury to the lowest level reasonably practicable.
- The measures should encompass the provision of suitable training and information to employees. Training should include the principles of correct handling, a safe system of work and the use of any risk reduction measures provided by the employer.
- Precise information should be provided on:
  - the weight of each load
  - the heaviest side of a load, where the centre of gravity is not positioned centrally, including loads where the centre of gravity is likely to shift during handling.
- Monitoring must take place to ensure the effectiveness of those measures, and reassessment must be carried out where necessary.

Employees' Duties

7.367

- Employees have a duty to take reasonable care of their own health and safety and that of other people who may be affected by their work under the Health and Safety at Work, etc Act 1974.
- The Manual Handling Operations Regulations 1992 require employees to make full and proper use of any system of work intended to reduce the risk of injury from manual handling activities.
- The Management of Health and Safety at Work Regulations 1999 require employees to:
  - use any machinery, equipment, transport, safety devices and means of production in accordance with any training and instructions provided by the employer
  - inform the employer of any serious and imminent dangers to health and safety
  - inform the employer of any shortcomings in the employer's health and safety arrangements.

In Practice

7.368 Manual handling is an activity involving the movement or support of a load by hand or by bodily force. Common examples include lifting pallets, throwing sacks into the back of a trailer or pushing a loaded trolley.

7.369 More than a quarter of all reportable accidents are associated with manual handling. The risks include musculoskeletal injuries such as back strain, injuries caused by the load falling onto or trapping part of the handler, or injuries caused by the handler falling.

7.370 The Manual Handling Operations Regulations 1992 contain a well defined sequence of steps that employers should follow to comply with legislation.

Preliminary Manual Handling Assessment

7.371 A detailed assessment of every manual handling operation could be a major undertaking — a number of handling operations may only have a negligible risk.

7.372 The purpose of a preliminary assessment is to determine which manual handling activities involve a significant risk of injury and therefore warrant a full risk assessment.

7.373 The starting point is to identify all manual handling activities undertaken by employees, eg by:

- referral to a generic risk assessment of work activities
- consultation with employees
- workplace inspections.

7.374 The identified manual handling activities must then be evaluated to determine those that present a significant risk. Certain information can be used in making a judgment as to whether a significant risk exists, including:

- accident and sickness records in relation to manual handling
- complaints and information from members of staff involved in manual handling
- guidance produced by the Health and Safety Executive (HSE) and organisations such as Back Care.

7.375 The assessment filter and manual handling assessment charts may also be useful guides.

Risk Assessment Filter

7.376 The HSE includes a guideline filter (as the graphic below illustrates) in its guidance on manual handling that provides a starting point for assessing manual handling tasks. The filter is intended to assist in the screening out of straightforward, low-risk manual handling operations. If the filter shows that the load is within the numerical guidelines, and it is easy to grasp, the handler is in a stable position and the working environment is good, it is not normally necessary to perform a full assessment unless:

- an individual is at significant risk, eg due to pregnancy, previous injury or health condition, etc
- the activity is complex and requires greater preliminary assessment
- there are other considerations to take into account, such as psychosocial factors, eg high workloads, tight deadlines and a lack of control over working practices.

7.377 Application of the guidelines should provide a reasonable level of protection to around 95% of working men and women.

7.378 The figures used in the filter are based upon scientific literature and practical experience. However, the intention is to set an *approximate* boundary within which the load is unlikely to create a risk. It is important to note the guidelines should not be considered as safe weight limits for lifting. There are no limits below which manual handling activities can be regarded as safe. If in doubt, a more detailed risk assessment should be conducted.

Manual Handling Assessment Charts

7.379 Manual handling assessment charts may assist in identifying high-risk manual handling operations. Based upon numerical guidelines and practical experience, they are intended to guide assessors through a logical process to identify any high-risk handling operation. They can be used to assess lifting operations, carrying operations and team-handling operations.

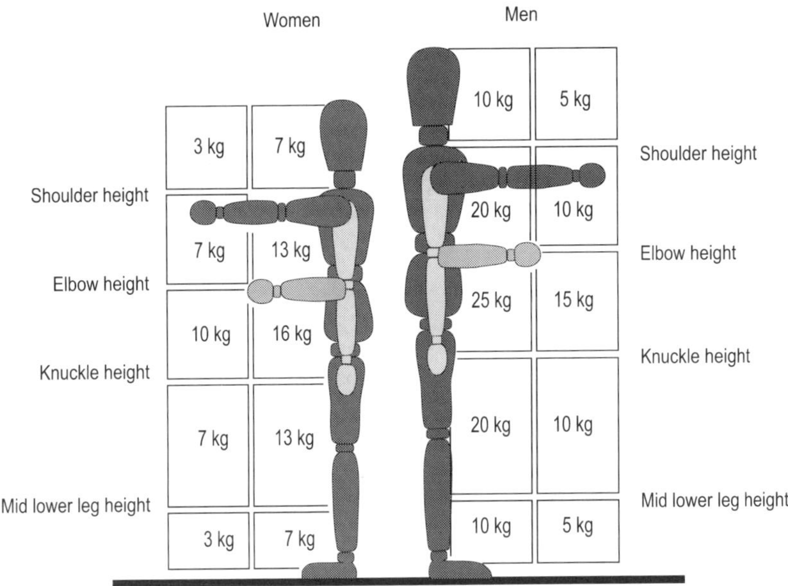

HSE Guidance on Lifting and Lowering

7.380 The chart considers seven basic factors, each of which is given a numerical value that can then be used to identify whether or not the activity is high-risk. Copies of the charts are available from *www.hse.gov.uk/msd/mac.*

Action if a Significant Risk Exists

7.381 Through observation and consultation with handlers, it may be helpful to complete a record. On the basis of this initial appraisal, a decision can be reached as to the level of risk associated with the operation.

7.382 If a significant risk exists, one of the following actions should be taken.

- Eliminate the manual handling activity altogether.
- Automate or mechanise the manual handling process.
- Carry out a full risk assessment of those tasks identified as being of significant risk that cannot be eliminated or automated.

Conducting a Full Manual Handling Risk Assessment

7.383 Employees at all levels of the organisation should be involved in the risk assessment process. This may include safety representatives, occupational health staff, ergonomic specialists, physiotherapists, etc. They may be able to provide valuable information on:

- the requirements of the regulations
- the nature of the handling operation
- human capabilities
- the identification of high-risk activities
- the practical steps required to reduce the risk.

7.384 A full risk assessment should take into account the:

- task
- load
- environment in which it takes place
- individual's capabilities.

7.385 Those carrying out the manual handling risk assessment must be competent to do so.

Task

7.386 The task refers to the specific actions, postures and movements that occur during the manual handling operation. The following features of the operation should be considered as part of the risk assessment.

Distance of Load from the Trunk

7.387 This is a crucial feature of manual handling operations that involve lifting, lowering or carrying. The further away a load is from the body, the greater the stress on the lower back.

7.388 The handler is also much more likely to topple over when the load is held away from the body. As a general rule, the weight that a handler can safely hold at arms' length is only one-fifth of that which can be safely held against the body. The following table provides some guidelines.

Distance from Body and Carrying Capacity

Distance from the Body	Percentage of Maximum Carrying Capacity
Close to body (less than 20cm)	100%
Elbows at 90° (about 35cm)	80%
Elbows at 135° (about 50cm)	50%
Elbows straight, arms straight out (about 70cm)	25%
Arms out, body bent forward (more than 70cm)	20%

Position of Load in Relation to Trunk

7.389 The position of the load influences the risk of back injury. It is best if the load is situated directly in front of the handler, rather than to one side. If the load is to one side, a sidewards bend is created. This puts uneven pressure on the discs in the back and means the centre of gravity is further away from the pivot point at the base of the spine, which increases the overall pressure.

Posture

7.390 Good posture during a manual handling operation minimises the risk of losing control of the load. It also enables the strain on different parts of the body to be better balanced. Bad postural features are high priority risks (eg weight is too far forward on the toes, heels are off the ground and feet are too close together). Measures should be taken to reduce them as soon as possible, regardless of their exact nature.

Distribution of Weight on Feet

7.391 The feet should be arranged so that the load can be evenly distributed. As a general principle, keeping the feet comfortably apart and, if possible, at right angles to each other, makes for the greatest stability.

Twisting the Trunk

7.392 This occurs frequently when picking up and moving off with items or when seated at a workbench. It increases the stress on the lower back and reduces the safe load, as the approximate figures outline.

Twist and Reduction in Safe Load

Degree of Twist	Reduction in Safe Load
30°	0%
60°	15%
90°	20%

Stooping

7.393 Stooping moves the centre of gravity of both the head and torso, and any objects that are being held, away from the pivot point at the base of the spine. The further away the centre of gravity from the pivot point, the greater the leverage effect upon the base of the spine and the greater the risk of injury. The following table provides an approximation of the reduction in safe load.

Stoop and Reduction in Safe Load

Angle of Stoop	Reduction in Safe Load
20°	25%
45°	35%
90°	50%

Precision

7.394 When a load needs to be positioned very precisely, eg when trying to relocate engine parts in a garage, a number of small adjustments will often have to be made to get the load in exactly the right place.

7.395 If these small adjustments have to be made at the same time as keeping the load static, the effect on the muscles is such that the safe weight will be much reduced. Where such operations cannot be eliminated, aids should be provided so that the handler is not bearing the full weight of the load at the same time as trying to move it, but is simply applying the manoeuvring forces.

Lifting Position and Lowering Distance

7.396 Stresses upon the back are minimised when the handling points (eg where the hands touch the load) are between the middle of the thigh and the waist. As the handling points move away from this area, stresses will increase. In addition, the further the vertical distance through which the load has to be moved, the greater the physical workload.

7.397 If the distance and load are such that the grip has to be changed, stresses are increased still further. An initial assessment should note the presence/absence of handling points and distances, while the full assessment may require more detailed measurements.

Carrying Distance

7.398 A load which is regarded as acceptable for lifting will not endanger the back if carried for a short distance. However, the fatigue associated with static work, such as holding an object, will build up if carrying over any great distance.

7.399 If the load is carried more than about 10m, then the safe weight limit (as shown in the figure *HSE Guidance on Lifting and Lowering*) will be reduced. The extent of reduction is not easily quantified and will depend as much on the grip possible on the load as on its weight. In the initial appraisal, distances over 10m should be noted. In the full assessment, exact distances and time taken to travel them should be noted.

Pushing and Pulling

7.400 Pushing and pulling exert forces across the spine rather than down it, and the safe force limit for optimum pushing is lower than it would be for lifting. Pushing and pulling with the hands much below waist height or above the shoulder increases the risks of injury. Where pushing and pulling are to be used, it is especially important to ensure that the loads are not too heavy or likely to either jam or run away.

7.401 The guidelines given in the table below assume that the distance involved is no more than 20m and that the force is applied between knuckle and shoulder height with the hands.

Guideline Load Figures for Pushing and Pulling

	Men	**Women**
Guideline figure for stopping/ starting load	20kg (200 newtons)	15kg (150 newtons)
Guideline figure for keeping load in motion	10kg (100 newtons)	7kg (70 newtons)

7.402 As a rough guide, the amount of force required to move a load over a flat surface is at least 2% of the load weight. On uneven surfaces, slopes or soft ground this figure will increase. A full assessment of the risks will be required in such conditions.

Sudden Movement of the Load

7.403 If an object being manipulated moves suddenly and unexpectedly, the forces generated in the spine are often large and unpredictable. Examples of this are the sudden freeing of a load that was jammed or encountering friction on the floor where a load is being pushed. When this happens the risk of injury is great, especially if the handler's posture is unstable.

Dynamic Activities Such as Throwing

7.404 Safe force limits for throwing activities, eg in delivering post bags or during refuse collection, are lower than those for lifting. The force required to impart momentum to a load, especially upward, is greater than for simply lifting it.

7.405 If the object fails to release properly, the unexpected stresses imposed on the spine can be increased considerably. Team throwing of objects, eg throwing sacks of grain, requires special precautions. Lack of co-ordination between the throwers can lead to the application of opposing forces, ie pushing or pulling against each other, or the total load being imposed upon one individual.

Frequency and Duration of Effort

7.406 Frequent movement of relatively light loads, or manual handling operations that require long periods in fixed postures, can lead to as much damage as one handling of a heavy load. This is because the muscles become fatigued, which means the ability to control the body accurately decreases and the risk of injury increases. This is more likely to occur with static work than a work activity that requires the handler to move somewhere with the load.

7.407 Any risk assessment should take into account the fact that, as frequency increases, the safe weight limit decreases. HSE guidance suggests the following.

Frequency of Repetition and Reduction in Safe Load

Frequency of Repetition	Reduction in Safe Load
Once or twice per minute	30%
Five to eight times per minute	50%
Twelve times per minute	80%

Rest and Recovery Periods

7.408 Objectivity in the assessment of the need for rest and recovery is difficult to achieve and will rely on observation and consultation with the handlers.

Seated Handling

7.409 The seated handler has to rely on the arms and torso for strength. The much stronger leg muscles normally used for lifting play no role. Moreover, the ability to use the body as a counterbalance is severely reduced. Consequently, the safe weight limit for seated work is much lower than for standing work.

7.410 The HSE guidance suggests that weights in excess of 5kg (3kg for women) should be regarded as risky for the purposes of an initial appraisal. Postural features of the seated activity, ie stretching, stooping and twisting, reduce the safe weight limit further. Lifting from below the level of the work surface will almost always create these risk factors.

Team Handling

7.411 Under ideal conditions, two people can lift about 1.3 times as much as one person, while three people can lift about 1.5 times as much. When climbing stairs, or carrying loads with very uneven weight distribution, the bulk of the load may fall on one individual and the safe load may be greatly reduced. Even on a level surface, a load with evenly distributed weight, eg a ladder or plank, can create risks if the team members are of different sizes. This is because the greater share of the weight would normally fall on the shorter individual.

Load

7.412 The load is the object, person or animal being handled, whether lifted, lowered, pushed, pulled or carried. The features of the load which should be taken into account are:

- the forces required to handle it
- its size
- how easy it is to grasp
- its stability
- external features that may create a hazard.

Heaviness of the Load

7.413 There is no single safe weight for lifting or safe force for pushing or pulling, as it depends on other features of the operation. Guidance for acceptable weights applies to very few operations, eg two-handed lifting in front of the body. The HSE has, however, provided guidance as to weights and forces *unlikely* to create a risk of injury such that a more detailed risk assessment is required.

7.414 The HSE states that the guidelines would give reasonable protection to about 95% of the population. The figures should not be viewed as absolute limits and may be exceeded provided a more detailed assessment shows that it is safe. Any employer continuing to require employees to use weights or forces greater than these will have contravened the Manual Handling Operations Regulations 1992, unless they can demonstrate they have made a detailed assessment.

Shape and Size of the Load

7.415 The further the centre of gravity of the load from the body, the greater the leverage effect on the spine, and the higher the risk of injury. If a load has an even weight distribution, the centre of gravity is in the centre of the load. The centre of gravity of a large load is always further away from the body than the centre of gravity of a small load. This is because the physical bulk of a large load separates the handler from the centre of gravity. Even if the two loads weigh the same, the larger load will cause more strain, due to the leverage effect.

7.416 When loads have weight unevenly distributed, the heaviest side should be nearest to the handler. When objects are contained within a carton or packing case, it may not be apparent which part is heaviest and the risk of injury is increased.

7.417 The regulations require information on weight distribution to be marked on loads. This is applicable to both the employer who requires employees to carry out handling and those who originate a load, eg a manufacturer.

Handling Points of the Load

7.418 Many loads are not particularly easy to grasp, as the load may be large, slippery or have sharp edges. In these cases, extra grip strength is needed, which fatigues the muscles quickly. As handlers tire, their grip will become weaker and they may have to change either the grip or posture to maintain control of the load. The risk of dropping the load is increased.

Stability of the Load

7.419 If a load is unstable or its contents are likely to shift during handling, eg containers of fluid, the risk of injury increases, since stresses on the spine are less predictable and the handler may not be prepared.

External Features of the Load

7.420 Assessing hazards arising from the external features of a load is largely about observation. Factors to look out for are:

- sharp or rough edges which may make holding a load difficult and cause other hazards, eg splinters of wood
- hot or cold objects, such as loads from a cold store
- chemical hazards that may expose the handler to risks from inhalation or absorption
- slippery loads, eg those which are wet, greasy or have a non-stick covering
- loads with damaged containers.

Working Environment

7.421 The working environment, for the purposes of the risk assessment, refers to:

- thermal conditions
- noise
- lighting
- chemicals
- space constraints
- the conditions of the floor and other surfaces
- standards of housekeeping.

7.422 An initial appraisal should only require observation of conditions.

Space Constraints on Posture

7.423 If the space available limits the posture a handler can adopt, the risk of injury will be increased. Restricted headroom causes stooping, which is a risk factor. Narrow walkways or tight gaps between machinery will increase the likelihood of having to twist while carrying an object. The risk of trapping the hands against a wall or fixed object will also increase.

Stability of Floor Surface

7.424 The risk of slipping, tripping or falling will be increased if the floor is slippery, uneven, unstable or cluttered. These features hinder smooth movement and, because they are unpredictable, slow the progress of the handler. This increases the length of time the load is carried and fatigue.

Levels of Floor and Work Surfaces

7.425 Steps and slopes increase the risks involved with handling loads. If handlers need to use their hands for stability, eg when going up a ladder, then the handling of loads can become much more difficult. The presence of steps or slopes should be regarded as risk factors.

7.426 When loads are pushed or pulled, eg using trolleys, it is important to ensure smooth transitions between levels. Steps require a trolley to be lifted, imposing a further load. Ramps should be used to assist movement of trolleys or wheeled objects. However, it should be remembered that ramps or other slopes will increase the forces required to push or pull the load. As an example, for a 1 in 12 slope, with a load of 400kg, an additional force of 33kg (330 newtons) will be required to move the load.

7.427 Floor surfaces should be properly maintained, as a poorly maintained floor is a risk factor.

7.428 Moving a load from one surface to another at a different height can increase the risk of injury, especially if lifting from floor level to above shoulder height. The optimum height for a work surface is normally around waist level, and will obviously vary from individual to individual.

Extremes of Temperature, Humidity or Air Movement

7.429 Extreme thermal conditions increase the risks of injury associated with manual handling. Very hot and humid conditions reduce the ability to carry out physical work, increasing the rate of perspiration and reducing the grip on the load.

7.430 Low temperatures impair the sensation of touch and the ability to control the muscles. In cooler temperatures, fatigue sets in earlier and recovery is slower. High wind speeds reduce temperature further because air warmed by natural body heat is blown away.

7.431 The initial appraisal should be concerned with answering simple questions, such as the following.

- Is the working environment governed by weather? If it is, risks to health may be increased. The nature of the risk will depend upon the season.
- Is work done in cold stores or refrigerated areas?
- Is work done near sources of heat, such as furnaces or in boiler rooms?
- Is work done in areas with high rates of airflow, such as outdoors or in ventilated tunnels?
- Is the workload intensive?

7.432 More detailed assessment may require the use of thermometers and hygrometers to accurately determine temperature and humidity.

Lighting

7.433 Poor lighting can increase the risk of injury: the governing factor is visibility. Correct levels of illumination are obviously relevant. However, other features such as glare or shadowing can also influence safety, as they might lead to misjudgment of distance or height. Too much contrast or glare can render objects invisible.

7.434 Visibility, rather than overall levels of illumination, should be the prime focus of the assessment.

Chemicals

7.435 Many of the chemicals used in industry, especially organic solvents, can have a narcotic influence similar to the effects of alcohol consumption, which may disturb balance and co-ordination. Exposure to chemicals may increase the risks of injury when carrying or lifting loads, as they can be inhaled or absorbed through the skin.

Noise

7.436 Excessive noise, in addition to damaging hearing, reduces the ability to notice sounds that may occur when lifting or adjusting a load. Noise may also obscure sounds that indicate when the contents of a heavy load are shifting.

Individual Capability

7.437 It is important to recognise that each individual is different in terms of build, strength, health, skills and knowledge. Employers should design manual handling operations to be safe for the majority of employees and protect individuals who are at an increased risk.

7.438 The assessment should identify those involved in manual handling operations that may be at risk. The assessor would be expected to know, or find out:

- whether the individuals are men or women
- if they are particularly small or large
- whether any have suffered a musculoskeletal injury recently
- what training in manual handling they have received
- if any of the female handlers are or have recently been pregnant.

7.439 The assessor should also make a note of any particular clothing or special equipment used by the handlers.

Strength and Height

7.440 There is great variation in individual ability to lift and in susceptibility to injury, and identification of susceptible individuals is very difficult. Effort expended in identifying such individuals is less valuable than time spent on improving handling methods. An improvement in a task has positive benefits for all employees. Selecting or de-selecting an individual leaves the manual handling problems unsolved.

7.441 Generally, the lifting strength of women is less than that of men but it should be remembered that the lifting range for both will vary considerably. Younger employees are likely to be stronger but may take greater risks. Older employees may be weaker, but will have more experience and a better understanding of their limitations.

Pregnancy and Health Problems

7.442 Women should not handle significant loads. Employers are required to assess particularly the risks to new and expectant mothers.

7.443 If individuals have suffered from injuries or prolonged illnesses that might reduce their strength, medical advice should be sought before allowing manual handling operations to be performed. Manual handling operations that have a very low risk for the majority of the population could still be excessive for an injured person.

Clothing and Equipment Worn by Workers

7.444 An individual's clothing, eg unsuitable shoes, can affect stability or reduce safety. Hand jewellery could interfere with the ability to grip a load.

Knowledge and Training

7.445 A lack of knowledge on how to lift loads or use any special equipment will increase the risk of injury. Similarly, a lack of understanding in management and supervisory staff can result in incorrect instructions or allocation of operations to staff who lack skills. It is important in carrying out an assessment to ask questions about general training and to become aware of any special equipment or needs.

Review of Assessment

7.446 A manual handling assessment is not a one-off exercise and reviews will be required if:

- there is reason to believe the assessment is no longer valid, eg following a change in personnel
- manual handling operations change
- an injury or accident occurs.

7.447 Assessments should be reviewed on a regular basis regardless, eg every three years, depending on the level of risk.

Control Measures

7.448 Consideration should be given as to whether it is reasonably practicable to eliminate the need to manually handle a load. Altering the task is a primary aim. The question is: "Does the load need to be moved?" If the operation cannot be eliminated, then automation or mechanisation should be considered. Any steps taken to avoid manual handling should be in line with current best practice and technology.

Automation

7.449 The automation of a process may appear expensive in the short-term but a cost-benefit analysis will show long-term gains. Fully automated robotic and conveyor systems eliminate the need for manual handling. Mobile, computer-controlled load-carrying systems that run on predetermined routes have been successfully used to transport heavy and unwieldy loads. Partial automation can also be a cost-effective method of risk control. Further examples of automation are as follows.

Vacuum Handling Systems

7.450 Vacuum handling systems can be mounted overhead or on a jib, with a variety of lifting heads suitable for drums, sacks and boxes.

Conveyors

7.451 Conveyor systems include belt systems, powered roller systems, slat systems (useful for uneven routes), overhead systems and enclosed dual screw systems (useful for transporting large quantities of powders or granules).

Mechanical Handling Aids

7.452 A large variety of simple but effective handling equipment is available, as the following examples illustrate. Maintenance of any handling equipment, particularly wheels on trolleys, is important. A sudden change in the weight of the load on a trolley can lead to tearing of muscles and other injuries. All equipment provided should be checked as part of a planned preventive maintenance programme, which includes a defect reporting and correction system. Employees should be made aware that they must report any defects in such equipment immediately. A system for recording maintenance undertaken should be in place.

Lifting Hooks

7.453 A hook attachment can be used in activities if a load would not suffer if punctured. Using long-handled hooks which hook around the bottom of the load, avoiding the need to stoop, may solve the problem of lifting and carrying large, thin sheet-like loads.

Hand-operated Trucks and Trolleys

7.454 These usually fall into the following categories.

- Standard type, which is useful for a variety of loads.
- Wheelbarrow-type trucks with large wheels for uneven ground.
- Trucks with detachable hand-tug units, allowing a number of platforms to be serviced by one unit.
- Trucks with a hand or power-operated raising and lowering platform. These are particularly helpful for stacking and unstacking because the employee does not have to stoop. They can also bring the platform height to the work surface height, making transfer of the load easier.
- Balance trucks, which have a central axle with swivel wheels at each end. This makes them easily manoeuvrable in workplaces with space constraints.
- Pallet trucks, which are moved manually but lift and lower loads hydraulically.
- Electrically powered, hand-steered trucks (like those used in hospitals).
- There are many versions of the simple sack trolley, eg one with star wheels (three small wheels on each side) for moving loads up and down stairs, and a version that has a hydraulically powered lifting mechanism. Trolleys designed for specific purposes include:
  - those for drum-shaped loads
  - those for loads that can hang on a rail (clothing)
  - those similar to a supermarket trolley but with several tiers for carrying small loads/components
  - shelf trolleys, as commonly seen in canteens.

7.455 The trolleys should have two swivel wheels on the side closest to the employee, to aid manoeuverability.

Lifting Devices and Chutes

7.456 Lifting devices include various types of hoists, which can be hand-operated, hydraulic or electric. Mobile hoists can be useful if heavy lifting is required at a variety of locations. Chutes are also an effective method for moving loads between levels.

7.457 Where automation and/or mechanisation is not practical, employers must consider other methods of reducing the risk of injury.

Reducing Risks by Modifying the Task

7.458 Having carried out an assessment, it is possible to devise task-oriented changes that will reduce the risks.

Task Involves Poor Handling Techniques and Posture

7.459 The distance and position of the load from the trunk, twisting of the trunk and poor posture can be controlled through changing the task, altering the layout of the task and effective use of the body.

7.460 Tasks that are designed to permit the load to be held closer to the body reduce risks of injury. The approach, grasping and placing of the load should be carefully controlled by:

- designing layout so the handler can get close to the load without twisting
- designing layout so the handler's feet can get as close as possible to the centre of gravity of the load
- storing objects so they can be picked up and carried without an immediate change of direction or twisting.

7.461 Simple aids may enable undesirable handling actions to be replaced by less stressful ones. For example, heavy cylinders could be transported using lifting equipment or trolleys instead of being carried by two people. Consideration must also be given to the information, instruction and training given to employees.

Task Involves Excessive Lifting, Lowering and Travel Distance

7.462 Stresses on the back during lifting, lowering and stooping, as well as having to travel distances with a load, can be controlled through change of task and use of the body. The following measures could be considered.

- Particular emphasis should be given to providing loads at waist height, the optimum height for lifting. Storage or delivery systems that allow for this can significantly reduce the risks of handling.
- Good design of storage racks, and more importantly, placing heavy or frequently handled items at waist height on racks, can reduce the risks. The need to stoop is minimised and, when stooping does occur, the distance through which the load has to be lifted or lowered is lessened. (Stooping, however, may be preferable to adopting a squatting posture, which can place excessive loads on knees and hip joints).
- Providing a suitable step, stool or platform can help handlers reach items on higher shelves, although it should be remembered that this might create a new risk of trips or stumbles.
- Heavier loads should never be placed on shelving which requires the employee to reach above head height.
- Shelves fitted with rollers make it easier to pull out or push loads, as do wheeled containers used to store loads underneath low shelving.
- Rollered tables or air beds can make it much easier for a machine operative to manoeuvre a load on a work bench or on the feed into a machine.

- The relocation of a goods receivable point closer to the point of delivery may remove the need to carry bulk deliveries.
- Handling aids, including levers for lifting objects off the ground and rollers, remove the need for stooping and lifting.
- Change the task by replacing the carrying with controlled pushing or pulling.
- Where appropriate, loads could be divided to enable lifting half with each hand. This means the weight is evenly distributed, avoiding the risks associated with one-handed lifting.
- Conveyor systems can be used to get the load as close to the handler as possible, which prevents overreaching and excessive travel.

Task Involves Excessive Effort, Fixed Posture and Insufficient Rest and Recovery

7.463 Changes to work routine can help reduce the stress on the handler's back. The following measures could be considered.

- If operations are relatively frequent, larger but less frequent loads may be possible in conjunction with a manual handling aid.
- When the same posture is maintained for a long period of time, stresses should be minimised by arranging for the weight to be borne by other means, such as a sling or hoist.
- Handling activities conducted at the pace of the individual (as opposed to the pace being dictated by a machine) are preferable wherever possible, as this gives the handler more control.
- A system which allows for flexible work breaks should be adopted if possible to prevent fatigue. If the work involves heavy static loads, handlers should be encouraged to take breaks and move around at convenient times.
- Job rotation is a particularly effective way of avoiding fatigue, especially if the different operations use different muscle groups. It also has the advantage of reducing monotony.

Seated Handling is Identified

7.464 Measures that can be taken to reduce the risks from seated handling are broadly similar to those for standing work, as follows.

- Lifting loads from the floor should be avoided. The load should be provided at working height and the worker should be able to grasp it without stretching.
- The work surface should be at the correct height relative to the seat. Since the correct height depends on the operator, adjustable seating is preferable.
- A swivel seat should be provided to reduce the need to twist.
- To prevent twisting, loads should be lifted forwards from the body.
- Seats should have a suitable backrest to reduce the load on the spine as well as undesirable movement.
- Castors on hard floors may also represent a risk and should be avoided.

Team Handling is Involved

7.465 The reduction of risks associated with team handling can be achieved by ensuring:

- enough space is available for the team to manoeuvre
- all team members have adequate access to the load
- sufficient handholds are present and, if not, stretchers or slings are used
- one person plans and takes control of the lifting task
- clear communication takes place between all team handlers.

Handling Aids and Personal Protective Equipment

7.466 Personal protective equipment (PPE) such as gloves, safety shoes, overalls and aprons, may have to be used during manual handling operations. Employers should take the following measures.

- Ensure that a range of sizes of PPE is available, as gloves which are too tight, too loose or made of inappropriate material could lead to loss of control of the load.
- Ensure that the PPE is appropriate to hazards present.
- Instruct employees in the appropriate use of the PPE provided.
- Emphasise the importance of PPE to staff, and monitor and enforce its use. A positive example from management is essential.

7.467 If equipment, including PPE, is used for manual handling, employers must ensure:

- it is appropriate
- it is maintained
- there is a defect reporting and correction system
- the equipment is supplied in sufficient quantities
- it is easily accessible
- employees use it correctly.

Reducing Risks by Modifying the Load

Heavy Loads

7.468 If weight guidelines are exceeded, or a detailed analysis shows weight to be a problem, possible solutions include:

- splitting the load into lighter containers
- instructing suppliers to provide items in lighter containers.

7.469 Making the load lighter is usually an effective way of reducing the risks of manual handling. However, making individual loads lighter can result in a greater number of journeys, which tends to increase the frequency of lifting. Other risks from high hand and/or arm repetition rates or energy expenditure may also arise.

7.470 Instead of reducing the weight of the load, consideration could be given to making the load a size such that it cannot be handled manually.

Awkward and/or Bulky Load

7.471 If the assessment shows size or bulk to be a risk factor, the load should be brought closer to the handler's body by:

- making the load smaller and less bulky, eg reducing the size of the packaging
- distributing the weight so the centre of gravity is nearer the handler's body.

7.472 If any dimension of the load exceeds 75cm, appropriate handling aids should be considered. If the load obscures vision, team handling may be a better option.

Load is Difficult to Grasp

7.473 If there is a risk of the load slipping, or undesirable postures (especially twisting) appear to be related to difficulty with holding the load, consider providing:

- handles, hand grips or indentations to improve the grasp on the load
- handling points towards the top of loads, to reduce stooping.

7.474 Ensure that handholds are wide enough to allow the whole palm to grip the object (preferably a power grip) and are deep enough to accommodate the knuckles and gloves, if worn. However, it should be noted that the use of gloves to increase grasp may, in fact, make the load more difficult to hold.

7.475 On loads to be pushed or pulled, suitable hand grips should be provided at a height of between 91 to 114cm.

Load is Unstable

7.476 If the assessment shows the load to be unstable, stability should be increased by:

- the use of packaging which will prevent objects moving around
- the use of slings or a stretcher
- ensuring containers of liquids or powders have only a small amount of free space inside, to minimise movement.

Loads with Difficult Physical Characteristics

7.477 If hazards arise from the nature of the load, consider:

- keeping the load clean and free of oils or contaminants
- placing hot or cold loads in an insulated container
- removing or covering sharp edges and corners, or providing suitable gloves.

Reducing Risks by Modifying the Environment

7.478 Once the assessment has been completed, a range of measures may be suggested to reduce risks posed by the working environment. The following changes are environment-oriented and many should already be in place if a safety policy exists.

Space Constraints

7.479 Space constraints can be removed by the following.

- Relocate equipment and/or machinery.
- Ensure gangways and other work areas are of sufficient width.
- Loads should be delivered to appropriate areas, removing the need to carry them at all or very far.
- General housekeeping should be of a high standard. Walking and carrying routes should be clearly marked to prevent objects being placed in these areas.
- When frequently moving loads through doors, automatic rather than manual operation should be considered.

Slippery or Uneven Floors

7.480 The following changes will reduce the risk of injury by improving floor conditions.

- Provide flat and well-drained floor surfaces.
- Ensure temporary work platforms are firm and stable.
- Promptly clear up spillages.
- Consider the use of slip-resistant flooring, particularly when floors can become wet.
- Carpet tiles must be securely fixed, without raised edges, especially where objects are wheeled on castors or where trolleys are used.
- Consider selecting protective footwear that is appropriate for the working environment.

Changes in Working Levels

7.481 If possible, changes in the working level should be avoided, eg by modifying the route taken by handlers and by ensuring working surfaces are of the same height to avoid the lowering or raising of loads. If changes in working level cannot be avoided, consider the following.

- Slopes should be provided in place of steps where wheeled equipment is used.
- Carrying loads up and down ladders should be avoided.
- Lifting aids, such as pulley systems and lifts, should be used where possible.
- Identify any changes in working level clearly, eg with signs, notices, and yellow and black tape or paint.

Extreme Thermal Environment

7.482 If the thermal environment may cause a risk of manual handling injuries, solutions will depend upon the exact nature of the risk. Consider:

- transfer of work to a more comfortable thermal environment
- alteration of the environment by providing heating, cooling or air-conditioning.

7.483 If thermal conditions cannot be sufficiently modified, then provision of PPE must be considered.

Poor Lighting

7.484

- Provide sufficient, well-directed light to enable visibility of handled objects and immediate surroundings.
- Avoid the need for handlers to travel through poorly lit areas and from bright to dim light.

Chemical Hazards

7.485 Risks posed by chemicals in the work environment can be dealt with by using the following standard hierarchy of measures.

- Change the chemicals used to a less hazardous type.
- Separate the chemicals from workers by using enclosures.
- Provide extraction systems to remove noxious chemicals from the environment.
- Supply suitable protective equipment, such as respirators.

Excessive Noise

7.486 Risks arising from noise can be dealt with by appropriate noise control methods such as:

- using quieter processes
- enclosing the noise sources.

Reducing Risks Related to Individuals

7.487 Once individual capability has been assessed, a range of changes may reduce the risks to health arising from these individual factors.

Task Requires Unusual Capabilities

7.488 If the task requires unusual levels of strength and height the employer should consider:

- modifying the task
- providing team lifting.

Individual at Risk Due to Health Problems

7.489 Pre-employment screening should be carried out. In most circumstances, a questionnaire covering previous musculoskeletal injuries should suffice. However, a medical examination may be justified if the intended handling activity is considered to carry significant risk and there is a history of injuries. Women should be asked whether they are or have recently been pregnant.

7.490 In addition, employers might want to consider the following measures.

- After sickness absence, employees should be screened and allocated light work if risk factors are identified.
- Supervisors and employees should be informed about individuals particularly at risk.
- Sufficient scope for job allocation should be possible to reduce risks for individuals.

New or Expectant Mothers

7.491 Reassess the manual handling operation to reduce risk and consider the improvements that can be made as follows.

- Allow more frequent rest breaks.
- Provide lighter duties temporarily, relocate, job-share or suspend manual handling duties.
- Liaise with the woman's GP over continuation of manual handling.
- Provide handling aids.
- Continue monitoring the situation, both during pregnancy and for three months following the return to work.

Task Requires Specialist Training

7.492 Training should be introduced as part of the effort to reduce manual handling risk. In itself, however, it is not a sufficient means of reducing risk. If an employer introduces a training programme without introducing any other control measures, this would not be sufficient to meet the requirements of the legislation.

7.493 Some manual handling activities in certain sectors may require more specialist training, eg the lifting of patients in a healthcare setting, or certain activities of the fire service.

Individual Requires Special Protective Clothing

7.494 If specialist clothing is required, consideration should be given to handling aids and personal protective equipment used.

7.495 Support belts, as used by competitive weightlifters, are marketed widely and are regarded as a form of PPE, but their use should be considered carefully. They should not be used to replace a proper assessment and should only be permitted if the assessment explicitly justifies their use. Under no circumstances should support belts be used as a justification for lifting heavier weights.

Training

7.496 Adequate information and training must be provided to all employees assessing and carrying out manual handling operations. This will include:

- identifying those who require training
- designing an appropriate training programme

- implementing the training programme, using the most effective methods
- monitoring the workplace to ensure the training has been effective
- keeping a record of any training employees have received
- providing refresher training at suitable intervals.

Who Needs Manual Handling Training?

7.497 Information and training will have to be provided to three main groups:

- those designated to carry out manual handling assessments, eg the competent persons
- those carrying out manual handling activities who have been identified through the assessment as requiring training
- line managers who are responsible for the supervision and monitoring of manual handling activities.

What Should Training Include?

Assessors of Manual Handling

7.498 It is important that those conducting the risk assessment have the necessary competence to carry out the assessment. Training for persons to perform assessments should include:

- an understanding of the Manual Handling Operations Regulations 1992
- knowledge of the operations/activities to be assessed
- the ability and experience to identify unsafe working practices, risk of injury and corrective actions required
- an understanding of safe manual handling techniques
- the ability to recognise when additional expert advice may be required (such as for lifting persons)
- how to effectively communicate relevant findings and implement necessary actions.

Manual Handlers and Supervisors

7.499 Any training for employees performing or supervising manual handling activities will have to be tailored to specific work activities. Training will usually be a combination of explanation, demonstration by the trainer and practice by the employees.

7.500 Training may take place during employee induction, after significant risks are identified or when work activities change. It may be necessary to carry out refresher training, eg after a review, an accident or because the manual handling activity is high risk. Typically, the programme should cover:

- the risks and injuries associated with manual handling
- the legal obligations of the employer and employee under the Manual Handling Operations Regulations 1992
- company rules and procedures
- how potentially hazardous loads may be recognised
- the weight and nature of the loads to be encountered
- how to deal with unfamiliar loads/non-standard operations
- proper use of handling aids
- proper use of personal protective equipment (PPE)

- relevant features of the work environment
- importance of good housekeeping
- factors influencing individual capability
- good handling techniques
- reporting of unsafe activities and accidents.

7.501 Line managers and supervisors will be expected to have sufficient knowledge of these factors to enable them to carry out their role in:

- supporting safe systems of work
- assessing manual handling risks.

List of Relevant Legislation

- Management of Health and Safety at Work Regulations 1999
- Manual Handling Operations Regulations 1992
- Health and Safety at Work, etc Act 1974

Further Information

Publications

HSE Publications

The following are available from *www.hsebooks.co.uk*.

- HSG115 *Manual Handling. Solutions You Can Handle*
- INDG143(L) (rev 2, 2006) *Getting to Grips with Manual Handling: A Short Guide for Employers*
- INDG398 *Are You Making the Best Use of Lifting and Handling Aids?*
- L23 (rev 2004) *Manual Handling. Manual Handling Operations Regulations 1992 (as amended). Guidance on Regulations*

Organisations

- BackCare
  Web: *www.backcare.org.uk*
  BackCare is a source of information about the causes, treatments and management of back pain and promotes best practice in the diagnosis, treatment and management of disorders of the spine and aims to prevent incapacity from back pain.
- Ergonomics Society
  Web: *www.ergonomics.org.uk*
  The Ergonomics Society aims to increase the general understanding of ergonomics and its importance.

Metal Working

- Metal work normally takes place in engineering workshops of various sizes and includes processes such as cutting, grinding, pressing and forging.
- Work with metals can be hazardous. Injuries and accidents arise from handling and carrying, metal shards, metal-working fluids (causing dermatitis, asthma, etc) and from trapping in, or contact with, moving parts of machinery.
- A thorough risk assessment must be carried out before beginning any work with metals.
- The risk assessment should consider the risk associated with each task undertaken, the potential hazards and those who may be harmed.
- After completing a risk assessment, the workplace environment should be made as safe as possible to minimise any risks identified during the assessment.
- The findings of the risk assessment must be recorded and periodically reviewed.
- It is important to maintain good housekeeping, ie the workshop should be kept clean and free from obstructions. Factors such as lighting, temperature and ventilation should all be considered.
- Many injuries sustained when working with metals occur through incorrect manual handling, eg when handling sharp edges. The risk of injury from sharp edges should be considered in the risk assessment.
- Control measures to combat the risk of injuries from sharp edges should include methods to remove sharp edges before manual handling of an item, or use of trays and baskets to move items around.
- Many metal-working operations and processes generate significant amounts of noise. Areas with high levels of noise (greater than 85dB(A)) should be clearly marked — including indications that ear defenders should be used — and efforts should be made to reduce noise at the source.
- Many metal-working operations involve the use of items of machinery that have common associated hazards. Guards should be fitted to protect operators. Employees should be fully trained and competent to use the machinery.
- Any metal-working fluids used may be irritants and therefore care should be taken to avoid skin contact or inhalation. Appropriate personal protective equipment should be worn.

7.502 There is no specific legislation in place for metal work operations. However, as a result of the varied nature of this type of work, various health and safety legislation applies.

7.503 As a wide range of activities is undertaken in engineering workshops, this guidance does not cover every hazard that may occur. It does, however, cover the major areas of concern and provides the basic tools to identify and control hazards related to metal-working activities undertaken in engineering workshops. It also examines the hazards relating to the use of metal-working fluids.

Employers' Duties

7.504

- Employers have a general duty to ensure, so far as is reasonably practicable, the health, safety and welfare at work of all employees under the Health and Safety at Work, etc Act 1974.
- Employers have a duty to carry out risk assessments in order to identify risks to the health and safety of employees and to put procedures in place to protect employees from serious and imminent danger, under the Management of Health and Safety at Work Regulations 1999.
- Employers also have a duty to assess the risks to health created by an activity in which a hazardous substance is used, produced or encountered, eg metal-working fluids, under the Control of Substances Hazardous to Health Regulations 2002.
- Under the Provision and Use of Work Equipment Regulations 1998, employers must ensure that:
  - all work equipment provided is suitable for the use and purpose for which it was provided
  - equipment is maintained in good working order and repair
  - information, instruction, training and supervision in relation to work equipment is adequate
  - equipment is safeguarded to prevent risks from mechanical and other hazards.
- Under the Manual Handling Operations Regulations 1992, employers must ensure that:
  - the need for hazardous manual lifting and handling is avoided if reasonably practicable
  - the risks of hazardous manual lifting and handling are assessed if it cannot be avoided and the risk reduced accordingly.
- Under the Control of Noise at Work Regulations 2005, employers must ensure that:
  - noise assessments are carried out which consider the persons at risk, enabling employers to reduce noise where reasonably practicable other than by the use of ear protection
  - ear protection is made available if the daily noise exposure exceeds 80dB(A) and that ear protection is worn if the daily noise exposure exceeds 85dB(A).
- Under the Workplace (Health, Safety and Welfare) Regulations 1992, the workplace should be maintained in an efficient state from the viewpoint of health, safety and welfare.
- Under the Personal Protective Equipment at Work Regulations 1992, suitable personal protective equipment must be provided free of charge to protect employees against risks which have not been controlled by other means. Employers must take all reasonable steps to ensure that the equipment is properly used.
- Any cases of skin cancer or oil acne must be reported under the Reporting of Injuries, Diseases and Dangerous Occurrences Regulations 1995.

Employees' Duties

7.505

- Employees have a general duty to take reasonable care of their own health and safety and that of other people who may be affected by their work under the Health and Safety at Work, etc Act 1974.
- The Management of Health and Safety at Work Regulations 1999 place a duty on all employees to follow health and safety instructions and to report danger.
- The Manual Handling Operations Regulations 1992 require that employees follow safe systems of work laid down by employers and use any mechanical aids that are provided.
- The Personal Protective Equipment at Work Regulations 1992 require employees to use any personal protective equipment (PPE) provided and to report any loss or defect to their employer.

In Practice

Main Causes of Injury in Metal Working

7.506 Activities associated with metal work are cutting, grinding, pressing and forging. Metal-working activities normally take place in engineering workshops of various sizes. The data published by the Health and Safety Executive (HSE) on accidents in engineering industries indicates that there are six main areas that account for the majority of accidents. These are (in order):

- handling and carrying
- being struck by moving plant and equipment
- slips and trips
- machinery, particularly trapping or contact with rotating parts of the machine
- falls from height
- workplace transport.

7.507 Many accidents also occur during maintenance operations on machines, due in part to failures of machine safeguarding, and there are hazards to health associated with metal-working fluids.

Risk Assessment of Metal Working

7.508 The assessment of risk involves the consideration of:

- each task that is to be undertaken
- the potential hazards associated with the task
- those who may be harmed.

7.509 Due to the large number and variation in metal-working tasks that occur within an engineering workshop, each task should be subjected to formal periodic risk assessments carried out by a competent person. In addition, employees should be provided with adequate information and training to allow them to assess the risk of each task undertaken during the working day.

The Five Basic Steps of a Risk Assessment

Identify the Hazards

7.510 Consider all metal-working activities undertaken and identify all potential hazards that may arise from the environment in which the task takes place. For example, there may be hazards from:

- the machine
- equipment and materials used
- other work activities taking place in the vicinity
- the ergonomics of the task itself.

Decide Who Might be Harmed

7.511 Consider not only the employee undertaking the task being assessed but also others who work in the vicinity or may pass through the work area.

Evaluate the Risk and Implement Any Control Measures Required

7.512 For each of the hazards identified, determine the associated risk, the probability that it may occur and the consequences. The risks could be prioritised as low, medium or high. Control measures should be implemented that reduce the risk to a level that is as low as reasonably practicable, beginning with those work activities identified as high risk.

Record the Findings of the Risk Assessment

7.513 There is a requirement to record the findings of the risk assessment. This is also good practice as it provides an auditable record of the process and information that can be used to brief employees on the hazards associated with specific tasks.

Periodically Review the Risk Assessment

7.514 The risk assessment should be reviewed immediately if there is reason to believe there have been major changes to the task or to the equipment used, or if it has been found to be inadequate.

A Safe Place to Work

7.515 The majority of metal-working tasks take place in an engineering workshop of some kind. It is therefore important that the hazards of the workshop itself are considered. While the work environment should be considered as part of the risk assessment of individual tasks, it is also important to look at the hazards related to the workshop itself in a separate assessment.

7.516 The aspects of the workshop environment to be considered in the risk assessment should include the following.

Workshop Hygiene

7.517 The workshop area should be in good repair and kept clean and free from obstructions and materials that present a hazard, eg that might cause slips, trips and falls. The workshop should also have suitable washing and toilet facilities: these are required to be kept clean and orderly. Where appropriate, separate toilet facilities should be provided for men and women.

Moving Around the Workshop

7.518 In workshops where both pedestrians and vehicles are present it may be necessary to clearly define and separate the routes to be used. Generally, the workshop floor should be even, not slippery. Handrails and ramps should be provided, as necessary.

Lighting

7.519 The lighting levels (luminance) within the workshop may vary depending on the tasks being undertaken. However, the luminance level should always be suitable and sufficient for the task. Where rotating machinery is used in the workshop, it is important to consider the strobe effects of fluorescent lighting. This is because rotating equipment can appear to be stopped. The effect can be overcome by feeding adjacent lights from different phases of the electrical supply or by the use of a high-frequency electrical supply.

Temperature and Ventilation

7.520 There is a legal minimum temperature of 16°C, with a lower limit of 13°C specified where physical effort takes place. However, there is no maximum temperature limit specified, only guidance on what is considered a reasonable temperature. There should be airflow through the workshop of a minimum of 5 litres per second per person, although a figure of 8 litres per second per person is the recommended comfort level. The velocity of the air through the workshop should be a minimum of 0.1 to 0.15 metres per second. For work activities that generate fumes, such as soldering/brazing/welding, local exhaust ventilation should be considered as part of the risk assessment for that task.

Manual Handling

7.521 In addition to those injuries normally associated with manual handling, such as injuries to the spine, hand, arm shoulders and neck, there are many injuries caused during metal work operations by the handling of material with sharp edges. The risk of injury from sharp edges should be considered as part of the assessment of risk for a specific task. Control measures to be considered include:

- implementing methods of removing sharp edges before manual handling of the item
- avoiding handling items with sharp edges by the use of trays, holders or baskets
- automating the process, to remove the requirement for manual handling
- use of the appropriate personal protective equipment, provided that its use does not introduce additional risks.

7.522 In assessing the risks associated with general manual handling tasks, the following areas should be considered:

- the task, eg height of the lift, time allowed, distance, number of repetitions, body movements required while lifting
- the load, eg the weight, size, centre of gravity, its temperature and shape
- task environment, eg lighting levels, floor surface, route of travel, obstructions, number of steps
- the employee, eg physical condition, training needs.

Noise

7.523 Many metal-working operations and processes generate significant amounts of noise. This problem is often made worse by the voluminous size of many engineering workshops. Noise is measured on a logarithmic scale known as the decibel, which is weighted to reflect the characteristics of the human ear and is normally written as dB(A). It is important to remember that due to the logarithmic scale, the noise level doubles every 3dB(A). There are two factors to consider in the risk assessment of noise:

- the noise level
- the duration of exposure to that noise level.

7.524 These two factors produce a combined figure that is referred to as the "daily personal noise exposure", expressed as L_{Epd}. However, because workers are exposed to varying levels of noise throughout the day a measure called "continuous equivalent" noise level, or L_{EQ}, is used. This measurement is normally directly available from modern integrating noise meters.

7.525 Noise should be reduced to as low a level as is reasonably practicable without the need for ear defenders. There are three action levels for noise — 80 and 85dB(A) and the exposure limit value of 87 dB(A). Employers are required to make ear defenders available to employees if the noise is equal to or greater than 80dB(A). If the noise is greater than 85dB(A), employers are required to provide employees with ear defenders and must ensure, as far as is reasonably practicable, that they use them. Areas where high levels of noise are likely to be present should be clearly marked and signs should indicate the use of ear defenders.

7.526 In line with the provisions of the Control of Noise at Work Regulations 2005, the employer must undertake a noise risk assessment if there is a risk to the health or safety of employees arising from workplace exposure to noise.

Machining

7.527 There are various types of machines used in metal-working operations, ranging from manually controlled machines to machines fully controlled by computer. There are hazards common to the operation of the majority of such machines.

7.528 When selecting new machinery, the purchaser/employer should give due consideration to the following.

- The machine design is such that hazards have been removed, where reasonably practicable.
- Where hazards exist, guards are fitted to protect operators and instructions are provided on how risks can be avoided.
- The noise generated by the machine is as low as is reasonably practicable.
- Clear instructions are available on the safe operation of the machine including how to start/stop in normal operation, how to stop in an emergency, and maintenance and inspection details.
- Emissions from the machine should be contained or provision made for the connection of ventilation equipment.
- If the machine uses metal-working fluids, details of that fluid should be provided, including a safety data sheet and instructions on the safe handling of the fluid.

7.529 All machinery should be checked at appropriate intervals, ie there should be both daily checks and thorough checks every 6 to 12 months, depending on the equipment and the likelihood of any defects or damage. These checks are to ensure that any deterioration can be detected and remedied in good time.

Metal-working Fluids

7.530 Metal-working fluids contain mineral oils or synthetic lubricants and may contain other substances such as emulsifiers, stabilisers, fragrances, etc. Metal-working fluids are hazardous to health and, as such, they are subject to the Control of Substances Hazardous to Health Regulations 2002.

7.531 Metal-working fluids are usually applied by continuous jet, spray, mist, etc and therefore care needs to be taken to avoid skin contact. All types of metal-working fluids may cause irritation or dermatitis. Any cases of folliculitis (oil acne) or skin cancer must be reported under the Reporting of Injuries, Diseases and Dangerous Occurrences Regulations 1995 (RIDDOR). Any fumes/aerosols that might contain metal-working fluids need to be avoided as these can also cause irritation or asthma.

7.532 Suitable eye protection should be worn when handling metal-working fluids or when at risk of being splashed. All personal protective equipment (PPE) should be used in accordance with the Personal Protective Equipment at Work Regulations 1992 and be suitable for the purpose.

Training

7.533 Employers should ensure that any employee or contractor who is involved in metal-work activities is competent and has received appropriate and adequate information and training.

7.534 For manual lifting and handling tasks, training should cover the following:
- how to recognise potentially harmful tasks
- the appropriate safe systems of work
- the use of mechanical aids
- good handling techniques.

7.535 Many metal-working operations involve the use of machinery. It is therefore important that operators are trained in the correct use of such equipment, including:
- the identification of the main dangers and safeguards of each machine
- how to start/stop the machine safely
- where appropriate, how to safely load and unload components, remove swarf (fine filings of metal) and adjust coolant flow
- the hazards of, and how to work safely with, metal-working fluids.

7.536 Those employees who work in noisy environments should be trained in the use of any noise protection.

List of Relevant Legislation

- Control of Noise at Work Regulations 2005
- Control of Vibration at Work Regulations 2005
- Control of Substances Hazardous to Health Regulations 2002
- Management of Health and Safety at Work Regulations 1999
- Provision and Use of Work Equipment Regulations 1998
- Reporting of Injuries, Diseases and Dangerous Occurrences Regulations 1995
- Manual Handling Operations Regulations 1992
- Personal Protective Equipment at Work Regulations 1992
- Workplace (Health, Safety and Welfare) Regulations 1992
- Health and Safety at Work, etc Act 1974

Further Information

Publications

HSE Publications

The following are available from *www.hsebooks.co.uk*

- EIS16 *Preventing Injuries from the Handling of Sharp Edges in the Engineering Industry*
- EIS21 *Immersion and Cold Cleaning of Engineering Components*
- HSG42 *Safety in the Use of Metal-cutting Guillotines and Shears*
- HSG129 (rev 1999) *Health and Safety in Engineering Workshops: New Edition Including Guidance on LOLER and PUWER 98*
- HSG231 *Health and Safety with Metal-working Fluids*
- HSG236 *Power Presses: Maintenance and Thorough Examination*
- HSG246 *Safety in the Storage and Handling of Steel and Other Metal Stock*
- INDG165(L) *Health Surveillance Programmes for Employees Exposed to Metal-working Fluids — Guidance for the Responsible Person*
- INDG365 *Working Safely with Metal-working Fluids: A Guide for Employers*
- L21 (rev 2000) *Management of Health and Safety at Work. Management of Health and Safety at Work Regulations 1999. Approved Code of Practice and Guidance*
- L22 (rev 1998) *Safe Use of Work Equipment. Provision and Use of Work Equipment Regulations 1998. Approved Code of Practice and Guidance*
- L24 *Workplace (Health, Safety and Welfare) Regulations 1992 — Approved Code of Practice*

Roofwork

- Each year, 28 workers are killed, typically by falling through fragile roofing materials or by falling from unprotected edges.
- Designers can eliminate hazards and make risks easier to manage "so as to provide a safer place of work on the roof", principally by avoiding, through good design practice, the foreseeable risks.
- It is likely that a large number of older buildings will not have appropriate access and edge protection and many may have fragile roofs — therefore extra care and planning needs to be taken when carrying out inspections or work on these roofs.
- The Work at Height Regulations 2005 (WAHR) set out a hierarchy of control for determining how to work at height safely.
- The WAHR contain specific requirements in relation to fragile surfaces, which are given the definition of "a surface which would be liable to fail if any reasonably foreseeable loading were to be applied to it".
- The Red Book defines a method for testing for non-fragility which gives consistent results when repeated or reproduced by different assessors.
- Contractors are expected to follow the general principles of the WAHR hierarchy of control for working at height, giving collective protective measures priority.
- It is recommended that workers do not consider going on any roof in poor weather conditions.
- If work on a fragile surface is unavoidable it is essential to identify all fragile materials and decide on and implement stringent precautions.
- Employers are required to ensure that no person engages in any activity, including organisation, planning and supervision, in relation to work at height or work equipment for use in such work unless they are competent to do so or if being trained or supervised by a competent person.

7.537 One of the main causes of deaths and injuries at work each year is falling from height, particularly through or from roofs. On average, 28 workers are killed each year by falling through fragile roofing materials or from unprotected edges. In addition, several hundred are seriously injured.

7.538 Employees and members of the public may also be killed and injured when hit by materials falling or thrown carelessly from roofs.

7.539 There may be many reasons for work to be carried out on or near roofs, not least the original construction. There is then the necessary maintenance, cleaning, repairs, refurbishment and finally stripping/dismantling of a roof. Managing the hazards connected with these work activities is just as important as managing any other hazards that people at work are exposed to.

7.540 Health and safety legislation such as the Construction (Design and Management) Regulations 2007 (CDM 2007) and the Work at Height Regulations 2005 set the framework for the use of suitable and sufficient protective and preventive measures to control the risks associated with roofwork.

Employers' Duties

7.541 Under the Work at Height Regulations 2005, duty holders must:

- avoid work at height where they can
- where work at height cannot be avoided, use work equipment to prevent falls
- where the risk of falls cannot be eliminated, take measures to minimise the distance and consequence of any falls.

7.542 In addition to these basic principles, duty holders must ensure that:

- work at height is properly planned and takes into account weather conditions
- all those who work, plan or supervise such work are competent and properly trained
- the place of height is safe
- equipment such as scaffolding and ladders are inspected in accordance with the specific requirements of the regulations
- the risks associated with fragile surfaces are controlled
- the risks from falling objects are properly controlled.

Employees' Duties

7.543

- Under the Health and Safety at Work, etc Act 1974 employees have a duty to:
  - take reasonable care of their own health and safety and that of other people who may be affected by their activities at work
  - co-operate with their employer to enable the employer to comply with health and safety duties.
- Under the Management of Health and Safety at Work Regulations 1999 employees are required to:
  - use any machinery, equipment, dangerous substances, transport, safety devices and means of production in accordance with any training and instructions provided by the employer
  - inform the employer of any serious and imminent dangers to health and safety
  - inform the employer of any shortcomings in the employer's health and safety arrangements.
- Under the Work at Height Regulations 2005, employees must:
  - report to their employer or person in control of the work, any activity or defect relating to work at height that may endanger the safety of themselves or others
  - use any equipment provided for their use in accordance with any training and/or instructions relating to its use so as to enable their employer or person in control of the work to comply with their own health and safety duties.

In Practice

7.544 To help meet legislative requirements and to improve safety in roofwork, the Advisory Committee for Roofwork (ACR) was established in 1998. Made up of nominees from trade associations and organisations involved in the industry, the ACR provides information that it hopes will make it easier to carry out roofwork safely. The committee also defines the UK's only test, recognised by the Health and Safety Executive (HSE), to determine the non-fragility status of roofing products.

The ACR "Orange Book"

7.545 The ACR's Orange Book, *Recommended Practice for Work on Profiled Sheeted Roofs*, emphasises that "failure of any roof while it is being worked on can have devastating effects" and that it is "essential to plan working on a roof carefully".

7.546 Although the Orange Book is specifically for profiled roofs, the general advice and principles contained within the publication can be applied to other types of roof design.

7.547 In particular, the Orange Book highlights key elements to ensuring safety for persons working on roofs, which include:

- designers designing-out risks
- clients employing competent practitioners
- proper planning and control of the work
- employing a competent roofing company
- ensuring that all workers on roofs are properly trained in the necessary skills and in safety procedures and are appropriately managed
- providing the building owner with good advice about the essential maintenance and advice about a safe system of work to be applied every time access to the roof is required.

New Buildings

7.548 Guidance from the HSE, HSG33: *Health and Safety in Roof Work*, notes that designers can eliminate hazards and make risks easier to manage "so as to provide a safer place of work on the roof", principally by avoiding, through good design practice, the foreseeable risks. This should include designing in fall-protection measures (both for the construction phase and any work thereafter such as maintenance and inspection work) and safe means of access to the roof.

Existing Buildings

7.549 For buildings built prior to the introduction of the Construction (Design and Management) Regulations 2007 (CDM 2007) and the formation of the ACR, it is likely that a large number of buildings will not have appropriate access and edge protection and many may have fragile roofs, or at least elements within it that are fragile. In addition, new build roofs may not remain non-fragile forever and can still represent a serious issue for those carrying out the work.

Work at Height Regulations 2005

7.550 Guidance from the ACR complements and reflects the legal obligations and goal-setting requirements of the Work at Height Regulations 2005 (WAHR). The regulations require employers to ensure that any work taking place at height is properly planned, appropriately supervised and "carried out in a manner which is as far as is reasonably practicable safe" (this includes planning for emergencies).

7.551 The regulations require the duty holder to:

- avoid work at height where they can
- use work equipment to prevent falls when work at height cannot be avoided
- when the risk of falls cannot be eliminated, use work equipment to minimise the distance and consequences of any fall.

7.552 In some circumstances, the need to work on a roof can be avoided, eg inspections may be possible from adjacent structures, from access equipment, or from below the roof. However, it is inevitable that some form of work will have to take place on roofs and it is essential that planning and assessing of any roofwork is undertaken. The WAHR set out a hierarchy of control for determining how to work at height safely.

7.553 The hierarchy should be followed systematically and only when one level is not reasonably practicable may the next level be considered. Where roofwork cannot be avoided under the WAHR, the duty holder must ensure that:

- all roofwork is properly planned and organised
- those involved in planning and carrying out the work are competent
- the risks from the work are assessed
- appropriate work equipment is selected, used, maintained and inspected
- the risks from fragile surfaces are properly controlled
- the risks from falling objects are properly controlled
- emergencies and rescue are planned for, eg if a fire occurs.

Risk Assessment

7.554 The WAHR do not contain a specific requirement to carry out a risk assessment but rather state that "every employer shall take account of a risk assessment under regulation 3 of the Management of Health and Safety at Work Regulations 1999".

7.555 The factors to be considered in a risk assessment should include the:

- work activity, ie the proposed roofwork
- equipment to be used, both for access and egress purposes, and in the actual work activity
- duration of the work
- location of the roof to determine the presence of hazards such as overhead power cables
- work environment, including weather conditions, lighting, space, etc
- type, condition and stability of the roof, including fragility, sloping, flat or industrial roof
- physical capabilities and competency of the workers
- emergency procedures required in the event of an incident or accident, eg a fire.

7.556 In addition to the above, when selecting work equipment the WAHR require consideration of the:

- working conditions and risks to the safety of persons at the place where the work equipment is to be used
- the distance to be negotiated (in the case of work equipment for access and egress)
- distance and consequences of a potential fall
- duration and frequency of use
- need for easy and timely evacuation and rescue in an emergency.

7.557 Any additional risk posed by the use, installation or removal of that work equipment or by evacuation and rescue from it.

Fragile Surfaces and the Legislation

7.558 Fragile surfaces are a particular issue when short-term maintenance or cleaning work is undertaken. It is often the case that little attention is paid to such work, and a full assessment may not necessarily be carried out. HSG33 *Health and*

Safety in Roof Work states that "everyone with responsibility for this type of work, at whatever level, should treat it as a priority".

7.559 The WAHR contain specific requirements in relation to a fragile surface, which is given the definition of "a surface which would be liable to fail if any reasonably foreseeable loading were to be applied to it". Regulation 9 states that "every employer shall ensure that no person at work passes across or near, or works on, from or near, a fragile surface where it is reasonably practicable to carry out work safely and under appropriate ergonomic conditions without his doing so". Where it is not reasonably practicable to carry out work safely and under appropriate ergonomic conditions, the employer should:

- ensure, so far as is reasonably practicable, that suitable and sufficient platforms, coverings, guard rails or similar means of support or protection are provided and used so that any foreseeable loading is supported
- take suitable and sufficient measures to minimise the distances and consequences of the fall where a risk of a person at work falling remains.

The "Green Book" and the "Red Book"

7.560 The ACR has so far published two guidance documents in relation to fragile surfaces: the Green Book, *Guidance Note for Safe Working on Fragile Roofs* and the Red Book, *Test for Non-Fragility of Profiled Sheeted Roofing Assemblies*. The Green Book has been prepared to reinforce the point that working on older roofs represents a serious hazard.

7.561 The preface to the Green Book highlights that in the future, "serious accident statistics associated with roofing will primarily arise out of maintenance work involving fragile roofs". It recommends that working on or inspecting an existing fragile roof should be treated as working on a roof with no covering with the following general requirements being applied.

- Any person carrying out an inspection or work on a fragile roof must have adequate information, training, instruction and management control to assess and carry out the task safely.
- A roof should always be treated as fragile until a competent person has agreed otherwise.
- No one should be allowed access to the area below a roof when it is being worked on unless there is an adequate protection system designed to prevent injury from and retain all falling objects.
- Only suitable persons having appropriate competence, training and physical fitness should be allowed on any roof.
- An inspection will always be needed prior to work on any roof commencing.
- All persons involved in roofwork need to be aware of the risks that can occur through a roof being fragile or becoming fragile over time.
- Existing asbestos cement sheets and old roof lights should always be treated as fragile.

Assessing Fragility

7.562 Since the introduction of the CDM Regulations, the details of any roof that has a fragile element should be noted in the health and safety file prepared for the building in question. However, fragility is not determined solely by the nature of the material. It may arise from the:

- general deterioration of the roof due to ageing, neglect and lack of maintenance
- corrosion of metal clad roofs and fixings
- quality of the original installation

- selection of original material, fixings and washers
- subsequent impact and thermal damage
- deterioration of the supporting structure, sheeting and fixings from below due to processes within the building or other causes
- damage from rain and storm water leading to random areas of weakness
- increased frequency of inspection (eg to support manufacturers' guarantees or regular maintenance of equipment on the roof) exposing the roof to excessive traffic.

7.563 Other factors that require consideration include the "element's" span, thickness, profile, fixings (type, location and number) as well as the design of the supporting structure.

7.564 Testing for fragility has always been a problematic area and both industry and the HSE have expressed concern about the lack of guidance on what is a fragile roof assembly. The Red Book defines a method for testing for non-fragility which gives consistent results when repeated or reproduced by different assessors.

7.565 It must be stressed that the Red Book can only be considered as giving information on a product's performance under test at the time of the test. It should be borne in mind that a product's properties may change during its service life as detailed above. For older roofs there may be little information available and it must be emphasised that a roof should always be treated as fragile until a competent person has agreed otherwise. To assess the non-fragility of a roof a competent person should have:

- sufficient knowledge of the mechanical and physical properties of the materials and assemblies involved
- practical experience of installation of the product, usage, behaviour and failure in service.

7.566 In addition, guidance on longer-term non-fragility for roofing assemblies should be obtained from recognised trade association publications and industry standards. For example, guidance on achieving up to 25 years non-fragility for in-plane rooflights is contained in Guidance Note 2004/1, from the National Association of Rooflight Manufacturers.

Safe Working on Roofs

7.567 HSG33 details many of the technical aspects of safe roofwork, while the ACR's Orange Book provides a source of essential practical information addressing the roles and responsibilities of all who may be concerned with working on roofs. It highlights that the proper management of the hazards associated with roofworking "can only be achieved if all those responsible for this work undertake their duties conscientiously". The recommendations in the document are intended to reduce the level of accidents by encouraging clients, designers and roofers to recognise their responsibilities and co-operate to make working on roofs a less hazardous occupation.

The Client

7.568 The Orange Book recommends a number of preliminary actions to be carried out. It highlights the need to determine if the roof is fragile or not. Where a non-fragile roof assembly is identified, the need to support this with documentary evidence is stressed. Where the roof is identified as fragile or suspected of being fragile, "systems of work, which protect the worker from the hazards of working on or close to fragile roof areas, must be put in place" .

7.569 Selecting a competent and capable contractor to carry out roofwork is essential and the Orange Book recommends that any contractor should be able to demonstrate:

- knowledge and understanding of the work and the health and safety laws covering roofwork
- that he/she can manage/eliminate the risks involved in working on roofs
- he/she has been assessed for competence and employs a trained workforce, preferably holders of a CSCS certificate (the Construction Skills Certificate Scheme)
- he/she understands the methods of use and mechanical properties of all the roof materials and systems involved in the installation
- he/she is a member of a relevant trade association.

7.570 Contractors undertaking work should be capable of providing a high quality service, ensure that work is carried out in accordance with current British or other accredited industry standards, and be able to carry out the work in a safe manner. Companies should preferably be registered by a reputable trade organisation that has a code of practice and complaints procedure, eg the National Federation of Roofing Contractors (NFRC), the Flat Roofing Alliance (FRA), the Single Ply Roofing Association (SPRA), Mastic Asphalt Council (MAC) and the Rural and Industrial Design and Building Association (RIDBA).

7.571 As with any client-contractor relationship, the flow of information is important and the client should assemble all existing information about the roof and pass this onto the prospective contractors to enable them to draw up an appropriate method statement. Contractors should also provide a risk assessment covering the work, which should include managing the risks to people who will be in the vicinity while the work is being carried out.

7.572 Requirements for power and ancillary structures associated with roofwork and delivery schedules should also be detailed where necessary. In addition, contractors should supply a list of those who will be working on the roof, accompanied by proof that they are competent to do so.

7.573 Other factors that the client should consider include:

- accommodation for the contractor and any necessary materials and equipment storage
- the planning of any shut down of the premises that may be required to enable the work to be carried out safely
- the provision of information to employees of the potential hazards while the work is being undertaken
- allowing adequate time for the work to be undertaken including any potential snagging time that may be required.

7.574 It is also recommended that a permit-to-work system be operated for anyone who will have to access the roof. The system should ensure that everyone:

- is competent to work on a roof
- is given safety induction training before commencing work
- is properly briefed about the hazards associated with a particular roof and safe access to it
- has access to and is competent to use suitable safety equipment.

The Contractor

7.575 The Orange Book highlights the potential hazards involved with working on roofs but also emphasises that the use of certain tools and materials may also represent a risk.

7.576 Contractors are expected to follow the general principles of the WAHR hierarchy of control for working at height, giving collective protective measures priority. The guidance contains a typical list of best practice that should be adopted when working on roofs. As an example, if walking on roofs is necessary, attempts to minimise the amount of walking on roofs should be undertaken, eg by ensuring that materials necessary for the work are deposited as close as possible to the point at which they will be required. In addition, dedicated walkways for carrying materials to their point of use should be provided.

Weather Conditions

7.577 As required by the WAHR, weather conditions should be considered.

7.578 It is recommended that workers do not consider going on any roof in poor weather conditions such as rain, ice, frost or strong winds (particularly gusting) or if slippery conditions exist on the roof. Winds in excess of 23mph (force 5) will affect a person's balance.

Inspections

7.579 The Green Book contains guidance on safe working methods for various activities involving fragile roofs. It also recommends that an inspection be carried out before any work commences. Guidance for this in the Green Book is comprehensive and the general principles can be applied to any pre-works inspection irrespective of whether the roof is fragile. Included in the list of inspection points are:

- the identification and location of all services on or near to the roof
- the available space for the siting of access equipment and consideration of how it will be moved around
- the steepness of roof, surface texture and weather conditions influencing the potential for sliding off and falling from or through the roof
- any old and broken roofing sheets, evidence of water leaks and staining, rusting end laps and peeling paint
- any missing or damaged ridge and verge cappings
- debris on the roof surface, blocked gutters or downspouts which may mask potential hazards on the roof surface
- any damaged roof lights
- the condition of fixings and washers, including evidence of wear and cracking around fixing points
- any evidence of external surface over coating treatments, which may have blacked out roof lights visible from within the building
- any evidence of asbestos identified within the asbestos management plan
- any other toxic/hazardous substances, eg gases emitted from ventilators on the roof
- the condition and suitability of access, safety systems and anchorages, and other hazards relevant to the future works.

7.580 Where it is necessary to access roofs, all reasonable precautions should be taken, having regard for the duration of the inspection, pitch and condition of the roof, weather conditions and the risks to other persons. The use of ladders and other such equipment must be justified through the assessment process and the use of crawling boards and roof ladders may be deemed to be reasonably practicable.

Precautions for Work on Fragile Surfaces

7.581 HSG33 states that if work on a fragile surface is unavoidable "it is essential to identify all fragile materials and decide on and implement stringent precautions". The Green Book contains guidance about safe working methods for various activities involving fragile roofs.

7.582 The principles of WAHR to prevent falls from height and avoid work near fragile surfaces are at the forefront of this guidance. In addition, guidance is provided for the client on matters to consider before undertaking any repair or refurbishment, again in an attempt to avoid the need for access to fragile roofs. Questions that clients should ask include the following.

- Can taking action within the building solve the consequences of the defect and avoid the need to access the roof?
- When considering the future use of the building, can the frequency of access be reduced through balancing the cost and life span of any repairs against an overall refurbishment?
- Is there a valid reason in the replacement policy why fragile materials should not always be replaced with non-fragile assemblies to remove the hazard at source for future work?

Roofwork Competence

7.583 The WAHR require employers to ensure that "no person engages in any activity, including organisation, planning and supervision, in relation to work at height or work equipment for use in such work" unless they are competent to do so or if being trained or supervised by a competent person.

7.584 Working on a roof can be an extremely hazardous activity. The HSE's accident statistics show that many accidents happen because the people carrying out the work or the inspection are not trained or competent to do so. The ACR Black Book, *Guidance Note for Competence and General Fitness Requirements to Work on Roofs*, states that "it is easy to overlook the ability of the people that are employed to go on to the roof to carry out inspections or perform work tasks".

7.585 The book also gives a general definition for competence as a "person who can demonstrate that they have sufficient professional or technical training, knowledge, actual experience and authority to enable them to:

- carry out their assigned duties at the level of responsibility allocated to them
- understand any potential hazards related to the work (or equipment) under consideration
- detect any technical defects or omissions in that work (or equipment), recognise any implications for health and safety caused by those defects or omissions, and be able to specify a remedial action to mitigate those implications".

Black Book Recommendations

7.586 The guidance provides checklists of required competences for those commissioning, planning, managing, supervising and carrying out roofwork. As an example, to organise and plan roofwork, the Black Book recommends that a person should have:

- a knowledge of health and safety legislation applicable to roofwork and how it should be applied to the task in hand
- an understanding of the hazards associated with the type of roof or system
- the ability to carry out and implement comprehensive risk assessments to deal with associated hazards

- the ability and authority to select the most appropriate working equipment for the task, regardless of financial constraints
- an understanding of method statements and what needs to be included within them
- an understanding of available techniques (including rescue) which can be employed to safely access and carry out work on a roof
- delegated authority to grant permission to access the roof.

Green Book Competence

7.587 Emphasis is also given in the Green Book to competence, and this states that competence to work on fragile roofs includes:

- recognising that this is a highly hazardous task
- sufficient training in the use of the equipment and how to deal with the hazards associated with the task involved
- an understanding of the need for and the ability to check the adequacy of the allocated safety equipment
- being able to state the correct procedure for the task and the emergency procedures in place for the work.

Working and Inspecting Roofs

7.588 Working on or inspecting a roof may involve:

- work at considerable heights for long periods of time
- work outdoors often in extreme weather conditions
- repetitive materials handling
- reaching, stretching and maintaining balance in awkward postures while carrying loads on varying roof terrains.

7.589 These activities will require physical exertion as well as good balance and mobility, therefore using competent workers with appropriate levels of fitness is vital in ensuring roofwork safety. The Black Book recommends that evidence be provided to show that individuals do not suffer from:

- any neurological condition likely to cause seizures, weakness of limbs, loss of balance including vertigo (dizziness from being at height)
- any heart or lung condition likely to be aggravated by strenuous work
- any disability/impairment of limb function
- any other disease, disability, medication, alcohol, drugs or effects of toxic substances (lead, etc) likely to impair mental or physical activity especially at a height
- temporary ailments such as influenza or other conditions that may affect judgment
- uncorrected sight problems
- a physique that would be unsuitable for the work environment.

Training

7.590 Employers must ensure that any person carrying out an inspection or work on a fragile roof must have adequate information, training, instruction and management control to assess and carry out the task safely.

List of Relevant Legislation

- Construction (Design and Management) Regulations 2007
- Work at Height Regulations 2005

Further Information

Publications

HSE Publications

The following is available from *www.hsebooks.co.uk*.

- HSG33 (rev 1998) *Health and Safety in Roof Work*

Advisory Committee for Roofwork Publications

The following are available from *www.roofworkadvice.info*.

- ACR[CP]001: 2007 Rev. 2 *Recommended Practice for Work on Profiled Sheeted Roofs*, (the Orange Book)
- ACR[CP]002: 2005 *Safe Working on Fragile Roofs*, (the Green Book)
- ACR[CP]005: 2006 *Guidance Note for Competence and General Fitness Requirements to Work on Roofs*
- ACR[M]001: 2005 *Test for Non-fragility of Roofing Assemblies* (the Red Book)

National Association of Rooflight Manufacturers Publications

The following is available from *www.narm.org.uk*.

- Guidance note 2004/1: *Test for Fragility of Roofing Assemblies*

National Federation of Roofing Contractors Publications

The following is available from *www.nfrc.co.uk*.

- *Roofing and Cladding in Windy Conditions*

Organisations

- Advisory Committee for Roofwork
  Web: *www.roofworkadvice.info*
  The Advisory Committee for Roofwork is a body dedicated to making working on roofs safer. It was established in 1998 and is made up of nominees from trade associations and organisations involved in roof work who provide the experience of many years' involvement in working on roofs in the advice given in their documents.
- Flat Roofing Alliance
  Web: *Flat Roofing Alliance*
  The Flat Roofing Alliance is the recognised trade association for the flat roofing

industry. It was formed in 1997 from the merger of two long-established bodies, the Roofing Contractors Advisory Board, and the Association of British Roofing Felt Manufacturers.

- Health and Safety Executive (HSE)
  Web: *www.hse.gov.uk*
  The HSE is responsible for the regulation of almost all the risks to health and safety arising from work activity in the UK.
- Mastic Asphalt Council
  Web: *www.masticasphaltcouncil.co.uk*
  The Mastic Asphalt Council (MAC) is the trade association for the UK mastic asphalt industry. MAC represents more than 90 firms — including mastic asphalt manufacturers, the contractors responsible for its installation, and associated suppliers of equipment and services.
- National Association of Rooflight Manufacturers
  Web: *www.narm.org.uk*
  The National Association of Rooflight Manufacturers represents a complete cross section of the rooflight design and material type manufacturers in the UK. The association has been formed to promote co-operation between member companies in order to develop and maintain standards and codes of practice — and to provide an authoritative information portal for rooflight specifiers.
- National Federation of Roofing Contractors
  Web: *www.nfrc.co.uk*
  The National Federation of Roofing Contractors is the UK's leading trade association for the roofing industry. Its mission statement is to promote trade members to gain more work and through this achievement to help its associate members to sell more products and lectively to achieve quality installation.
- Rural and Industrial Design and Building Association (RIBDA)
  Web: *www.ribda.org.uk*
  RIBDA is the only association in the UK having the detailed knowledge of the function and environmental requirements of modern agricultural and industrial buildings, together with the breadth of expertise in their siting, planning, design and construction. This expertise also extends to conversion for diversification and other rural building design issues.
- Single Ply Roofing Association (SPRA)
  Web: *www.spra.co.uk*
  SPRA aims to ensure that its clients obtain high-quality polymer-based single layer roofing through a partnership of quality assured manufacturers and contractors.

Slips, Trips and Falls

- One in three serious injuries at work involves slips and trips on the same level, and a quarter of all workplace fatalities involve falls from heights of 2m or more.
- A number of factors associated with the individual, the activity and the workplace environment can cause slips, trips and falls.
- An individual's balance and control of movement can be influenced by a wide range of factors, including vision, muscle endurance, footwear type and the friction level of a surface.
- A preliminary assessment can identify which areas and activities present a significant risk of slips, trips and falls.
- Assessments of the risks of slips, trips and falls should concentrate on the individual, the task, the traffic route and the immediate environment.
- Control measures include regular cleaning and maintenance, sufficient storage space, adequate lighting, suitable footwear for employees and non-slip flooring in high-risk areas.
- Effective supervision and disciplinary procedures can help ensure that appropriate footwear is worn and spillages and wet areas are dealt with quickly.

7.591 The danger with slip, trip and fall accidents is that they are so commonplace that people begin to believe that they are inevitable, and can only be accepted — not controlled. A third of all serious injuries in the workplace involve slips and trips on the same level, and a quarter of workplace fatalities involve falls from over 2m in height. However, a rigorous risk management programme, which focuses on factors related to the individual, the activity or task being performed and the workplace environment, can dramatically reduce both the likelihood of such accidents taking place and the severity of those that do occur.

Employers' Duties

7.592 The legal requirements imposed on employers in relation to slips and trips can be sub-divided into those covering the employer's duty to employees, and those covering the employer's duty to others (non-employees), eg visitors, contractors and members of the public. (In many instances, these requirements overlap in terms of the practical action required to reduce the risk of slip, trip or fall accidents.)

- Under the Health and Safety at Work, etc Act 1974 employers have a general duty to:
  - ensure, so far as is reasonably practicable, the health, safety and welfare at work of all employees
  - carry out their activities in such a way as to ensure, so far as is reasonably practicable, that persons not in their employment are not exposed to risks to their health and safety.
- Under the Management of Health and Safety at Work Regulations 1999, employers must make an assessment of the risks to the health and safety of their employees while at work.
- The Workplace (Health, Safety and Welfare) Regulations 1992 require that:
  - the workplace is maintained in an efficient state, in efficient working order and in good repair

- the workplace and all furniture, furnishing and fittings within it to be kept sufficiently clean
- it must be possible to keep the surfaces of the floors, walls and ceilings of all workplaces inside buildings sufficiently clean
- so far as is reasonably practicable, waste materials should not be allowed to accumulate in a workplace except in suitable receptacles
- every workplace has suitable and sufficient lighting
- every room in which persons work must have sufficient floor area, height and unoccupied space for purposes of health, safety and welfare
- outdoor workstations must be arranged so that persons at the workstations are not likely to slip or fall
- surfaces of traffic routes must be suitable for their purposes and suitable for the persons or vehicles using them (sufficient in number, positions and size)
- all traffic routes must be suitably indicated where necessary
- so far as is reasonably practicable, suitable and effective measures must be taken to prevent "any person falling a distance likely to cause personal injury"
- all windows and skylights in a workplace must be designed so that they can be cleaned safely.

- Although construction sites are excluded from the provisions of the Workplace (Health, Safety and Welfare) Regulations 1992, they are covered by the Construction (Design and Management) Regulations 2007, which contain, among other matters, requirements relating to safe places of work, traffic routes, lighting and good order.
- The Occupiers' Liability Act 1957 — purely a civil law statute — states that a common duty of care is owed by an occupier to all visitors, except where the occupier has extended, restricted, excluded or modified this duty. The common duty of care is the duty to take reasonable care to ensure the safety of visitors using the premises (within the limits of their reasons for having been invited).

Employees' Duties

7.593

- Under the Health and Safety at Work, etc Act 1974, employees have a duty to:
  - take reasonable care of their own health and safety and that of other people who may be affected by their activities at work (this may extend to matters such as wearing appropriate footwear, etc)
  - co-operate with employers to enable employers to comply with their health and safety duties. (This includes co-operating with employers in relation to controls introduced to either eliminate or reduce the likelihood of slip, trip and fall accidents.)
- Under the Management of Health and Safety at Work Regulations 1999 employees must, among other duties, inform employers of shortcomings in their protection arrangements and of situations that they reasonably consider represent a serious and immediate danger to health and safety. This applies to slip, trip and fall hazards as well as to other hazards.

In Practice

7.594 Slip and trip accidents generate many costly civil claims, either directly or indirectly. Other cost factors include:

- lost income, pain suffered and reduced quality of life for the individual involved
- damages, administration and insurance costs, lost production time and temporary absences from work for the employer
- loss of potential output, medical costs and social security costs for society.

Main Causes of Slips, Trips and Falls

7.595 Slips often take place when there is not effective contact between a shoe sole and the floor surface. This can be due to:

- faults with the floor surface
- an inappropriate floor surface
- inappropriate footwear or contamination between the shoe sole and the floor surface such as oils, greases and water.

7.596 All of these factors, either individually or in combination, affect slip resistance. Trips, on the other hand, occur when an obstruction prevents the normal movement of the foot and this results in a loss of balance, eg objects left on the floor, uneven floor surfaces and poorly maintained floor surfaces.

Secondary Events Arising from Slips and Trips

7.597 A slip or trip can begin a chain of events that results in a serious accident.

- An employee died when she slipped on the kitchen floor of the residential care home where she worked and severed an artery in her neck with the large knife she was carrying.
- A 16-year-old girl slipped on water leaking from an ice-making machine and instinctively put out her hand to break her fall. Unfortunately her hand and forearm went into a deep fat fryer.

7.598 Many slip and trip accidents lead to a fall. According to a study by the Health and Safety Executive (HSE), slips and trips account for 30% of falls from heights.

Preliminary Assessment

7.599 The purpose of a preliminary assessment is to determine what areas and activities involve a significant risk of slip, trip and fall injury and therefore warrant a full risk assessment. Often such an approach involves direct observation during safety sampling exercises or inspections. Also, analysis of accident and incident records may help identify areas and activities which present particular problems related to slips, trips and falls and that may require more detailed assessment.

Conducting a Full Slips, Trips and Falls Assessment

7.600 When carrying out a detailed assessment of the risks of slips, trips and falls it is important to consider the individual, the task being performed and the immediate environment. When considering the immediate environment, it is useful to compare the physical layout of the workplace with recognised standards.

The Individual

7.601

- What are the ages of the employees involved in carrying out the tasks?
- Will the employees involved have difficulty in lifting their feet when walking?
- Do any of the employees suffer from health problems that may affect their sense of balance?
- Will the footwear worn provide adequate grip?
- Will any individuals be rushing or hurrying or have they been informed not to rush to complete tasks?
- Does any individual suffer from a disability that may place them at greater risk of slips, trips and falls?
- Are any employees, or others, at greater risk should they suffer a slip, trip or fall (eg pregnant women, those with health problems, etc)?
- Will any employees require specific training in order to reduce the risks of slipping, tripping or falling?

The Task

7.602

- Will the employees involved be carrying difficult loads or carrying out activities which may affect their balance?
- Will the task affect visibility in such a way as to increase the risk of slips, trips and falls?
- Will employees be moving from wet to dry areas or will they be moving from outdoor to indoor areas, which may result in the soles of their shoes becoming wet?
- Will employees be changing levels using stairs, steps, slopes or ramps, etc?
- Are there any distractions present that might affect concentration?
- Will any equipment or substances be used that could result in a more serious injury should a slip, trip or fall occur?

The Floor

7.603

- Is the floor level, dry and free from spillages and obstructions?
- Does the floor provide sufficient grip? (The cleaning materials used may affect grip.)

The Steps and Stairs

7.604

- Are the stairs in good condition?
- Are the treads level and even?
- Are all the treads of similar lengths (goers) and dimensions?
- Are all the heights between treads (risers) of similar dimensions?
- Is the slope (pitch) suitable for safe access and not too steep or too shallow?
- Is a handrail required for safe access and, if so, is it suitable and in good condition?
- Are the edges of the treads (nosings) in good condition?
- Where coverings, such as carpets, are used, are they suitable and in good condition?
- What materials and coverings are used for landing places, and will they help to lessen the impact of a fall?

Lighting

7.605

- Is there enough lighting in possible slip, trip and fall areas, such as stairways and steps?
- Are there areas of shadow or significant changes in levels of lighting?
- Are there areas affected by glare?
- Is the lighting behind the person, which may cast a shadow (this is particularly significant in possible slip, trip and fall areas, such as stairways and steps)?
- Is the lighting well maintained?
- Is sufficient emergency lighting provided, particularly on traffic routes?

Wet Areas

7.606

- Is the area generally wet (eg this is a possibility in the kitchen or around a swimming pool)?
- Is non-slip flooring provided?
- Are mats, etc provided to dry shoes when moving from wet to dry areas or from outdoor to indoor areas?

Cleanliness and Housekeeping

7.607

- Are areas and traffic routes cleaned regularly and kept free from spillages and sources of obstruction?
- Are areas and traffic routes inspected regularly and are records of these inspections kept?

Equipment

7.608

- Are there any substances or equipment in the area that a person could slip, trip or fall into, and which would cause a more serious injury to occur?

Control Measures

7.609 Consideration should be given firstly to whether it is reasonably practicable to eliminate slip, trip and fall hazards, and secondly to reducing the remaining risks to an acceptable level. Often the best approach is to incorporate the control of slip, trip and fall hazards in the overall risk management system, rather than trying to deal with individual issues in an *ad hoc* manner.

7.610 A risk management approach to slip, trip and fall accidents requires the identification and control of factors relating to the individual, work activity and workplace environment which could give rise to slips, trips and falls. In terms of the workplace environment, this often involves comparison against a recognised standard for what should exist. This "gap analysis" will produce recommendations to bring the workplace up to the required standard.

7.611 Measures to control slip, trip and fall hazards include the following.

- Cleaning and maintenance, especially of floor surfaces, should take place regularly. Maintenance schedules and procedures should ensure that the building fabric, traffic routes and lighting, etc remain effective and in good order. If wet floor cleaning is necessary, it should be scheduled to take place outside of normal working hours, wherever possible.
- Equipment should be maintained in order to reduce the leakage of liquids, and containers holding substances need to be regularly inspected for leaks.
- Adequate storage facilities must be provided, and rigorous housekeeping regimes should be implemented in order to reduce clutter that might cause trips.
- Adequate lighting levels must be provided so that people can spot obstructions and slippery areas, etc. Additional lighting may be required at changes in level.
- Floor surfaces must be checked regularly for loose finishes, holes and cracks, and worn coverings, and must be suitable for wet or dusty conditions if these are likely to arise.
- Obstructions and spills must be removed immediately, and work areas and means of access and egress must be kept in a good condition generally.
- Where obstructions and spills cannot be removed immediately, warning signs and barriers should be erected.
- Non-slip flooring should be provided in wet and other high-risk areas.
- Effective cable-management procedures should be implemented to prevent trailing leads, etc in walkways and other traffic routes.
- Employees should wear suitable footwear. There should be means to dry footwear where it is likely to become wet. It is important to provide signs, etc where a change from a dry to a wet floor surface takes place.
- Workplace designs and layouts should reduce risk, and appropriate methods of work should be introduced and followed.

7.612 Slip and trip accidents are so commonplace that it is almost accepted that there are some areas of risk over which full control can be exerted. However, an effective risk management programme can drastically reduce the numbers of such accidents. This will result in both a cost saving for the employer and a reduction in the pain and suffering caused by such accidents. Therefore it is in everyone's interest to work together in order to bring such accidents under control.

Inspection and Monitoring

7.613 Workplaces should be subject to regular inspection and random monitoring using techniques such as safety sampling. Safety sampling involves dividing the workplace into routes that can be walked briskly in 5–10 minutes. A route is selected at random and walked by someone independent of the area, who makes a note of any hazards (including slip, trip and fall hazards) spotted.

7.614 Inspections involve observation of physical factors in an area, making use of a standard checklist (which will include reference to slip, trip and fall hazards). These techniques help to ensure that traffic routes remain clear and unobstructed, that no trailing leads obstruct walkways and that stairways and passageways remain well lit, free from obstruction and in good repair.

Disciplinary Procedures and Systems

7.615 Supervision and disciplinary procedures should help to ensure that appropriate footwear is worn at all times and that spillages and wet areas are dealt with quickly and effectively.

Recipe for Safety Initiative

7.616 The effectiveness of control measures is demonstrated by the success of the HSE's *Recipe for Safety* initiative in the food and drink industry. Since 1989, it has resulted in a 15% reduction in the total number of injuries, and a 10% reduction in the injury incidence rate per 10,000 employees. The prevention of slips is a major focus of the initiative.

Standards

7.617 Accidents on steps and stairs often involve two distinct hazard areas: the primary hazard leading to the initiating event, and the secondary hazard, which affects the nature of the harm. In *Building Regulations and Safety: Building Research Establishment Report 1995*, the primary and secondary hazards are described as follows.

Primary and Secondary Hazards

Primary	Secondary
Obstructions	Non-cushioned surfaces
Worn treads and nosings	Obstructions
Lack of/inappropriate rails	Dangerous collision surfaces
Surface covering	Nearby sharp corners
Inappropriate tread dimensions	Distance to fall
Unusual design features	
Insecure fixings	
Absence of/inappropriate guards	
Open risers	
Poor lighting	
Inappropriate landings	

7.618 The Building Regulations have included requirements relating to stairs since 1965, and are the subject of *Approved Document K: Protection from Falling, Collision and Impact*. The Building Regulations are also supported by British Standard BS 5395: *Stairs, Ladders and Walkways*.

Building Regulations

7.619 Section K1 of Approved Document K to the Building Regulations 2000 deals specifically with stairs, ladders and ramps. Some of the key requirements of Section K1 are outlined below.

- In a flight, the steps should all have the same rise and the same going. The dimensions for the rise must be between 150mm and 190mm, and the dimensions for the going must be between 250mm and 320mm. The normal relationship between the dimensions of the rise and going is twice that of the rise, plus the going (2R+G) and should be between 550mm and 700mm.
- Headroom of 2m is adequate on the access between levels.
- A door may swing across a landing at the bottom of a flight, but only if it will leave a clear space of at least 400mm across the full width of the flight.

Design of Industrial Stairs and Walkways

7.620 British Standard BS 5395–3: 1985, *Stairs, Ladders and Walkways — Code of Practice for the Design of Industrial Type Stairs, Permanent Ladders and Walkways* offers guidance to designers on industrial buildings, plants and installations.

7.621 BS 5395–3 contains the following key requirements.

- It is essential to make all risers in a flight uniform — consistency of risers and goings are of primary importance for user confidence and safety. The relationship between riser and going for a stair should not change along the walking line.
- The minimum clear width should be 600mm for occasional one-way traffic and 800mm for regular one-way traffic.
- The minimum pitch for straight stairs should be 30° and the maximum pitch for occasional access should be 42°.

Studies on Safe Stair Design

7.622 Studies on the technical features of stairs have considered several issues, including:

- the dimensions of risers and goings
- nosing projections and materials
- fixing of nose coverings
- stair and step width
- length of flight
- tapered treads
- alternate tread stairs
- changes of direction
- landings
- handrails
- guarding.

7.623 Studies concerning the behaviour of stair users suggest that more serious accidents occur in stair descent.

7.624 According to the *Building Research Establishment Report 1995*, accident research demonstrates that the following issues are important in achieving an acceptable level of safety on stairs.

- Step geometry should meet the minimum requirements to ensure there is sufficient room for the foot and to discourage tripping and slipping.
- Steps and floor surfaces should not be slippery and should be detectable by tactile means by all users.
- Handrails and guarding (banisters) should be correctly dimensioned, shaped and located.
- There should be adequate lighting to give stair users clear visibility.
- Consideration of secondary safety is important. If an accident does occur, the avoidance of projections, sharp surfaces and obstacles may reduce the level of harm.

Training

7.625 Adequate information and training must be provided to all employees. Such training must cover slip, trip and fall risks, and the necessary controls relating to both the task being performed and the environment in which it is being performed. Additionally, individuals who are exposed to a particularly high risk of slip, trips and falls may require specialist training in methods such as the Alexander Technique, which has been shown to improve posture and stability.

7.626 If slip and trip accidents are to be effectively controlled, however, efforts must be made to eliminate or reduce the risks associated with the task and the environment, as well as to improve the individual's ability to carry out the tasks and to better withstand the environment.

List of Relevant Legislation

- Building Regulations 2000
- Management of Health and Safety at Work Regulations 1999
- Reporting of Injuries, Diseases and Dangerous Occurrences Regulations 1995
- Workplace (Health, Safety and Welfare) Regulations 1992
- Health and Safety at Work, etc Act 1974

Further Information

Publications

HSE Publications

The following are available from *www.hsebooks.co.uk.*

- HSG155 *Slips and Trips: Guidance for Employers on Identifying Hazards and Controlling Risks*
- INDG225(L) (rev 2003) *Preventing Trips, Slips and Falls at Work*
- L24 *Workplace (Health, Safety and Welfare) Regulations 1992 — Approved Code of Practice*

British Standards

The following are available from *www.bsi-global.com.*

- BS 5395–1: 2000 *Stairs, Ladders and Walkways. Code of Practice for the Design, Construction and Maintenance of Straight Stairs and Winders*
- BS 5395–3: 1985 *Stairs, Ladders and Walkways. Code of Practice for the Design of Industrial Type Stairs, Permanent Ladders and Walkways*

Other Publications

- *Building Regulations and Safety: Building Research Establishment Report 1995*, Cox, S J, O'Sullivan, E F, Building Research Establishment, 1995
- *Contract Flooring Journal*, No. 16, Health and Safety Laboratory, 2003
- *The Building Regulations 2000 Approved Document K: Protection from Falling, Collision and Impact*
- *The Safety and Health Practitioner*, September 2000

Organisations

- British Standards Institution (BSI)
  Web: *www.bsi-global.com*
  Founded in 1901, the BSI develops and provides supporting information on national and international standards in a wide range of areas.
- Building Research Establishment (BRE)
  Web: *www.bre.co.uk*
  The BRE provides consultancy, testing and commissioned research services covering all aspects of the built environment and associated industries.
- Health and Safety Laboratory (HSL)
  Web: *www.hsl.gov.uk*
  The HSL operates as an agency of the HSE and is the UK's leading industrial health and safety facility.
- Rapra Technology
  Web: *www.rapra.net*
  Rapra Technology, formerly the Rubber and Plastics Research Association, is an independent plastics and rubber consultancy, which provides consultancy, technology and information services for the polymer industry and industries using plastics and rubber in any component, product or production process.
- Ergonomics Society
  Web: *www.ergonomics.org.uk*
  The Ergonomics Society aims to increase the general understanding of ergonomics and its importance.

Working at Height

- Working at height can mean any activity that requires the employee to work at a distance from which a fall is likely to cause personal injury.
- The two main hazards are falls of persons from height and falls of objects from height.
- There is no minimum height that should be considered safe. The most important consideration is the potential for harm rather than the distance through which someone or something might fall.
- Employers have a general duty to ensure, as far as reasonably practicable, the health and safety of all employees, including those working at height.
- Employers should identify all activities that require working from height.
- Employers should undertake an initial assessment to determine those activities that involve significant risk.
- If it is possible to do so, employers should consider removing the need to work at height altogether.
- All work at height activities that pose a significant risk should be subject to a detailed risk assessment.
- The control of risks from height work is based upon a hierarchy. The hierarchy has to be followed systematically and only when one level is not reasonably practicable may the next level be considered.
- All risk assessment and control measures must be recorded.
- Employers should undertake regular inspections and monitor control measures to ensure they are working.
- Adequate information, instruction and training should be provided to everyone involved.
- Falls from height are the most common cause of fatal injury and the second most common cause of major injury to employees.

7.627 Working at height is defined in the Work at Height Regulations 2005 (WAHR) as work in any place, including a place at or below ground level, from which a person could fall a distance liable to cause personal injury if the precautions required by the regulations were not taken. This includes obtaining access and egress to or from a place of work but not a staircase in a permanent workplace.

7.628 The regulations, which cover almost all work situations, set out specific precautions for all such work regardless of the distance a person may fall. They also set out a hierarchy of control for determining how to work at height safely. The hierarchy has to be followed systematically and only when one level is not reasonably practicable may the next level be considered. Under WAHR, it is not acceptable to select work equipment from lower down the hierarchy in the first instance. According to the hierarchy, duty holders must act as follows.

1. Avoid work at height where possible.
2. When work at height cannot be avoided, use work equipment to prevent falls.
3. When the risk of falls cannot be eliminated, use work equipment to minimise the distance and consequences of any fall.

7.629 Hence, when considering work at height, it is important to note that it is the potential to cause harm that needs to be considered and not the distance through which someone or something might fall.

7.630 Falls from height are the most common cause of fatal injury and the third most common cause of major injury to employees. During the period 2005/06, a total of 46 workers died as a result of a fall from height. This equates to 22% of the total number of workplace fatalities for the period. In addition, 3351 workers suffered major injuries during the period as a result of a fall from height, 66% as a result of falling less than 2m. All industry sectors are exposed to the risks and therefore all employers need to give priority to the reduction of risks arising from working at height.

Employers' Duties

7.631

- Under the Work at Height Regulations 2005, duty holders must:
  - avoid work at height where they can
  - where work at height cannot be avoided, use work equipment to prevent falls
  - where the risk of falls cannot be eliminated, take measures to minimise the distance and consequence of any falls.

  In addition to adherence to these basic principles, duty holders must ensure that:

  - work at height is properly planned and takes into account weather conditions
  - all those who work, plan or supervise such work are competent and properly trained
  - the place of work at height is safe
  - equipment such as scaffolding and ladders are inspected in accordance with the specific requirements of the regulations
  - the risks associated with fragile surfaces are controlled
  - the risks from falling objects are properly controlled.
- Under the Health and Safety at Work, etc Act 1974 employers have a general duty to ensure, so far as is reasonably practicable, the health and safety of all employees at work.
- Under the Management of Health and Safety at Work Regulations 1999 employers have a duty to undertake suitable and sufficient assessments of risks, including those associated with working at height.
- Under the Safety Representatives and Safety Committees Regulations 1977 and the Health and Safety (Consultation with Employees) Regulations 1996 employers are required to consult with employees.
- Under the Construction (Design and Management) Regulations 2007 (CDM) designers are required to critically assess their design proposals to identify health and safety issues and to take steps to reduce the risks, including for example, the need to work at height.

Employees' Duties

7.632

- Under the Health and Safety at Work, etc Act 1974 employees have a duty to:
  - take reasonable care of their own health and safety and that of other people who may be affected by their activities at work
  - co-operate with their employer to enable the employer to comply with health and safety duties.
- Under the Management of Health and Safety at Work Regulations 1999 employees are required to:

  - use any machinery, equipment, dangerous substances, transport, safety devices and means of production in accordance with any training and instructions provided by the employer
  - inform the employer of any serious and imminent dangers to health and safety
  - inform the employer of any shortcomings in the employer's health and safety arrangements.
- Under the Work at Height Regulations 2005 employees must:
  - report to their employer or person in control of the work, any activity or defect relating to work at height that may endanger the safety of themselves or others
  - use any equipment provided for their use in accordance with any training and/or instructions relating to its use so as to enable their employer or person in control of the work to comply with their own health and safety duties.

In Practice

Identifying Work at Height Activities

7.633 When working at height, the height at which the work is being undertaken is not the primary factor to be taken into account. Instead, to satisfy the general duty to ensure safety, the employer needs to consider the potential a fall from height has to cause personal injury. It therefore follows that persons could be considered to be working at height where the potential to fall a distance is small but where the consequences of such a fall are great.

7.634 This brings into consideration a broad range of activities including:

- undertaking maintenance work on overhead systems such as conveyors
- installing lighting systems
- work on ladders, eg cleaning of windows and other structures
- work on staging or trestles, eg erecting bill posters and painting and decorating
- most construction work
- erection of scaffolding and other steelwork
- any activity undertaken from a scaffold or from a mobile-elevated work platform
- work on permanent structures such as gantries, pylons and street lighting
- working on the back of a lorry, eg sheeting a load
- using a ladder/stepladder or kick stool to store or retrieve goods at height, eg in a library
- work performed in trees, such as cutting branches
- working close to excavations or other voids.

Initial Assessment

7.635 The purpose of the initial assessment is to determine whether there are work activities that involve a risk of personal injury from falls or falling objects.

7.636 Care should be taken to avoid dismissing work activities just because the distance through which someone or something might fall seems insignificant. Not only are there records of fatalities involving falls of less than 2m, but also of badly torn muscles, tendons and ligaments as a consequence of someone falling through quite short distances, eg from a step stool.

Eliminating Working at Height

7.637 It is a statutory duty to consider whether a task could be undertaken without working at height. This should be undertaken at the earliest opportunity in the process, eg:

- designers may specify windows for a tall building that can be cleaned from the inside
- an office manager might increase storage capacity by selecting mobile shelving instead of higher storage units
- managers may install a permanent means of access for maintenance workers who need to gain frequent access to an overhead conveyor system
- street lighting systems could be designed so that they can be lowered to the ground for work such as cleaning or replacing lamps.

Detailed Risk Assessment of Working at Height

7.638 Where initial assessment has found that working at height cannot be eliminated, employers are required to undertake a detailed risk assessment. The assessment should include a careful examination of what harm could be caused from working at height and the necessary steps to reduce the likelihood of this harm occurring. The level of risk should be evaluated whilst taking into account all existing control measures. The complexity of the risk assessment will vary according to the work activity.

7.639 The risk assessment should address:

- those activities that involve working at height
- the level of risk associated with working at height
- the need to introduce new or improve existing control measures.

7.640 When assessing the work activity, it may be helpful to consider:

- whether the hazards are associated with the working platform, the means of access to and from the working platform, or both
- the potential for persons to fall through a distance — this could include those working on the ground who are at risk of falling into a tank, pit, excavation or similar structure
- the potential for tools, materials and equipment to fall through a distance
- the need to effect an evacuation from the working area in case of emergency and/or injury.

7.641 When evaluating the level of risk, the following should be considered:

- existing control measures
- persons working at height and their level of skills, knowledge, experience and physical health
- persons not working at height but whom falling objects might otherwise affect
- the nature of the activity
- the load imposed on a working platform by the number of persons
- additional loads imposed on the working platform by any materials, plant or equipment
- frequency and duration of the activity
- proximity of fragile surfaces, eg glass planes
- the potential influence of changing weather conditions
- the need to consider supplementary lighting
- effects on and of the surrounding area such as pedestrian and vehicular traffic

- the distance of the potential fall that ultimately influences the speed of impact (eg the impact speed when falling from the average height ladder is around 20mph)
- physical state of the impact site
- in the case of falling into a tank or pit, its contents, as these will greatly influence the potential outcome.

Control Measures

7.642 When considering risk reduction strategies, employers will need to ensure that any controls:

- are proportionate to the potential for harm, ie satisfy the measure of "so far as is reasonably practicable"
- meet or exceed the standards specified in the relevant legislation and official guidance.

7.643 The fundamental principle for work at height is that it should be undertaken according to the staged process outlined in the Work at Height Regulations 2005 (the "hierarchy"). Thus, where it is not reasonably practicable to avoid working at height:

- safe systems of work are established
- proper planning and organisation takes place
- competent staff are engaged in the work
- appropriate equipment is chosen and used correctly.

7.644 The following is a detailed explanation of the hierarchy.

1. Avoid work at height.
2. Prevent falls by using existing places of work.
3. Prevent falls by using "collective" work equipment (ie equipment with guardrails, working platforms, tower scaffolds, etc).
4. Prevent falls by using personal work equipment (ie fall-restraint systems).
5. Use "collective" work equipment to minimise the distance and consequences of falls (ie safety nets).
6. Use "personal" work equipment to minimise the distance and consequences of falls (ie fall-arrest systems).
7. Use "collective" work equipment that minimises consequences of falls (ie nets at a low level).
8. Use "personal" work equipment that minimises consequences of falls (ie personal injury prevention equipment such as life jackets if working next to unguarded water).
9. Mitigate occurrence of falls through training and instruction (ie safe use of ladders).

7.645 Collective measures are given priority over personal protection measures as they protect more than one person and are usually passive. Personal control measures (such as fall-arrest equipment) are usually active in that they require the user to do something to make them work effectively.

7.646 Where it is reasonably practicable, control systems that provide collective preventive measures should always take priority over those systems that provide personal protection. For example, scaffolding with guardrails and toe boards will usually take preference over lanyard and harness systems.

7.647 The means of securing a safe system of work for temporary work at height activities may include considering the use of any or a combination of the systems set out below.

7.648 Where the employer determines that the control measures have reduced the level of risk to as low a level as is reasonably practicable, he or she should:

- record the assessment
- maintain those systems
- periodically review the assessment.

Existing Places of Work

7.649 Reference is made in the Work at Height Regulations 2005 to "existing places of work"; such places should be the first option if height work cannot be avoided. No definition is given in the regulations for an existing place of work. However, it is best thought of as being a place where the use of extra work equipment to remove the risk of a fall from height is not required. For example, if a flat roof has permanent edge protection around the perimeter, then it is unlikely that further equipment will be necessary as the risk of falling is already controlled. When work is carried out from such a place, that place should meet the requirements of the Work at Height Regulations. These include requirements that existing places:

- are stable
- are of sufficient strength and dimensions
- have a surface with no gaps
- are free from slipping or tripping risks (which have been eliminated).

Working Platforms

7.650 In the prevention of falls, collective protection measures should always take priority over personal protection measures. The use of collective work equipment will normally entail the use of some form of access equipment that has a working platform.

7.651 Under the Work at Height Regulations 2005, working platforms are defined as any platform used as a workplace or as a means of access to or egress from such a place. The definition includes:

- scaffolds
- suspended scaffolds
- cradles
- mobile platforms
- trestles
- gangways
- runs
- gantries
- stairways
- crawling ladders.

7.652 The structures that support a working platform must be stable and strong enough to support the loads that are placed on it. These loads include:

- people
- plant
- materials
- any extraneous loads, such as high winds.

7.653 The size of the working platform is important and its dimensions must allow for the safe passage of persons, the safe use of plant and equipment and the storage of materials.

7.654 The selection of equipment for the means of access and egress to scaffolds and other height work equipment requires consideration. If frequent access is required and the place to be used at height is likely to be there for "some time", the Health and Safety Executive (HSE) now expects consideration to be given to the use of equipment other than ladders, for example by:

- using a staircase with handrails
- the use of hoists to raise loads and equipment.

7.655 Ladders should now only be used as a means of access and egress "where a risk assessment shows that the use of other work equipment is not justified because of the low risk and short duration of the job or unalterable features of the worksite".

7.656 Falls, of both persons and materials, are prevented by the provision of guardrails and toe boards and by ensuring that there are no gaps in the working surface. Systems must be in place to prevent access by unauthorised persons to a partially complete or unstable working platform, eg warning notices.

Ladders

7.657 The Work at Height Regulations 2005 lay down general requirements relating to the use of ladders.

7.658 Ladders should only be used for work at height when the risk assessment has determined that the use of more suitable means, such as tower scaffolds or mobile elevated working platforms, are not justified because of the low risk. The assessment should also demonstrate that their use is of short duration and that existing features of the site cannot be altered.

7.659 In addition, the schedule requires that ladders:

- rest on a stable and firm surface of sufficient strength to support the ladder and keep the rungs horizontal
- are positioned so as to ensure stability in use
- (in the case of suspended ladders) are attached in a secure manner
- (in the case of portable ladders) are prevented from slipping by:
  - securing the stiles near or at the top
  - the use of anti-slip or other stability devices
  - any other equivalently effective arrangement
- (where used for access) are long enough to protrude above the place of landing unless other means of securing a handhold are available
- (in the case of interlocking or extension ladders) are not used until the sections are prevented from moving relative to each other in use
- (in the case of a ladder or a series of ladders which rise a vertical distance of 9m or more) include, at suitable intervals, safe landing areas or rest platforms where reasonably practicable
- are used in such a way that:
  - a secure handhold and secure support are always available
  - the user can maintain a safe handhold when carrying a load (unless, in the case of a step ladder, when the maintenance of a handhold is not practicable and a risk assessment has found that their use is justified because of the low risk and short duration of use).

Guardrails and Toe Boards

7.660 These have the great advantage of facilitating movement because they do not require any physical connection between the worker and rail.

7.661 In connection with guardrails and toe boards used for construction work, the Work at Height Regulations 2005 require a top guardrail to be at least 950mm (910mm for rails already fitted when the regulations were introduced) above the edge from which a person could fall. Intermediate guardrails must be fitted which leave a gap of no more than 470mm. Toe boards are required to be "suitable and sufficient".

7.662 Workers may still be at a risk of falling if they lean over or through the rail to gain access. Falling objects may require the provision of a more substantial barrier.

Safety Nets, etc

7.663 Safety nets, and other soft-landing systems, facilitate mobility because they have no connection between worker and net. However, unlike guardrails, they are designed to limit a fall from height as opposed to preventing a fall from height.

7.664 Of course the erection of safety nets is a hazardous process in itself and should only be undertaken by competent personnel, usually specialists. Ideally, to minimise the distance a worker could fall, nets should be installed as close to the working level as possible.

7.665 Special consideration needs to be given to the arrangements for rescuing a worker after a fall. Further information in respect of safety nets can be found in BS EN 1263–1: 2002 *Safety Nets. Safety Requirements, Test Methods*, and BS EN 1263–2: 2002 *Safety Nets. Safety Requirements for the Positioning Limits*.

Fall-restraint Systems

7.666 There will be occasions when collective measures (such as working platforms with guardrails or safety nets) may not be reasonably practicable. On such occasions, the duty holder will have to select personal protection measures that may prevent or mitigate the consequences and distance of a fall.

7.667 The idea behind this approach is to prevent the worker getting into a position where they could fall from a height. This is achieved by tethering the worker to the structure in such a way that the length of the lanyard prevents them straying from a place of safety. Because the worker is prevented from exposure to the risk of falling, the specifications for the equipment are less demanding than those for fall-arrest systems. BS EN 358: 2002 gives full details of personal equipment for work positioning and prevention of falls from a height.

Note: This control measure will not prevent falling objects.

Work Positioning Systems

7.668 Work positioning systems are fall-protection systems that include a harness connected to a reliable anchor point. This will support the user in tension or suspension in such a way that a fall is prevented or restricted. An example is a boatswain's chair. This equipment in its own right is not intended for use as fall-arrest equipment but rather it is normally incorporated in or used in conjunction with a fall-arrest system. The Work at Height Regulations 2005 require these systems to include a suitable back-up system to prevent or arrest falls (which, if another line, the user should be connected to).

Rope Access and Positioning Systems

7.669 Rope access and positioning systems use two ropes each secured to different anchor points. One rope is connected to a harness and the other acts as a safety back-up. Rope access is often used to access the sides of tall premises when the use of cradles is not a suitable or reasonably practicable option. As well as making requirements for the anchor points and connection to a harness, the Work at Height Regulations 2005 require a seat with appropriate accessories to be provided (taking into account the risk assessment). Single rope access can be employed only when:

- the use of a second line would entail a higher risk
- appropriate measures have been taken to ensure safety.

Fall-arrest Systems

7.670 The first consideration must be to prevent persons from being in, or reaching a position, where they could fall from height. Where this is not reasonably practicable, personal protective equipment (PPE) in the form of a personal fall-arrest system may be considered as a last resort. This provides a secure connection between the worker and workplace structure and comprises:

- a full-body safety harness conforming to BS EN 361: 2002
- a safety lanyard and energy absorber conforming to BS EN 354: 2002 and BS 355: 2002
- the anchorage conforming to BS EN 795: 1997
- connectors that link the parts of the system together conforming to BS EN 362: 1993.

7.671 Various European Directives require only the use of full-body safety harnesses. The lanyard length should be kept as short as possible whilst still retaining a sufficient degree of free movement for the worker. It must not exceed 2m in length and must have an integral energy absorption system.

7.672 An integral energy absorption system is necessary because a fall of 2m that ends in a sudden stop can still result in sudden impact injuries to the body. A person weighing 100kg who falls through 2m would be subject to twice the limit of 6 kilonewtons (unit of force) imposed by current standards. The absorption system may be built into the design of the webbing, or ropes may be used in conjunction with fall-arrest blocks. In both cases, the systems work by decelerating the speed of decent over time.

7.673 Increased freedom of movement can be obtained by the use of vertical or horizontal sliding devices that move up and down or along a fixed rope or rail in response to movement.

7.674 Care should be taken when choosing the location of the anchorage point as this can greatly increase the unprotected fall distance. Wherever possible and always when using fall-arrest blocks, the anchor point should be overhead. Sufficient space must always be maintained to ensure the length of any fall is not shorter than that for which the system is designed. This is particularly important as space can be reduced over time by the build-up of debris and/or materials.

7.675 Connectors include devices such as hooks and karabiners (a metal clip with a spring) and compatibility between the various parts of the system should be ensured. The responsibilities of the manufacturer, the purchaser and the assembler are given in BS EN 363: 2002.

7.676 It should be remembered that fall-arrest systems will severely limit a worker's freedom to move around and this may bring with it a consequential increase in motivation to defeat the system. For this reason, as with any control system that relies on the use of PPE, effective utilisation of such a system will depend on comfort, compatibility, training, inspection, maintenance and supervision.

Note: This control measure will not prevent falling objects.

Falling Objects and Danger Areas

7.677 The Work at Height Regulations 2005 require the employer to take suitable and sufficient steps to prevent, so far as is reasonably practicable, the fall of any material or object. Where it is not reasonably practicable to do this, suitable and sufficient steps to prevent any person being struck by any falling material or object must be taken. Steps must also be taken to ensure that no material or object is thrown or tipped from height in circumstances where it is liable to cause injury to any person.

7.678 All materials and objects must be stored in a way which prevents risk to any person arising from the collapse, overturning or unintended movement of those materials or objects.

7.679 Where a workplace contains an area in which, owing to the nature of the work, there is a risk of any person falling a distance or being struck by a falling object, which is liable to cause personal injury, that workplace should (so far as is reasonably practicable) be equipped with devices preventing unauthorised entry. Dangerous areas should also be clearly indicated.

Fragile Materials

7.680 Where it is reasonably practicable to carry out work safely by other means, employers are required to ensure that no person at work:

- passes across or near a fragile surface
- works on, from or near a fragile surface.

7.681 Where this is not reasonably practicable, suitable and sufficient platforms, coverings, guardrails or similar means of support or protection must be provided. Where a risk of falling still remains, suitable and sufficient measures to minimise the distances and consequences of falling must be taken. In addition, prominent warning notices should be affixed at the approach to the place where the fragile surface is situated.

Inspections

7.682 All temporary working platforms and associated equipment including scaffolds, ladders, personal suspension equipment and nets, etc must be inspected:

- before first use
- after substantial additions, dismantling or other alteration
- after any event liable to affect strength or stability.

7.683 In the case of working platforms used for construction work and from which a person could fall a distance of 2m or more, in addition to the inspection requirements above, they must not be used in any position unless:

- it has been inspected in that position within the previous seven days
- in the case of a mobile working platform inspected on-site within the previous seven days.

Reports

7.684 Inspections must be carried out by a competent person. In the case of an inspection carried out on working platforms used for construction activities and from which a person may fall 2m or more, a written report must be made that includes:

- the name and address of the person on whose behalf the report is made
- the location
- a description (including any relevant plant, equipment and materials)
- the date and time
- a description of defects which could affect health and safety
- details of remedial action
- details of further action considered necessary
- the name and position of the person completing the report.

Consulting on Proposed Control Measures

7.685 Regulations require employers to consult with employees on:

- the introduction of any measure that will affect the health and safety of employees
- the persons nominated to provide health and safety assistance, and assist in emergency procedures
- the health and safety consequences of the planning and introduction of new technologies into the workplace.

7.686 Health and safety training and information must be provided to employees directly or via a safety representative.

7.687 Potentially, there are major benefits to be gained through the consultation process. Not only is it an opportunity to learn from the skills and experience of the workforce, but control measures are also more effective when individuals who have responsibility for using them own those measures. Individuals who are involved in the design and selection of control measures are more likely to use such controls.

Implementing the Control Measures

7.688 The risk assessment process is all about protecting people. It may seem obvious but unless the outcomes of the risk assessment are effectively implemented the effort in developing the control measures will have been wasted.

7.689 One way to ensure effective implementation is to follow the requirements of s.2(2)(c) of the Health and Safety at Work, etc Act 1974, which requires employers to provide information, instruction, training and supervision. Using the risk assessment as the basis for this ensures that employees are:

- given information about the hazards of working at height
- given instructions about the control measures devised to reduce risk and the role employees play in maintaining the integrity of those control measures
- trained so that they have the requisite skills and knowledge required to undertake their role in accordance with instructions
- aware that they will be supervised to ensure their continued compliance with safe systems of work.

7.690 Where there is a possibility of both falls from height and falling objects, it will be necessary to widen the scope of persons who need to be subject to this process. The scope will need to include all those who might reasonably be foreseen as being affected by a fall of materials. Depending on the location, scheduling and duration of the work, this might include:

- other employees
- contractors
- visitors
- members of the public
- trespassers.

7.691 Special care will be needed to ensure that the above is provided in a form that is comprehensible to employees. Simply getting someone to sign to say they have received such information will not be enough to discharge the employer's duty. For example, there may be employees who have special learning needs or literacy problems. For others, English may not be their first language.

7.692 When operating in public places, the strength and reliability of barriers, the effectiveness of any warning signs and the needs of special groups, such as the very young, should be considered.

Monitoring Strategies

7.693 Monitoring is an integral part of successful health and safety management. It is proactive and should comprise a systematic approach to routine inspections of plant, equipment and behaviour. In other words employers should actively check to see that:

- the correct plant and equipment, such as ladders, scaffolding, mobile-elevated work platforms (MEWP) have been provided
- the equipment is used for the purpose for which it was provided
- the equipment is used in accordance with instruction and training
- the equipment is well maintained and free from defects.

7.694 Monitoring should also look at supporting systems, such as:

- arrangements for reporting defects and taking remedial action
- arrangements in case of serious and imminent danger, eg following a collision with a temporary working platform or when a scaffolding has been subject to adverse weather conditions such as high winds.

Reviewing Assessments

7.695 Under the Management of Health and Safety at Work Regulations 1999, employers are required to review and if necessary to modify their risk assessment where:

- there is reason to believe that the assessment is no longer valid, eg where there has been an accident
- the work at height operations have changed, eg as the work progresses, additional hazards may have been identified.

7.696 The clear indication here is that risk assessment is not a one-off process but an integral part of the ongoing management of the working environment. It is good practice to review risk assessments on a regular basis. With some working at height operations where the level of risk is more prominent, it might be appropriate to review the assessment each time the activity is undertaken.

Training

7.697 Adequate information and training must be provided to all employees involved with working at height. This will entail:

- identifying those who will require training
- designing an appropriate programme of training
- implementing the training programme using the most effective methods
- monitoring the workplace to ensure that the training has been effective
- keeping a record of any training employees receive.

Who Needs Training?

7.698 All those who:

- design the safe system of work
- supervise working at height
- work at height
- erect, dismantle or inspect the working platform and/or means of access.

What Should Training Include?

Designers

7.699 For this group training should include:

- knowledge of work at height hazards
- knowledge of how to eliminate or reduce work at height hazards
- knowledge of the Work at Height Regulations 2005
- knowledge of the Construction (Design and Management) Regulations 2007 (CDM) requirements for designers.

Supervisors

7.700 For this group training should include:

- responsibilities for self and others
- active and reactive methods of monitoring
- action in case of non-compliance
- action in case of emergency
- competence requirements for working at height
- when and how to use PPE
- inspection requirements.

Those Who Work at Height

7.701 For this group training should include:

- responsibilities for self and others
- how to move and work safely
- dangers of overreaching
- safe means of access and egress

- means of moving materials
- how to use PPE, such as harness systems
- how to operate equipment, such as mobile access platforms
- indications of faults and how to report them
- precautions to prevent materials falling
- how to spread loads and the dangers of overloading
- dangers of fragile materials
- arrangements for making changes to the working platform and/or means of access/egress.

Those Who Erect, Dismantle or Inspect

7.702 In addition to the training requirements for those who work at height, this group will require specific technical competencies relating to the type of access/egress and working platform being used. Their training will need to cover:

- practical and theoretical knowledge of the system employed, including the plan for the assembly, dismantling or alteration of any work at height equipment
- a full understanding of potential risks related to their work
- how to detect defects and their implications for health and safety
- an ability to specify appropriate remedial actions when defects are detected
- appropriate measures in the event of weather conditions which adversely affect the safety of the equipment.

Work at Height (Amendment) Regulations 2007

7.703 The Work at Height (Amendment) Regulations 2007 came into effect on 6 April 2007. These amendment regulations bring those paid to lead or train others in climbing or caving activities in the adventure activity sector within the scope of the Work at Height Regulations 2005.

List of Relevant Legislation

- Construction (Design and Management) Regulations 2007
- Work at Height (Amendment) Regulations 2007
- Work at Height Regulations 2005
- Management of Health and Safety at Work Regulations 1999
- Provision and Use of Work Equipment Regulations 1998
- Lifting Operations and Lifting Equipment Regulations 1998
- Health and Safety at Work, etc Act 1974

Further Information

Publications

HSE Publications

The following are available from *www.hsebooks.co.uk*.

- CIS10 (rev 4) *Tower Scaffolds*
- CIS49 (rev 2003) *General Access Scaffolds and Ladders*
- CIS56 *Safe Erection, Use and Dismantling of Falsework*
- HSG33 (rev 1998) *Health and Safety in Roof Work*
- HSG150 (3rd ed, 2006) *Health and Safety in Construction*
- INDG244(L) (rev 1, 2006) *Workplace Health, Safety and Welfare: A Short Guide for Managers*
- INDG278 *Tree Work Accidents: An Analysis of Fatal and Serious Accidents*
- INDG284 *Working on Roofs*
- INDG367 *Inspecting Fall Arrest Equipment Made From Webbing or Rope*
- INDG369 *Why Fall For It?*
- INDG384 *The High 5: Five Ways to Reduce Risk on Site*
- INDG401 *The Work at Height Regulations 2005 (as amended): a Brief Guide*
- INDG402 *Safe Use of Ladders and Stepladders: An Employers' Guide*
- INDG403 *A Toolbox Talk on Leaning Ladder and Stepladder Safety*
- INDG405 *Top Tips for Ladder and Stepladder Safety*
- RR116 *Falls From Height: Prevention and Risk Control Effectiveness*
- RR205 *Evaluating the Performance and Effectiveness of Ladder Stability Devices*
- HSE Video *A Head for Heights: Guidance for Working at Height in Construction*
- HSE Video *High Designs*

British Standards Publications

The following are available from *www.bsi-global.com*.

- BS EN 131–1: 1993 *Ladders. Specification for Terms, Types, Functional Sizes*
- BS EN 354: 2002 *Personal Protective Equipment Against Falls From a Height: Lanyards*
- BS EN 355: 2002 *Personal Protective Equipment Against Falls From a Height: Energy Absorbers*
- BS EN 358: 2000 *Personal Protective Equipment for Work Positioning and Prevention of Falls From a Height: Belts for Work Positioning and Restraint and Work Positioning Lanyards*
- BS EN 361: 2002 *Personal Protective Equipment Against Falls From a Height: Full Body Harnesses*
- BS EN 362: 2004 *Personal Protective Equipment Against Falls From a Height: Connectors*
- BS EN 365: 2004 *Personal Protective Equipment Against Falls from a Height. General Requirements for Instructions for Use, Maintenance, Periodic Examination, Repair, Marking and Packaging*

- BS EN 795: 1997 *Protection Against Falls From a Height: Anchor Devices: Requirements and Testing*
- BS EN 1004: 2004 *Mobile Access and Working Towers Made of Prefabricated Elements. Materials, Dimensions, Design Loads, Safety and Performance Requirements*
- BS 1129: 1990 *Specification for Portable Timber Ladders, Steps, Trestles and Lightweight Stagings*
- BS EN 1263–1: 2002 *Safety Nets. Safety Requirements, Test Methods*
- BS EN 1263–2: 2002 *Safety Nets. Safety Requirements for the Positioning Limits*
- BS 2037: 1994 *Specification for Portable Aluminium Ladders, Steps, Trestles and Lightweight Stagings*
- BS 4211: 2005 *Specification for Permanently Fixed Ladders*
- BS 8454: 2006 *Code of Practice for the Delivery of Training and Education for Work at Height and Rescue*
- BS EN 12811–1: 2003 *Temporary Works Equipment. Scaffolds. Performance Requirements and General Design*

Organisations

- Prefabricated Access Suppliers' and Manufacturers' Association
  Web: *www.pasma.co.uk*
  The association provides training and strives to advance safety standards in the UK access industry.
- International Powered Access Federation (IPAF)
  Web: *www.ipaf.org*
  The IPAF works to promote the combination of the capabilities of modern platforms and training for safe and effective work at height.
- Fall Arrest Safety Equipment Training (FASET)
  Web: *www.faset.org.uk*
  FASET is a trade association and a training body in the fall arrest industry.
- Construction Industry Training Board (CITB) and Construction Skills
  Web: *www.citb.org.uk*
  The CITB and Construction Skills provide assistance in all aspects of recruiting, training and qualifying the construction workforce.
- National Access and Scaffolding Confederation Ltd (NASC)
  Web: *www.nasc.org.uk*
  The National Access and Scaffolding Confederation is the national representative employers' organisation for the access and scaffolding industry.

Working on Live Roads

- A live road is one which is open to day-to-day traffic. It may be a minor road through a small village or a major trunk road. All roads and the services that run along and across them will from time to time require maintenance or repair.
- Any works carried out on a road where there is no physical barrier between the works and the live traffic would be deemed working on live roads.
- The scale of the works can vary from minor maintenance on a small road to the reconstruction of a major national trunk road.
- Consider if there are any alternatives to road closures when planning work on live roads.
- Before mobilisation give adequate warning to the travelling public — display notices, announce on local radio, etc and notify householders who are likely to be affected (letter drop).
- Risk assessments and safety provisions must be regularly reviewed during work.
- There are a variety of safety provisions that must be adhered to when carrying out work on live roads.
- Provisions must always include adequate safety clearances between operatives and the passing traffic.

7.704 A live road is a public road on which traffic flows. It may comprise from one to eight lanes wide, have height or weight restrictions and have a speed limit of between 20 and 70mph. Traffic may vary from a two-wheeled bicycle to lorries in excess of 40 tonnes. Roads include pedestrian footpaths.

7.705 A large number of essential services are buried in roads, including water, gas, electricity, telephone and communication cables.

7.706 From time to time it becomes necessary to maintain or repair the road or one of the services buried within it or to connect a new property to an existing service. In order to do so it becomes necessary to close the road or part of it to carry out the works.

In Practice

Live Roads

7.707 In most instances it is not possible to close the road to traffic and the works must be carried out while maintaining the traffic flow. Even where it is possible to close a road and divert the traffic it is likely to be necessary to accommodate residents or the users of premises on that road and will require access through the works.

7.708 Any works carried out on a road where there is no physical barrier between the works and the live traffic would be deemed working on live roads.

Scale

7.709 The scale of the works can vary from minor maintenance on a small road to the reconstruction of a major national trunk road. A competent risk assessment will identify specific requirements for individual sites.

Planning Works

7.710 Before carrying out work on live roads it is necessary to consider the following.

- Is there an alternative to closing the road (ie thrust boring or "mole")?
- Can closure be delayed to a more convenient time, ie avoiding peak periods.
- Are alternative routes already closed for works? Prioritise.

7.711 When planning the works the following should be taken into account.

- Consider the best work scheme — single shift or 24-hour working. It will be necessary to consider the following points before selecting 24-hour working.
  - Is there sufficient skilled labour available to work two shifts?
  - Is the supply chain available during the night? Concrete, tarmac, etc.
  - Is it a noise-sensitive area and are the operations noisy?
  - Night shifts are prone to more accidents and are less productive.
  - It is a premium shift, leading to greater costs.
  - It will be necessary to hire more equipment, eg lighting.
  - With less traffic on the road passing vehicles tend to travel faster, increasing the danger.
- Plan the works to progress in the same direction as the traffic flows. This will better accommodate single directional traffic flow through the works.
- The longer notice period requirement before commencing work is to allow other services to carry out essential work at the same time in the same work space. These works will need co-ordinating.

Pre-work

7.712 Before mobilisation:

- give adequate warning to the travelling public — display notices, announcements on local radio, etc
- notify householders who are likely to be affected (eg by letter drop)
- where necessary inform the police, fire brigade and ambulance service
- for medium or long-term works select a suitable location for the site set-up, taking into account connection to services, ie water, sewerage, electricity
- obtain service drawings and locate and mark services
- select and appoint suitable subcontractors
- plan traffic routes to and from site, taking into account the suitability of the roads and roads from which construction traffic is banned — pass this information to contractors, material and service suppliers
- visit the site and prepare risk assessments and method statements
- plan the site set-up, including offices and toilets
- arrange connection of telephone, gas, electricity, water, sewerage
- locate suitable disposal site for waste and redundant materials
- arrange suitable briefing for site personnel. This may vary from a briefing for small works to full induction for major works
- prepare a programme and select suitable personnel. Pass programme details to subcontractors
- plan delivery of materials and place orders
- arrange storage facilities for flammable liquids and LPG (liquid petroleum gas)
- arrange weekly site meetings where necessary.

Record Keeping

7.713

- Commence site records to include:
  - full site address, including post code and telephone number
  - identity and contact details of first aiders
  - location of first-aid equipment
  - contact details for emergency services, police, fire, ambulance, hospital accident and emergency, including route to hospital
  - contact details for subcontractors and suppliers
  - copies of plant registers
  - copies of operator competences
  - risk assessments and COSHH (Control of Substances Hazardous to Health Regulations 2002) assessments
  - subcontractors' risk assessments
  - register for induction training and toolbox talks
  - register of inspections for excavations, scaffolds, falsework including permits to load
  - certificates of inspection for electrical installations
  - fire extinguisher inspection certificates
  - on larger sites fire risk assessment, fire drills, etc
  - accident register
  - waste disposal notices.
- Post statutory notices including health and safety notices and employers' liability insurance.

Signage

7.714 Checks on signage relating to live road works should include the following.

- Check, and if necessary, renew signs on any LPG store and ensure a suitable fire extinguisher is provided.
- Post site speed limit signs and other safety notices (eg no reversing without banksman, overhead cables, etc).
- Ensure that high-visibility clothing of the correct standard is being worn by everyone on-site.
- Ensure all signs, cones barriers and lighting are correctly placed.
- Check that signs are not obscured by bends, hills or dips in the road.
- Check that there is adequate width for traffic to use.
- Ensure that the company name and contact details are displayed on the site.
- Ensure that the pedestrian arrangements cater for blind and disabled users, and are clearly marked. Ensure ramps are installed where pedestrians have to mount the kerb.

During Work

7.715 As work progresses:

- review risk assessments
- set up regular formal site safety inspections
- provide information to residents on an ongoing basis
- review safety at weekly progress meetings
- ensure that all signs and lamps, cones are clean
- where necessary ensure permanent traffic signs are covered
- review arrangements to see if delays can be reduced
- check that arrangements cover the needs of cyclists and horse riders.

7.716 If work is suspended ensure that arrangements are in place for checking signage, lighting and guarding.

Safety Provisions

7.717 There are many safety provisions that must be taken into account when working on live roads as follows.

- Where it is possible for a workman to inadvertently walk into the path of passing traffic, provision must be made to remove this risk, eg a lifeline of high visibility cord or tape or close coning along the inside of the safety margin.
- The maintenance of road signs and cones must be undertaken on a regular basis. Records must be kept of such maintenance and will be required in the event of an accident. Revellers in particular have a penchant for moving road signs and cones so greater attention must be paid to maintenance in areas known to be frequented by such persons.
- Workmen must never cross a live lane.
- All personnel working at roadworks on a live road must wear distinctive high-visibility clothing in addition to any other personal protective clothing required for the specific operation.
- All vehicles entering, travelling along or leaving the works must be equipped with a distinctive roadworks label and an amber flashing lamp. Hazard warning lights must not be used for this purpose.
- Vehicles must enter and leave site only by the correctly signed access points.
- Reversing vehicles must be under the direction of a banksman.
- Where work extends beyond dusk adequate lighting must be provided. Lighting must not dazzle motorists.
- On motorways and dual carriageways where traffic is diverted crossovers must be illuminated.
- The placing of signs and cones must be carried out under the supervision of a suitably trained person.
- On major roadworks suitable arrangements must be made for the prompt recovery of broken down vehicles. This is normally CCTV cameras and stand-by recovery vehicles.
- It is important to inform the travelling public of delays. This is now possible using Roadmaster cameras on the road and signs approximately 10 miles in advance of the roadworks. Technology enables the cameras to time a selected vehicle through the works and feed the information to the sign. The time is set against a predetermined time on the server and the delay is displayed on the sign.
- Adequate welfare facilities must be made available for the personnel on-site. For small works they may be public toilets or facilities at a local supermarket by arrangement provided they are "reasonably accessible".

Safety Clearances

7.718 Provisions must always include adequate safety clearances between operatives and passing traffic. Clearances will vary with the speed limit in force at that time.

7.719 If adequate safety clearances are not available for the proposed speed it will be necessary to apply for a lower speed limit on the grounds of safety.

Safety Clearances

Speed (mph)	Clearance/distance (m)
<40	0.5
≥50	1.2

Completion of Work

7.720 On completion of works:

- preserve and archive site records
- ensure that all temporary signs have been removed
- ensure all permanent signs have been reinstated
- notify the authorities that work has been completed.

List of Relevant Legislation

- Control of Substances Hazardous to Health Regulations 2002
- New Roads and Street Works Act 1991

Further Information

Publications

British Standard Publications

The following are available from *www.bsi-global.com*.

- BS EN 471: *2003 High-visibility Warning Clothing for Professional Use. Test Methods and Requirements*
- PAS 43: 2006 *Safe Working of Vehicle Breakdown, Recovery and Removal Operations — Management System Specification*

HSE Publications

The following are available from *www.hsebooks.co.uk*.

- CIS53: *Crossing High-speed Roads on Foot During Temporary Traffic-management Works*, 2000

Highways Agency Publications

The following are available from *www.highways.gov.uk*.

- *Guidance for Safer Temporary Traffic Management*, 2002
- National Highway Sector Scheme 12A — Sector Scheme Document, *Installing, Maintaining and Removing Static Temporary Traffic Management on Motorways and High Speed Dual Carriageways for Schemes Incorporating Contraflow Operations and/or Temporary Road Markings*, 2005

- National Highway Sector Scheme 12B — Sector Scheme Document, *Static Temporary Traffic Management on Motorways and High Speed Dual Carriageways for Schemes not Incorporating Contraflow Operations and/or Temporary Road Markings*, 2005
- National Highway Sector Scheme 12C — Sector Scheme Document, *Mobile Lane Closure Traffic Management on Motorways and Dual Carriageways*, 2005
- National Highway Sector Scheme 12D — Sector Scheme Document, *Installing, Maintaining and Removing Temporary Traffic Management on Urban Roads*, 2005
- National Highway Sector Scheme 17 — Sector Scheme Document, *Vehicle Recovery on Highway Construction Sites*, 2005
- *Temporary Traffic Management on High Speed Roads — Good Working Practice*

Department for Transport Publications

The following is available from *www.dft.gov.uk*.

- *Traffic Signs Manual*

Organisations

- Department for Transport (DfT)
  Web: *www.dft.gov.uk*
  The DfT oversees the delivery of a reliable, safe and secure transport system that responds efficiently to the needs of individuals and business while safeguarding the environment.
- Health and Safety Executive (HSE)
  Web: *www.hse.gov.uk*
  The HSE is responsible for the regulation of almost all the risks to health and safety arising from work activity in the UK.
- The Highways Agency
  Web: *www.highways.gov.uk*
  The Highways Agency is responsible for the construction, maintenance and improvement of the strategic road network (motorways and trunk roads) in England. All other roads in England are the responsibility of local authorities.

Working On or Over Water

- Working on, near or over water can give rise to a number of risks to the safety and health or workers.
- Water can be a danger if it is fast flowing as a river or in man-made aqueducts or sewers, or if it is not flowing as in a reservoir, dock or lake.
- Suitable precautions to prevent falls from height into water must be taken in compliance with the Work at Height Regulations 2005.
- Life jackets significantly increase the possibility of survival if they are of the correct design, correctly maintained and worn properly.
- Some clothing may help reduce the speed of heat loss when immersed in cold water and special clothing for this purpose is available.
- Ropes slung across fast flowing water can "catch" someone in the water and help with rapid rescue if installed correctly.
- Rescue boats may need to be available for rapid rescue of anyone falling into water.
- Hypothermia is the condition resulting from the loss of core body heat and causes loss of awareness, strength and feeling, sleepiness, inability of think clearly, unconsciousness and eventually death.
- Infections are common where the correct precautions are not taken: leptospirosis is a disease that can develop from contact with rat's urine; tetanus is a disease resulting from contact with contaminated soil.
- All those involved with working over, near or on water must understand the hazards and risks and have received appropriate information, instruction and training

7.721 Working on, near or over water can give rise to a number of risks to safety and health. Such work can include:

- working from a boat, floating dock, barge or other structure on the water
- working on the construction, repair or maintenance of a structure that is near or over water
- working near to water, eg cutting weeds on a riverbank.

7.722 Some work can involve intimate contact with water such as sewer repair and maintenance and work on sewage treatment sites.

7.723 Work under water is a highly specialised category and not covered in this topic. Other types of work may take place below a water table in an excavation, in a tunnel below water or in pressurised workings or caissons. These may be undertaken by construction workers rather than trained divers but in each case the people involved must be made aware of the specific additional risks. Long-term work under pressure for instance, can involve bone necrosis and sudden decompression can result in "caisson disease".

7.724 Some people may have to cross water by boat to attend their place of work. Some of the information in this section may be of relevance to making an assessment of risks for that situation.

Employers' Duties

7.725 There are no specific regulations placing duties on an employer (or self-employed person) organising or carrying out work on or near water. However, general health and safety legislation applies to all cases of work on or near water as follows.

- Under the Health and Safety at Work, etc Act 1974, employers have a general duty in respect of protection of employees at work. Section 3 places duties on an employer regarding protection of other persons who may be affected by the work and also on self-employed persons to look after their own health and safety and not to affect any other persons' health or safety.
- The Management of Health and Safety at Work Regulations 1999 include the specific duty on an employer to carry out suitable and sufficient assessment of risk and this is the keystone to avoiding risks of injury or ill health with work on or near water.
- The Construction (Design and Management) Regulations 2007 include both broad and specific duties which cover the wide range of work likely on or near water. These include:
  - welfare requirements such as toilets, rest areas and washing facilities (possibly of significant importance for personal hygiene)
  - the creation, support, working in and the inspection of excavations
  - traffic routes and the segregation of pedestrians and vehicles
  - higher risk activities such as demolition
  - emergency and fire provisions and arrangements.
- The Work at Height Regulations 2005 include specific duties regarding any work where a person may fall any height and be injured. The key provisions of the regulations are that duty holders should:
  - avoid work at height where reasonably practicable
  - (where work at height is not avoidable) undertake a suitable and sufficient assessment of risks and put in place adequate control measures
  - use suitable equipment, such as scaffolds or mobile elevating work platforms to prevent falls from height
  - use suitable equipment to minimise the distance and consequences of a fall from height.

Employees' Duties

7.726 Employees have a general duty under the Health and Safety at Work, etc Act 1974 to ensure their own safety and the safety of others who may be affected by their acts or omissions.

7.727 Employees must co-operate with their employers to ensure compliance with any relevant statutory provisions.

7.728 Employees are required under the Management of Health and Safety at Work Regulations 1999 to:

- use "any machinery, equipment, dangerous substance, transport equipment, means of production or safety device" in accordance with any training or instruction received

- inform the employer or an employee who has safety responsibilities of:
  - any situation that may present a danger to health and safety
  - any omissions or shortcomings in the employer's health and safety protection arrangements.

In Practice

The Range of Risks

7.729 The obvious risk from working over water is drowning, but few accidents of this type occur. Falls from height into water can result in serious injury either from striking fixed structures on the way down, or from hitting the water itself should the height be great enough — accurate statistics for this are not available. A possible consequence of a fall into water may be the survivor suffering subsequent psychological problems and stress.

7.730 Risks to health when working near water can be serious. Work on or near rivers, sewers or any body of water where rats exist can give rise to serious hazards such as leptospirosis (Weil's Disease).

7.731 Contact with untreated sewage leads to health risks from infection through unprotected cuts or lesions, the eyes or through inhalation or swallowing contaminated water.

7.732 As with most manual work, many more commonly thought of risks should be considered, eg vibration, dermatitis, electricity, repetitive work, manual handling, etc.

Identifying Relevant Activities

7.733 In order to identify risks and precautions to be taken it is necessary to understand the type of work being undertaken. Work on, near or over water would include:

- any activity where a person, either employee or member of the public, could fall into water as a result of the work, eg:
  - construction work, repair, alterations to a structure
  - work from a boat or other floating structure
  - work from a work platform which may project over water including a mobile elevating work platform or platform suspended from a crane
- work near to open water
- work in a trench or other workings where there could be a sudden inrush of water from a nearby river, reservoir or similar body of water
- work in tidal areas where there may be a rapid inrush of water
- work in a sewer, culvert (a conduit used to enclose a flowing body of water, ie under a road or railway) or similar structure where water levels can vary quickly and there may be a sudden inrush of water
- work near a substantial storage tank with an open top or work near to an opening in the top through which a person may fall.

Risks from Falling into Water

7.734 The most obvious and immediate risk is that of drowning. Contributory factors include:

- inability of the individual to swim
- incapacity of the individual due to striking an object during the fall
- incapacity of the individual due to striking an object in the water
- incapacity of the individual due to striking the water after a high fall (it is quite possible to sustain very serious injuries to the arms, legs, spine, neck from a high fall into water where entry into the water is uncontrolled
- shock of sudden immersion into cold water which may affect breathing, cause panic or result in unintentional inhalation of water
- waterlogged clothing dragging the individual below the water
- fast moving/rough water making staying on the surface difficult
- fatigue and/or hypothermia
- lack of rescue facilities.

7.735 Other risks that may occur include:

- infection from untreated sewage
- infection from rat infestation and rat urine
- unseen underwater obstructions
- unseen underwater traps such as nets, ropes, etc
- water draw-off points or drains which may result in fingers, arms, legs being drawn in and then held due to water pressure (eg, a water draw-off point in a reservoir)
- being hit by other vessels where rescue is not immediate.

Risk Assessment

7.736 Any risk assessment for work on or over water should look at the work to be carried out and consider the following.

- Can work over water be avoided?
- Can work near water be avoided?
- Is the water flowing or stationary?
- Is the water deep or cold; does it have other characteristics likely to increase risk?
- Can there be a sudden change in water levels?
- Do the planners and supervisors have sufficient knowledge and experience to adequately assess risks and plan the work for safety?
- What knowledge and experience do the workers involved have?
- Can members of the public gain access to the area where work is carried out?
- Could children gain access to the area where work is carried out?
- Are rescue facilities readily available?
- Are medical facilities readily available?
- Will any form of mobile elevating work platform (MEWP) be used near/over water? (If this is the case the normal rules for wearing a harness attached to the MEWP cage are ignored due to the risk of drowning should the equipment overturn and fall into the water. A life jacket must be worn.)
- Is access to the water easy or difficult? Could a person in the water climb from the water unaided?

Control Measures

Selection of People

7.737 Anyone working over, near or in water must understand the additional risks to safety or health this type of work can give rise to. The provision of information, instruction and training will depend on:

- the location of the work
- the activities to be carried out (which may have their own serious risks)
- any other factors identified during the early assessment of risks.

7.738 Assessment of risks must take account of an individual's capability and also the possible lack of maturity, experience and knowledge if under 18 years of age. If the risk of falling into water is high then careful consideration should be given to a number of factors as follows.

- Can the individual swim? This may be irrelevant where any fall from a substantial height could be involved as the individual may be badly injured or unconscious. In practice, it seems that for incidents involving people in cold water immersion, about the same proportion of swimmers and non-swimmers drown.
- Does the individual suffer from any medical condition that makes them more vulnerable from sudden immersion into cold water or from prolonged exposure during rescue efforts?
- Will a lack of maturity, or lack of appreciation of consequences, lead to risk-taking behaviour?
- Have all possible infection risks been identified and does each individual have the appropriate immunisation where this is available?
- Do all individuals know how to use ALL the safety equipment provided and understand how ALL the rescue procedures work?

Falls from Height

7.739 This is a major concern as immersion in water is rapid, possibly with serious injury already inflicted. Compliance with the Working at Height Regulations 2005 (WAHRR) should be sufficient to prevent falls and ensure suitable and safe access.

7.740 The WAHRR make no distinction between falls from high or low levels. Care must be taken to provide adequate edge protection and safety systems where a fall could occur from a low level that may be ignored in "dry" circumstances.

7.741 When working near to water, slips can be a higher risk than in normal circumstances. For example, a slip on a fast-flowing river bank resulting in immersion may have serious consequences compared to a similar slip elsewhere.

Control of Water

7.742 Control of the water may or may not be physically and technically possible. The ideal option would be to have no water under/near the work to be carried out. If work has to take place near a fast-flowing water channel it may be possible to divert water to reduce the depth, volume and speed of water.

7.743 Where there is static or slow-moving water it may be possible to reduce the depth. However, consideration must be given to the bed of the water course or structure and whether it is suitable for walking on. For example, there may be:

- obstacles underwater creating trip hazards
- sudden depressions creating deep water hazards
- soft layers unable to support an individual's weight, etc.

7.744 Where control of water flow may suddenly fail or where water volume may change through natural influences (eg rain) this MUST be taken into account. It may be appropriate to have water level monitoring and/or automatic or manual alarm systems to warn workers to evacuate.

Personal Protective Equipment

7.745 Any personal protective equipment supplied should comply with the Personal Protective Equipment at Work Regulations 1992 and all users should be fully trained in accordance with those regulations.

Flotation Devices

7.746 Buoyancy aids are intended to help keep a conscious person afloat. They may be suitable for use by people using boat transfer to a place of work, or for working on a boat. They are unsuitable in situations where a person in the water is unconscious as they may not turn them over from a face-in-the-water position.

7.747 Life jackets can be of two basic types as follows.

- *Manual* — manual inflation will only work where the person in the water is able to operate the inflation system which is a stored compressed gas released by valve or pull cord.
- *Auto-inflation* — auto-inflation systems operate on immersion in water and will inflate and support an unconscious person. If the jacket is the right size and correctly fitted it should also turn an unconscious person over from a face-in-the-water position and support them with the head slightly back.

7.748 Life jackets should be fitted with manual inflation for a person to be able to top up inflation if necessary (where immersion is for a longer period). Life jackets should also be fitted with a whistle for attracting attention. If immersion in the dark is a possibility, a strobe beacon should be fitted. In higher risk situations, the fitting of an emergency radio transponder may be appropriate.

7.749 Life jackets and buoyancy aids must be inspected at the intervals specified by the manufacturer. Any records of issue, use, inspection and repair should be kept.

Immersion and Flotation Suits

7.750 Immersion suits are used extensively where workers are transported across open water and there is a risk of immersion. They are also used where evacuation from a structure, such as an oil production platform, may be necessary. They are usually one-piece insulated suits with a hood which are designed to minimise water ingress and reduce heat loss.

7.751 Flotation suits are similar to an immersion suit but also incorporate flotation aids. An immersion suit requires a separate flotation device to be donned.

7.752 Because of the bulky nature of both types of suit they are not likely to be of use to workers except for:

- transportation purposes
- cases where evacuation from the worksite may be necessary and may involve entry in the water.

7.753 They may be useful for work in extreme weather conditions as they do provide waterproofing, windproofing and insulating protection. (Flotation suits are often used by fishermen, etc.) However, their use in normal weather conditions should be judged very carefully. This can lead to other problems, eg heat stress and exhaustion, due to the inability of the body to lose heat during exertion.

Catch and Rescue Systems

7.754 Where work is over flowing water, there must be some system in place to either prevent them being carried away:

- from the point of immersion too quickly
- too far to allow the alarm to be raised and rescue to be initiated.

Catch Lines

7.755 Catch lines may be strung downstream across flowing water courses at intervals to allow a worker to "catch" one and prevent them being carried further.

7.756 Catch lines should be tensioned across the water at an angle of 45 degrees. The most downstream end should be on the river bank where easiest access to and egress from the water and rescue is possible. It must not be placed directly across the water at 90 degrees.

7.757 The line must have floats at intervals to make it more visible and should be a man-made fibre which floats itself and is also a bright colour.

Throw Lines

7.758 Throw lines are used by a second person to give a possible lifeline to the person in the water.

7.759 A throw line should not be tied to anything, but the "thrower" must ensure that they have a firm grip on the line end. A throw line should be a brightly coloured floating rope with a diameter of 8–12mm. The "float" should be of an acceptable design.

7.760 It is possible, in fast-flowing water, that the rescuer may not be able to hold position once the rescuee has grabbed the line. They should then walk along the bank to avoid themselves being pulled into the water. A snagged or tied line may result in the person in the water becoming submerged.

7.761 The thrower must have been trained to use a throw line.

Rescue Boats

7.762 Rescue boats with competent operators may need to be provided for both flowing and non-flowing water situations.

7.763 The rescue boat must be:

- large enough to accommodate at least two operators plus the person to be rescued
- powerful enough to overcome any fast water flow and stable in all anticipated conditions
- low in the water to facilitate rescue.

7.764 If the boat is a larger design, it should have additional facilities such as a "swim board" to allow rescue to be carried out easily. It may be appropriate for the boat crew to be trained in first aid including the use of resuscitation equipment.

7.765 Where work is over fast-flowing water, such as a tidal estuary, boats may have to be on permanent patrol. In such water, an immersed person may travel a considerable distance very quickly. In other situations it may be satisfactory to have a boat moored or at a launch site ready for action should the need arise.

7.766 In this second scenario it is important that regular checks are made on the serviceability of the boat. The assessment may show that two boats or a twin engine boat are required to ensure rapid deployment should a mechanical failure take place.

Provision for Emergency Services

7.767 Where rescue has taken place, transfer to the emergency services may be necessary. Access for a two-wheel drive ambulance to the rescue boat launch and recovery site should be provided. However, this is not always possible and other means of transfer must be considered.

7.768 Remoteness from emergency medical services should also be considered. For example, on some larger construction projects a temporary helicopter landing site is provided to ensure that a serious emergency could be adequately dealt with.

Hypothermia

7.769 Water temperatures below 26.5°C (80°F) will have an adverse effect on survival for a person not rescued quickly. Inland water is generally colder than the sea and most inland water in the UK probably remains at temperatures below 10°C throughout most of the year.

7.770 The life threatening initial cold shock response begins at water temperatures below 25°C. Cold shock is an increased respiratory response to cold water immersion. At first there is an involuntary gasp which is followed by hyperventilation. There is usually an associated degree of disorientation, where sufferers are not sure which way is up, or where they are in relation to the boat, the bank, etc.

7.771 Effects of cold shock are proportional to reduction in water temperature with the maximum effect being at 10–15°C. Ability to hold breath is proportionally reduced the colder the water.

7.772 "Predicted survival curves", giving an expected survival time when immersed in water at various temperatures, are of limited use. They are based on rates of body core cooling, but the early localised effects of hypothermia may be fatal long before body core temperature reaches life-threatening levels. For example, manual dexterity is rapidly and severely degraded in water below 15°C. This would badly hamper the ability to carry out essential survival tasks, ie trying to catch or hold a throw line.

Infectious Diseases

Leptospirosis

7.773 Leptospirosis, or Weil's disease, is fatal if untreated. It is an acute infection caused by bacteria known as *Leptospira*. These bacteria infect wild and domestic animals, particularly rats, cattle, pigs and dogs and are excreted in their urine.

7.774 Human infection results from direct contact with the urine or tissues of an infected animal or indirectly from contaminated water, soil or sewage. Infection occurs through breaks in the skin or the mucous membranes of the mouth and nose.

7.775 People who work in and around water and waterways, where rats can be present and where urine from animals can be a contaminant have an increased risk of contracting this disease.

Tetanus

7.776 Tetanus is a risk always present for anyone working outside, whether near water or not. Tetanus is a dangerous nerve ailment caused by the toxin of a common bacterium, *Clostridium tetani.*

7.777 Bacterial spores are found in soil — most frequently in cultivated soil, less frequently in virgin soil. The spores can remain infectious for more than 40 years in soil. They also exist in environments as diverse as animal excrement, house dust and the human colon. If the spores enter a wound that penetrates the skin and extends deeper than oxygen can reach, they germinate and produce a toxin that enters the bloodstream.

7.778 Cuts, etc should be covered with a waterproof plaster and it is essential to keep tetanus immunisation up-to-date with boosters at intervals of 6–10 years or as advised by an individual's GP. Some medical opinion is that immunisation can last for most of an adult lifetime if given at school age and that a booster is necessary after a possible infection. Medical advice should be sought in each case.

Raw Sewage and other Contamination

7.779 Workers likely to come into contact with water contaminated with raw sewage should follow the precautions mentioned above. There may also be contamination from chemical or other toxic waste disposal which has taken place illegally. Even years later there may be ground contamination near to previously contaminated water.

7.780 Whether working in an inland waterway or open water with a sewage outlet, the ingestion or inhalation of contaminated water is a very serious risk. Inhalation is particularly serious and may give rise to the rapid development of pulmonary or lung infection with potentially very serious consequences.

Precautions

7.781 Precautions against infection are quite simple. Any cuts, grazes or skin lesions should be covered with a waterproof plaster and suitable gloves worn when handling any material that may be contaminated.

7.782 Personal hygiene is important and all workers must receive adequate information, instruction and training so they fully understand the risks and the precautions to take. With any of the health risks described here, any worker who becomes ill MUST tell their GP or treating medical practitioner that they work near water, raw sewage, etc. This can influence the diagnosis. For example, Weil's disease in the early stages is easily confused with severe flu symptoms and wrong treatment may result unless the likelihood of this infection is known.

Accident investigation and Reporting

7.783 Normal accident investigation, recording and reporting procedures to comply with the Reporting of Injuries, Diseases and Dangerous Occurrences Regulations 1996 (RIDDOR) should be in place. Any relevant disease as specified in RIDDOR should be reported after diagnosis by a medical practitioner using report form F2508A or the alternative internet reporting options.

Further Information

Publications

HSE Publications

The following are available from *www.hsebooks.co.uk*.

- HSG150 (3rd ed, 2006) *Health and Safety in Construction*
- L144 *Managing Health and Safety in Construction. Construction (Design and Management) Regulations 2007. Approved Code of Practice*

The following free leaflets are available from *www.hse.gov.uk*.

- AIS1 *Personal Buoyancy Equipment on Inland and Inshore Waters*
- CIS47 *Inspections and Reports*
- CIS56 *Safe Erection, Use and Dismantling of Falsework*
- INDG367 *Inspecting Fall Arrest Equipment Made From Webbing or Rope*
- MISC614 *Preventing Falls from Boom-type Mobile Elevating Work Platforms*
- *Height Safe — Absolutely Essential Health and Safety Information for People who Work at Height*

Workplace Transport

- The HSE defines workplace transport as any vehicle or piece of mobile equipment that is used by an employer, employees, self-employed people or visitors in any work setting, excluding travelling on public roads.
- Workplace transport covers a wide range of vehicles and the many forms of industrial trucks, eg dumpers. It also covers less common vehicles and plant, such as straddle carriers, rubber-tyred gantries and self-propelled machinery.
- In managing the risks associated with workplace transport, the first step is to identify the hazards, including vehicle arrival and departure and actions of the driver, as well as vehicle movement around the site.
- Identify who might be harmed and how, taking note of any hazards to the general public.
- Evaluate the risks and assess whether existing precautions are adequate or whether more precautions are required.
- Record the significant findings of the assessment and periodically review it.

7.784 In the past two years, there have been just over 100 fatal injuries to workers resulting from workplace transport accidents and just under 4000 major injuries reported under the Reporting of Injuries, Diseases and Dangerous Occurrences Regulations 1995 (RIDDOR). The majority of accidents related to workplace transport involve people:

- being hit or run over by moving vehicles
- falling from vehicles
- being struck by objects falling from vehicles
- being injured as a result of vehicles collapsing or overturning.

Employers' Duties

7.785

- Employers have a general duty to ensure, so far as is reasonably practicable, the health, safety and welfare at work of all employees under the Health and Safety at Work, etc Act 1974.
- Employers are required to assess the risks to employees and anyone who may be affected by their operations under the Management of Health and Safety at Work Regulations 1999.
- Mobile work equipment is governed by the Provision and Use of Workplace Equipment Regulations 1998. Its main requirements include the following.
  - Employees must not be carried on mobile work equipment unless it is suitable.
  - Risks to employees riding on mobile work equipment must be minimised by stabilising the equipment or by structural design.
  - Fork-lift trucks which carry employees must be adapted or equipped to reduce risks from overturning to as low a level as reasonably practicable.
  - If self-propelled work equipment may involve risks to safety due to its motion, it should not be able to be started by unauthorised people.
  - If appropriate to ensure safety, remote-controlled self-propelled work equipment should stop automatically after it leaves its control range.
  - The drive shaft on mobile equipment should be safeguarded if it could become soiled or damaged by contact with the ground while uncoupled.

- The Workplace (Health, Safety and Welfare) Regulations 1992 place duties on employers. Its requirements include the following.
  - Every workplace should be organised so that pedestrians and vehicles can circulate in a safe manner.
  - Traffic routes in a workplace should be suitable for the persons or vehicles using them, sufficient in number, in suitable positions and of sufficient size
  - All traffic routes within a workplace should be suitably indicated if necessary.

Employees' Duties

7.786

- Employees have a duty to take reasonable care of their own health and safety and that of other people who may be affected by their work under the Health and Safety at Work, etc Act 1974.
- Employees have a duty to co-operate with their employers on health and safety matters under the Health and Safety at Work, etc Act 1974.
- Employees are required to follow health and safety instructions and to report danger under the Management of Health and Safety at Work Regulations 1999.

In Practice

Identify the Hazards

7.787 The hazards of workplace transport need to be considered against the activities undertaken within the particular workplace and should involve the consideration of site-based and visiting vehicles.

7.788 There is a wide range of activities that should be considered in identifying the hazards.

Vehicle Arrival and Departure

7.789 How vehicles arrive and depart the business must be dealt with. The following questions should be considered.

- How are they controlled?
- How is information on workplace layout and precautions communicated to visitors?

7.790 Anyone entering a site needs to be aware of transport hazards and safety rules. Typical types of information include:

- verbal instructions on arrival
- site induction
- issue of site maps
- delivery instructions
- displaying site maps and rules around site.

Vehicles

7.791 The vehicles themselves should also be considered, ie are they:

- suitable for the task?
- properly maintained?

7.792 The vehicle selected needs to be capable of completing its tasks. Areas to be considered include:

- stability under all foreseeable operating conditions
- safe access to and from the cab
- effective handling systems
- adequate visibility for the driver
- headlights, horn, windscreen wipers, reversing alarms, etc
- physical guards to protect dangerous parts
- driver protection from falling objects/vehicle overturning
- driver protection from weather, noise, fumes, vibration, etc.

7.793 A vehicle inspection program, outlining the frequency and specifications of checks should be devised, according to the anticipated risks. A maintenance log should be kept of all checks. Examples of checks to make include:

- braking systems
- seat belts
- tyres
- steering
- mirrors, CCTV, etc
- lights and indicators
- fire-fighting equipment
- warning signals, alarms.

Travel Routes Within Workplaces

7.794 Travel routes within workplaces are another point to be considered. Employers should consider nearby hazards and obstructions and whether travel routes are suitable for the vehicles that use them. The risks from the movement of vehicles around the site need to be assessed and suitable control measures must be introduced.

Actions of Drivers

7.795 The actions of drivers obviously impact on health and safety. Employers should consider whether drivers are using safe working practices when using the vehicles, especially in reversing manoeuvres and when carrying out loading and unloading activities.

7.796 When planning and controlling site vehicle operations, the following hierarchy of control measures for reversing operations should be applied.

- Eliminate need to reverse — introduce one-way systems, designated turning areas, etc.
- Reduce reversing operations — reduce number of vehicle movements.
- Ensure adequate visibility for drivers — CCTV, mirrors, etc.
- Follow safe systems of work — allow adequate space for reversing, prohibit pedestrian access to area, signs or physical stops for drivers reaching reversing limit, etc.

7.797 Vehicles should be loaded and unloaded on level ground, away from passing traffic, pedestrians and overhead hazards.

7.798 Vehicles should not be loaded above capacity in terms of weight and dimensions. All loads should be secured to the vehicle and should be distributed as evenly and as close to the centre of gravity as possible.

Activities of Others

7.799 Employers need to consider the activities of others. Is there a separation between workplace vehicles and others who may be harmed by their activities, such as other employees, visitors and the general public?

Shared Workplaces

7.800 The Management of Health and Safety at Work Regulations 1999 require all employers and self-employed people who share a workplace to co-operate in health and safety matters. They must:

- co-operate so everyone can comply with their health and safety duties
- take all reasonable steps to co-ordinate the measures they take with those taken by other employers or self-employed people
- take all reasonable steps to tell other employers and self-employed people about risks to their employees' health and safety as a result of their work activities.

7.801 It is normal for the site operator (or main employer) to take control of the site. This person should also take responsibility for co-ordinating health and safety measures. They will need to hold discussions with other employers, obtain health and safety information from them and seek their agreement to (and ensure their understanding of) site-wide arrangements.

7.802 Vehicles on which employees of more than one company are working are considered shared workplaces. This is the case even if the activity takes place for only a brief period (such as loading, unloading or sheeting operations). All the employers concerned are responsible for the safety of their own employees and those of the other employers involved. Those involved should agree at the start who will be responsible for what, including safety.

Loading and Unloading of Vehicles

7.803 Consider the siting and layout of loading/unloading areas carefully. They should be sited to minimise vehicle movements, including those of fork-lift trucks and other loading vehicles, and should be sited away from overhead power cables, etc.

7.804 Avoid excessive gradients, especially in areas where lift trucks operate, and consider the use of one-way systems and drive-through loading/unloading areas.

7.805 Ensure that the loading areas have at least one safe pedestrian exit route. Arrangements should be made to exclude non-essential personnel from the loading/unloading area, and essential personnel should wear suitable high-visibility clothing, including suitable waterproof and bad weather clothing.

7.806 In order to prevent unexpected vehicle movements during loading/unloading operations (such as drive offs), appropriate control measures should be introduced, including:

- ensuring that the engine is stopped and the handbrake is on
- ensuring the vehicle is left in gear
- the driver should remove the key from the ignition and (if appropriate) pass it to a responsible person.

Vehicle Maintenance

7.807 Vehicle maintenance activities should be covered by safe working practices.

Identify Who Might be Harmed and How

7.808 The identified hazards should be considered individually to determine who might be harmed and how.

7.809 The people at risk of harm from workplace transport are likely to include any employee, contractor, visitor or member of the public.

Evaluate the Risks

7.810 In evaluating the risks, it is important to consider the likelihood that harm will occur and the severity of consequences.

7.811 The evaluation of risk should consider whether any existing precautions are adequate and whether the risk has been reduced to as low a level as reasonably practicable.

7.812 If further measures are deemed necessary to reduce the risk to as low a level as reasonably practicable, the following hierarchy of control measures should be considered.

- Eliminate the hazard — this could include the restriction of routes.
- If possible, modify the workplace — separate vehicles from walkways, use vehicles with appropriate safety features.
- Use safe systems of work — enforce speed limits within the workplace, ensure that all users of workplace transport have sufficient information and training in their proper use.
- Use personal protective equipment (PPE) — as a last resort, train people to use and take care of the appropriate PPE.
- Use appropriate signage, etc.

Record the Findings

7.813 The Management of Health and Safety at Work Regulations 1999 require organisations with five or more employees to record the significant findings of risk assessments, as follows.

- Record in writing significant hazards identified. It is good practice to record all the hazards identified.
- Record any control measures that have been considered or are to be introduced, along with the name of the person responsible for implementing them.
- Any actions to be taken to mitigate an identified hazard should be prioritised and included in the records.

Review the Assessment

7.814 The risk assessment should be reviewed at regular intervals or if there is reason to suspect that the risk assessment is no longer valid, for example if:

- there has been a significant change in the work
- the results of monitoring show it to be necessary
- new vehicles or workplace equipment have been introduced.

Training

7.815 The Health and Safety at Work, etc Act 1974 requires all employees to receive suitable information and training relevant to their work.

7.816 The aim is to ensure that all employees are competent to carry out their duties. In relation to workplace transport, employers should ensure that suitable information and training is provided to all those who are at risk. This will include:

- identifying all those who require training
- designing an appropriate training programme
- implementing the training programme
- monitoring the effectiveness of the training programme
- keeping records of any training provided.

7.817 It is important that training is not seen as a "one off" exercise. The training programme should take account of the output from the monitoring and review of the risk management system addressing any specific risks identified.

List of Relevant Legislation

- Traffic Signs Regulations and General Directions 2002
- Management of Health and Safety at Work Regulations 1999
- Provision and Use of Work Equipment Regulations 1998
- Reporting of Injuries, Diseases and Dangerous Occurrences Regulations 1995
- Workplace (Health, Safety and Welfare) Regulations 1992
- Health and Safety at Work, etc Act 1974

Further Information

Publications

HSE Publications

The following are available from *www.hsebooks.co.uk*.

- HSG6 (rev 2000) *Safety in Working with Lift Trucks*
- HSG136 *Workplace Transport Safety: An Employers Guide*
- INDG199 (rev 1) *Workplace Transport Safety: An Overview*
- INDG279 *Fatal Traction: Practical Advice on Avoiding Agricultural Transport Accidents*
- WPT02 *Safe Access to Road-going Vehicles: Specifying the Right Equipment*
- WPT05 *Managing Work to Avoid Falls from Vehicles*
- WPT06 *Co-operating to Prevent Workplace Vehicle Accidents*

Chapter 8

Legal Requirements and Management Responsibilities

Accident Reporting

- The Reporting of Injuries, Diseases and Dangerous Occurrences Regulations 1995 (RIDDOR) require a "responsible person" to report the following instances to the relevant enforcing authority:
  - deaths
  - certain major injuries
  - injuries that cause absence for more than three days (including days that are not normally worked, eg weekends)
  - certain diseases and dangerous occurrences.
- The "responsible person" will either be an employer, a self-employed person or someone in control of the premises.
- Reports under RIDDOR can be made to the Incident Contact Centre or the local office of the enforcing authority.
- Reports must be made by the quickest practicable means, which is usually a phone call.
- F2508 is the official form that should be used for the reporting of injuries, diseases and dangerous occurrences.
- All employees must give notice to employers of any personal injury caused by an accident at work. This is usually done by entry of the details into an accident book.
- The Safety Representatives and Safety Committees Regulations 1977 require the employer to inform an appointed union safety representative of a notifiable accident to allow the representative to conduct an investigation.
- An accident reporting policy should be in place to ensure that all procedures are followed correctly.
- Employees and managers should receive training to raise their awareness of the need to report all incidents and the procedures to follow.
- A monitoring system should be in place to ensure that accidents, diseases, dangerous occurrences and near misses are reported.

8.1 It is essential that employers introduce a reporting system in the workplace for all:

- accidents and incidents
- dangerous occurrences
- diseases
- near misses.

8.2 This system should apply equally to employees, contractors and visitors. It will allow employers to:

- comply with statutory provisions under the Reporting of Injuries, Diseases and Dangerous Occurrences Regulations 1995 (RIDDOR)
- carry out any necessary investigations
- measure their safety performance and benchmark against national statistics.

8.3 Incident/accident reporting is an important aspect of monitoring or measuring safety performance and allows an organisation, through investigations, to learn from mistakes and improve safety.

8.4 Accident reports made to the enforcing authority are used to compile national statistics on health and safety performance in specific work sectors, such as construction, agriculture and the office environment.

8.5 An accident report may also prompt an investigation by the enforcing authority. Enforcement officers have the power to investigate every incident notification that they receive. However, investigations are normally prioritised in terms of the severity of injury that has either occurred or had the potential to occur.

Employers' Duties

8.6

- The Reporting of Injuries, Diseases and Dangerous Occurrences Regulations 1995 require the specified responsible person, usually an employer, to report certain defined work-related events to the enforcing authority. The enforcing authority will either be the Health and Safety Executive (HSE) or the local authority. Accidents or incidents to be reported include:
  - all fatalities
  - accidents resulting in any of the specified major injuries
  - certain defined, work-related diseases
  - accidents resulting in employees being absent from work for more than three days
  - certain dangerous occurrences, such as when a building collapses or a gas explosion occurs.
- The Social Security (Claims and Payments) Regulations 1979 require that every owner or occupier of premises to which any of the provisions of the Factories Act 1961 apply, and every other employer with 10 or more employees, keep an accident book or books in an approved format.
- The Safety Representatives and Safety Committees Regulations 1977 require that the employer informs an appointed union safety representative after the occurrence of a notifiable accident in order to allow that representative to conduct an investigation.

Employees' Duties

8.7

- The Health and Safety at Work, etc Act 1974 requires employees to co-operate with employers to enable them to fulfil their statutory duties. This includes reporting all dangerous occurrences, near misses and accidents whether or not they resulted in injury, damage or disease.
- The Social Security (Claims and Payments) Regulations 1979 require employees to give notice to their employers, either verbally or in writing, of any personal accident in respect of which a benefit may be payable. The duty of the employee to report accidents is fulfilled if the employee, or his or her representative, records the details of the accident in a book or forms containing the same information.

In Practice

Internal Reporting Procedures

8.8 It is important for employers to implement an effective internal reporting procedure, which includes the provision of an accident book, in addition to reporting procedures under the Reporting of Injuries, Diseases and Dangerous Occurrences Regulations 1995 (RIDDOR).

8.9 All reporting procedures should be detailed within a policy. Management should encourage the reporting of all incidents.

8.10 Internal reporting procedures will depend upon:

- organisational factors
- the type of incident involved
- who is involved
- the subsequent legal requirements.

8.11 The following is a sample accident/incident reporting procedure for employees to follow. All staff should receive suitable training in the procedures adopted.

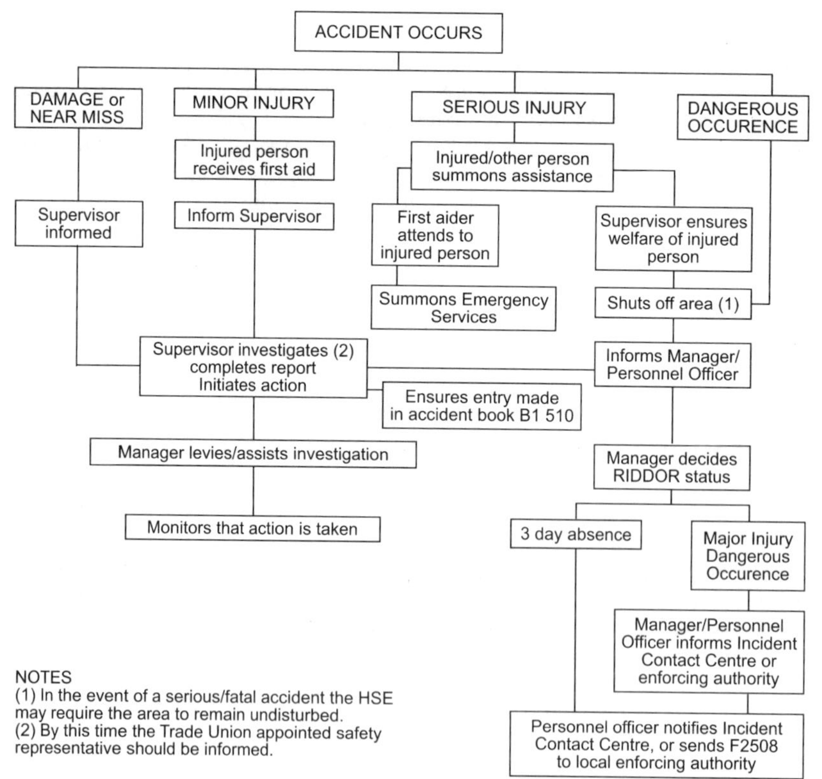

8.12 Accidents involving non-employees should also be entered either into the accident book or onto a suitably designed form to ensure that suitable records are maintained.

8.13 When accidents involving contractors occur, they should be encouraged to report any such accidents to their own employer to ensure that the accident is suitably recorded.

8.14 It is important to establish who is responsible for reporting an incident, including when a non-employee suffers an injury.

Accident Book

8.15 An employee must give notice, either orally or in writing, of any personal injury caused by an accident at work.

8.16 This notice must be given as soon as is practicable after the accident. An entry of the appropriate particulars in an accident book will be sufficient notice of the accident. This may be done by the injured person or by another person acting on his or her behalf.

8.17 An employer is required to take reasonable steps to investigate the circumstances of every accident that is reported. If there appears to be a discrepancy between the circumstances found by the investigation and those reported, there is a requirement to record those circumstances.

8.18 The prescribed form of record required is BI510: *Accident Book*, available from HSE Books.

8.19 The use of this book is not compulsory where there are fewer than 10 employees or where the Factories Act 1961 does not apply. It is left to the responsible person to devise a form of record which best suits the individual circumstances. However, the following information must be contained in either record.

- Full name, address and occupation of injured person.
- Date and time of accident.
- Place where accident happened.
- Cause and nature of injury.
- Name, address and occupation of person entering the details, if it is someone other than the injured person.

8.20 All accident books must now meet the requirements of the Data Protection Act 1998. This means that other members of staff, except for safety staff and the employee's manager, must not view accident records.

8.21 The accident book must be retained for at least three years from the last date of entry.

8.22 The obligation to maintain an accident book does not relieve an employer of the similar, but unrelated, duty to report certain accidents and industrial diseases to the health and safety authorities under the RIDDOR Regulations.

Near Misses

8.23 These are incidents where an unplanned event occurs for which there is no resultant injury or damage, but the potential exists for harm to result if the event occurred under different circumstances. Due to the potential for harm, all of these incidents should be reported internally in the same way as an accident.

8.24 All near misses should be recorded and, if appropriate, an investigation should be carried out.

The RIDDOR Regulations Reporting Procedures

8.25 Where an injury, disease or dangerous occurrence is reportable under the RIDDOR Regulations, the responsible person has the following two options.

1. Incident Contact Centre

8.26 Notify the Incident Contact Centre (ICC). Reports can be made by:

- tel: 0845 300 9923
- fax: 0845 300 9924
- website: *www.riddor.gov.uk*
- post: Incident Contact Centre, Caerphilly Business Park, Caerphilly, CF83 3GG.

8.27 The report will then be forwarded on to the appropriate enforcing authority. A copy of the report made to the ICC will be sent back to the duty holder for correction, if necessary. A record of reported incidents must be kept for inspection by visiting officers.

2. Enforcing Authority

8.28 Notify the local office of the enforcement authority (either the HSE or the local authority) by the quickest practicable means and/or on the approved statutory form (F2508), according to the category of the incident.

8.29 The enforcing authority is the body responsible for enforcing the Health and Safety at Work, etc Act 1974 (and other relevant statutory provisions) in the premises where, or in connection with the work at which, the reportable event happened.

8.30 The split of enforcement responsibility between HSE and local authorities is determined by the Health and Safety (Enforcing Authority) Regulations 1998.

8.31 Local authorities are responsible for enforcing health and safety legislation in:

- retail premises
- some warehouses
- most offices
- hotels and catering establishments
- sports and leisure facilities
- consumer services
- places of worship
- residential accommodation.

8.32 In practice, notifying "by the quickest practicable means" will normally entail a telephone call to the enforcing authority during normal office hours. It is advisable to keep a note of telephone notifications, including:

- the time of the call
- the name of the caller
- the details given of the notified event.

8.33 The approved form for making reports is F2508. This may be freely photocopied in order to make reports.

What Should be Reported?

8.34 The following are notifiable to the relevant authority.

- Deaths or major injuries.

- Over-three-day injuries.
- Violent incidents (in certain circumstances).
- Dangerous occurrences.
- Certain diseases.

Reporting Death or Major Injury

8.35 A report must be made to the appropriate enforcing authority by the quickest practicable means in the event of an accident arising out of a work activity which results in:

- the death of, or major injury to, an employee or self-employed person on work premises
- the death of a member of the public
- a member of the public being taken to hospital.

8.36 The report should be made either to the local enforcing authority or the ICC. Where a verbal report is given to the local enforcing authority, this must be followed, within 10 days, by a written report on a prescribed form (F2508). Where the ICC has been notified, no follow-up action in respect of reporting will be required.

8.37 The following should be kept in mind.

- Where the nature and severity of an injury is not immediately apparent, the report required shall be submitted as soon as the nature of the condition is confirmed.
- Deaths to be reported include those where an employee dies within one year as a result of an accident at work, whether or not this was reported at the time of the original accident.
- Although the police would notify the HSE in cases of accidental fatalities at work, this does not relieve the responsible person of the duty to report the fatality to the HSE.

8.38 Major injuries are defined in the RIDDOR Regulations as:

- fractures other than fingers, thumbs and toes
- amputation (including surgical amputation following an accident)
- the dislocation of a shoulder, hip, knee or spine
- an eye injury, by chemical or hot metal burn, or penetration, resulting in temporary or permanent loss of sight
- unconsciousness caused by electric shock, exposure to a hazardous substance, biological agent or asphyxia
- any acute condition or illness resulting in loss of consciousness or requiring resuscitation or admission to hospital for more than 24 hours
- an illness requiring medical treatment because of exposure to a hazardous substance.

Reporting Lost-time Injuries

8.39 A report must be made within 10 days following an accident caused by a work activity that resulted in the incapacity of an employee (or self-employed person working on the premises) for more than three consecutive days.

8.40 An over-three-day injury is one that is not major but results in the injured person being away from work or unable to perform the full range of his or her normal duties for more than three days.

8.41 In the calculation of "more than three consecutive days", the day of the accident should not be counted, only the period after it. Any other days the injured person would not have been expected to work normally, such as weekends, rest days or holidays, must be included.

8.42 A degree of judgment is required in determining whether the injured person would have been unable to perform his or her normal range of duties for more than three consecutive days. It might be necessary to ask the injured person if he or she would have been able to carry out all duties if he or she had been at work.

Examples of Lost-time Injuries

8.43 The following are two examples of lost-time injuries.

- A trainee who normally works Monday to Friday is injured at work on Thursday and left unable to do his or her job. The trainee returns to work the following Tuesday. The days counted would be Friday, Saturday, Sunday and Monday. This makes a total of four days when the trainee would have been unable to work because of the injury. In this case the injury must be reported.
- A shift worker, who normally works five days on and five days off, is injured at work on the third day of his or her shift and left unable to work. By the fourth day of the five-day rest period, the worker is fit enough to do the full range of normal duties. The worker returns to work at the start of the next shift. The days counted would be the last two days of the shift, plus the first three days of the rest period, making a total of five days of incapacity. This injury must be reported.

Violence

8.44 The definition of an accident includes "an act of non-consensual physical violence done to a person at work".

8.45 The RIDDOR Regulations would apply if a person died or suffered a major or over-three-day physical injury caused by a non-consensual act of physical violence whilst they were at work. This means that the death or injury must be reported.

8.46 In the case of an over-three-day injury, the incapacity must arise from the physical injury. It must not be the result of a psychological reaction to the act of violence alone.

Examples of Reportable Violent Incidents

8.47 The following incidents are reportable.

- A supervisor is hit by an employee whilst giving an instruction to carry out a work-related task.
- A new employee is injured when being forced to take part in an initiation ceremony.

Examples of Non-reportable Violent Incidents

8.48 The following incidents would not be reportable.

- An employee working on a factory production line hits another employee during an argument over a personal matter.
- An employee at work at a public enquiry desk is hit by one of his or her relatives who comes in to discuss a domestic matter.

8.49 Incidents involving acts of violence may need to be reported to the police, whether or not they are reportable to the HSE or local authorities.

Reporting Dangerous Occurrences

8.50 If there is an accident because of a work activity that does not result in a reportable injury, it may still be a "dangerous occurrence". A dangerous occurrence must be immediately reported to the enforcing authority or the ICC.

8.51 A dangerous occurrence includes, for all industries and occupations:

- the collapse of a load-bearing structure (eg building, freight container, crane, scaffolds over 5m high or near water and fairground structures)
- the explosion of a closed vessel or pipework
- contact with overhead lines or other electrical fault causing fire or explosion resulting in plant stoppage for more than 24 hours
- other fires or explosions causing normal work to stop for more than 24 hours
- the unintended operation of an explosive device
- the accidental release of a hazardous biological agent
- the failure of radiography equipment
- the malfunction of breathing apparatus in use or during pre-use testing
- the failure of diving equipment causing danger or uncontrolled ascent
- collision between trains and other vehicles
- dangerous failures of fairground equipment, such as derailment or collisions of cars and trains or failure of passenger restraints, whether in use or under test
- incidents at wells (but not water wells) and pipelines that may endanger life
- the uncontrolled release of, or fires involving, a dangerous substance being moved by road.

8.52 Additional dangerous occurrences are defined for incidents in mines and quarries, railway transport systems and workplaces offshore.

8.53 A full description of all reportable dangerous occurrences is given in schedule 2 of the RIDDOR Regulations.

Reporting Cases of Disease

8.54 Where a doctor sends notification that an employee is suffering from a reportable work-related disease, a report must be made to the ICC or the appropriate enforcing authority by submission of a completed form (F2508).

8.55 Reportable diseases are related to particular occupations and include:

- poisoning by specified chemicals and chemical compounds
- skin diseases such as dermatitis, oil acne, chrome ulcers and skin cancers
- lung diseases such as occupational asthma, farmer's lung, pneumoconiosis, asbestosis, mesothelioma
- infections due to biological agents such as leptospirosis, hepatitis, tuberculosis, anthrax, tetanus and legionnaires' disease
- occupational cancers
- conditions related to physical agents and demands of work such as:
  - inflammation or disease from exposure to heat or ionising radiation
  - musculoskeletal disorders due to prolonged or repeated physical effort
  - decompression illness and barotrauma when working at increased air pressure
  - hand-arm vibration syndrome.

8.56 A full description of all reportable diseases is given in schedule 3 of the RIDDOR Regulations.

Special Provisions

Mines and Quarries

8.57 Schedule 5 of the RIDDOR Regulations extends the definition of a duty holder to include:

- a person appointed under the Management and Administration of Safety and Health at Mines Regulations 1993
- a person treated as a manager for the purpose of the Mines and Quarries Act 1954.

8.58 Special provisions require that the site of any accident resulting in death, major injury or a dangerous occurrence is left undisturbed for three days following the making of a report or until the site has been visited by an inspector and workmen's inspectors.

Offshore Workplaces

8.59 Schedule 6 of the Reporting of Injuries, Diseases and Dangerous Occurrences Regulations 1995 makes additional requirements that the site of any accident resulting in death, major injury or a dangerous occurrence is left undisturbed either for three days following notification in accordance with the regulations or until the site has been visited by an inspector.

Exceptions

8.60 The Reporting of Injuries, Diseases and Dangerous Occurrences Regulations 1995 (RIDDOR) do not apply to accidents involving vehicles moving on public roads unless they involve or are connected with:

- exposure to any substance being conveyed by road
- vehicle loading and unloading activities such as those performed by refuse collectors or furniture removers
- the specified construction, demolition, alteration, repair or maintenance activities on or alongside public roads
- an accident involving a train where a person is killed or injured.

8.61 In the case of the first three items mentioned above, RIDDOR applies whether the injured person is engaged in one of the listed activities or whether they are injured as a result of the work of someone else who is engaged in such activities. For example, the following incidents would be reportable.

- An employee of a furniture remover dies as a result of being struck by a passing car whilst unloading furniture from a lorry.
- Falling scaffolding injures a motorist driving past a building site alongside the road.
- The driver of a road tanker suffers gassing and acute illness as a result of exposure to a toxic substance spilled from the tanker or, as a result of the same spillage, a member of the public is taken to hospital for treatment because of exposure to the substance.
- An employee painting road markings is hit by a car and does not suffer a major injury as a result, but is unable to do the full range of normal duties for four days.

Who Should Make Reports under RIDDOR?

8.62 The person with responsibility for notifying and reporting deaths, injuries, diseases and dangerous occurrences to the enforcing authority under the Reporting of Injuries, Diseases and Dangerous Occurrences Regulations 1995 (RIDDOR) depends upon the circumstances involved.

8.63 The responsible person may be:

- the employer of the individual involved
- a self-employed person
- someone in control of premises where work is carried out.

8.64 The following table gives details of the responsible person.

Responsible Persons under RIDDOR

Reportable Incident	Responsible Person
Death, major injury or over-three-day injury to: • an employee • a self-employed person at work in premises under the control of some other person.	**Person responsible in this instance:** • the employee's employer • the person in control of the premises at the time of the event and connected to the trade, business or undertaking.
Major injury, over-three-day injury to: • self-employed person at premises under his or her control.	**Person responsible in this instance:** • the self-employed person or someone acting on his or her behalf.
Death or injury to a person not at work that results in that person being taken to hospital for treatment.	The person in control of the premises at which the accident occurred at the time of the event and connected to the trade, business or undertaking.
A reportable disease contracted by: • an employee • a self-employed person.	**Person responsible in this instance:** • the employee's employer • the self-employed person.
A specified dangerous occurrence **Note:** some exceptions exist for occurrences at mines, wells, pipelines and with vehicles carrying a dangerous substance	The person in control of the premises where the dangerous occurrence happened at the time it occurred and connected to the trade, business or undertaking.
Special cases — all reportable events at: • mines • quarries • offshore installations • diving operations.	**Person responsible in this instance:** • mine manager • the owner • the owner (mobile installations), the operator (fixed installations) • the diving contractor.

8.65 This responsible person does not have to be a named individual. In practice though, an individual, such as a safety advisor, with the knowledge of the procedures and requirements of reporting under the Reporting of Injuries, Diseases and Dangerous Occurrences Regulations 1995 (RIDDOR) will often be expected to perform the function of reporting all necessary incidents.

Training

Employees

8.66 Training should concentrate on the legal and moral obligation to report all accidents and near-miss incidents, however minor, and the consequences of failing to do this.

8.67 The purpose of accident investigation and the subsequent benefits of this action should also be covered.

8.68 Employees will need to understand:

- the definition of an accident and near-miss incident and what should be reported

- the benefits to be achieved by reporting accidents and near misses
- the legal duties of employees and employers in accident reporting
- how and to whom to report an accident/incident under the company's own procedure
- the location and availability of the accident book.

Managers and Supervisors

8.69 In addition to the above, managers and supervisors must understand:

- the Reporting of Injuries, Diseases and Dangerous Occurrences Regulations 1995 (RIDDOR) and the incidents that fall within its scope
- the procedure for notifying the enforcing authority
- the information that must be recorded following an incident
- statutory record-keeping requirements
- any possible follow-up action by the enforcing authority.

8.70 As a matter of policy, all records of training given in accident reporting should be kept.

List of Relevant Legislation

- Health and Safety (Enforcing Authority) Regulations 1998
- Health and Safety (Consultation with Employees) Regulations 1996
- Reporting of Injuries, Diseases and Dangerous Occurrences Regulations 1995 (RIDDOR)
- Social Security (Industrial Injuries) (Prescribed Diseases) Regulations 1985
- Health and Safety (First-aid) Regulations 1981
- Social Security (Claims and Payments) Regulations 1979
- Safety Representatives and Safety Committees Regulations 1977
- Health and Safety Inquiries (Procedure) Regulations 1975
- Health and Safety at Work, etc Act 1974

Further Information

Publications

HSE Publications

The following are available from *www.hsebooks.co.uk*.

- HSG96 (rev 1997) *The Costs of Accidents at Work*
- L73 (rev 1999) *A Guide to the Reporting of Injuries, Diseases and Dangerous Occurrences Regulations 1995*
- MISC769 *Incident at Work?*
- *Accident Book: BI510 Log Book*

Building Regulations

- The Building Regulations exist in order to ensure the health and safety of people in and around buildings by providing functional requirements for design and construction.
- They also cover areas such as improving energy efficiency and meeting the needs of disabled people.
- The Building Regulations 2000 will apply to various building operations, with some exceptions.
- If the Building Regulations do apply to the proposed building operations then it is a legal requirement to obtain Building Control Approval.
- Building Control Approval is an independent check to show that the Building Regulations have been complied with — either by:
  - the local authority
  - an Approved Inspector.
- Building work that is subject to the Building Regulations 2000 must comply with schedule 1 of the regulations.
- The Building Regulations rely on a series of Approved Documents — approved by the Secretary of State — that contain practical guidance on the requirements of the regulations.
- Under s.35 of the Building Act 1984, it is an offence to contravene any of the provisions contained in the Building Regulations.
- Equivalent Scottish legislation is predominantly contained in the Building Standards (Scotland) Regulations 1990.

8.71 The Building Act 1984 is the primary legislation which introduces the Building Regulations. It allows the Secretary of State to make regulations with respect to the design and construction of buildings, the demolition of buildings and the provision of services, fittings and equipment in or in connection with buildings. Any such regulations are for the purposes of:

- securing the health, safety, welfare and convenience of persons in or about buildings and of others who may be affected by buildings or matters connected with buildings
- furthering the conservation of fuel and power
- preventing waste, undue consumption, misuse or contamination of water
- furthering the protection or enhancement of the environment
- facilitating sustainable development
- furthering the prevention or detection of crime.

8.72 The Building Regulations 2000 came into force on 1 January 2001, and are regularly amended and updated. Therefore it is essential to check when making reference to documents on the regulations that the latest edition is being used.

8.73 The Building Regulations rely on a series of Approved Documents — approved by the Secretary of State — that contain practical guidance on the requirements of the regulations. Evidence of compliance or non-compliance with the Approved Documents is evidence of compliance with, or contravention of, the regulations. However, there is no obligation to adopt one particular solution if the requirements can be met in an alternative way. These Approved Documents are subject to ongoing revision.

8.74 The Building Regulations exist in order to ensure the health and safety of people in and around buildings by providing functional requirements for design and construction. They also cover areas such as improving energy efficiency and meeting the needs of disabled people. The documents make many references to other relevant governing legislation that needs to be complied with. Of topical note is the Disability Discrimination Act 1995 and the Regulatory Reform (Fire Safety) Order 2005.

8.75 From 1 January 2005, the term building work was extended to include work on domestic electrical installations, and from 6 April 2006, amendments were made to part L to incorporate some of the consequences of the EU Energy Performance of Buildings Directive. The scope again changed to include space cooling as well as space heating elements. Parts L1 and L2 now include requirements to provide calculations in support of carbon emissions and other energy usage targets.

In Practice

When do the Building Regulations Apply?

8.76 As a general rule, the Building Regulations 2000 will apply to the following building operations.

- The erection of a building except for:
  - buildings controlled by other specialist legislation
  - agricultural buildings, greenhouses and some plant rooms
  - temporary buildings (erected for less than 28 days)
  - small detached buildings of less than $30m^2$ (although any fixed electrical installation may still need to comply with part P).
- The extension of a building except for ground level extensions of less than $30m^2$, eg conservatories and porches (any fixed electrical installation may still need to comply with part P).
- The material alteration of a building; this term does not cover any other alteration or repairs to a building.
- The provision, extension or material alteration of:
  - sanitary equipment drainage for a building
  - unvented hot water systems
  - fixed heating appliances in which fuel is burned except in small buildings, ie those with a total floor area of less than $30m^2$ (any fixed electrical installation may still need to comply with part P).
- The provision, extension or material alteration of insulation and energy-saving controls for space heating and hot water systems, hot water pipes and warm air ducts, except in dwellings or for industrial purposes.
- A change of use of an existing building.

8.77 It is advisable to contact the local authority prior to commencing work to ensure that any exemptions are applicable.

8.78 Application may be made to the local planning authority for a particular requirement of the regulations to be relaxed or dispensed with if the developer thinks that the circumstances justify it. Even if the Building Regulations do not apply to the proposal, there may be legislation that does, eg planning.

8.79 Building regulations are different to planning permissions and separate permission will be required for many types of building work.

Procedures to Follow if the Regulations Apply

8.80 If the Building Regulations do apply to the proposed building operations, then it is a legal requirement to obtain Building Control Approval. This is an independent check — either by the local authority or an Approved Inspector — to show that the Building Regulations have been complied with.

Supervision by the Local Authority Building Control

8.81 If this option is chosen, either:

- a building notice will need to be given (this may not be done if shops or offices are being built)
- full plans will need to be deposited.

8.82 In both cases, a fee will need to be paid to the local authority. At this stage, work may begin provided that the local authority is given 48 hours' notice.

8.83 In the case of a building notice, the local authority may ask for more plans to help with inspection of the work. Where full plans have been deposited with the local authority, they must pass or reject them within five weeks. This time limit may be extended to two months if the person proposing the building operations agrees. If the plans are rejected, the Secretary of State for the Environment can be approached for a determination, but a fee is payable.

8.84 If the local authority considers that the work done contravenes the Building Regulations, a notice may be served requiring that the structure be taken down or altered. However, the local authority may not do this if the building work is in accordance with plans that it passed.

8.85 The local authority is not required to issue a completion certificate unless one is requested in writing at the time of application. It is always advisable to ask for this, together with written confirmation that any conditions have been complied with.

Supervision by an Approved Inspector

8.86 If this option is chosen, an initial notice must be given to the local authority jointly by the Inspector and the person proposing the building works. Certain plans and evidence of insurance must also be submitted.

8.87 The local authority must accept or reject the initial notice within 10 working days. Once the notice has been accepted, the local authority's powers to enforce the Building Regulations are suspended. The Approved Inspector may be asked for a plans certificate before the building work commences. If the Inspector is unable to provide such a certificate, the Secretary of State for the Environment may be approached for a determination. A fee will be payable. The fee payable to an Approved Inspector is negotiable between the parties.

8.88 The building work should not be commenced until the initial notice has been accepted by the local authority or 10 days have gone by without it being rejected. The Inspector may have to be notified of the date of commencement of the work or at various stages of the work.

8.89 If the Approved Inspector considers that the building work contravenes the Building Regulations, they must notify the person responsible for the work. If the contravention is not remedied within three months, the Inspector must cancel the initial notice. Another Approved Inspector may then be engaged or the local authority can take over responsibility for the project.

8.90 When the work is completed to the Inspector's satisfaction, they will give a final certificate to the local authority and to the developer.

The Building Notice Procedure

8.91 The building notice procedure does not involve the passing or rejecting of plans. It therefore avoids the preparation of detailed "full plans", and is designed to enable some types of building work to get under way quickly. It is perhaps best suited to small work however. There are also specific exclusions in the regulations as to when building notices cannot be used.

8.92 If this procedure is used, it is important to be confident that work will comply with the Building Regulations otherwise the local authority may request correction work to be carried out.

Compliance with Schedule 1 of the Building Regulations

8.93 Building work that is subject to the Building Regulations 2000 must comply with schedule 1 of the regulations. The various requirements of the schedule are covered in detail in the Approved Documents.

Energy Certificates

8.94 The Energy Performance of Buildings (Certificates and Inspections) (England and Wales) Regulations 2007 implement some of the requirements of the EU Energy performance of buildings directive (2002/91/EC). This requires the production of energy certificates for certain classes of buildings.

8.95 Energy certificates show the energy performance of a building. They are intended to be similar in appearance to the well-established energy labels attached to fridges and washing machines.

8.96 There are two types of certificate.

- Energy performance certificates (EPCs).
- Display energy certificates (DECs).

Energy Performance Certificates

8.97 EPCs apply to all buildings and will be required whenever a building is constructed, rented or sold.

8.98 The certificate will provide a rating of the energy efficiency and carbon emissions of a building from A to G, where A is very efficient and G is very inefficient.

8.99 EPCs are produced using standard methods with standard assumptions about energy usage, so that the energy efficiency of one building can be easily compared with another building of the same type. This is intended to allow prospective buyers, tenants, owners, occupiers and purchasers to consider energy efficiency and fuel costs as part of their investment.

Display Energy Certificates

8.100 DECs are intended to show the actual energy usage of a building, the operational rating, and help the public be aware of the energy efficiency of a building. They will show the energy consumption of the building as recorded by gas, electricity and other meters. The DEC should be clearly displayed at all times in a prominent place clearly visible to the public.

8.101 DECs are only required for buildings that are occupied by a public authority or an institution providing a public service to a large number of persons, with a total useful area greater than 1000m^2.

8.102 The certificates are valid for one year and the accompanying Advisory Report is valid for seven years.

8.103 In the longer term, the Government has announced its intention to consult on whether this requirement should be extended to include private sector buildings occupied by commercial organisations where large numbers of members of the public regularly visit the building. This measure would require new legislation.

8.104 The current timetable for the introduction of DECs is October 2008.

Approved Documents

8.105 Approved documents are issued by the Secretary of State to provide practical guidance on ways of meeting the requirements. These documents are subject to regular review and updating. The Approved Documents are as follows.

A	—	Structure (2004 edition): loading, ground movement and disproportionate collapse.
B	—	Fire Safety (2000 edition, amended 2002 and 2007): means of escape from fire, internal fire spread (surfaces), internal fire spread (structure), external fire spread, access and facilities for the fire brigade.
C	—	Site Preparation and Resistance to Contaminants and Moisture (2004 edition): preparation of site, dangerous and offensive substances, sub-soil drainage, resistance to weather and ground moisture.
D	—	Toxic Substances (1992 edition, amended 2002): cavity insulation.
E	—	Resistance to the Passage of Sound (2003 edition, amended 2004): airborne sound (walls), airborne sound (floors and stairs), impact sound (floors and stairs).
F	—	Ventilation (2006 edition): means of ventilation, condensation in roofs.
G	—	Hygiene (1992 edition, amended 2002): bathrooms, hot water storage, sanitary conveniences and washing facilities.
H	—	Drainage and Waste Disposal (2002 edition): sanitary pipework and drainage, cesspools and tanks, rainwater drainage, solid waste discharge.
J	—	Combustion Appliances and Fuel Storage Systems (2002 edition, 2nd ed. 2004): air supply, discharge of products of combustion, protection of buildings.
K	—	Protection from Falling, Collision and Impact (1998 edition, amended 2000): stairs, ladders and ramps, protection from falling, vehicle barriers.
L1A	—	Conservation of Fuel and Power in New Dwellings (20062 edition): Part L1B – Conservation of Fuel and Power in Existing Dwellings (2006).
L2A	—	Conservation of Fuel and Power in New Buildings Other than Dwellings (2006 edition): Part L2B Conservation of Fuel and Power in Existing Buildings Other than Dwellings (2006).
M	—	Access to and Use of Buildings (2004 edition): means of access, sanitary conveniences, audience or spectator seating.
N	—	Glazing — Safety in Relation to Impact, Opening and Cleaning (1998 edition): safe access for cleaning, critical locations, safe operation.
P	—	Electrical Safety (2006 edition): electrical installations in dwellings. Approved document to support regulation 7 — Materials and workmanship (1999 edition, amended 2000).

Contravening the Building Regulations — Consequences

8.106 Under s.35 of the Building Act 1984, it is an offence to contravene any of the provisions contained in the Building Regulations. Any person who does so is liable to a fine of up to £5000 and to a further daily fine for each day the default continues after conviction.

8.107 In addition, or as an alternative, the local authority may, by notice under s.36, require the owner to pull down or remove offending work and later work so that it complies. Breach of a duty imposed by the Building Regulations 2000 is actionable in civil proceedings unless the regulations provide otherwise, so far as it causes damage.

8.108 Fire safety breaches under the Regulatory Reform (Fire Safety) Order 2005 are criminal offences, and penalties can include prison sentences.

Building Control in Scotland

8.109 The Scottish Building Standards Agency (SBSA) was set up on 21 June 2004 as an executive agency of the Scottish Executive. It carries out the duties of the Scottish Ministers set out in the Building (Scotland) Act 2003.

8.110 The Act gave ministers the power to make building regulations to maintain the health, safety, welfare and convenience of persons in and about buildings and others who may be affected by buildings or matters connected with buildings, to encourage the conservation of fuel and power and to continue the progress of sustainable development. The Act introduced the new Scottish building standards system, which was introduced on 1 May 2005.

8.111 The system is designed to protect the public interest in the design, construction, conversion and demolition of buildings. Before work begins, plans must be 'verified' as meeting the standards set in the Scottish building regulations. The process is overseen by the Scottish Building Standards Agency (SBSA).

8.112 The functions of the SBSA are to:

- prepare the building regulations and write guidance on how to meet the regulations
- provide views on compliance to help verifiers make decisions
- grant relaxations of the regulations in exceptional cases
- maintain a register of Approved Certifiers
- monitor and audit the certification system
- monitor and audit the performance of verifiers
- verify Crown building work.

Procedural Handbook

8.113 The Procedural Handbook explains the procedures set up by the Building (Scotland) Act 2003 and the Building (Procedure) (Scotland) Regulations 2004.

8.114 The purpose of the Handbook is to clarify the intent of the procedure regulations and expand upon the procedures set out by the Act. Unlike the Technical Handbooks and other guidance documents issued by Scottish Ministers to support the Building Regulations, this handbook has no specific legal status, but is designed to aid the practical operation of the procedures.

2007 Technical Handbooks

8.115 A revised edition of the Technical Handbooks came into force on 1 May

2007. The revisions cover sections 0, 1, 3, 4 and 6. The Technical Handbooks are available in hard copy from The Stationery Office (TSO).

Domestic Handbook 2007

Full Publication
Section 0 – General
Section 1 – Structure
Section 2 – Fire
Section 3 – Environment
Section 4 – Safety
Section 5 – Noise
Section 6 – Energy
Appendix A
Appendix B

Non-domestic Handbook 2007

Full Publication
Section 0 – General
Section 1 – Structure
Section 2 – Fire
Section 3 – Environment
Section 4 – Safety
Section 5 – Noise
Section 6 – Energy
Appendix A
Appendix B

List of Relevant Legislation

- Energy Performance of Buildings (Certificates and Inspections) (England and Wales) Regulations 2007
- Building (Scotland) Amendment Regulations 2006
- Building (Forms) (Scotland) Amendment Regulations 2006
- Building (Forms) (Scotland) Regulations 2005
- Building (Amendment) Regulations 2004
- Building (Approved Inspectors, etc) (Amendment) Regulations 2004
- Building (Amendment) (No. 2) Regulations 2004
- Building (Fees) (Scotland) Regulations 2004
- Building (Procedure) (Scotland) Regulations 2004
- Building (Scotland) Regulations 2004
- Building Standards Advisory Committee (Scotland) Regulations 2004
- Building (Amendment) Regulations 2003
- Building and Building (Approved Inspectors, etc) (Amendment) Regulations 2003
- Building (Amendment) Regulations 2002
- Building (Amendment) (No. 2) Regulations 2002
- Building (Approved Inspectors, etc) (Amendment) Regulations 2002
- Building (Amendment) Regulations 2001
- Building (Approved Inspectors, etc) (Amendment) Regulations 2001
- Building (Approved Inspectors, etc) Regulations 2000

- Building Regulations 2000
- Building (Local Authority Charges) Regulations 1998
- Building (Inner London) Regulations 1987
- Building (Approved Inspectors, etc) Regulations 1985
- Building (Prescribed Fees, etc) Regulations 1985
- Building (Scotland) Act 2003
- Building (Scotland) Act 2003 (Commencement No. 1, Transitional Provisions and Savings) Order 2004
- Building Act 1984

Construction Risk Assessment

- Risk assessment provides a systematic approach to the control and elimination of accidents at work.
- Risk assessments should be carried out before the work activity begins.
- Risk assessments should be carried out by competent persons with appropriate knowledge and experience of the activities to be assessed.
- The assessment must be suitable and sufficient. There are several types of risk assessment, and the choice will depend on what is most appropriate for the work situation.
- All aspects of the task should be evaluated for potential hazards.
- Persons at particular risk should be identified when considering the hazards.
- Even after all precautions have been taken, some risk is likely to remain. The key is to decide, for each significant hazard, whether this remaining risk is high, medium or low.
- Any existing control measures and further control measures that are required to be introduced should be assessed to evaluate their effectiveness in preventing harm.
- Safe place measures should be introduced via a process of elimination, reduction of the risk at source, and control of exposure to the hazard.
- Personal protective equipment (PPE) may be used if control measures do not adequately control the hazard.
- The significant findings of a risk assessment should be recorded where the employer employs five or more employees.

8.116 Organisations that have high standards of health and safety are often the most successful, irrespective of size or industry. Accidents at work can prove damaging to a company, both financially and in terms of reputation The potential for serious injury and death on a construction site is considerable. The management of health and safety is of paramount importance in construction work, and risk assessments are an essential component to the achievement of a safe working environment.

8.117 Risk assessment comprises the identification of risks, the evaluation of associated risks and the determination of appropriate control measures. It provides a systematic approach to the control and elimination of accidents at work. It applies to every aspect of an employer's business, however, in many cases risk assessment may be no more than a check to make sure nothing has been missed, and a record that an activity has been assessed. A full risk assessment will be required for many activities on a construction site because of the hazardous nature of the work involved.

8.118 Hazards are defined as anything with the potential to cause harm (substances, equipment, activities, etc). Risk is the likelihood of the harm occurring. Most risk assessments take account of the following.

- Likelihood — the chance of the harm occurring.
- Frequency — how often the chance of the harm happening occurs during the process or activity being assessed.
- Duration — the length of time that the likelihood of harm exists.
- Severity — the likely severity of the harm should it occur. Often this makes use of categories such as those contained in the Reporting of Injuries, Diseases and Dangerous Occurrences Regulations 1995.
- Extent — the extent of the risk depends on those who might be affected by the hazard.

8.119 By considering the extent of each risk, the level and rating of the risk can be determined. Risks can then be ranked into a prioritised hierarchy.

Employers' Duties

8.120

- Under the Health and Safety at Work, etc Act 1974 (HSWA), the main duties of employers are to:
  - provide and maintain safe plant, safe systems of work and methods of using, handling, storing, and transporting articles and substances
  - provide information, instruction, training and supervision
  - provide a safe place of work and ways of getting to and leaving that place
  - provide welfare and a safe working environment
  - make arrangements to ensure others who are affected by work are also kept safe.

Suitable and Sufficient Assessment

8.121 Under the Management of Health and Safety at Work Regulations 1999 (MHSWR), employers must carry out a suitable and sufficient assessment of risks to health and safety of:

- employees at work
- non-employees affected by the business.

8.122 The assessment must identify measures which need to be taken to comply with statutory provisions. If more than four people are employed, the assessment must be recorded in writing. The assessment must also be reviewed if there is a significant change or reason to believe it is no longer valid.

Preventive and Protective Measures

8.123 Under regulation 4 of the MHSWR , employers must ensure preventive and protective measures are implemented. After either avoiding the risk altogether or evaluating any remaining risk which cannot be avoided, the following principles should be adopted.

- Combating the risk at source.
- Adapting the work to the individual.
- Adapting to technical progress.
- Replacing the dangerous with the non-dangerous or the less dangerous.
- Developing a coherent prevention policy which covers technology, organisation of work, working conditions, social relationships and the influence of factors relating to the working environment.
- Giving collective protective measures priority over individual protective measures.
- Delivering appropriate instructions to the employees involved.

Effective Planning, Organisation, Control, Monitoring and Review

8.124 Under regulation 5 of the MHWSR , arrangements should be made for the effective planning, organisation, control, monitoring and review of the preventive and protective measures introduced.

8.125 If more than four people are employed, the arrangements must be recorded in writing. It is important to clearly distinguish between workplace precautions and risk control systems.

8.126 Risk control systems provide the basis for ensuring adequate workplace precautions are provided and maintained. The three basic stages are:

- hazard identification
- risk assessment
- risk control.

8.127 Following the assessment, measures must be introduced to reduce the risk of injury to the lowest level reasonably practicable. For risk control, the following hierarchy is suggested.

- Eliminate risks by:
  - substituting use of less hazardous substances
  - using a better-guarded machine
  - avoiding certain activities and processes, perhaps by buying in from sub-contractors.
- Combat risks at source by engineering controls and giving collective protective measures priority by:
  - separating operator from hazard through enclosure
  - protecting dangerous parts through guarding
  - designing processes, machinery and activities to minimise the release of airborne hazards or to suppress or contain them
  - design machinery which can be operated remotely and which features automatic feeding.
- Minimise risks by:
  - designing suitable safe systems of work
  - providing, as a last resort, personal protective equipment and ensuring it is used.

Provide Comprehensible and Relevant Information

8.128 Under regulation 10 of the MHSWR, employers must provide employees with comprehensible and relevant information on:

- the risks to their health and safety identified by the assessment
- the preventive and protective measures introduced.

8.129 If an employee is below the minimum school-leaving age, this information must be provided to the parents or guardian before the child starts work.

8.130 Information about and consultation on risk assessments, introduction of control measures and training can be given to employees directly or through union appointed safety representatives, under the Safety Representatives and Safety Committees Regulations 1977. The information can also be delivered through non-union representatives of employee safety, under the Health and Safety (Consultation with Employees) Regulations 1996.

Employees' Capabilities

8.131 Under regulation 13 of the Management of Health and Safety at Work Regulations 1999, employers must, in entrusting tasks to their employees, take into account their capabilities.

8.132 This applies to employees entrusted to carry out risk assessments and others. Regulation 13 also stipulates that employers must ensure employees are provided with adequate health and safety training:

- on being recruited into the employer's undertaking
- on being exposed to new or increased risks because of:
  - a transfer or change of responsibilities
  - the introduction of new work equipment
  - a change in work equipment already in use
  - the introduction of new technology
  - the introduction of a new system of work
  - a change in a system of work.

8.133 Those carrying out risk assessments will require training, which reflects a change in job responsibility. Those affected by changes will also require training.

Monitoring

8.134 Monitoring must take place to ensure the effectiveness of measures, and reassessment must be carried out where necessary.

8.135 The Construction (Design and Management) Regulations 2007 do not include any specific requirement to carry out and record risk assessments, however, this is still an essential basis for compliance for all the duty holders.

8.136 Contractors are required to identify significant hazards and eliminate or control significant risks. The risks involved must be considered, and risk assessments under other legislation, eg the MHSWR, or the Control of Asbestos Regulations 2006, must be carried out and included in the health and safety plan.

Employees' Duties

8.137 The main duties of employees are to:

- obey all safety instructions
- ensure that actions at work do not put the employer or others at risk
- care for safety equipment.

8.138 Under the MHSWR, employees are also required to participate in safety training and to inform the employer of any shortcomings in safety.

In Practice

When to Carry Out Risk Assessments

8.139 Risk assessments should be carried out before the work activity begins. Points to consider include:

- the nature of any hazards present
- the severity of the harm that could result
- the frequency of exposure to the hazards of both persons or property.

8.140 In the construction industry, risk assessments apply to a wide range of activities, plant and equipment; from the very simple, such as standing on a ladder to painting a window frame, to the more complex, such as excavation work. The number of people involved in the activity itself or carrying out other activities on the site (sometimes under the direction of other contractors), and members of the public in the vicinity varies greatly.

8.141 The time and effort taken to carry out the risk assessment and implement required controls should reflect the danger of the activity to be undertaken.

8.142 The assessment should take into account any other activity in the area which may be affected by the work, and those carrying out the risk assessment should be familiar with other hazards in the vicinity.

Who Should Carry Out Risk Assessments

8.143 Risk assessments should be carried out by competent persons with appropriate knowledge and experience of the activities to be assessed.

8.144 In certain circumstances it may be necessary to call in specialists to carry out the risk assessment, eg when working with asbestos or when a noise assessment is required. It should be a team exercise involving as many people as possible to ensure that all aspects of the work are considered. Making people feel involved in this way will help develop a positive safety culture.

8.145 The assessor should have a basic understanding of:

- how accidents are caused
- health and safety management
- the principles of accident prevention and hazard management
- basic health and safety law.

Preparation

Information Required

8.146 An essential requirement for risk assessment is access to up-to-date information. In order to identify hazards, analyse risks and arrive at an evaluation of the adequacy of controls for a particular hazard, risk assessments should be made with reference to standards and advice as set out in:

- legislative requirements
- Approved Codes of Practice (ACOPs)
- guidance from the Health and Safety Executive
- British Standards
- industry guidance
- manufacturers' literature and manuals
- in-house safe working procedures.

8.147 For topics that are very relevant to an organisation, a copy of the actual guidance information should be obtained for ongoing reference. For example, if joinery work is to take place the company should purchase copies of the relevant Guidance Notes from the HSE, such as:

- WIS15 *Safe Working at Woodworking Machines*
- L114 *Safe Use of Woodworking Machinery. Provision and Use of Work Equipment Regulations 1998 as Applied to Woodworking Machinery. Approved Code of Practice and Guidance.*

Types of Risk Assessments

8.148 Although the law requires employers to undertake a suitable and sufficient risk assessment, the Approved Code of Practice accompanying the Management of Health and Safety at Work Regulations 1999 outlines what counts as such a risk assessment in a variety of situations.

Preliminary Risk Assessment

8.149 An initial, informal risk assessment has to be completed to determine the sophistication and level of detail that will be required in a full risk assessment.

8.150 In general, insignificant risks can be ignored, along with risks arising from routine activities associated with life in general, unless the work either compounds or significantly alters the risk. The purpose of a preliminary assessment is to determine which work activities involve a significant risk of injury and warrant a full assessment.

Qualitative Assessments

8.151 Undertakings involving a small range of simple risks can utilise a straightforward assessment process based on informed judgment and reference to any appropriate guidance material. Such a qualitative process may be based solely on the subjective judgment of the assessor.

Semi-quantitative Assessments

8.152 In intermediate cases, such as those involving medium-sized undertakings dealing with more complex risks, the risk assessments will need to be more sophisticated. Specialist advice may be needed and analytical techniques relating to monitoring may need to be considered.

8.153 Often semi-quantitative processes, where scoring schemes are ascribed to likelihood and potential severity along with guide words to aid the assessor's subjective judgment in scoring, are used. While these techniques limit the range of the assessor's subjective input, they are not fully objective, as a judgment still has to be made with regard to scoring for likelihood and severity.

Quantitative Assessments

8.154 Large and hazardous undertakings require more developed and sophisticated processes, such as fault tree analysis, event tree analysis and quantitative assessments based on historical data. There are many techniques for high-risk activities, which are applied when complex or novel processes are involved.

8.155 The intention of such assessment processes is to produce objective assessments which are not reliant on the assessor's subjective judgment. However, subjective elements can still creep in through choices made in the construction of the models used and the historical data to be applied.

Seven-steps Approach to Risk Assessment

8.156 The following seven-steps approach has been successfully adopted by many organisations.

1. Complete an inventory of all sources of hazards. This will form the bedrock of the safety management system, as it will list all equipment, substances and activities that need to be managed. The inventory should be subject to regular review and revision.
2. Identify all hazards. Often each source on the inventory will involve more than one hazard. A fork-lift truck, for example, can be a source of many hazards. A useful aid to identifying hazards is to classify them as physical, chemical, biological, ergonomic or psychological.
3. Rate the risks. By rating the risks associated with each hazard, it is possible to prioritise which needs to be dealt with first. In a simple system, the rating can be in terms of high, medium and low. In more sophisticated systems, a figure will be arrived at through ascribing scores for likelihood and severity, and multiplying them.
4. Determine whether the risk is acceptable or not. This can only be done once the risks have been rated. In general, low risks are equivalent to the risks in everyday life and are therefore acceptable, but high risks require urgent action, as employees could suffer serious injury. Medium risks obviously fall between high and low and, while being above the risks in everyday life, do not have the urgency of high risks. The time-frame for dealing with them is therefore longer.
5. Select appropriate control measures. Workplace precautions deal with the actual risk, whereas risk control systems ensure the workplace precautions remain effective. The arrangements for the protective and preventive measures need to be:
   - planned
   - organised
   - controlled
   - monitored
   - reviewed
   - recorded.
6. Record assessments. This is always good practice and may be required by clients. By law, the assessment has to be recorded in writing if more than four people are employed.
7. Review assessments. The safety management system should specify how often reviews are held. By law, assessments must be reviewed following a significant change or if there is reason to believe they are no longer valid.

8.157 The more effort put into the risk assessment process, the greater the benefits to the organisation.

Hazard Identification

8.158 Risk assessors should walk around the workplace and "look afresh at what could reasonably be expected to cause harm".

8.159 A review of accident books, copies of forms kept under the Reporting of Injuries, Diseases and Dangerous Occurrences Regulations 1995, records of first-aid treatments, and sickness and absence records may also reveal information about hazards and risks.

8.160 Ultimately, low-risk hazards can be ignored and concentration should be focused on significant hazards which could result in serious harm or affect several people.

8.161 Typical hazards to be found on a construction site include the following.

Typical Hazards on a Construction Site

Hazard	Explanation
Slips and trips	A third of all serious injuries result from slipping on a level surface.
Working at height	A quarter of all fatalities result from falls of over 2m.
Moving parts of machinery	The control of dangerous parts is subject to the requirements of the Provision and Use of Work Equipment Regulations 1998. Access to a dangerous part has to be prevented while it is in motion. Where access is possible, fixed guards should be in place. Otherwise, use of other guards, protective devices, protective appliances or information, instruction, training and supervision need to be introduced.
Vehicles	Hazards from vehicles can come from vehicles used in-house and on public roads for business, even if these vehicles are privately owned.
Electricity	Virtually all workplaces have electrical systems and use electrical equipment, including portable equipment. It should be remembered that portable electrical equipment forms part of the electrical system once connected to it, therefore portable electrical equipment brought into the workplace by employees, visitors and contractors also has to be considered.
Dust	Even "nuisance" dust presents a risk of lung injury when present in significant concentrations and therefore has to be considered along with toxic and harmful dusts.
Manual handling	Over a third of three-day or more reportable injuries result from manual handling operations and the largest claim area involves musculoskeletal disorders.
Noise	If a normal conversation cannot be carried out in a workplace, it is likely the noise level is around 80dB(A) which, as the first action level, triggers the need for a noise survey.
Vibration	Both whole-body vibration and hand-arm vibration should be considered as hazards within the workplace.
Extreme temperature	Tasks and activities in cold stores and near furnaces must be considered.

8.162 It is important to spend adequate time on preparation. A systematic approach should be adopted which starts with the preparation of a preliminary list of identified potential hazards. This list can then be confirmed or modified through a tour of the site.

8.163 Certain hazards, including manual handling, electricity and electrical equipment, fire, hazardous substances and slips, trips and falls, can be added to the preliminary list for virtually all workplaces, tasks and activities.

8.164 Specific hazards, such as machinery, noise, welding, compressed air, lift truck operation, traffic movements and falling objects, can be added to the generic risks. Such an approach will help ensure all significant hazards have been identified.

8.165 Observation of workplaces and activities is an important part of hazard identification, but it has limitations. Some hazards, such as non-visible radiation, odourless vapours, micro-organisms, etc are not directly perceptible, and even when they are, often a certain level of knowledge is required for them to be identified as hazards.

8.166 Observations need to be supported by a structured checklist of questions. In a more formal situation, it might be developed into a hazard and operability study.

8.167 If task analysis and method study is used in a manufacturing or production process, then job safety analysis may play an important role in hazard identification at each step in the process.

Who is at Risk?

8.168 Having identified the hazards, it is necessary to identify who is at risk. It is not necessary to identify individuals by name, rather specific groups of employees and others can be resorted to, eg:

- security staff
- managers
- supervisors
- clients
- designers
- construction workers
- principal contractors
- visitors
- the general public.

8.169 Particular attention needs to be given when vulnerable people are identified, eg young workers or new and expectant mothers. Additional controls may be required to reduce the risk to an acceptable level for such employees.

What is the Harm?

8.170 Having identified the hazards and those exposed to them, it is necessary to determine the likely harm that could result from a hazard. Reference to the type of injuries treated by first aiders may prove useful, as may analysis of accident and ill-health records.

8.171 It is important to consider the potential severity of injury in each case rather than the actual injury, which may have been a matter of luck.

Risk Evaluation

8.172 Even after all precautions have been taken, some risk is likely to remain. The key is to decide, for each significant hazard, whether this remaining risk is high, medium or low.

8.173 This calculation should take into account answers to the following questions.

- Have all legal requirements been satisfied?
- Are generally accepted industry standards in place?
- Have all reasonably practicable measures been taken to keep the workplace safe?

8.174 Another approach to risk analysis involves the ranking of risk according to the perception of the likelihood of a range of incidents actually occurring.

8.175 It is important to remember such an analysis is based on perception and subjective judgment.

Incident and Perceived Rank

Incident	Perceived Rank
Fall from height	1
Striking an object	2
Being struck by an object	3
Fall or trip on level ground	4

8.176 If the incidents were ranked in accordance to how often they had resulted in injury, utilising accident book entries, records of reportable incidents, sickness records and first-aid treatment records, a different ordering might result.

8.177 The various types of risk rating techniques operate on a basic assumption that risk equates to the likelihood (of the hazard occurring) times the severity (of the loss). Or, risk = likelihood x severity.

8.178 With less sophisticated systems, this calculation is implied in the assessor's subjective judgment of whether the risk is high, medium or low. More sophisticated systems ascribe numbers and levels of probability to likelihood and severity to decide a final risk rating.

8.179 The following points be adopted when calculating likelihood and severity.

- Likelihood:
  - high — where it is certain, or near certain, harm will occur
  - medium — where harm will often occur
  - low — where harm will seldom occur.
- Severity:
  - major — death, or major injury or illness, causing long-term disability
  - serious — injuries or illness causing short-term disability
  - slight — all other injuries or illness.

8.180 The results from such an approach can be used as part of a risk level estimator system.

Risk Level Estimator

Likelihood of Harm		**Severity of Harm**	
	Slight harm	*Moderate harm*	*Extreme harm*
Very unlikely	Very low risk	Very low risk	High risk
Unlikely	Very low risk	Medium risk	Very high risk
Likely	Low risk	High risk	Very high risk
Very likely	Low risk	Very high risk	Very high risk

8.181 It should be noted that the use of "low risk, very high risk", etc in the estimator is suggestive of perception of risk, rather than an exact measurement of risk. The perception of risk can be modified by many factors including media coverage and trust in those in control of the activity.

8.182 Scoring systems and guide words are often introduced in the assessment process, to curb any subjective bias by the assessor when calculations of likelihood and severity are made, as shown below.

Likelihood and Severity Guide Words

Likelihood		**Severity**	
Rating	*Guide words*	*Rating*	*Guide words*
0	Almost impossible	0	No harm
1	Extremely unlikely	1	Minor harm
2	Unlikely	2	Moderate harm
3	Likely	3	Serious harm
4	Extremely likely	4	Major harm
5	Almost certain	5	Catastrophic

8.183 This model would result in an overall risk rating ranging from 0 to 25. The following points must also be considered.

- Scoring a 0 for either likelihood or severity results in a final 0 outcome, and this may not always appear to be intuitively correct.
- Not all the numbers between 0–25 are achievable and a gap exists between 20 and 25, which is at the high-risk end of the scale, where more discrimination may be required.

8.184 Some systems dispense with the 0 scores and reverse the remaining scores in order to make 1 represent certainty and catastrophic harm and 5 represent extreme unlikelihood and minor harm. Such a system also has the advantage that the risks can then be dealt with in priority order, ie 1, 2, 3, etc.

Risk Level

8.185 Whether or not a number system is used to rank risks, the end result should be a system that clearly indicates the risk level determined and the action that needs to follow. According to BS 8800: 2004, the following approach can be adopted.

Risk Level and Action

Risk Level	**Tolerability: Guidance on Action**
Very low	These risks are considered acceptable. No further action is necessary other than to ensure that the controls are maintained.
Low	No additional controls are required unless they can be implemented at very low cost (in terms of time, money and effort). Actions to further reduce these risks are assigned low priority.
Medium	Consideration should be given as to whether the risks can be lowered, where applicable, to a tolerable level, and preferably to an acceptable level, but the costs of additional risk reduction measures should be taken into account. The risk reduction measures should be implemented within a defined time period.
High	Substantial efforts should be made to reduce the risk. Risk reduction measures should be implemented urgently within a defined time period and it might be necessary to consider suspending or restricting the activity or to apply interim risk control measures until this has been completed.
Very high	These risks are unacceptable. Substantial improvements in risk controls are necessary so that the risk is reduced to a tolerable or acceptable level. The work activity should be halted until risk controls are implemented that reduces the risk so that it is no longer very high.

Note: Arrangements should always be made to ensure that the controls are maintained.

Risk Assessment Techniques for High-risk Activities

8.186 Hazard identification and risk assessment techniques need to be applied by designers, manufacturers, suppliers, installers and users of articles and substances as well as by employers in relation to the work activities being carried out.

8.187 There are a number of risk assessment techniques, depending on the nature and complexity of the hazards and risks. The techniques range from qualitative, semi-quantitative or quantitative assessments, to more specialist methods, such as job safety analysis, hazard and operability studies, failure mode and effect analysis, fault tree analysis and event tree analysis.

Hazard and Operability Studies

8.188 Hazard and operability (HAZOP) studies adopt a qualitative approach to hazard identification and are often employed at the planning or design stage of a project. Although the technique is strictly a critical enquiry into the operation of a plant, it can be extended to investigate foreseeable risks related to products in the development stage.

8.189 The HAZOP approach includes asking questions at three vital stages.

- Development.
- Operation.
- Maintenance.

8.190 The questions relate to:

- the intention of the product or operation
- any possible deviations that could arise from the declared intention
- the causes of possible deviations from the declared intention
- the consequences of the deviations.

8.191 In effect, the technique comprises of a structured brainstorming approach to hazard identification. It is important a team takes part in the HAZOP activity rather than an individual, as all foreseeable deviations from original intention are more likely to be considered. Individuals may focus upon certain deviations at the expense of others, perhaps due to experience. The results from the exercise can be tabulated under the following headings.

Intention	Guide Words	Foreseeable Deviation from Intention	Causes	Consequences	Action Required

8.192 However, the core of the HAZOP approach involves the use of guide words, which alert users to foreseeable deviations from the intention. Typical guide words include: *no*, *not*, *more*, *less*, *as well as*, *part of*, *reverse* and *other than*.

Fault Tree Analysis

8.193 Fault trees are diagrammatic representations of all the events which may give rise to a major event — the way in which individual events can combine to produce dangerous situations.

8.194 The first step is to identify what failure must be avoided, labelled in the analysis as the top event. The combination of events required to produce the top event are then considered. If two or more events must occur at the same time, they are combined and labelled an "and" gate. If only one of a number of events needs to occur at the same time, it is labelled an "or" gate.

8.195 Once the full fault tree is completed, the probability of the bottom events occurring, according to statistical analysis, can be ascribed. For example, a certain event scores 1, a 50/50 chance scores 0.5 and an impossibility scores 0.

8.196 If an "and" gate is involved, the probabilities of the inputs are multiplied to calculate the probability of the output. If an "or" gate is involved, the probabilities of the inputs are added to calculate the probability of the output. In this way, the probability of the top event can eventually be calculated.

8.197 Given the likelihood of the top event occurring, and the severity of the consequences should it occur, the necessity of remedial action can be considered, including redesign, use of better quality components, back-up systems and fail-safe devices.

Event Tree Analysis

8.198 Event tree analysis complements fault tree analysis, through determining the consequences that flow from a failure or undesired event or top event. If the fault tree illustrates how the top event could occur, the event tree illustrates what will result from the top event, taking the top event as the initiating event.

Failure Mode and Effect Analysis

8.199 Failure mode and effect analysis is a hazard identification and risk assessment technique which explores the effects of failures or malfunctions of individual components within a system. Basically, the question is asked: "if this part fails in this manner, what will be the result?"

8.200 At the design stage, a team considers all the possible failure modes of every component of the final product. Each component can have several failure modes, eg a thread can shear completely or the thread can become partly worn. For each failure mode, scores are ascribed (often between 1-5) for:

- likelihood
- severity of consequence
- inability to detect prior to failure, eg a part being located in an inaccessible location.

8.201 A risk rating is determined for each failure mode by multiplying the three scores together and a course of remedial action suggested. The results are often tabulated under the following headings.

Component	Failure Modes	Likelihood	Severity	Detection	Risk Rating	Action

Job Safety Analysis

8.202 Job safety analysis adopts techniques used in method study and task analysis. Each job or task is divided into component steps and the health and safety implications of each step are carefully considered. Usually, the following four-stage approach is adopted.

1. Identify the key steps in the job or operation.
2. Analyse and assess the risks associated with each stage.
3. Define the precautions and controls that need to be taken for the operation to proceed safely and without danger to the operator or anyone else.
4. Draw up safe operating procedures and provide job safety instructions to those carrying out the operation.

Integrated Strategy

8.203 To fully investigate all foreseeable risks at the development, installation, use and maintenance stages, a combination of the various techniques is often required.

8.204 An integrated strategy can apply failure mode and effect analysis to identify any potential hardware component failure modes and to analyse effects, fault tree analysis to identify hardware failure modes and human error that can lead to a particularly undesired outcome, and event tree analysis to investigate the chain of events following a component failure mode. The HAZOP approach is used to ask the often unasked questions — to guide and focus the other, more analytical techniques.

Considering and Implementing Control Measures

8.205 Any existing control measures and further control measures that are required to be introduced should be assessed to evaluate their effectiveness in preventing harm. The risk assessment should identify whether there is a need for further control measures, and the assessor should consider the most suitable control measures to be implemented.

8.206 It may also be necessary to introduce interim controls to deal with high risks until longer term risk control measures can be introduced.

8.207 The accompanying Approved Code of Practice to the Management of Health and Safety at Work Regulations 1999 states: "where risks are already controlled in some way, the effectiveness of those controls needs to be considered when assessing the extent of risk which remains. You also need to:

- observe the actual practice; this may differ from the works manual, and the employees concerned or their safety representatives should be consulted
- address what actually happens in the workplace or during the work activity
- take account of existing preventive or precautionary measures; if existing measures are not adequate, ask yourself what more should be done to reduce risk sufficiently."

8.208 Risk assessors should first ask whether everything has been done that the law requires and secondly, whether generally accepted industry standards and good practice have been reached.

8.209 Once this minimum standard has been reached, it is necessary to go further and to reduce any remaining risks, so far as is reasonably practicable. An action list needs to be created, which prioritises remaining risks, based on their level and the number of persons exposed to them.

8.210 If additional controls are required, then the following principles, in priority order, need to be adopted wherever possible.

- Select another option, so the work has a lower risk.
- Prevent access to the hazard, through the use of fencing, barriers and guards, etc.
- Organise the work so exposure to the hazard is reduced.
- Issue personal protective equipment to those involved in the work, or those affected.
- Provide adequate welfare facilities, including washing facilitates and first-aid facilities.

8.211 It is not enough to introduce these additional controls without also providing effective systems and procedures, and giving adequate information, instruction and training to relevant people.

8.212 The design of workplace precautions should also consider human factors. Precautions should be fully integrated into plant and work design procedures so they satisfy output, quality and health and safety requirements, rather than being perceived as "bolt on" safety requirements.

8.213 The level of risk evaluated by the assessment implies the effectiveness of the existing controls. If additional controls are introduced as a result of a risk assessment, it is important that a reassessment is done to determine the new level of risk and the effectiveness of the new controls.

8.214 It is also important that reassessment is carried out logically and it is not just assumed that the level or risk is reduced. As risk is calculated through multiplying likelihood by severity, the new controls must either reduce likelihood, severity or both if a reduction in risk is to be achieved. Placing barriers or fences to prevent access to dangerous areas, for example, reduces the likelihood of accidents happening, but does nothing to reduce their severity should they occur. Wearing personal protective clothing does nothing to reduce the likelihood of an accident occurring, but reduces the severity of the injury suffered should the accident happen.

8.215 Prohibiting work with certain electrical equipment in wet conditions may, however, reduce both the likelihood of an accident occurring and the resulting severity.

8.216 A work activity should not begin until it is decided that any risk present is adequately controlled. Therefore control measures should not be implemented until it is decided that they do adequately control the risk.

8.217 Safe place measures should be considered to control a hazard before considering which safe person measures will be needed. For certain activities there are specific legal requirements for particular methods of risk control, eg noise.

8.218 Risk control is the process of designing, implementing and maintaining measures which reduce a particular risk, and specific risk control measures for industry must be considered. Most risk control measures tend to adopt one or more of the following.

- Hazard reduction methods — elimination, substitution and redesign, etc.
- Separation methods — height, distance and remote locations, etc.
- Physical barrier methods — around the hazard, around the person or around the location.
- Dose-limitation methods — reduce the dose, number of people exposed to the dose or time exposed to the dose.

Safe Place Measures

8.219 To incorporate safe place measures the following hierarchy of control should be considered.

1. Eliminate the hazard.
2. Reduce the risk of the hazard at source.
3. Control exposure to the hazard by means of physical measures, preferably those which protect everyone rather than individuals.

8.220 The measures that reduce risk most effectively should be considered where possible.

Elimination

8.221 Elimination involves the complete removal of the hazard, eg if the hazard is a hole in the floor it can be eliminated by filling it in. Elimination is the most effective method of dealing with a hazard. However, it should be noted that if a hazard is eliminated by providing an alternative method of work, this alternative method of work may also introduce new and different hazards, eg the use of compressed air tools, instead of electrical tools, may increase hand-arm vibration.

Reduction

8.222 The option of reducing the risk at source or using a safer alternative should be considered, eg providing a water spray to reduce the amount of dust produced during cutting or grinding.

Control of Exposure

8.223 Controlling exposure to the hazard is not as effective as elimination as it does nothing to the hazard itself. It does not reduce the severity factor of the hazard but it does reduce the likelihood of the hazard occurring. Controls can be introduced, such as the following.

- Full enclosure, eg a noise-reducing enclosure around a machine.
- Part enclosure, eg local exhaust ventilation.
- Remove persons from the hazard or provide a barrier between the persons and the hazard if it cannot be enclosed or covered, eg a guard-rail fitted to a scaffold.
- Reduce the time of contact or quantity of contact by some means, eg a device that cuts out power to a machine if the safe noise level is exceeded.

Safe Person Controls

8.224 Safe person controls will always be required to:

- prevent accidents
- ensure that the physical controls remain effective
- raise awareness
- increase protection of the individual.

Protective Devices

8.225 Protective devices can be provided to distance or protect a person from a hazard, eg a rubber hand guard fitted to a chisel.

Safe Systems of Work

8.226 A safe system of work is a definition, usually in writing, of the correct methods, plant, equipment, processes, etc to follow for a certain work activity. High-risk activities, eg demolition or working with asbestos, will require very detailed written safe systems of work. A safe system of work would normally involve the use of method statements and, in some cases, permits to work.

Permits to Work

8.227 A permit to work is a control procedure which may be issued for virtually any work activity. However, historically, it has been reserved for what would be classified as a high-risk activity, task or process. The procedure is a formalisation of assessed controls and limitations, designed to provide a safe system of work. The permit is issued by, and returned to, an authorised (usually senior) manager.

8.228 There are a number of key points relating to the permit-to-work procedures that need to be considered.

- Specialised training is usually required for permit to work tasks.
- The work is usually classified as being high risk by risk assessments.
- The work or task may be complex.

8.229 The nature of tasks for which permits to work may be used include:

- electrical work
- work in confined spaces
- pressure testing
- excavation work
- scaffolding design
- radiography (X-ray)
- mechanical (maintenance)
- welding or other hot work.

8.230 Safe working procedures or health and safety plans should identify circumstances, locations and activities subject to permit-to-work procedures and ensure that formal requests are made for permits to work. This, in turn, should concentrate the minds of managers responsible for co-ordinating safety on the fact that a safety critical task is scheduled to take place.

8.231 The request for a permit to work should be given careful consideration, and should identify:

- the reference number of the request
- to whom the request is made
- the name and status of the individual making the request
- what the proposed work involves
- where the proposed work will be carried out
- the date and time of the proposed work
- the planned duration of the work
- the safety critical element(s).

8.232 The safety management co-ordinator will, at this stage, begin to prepare the notice of permit to work, which will include details of:

- the work process
- the location
- issuing procedures for the formal permit to work
- notifications to third parties about limitations or restrictions of access.

8.233 The next stage is the appointment and recording of the authorised person to control the permit to work. This individual will carry a considerable responsibility and must therefore be fully conversant with, and competent and qualified to discharge the delegated duties associated with the procedure.

8.234 Permit-to-work formats vary from task to task, but there are some common points, including:

- a unique permit number
- date and time of issue
- duration of the permit to work
- identification of the precise work location
- description of the hazard, the foreseeable risks, and the potential consequences should the risk be realised
- list of precautionary measures to be taken
- testing and proving procedures
- details of emergency procedures and signals
- acknowledgement and confirmation of understanding of the work, the hazard and precautions, signed by the person in charge of the work

- signing off the completed or partly completed work by the person in charge of the work
- debriefing
- cancellation of permit-to-work by authorised person.

Personal Protective Equipment

8.235 Personal protective equipment (PPE) may be used if control measures do not adequately control the hazard. The use of PPE should be carefully considered. PPE may in some circumstances increase the risks from other hazards, eg heat stroke.

Monitoring Performance

8.236 Supervision or monitoring of work activities is always required even if the activity is low risk or the people involved are very competent. Having analysed the risks related to work activities, assessors should determine the level of supervision required to check on control measures.

8.237 Monitoring involves checking conditions, plant and systems to ascertain whether performance is being maintained to the defined standards, using the techniques of safety inspections and safety audits.

8.238 Control measures must be monitored to ensure their effectiveness. This includes:

- checking the maintenance procedures of control measures
- ensuring that the selected control measures are properly and consistently used by workers.

8.239 The maintenance of control measures is essential to the safety of users, and a system of specific checks designed to ensure that the equipment remains in a safe and serviceable condition should be introduced. The nature and extent of any routine inspections should be decided by the risk assessor, in consultation with a specialist company if required.

8.240 Any relevant legislation or guidance on the frequency of routine inspections should be consulted and conformed with.

8.241 Employees are required to make correct use of control measures and report any defects that may occur in such measures. Employers should ensure that the control measures selected for use are those that are strictly necessary and employees should be trained in the correct use of the protective equipment.

8.242 Health surveillance is required under certain circumstances to identify and protect individuals at increased risk.

Review

8.243 The review is the final stage in the risk management process. A review of the assessment must be undertaken if:

- there is reason to suspect that it is no longer valid
- there has been a significant change in the matters to which it relates.

8.244 The review process should be built into risk assessment as a regular event. Every risk assessment should have a date for review, which should be based on the level of risk. The review should look at the whole of the risk management process to re-evaluate each stage of the process. This is particularly important with respect to generic risk assessments.

Recording Risk Assessments

8.245 The significant findings of a risk assessment should be recorded where the employer employs five or more employees. This includes the recording of the arrangements for:

- planning, organisation, control, monitoring and review of the preventive and protective measures
- the establishment of procedures for serious and imminent danger and for danger areas.

8.246 As well as satisfying legal requirements, it is also necessary to record the risk assessments in order to demonstrate legislative compliance and for use in legal proceedings and in civil claim cases. The record may be handwritten, printed or stored electronically.

8.247 The HSE has not produced a standard format document for risk assessments; it has been left to individual employers to decide how they will record their own risk assessments. However, the supporting ACOP to the Management of Health and Safety at Work Regulations 1999 (L21) does intimate the contents of the risk assessment format regarding the "significant findings", ie:

- the significant hazards identified
- the existing control measures in place and the extent to which they control risk
- the persons who may be affected by these significant risks, including those especially at risk.

Training

8.248 Risk assessments must be carried out by competent persons. The need for training and the communication of information related to the risk assessment must be considered. Suggested training courses for risk assessors are the IOSH Managing Safety course, or the more advanced NEBOSH General Certificate in Safety and Health.

Information, Instruction and Training

8.249 Training is an important method of achieving competence and is required at all levels. Training is required when:

- an employee is first recruited, eg induction training
- there are new or increased risks as a result of:
  - transfer or new responsibilities
  - new or changed work equipment
  - new technology
  - new or changed systems of work.

8.250 The skills required for particular tasks should be assessed and any workers required to carry out the work should be trained to the required standard.

Provision of Information

8.251 Information about risks and control measures must be supplied to workers (and to others if necessary). The legislation requires that employers issue employees with "comprehensible and relevant information" on risks identified by the assessment, the preventive and protective measures, emergency procedures and the staff involved in them, and any risks notified to the employer by others.

8.252 Employers are also required to provide information to other employers with employees working on the same premises or involved in the same undertaking. If these employees do not have sufficient understanding of English, then for the relevant information to be comprehensible to them it will need to be translated. If they cannot read, it will need to be explained to them in a form which they can understand. For the average worker, toolbox talks are an effective means of providing information, as is a pre-start briefing.

Instructions

8.253 Instructions, written or verbal, should be issued to workers on what they must or must not do to deal with a specific hazard. The degree to which instructions need to be provided and the detail they contain will depend on the level of risk involved, and the complexity of the controls.

List of Relevant Legislation

- Construction (Design and Management) Regulations 2007
- Control of Asbestos at Work Regulations 2006
- Work at Height Regulations 2005
- Control of Noise at Work Regulations 2005
- Control of Vibration at Work Regulations 2005
- Control of Substances Hazardous to Health Regulations 2002
- Control of Lead at Work Regulations 2002
- Dangerous Substances and Explosive Atmospheres Regulations 2002
- Pressure Systems and Transportable Gas Containers Regulations 2000
- Management of Health and Safety at Work Regulations 1999
- Ionising Radiations Regulations 1999
- Provision and Use of Work Equipment Regulations 1998
- Lifting Operations and Lifting Equipment Regulations 1998
- Confined Spaces Regulations 1997
- Work in Compressed Air Regulations 1996
- Health and Safety (Display Screen Equipment) Regulations 1992
- Manual Handling Operations Regulations 1992
- Personal Protective Equipment at Work Regulations 1992
- Supply of Machinery (Safety) Regulations 1992
- Construction (Head Protection) Regulations 1989
- Electricity at Work Regulations 1989
- Health and Safety (First-aid) Regulations 1981
- Health and Safety at Work, etc Act 1974
- Regulatory Reform (Fire Safety) Order 2005

Further Information

Publications

HSE Publications

The following are available from *www.hsebooks.co.uk*.

- HSG65 (rev 1997) *Successful Health and Safety Management*
- HSG151 *Protecting the Public — Your Next Move*
- INDG163(L) *5 Steps to Risk Assessment*
- L21 (rev 2000) *Management of Health and Safety at Work. Management of Health and Safety at Work Regulations 1999. Approved Code of Practice and Guidance*

British Standards

The following is available from *www.bsi-global.com*.

- BS 8800: 2004 *Occupational Health and Safety Management Systems — Guide*

Other Publications

- *Health and Safety Risk Assessment*, Dr Tony Boyle. IOSH Services, Leicester, 2nd Edition.

Organisations

- Health and Safety Executive (HSE)
  Web: *www.hse.gov.uk*
  The HSE is responsible for the regulation of almost all the risks to health and safety arising from work activity in the UK.
- Institution of Occupational Safety and Health (IOSH)
  Web: *www.iosh.co.uk*
  IOSH is Europe's leading body for health and safety professionals and provides guidance on health and safety issues.
- Loss Prevention Council
  Web: *www.redbooklive.com*
  Owned by the Association of British Insurers and Lloyds of London, the council provides a variety of technical services in the field of loss prevention, including research, standard setting, product testing, certification and training.
- National Examination Board in Occupational Safety and Health (NEBOSH)
  Web: *www.nebosh.org.uk*
  Founded in 1979, NEBOSH is an independent awarding body.

Construction Training Schemes

- UK construction output growth is forecast to average 3% annually between 2008–2010, and all new recruits will need to be trained to a competent level.
- Only registered workers may be on-site from 2003 and only qualified workers may be on-site from 2003 in the case of major contractors and from 2010 for all other contractors.
- There are obvious advantages to passport schemes: they provide assurance that workers are appropriately trained and competent in the job they are being employed to do, particularly in the area of health and safety.
- The most well-known of these schemes currently in use in the UK is the one administered by CITB-ConstructionSkills, known as the CSCS (Construction Skills Certification Scheme).
- If an employee (prospective or otherwise) holds themselves out to have a particular skill, trade or qualification, they should be able to prove this to prospective employers.
- To ensure that contractors have completed prescribed health and safety awareness training, entry to construction sites should be restricted to workers who can demonstrate they have the necessary skills and training by means of a safety passport.

8.254 According to the Construction Skills Network publication, *Blueprint for UK Construction Skills 2008–2012*, "UK construction output growth is forecast to achieve 1.7% annually between 2008–2012 — lower than the average 2.6% increase over 2007–2011. This is probably due to the completion of the buildings for the Olympic Games and other large projects in the next 4 years. It is anticipated that growth will tail off in 2012 and further estimated that 2.8 million construction workers will be needed in the construction industry by 2012.

8.255 Therefore, to deliver forecast growth between 2008 and 2012, the number of construction workers needed is likely to increase by around 180,000 across the whole of the UK. This translates into a need for an additional 88,400 new recruits a year on average to fulfil the requirement created by additional demand and to take account of those who will leave the industry during the same period.

8.256 ConstructionSkills state that: "In order to ensure the construction industry is on the right track for the future, it is vital that we identify exactly where skills shortages lie and that we understand the skills needs of the future."

In Practice

Passport Schemes

8.257 A number of schemes have been set up to allow workers in many industries, including construction, to demonstrate their capabilities to prospective employers. Most of these schemes are now independently accredited by UKAS.

8.258 One of the more popular methods adopted so far are the so-called passport schemes, whereby workers carry a simple card showing their identity and the level of qualification they have in a particular field. Participating contractors and the controllers of building sites are supposed to allow admittance to their sites to card-carrying workers.

8.259 There are obvious advantages to passport schemes: they provide assurance that workers are appropriately trained and competent in the job they are being employed to do, particularly in the area of health and safety. However, the sheer number of different schemes in operation in the UK and Europe has led to accusations that schemes are being misused, and in some cases the scheme itself is being abused.

8.260 Schemes are operated, among others, by the:

- Safety Pass Alliance (SPA)
- Client/Contractor National Safety Group
- Airport Construction Training Alliance
- Energy and Utilities Skills Limited.

8.261 The most well-known of these schemes currently in use in the UK is the one administered by CITB-ConstructionSkills, known as the CSCS (Construction Skills Certification Scheme).

CSCS

8.262 The CSCS offers a wide variety of skills cards. Their use is promoted by most of the industry bodies, and in the case of standard building contracts are issued by the Joint Contracts Tribunal (JCT), where their use is written into the contract itself as the preferred option.

8.263 The CSCS was originally set up in the mid-1990s as a separate organisation owned and managed by CSCS Limited and controlled by a management board whose members are from the:

- Construction Confederation
- Federation of Master Builders
- GMB Union
- National Specialist Contractors Council
- Transport and General Workers Union (now part of the UNITE trade union)
- Union of Construction Allied Trades and Technicians
- Confederation of Construction Clients
- Construction Industry Council.

8.264 ConstructionSkills is an industry-led Sector Skills Council for the construction sector. This employer-led body is responsible for ensuring the skills and productivity needs of the construction sector are addressed. It is a partnership between CITB-ConstructionSkills, the Construction Industry Council (CIC) and CITB Northern Ireland and as such covers the whole industry and the whole of the UK.

8.265 CITB-ConstructionSkills helps industry in England, Scotland and Wales in all aspects of recruiting, training and qualifying the construction workforce. It also works with government to improve the competitiveness of the industry as a whole.

8.266 The CSCS card is the industry's largest scheme, and at present covers almost 250 organisations (including trades, technical, supervisory and management) for which cards are approved by the major contractors. Around 1.2 million cards have been presented by the CSCS scheme, meaning that the holders have formally proved their work competency. Various cards are issued depending on the skills level of the individual.

8.267 CSCS cards list the holder's qualifications and are valid for either three or five years. They also prove health and safety awareness as all cardholders have to pass the appropriate CITB-ConstructionSkills health and safety test, which currently costs £17.50. Most elements of the test may now be taken online. The CSCS card costs £25.

Qualified Workers

8.268 If an employee (prospective or otherwise) holds themselves out to have a particular skill, trade or qualification, they should be able to prove this to prospective employers.

8.269 There are a number of ways of demonstrating this, eg:

- tradespeople may gain formal qualifications such as those awarded by Oxford, Cambridge and the RSA (OCR), or Edexcel in a vocational field
- professional consultants can join specialist institutions such as the Royal Institute of British Architects (RIBA) or the Royal Institution of Chartered Surveyors (RICS), which have education and training-based entry requirements.

8.270 By 2010, the industry aims to have all construction workers qualified. This means that over 70,000 construction workers will require NVQ level 2 qualification. ConstructionSkills is co-ordinating efforts to ensure this target is met. A major part of their effort is working through 200 on-site assessment training centres which assist in assessing the competence of construction workers.

Safety Passports

8.271 In recent years there has been an increasing requirement for contractors to undergo prescribed health and safety awareness training, as well as having relevant practical experience.

8.272 To ensure contractors have completed this training, entry to construction sites should be restricted to workers who can demonstrate they have the necessary skills and training. This is achieved by means of a "safety passport".

8.273 These are an increasingly important way for employers and companies who employ contractors to establish health and safety assurances among their workforce. The safety passport is similar to any other passport in that it allows the holder access to a passport-controlled environment.

8.274 As well as containing personal details such as the holder's name and photograph, the safety passport provides details of the health and safety training that the holder has received and is used as evidence that the bearer is qualified to an acceptable and recognised level of health and safety awareness and skill level.

8.275 There are a number of different schemes, including the following.

- Safety Pass Alliance (SPA). Established in response to the needs identified within industry and has undertaken to develop and extend the safety passport culture across all interested sectors. Training includes:
  - introduction to health and safety, environmental considerations and safe systems of work
  - workplace safety access, egress, emergencies, vehicles, equipment and machinery
  - accidents: prevention and reporting, first aid
  - hazardous substances: COSHH
  - manual handling
  - noise at work.

- The Institution of Occupational Safety and Health (IOSH) provide passport programmes for larger organisations.
- Client/Contractor National Safety Group (CCNSG) safety passport. The scheme, which results in the award of a "safety passport" to contractor site personnel is monitored and controlled by CCNSG, although there is also representation from contractors, training providers and trade unions. The CCNSG meets regularly to ensure that standards are maintained and that the content of the training scheme continues to meet current safety legislation and best practices.
- Airport Construction Training Alliance (ACTA). ACTA is a non-profit making organisation which operates as a strategic alliance between BAA and its construction partners. It aims to meet head-on the challenges of attracting, recruiting and retaining skilled and motivated people to work in airport construction. The ACTA safety passport is linked to a database of training and competence-based qualifications.
- The Energy and Utility Skills organisation provides a Health, Safety and Environment Passport programme for specific utilities. It is the sector skills council for the electricity, gas, waste management, and water industries. Employer-led, its purpose is to identify employers skills needs and provide effective solutions to improve business performance.

Further Information

Publications

- *Blueprint for UK Construction Skills 2008–2012*, ConstructionSkills
- *The Sector Skills Agreement for Construction: England 2005–2010*, ConstructionSkills
- HSL/2003/10 *A Review of Safety Passport Training Schemes*, Health & Safety Laboratories

Organisations

- Airport Construction Training Alliance
  Web: *http://web.acta.easitrack.co.uk*
  ACTA is a non-profit-making organisation operating as a strategic alliance between BAA and its construction partners. It aims to meet the challenges of attracting, recruiting and retaining skilled and motivated people to work in airport construction.
- Client Contractor National Safety Group (CCNSG)
  Web: *www.ccnsg.com*
  The Client Contractor National Safety Group run the CCNSG Safety Passport Scheme, to ensure a basic knowledge of health and safety for all site personnel and enable them, after appropriate site induction, to work on-site more safely with lower risk to themselves and others.
- Confederation of Construction Clients
  Web: *www.clientsuccess.org.uk*
  Confederation of Construction Clients' mission is to represent the interests of construction industry clients lectively by encouraging clients to achieve value for money through best practice, securing major and measurable improvement in the performance of supply, and promoting policies which can achieve a safe, stable and skilled industry.
- Construction Confederation
  Web: *www.constructionconfederation.co.uk*

The leading representative body for contractors, representing some 5000 companies who in turn are responsible for over 75% of the industry's turnover.

- Construction Industry Council (CIC)
  Web: *www.cic.org.uk*
  The CIC is the representative forum for the professional bodies, research organisations and specialist business associations in the construction industry.
- Construction Industry Training Board (Northern Ireland) (CITB)
  Web: *www.citbni.org.uk*
  The CITB works in partnership with CITB-ConstructionSkills in Great Britain and the Construction Industry Council (CIC), as the Sector Skills Council for Construction. The organisation is called ConstructionSkills, responsible for representing employers' skills needs and for tackling the skills and productivity needs of the sector.
- Energy and Utility Skills
  Web: *www.euskills.co.uk*
  Working under the governance of the Sector Skills Development Agency (SSDA), and within the Skills for Business network (SfBn), Energy and Utility Skills'employer-led sector skills council tackles the skills debate, qualifications and skills delivery.
- Federation of Master Builders (FMB)
  Web: *www.fmb.org.uk*
  The FMB is a the largest trade association in the UK building industry, protecting the interests of small and medium-sized building firms.
- National Specialist Contractors Council (NSCC)
  Web: *www.nscc.org.uk*
  The NSCC brings together the common aims of specialist trade organisations within the construction industry and is the authoritative voice of Specialist Contractors in the UK.
- Oxford, Cambridge and the RSA (OCR)
  Web: *www.ocr.org.uk*
  The OCR is a leading UK awarding body, dedicated to supporting education providers by offering respected qualifications and comprehensive support.
- Safety Pass Alliance (SPA)
  Web: *www.safetypassports.co.uk*
  The Safety Pass Alliance has undertaken to develop and extend the safety passport culture across all interested sectors within industry.
- Transport and General Workers Union
  Web: *www.tgwu.org.uk*
  The Transport and General Workers Union amalgamated with AMICUS to form the UNITE union, representing members in the workplace.
- Union of Construction Allied Trades and Technicians (UCATT)
  Web: *www.ucatt.org.uk*
  UCATT is Britain's only specialist construction workers union with over 120,000 members employed in all building trades, in both the private and public sectors. The union organises workers in the UK and the Republic of Ireland, and has an extensive network of regional offices.

Corporate Manslaughter and Corporate Homicide Act 2007

- Prosecution for corporate manslaughter is the prosecution of a company as a result of a death or deaths for which the company is held responsible.
- Corporate manslaughter is a type of involuntary manslaughter (killing by gross negligence).
- Successful corporate manslaughter prosecutions are extremely rare because of the need to identify a "directing mind" of the company who is also guilty.
- A new piece of legislation which came into force on 6 April 2008 — the Corporate Manslaughter and Corporate Homicide Act 2007 — introduces a new offence for prosecuting companies and other organisations where there has been a gross failing in the management of health and safety with fatal consequences.
- Unlimited fines may be issued, and the courts may force companies to publicise their convictions through a "publicity order", leading to severe damage to reputations under the new legislation.
- Under the new legislation, individual directors will not be liable for any deaths due to a general breach of the duty of care by the firm.
- Employers should take steps now to review their management structures and health and safety policies.

8.276 Prior to the coming into force of the Corporate Manslaughter and Corporate Homicide Act 2007, the state of the law surrounding the criminal liability of companies for corporate manslaughter was unclear and confusing, and successful prosecutions of companies for manslaughter were rare. These facts have been highlighted by the collapse or failure of manslaughter prosecutions against companies brought in the wake of high profile public disasters such as the sinking of the Herald of Free Enterprise.

In Practice

The Law Prior to the New Act

Background — the Law of Manslaughter

8.277 In categorising criminal behaviour, English law required two elements to be present.

- There must be a criminal act, known to lawyers as the *actus reus*. In general, no crime can be committed unless there is a voluntary act or omission.
- There must be criminal intent, known to lawyers as the *mens rea*. Recklessness and gross negligence are sufficient to be included within the definition of criminal intent for manslaughter.

8.278 Manslaughter is divided into two types: voluntary and involuntary. It is the latter type which relates to corporate manslaughter.

8.279 Involuntary manslaughter occurs when there is an unlawful killing but there is no intent to kill or cause grievous bodily harm. Involuntary manslaughter is further sub-divided into unlawful act manslaughter and gross negligence manslaughter.

Unlawful Act Manslaughter

8.280 Unlawful act manslaughter occurs when the killing is the result of an unlawful or criminal act that all sober and reasonable people would realise would expose the victim to the risk of physical harm. It is irrelevant whether or not the accused knows the act is unlawful and criminal and whether or not he or she intends to do harm.

Gross Negligence Manslaughter

8.281 In gross negligence manslaughter, the following elements must be present.

- The accused must owe a duty of care to the victim. The ordinary principles of the law of negligence will apply in ascertaining whether such a duty exists.
- The accused must have breached that duty. A person may be liable for manslaughter by neglect of a positive duty arising from the nature of his occupation.
- The breach must have caused the death of the victim. This is a question of fact that the jury has to decide.
- The breach must be characterised as gross negligence or recklessness. In the words of Land J in *R v Stone and Dobson* (1997): "It is clear ... that the indifference to an obvious risk and appreciation of such risk coupled with a determination nevertheless to run it, are both examples of recklessness. What the prosecution have to prove is a breach of ... duty in circumstances that the jury feel convinced that the defendant's conduct can properly be described as reckless. That is to so say a reckless disregard of danger to the health and welfare of the infirm person. Mere inadvertence is not enough. The defendant must have been proved to have been indifferent to an obvious risk of injury to health, or actually to have foreseen the risk but to have determined nonetheless to have run it."
- Failure to appreciate that the risk exists does not of itself amount to recklessness.

8.282 With effect from 6 April 2008, the common law offence of manslaughter by gross negligence in its application to corporations and other organisations is abolished.

Corporate Manslaughter

8.283 In this context, the most common entity is a company or corporation but the law also applies to other abstract entities such as councils. For example, Barrow Borough Council was charged with manslaughter in February 2004 following the death of seven people from legionnaires' disease.

8.284 A company is a separate legal entity in its own right but it is not a "real" person. Because it is not a real person, a company cannot commit voluntary manslaughter, nor can it commit unlawful act manslaughter. It can, however, commit manslaughter by gross negligence or recklessness.

8.285 The law of corporate manslaughter is separate from health and safety offences for which companies are liable and in respect of which companies are regularly prosecuted and fined. For example, Thames Trains was fined a record £2 million for breaches of health and safety law in respect of the Paddington rail crash.

The Identification Principle

8.286 A company could only be prosecuted successfully for manslaughter if the guilt of an identified human individual could be attributed to the company. In other words, the company must have a criminal intent, or *mens rea*. In the words of Rose LJ in Attorney General's Reference (No 2 of 1999) (2000): "Unless an identified individual's conduct, characterised as gross criminal negligence, can be attributed to the company, the company is not, in the present state of the common law, liable for manslaughter."

8.287 This is known as the "identification principle" of corporate liability and was expounded in a ruling arising out of the failed prosecution of Great Western Trains for the Southall train crash.

8.288 The effect of this is that the relevant person must be so closely identified with the company as to amount to its alter ego and must be the company's "directing mind".

8.289 In practice, for there to be a successful manslaughter prosecution of a company it was necessary for there to be a successful prosecution of a company director or similar for manslaughter.

8.290 In small companies, where there are one or two directors, identifying an individual who is a directing mind of the company is relatively straightforward. Indeed, the five successful corporate manslaughter prosecutions to date have all been of small companies.

8.291 However, in a large company, where structures can be very complex and responsibilities are divided, identifying specific individuals who can be characterised as the embodiment of the company and who have been grossly negligent or reckless is often impossible. This is why prosecutions for corporate manslaughter are rarely brought and often fail, even in the highest profile cases.

8.292 The first successful prosecution of a company for manslaughter occurred as recently as 1994, when the company OLL Ltd was found guilty in relation to the deaths of four teenagers who drowned during a canoeing trip off Lyme Regis. The managing director was also found guilty of manslaughter and imprisoned.

8.293 The most high-profile failed corporate manslaughter prosecution was of P&O Ferries in relation to the sinking of the Herald of Free Enterprise. Although the judge commented that P&O was "infected from top to bottom with the disease of sloppiness" and was guilty of "staggering complacency", no controlling mind could be identified and so no criminal guilt could be attributed to the company.

Manslaughter Penalties

8.294 Although the courts can impose unlimited fines, in practice the fines imposed upon companies convicted of manslaughter have been low, possibly reflecting the size of the companies.

Corporate Manslaughter Legislation

8.295 The Corporate Manslaughter and Corporate Homicide Act 2007 introduces a new offence across the UK for prosecuting companies and other organisations where there has been a gross failing, throughout the organisation, in the management of health and safety with fatal consequences (known as corporate manslaughter in England and Wales, and corporate homicide in Scotland).

8.296 The law applies to all organisations that employ people, including partnerships, clubs, trade unions, schools and other educational institutions, local authorities, hospital trusts and individuals who employ others in small businesses. Its applicability is very wide.

8.297 Under the new law, companies, organisations and, for the first time, government bodies, face an unlimited fine if they are found to have caused death due to their gross corporate health and safety failures.

8.298 The new legislation:

- makes it easier to prosecute companies and other large organisations when gross failures in the management of health and safety led to death by delivering a new, more effective basis for corporate liability
- allows unlimited fines to be issued and enables the courts to force companies to publicise their convictions through a "publicity order", leading to severe damage to reputations
- reformed the law so that a key obstacle to successful prosecutions was removed because both small and large companies can be held liable for manslaughter where gross failures in the management of health and safety cause death, not just health and safety violations
- complements the law under which individuals can be prosecuted for gross negligence manslaughter and health and safety offences, where there is direct evidence of their culpability
- is designed to complement existing health and safety legislation so the new offence does not impose new regulations on business
- lifts Crown immunity to prosecution meaning that Crown bodies — such as government departments — can be liable to prosecution
- applies to companies and other corporate bodies, in the public and private sector, government departments, police forces, and certain unincorporated bodies — such as partnerships — where these are employers.

8.299 Safety culture will be a key factor for juries to consider in deciding whether or not there has been a gross breach in health and safety management.

8.300 The individuals within the company are not liable to prosecution under the Act.

Who Will Prosecute?

8.301 The police and the health and safety enforcing authority, eg the Health and Safety Executive, will investigate most workplace fatalities and such investigations may lead to corporate manslaughter prosecutions.

Penalties for Corporate Killing

8.302 Under the Corporate Manslaughter and Corporate Homicide Act 2007, unlimited fines may be issued, and the courts may force companies to publicise their convictions through a "publicity order", leading to severe damage to reputations.

Evasion of Liability

8.303 Most prosecutions for the new offence are likely to be against companies, which gives rise to the real risk that corporate structures could be exploited to try to evade liability, eg by establishing financially weak subsidiary companies to carry out risky activities or by dissolution or insolvency of a guilty company. The Government is anxious that enforcement action should be a real deterrent.

8.304 The law applies to foreign companies doing business in the UK so that it will not be possible for a company incorporated overseas but carrying out activities here to escape prosecution. The Government recognises that there may be practical problems in taking enforcement action against such companies.

8.305 Criminal proceedings may be brought against parent companies or other companies within a group if their management failings contributed to the death. This will prevent group structures from being used for evasion, eg for establishing subsidiary companies to carry out risky business or by establishing financially weak subsidiary companies with insufficient assets to pay any fine.

8.306 The Act makes provisions against criminal liability evasion by dissolving a company or making it insolvent before the prosecution takes place. Companies could be subject to legal proceedings to freeze their property and assets and that such proceedings could be taken before a prosecution commences, to prevent assets being transferred or dissipated.

The Individual

8.307 The law makes companies, not directors, liable for any deaths due to a general breach of the duty of care by the firm. Individual directors cannot be personally liable under the Act, as the manslaughter offence will apply only to corporations, including public bodies.

Territorial Extent

8.308 With one or two exceptions, English criminal law applies only to England and Wales so that the courts do not have jurisdiction over criminal acts committed outside England and Wales. One of the exceptions to this rule is that the courts have jurisdiction over homicide offences, which include manslaughter, committed abroad.

8.309 However, the corporate manslaughter law applies only to offences of corporate killing committed in the UK and if a corporate killing by a UK company takes place overseas it will be a matter for the courts of the particular country concerned.

Crown Immunity

8.310 Under the law, Crown immunity was lifted. This means that Government bodies are within the remit of the law, and will face an unlimited fine if they are found to have caused death due to their gross corporate health and safety failures.

Private Prosecutions for Corporate Manslaughter

8.311 The consent of the Director of Public Prosecutions will be required for all prosecutions brought under the legislation.

Reckless or Grossly Careless Transmission of a Disease

8.312 There is increasing pressure generally on companies and employers to increase the safety of the workforce in the wake of the string of public diseases in recent years, where complacency and neglect have been seen to be contributory factors to fatalities.

8.313 If a company's management failure leads to the transmission of a fatal disease, the company may be guilty of corporate manslaughter.

List of Relevant Legislation

- Corporate Manslaughter and Corporate Homicide Act 2007
- Local Employment Act 1960

Further Information

Publications

- *Legislating the Criminal Code: Involuntary Manslaughter*, The Law Commission, Report No. 237
- *Reforming the Law on Involuntary Manslaughter: The Government's Proposals*, The Home Office

Organisations

- Centre for Corporate Accountability
  Web: *www.corporateaccountability.org*
  The Centre for Corporate Accountability is a charitable organisation that monitors worker and public safety, and undertakes research and provides advice on law enforcement and corporate criminal accountability.
- Home Office
  Web: *www.homeoffice.gov.uk*
  The Home Office is the government department responsible for internal affairs in England and Wales.
- Crown Prosecution Service (CPS)
  Web: *www.cps.gov.uk*
  The CPS is responsible for prosecuting people in England and Wales charged with a criminal offence.
- Health and Safety Executive (HSE)
  Web: *www.hse.gov.uk*
  The HSE is responsible for the regulation of almost all the risks to health and safety arising from work activity in the UK.

Demonstrating an Effective Health and Safety Management System

- To avoid prosecution under the Corporate Manslaughter and Corporate Homicide Act 2007, a demonstrable and effective health and safety management system should be in place.
- When considering whether or not a gross breach has taken place, the Act requires juries to consider, among other matters, the attitudes, policies, systems or accepted practices likely to have encouraged the failure.
- The health and safety management system should be based on effective risk assessment and be subject to robust systems of internal control and review.
- The health and safety roles and responsibilities of senior managers needs to be clearly outlined, and their performance should be measured and evaluated to ensure they have the necessary competencies to carry out their roles and responsibilities effectively.
- The attitude of those at director level plays an important part in developing the organisation's culture and its tolerance of breaches of safety legislation.
- The effective management of health and safety will depend on a suitable and sufficient risk assessment being carried out and the findings being effectively used.
- The organisation needs to establish an effective health and safety management system to implement their health and safety policy. It needs to be proportionate to the hazards and risks they need to combat.
- Managers need to ensure that all employees, at all levels, are motivated and empowered to work safely.
- A systematic review of performance, based on the information obtained from monitoring and auditing, is required if an organisation is to learn from all relevant experience.
- Only when directors and senior executives give a positive commitment to any changes that need to be implemented will those occupying other managerial levels follow suit.

8.314 In order to ensure that an organisation's health and safety management system operates efficiently and in a way that provides a defence under the requirements of the Corporate Manslaughter and Corporate Homicide Act 2007 (the Act), it is essential that evidence can be obtained, when required, that demonstrates that effective management systems are in place and that those occupying senior management positions are both competent and fully committed to ensuring the effectiveness of the health and safety management system.

Employers' Duties

Corporate Manslaughter and Corporate Homicide Act 2007

8.315 The Corporate Manslaughter and Corporate Homicide Act 2007 places senior managers and organisations at risk of prosecution if their gross management failings lead to the death of an employee or a member of the public to whom a duty of care is owed.

8.316 The legislation states that there has to be a relevant duty of care owed by the organisation in the law of negligence to the deceased and a gross breach of that relevant duty of care leading to that person's death.

8.317 In order to ensure that the organisation's health and safety management system operates efficiently and in a way that provides a defence under the requirements of the Act, it is essential that evidence can be obtained, when required, that demonstrates that effective management systems are in place and that those occupying senior management positions are both competent and fully committed to ensuring the effectiveness of the health and safety management system.

Management of Health and Safety at Work Regulations 1999

8.318 The Management of Health and Safety at Work Regulations 1999 (MHSWR) provide the legal framework for a health and safety management system.

8.319 Employers are required to make and give effect to such arrangements as are appropriate — having regard to the nature of their activities and the size of the undertaking — for the effective planning, organisation, control, monitoring and review of the preventive and protective measures.

8.320 Where the organisation employs five or more employees, these arrangements must be recorded.

Health and Safety at Work, etc Act 1974

8.321 The Health and Safety at Work, etc Act 1974 requires employers to ensure "so far as is reasonably practicable" the health, safety and welfare of their employees and other persons who may be affected by their work. This includes sub-contractors and the general public.

Employees' Duties

8.322 Under the MHSWR, employees must inform their employer, or other employees with a responsibility for health and safety, of any work situation where there is a serious and immediate risk to health and safety. Similarly, they must inform employers of any shortcomings in the employer's protection arrangements.

In Practice

8.323 The emphasis in the Corporate Manslaughter and Corporate Homicide Act 2007 is on effective management of health and safety and therefore it is vital that senior management can demonstrate that they have an effective system in place.

Gross Breach of Care

8.324 To be guilty of an offence under the Act, an organisation's activities must be managed or organised in such a way as to amount to a gross breach of a relevant duty of care owed by the organisation to the deceased.

8.325 A "gross" breach involves conduct that falls far below what can reasonably be expected of the organisation in the circumstances. Furthermore, the way in which the organisation's activities are managed or organised by its senior management must form a substantial element in the breach.

8.326 "Senior management" is defined as the persons who play significant roles in the making of decisions about how the whole, or a substantial part, of an organisation's activities are to be managed or organised, or the actual managing or organising of the whole, or a substantial part, of those activities.

8.327 When considering whether or not a gross breach has taken place, the Act requires juries to consider, among other matters:

- failure to comply with any relevant health and safety legislation and, where this occurs, the seriousness of the failure and the risk of death it posed
- attitudes, policies, systems or accepted practices likely to have encouraged the failure to comply with the relevant health and safety legislation, or led to a tolerance of this failure
- any relevant health and safety guidance
- any other matters they consider relevant.

Relevant Duty of Care

8.328 A "relevant duty of care" is taken to mean any of the following duties owed by an organisation under the law of negligence.

- The duty owed to an employee or to other persons working for the organisation or performing services for it.
- The duty owed as an occupier of premises.
- The duty owed in connection with the supply of goods or services, the carrying on of construction or maintenance operations, the carrying out of other activities on a commercial basis or the keeping of any plant, vehicle or other item.

8.329 Therefore, in order to defend itself against such a charge, it is necessary for senior managers to ensure that they have evidence to demonstrate a transparent and effective health and safety management system is in place and that its policies and procedures are translated into positive action on a day-by-day basis.

8.330 In line with the requirements of the Turnbull Report — *Internal Control: Guidance for Directors on the Combined Code* — it is essential that this health and safety management system is based on effective risk assessment and is subject to robust systems of internal control and review. It is also essential for senior managers to ensure that the health and safety management system covers the full range of duty of care situations outlined and, as a minimum, ensures compliance with all relevant health and safety legislation and guidance.

The Role, Responsibilities and Competency of Senior Managers

Combined Negligence

8.331 It is important to remember that a "gross" breach under the Act need not purely relate to the failures of one individual, rather the negligence of several individuals may add up to gross negligence, particularly where the individuals occupy senior management positions.

8.332 Therefore, it is essential that the health and safety roles and responsibilities of senior managers are clearly outlined, that their performance is measured and evaluated and that they have the necessary competencies to carry out their roles and responsibilities effectively.

Directors' Responsibilities

8.333 Those operating at director level in organisations (titles may differ depending on whether the organisation is in the private or public sector and on its legal status) have a critical part to play in ensuring that health and safety is effectively managed. For example, they are responsible for ensuring robust internal monitoring, that reviews take place and that effective risk assessments are undertaken.

8.334 Additionally, the attitude of those at director level plays an important part in developing the organisation's culture and its tolerance of breaches of safety legislation and organisational rules and procedures.

Impact of Health and Safety Director Role

8.335 Some organisations may choose to appoint a specific director to carry the health and safety responsibilities. This has the advantage of providing a focal point for health and safety within the organisation and allows that person to provide authority and credence to health and safety initiatives.

8.336 However, it may also lead to others at director level feeling that they have no health and safety responsibilities. In reality, health and safety factors need to be taken into account when all business decisions are made and they cannot be considered in isolation. For example, when deciding on the purchase of equipment, the health and safety issues related to the various items of equipment under consideration need to be taken into account along with factors relating to purchase costs, maintenance requirements, training needs and installation costs, etc.

8.337 In fact, health and safety impacts on the decisions made by directors with a diverse range of responsibilities and not merely on the director allocated health and safety responsibilities.

No Directing Mind, No Excuse

8.338 In order for any health and safety management system to be effective, it is essential that those at senior management level take the lead and provide visible and active support, strong leadership and commitment.

8.339 Where this is lacking, line managers will perceive that health and safety is a "cosmetic" rather than a genuine concern for the organisation and a sloppy culture, which tolerates rule breaking, will quickly be generated.

8.340 This was highlighted by the Herald of Free Enterprise disaster where it was found that all those concerned in management, from the members of the board of directors down to the junior superintendents, were guilty of fault. From the top to the bottom, the body corporate was infected with the "disease of sloppiness".

8.341 In the past, the need to identify a particular individual at the directing mind level, whose negligence directly caused death, meant that this combined negligence was not enough to successfully prosecute an organisation for manslaughter. However, under the new Act, the total of the various examples of individual negligence may, taken together, count as gross negligence. This is particularly true where senior managers are involved.

8.342 Put bluntly, the Act will make it easier to successfully prosecute in cases such as the Herald of Free Enterprise. Therefore, it is essential that those at senior management level set positive examples and ensure that those beneath them in the management chain do likewise.

Health and Safety Commission's Guidance

8.343 The Health and Executive's (HSE) guidance INDG417 advises that:

- directors formally and publicly accept a role in providing health and safety leadership
- directors accept their individual roles in relation to health and safety
- all board decisions reflect the board's health and safety intentions (as stated in policy)
- directors accept a role in engaging active participation of their staff in improving health and safety
- directors ensure that they are kept informed of, and alert to, relevant health and safety issues.

8.344 The HSE also recommends the appointment of a health and safety director.

The Health and Safety Management System

8.345 As juries will need to consider the extent of health and safety legislative breaches when deciding if gross negligence led to the death, it is clear that no new or additional standards have been introduced by the Act.

8.346 Senior managers need to ensure, and to be able to demonstrate to enforcement authorities and the courts, that the health and safety management system in operation meets the requirements of current health and safety standards and legislation.

8.347 Organisations with a positive attitude towards health and safety will, of course, be doing this already. However, they will still need to ensure that they have evidence available to demonstrate that this is the case. In organisations lacking such a positive attitude a lot more work will be required.

Health and Safety Legislation

8.348 There is a plethora of specific health and safety legislation relating to, eg the control of hazards — such as chemicals, noise, vibration, work at height, electricity and equipment, etc — which must be adhered to. However, the main requirements of the Management of Health and Safety at Work Regulations 1999 apply to health and safety management systems. These include the following.

- Assessments of the risks to employees and others (Regulation 3).
- Principles of prevention (Regulation 4).
- Health and safety arrangements (Regulation 5).
- Health and safety assistance (Regulation 7).
- Procedures for serious and imminent danger (Regulations 8 and 9).
- Information for employees (Regulation 10).
- Co-operation and co-ordination between employers (Regulation 11 and 12).
- Capabilities and training of employees (Regulation 13).

8.349 Employers are required to make and give effect to such arrangements as are appropriate — having regard to the nature of their activities and the size of the undertaking — for the effective planning, organisation, control, monitoring and review of the preventive and protective measures. Where the organisation employs five or more employees, these arrangements must be recorded. This regulation provides the legal framework for a health and safety management system.

8.350 According to the accompanying Approved Code of Practice (ACOP), the effective management of health and safety will depend on a suitable and sufficient risk assessment being carried out and the findings being effectively used.

Planning

8.351 The organisation needs to establish an effective health and safety management system to implement their health and safety policy and which is proportionate to the hazards and risks they need to combat.

8.352 This will involve:

- adopting a systematic approach to risk assessment with deadlines for completion
- using risk assessment methods to set priorities and objectives for hazard elimination and risk reduction
- selecting appropriate methods of risk control
- setting deadlines for the design and implementation of the preventive and protective measures
- developing performance standards for the completion of risk assessments and the implementation of the preventive and protective measures.

Organisation

Consultation and Communication

8.353 The organisation needs to involve employees and, where appropriate, their representatives in the carrying out of risk assessments and in the selection and implementation of the preventive and protective measures. It also needs to establish effective means of communication and consultation that makes clear the commitment to health and safety.

8.354 This communication will ensure that employees receive sufficient information to implement effectively the control measures that have been introduced.

Provision of Information and Training

8.355 The provision and evaluation of information, instruction and training should also ensure the competence of employees, particularly those involved in risk assessment and the selection and implementation of the preventive and protective measures.

8.356 Where necessary this competence will need to be supported by the provision of health and safety assistance or advice.

Control

8.357 In order to establish effective control, an organisation needs to:

- clarify health and safety responsibilities and ensure those with such responsibilities clearly understand what they need to do and have the time and resources needed to discharge them
- co-ordinate the activities of various individuals
- set standards to judge the performance of those with health and safety responsibilities
- ensure adequate supervision is provided.

Monitoring

8.358 Organisations need to monitor:

- how well their health and safety policy is being implemented
- how well identified risks are being controlled
- how successfully a positive safety culture is being developed.

8.359 This will include both proactive and reactive monitoring, eg:

- planning and carrying out routine inspections to check the effectiveness of the preventive and protective measures
- investigating the immediate, underlying and root causes of incidents and accidents to ensure that remedial action is taken
- recording and analysing the findings of proactive and active monitoring in order to identify themes or trends.

Review

8.360 Organisations should carry out reviews in the following circumstances.

- In order to establish priorities for any remedial action identified as a result of proactive or reactive monitoring.
- In order to periodically check the effectiveness of the entire health and safety management system.

A Model for a Health and Safety Management System

8.361 The HSE's HSG65 *Successful Health and Safety Management* suggests that the following key elements need to be incorporated into an effective health and safety management system.

- Policy. Setting a clear direction for the organisation to follow in terms of health and safety.
- Organising. Ensuring that all employees, at all levels, are motivated and empowered to work safely. This relies on effective involvement and participation and is sustained by effective communication and the promotion of competence. A positive health and safety culture needs to be fostered through the visible and active leadership of senior managers.
- Planning and Implementing. Risk assessment methods should be used to decide priorities and to set objectives for eliminating hazards and reducing risks. Performance standards need to be established and used for measuring achievement.
- Measuring Performance. Performance needs to be measured against agreed standards through the use of proactive and reactive monitoring techniques. The monitoring needs to consider hardware, in terms of premises, plant and substances, and software, in terms of people, procedures and systems.

Additionally, the monitoring needs to consider individual behaviour and performance, including that of senior managers in respect of health and safety. The objectives of the monitoring are to determine the immediate causes of sub-standard performance, and to identify the underlying causes and their implications for the design and operation of the health and safety management system.

- Auditing and Reviewing Performance. A systematic review of performance, based on the information obtained from monitoring and auditing, is required if an organisation is to learn from all relevant experience. This not only ensures that the organisation adopts a philosophy of continuous improvement but also forms the basis of self-regulation implied in the Health and Safety at Work, etc Act 1974. The results of the review can be incorporated into any annual reports that are required to be produced. This is also in line with the requirements of the Turnbull Report.

Securing Commitment at Board and Senior Executive Level

8.362 In order for an organisation to effectively manage health and safety in a way that will offer a defence against a possible prosecution under the Act, it is essential that those at the top give a positive commitment to any changes that need to be implemented. Only then will those occupying other managerial levels follow suit.

8.363 In order to secure such a commitment from those at board or senior executive level, the following approach could be adopted.

- The preparation and delivery of a presentation, at board or senior executive level, outlining the implications of the Act and providing detailed recommendations concerning the action that needs to be taken to defend the organisation.
- Preparation of a brief for a board or senior executives meeting that, in order to focus their attention and interest, outlines the likely outcomes of a failure to defend a prosecution under the Act, such as the following.
  - Fines may be in the order of 2.5% to 10% of annual turnover if the recommendations of the Sentencing Advisory Panel are followed. Where the organisation pleads not guilty, the starting point may be 5% for a first offence. Turnover is calculated from the average of three years. The panel suggest that the fine imposed needs to be set at a level significantly higher than for an offence involving a death, under the Health and Safety at Work, etc Act 1974, because the new offence involves more serious instances of management failure. The sentencing guidelines will lead to far more significant fines as the largest current fine under health and safety legislation is the £15 million awarded against Transco (in 2005, for safety breaches that led to the deaths of four members of a family), which represented less than 1% of the organisation's annual turnover.
  - Brand image could be seriously damaged if the courts impose a publicity order on every organisation convicted as has been suggested. This might require publicity on television, in local and national newspapers, in the trade press and in notices to shareholders or letters to customers.
- Preparation of a report for consumption at board or senior executive level that reviews the effectiveness of the existing health and safety management system. This should consider the following.
  - Whether the policy is up-to-date and translated effectively into action.
  - Whether health and safety roles and responsibilities are clearly identified at all levels and whether those allocated with these roles and responsibilities understand them and have the competencies to discharge them effectively.
  - All relevant health and safety legislation, guidance and standards and how well they are being complied with.
  - The status of risk assessments — have they been completed and are the reviews up-to-date?

- The disciplinary records relating to rule-breaking, etc.
- Whether key performance indicators are established and are being met.
- Whether training needs analysis has been undertaken, training provided and training records kept.
- The effectiveness of the proactive and reactive monitoring regimes with regard to inspections, safety sampling, accident and incident investigations, etc.
- Whether all identified remedial work has been actioned and follow-up monitoring instigated.
- The effectiveness of the auditing and review process.
- Whether current insurance arrangements are sufficient to cover possible defence costs.

- Preparation of costed and prioritised recommendations for action that needs to be taken, for consideration at a board or senior executives meeting. These recommendations are based on a gap analysis that compares what evidence will be required to successfully defend a prosecution under the Act with the evidence that can be provided by the present health and safety management system. In this way, cost-effective decisions can be taken based on analysis of both the costs and benefits involved.

Training

8.364 The provision and evaluation of information, instruction and training should also ensure the competence of employees, particularly those involved in risk assessment, and the selection and implementation of the preventive and protective measures.

8.365 Where necessary, this competence will need to be supported by the provision of health and safety assistance or advice.

List of Relevant Legislation

- Management of Health and Safety at Work Regulations 1999
- Corporate Manslaughter and Corporate Homicide Act 2007
- Health and Safety at Work, etc Act 1974

Further Information

Publications

HSE Publications

The following are available from *www.hsebooks.com*.

- HSG65 (rev 1997) *Successful Health and Safety Management*
- INDG417 *Leading Health and Safety at Work. Leadership Actions for Directors and Board Members*

- L21 (rev 2000) *Management of Health and Safety at Work. Management of Health and Safety at Work Regulations 1999. Approved Code of Practice and Guidance*

Other Publications

- *Internal Control: Guidance for Directors on the Combined Code*, Institute of Chartered Accountants for England and Wales

Organisations

- Centre for Corporate Accountability
  Web: *www.corporateaccountability.org*
  The Centre for Corporate Accountability is a charitable organisation that monitors worker and public safety, and undertakes research and provides advice on law enforcement and corporate criminal accountability.
- Health and Safety Executive (HSE)
  Web: *www.hse.gov.uk*
  The HSE is responsible for the regulation of almost all the risks to health and safety arising from work activity in the UK.
- Institute of Directors (IoD)
  Web: *www.iod.com*
  The IoD provides a professional network for the business community.

Directors' Responsibilities

- Directors have legal responsibilities for health and safety under the Health and Safety at Work, etc Act 1974, the Fire Precautions Act 1971 and the Companies Act 1985.
- Directors can be prosecuted under these Acts, under relevant subsidiary legislation and under common law for manslaughter.
- Directors should ensure that health and safety is included in the overall corporate social responsibility programme.
- Directors should provide health and safety leadership and accept their individual roles in providing this leadership.
- Directors should actively support the participation of employees in improving health and safety and be kept informed of health and safety risk management issues.
- An organisation's competent person can use a number of arguments to ensure that directors are committed to health and safety.

8.366 Directors have a critical role to play in ensuring that health and safety is properly managed and that risks are controlled. The Health and Safety Commission (now the HSE) has been taking action to motivate boardrooms to improve health and safety for some time. It is now recognised that boardrooms and directors need to take greater responsibility for health and safety within their organisations.

8.367 The most fundamental factor affecting health and safety performance is what is commonly referred to as "safety culture". There is little doubt that the strongest influence on safety culture within most businesses is "the message from the top". The directors of a company send out signals, either consciously or subconsciously, of what they expect of their employees. Strong leadership from the top is vital in delivering effective control of risks to health and safety. Directors need to be made aware of the need to ensure that health and safety is taken into account when business decisions are taken.

In Practice

Legal Responsibilities

Health and Safety at Work, etc Act 1974

8.368 Under s.37, if an offence is committed with the consent or connivance of, or if an offence is attributable to any neglect on the part of, any:

- director
- manager
- company secretary
- other similar officer of the body corporate
- any person acting in one of these capacities
- they, as well as the body corporate, shall be guilty of the offence and can be proceeded against and punished accordingly.

8.369 This also applies where the affairs of a body corporate are managed by its members. The acts and defaults of a member in connection with his or her functions of management can be treated as if he or she were a director of the body corporate.

Companies Act 1985

8.370 According to the Companies Act 1985 (as amended by the Companies Act 2006), the Secretary of State may prescribe cases whereby directors' reports will contain information about the arrangements in force for that year for:

- securing the health, safety and welfare at work of the employees of that company (and any subsidiary company)
- protecting other persons against risks to health resulting from the activities at work of the employees.

Company Directors Disqualification Act 1986

8.371 The court may disqualify a person from being a director of a company if he or she is convicted on an indictable offence, whether on indictment or summarily, if the offence was in connection with the management of the company.

8.372 A magistrates' court may impose a disqualification for up to 5 years, whilst higher courts may disqualify for up to 15 years.
Common Law: under common law, a director may be prosecuted for manslaughter.

Corporate Social Responsibility

8.373 There is increasing pressure on businesses to:

- promote greater transparency in corporate reporting and in the marketplace
- answer public concern about directors' and organisations' behaviour
- stimulate good practice.

8.374 The Government's strategy document, *Revitalising Health and Safety*, recognises that health and safety management needs to be set in the wider context of corporate social responsibility. Thereby it is seen as one of the indicators of a company's performance. This has been seen as an opportunity to work with investors and stakeholders to promote health and safety and to establish it as an important factor in corporate social responsibility. This has potential for influencing health and safety performance.

8.375 Corporate social responsibility and investors'/stakeholders' influences now impact on health and safety. Thus those with corporate responsibility (ie directors) should be aware of how the safety performance of the organisation and the commitment given to health and safety by directors will impact on their business functions.

8.376 As part of this process, the Health and Safety Executive launched the Corporate Health and Safety Performance Indicator (CHaSPI), which aims to help in the assessment of how well an organisation manages its risks and responsibilities towards its workers, the public and other stakeholders.

8.377 There is a clear opportunity for companies and directors to demonstrate their commitment to corporate social responsibility through proactive occupational safety and health policies. There is also an opportunity to win over consumers at the same time.

Prosecution of Directors

8.378 Directors, managers and company secretaries are personally under a duty to ensure that the company's statutory duties are performed. Prosecution will normally occur under s.37 of the Health and Safety at Work, etc Act 1974 but other legislation may be employed. Directors may also be prosecuted under common law for manslaughter.

8.379 It should be noted that anyone who acts in a managerial capacity may be held liable under s.37 of the Health and Safety at Work, etc Act 1974 whatever title he or she may have. For example, if the affairs of the body corporate are being managed by its members (eg a workers' co-operative) the acts of a member which are in connection with his or her managerial functions are within the meaning of s.37.

8.380 Persons who purport to act as directors, managers, secretaries or similar officers are also equally liable. Thus if a person acts as a director even though he or she has been disqualified from doing so under the Companies Act 1985, he or she is purporting to act as such.

Appointment of Third Parties

8.381 A director does not escape responsibility for health and safety matters by appointing a third party to carry out the necessary duties. The director must make sure that an independent third party is competent and properly supervised. However, the director still has to accept greater responsibility for the safety of employees and others than, eg those who are engaged on a part-time basis to assist.

Offences

8.382 An offence will be committed if:

- a director consents to the commission of an offence, ie he or she is well aware of what is going on and agrees to it
- a director connives to an offence, ie he or she is well aware of what is going on and his or her agreement is tacit, not actively encouraging of what happens but letting it continue and saying nothing about it
- a director is under a duty to do something and fails to do it, or does it in a negligent manner. In this case an act of neglect is committed.

8.383 Directors may also be disqualified for breaches of safety legislation (eg a failure to comply with a prohibition notice). The disqualification can be in addition to any other penalty imposed.

HSE Guidance on Directors' Responsibilities

8.384 The HSE's publication *Successful Health and Safety Management* (HSG65) highlights the need for senior management to take the lead. It states that "visible and active support, strong leadership and commitment of senior managers and directors are fundamental to the success of health and safety management".

8.385 Boardrooms and directors are expected to take a number of actions in respect of health and safety.

- The board is expected to accept its collective role in providing health and safety leadership within the organisation. This should be done formally and in public.
- The individual members of the board need to accept individual roles in providing health and safety leadership.

- Board decisions need to reflect health and safety intentions, as articulated in the health and safety policy statement.
- The active participation of employees in improving health and safety must be recognised.
- The board should ensure that it is kept informed of, and alert to, health and safety risk management issues.

8.386 In support of the guidance *Directors' Responsibilities for Health and Safety*, the HSE has also produced *Director Leadership — Case Studies* highlighting the vital role that directors play in health and safety. These can be accessed via the HSE website.

Importance of Strong Leadership

8.387 Strong leadership is vital in delivering effective health and safety risk control. Everyone should know and believe that directors are committed to continuous improvements in health and safety performance. Expectations must be explained along with the procedures to deliver them.

8.388 Board members need to ensure that actions and decisions at work always reinforce the messages in the health and safety policy statement. Any mismatch between individual attitudes, behaviour and the organisation's health and safety policy will undermine good health and safety practice and workers' beliefs. A director's personal responsibilities and liabilities under health and safety law must be recognised.

New Plant, Premises, Processes or Products

8.389 Many business decisions will have health and safety implications. It is particularly important that the health and safety ramifications of investment in new plant, premises, processes or products are taken into account as decisions are made.

Legal Responsibility Rests with the Employer

8.390 It is important for boards to remember that, although health and safety functions can (and should) be delegated, legal responsibility for health and safety rests with the employer.

Active Participation of Workers

8.391 Effective health and safety risk management requires the active participation of workers. Directors should encourage workers at all levels to become actively involved in all aspects of the health and safety management system. Worker involvement supports a positive health and safety culture where health and safety is everyone's business. The best form of participation is a partnership for prevention, where workers and their representatives are involved in identifying and tackling potential or actual problems. This is much better than consulting workers and their representatives only after decisions have already been taken.

Appointing a Health and Safety Director

8.392 Appointing a "health and safety director" creates a board member who can ensure that health and safety risk management issues are properly addressed, both by the board and by the organisation as a whole.

8.393 The chairman and/or chief executive have a critical role to play in ensuring that risks are properly managed and that the health and safety director has the necessary competence, resources and support of other board members to carry out his or her functions.

8.394 It is important that the role of the health and safety director should not detract either from the responsibilities of other directors for specific areas of health and safety risk management or from the health and safety responsibilities of the board as a whole.

Health and Safety Policy

8.395 The health and safety responsibilities of all board members should be clearly articulated in the organisation's statement of health and safety policy and arrangements.

8.396 The Health and Safety Commission suggests that any board and directors will need to:

- review the health and safety performance regularly
- ensure that health and safety policy statements reflect current board priorities
- ensure that management systems provide effective monitoring and reporting procedures
- be kept informed about significant health and safety failures and of the outcome of the investigations into their causes
- ensure that implications in respect of health and safety are addressed in all decisions
- ensure that risk management systems for health and safety are in place and effective. Periodic audits can provide information on their operation and effectiveness.

Getting Directors' Commitment

8.397 One of the challenges to be faced is convincing staff at a senior level that health and safety is a boardroom issue and that those with corporate roles do have a responsibility for health and safety both collectively and as individuals.

8.398 Senior staff need to ensure that the actions and decisions they take reinforce the safety message. An expressed attitude needs to be supported by observed action. A mismatch will undermine employees' trust and belief and thus make the task of implementing an effective safety management system that much harder. Visible management and good communication are essential.

8.399 Commitment by directors will increase motivation and concern throughout the organisation for health and safety. It is best indicated by the proportion of resources and support allocated, including time, staff and finances. Typically this will be done by:

- arranging for a director or the chairman to sign the health and safety policy
- arranging for a senior director to chair the committee on safety
- ensuring that sufficient financial resources are allocated to health and safety
- ensuring that staff are fully involved in health and safety
- appointing and training sufficient competent persons.

8.400 For success to be achieved, it is crucial that the competent person within the organisation must:

- wield control and authority, with the support of the board of directors
- have a direct line of communication to directors on matters of policy.

Arguments for Health and Safety

8.401 Those with health and safety responsibilities will be required to support their case for safety and the application of risk reduction measures. Highlighting legal responsibilities and subsequent case law can be a convincing tool to utilise when striving to make health and safety a corporate issue. Clearly a balance must be drawn between health and safety and business needs. Health and safety should be seen as a business asset by directors. The following arguments can be used for convincing directors to take responsibility for health and safety. They can also be used to show how health and safety can be of benefit to the business.

8.402 Positive arguments include the following.

- The reduction in exposure of the company and key personnel to criminal liability.
- The minimising of the likelihood of prosecution and consequent penalties.
- Reduction in the risk of directors' disqualification from office.
- Reduction in exposure of the company and senior managers to civil liability for death or injury.
- Control over insured losses to the satisfaction of insurers to ensure that policy cover remains valid.
- Reduction in financial risks and improved control over uninsured losses (such as lost production time, repair costs to plant and equipment, replacement costs for spoiled materials, rework costs for damaged products).
- Control over risks of loss in market share.
- Improvement of the organisation's reputation in the eyes of customers, competitors, suppliers, other stakeholders and the wider community.
- Reduction in risks to company share values, etc.

Training

8.403 Training for directors should concentrate on the following topics.

- The legal responsibilities of directors.
- The potential outcomes for the individual if prosecution occurs (to include case law).
- The need to integrate health and safety in all business decisions.
- The benefits of health and safety to the overall business function.
- The actions expected of directors in respect of health and safety.
- How directors can show commitment to health and safety.

List of Relevant Legislation

- Regulatory Reform (Fire Safety) Order 2005
- Management of Health and Safety at Work Regulations 1999
- Health and Safety (Consultation with Employees) Regulations 1996
- Safety Representatives and Safety Committees Regulations 1977
- Company Directors Disqualification Act 1986
- Companies Act 1985
- Health and Safety at Work, etc Act 1974

Further Information

Publications

HSE Publications

The following are available from *www.hsebooks.co.uk*.

- HSG65 (rev 1997) *Successful Health and Safety Management*
- INDG343 *Directors' Responsibilities for Health and Safety*

Other Publications

- *Internal Control: Guidance for Directors on the Combined Code*, Institute of Chartered Accountants for England and Wales
- The Corporate Health and Safety Performance Indicator can be accessed at *www.chaspi.info-exchange.com/default.asp.*
- *Director Leadership — Case Studies* can be accessed at *www.hse.gov.uk/corporateresponsibility/casestudies.*

Emergency Planning

- How an organisation responds to an emergency will hold sway with a jury when deciding whether or not a gross management failing has taken place.
- The Corporate Manslaughter and Corporate Homicide Act does not impose a duty of care on the emergency services to the victim when they are responding to a life-threatening incident.
- The King's Cross Underground station fire, the capsizing of the Herald of Free Enterprise at Zeebrugge, and the fire on the Piper Alpha North Sea platform illustrate how emergency procedures have been found wanting during fatal incidents.
- If an organisation's emergency procedures are good enough to prevent an incident becoming a fatality, the company cannot be prosecuted under the Act.
- For the vast majority of businesses, their emergency procedures need mainly to address fire precautions and first aid.
- Planning for emergencies should encompass a wide range of potential accidents, incidents and scenarios.
- Identifying hazards in the workplace is vital to implementing successful emergency procedures.
- Specific hazards will need specific procedures in place in the event of an emergency.
- Chemical hazards are covered by specific legislation.
- Planning should encompass consideration of what internal and external resources are needed.
- All emergency procedures should be integrated into an overall emergency plan.
- Emergency planning should consider the aftermath of an emergency, eg measures to achieve a swift return to normal working.
- Once the plan has been prepared, informing and training staff and others who may come onto the premises is vital.

8.404 The lack of suitable emergency plans has frequently caused companies loss of competent staff and/or damage to property with costs of restitution. In addition to the threat of prosecution for corporate manslaughter, the disruption caused is likely to have had impacts on productivity and profitability and any adverse publicity may have also taken its toll on demand for their product.

Employers' Duties

Corporate Manslaughter and Corporate Homicide Act 2007

8.405 The new Corporate Manslaughter and Corporate Homicide Act 2007 places senior managers and organisations at risk of prosecution if their gross management failings lead to the death of an employee or a member of the public to whom a duty of care is owed.

8.406 The legislation states that there has to be a relevant duty of care owed by the organisation in the law of negligence to the deceased and a gross breach of that relevant duty of care leading to that person's death.

8.407 When considering whether or not a gross breach has taken place, the Act requires juries to consider, among other matters:

- failure to comply with any relevant health and safety legislation and, where this occurs, the seriousness of the failure and the risk of death it posed
- attitudes, policies, systems or accepted practices likely to have encouraged the failure to comply with the relevant health and safety legislation, or led to a tolerance of this failure
- any relevant health and safety guidance
- any other matters they consider relevant.

8.408 Thus, how an organisation prepares for an emergency can directly impact on how liable it is under the Act.

Specific Health and Safety Legislation

8.409 The Health and Safety at Work, etc Act 1974 requires employers to ensure "so far as is reasonably practicable" the health, safety and welfare of their employees and other persons who may be affected by their work. This includes sub-contractors and the general public.

8.410 Under the Management of Health and Safety at Work Regulations 1999, employers must:

- undertake suitable and sufficient assessments of the risks to employees and non-employees
- implement procedures for serious and imminent danger and for danger areas
- establish and implement procedures
- nominate competent persons to implement evacuation procedures (eg fire marshals)
- liaise with external services for first aid, emergency medical care and rescue work.

8.411 Under the Regulatory Reform (Fire Safety) Order 2005, employers must:

- carry out a fire risk assessment identifying possible dangers and risks
- consider who may be especially at risk
- eliminate or reduce the risk from fire as far as is reasonably possible and provide general fire precautions (including fire detection and fire alarms, if appropriate) to deal with any possible risk left
- take other measures to make sure there is protection if flammable or explosive materials are used or stored
- create a plan to deal with any emergency and, in most cases, keep a record of the findings
- review the findings when necessary.

8.412 Under the Health and Safety (First-aid) Regulations 1981, employers must:

- provide equipment and facilities for enabling first aid to be rendered to employees; in most workplaces, this includes provision of trained first aiders
- inform employees of the arrangements that have been made.

8.413 The Control of Major Accident Hazards Regulations 1999 (COMAH) require that operators "take all measures necessary to prevent major accidents and limit their consequences to persons and the environment".

Employees' Duties

8.414 The Health and Safety at Work, etc Act 1974 states that employees must take reasonable care for the safety of themselves and of other persons who may be affected by their acts or omissions. They should co-operate with their employers and others in carrying out their statutory obligations.

8.415 There are no specific duties on employees in respect of emergency planning.

8.416 Self-employed persons should provide appropriate equipment to render first aid to themselves while at work.

In Practice

Reasons for Emergency Preparedness

8.417 The new Corporate Manslaughter and Corporate Homicide Act 2007 (the Act) places organisations at risk of prosecution if gross management failings lead to the death of an employee or a member of the public to whom a duty of care is owed. The legislation states that there has to be a relevant duty of care owed by the organisation in the law of negligence to the deceased and a gross breach of that relevant duty of care leading to that person's death.

8.418 When considering whether or not a gross breach has taken place, the Act requires juries to consider, among other matters:

- failure to comply with any relevant health and safety legislation and, where this occurs, the seriousness of the failure and the risk of death it posed
- attitudes, policies, systems or accepted practices likely to have encouraged the failure to comply with the relevant health and safety legislation, or led to a tolerance of this failure
- any relevant health and safety guidance
- any other matters they consider relevant.

8.419 How an organisation responds to an emergency will hold sway with a jury when deciding whether or not a gross management failing has taken place.

8.420 Further, as with most occupational health and safety matters, the reasons for taking action are combinations of both financial and moral factors, together with statutory duties. Planning for emergencies and training staff in the implementation of a plan and procedures will:

- reduce the time taken to respond to an incident
- assist the organisation in meeting its legal and moral responsibilities
- enable rapid recovery from any incident and assist with business continuity
- help mitigate any losses such as lost business, insurance premium increases, damage to property and goods, etc
- reduce exposure to liability, both criminal and civil
- improve the public image of the organisation.

Ambit of the Legislation

Emergency Services and the Act

8.421 Section 6 of the Act does not impose a duty of care on the emergency services to the victim when they are responding to a life-threatening incident. For example, if someone is trapped in a building by a fire and the response of the fire and rescue authority fails to save the person, that incident lies outside the scope of this legislation. Rescue operations involving NHS personnel, the coastguard and the Royal National Lifeboat Institution are similarly exempted.

8.422 However, if the fire engine responding to an incident had not been adequately maintained, and was shown to have caused the vehicle to be involved in a road traffic accident killing one of the crew, then a charge of corporate manslaughter might be brought against the fire authority.

8.423 However, since the implementation date of most provisions of the Corporate Manslaughter and Corporate Homicide Act 2007 was 6 April 2008, the Act cannot be used in the case of the deaths of four part-time firefighters in a warehouse fire in Warwickshire in 2007, where questions were raised about the wisdom of the decision to send the men into the building.

8.424 Also, while matters relating to the organisation and management of medical services fall within the scope of the Act, the eventual interpretation of this by the courts is difficult to predict.

Examples of Poor Contingency Planning

8.425 The King's Cross Underground station fire, the capsizing of the Herald of Free Enterprise at Zeebrugge, and the fire on the Piper Alpha North Sea platform are all cases where better emergency response could have, to some extent, mitigated the death toll.

8.426 While the initiating events, and the responsibility for them, were probably of greater significance, the purpose of contingency planning is to recognise that sometimes things go wrong (for whatever reason) and to put in place procedures to minimise the chance of fatalities.

8.427 The following examples illustrate how emergency procedures have been found wanting during fatal incidents. The emergencies have all arisen due to shortcomings in the management of the companies' normal activities, although deficiencies in their emergency procedures have also been identified.

King's Cross Underground Station Fire

8.428 At King's Cross, the probable cause of the fire was a discarded cigarette falling onto detritus that had accumulated under an escalator. In response to this, smoking on the Underground was banned (long before the smoking ban was introduced for health reasons in all public buildings) and a programme of improvements to escalators at many stations was also carried out.

8.429 The official report on the incident also found many deficiencies in the emergency procedures. A fire was a foreseeable occurrence (in fact there had been several), and adequate plans should have been developed to evacuate passengers safely. This in particular needed trained staff to implement.

8.430 In total there were 157 recommendations, which included the following.

- Station instructions for emergencies and closure must be agreed with the London Fire Brigade and used in training station staff.

- Fire hydrants and cabinets must be marked with outrigger signs.
- A rendezvous point for the emergency services and a staff assembly point must be agreed and marked.
- Station evacuation plans should include evacuation by train.
- Water fog equipment must be regularly tested and staff trained in its use.
- In agreement with the London Fire Brigade, London Underground shall produce and maintain up-to-date station plans, and place them in boxes it has provided, at locations agreed or specified by the London Fire Brigade.
- Every two years, all management and supervisory staff shall receive refresher training in controlling station emergencies and the use of fire and communications equipment.
- Every six months, fire and safety training must be provided for non-supervisory staff and booking clerks. Staff must be given site familiarisation training before they are permitted to take part in the running of the station.
- Instructions to the staff as to the calling of the fire brigade shall be re-drafted in plain English.

8.431 There were many shortcomings in emergency preparedness. Further, under legislation current at the time, a "controlling mind" had to be identified for a charge of manslaughter to be brought. However, the complications of a large organisation meant that this was impossible. Under the new Act, this would not be necessary, and the organisation rather than an individual could be prosecuted.

Herald of Free Enterprise

8.432 The case of the Herald of Free Enterprise is a sad one: there was a huge death toll despite the fact that the ship capsized within sight of the harbour and lay on her side rather than sinking.

8.433 The cause of the disaster was that the ship sailed with the bow doors open, and the reasons behind that would have been sufficient for a charge of corporate manslaughter had the new Act been in place at the time.

8.434 The facts of the case are that plans for abandoning ship should always be in place and rehearsed. Further, while their total effectiveness cannot be guaranteed, particularly in bad weather, the official report stated that not a single lifeboat was launched on the night of the disaster.

8.435 The scale of the disaster was due to rapid capsizing as soon as the sea flooded onto the car deck as a consequence of the ship's design. This flaw had never been properly considered, and even the best evacuation plans would have been severely tested under these circumstances.

Piper Alpha Platform

8.436 The Cullen report investigating the Piper Alpha platform fire concluded that the immediate cause of the fire was a leakage of flammable gas, as a result in major failures in the design and operation of safety systems. The automatic deluge system had been deactivated and no message was given over the public address system.

8.437 The limitations of the emergency plans may also owe much to the design of the rig rather than their execution, with men trapped in the accommodation block. The report drew attention to basic failures to show all newcomers the location of their lifeboats, and that evacuation drills and exercises were not carried out with the required frequency or in the required way.

Industrial Accidents

8.438 There are also various Health and Safety Executive (HSE) reports on a number of industrial accidents in the chemical industry indicating shortcomings in emergency arrangements.

8.439 An explosion causing fatalities occurred at the site of Hickson and Welsh in Castleford, killing four workers in the control room. A fifth victim was a young officer worker. The incident occurred at lunchtime when it was unclear which members of staff had left the site. There was a delay before it became known to the emergency services precisely where to search for the fifth person, who, when located, had been overcome by smoke. The smoke had spread above a false ceiling and through breaches in the fire-resisting construction. She died later in hospital.

8.440 One of the "lessons" identified in the report produced by the HSE states: "When exercising their on-site emergency plans companies should ensure that roll call information on missing persons is passed immediately, accurately and directly to the senior fire officer in charge. Roll call procedures should be practised routinely to ensure that they are effective when carried out at all periods of the working day."

8.441 The company was convicted of an offence under the Health and Safety at Work, etc Act 1974.

Developing Emergency Procedures under the Act

8.442 Developing emergency procedures is a double-edged sword so far as the Corporate Manslaughter and Corporate Homicide Act 2007 is concerned.

8.443 Firstly, no matter what regulations have been contravened leading to an incident, if an organisation's emergency procedures are good enough to prevent an incident becoming a fatality, the company cannot be prosecuted under the Act.

8.444 Secondly, if a fatality does occur, the organisation's procedures will be put under the microscope, and any positives might just swing the balance when the decision on whether to prosecute under this new Act (or alternatively under some less serious regulation) is made.

Management of Health and Safety at Work Regulations

8.445 The general requirement for emergency procedures is found in the Management of Health and Safety at Work Regulations 1999 (MHSWR), Regulation 8: procedures for serious and imminent danger and for danger areas. This requires employers to establish and implement procedures and to nominate competent persons to implement evacuation procedures (eg fire marshals).

8.446 Regulation 9 requires employers to liaise with external services for first aid, emergency medical care and rescue work. The accompanying Approved Code of Practice (ACOP) indicates the content of the procedures including:

- the nature of the risk and how to respond to it
- the additional responsibilities of any employees who have specific tasks to perform in the event of an emergency
- the role, responsibilities and authority of the competent people nominated to implement the detailed actions
- any requirements laid on employers by health and safety regulations covering specific emergency situations
- details of when and how the procedures are to be activated so that employees can proceed in good time to a place of safety.

Developing an Emergency Plan

8.447 In the smallest organisations, the emergency plan can be very simple and prepared internally, using common sense with observations of what is done elsewhere, readily available advice from various organisations, and perhaps supplemented by some basic health and safety training.

8.448 On the other hand, in high-risk organisations, the planning may require expert consultants, and liaison with not only emergency services but also neighbours. Even so, simplicity and user-friendliness should be retained; in an emergency nobody has time to study complex procedures.

8.449 Consideration will have to be given to setting up a control centre, and who will take charge — including during shift work when senior management must first be called from home. Equipment may be needed, not only for fire fighting, but also for protecting people and the environment.

8.450 The stages in developing emergency plans are as follows.

- Carry out a risk assessment to identify threats.
- Develop a general strategy allowing for what resources can be made available.
- Prepare specific procedures for principal scenarios.
- Co-ordinate procedures into a common framework that will be adaptable to all situations.
- Validate and test the plans.
- Review and revise the plans.

8.451 When developing emergency plans, several principles should be kept in mind. Namely, the plan should:

- respond to the incident and not the cause of the incident
- be integrated into the organisational structure and management systems
- be flexible to handle any incident
- include co-operation and co-ordination with other organisations and agencies
- identify and provide the necessary resources to ensure recovery and give an estimate of recovery time
- take account of the core and critical functions required to ensure recovery
- have the support of senior management.

Risk Assessment

8.452 The range of events that may be taken into account by contingency planning is vast. The starting point for any planning process will be the risk assessment, to determine the most foreseeable and likely events that may blight the organisation. Risk assessment will certainly aid in the planning process.

8.453 Anticipating the effect of emergencies on an organisation and devising a response may also include other risk management techniques, eg:

- impact analysis to estimate the costs to the organisation
- a vulnerability study to determine the parts of the organisation that are most vulnerable to disaster
- a threat scenario to assess the type of contingency most likely to effect the organisation.

8.454 Procedures should be developed to deal with a full range of scenarios, but they should focus on those scenarios that are assessed as of the highest risk.

Identifying Hazards

8.455 The first stage of any risk assessment is the identification of hazards. In the case of emergency planning, this is often expressed in terms of an analysis of threats. Factors which lead to an incident might be internal or external. *Table 1 — Emergency Threats* lists some possible hazards and causes.

Table 1 — Emergency Threats

Scenario	Internal Cause	External Cause
Fire	Faulty electrical equipment	Lightning strike
Flood	Overflow of tank	Heavy rain
Chemical pollution	Contamination by leakage of liquid chemical from storage tank	Toxic chemical gas cloud from adjoining premises
Militant action	Industrial action by employees	Protest action by pressure group
Disease	Legionella bacteria in water system	Introduced by livestock

8.456 The response to a scenario might well be the same, irrespective of whether the causative factor is internal or external. However, to avoid having to make the responses in the first place, emergency planners should always consider whether everything possible has been done to prevent the incident happening. For example, while procedures will be necessary in case of fire, adequate maintenance procedures of electrical equipment and lightning conductors should be in place to reduce the likelihood of fire.

Risk Management

8.457 When risks have been assessed there will be choices to be made as to the management strategy to deal with them.

8.458 Insurance will play a large part in the plans of many organisations, complementing the emergency procedures. Insurance might go beyond cover for property damage and personal injury and incorporate cover for loss of productivity. Some organisations also subscribe to schemes whereby they have an option on a suite of offices to which they can relocate if necessary, or schemes that provide them with alternative information technology (IT) facilities if their own fail.

Procedures for Principal Scenarios

Business Crises

8.459 For the vast majority of businesses, their emergency procedures need mainly to address fire precautions and first aid. However, emergency planning is not simply about evacuating premises if a fire breaks out.

8.460 The range of scenarios that have affected businesses in recent years include:

- terrorist incidents
- computer viruses
- petrol shortages
- transport accidents
- severe weather causing flooding, power loss, etc

- war
- protests such as public marches and demonstrations
- foot and mouth disease.

8.461 Some of these occurrences fall into what might be described as business crises. Many of the actions needed to recover from them are similar to the actions needed when handling the aftermath of a fire or flood.

8.462 Workplace emergency planning should focus primarily on emergencies that:
- threaten the safety of employees
- can cause damage to the workplace
- despoil the environment.

8.463 In all cases, plans should deal with restoring normality as well as "putting out the fire".

First Aid

8.464 The Health and Safety (First-aid) Regulations 1981 and accompanying Approved Code of Practice (ACOP) require that a risk assessment is carried out to determine what first-aid facilities are appropriate in the circumstances.

8.465 Some small premises, especially if near to medical facilities, may need nothing more than an appointed person to take control of the situation if someone needs assistance. Larger premises, and high-risk activities are likely to require at least one trained first aider present at all times, including cover for absences, shift-working, etc.

8.466 Where specific risks are identified, first aiders may need additional training, eg in the use of antidotes for phenol or hydrogen fluoride contamination.

8.467 On the basis of this, the appropriate resources should be provided. Selected personnel should be trained and first-aid boxes (as a minimum measure) provided. Locations of first-aid boxes should be clearly marked. All employees must be informed of the arrangements made.

8.468 All incidents and illnesses requiring first-aid treatment should be recorded in a book, indicating:
- the date, time and place of incident
- the name of person treated
- the work being undertaken by the victim at the time of injury
- the nature of the injury
- the treatment applied
- whether the person returned to work/was sent home/was sent to hospital/was advised to consult a doctor
- the name and signature of the first aider.

8.469 This record might be combined with the record to be kept in the statutory accident book. Procedures should make clear how it is made known which first aiders are on duty and how they can be contacted.

Fire Procedures

8.470 Since the introduction of the Regulatory Reform (Fire Safety) Order 2005, there is a responsibility on employers to undertake a fire risk assessment. For some higher-risk premises, this was previously done with the assistance of the fire authority. The fire risk assessment is a good basis for reviewing or preparing emergency procedures.

8.471 The basic elements of a fire evacuation procedure are as follows.

- In larger premises the means of raising alarm will be by an installed system, perhaps triggered by smoke or heat detectors or manually activated. In some workplaces this system will include a warning system for deaf and hard of hearing persons. In smaller organisations, the alarm might be raised by shouting a warning or using a horn on a canister of compressed air.
- The fire brigade must be contacted. Some alarm systems will do this automatically, otherwise procedures must specify who is responsible for this action.
- Only persons who have been trained in the use of hand-held fire extinguishers should attempt to tackle the fire and even then only if the fire is small and safe to tackle.
- All employees, contractors and visitors should leave the premises by designated escape routes, without using lifts, turnstiles or revolving doors. These persons may be helped by the fire wardens and may be asked to assist in the evacuation of disabled people.
- Industrial processes might need to be shut down, if it is possible to do so without jeopardising personal safety. What to do should be stipulated in procedures and covered in training.
- Nobody should stop to collect personal belongings.
- Windows and doors should be closed. In particular, fire doors delay the spread of smoke and fire and are vital for maintaining a safe means of escape from the building.
- Employees and contractors should quickly make their way to their designated assembly points for registration. The procedures should define how registration is to be carried out and by whom. There should be some list against which to check names. It will be possible to account for visitors in premises where visitors have been signed in. However, in buildings to which the public has free access, reliance for this must be placed on checking by fire wardens and the searching of the building by the fire brigade.
- Details of any missing persons should be reported to the employer and to the fire brigade.
- If an incident is likely to be protracted, evacuees should be allowed to go to an alternative building and consideration should be given to their personal needs.
- Re-entry to the premises should only take place when authorised.

Bomb Threats

8.472 The increase in terrorism in recent years has dictated that most organisations now require procedures to deal with the possibility of explosive devices being planted on their premises, or of packages being delivered that contain explosive devices or chemical/biological agents.

8.473 If a telephone call is received indicating that a bomb has been planted, every effort should be made to obtain as much information as possible for a decision to be made on what action to take. A standard form should be circulated to anyone who receives outside calls to their workstation, indicating how to proceed.

8.474 Any such call should be immediately reported to the police. How the police respond will depend on a number of factors, such as the current level of alert and the type of premises involved. It should not be expected that if a building has been evacuated, the police will go in to see whether everything is safe. They will contact the army bomb disposal team if they perceive the risk as high, eg if the warning message contains a pre-notified code word.

8.475 If a telephoned bomb threat is received, there are three options.

- Evacuate the premises.
- Search the premises, but only evacuate if something suspicious is found.
- Do nothing.

Evacuating the Premises

8.476 Immediate evacuation is rarely appropriate unless the location of the device is specified and there is an indication that it will explode imminently.

8.477 However, if the employer decides that this option is appropriate, the list below suggests a typical sequence of reasonable actions. This list also indicates some differences between bomb threat and fire evacuations.

- The alarm should be sounded. Ideally this should be distinguishable from the fire alarm as different actions are required.
- The police should be informed of the bomb threat.
- All persons should leave the premises, taking their personal belongings with them, eg bags, cases and clothes.
- All doors and windows should be closed for fire evacuations, but should be left open for bomb threat evacuations. Closing the windows restricts the supply of oxygen to the fire, whereas opening them dissipates the blast effect from an explosion.
- Lifts must not be used in fire evacuations but may be used in bomb threat evacuations.
- The safe distances from danger must be greater in the event of bomb threat evacuations than in fire evacuations. The assembly point should be behind another building. It is also possible that refuges (safe havens) within other buildings are designated as an alternative to external assembly points.
- The designated bomb wardens should carry out a search of their zones for any suspect packages and should report as each area is clear. If a suspected item is discovered, the police will notify the bomb disposal team.

Searching the Premises

8.478 One reason for not evacuating a building is that if the location of the explosive device is not known, persons outside the building could be at greater risk than those indoors, eg if there is a car bomb.

8.479 Furthermore, if the building is evacuated, nobody can search for the bomb, whereas the most effective check is by the regular occupants of the building, as they are most likely to spot something unusual.

8.480 Obtaining co-operation in these circumstances is only likely to be successful if the procedure has been determined in advance with full consultation. Often "bomb wardens" are appointed from willing volunteers.

Taking No Action

8.481 The only justification for doing nothing is if there is very good reason for believing that the bomb threat is a hoax.

Other Suspect Packages

8.482 Warnings are not always given when bombs have been planted by terrorists. Therefore procedures are required for dealing with any suspect item that is discovered. Abandoned items of luggage at locations such as airports are a frequent source of alerts. Depending on the circumstances, threat level, etc, it might be appropriate to isolate the item whilst an attempt is made to trace its owner rather than evacuate the building immediately.

Mail Rooms

8.483 Mail rooms are a particular concern, being the likely point of entry for packages containing explosive devices or chemical/biological agents. Such devices are usually triggered only on opening the item, so isolating the item is usually the most appropriate initial step.

8.484 Items should not be immersed in water, as this may well short out an electrical triggering device.

8.485 Organisations perceived as being at high risk will probably acquire some screening equipment and check all incoming mail. Otherwise staff dealing with mail should be advised to be alert for:

- protruding wires
- grease marks
- a smell similar to marzipan
- heaviness relative to size/uneven weight distribution
- excessive fastening of the package
- packages addressed to a specific person, crudely hand-written
- unfamiliar postmark/no name of sender.

Chemical Incidents

8.486 Companies working with chemicals will need emergency procedures to deal with appropriate scenarios, such as shelter to avoid exposure to escapes of toxic gases. The Control of Substances Hazardous to Health Regulations 2002 (COSHH) set out the requirements for dealing with accidents, incidents and emergencies, expanding on the general requirements in the MHSWR.

8.487 Similar requirements are also found in the Dangerous Substances and Explosive Atmospheres Regulations 2002, which are concerned with the flammability and explosivity of chemicals and combustible dusts.

8.488 For those companies with inventories exceeding the thresholds in the Control of Major Accident Hazards Regulations 1999 (COMAH), the emergency planning requirements are even more onerous, including, for top tier sites, an off-site emergency plan prepared by the local emergency planning authority.

8.489 Guidance produced by the HSE for COMAH sites may also prove useful to other companies, as it describes the general issues and topics such as training and testing, initiation and review, in addition to the specific requirements of COMAH.

8.490 In general, procedures for dealing with chemical emergencies need to address the following.

- Appropriate types of fire extinguishants. Most fire brigades have only limited stocks of foam, so companies storing flammable liquids in bulk might need to hold some foam concentrate themselves.
- Availability of containment measures (absorbents, bunds, booms, etc) to protect the environment. This should include containment for water run-off from fire-fighting if it is contaminated with chemicals.
- Air monitoring to determine level of contamination.
- Personal protective equipment (PPE), including respiratory protective equipment, required by responders.
- Means of raising an alert, including whether there is a need for a different sounding alarm from those for fire or bomb threat. For sites with major accident hazard potential, a public warning system might be required to alert the neighbourhood, in which case residents will need to be informed in advance about the response required.

- Assembly points up-wind of any release of toxic materials, or safe havens within buildings.
- Access to safety data sheets.
- Treatment of affected persons. Local hospitals might need to be advised of the chemicals present and given data on antidotes, etc.
- Inventories and location plans for chemical storage.

8.491 Companies despatching dangerous chemicals will also need to have a procedure for dealing with reports of incidents during transport.

Flooding

8.492 Not all emergency situations are man-made, and flooding (though it can occur from overflowing tanks or burst pipes) is most often a consequence of inclement weather. Naturally, premises in low-lying areas are predominantly those at risk.

8.493 When determining the likelihood of flooding it is recommended that the history of flooding in the area is researched along with the geographical details of the area. A number of questions can be asked.

- Has the property or surrounding land ever flooded in the past?
- Have neighbouring properties ever flooded?
- Is the property in a floodplain?
- Have flood warnings ever been issued for the area in which the property is located?
- Is the property close to streams, rivers, drainage ditches?
- Are there any major water mains near to or under the premise?
- Is the property in a hollow or low-lying ground?
- Is the property protected by any flood defences?
- Are there issues in relation to groundwater overflow?

8.494 Measures that might be taken to avoid ingress of floodwater include:

- barriers created with sandbags, etc
- sealing gaps around doors and blocking air vents
- preventing backflow from sewers by blocking drains (a sandbag in a toilet is particularly effective).

8.495 Measures to minimise damage include:

- isolating electricity supply (and not restoring it until after it has been checked by an electrician)
- moving items upstairs or to higher ground
- weighting down articles that could float.

Integration of Emergency Procedures

8.496 The procedures required for dealing with various types of incidents should be brought together within an overall emergency plan. This ensures a co-ordinated and effective response irrespective of the nature of an event. A structure for command and control needs to be established. Also, the procedures for various incidents should be made to dovetail with the response of the public emergency services and other organisations that might be involved in the response to an incident.

8.497 For the plan to be effective, it is important that the organisation knows when and how to implement it. To ensure flexibility, criteria should be set to state when implementation needs to take place. It must be remembered that emergencies can occur at any time, not just in working hours, and as such consideration must be given to who can implement the plan.

8.498 The type of action taken will also be influenced by the amount of time given between warnings and an actual emergency occurring. The correct response to the emergency in the first instance can greatly help to mitigate the long-term consequences of the incident.

8.499 Planning for possible losses and then having in place procedures to carry out salvage work, damage control and assessing the organisational requirements to ensure recovery are vital.

8.500 The overall emergency plan should designate who has the authority to activate various steps and who from the organisation will have control of the response. It is helpful if tabards are worn by persons taking certain roles so that they are easily recognisable, eg the person delegated to conduct the roll call at the assembly point.

8.501 Larger organisations will have an emergency centre, from where the response will be co-ordinated by the person described in *Table 2 — Respective Duties of Incident Controllers* as the "site main controller". This will reduce the burden on the person at the forward control point (the "site incident controller"). The location of the control centre should be chosen to be of low vulnerability.

Table 2 — Respective Duties of Incident Controllers

Site Main Controller:	Site Incident Controller:
• based in emergency control centre • receives situation reports from site incident controller (or nominee) • calls up external resources • activates crisis management procedure • informs company headquarters • contacts relatives of injured persons • issues press statements and information to neighbours • arranges relief of personnel if an incident is protracted. In extreme cases it might be necessary to arrange for some senior staff to be sent home to rest so that they can return later and take over from their colleagues.	• located near scene of incident • assesses situation • orders evacuation • calls up internal resources • keeps control centre informed of situation (probably through a designated nominee) • ensures that casualties are receiving assistance • confirms all personnel are accounted for • liaises with emergency services officers at forward control.

Note: These lists are not exhaustive, and need to be adapted to the requirements of specific organisations.

Importance of Communication

8.502 Good communication is always critical in handling emergencies. This applies to communication at the time of the emergency and to communication beforehand.

8.503 Consultation should take place with organisations such as the emergency services. It is important that site emergency procedures dovetail with those of external responders. When the fire or ambulance service is called, someone should be designated to intercept them on arrival, brief them and direct them to the appropriate location.

8.504 Emergency procedures should take into account the different levels of manning that apply during shift work. There should be arrangements for contacting key personnel out of normal working hours.

Consideration of What Resources are Available

8.505 Good emergency procedures are also the product of consideration as to what resources are needed and where these are to be provided.

8.506 Several resources should be made available, as follows.

- Plans of site and buildings showing utility supplies with isolation points, drains, etc.
- Telephones: it might be necessary to have dedicated lines for certain purposes as switchboards can easily become overloaded.
- Other communication equipment.
- Fire-fighting equipment.
- First-aid supplies, stretcher.
- Chair for the evacuation of physically disabled persons.
- Chemical containment equipment, eg absorbent, drain covers, bunds, booms, drums.
- Means to cordon off an area, warning signs.
- PPE, breathing apparatus.
- Equipment for flooding, eg sandbags, pumps.
- Tabards or arm bands for key personnel so they are readily recognised.

When the Emergency is Over

8.507 The emergency plan should include dealing with the aftermath of an incident and may include:

- what checks have to be carried out before a building can be reoccupied
- preservation of evidence
- clean-up, salvage, waste disposal
- stand-down of responders, debriefing
- damage assessment
- information to staff (those present in the workplace, those deployed elsewhere and those off-duty at the time) and safety representatives
- reporting to authorities such as the HSE or Environment Agency
- advising insurers
- counselling of affected persons
- investigation of the incident
- public relations
- environmental restoration.

Is the Emergency Plan Easy to Find and Use?

8.508 Consideration should be given regarding how the plan is to be presented. Though it needs to be comprehensive, the plan should be kept as simple as possible. A contents list will help readers to find the appropriate section quickly. If the plan is only available in electronic format, it might not be accessible in an emergency if the power supply fails. If issued in hard copy, consideration should be given as to how it will be updated. Summaries or checklists for quick reference are useful, eg laminated cards that people can keep in their pockets or can be kept with emergency equipment.

8.509 The plan should incorporate information needed quickly in an emergency, such as telephone numbers for key contacts. These are usually in an appendix that can be easily updated without re-issuing the entire plan.

Information and Training

8.510 Once the plan has been prepared, informing and training staff and others who may come on the premises is vital. Drills and exercises must be conducted, initially to validate the plan and to familiarise everyone with it, but thereafter a regular schedule of tests should be carried out to ensure everything is kept up to date. The incident reports investigating disasters link to Examples of Poor Contingency Planning page 5 illustrate how often this is lacking.

Training

8.511 All employees should be given sufficient information and be trained in what to do in the event of an emergency. Drills and exercises are useful in this respect to ensure that everyone is familiar with the procedures in place.

8.512 Many employees and contractors will only require familiarisation with evacuation procedures and first-aid arrangements. They should be informed about these in an Induction course. They should participate in a fire drill at least once annually.

8.513 Certain employees might receive specific training (eg as first aiders) in using fire extinguishers, or on how to wear chemical protective suits. The training needs for all responders in the emergency procedures should be analysed and a programme set up that will give them opportunity to participate in exercises from time to time. These will involve a certain amount of simulation, or perhaps "table-top" exercises. Virtual reality exercises are used for simulating the stressful conditions that can be encountered by senior personnel managing a major incident.

8.514 Training should also be considered for any person who might be required as a spokesperson for the organisation and have to face questions in front of a television camera. Representatives of the local media might be invited to participate in an emergency exercise and carry out a mock interview.

8.515 Well-run training exercises build the confidence of the participants and develop their teamwork skills.

List of Relevant Legislation

- Regulatory Reform (Fire Safety) Order 2005
- Control of Substances Hazardous to Health Regulations 2002
- Dangerous Substances and Explosive Atmospheres Regulations 2002
- Control of Major Accident Hazards Regulations 1999
- Management of Health and Safety at Work Regulations 1999
- Health and Safety (First-aid) Regulations 1981
- Corporate Manslaughter and Corporate Homicide Act 2007
- Health and Safety at Work, etc Act 1974

Further Information

Publications

HSE Publications

The following are available from *www.hsebooks.co.uk*.

- HSG191 *Emergency Planning for Major Accidents*
- L5 (rev 2005) *The Control of Substances Hazardous to Health Regulations 2002 (as amended). Approved Code of Practice and Guidance*
- L21 (rev 2000) *Management of Health and Safety at Work. Management of Health and Safety at Work Regulations 1999. Approved Code of Practice and Guidance*
- L74 *First Aid at Work: The Health and Safety (First-aid) Regulations 1981: Approved Code of Practice*
- L138 *Dangerous Substances and Explosive Atmospheres: Approved Code of Practice and Guidance. Dangerous Substances and Explosive Atmospheres Regulations 2002*

Stationery Office Publications

The following is available from *www.tso.co.uk*.

- *Fire Safety — An Employer's Guide*
- *Investigation into the King's Cross Underground Fire*, 1988, D Fennell
- *The Public Inquiry into the Piper Alpha Disaster*, 1990, WD Cullen
- *The Fire at Hickson and Welsh Ltd*, 1994

Other Publications

- *The Contingency Planning Disaster Recovery Guide*

Organisations

- Centre for Corporate Accountability
  Web: *www.corporateaccountability.org*
  The Centre for Corporate Accountability is a charitable organisation that monitors worker and public safety, and undertakes research and provides advice on law enforcement and corporate criminal accountability.
- Civil Contingencies Secretariat
  Web: *www.ukresilience.info*
  The Civil Contingencies Secretariat supports the Home Secretary and works to build the UK's resilience and ability to manage the consequences of major emergencies.
- Emergency Planning Society
  Web: *www.the-eps.org*
  The Emergency Planning Society is a member-led organisation for professionals dealing with emergency planning and crisis and disaster management.
- Health and Safety Executive (HSE)
  Web: *www.hse.gov.uk*
  The HSE is responsible for the regulation of almost all the risks to health and safety arising from work activity in the UK.
- National Chemical Emergency Centre (NCEC)
  Web: *www.the-ncec.com*
  The NCEC provides advice for chemical health and safety compliance and emergencies.
- Society of Industrial Emergency Service Officers (SIESO)
  Web: *www.sieso.org.uk*
  The SIESO provides a forum for sharing ideas, experience and practices for avoiding industrial and commercial accidents and improving the planning and management of responses.

First Aid

- Employers are required to provide adequate and suitable first-aid equipment and facilities for employees under the Health and Safety (First-aid) Regulations 1981.
- Employers are not legally obliged to provide first aid to non-employees, eg customers, although many organisations do so.
- An assessment must be made of the first-aid needs that are appropriate, taking into account factors such as the size and location of the workplace, and how hazardous work activities are.
- Employers must provide a suitable number of first aiders and/or appointed persons.
- First aiders must hold a certificate from an HSE-approved organisation.
- First-aid equipment should include a suitably stocked first-aid box or boxes and possibly portable first-aid kits for travelling employees.
- Analgesics, eg paracetamol, should not be included in first-aid boxes.
- First-aid rooms may be necessary if the workplace is large, includes higher risk activities or access to emergency medicine is limited.
- Employees must be informed of the first-aid arrangements in the workplace.
- Appointed persons and certified first aiders must attend appropriate training which includes initial training, annual updates and re-certification courses.
- The First Aid at Work Council was formed from representatives of providers of first aid training as the lead industry body responsible for regulating training provision.

8.516 First aid is administered to minimise the consequences of injury and illness until qualified medical assistance is available. It also includes the treatment of minor injuries which would otherwise receive no treatment.

8.517 The Health and Safety (First-aid) Regulations 1981 impose duties on the employer and the self-employed to provide facilities that enable first-aid to be given to employees if they are injured or become ill at work. In the majority of cases, trained and qualified first-aiders must administer first-aid. The regulations also require employers to inform employees of first-aid arrangements.

8.518 The treatment of minor illnesses with medication such as analgesics for pain relief and antacids is not considered first-aid. The training of first-aiders does not include such treatments and first-aid supplies must not contain such medications.

Employers' Duties

8.519

- Employers have a general duty to ensure, so far as is reasonably practicable, the health, safety and welfare at work of all employees under the Health and Safety at Work, etc Act 1974.
- Employers are required to make provision for employees to receive first-aid treatment under the Health and Safety (First-aid) Regulations 1981.
- Employers must ensure equipment and facilities provided are adequate and appropriate.
- Employers must provide an adequate and appropriate number of suitable persons to provide first-aid to employees who are injured or become ill at work.

- A person must be appointed to take charge of first-aid facilities and equipment if the first-aider is absent.
- Depending on the nature of the undertaking, number of employees at work and the location of the establishment, the employer may appoint a person to take charge of first-aid, rather than providing a qualified first-aider.
- Employers are required to inform employees of first-aid provisions, including the location of equipment, facilities and personnel.
- A self-employed person must provide, or ensure there is provided, adequate and appropriate equipment to enable self-administered first-aid while at work.

Employees' Duties

8.520

- Employees have a duty to take reasonable care of their own health and safety and that of other people who may be affected by their work under the Health and Safety at Work, etc Act 1974.
- Employees have a duty to co-operate with the employer's health and safety arrangements.

In Practice

Assessment of First-aid Needs

8.521 The level of first-aid provision depends on workplace circumstances. No fixed level exists, but each employer needs to assess what facilities and personnel are appropriate. Employers may delegate the responsibility for carrying out the assessment and advising on first-aid to an occupational health service. When assessing what is adequate and appropriate, the employer must take account of a number of factors.

Workplace Hazards and Risks

8.522 Employers need to consider the different risks in different parts of the establishment. Identifying the likely nature of an accident or injury will help the employer work out the most appropriate type, quantity and location of first-aid facilities and personnel. Greater provision of first-aid will be needed in workplaces with higher risk activities. For example, a construction site, than in those with lower risk activities, such as an office. Consideration should be given to any special arrangements that may be needed for dealing with injuries arising from exposure to hazardous substances or dangerous machinery.

Size of the Organisation

8.523 As a rule, the larger the workforce, the more first-aid provision is needed, but employee numbers should never be the sole basis for determining first-aid needs. Greater risks may exist when fewer people are at work, eg during maintenance.

Organisation's History of Accidents

8.524 This information could be helpful in determining what first-aid materials and equipment are necessary, where first-aiders should be located and what geographical area they should cover.

Nature and Distribution of Workforce

8.525 The needs of employees potentially at greater risk, such as young workers, trainees and people with certain disabilities, should be addressed. Employers should consider how many first-aiders or appointed persons will be required to give adequate provision to each floor in a multi-level establishment. It is important that sufficient first-aid provision is always available when employees are at work, and separate arrangements may have to be made for each area of work or shift.

Remoteness of Site from Emergency Medical Services

8.526 When a site is remote from emergency medical services, employers may need to make special arrangements to ensure appropriate transport is available. At the very least, employers should inform the local emergency services in writing of the site's location and any particular circumstances, including specific hazards.

Needs of Travelling, Remote and Lone Workers

8.527 The assessment should determine whether those who travel long distances or are continuously mobile should carry a personal first-aid kit. Organisations with employees who work in remote areas should consider making special arrangements, such as issuing personal communicators, providing special training and organising emergency transport facilities. When employees work alone, other means of summoning help, such as a mobile telephone, may be useful.

Employees Working on Shared or Multi-occupancy Sites

8.528 On a shared or multi-occupancy site, one employer can take responsibility for providing first-aid cover for all workers. In these cases, a full exchange of information about the risks and hazards involved should help ensure the shared provision is suitable and sufficient.

Employment Agency Employees

8.529 Where an employment business contracts out employees, the employment business should ensure, by arrangement with the user employer, that employees have access to first-aid.

Annual Leave and Other Absences of First Aiders and Appointed Persons

8.530 Employers must cover annual leave and other planned absences of first-aiders or appointed persons. Employers should also consider what cover is needed for unplanned and exceptional absences, such as sick leave or special leave due to bereavement.

Non-employees

8.531 The regulations do not oblige employers to provide first-aid for anyone other than employees. Many undertakings, eg educational establishments, health authority premises, places of entertainment, fairgrounds and shops, do provide services for others, and employers may wish to include them in the assessment of needs and make provision. Employers should be aware that the compulsory element of employers' liability insurance does not cover litigation resulting from first-aid to non-employees. However, many public liability insurance policies do cover this aspect and employers may wish to check such policies.

Reviewing First-aid Provision

8.532 Employers should periodically review first-aid needs to ensure the provision remains appropriate. As with many health and safety requirements, there are no set review periods, but one may be necessary:

- when processes change or are altered, creating different risks
- when personnel change, eg if contractor staff are employed
- after an accident or incident that may highlight first-aid deficiencies
- when the workplace location changes
- when working patterns change, eg if shift work is introduced.

First-aid Equipment

8.533 First-aid equipment must, as a minimum, be equivalent to the standard stated in the Approved Code of Practice. This includes a suitably stocked first-aid container identified by a white cross on a green background. There is no mandatory list of items. Employers should decide what to include from information gathered during their assessment of first-aid needs.

8.534 As a guide, where no special risk arises in the workplace, the Approved Code of Practice suggests a minimum stock would normally be:

- a guidance card or leaflet on first-aid, eg HSE leaflet INDG215L, which provides basic first-aid advice
- 20 individually wrapped, sterile adhesive dressings (assorted sizes) appropriate to the work environment
- two sterile eye pads, with attachments
- four individually wrapped triangular bandages
- six safety pins
- six medium, individually wrapped, sterile, unmedicated wound dressings (approximately 12cm x 12cm)
- two large, individually wrapped, sterile, unmedicated wound dressings (approximately 18cm x 18cm)
- a pair of disposable gloves.

8.535 When mains water is not readily available for eye irrigation, sterile water or sterile normal saline solution (0.9%) in sealed disposable containers should be provided.

8.536 This is only a suggested contents list and equivalent but different items will be considered acceptable. In addition, the assessment may find a need for additional materials and equipment, eg scissors, adhesive tape, disposable aprons or individually wrapped moist wipes. These may be kept in the first-aid container or stored separately, as long as they are available for use.

8.537 The contents of a portable first-aid kit should reflect the circumstances in which it may be used, but should include as a minimum:

- a guidance card or leaflet on first-aid
- six individually wrapped, sterile adhesive dressings
- one large, sterile, unmedicated dressing (approximately 18cm x 18cm)
- two triangular bandages
- two safety pins
- individually wrapped moist cleansing wipes
- a pair of disposable gloves.

8.538 The contents of first-aid boxes and portable first-aid kits should be replenished as soon as possible after use. All first-aid boxes should be checked regularly to ensure the contents are not used after the expiry date.

First-aid Rooms

8.539 The employer must assess the need for a first-aid room. Establishing a first-aid room should be considered where the workplace presents a high risk from hazards, access to outside accident and emergency facilities is difficult or when there are large numbers of employees on-site at one time. A suitable person should be responsible for the room and its contents at all times when employees are at work.

8.540 To be effective, first-aid rooms should:

- be large enough to hold a couch, and have enough space at each side for people to work, a desk, a chair and any necessary additional equipment
- have washable surfaces and adequate heating, ventilation and lighting
- be kept clean, tidy, accessible and available for use at all times when employees are at work
- be positioned as close as possible to a point of access for transport to hospital
- include a notice on the door advising of the names, locations and contact details of first-aiders.

8.541 Facilities and equipment that should be provided in first-aid rooms include:

- a sink with running hot and cold water (which should always be available)
- drinking water when not available on tap, and disposable cups
- soap
- paper towels
- smooth-topped working surfaces
- a suitable store for first-aid materials
- a range of first-aid equipment, at least to the standard required in first-aid boxes
- suitable, foot-operated refuse container, lined with a disposable plastic bag
- a couch (with a waterproof surface), and frequently cleaned pillow and blankets
- clean, protective garments for use by first-aiders
- a chair
- an appropriate record-keeping book
- a bowl
- a telephone or other suitable means of communication.

8.542 The room should be used solely for first-aid treatment (although it may also be used for health screening). The first-aid room should be clearly marked by a sign complying with the Health and Safety (Safety Signs and Signals) Regulations 1996. Specialised first-aid equipment should also be kept in the first-aid room.

First Aiders

8.543 An adequate and appropriate number of suitable persons must be provided to deliver first-aid. Deciding what is adequate and appropriate should be based on the risk assessment.

8.544 The selection of first-aiders depends on an individual's:

- reliability, disposition and communication skills
- aptitude and ability to absorb new knowledge and learn skills
- ability to cope with stressful and physically demanding emergency procedures
- normal duties, which that person should be able to leave to go to an emergency.

8.545 Doctors registered with the General Medical Council and nurses registered on part 1, 2, 10 or 11 of the UK's Central Council for Nursing, Midwifery and Health Visiting's Single Professional Register, can provide first-aid at work. When doctors or nurses are employed, the employer may take that into account in determining first-aid provision.

8.546 The regulations also stipulate that a person be appointed to provide emergency cover in the absence of first-aiders, but only when the absence is due to exceptional, unforeseen and temporary circumstances. Absences such as annual leave do not count.

Appointed Persons

8.547 When an assessment identifies that a first-aider is not necessary, the minimum requirement is to appoint a person to take charge of first-aid arrangements, including looking after equipment and calling the emergency services. An appointed person must be available to complete such duties when people are at work. Appointed persons are not necessary if there is an adequate number of first-aiders.

8.548 Appointed persons are not first-aiders and so should not attempt to give first-aid without training. However, as the appointed person is required to look after the first-aid equipment and should ideally know how to use it, employers are advised to consider training appointed persons.

Number of First-aid Personnel

8.549 It is impossible to identify the precise ratio for the number of first aiders to employees. Employers must decide how many first-aiders are needed.

8.550 Guidance notes which accompany the regulations give the following suggestion (see table), according to the number of employees, on the assumption that the more employees there are, the greater the probability of injury or illness.

8.551 It can be difficult for some organisations to find sufficient staff to become first-aiders on a voluntary basis. It may be necessary to provide a form of incentive, such as additional pay, or to make it a condition of employment that an individual becomes a first-aider.

8.552 It is good practice for employers to provide first-aiders and appointed persons with a form to record incidents which require attendance. When there are a number of first-aiders working for a single employer, it would be advisable for one central log to be used, though this would not be practical on larger, spread-out sites.

Determining the Number of First-aid Personnel Needed

Category of Risk	Number of Employees	Suggested Number of First-aid Personnel
Lower risk — shops, offices, libraries	Fewer than 50	At least 1 appointed person
	50 to 100	At least 1 first-aider
	More than 100	An additional first-aider for every 100 employed
Medium risk — light engineering, assembly work, warehousing	Fewer than 20	At least 1 appointed person
	20 to 100	At least 1 first-aider for every 50 employed
	More than 100	An additional first-aider for every 100 employed
Higher risk — construction, chemical, dangerous machinery, sharp instruments	Fewer than 5	At least 1 appointed person
	5 to 50	At least 1 first-aider
	More than 50	An additional first-aider for every 50 employed
	Hazards for which additional first-aid skills are necessary	In addition, at least one first-aider trained in the specific emergency action

Duty of Care of First Aiders

8.553 First aiders have the normal responsibilities as laid down by ss.7 and 8 of the Health and Safety at Work, etc Act 1974. Additionally, it is considered that they will have a civil duty of care to patients.

8.554 However, since the first-aider is appointed by the employer in most instances, it is considered that the normal laws regarding duty of care would apply, eg the first-aider is still an employee and the employer may be vicariously liable for the employee's actions while at work. Obviously, if the first-aider were to use their skills outside of the workplace, the duty of care would pass to the individual. There has not been a civil suit against a first-aider for negligence.

Information for Employees

8.555 First-aid arrangements operate efficiently in an emergency only where they are known, understood and accepted by all in the workplace. One way to achieve this is to set up procedures for informing staff, in consultation with employees or safety representatives.

8.556 The procedures should detail first-aid provision and explain how employees will be told the location of first-aid equipment, facilities and personnel. The procedures should also identify who will provide the relevant information to new and transferred employees.

8.557 A simple method of keeping employees informed is by using first-aid notices. The information needs to be clear and easily understood by all employees; those with reading and language difficulties should also be kept informed.

8.558 Notices must be designed and worded carefully to ensure information is effectively communicated to employees. At least one notice in a prominent position at each site, including the base for travelling employees, should give enough opportunity for employees to see the information.

8.559 The inclusion of first-aid information in induction training will help ensure new employees are made aware of first-aid arrangements. Employees should be made aware of where first-aiders work as well as their names.

Training

First Aiders

8.560 One of the outcomes of the 2004 review of the Health and Safety (First-aid) Regulations 1981 was that the Health and Safety Commission (HSC) agreed with the recommendation to make changes to first-aid training courses. The HSE has therefore developed new training courses in consultation with relevant stakeholders and with the First Aid at Work Council.

8.561 A person appointed as a first-aider must attend a minimum of 18 hours' First Aid at Work (FAW) training given by an HSE-approved organisation. At the end of training, the individual must pass an examination conducted by two examiners who are qualified first-aid trainers, one of whom must be a nurse or qualified medical practitioner.

8.562 The qualification is only valid for three years. In the first and second years after qualification, first-aiders should attend annual update training covering basic life support skills. These sessions should each last for at least three hours. After three years, first-aiders will need to complete a 12-hour FAW requalification course to obtain a new certificate.

8.563 An organisation or individual employer may conduct its own training, providing it gains approval from the HSE. This approval will be subject to a number of criteria, including the qualifications of trainers, the syllabus, the equipment and arrangements of examinations. Records of qualified first-aiders and appointed persons should be kept, with details of when refresher training will be required.

8.564 On completion of training, successful candidates should have the ability to:
- act safely, promptly and effectively when an emergency occurs
- administer cardio-pulmonary resuscitation (CPR) promptly and effectively
- administer first-aid safely, promptly and effectively to a casualty who is unconscious, wounded or bleeding
- administer first-aid safely, promptly and effectively to a casualty who has been burned or scalded, is suffering from an injury to the bones, muscles or joints, is suffering from shock, has an eye injury, may be poisoned or has been overcome by gas or fumes
- transport a casualty safely
- recognise common major illnesses and take action
- recognise minor illnesses and take action
- maintain simple, factual records and provide written information to a doctor or hospital if required.

8.565 Where work presents special or unusual hazards, the employer should provide relevant training. This may be required when there is a:
- danger of poisoning by certain cyanides and related compounds
- danger of burns from hydrofluoric acid
- need for oxygen as an adjunct to resuscitation.

Appointed Persons

8.566 Following the 2004 review, the HSE concluded that its focus should be on the structure and content of first-aid at work courses rather than the approval of training providers.

8.567 One of the outcomes of this review was that the HSC agreed with the recommendation to make changes to first-aid training courses and the HSE has now developed new courses in consultation with other relevant stakeholders.

8.568 The intention is that, in future, employers will be able to send suitable employees on either a 6-hour (minimum) emergency first-aid at work course or an 18-hour (minimum) first-aid at work course, based on the findings of their first-aid needs assessment.

8.569 After three years, first-aiders will be required to complete another course (either a 6-hour or a 12-hour re-qualification course, as appropriate) to obtain a new certificate. Within any three-year certification period, first-aiders should complete two annual refresher courses covering basic life support/skills updates, which last for at least three hours.

8.570 A date for the introduction of these courses has not been set but the HSE is aware that training providers require an adequate lead-in period.

Training Needs Flowchart

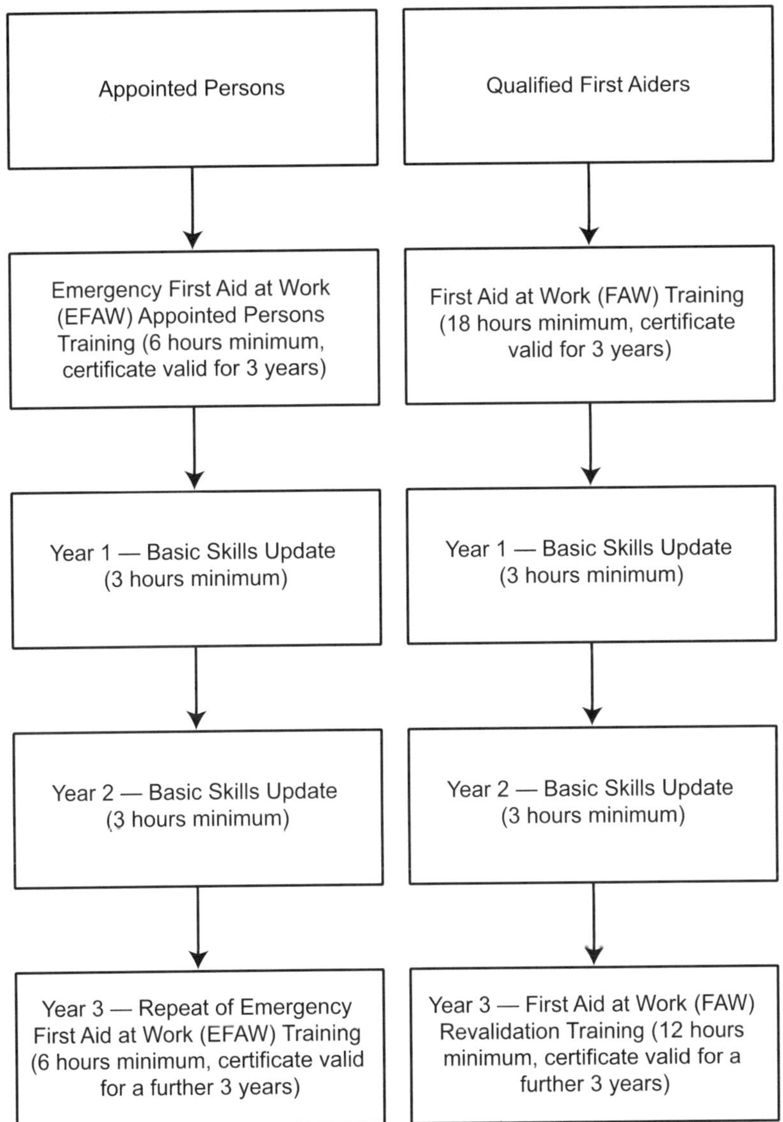

The First Aid at Work Council

8.571 Following the 2004 review of the Health and Safety (First-aid) Regulations 1981, the First Aid at Work Council (FWC) was subsequently formed from representatives of providers of first-aid training as the lead industry body.

8.572 The FWC was formed to:

- work together to form a representative industry lead body that will encompass all aspects of the first-aid at work industry, thus providing a focal point for ongoing first-aid development
- act as a national contact point for all those wishing to seek information and guidance with regard to first-aid at work issues
- develop, define, monitor and regulate UK first-aid at work training and act as provider of operational standards
- develop and publish a national register of all registered and approved first-aid at work training providers and trainers
- act as an advisory and representative voice for the first-aid at work industry in its dealings with UK businesses and associated industry and national government bodies.

Employees

8.573 First-aid arrangements operate efficiently in an emergency only where they are known, understood, and accepted by everyone in the workplace. All employees should be aware of:

- who the first-aiders and/or appointed persons are
- where they are located in the workplace
- how they can be contacted
- where first-aid equipment is situated
- where the first-aid room is situated
- the procedures to be followed if outside medical services are required (ie calling an ambulance)
- where in the workplace such information is displayed (such as notice boards and first-aid signs).

8.574 The procedures should also identify who will provide relevant first-aid information to new and transferred employees. A simple method of keeping employees informed is by displaying first-aid notices. The information needs to be clear and easily understood by all employees. Employers should ensure that those with reading and language difficulties are also kept informed.

8.575 Notices must be designed and worded carefully to ensure that the information is put across effectively to employees. At least one notice in a prominent position at each site, including the base for travelling employees, should give enough opportunity for employees to see the information. Clearly, all information provided must be kept up-to-date.

List of Relevant Legislation

- Health and Safety (Miscellaneous Amendments) Regulations 2002
- Management of Health and Safety at Work Regulations 1999
- Health and Safety (Safety Signs and Signals) Regulations 1996
- Health and Safety (First-aid) Regulations 1981

Further Information

Publications

HSE Publications

The following are available from *www.hsebooks.co.uk*.

- HSG212 *The Training of First Aid at Work: A Guide to Gaining and Maintaining HSE Approval*
- INDG214 *First Aid at Work: Your Questions Answered*
- INDG347 (rev 1, 08/06) *Basic Advice on First Aid at Work*
- L74 *First Aid at Work: The Health and Safety (First-aid) Regulations 1981: Approved Code of Practice*

Other Publications

- *First Aid Manual*, Dorling Kindersley, London, 8th edition, 2006.

Organisations

- British Red Cross Society
  Web: *www.redcross.org.uk*
  The British Red Cross is one of the UK's leading providers of first-aid training and support.
- First Aid at Work Council
  Web: *www.fwc–uk.co*
  The First Aid at Work Council is the leading industry body on first aid at work issues.
- St Andrew's Ambulance Association
  Web: *www.firstaid.org.uk*
  The St Andrew's Ambulance Association is one of Scotland's leading charities and the country's premier provider of first-aid training and services.
- St John Ambulance Association
  Web: *www.sja.org.uk*
  The mission of the St John Ambulance Association is to provide first-aid and medical support services, caring services in support of community needs and education, training and personal development to young people.

General Health and Safety Policy

- A health and safety policy is a document that outlines an organisation's commitment to, and procedures for, health and safety.
- Employers with five or more employees are required to prepare a written statement of their health and safety policy under the Health and Safety at Work, etc Act 1974.
- Employers who have fewer than five employees are exempt from the requirement to have a written policy.
- The health and safety policy should be tailored to the individual company.
- The policy should include a general statement that expresses a commitment from the organisation to their employees' health and safety.
- It should clearly identify who is responsible for what.
- The practical arrangements for health and safety should be included within the policy, with reference to other documentation, eg risk assessments and safe systems of work. The policy must specify how these matters are to be managed.
- Employees should be consulted on the organisation's health and safety policy, rather than just be informed of it.
- The most senior person in the organisation should sign the statement of commitment.
- The policy must be effectively implemented.
- The policy must be communicated to employees in a comprehensible form. The employer must take employees with learning difficulties, and foreign employees for whom English may be a second language, into account when deciding on the form of the policy.
- Ongoing commitment to the health and safety policy can be secured by promoting it, eg through poster campaigns and suggestion schemes.
- Employers must monitor the effectiveness of the policy and review it, where necessary.
- Adequate training in health and safety must be provided to all employees.
- The health and safety policy should be covered during induction training.
- Records must be kept to document the existence of the policy as part of an overall health and safety management system, in order to provide evidence of legal compliance.

8.576 The preparation of a company health and safety policy is the starting point for managing health and safety in the workplace. Such a policy sets out how the organisation will approach and discharge its duties in relation to the management of occupational health, safety and welfare. Where an employer has five or more employees, the policy must be in writing.

Employers' Duties

8.577

- Employers have a duty under the Health and Safety at Work, etc Act 1974 to prepare, implement and revise, as necessary, a health and safety policy. Employees must be informed of the policy and of any amendments made to it.

- Employers with fewer than five employees are exempt from the requirement to have a written health and safety policy under the Employers' Health and Safety Policy Statements (Exception) Regulations 1975. However, if an employer operates a number of small establishments, where each establishment employs fewer than five employees but where all form part of the same undertaking, the employer is obliged to prepare a written health and safety policy for the whole undertaking.
- Under regulation 4 of the Management of Health and Safety at Work Regulations 1999, employers have a duty to implement preventative and protective control measures on the basis of the principles specified in schedule 1 to the regulations. According to schedule 1, the policy must incorporate:
  - technology
  - organisation of work
  - working conditions
  - social relationships
  - influence of factors relating to the working environment.
- Under regulation 5 of the Management of Health and Safety at Work Regulations 1999, the employer must implement appropriate health and safety arrangements. The management system (health and safety policy) will need to reflect the complexity of the organisation's activities and working environment. The Approved Code of Practice (ACOP) to the Management of Health and Safety at Work Regulations 1999 refers to the approach taken in the British Standard for health and safety: BS 8800. A successful health and safety management system should include the five elements: planning, organisation, control, monitoring and review.
- For employers with five or more employees, these arrangements must be recorded.
- The Safety Representatives and Safety Committees Regulations 1977 require employers to consult with safety representatives on matters relating to health and safety, including the development of the arrangements section of a company health and safety policy.
- The Health and Safety (Consultation with Employees) Regulations 1996 require employers to consult directly with employees, who are not otherwise represented, on matters relating to health and safety, which includes the development of the arrangements section of a company health and safety policy.
- Regulation 5 of the Control of Major Accident Hazards Regulations 1999 requires operators to prepare, retain and implement a policy document for preventing major accidents.

Employees' Duties

8.578

- Employees have a duty to take reasonable care of their own health and safety and that of other people who may be affected by their work under the Health and Safety at Work, etc Act 1974.
- Under the Management of Health and Safety at Work Regulations 1999, employees must inform their employer of any danger to health and safety posed by a work activity. They must also inform their employer of any shortcomings in the employer's protection arrangements.
- Employees also have a duty to co-operate with the employer's health and safety arrangement under the Management of Health and Safety at Work Regulations 1999.

In Practice

Benefits of an Effective Health and Safety Policy

8.579 The benefits of having an effective company health and safety policy include the following.

- The company complies with the relevant legal requirements.
- A commitment to, and competence in, managing health and safety issues is demonstrated. This, in turn, can have a positive influence on:
  - employment and retention strategies
  - customer and client perceptions
  - the running costs of the business.

Consequences of a Failure to Have a Health and Safety Policy

8.580 It is an offence for an employer to fail to:

- prepare a written statement outlining the organisation's general policy on health and safety
- distribute that statement to employees
- review the organisation's general policy on health and safety.

8.581 Such an offence makes the employer liable, on summary conviction, to a fine not exceeding £20,000.

Tailoring a Health and Safety Policy to an Organisation

8.582 Policy statements do not necessarily need to be long or elaborate. Instead, they must be relevant to the organisation and its circumstances.

8.583 To be effective, whatever is written in the policy must be put into practice. This means that the policy must reflect the company's:

- culture and commitment to health and safety
- organisational structure, roles and responsibilities
- actual systems and procedures that it uses to discharge its responsibilities.

8.584 The true test of a health and safety policy is the actual conditions in the workplace, not how large the policy is or how well it is written. In other words, if you cannot recognise the company when reading the policy, it is highly likely that the policy is not relevant to that organisation.

8.585 Employers should, therefore, avoid the tendency to "clone" policy documents prepared for other organisations, as these will prove to be of little value to the company, its employees, its clients and suppliers. Indeed, such an approach might prove to be a significant impediment should an enforcing authority, or the courts, have cause to compare the policy with the organisation's practice.

8.586 Only a document that is unique to an organisation can truly reflect its commitment and practical approach to health and safety. This is key in achieving acceptable health and safety standards and reducing accidents and cases of work-related ill health.

Components of a Health and Safety Policy

8.587 Health and safety policy statements usually consist of three parts, namely:

- a general statement of intent, which is a declaration of the employer's commitment to providing a safe and healthy workplace and a clear direction for the organisation to follow
- details of responsibilities for health and safety throughout the organisation
- details of safe systems of work and/or safe working practices for all work activities.

8.588 While the organisation can make its own choice about the size and format of the health and safety policy, this will largely depend on the size and complexity of the organisation, the nature of its business and the extent of the risks associated with its undertakings.

Preparing the General Statement

8.589 The general statement of policy is usually quite short, often taking up no more that one sheet of A4 paper. Nevertheless, it should meet a number of criteria. The statement should:

- express a clear commitment from the organisation to minimise adverse effects to the health and safety of employees and others
- commit the organisation to complying with all relevant legal requirements
- stipulate compliance by all management personnel with the legal duties and any further requirements relating to occupational health and safety specified in the policy document
- require the compliance of all employees and the commitment to take an active part in maintaining a safe and health-conscious work environment
- specify a commitment to regularly review and, where necessary, revise the policy
- become effected by being signed and dated by senior management
- be communicated to all employees.

8.590 It could further:

- commit to the continuous improvement of the health and safety of employees and others
- oblige the organisation to comply with best practice
- convey the conviction that safety and health protection are not inconsistent with profitability.

Setting Out the Organisational Structure

8.591 It is the responsibility of senior management to specify the organisational and operational structures required for the implementation of its health and safety policy and strategy, as well as the authorities and the associated responsibilities of both line managers and staff functions.

8.592 The organisational structure should identify, in clear terms, the personnel to whom managerial and advisory functions are to be assigned. The need to constantly revise the policy following staff turnover can be minimised by referring responsibilities to job titles, rather than jobholders by name. However, where personnel have a primary responsibility — such as occupational health and safety specialists, industrial physicians or safety representatives — it may be more appropriate to refer to them by name.

8.593 The structure must be conducive to implementing the policy and for achieving the targets stated in the policy. Along with the rest of the health and safety policy, the structure should not be seen as set in concrete but subject to review. It should be revised from time to time, where necessary, to account for changes in the structure of the organisation, or the way in which it seeks to secure its objectives.

8.594 The person with day-to-day responsibility for the health and safety risk management system should, ideally, report directly to senior management within the organisation on the performance of systems. This will enable senior management to make its own assessment on the policy's effectiveness.

8.595 Regulation 13 of the Management of Health and Safety at Work Regulations 1999 requires specific considerations to be made when an employer entrusts tasks with health and safety implications to employees. In particular, the employer should ensure that those tasks are within the employees' ability, so as not to put themselves, or others, at risk. In other words, employers should take steps to ensure that their employees are competent.

8.596 Competence should be ascertained on the basis of professional qualifications, experience and the physical and mental characteristics that are suited to fulfil the tasks. If not already present when the employee is engaged, training may be required before a jobholder can assume full responsibility for his or her role.

Developing the Arrangements Section

8.597 This part of the policy covers the systems and procedures in place for ensuring employees' health and safety. The health and safety policy statement does not need to record the full details of all procedures. It can refer to other documentation, such as risk assessments, training programmes, emergency instructions, etc. However, the policy statement should record the arrangements and procedures for how these matters are to be managed.

8.598 Core elements to be addressed within the arrangements section of a policy include arrangements for:

- risk assessments
- consultation with employees
- information and instruction to be provided to employees
- supervision of employees
- training, including arrangements for promoting health and safety awareness
- accidents, first-aid and ill-health issues
- cases of emergency (eg fire)
- providing personal protective equipment (PPE)
- maintaining plant and equipment
- safe handling and use of substances
- monitoring
- machinery maintenance and breakdowns.

8.599 Other elements, such as those arrangements for the management of display screen equipment, noise or working at height, should be added as required, depending on the particular hazards an organisation faces.

Consulting Employees on the Policy

8.600 Acceptance of the policy by employees is crucial to its success. Not only does consultation help to overcome a natural resistance to change, it taps into a potential wealth of experience and knowledge that can improve health and safety arrangements.

8.601 By enabling staff to become architects in the design of solutions, they are empowered to take control and, hence, are more likely to "own" the control measures. This approach has the added benefit of increasing peer pressure to comply, which can be an important factor if the policy is to be a success in the long-term.

8.602 The law requires that employees be consulted — not just informed — on matters relating to health and safety. The consultation process can be simplified, encouraged and supported by training employees and help them to become organised through the election of safety representatives and the formation of a safety committee.

Securing Senior Level Commitment to the Policy

8.603 The health and safety policy will not be effective without genuine commitment from the most senior person in the organisation. This person should take steps to ensure that health and safety are taken seriously at all levels in the organisation and that the policy is signed and dated to demonstrate management commitment to health and safety in the workplace.

Implementing and Promoting the Policy

8.604 Occupational health and safety is all about protecting people. Unless a health and safety policy is implemented, the effort in developing it will be wasted. Successful implementation of the policy should be seen as a priority for management, in the same way as the launch of a new product or service is a priority. In the absence of sufficient resources, adequate planning and proper management, the implementation will fail. Employers must develop an action plan, assign responsibilities to individuals, agree to deadlines and check progress.

8.605 Special care will need to be taken to ensure that the policy is presented in a way that is understood by employees. While the law requires that the policy be brought to the attention of all employees, simply getting someone to say they have received such information will not be enough to discharge the employer's duty. For example, there may be employees who have a special learning need, such as a difficulty with literacy. For other employees, English might not be their first language.

8.606 Where a need has been identified, further consideration may have to be given to convey parts of the policy to others not directly connected with the organisation. Such persons might include employees of other organisations who may be sharing common workplaces (eg contractors), external bodies (eg public authorities, insurers, visitors) and the general public.

8.607 In large, complex organisations, the health and safety policy and supporting documents can consist of documents of considerable volume. Employees who perceive that there is too much to take in, or that they are being bombarded by irrelevant information, will "turn off" very quickly. To avoid this happening, employers should consider the following points.

- While all employees must see the general statement of policy itself, they may not need to see all of the other documents, especially if they are not relevant to their health, safety or welfare.
- Some hazards carry risks that are higher than others. Introducing the arrangements in a more strategic manner, with priority based on the level of risk, might be a way of reducing overall risk in the short-term.

8.608 Wider participation and ongoing commitment can be secured through the development of an employee handbook, poster campaigns, suggestions schemes, employee reporting systems for hazards and incidents and incentives for exemplary conduct.

Monitoring the Effectiveness of the Policy

8.609 Monitoring is an integral part of successful health and safety management; it is the principal means by which employers measure their health and safety performance. The arrangements section of the policy should describe the organisation's approach to this process.

8.610 Performance can be measured both reactively and actively.

8.611 Reactive measures will include examination of loss data, such as accident records, claims history, plant and machinery breakdown and repairs. These give an historical perspective and can show, not only trends, but also "hot spots" for incidents.

8.612 Active monitoring (or checking compliance with arrangements, eg by a compliance audit) can be more productive, for while this approach can also show trends and "hot spots", it has the additional benefits of:

- creating the opportunity to reward success
- being highly visible and therefore a means by which management can "make things matter".

Reviewing the Policy

8.613 Under s.2(3) of the Health and Safety at Work, etc Act 1974 and regulation 3 of the Management of Health and Safety at Work Regulations 1999, employers are required to review and, if necessary, modify their arrangements for managing health and safety. This should occur:

- when there has been significant change in the organisation or its undertakings
- after serious accidents, or cases of ill health, or damage to plant or equipment, that are attributable to deficiencies in the system
- where the monitoring reports indicate that a review is necessary
- at intervals of not more than one year.

8.614 Employees must be informed of the changes and receive as much information, instruction and training in them as may be necessary to avoid risks to health and safety.

8.615 The review should consider the:

- adequacy of the policy and strategy specified in connection with the health and risk management system
- relevance of the strategic objectives derived from them and of the necessary measures in the operational area
- organisation and performance of the management system
- issues of compliance with statutory requirements and other voluntary codes.

8.616 The review must indicate whether any action needs to be taken to improve the system and this serves as the basis for improvement. The result is a dynamic rather than static approach to the management of health and safety.

Training

8.617 Adequate training must be provided to all employees where this is required to ensure their own health and safety or the health and safety of others who may be affected by the undertakings of the organisation. Where an employee is deemed at the outset to be competent to perform the tasks assigned to him or her, there might be no need for further training.

Who Might Need Training?

8.618 All those who:

- advise on the content and format of the health and safety policy
- devise organisational structures and arrangements contained in, or referred to by, the policy
- participate in the consultation process
- monitor compliance with the policy
- have day-to-day responsibility for the operation of the health and safety management system.

What Should Training Include?

8.619 Any training should include the:

- importance of the health and safety policy
- strategies for the organisation as a whole and for each individual employee
- relevance of the tasks, and responsibility assigned to employees
- potential and real consequences of employees' behaviour for their own health and safety and that of others, and the safety of plant and equipment.

When Should Training Take Place?

8.620 Training should take place when:

- employees commence employment (eg as part of induction training), are transferred, have their job duties or working environment changed
- equipment, machinery and plant, or procedures and processes, are introduced or altered.

8.621 Training should be repeated at the intervals prescribed by law or at appropriate intervals specified by the organisation.

List of Relevant Legislation

- Railways (Safety Case) Regulations 2000
- Control of Major Accident Hazards Regulations 1999
- Management of Health and Safety at Work Regulations 1999
- Health and Safety (Consultation with Employees) Regulations 1996
- Safety Representatives and Safety Committees Regulations 1977
- Employers' Health and Safety Policy Statements (Exception) Regulations 1975
- Health and Safety at Work, etc Act 1974

Further Information

Publications

HSE Publications

The following are available from *www.hsebooks.co.uk.*

- HSG65 (rev 1997) *Successful Health and Safety Management*
- INDG275 *Managing Health and Safety: Five Steps to Success*

Health and Safety Induction Training

- The Management of Health and Safety at Work Regulations 1999 require employers to provide adequate health and safety training when an employee is recruited to an organisation. This is in addition to other legal requirements for health and safety training.
- Many criminal and civil cases have revolved around an employer's failure to provide adequate training.
- Requirements for induction training are determined via analysis of the organisation's training needs and as a result of the risk assessment process.
- Induction training generally presents information at three levels:
  - general information regarding health and safety in the organisation
  - local health and safety information, such as fire arrangements, names of fire wardens and first-aiders, etc
  - job-specific information as is necessary to allow the person to begin working safely.
- It is not only new employees who require induction training — existing employees moving to a new site, or workplace, might also require this form of training.
- An employer's duty to provide information to employees extends to employees on fixed-term contracts, employees of other employers and the self-employed working within the employer's undertaking. A principal method of providing information to these groups is via induction training.
- Methods of induction training often include:
  - lectures/talks
  - one-to-one training
  - tours of the work site
  - videos
  - computer-based training
  - booklets and literature.
- A single method, or a combination of methods, may be used. It is essential that the chosen method is appropriate to the person and the situation.
- In high-risk situations, it might be necessary to test understanding of the induction training material before allowing new employees to begin work.
- Induction training should be carried out as soon as possible after the employee joins the organisation. This is particularly so for employees working in hazardous situations.

8.622 One of the key areas to be addressed by an organisation when it is planning training is the provicion of training to people who are new to either the organisation or to a particular part of that organisation. This form of training is commonly referred to as induction training.

8.623 Training is a key process in the management of risk in the workplace and, although it cannot completely remove risk on its own, it is an essential step in the control of workplace hazards. There is much evidence to suggest that people are most at risk when first entering a new work environment. Induction training not only reduces the risk to people during this most vulnerable period, but also exposes them to the culture of the organisation. This is important, since the first impressions that people have of an organisation will shape their subsequent attitudes and behaviour.

Employers' Duties

Duties to Own Employees

8.624 The Health and Safety at Work, etc Act 1974 places a duty on employers, so far as is reasonably practicable, to provide such information, instruction, training and supervision as is necessary to ensure the health and safety of employees at work.

8.625 More specifically, regulation 13 of the Management of Health and Safety at Work Regulations 1999 requires employees to be provided with "adequate health and safety training on their being recruited to the employer's undertaking".

8.626 The Management of Health and Safety at Work Regulations 1999 also require that the information provided to employees includes any risks to their health identified by the risk assessment and the preventive and protective measures applied to control these risks.

8.627 Article 21 of the Regulatory Reform (Fire Safety) Order 2005 creates a duty on the responsible person to ensure that employees are provided with adequate fire safety training, including induction training.

8.628 Additionally, much of the body of UK health and safety legislation requires employees to receive information, instruction and training in relation to the specific hazards to which they may be exposed.

Duties to Others

8.629 In addition to providing induction training to their own employees, regulation 12 of the Management of Health and Safety at Work Regulations 1999 requires employers to provide information to:

- employees on fixed-term contracts
- employees of other employers and self-employed people who are working on the premises (this includes contractors and agency staff who are actually employees of the agency).

8.630 This information must be comprehensible and must cover the:

- risks to employees' health and safety arising out of the host employer's undertaking
- measures taken by the host employer to control the risks, so far as they relate to those people.

8.631 The method of providing this information will normally be via the induction training process, but may have to be adapted from the material provided to permanent employees.

Employees' Duties

8.632 Employees have a duty under the Health and Safety at Work, etc Act 1974 to co-operate with their employer on matters relating to health and safety. This includes attending, and acting in accordance with, induction training.

8.633 Under the Management of Health and Safety at Work Regulations 1999, employees shall use any item or piece of equipment provided to them by their employer in accordance with any:

- training provided by the employer
- instructions given by the employer.

8.634 Therefore, employees have a general duty to work according to any training they have received, whether it is induction training or training provided at a later date.

In Practice

Content of Induction Training

8.635 Safety induction training is often included as part of a general induction programme targeted at new employees, or existing employees who are new to a site or location. At this stage of employment, the employee may have large amounts of information to absorb and, as a result, may not necessarily attach an appropriate level of importance to the health and safety content.

8.636 It is important, therefore, that the safety induction training concentrates on the core elements that allow employees to work safely. The following key points should be considered.

- Training should not overload new employees with information. This can be both confusing and demotivating.
- The major issues that new employees will encounter during their early days with the organisation should be covered in the training.
- The level of detail in the training should be limited, but signposts to more detailed information should be provided.
- Safety training should be incorporated into any general induction training, as opposed to being "tacked onto" the end of it, as is often the case. If safety is not given equal prominence in a general induction programme, the message that the organisation does not rate safety as highly as other issues will be sent to new employees.

8.637 Generally, safety induction training presents information at three levels:

- general information regarding health and safety in the organisation
- local health and safety information
- job-specific information necessary to allow the person to begin working safely.

General Information Regarding Health and Safety in the Organisation

8.638 This should include the following.

- Details of the health and safety policy of the organisation — employees should read the policy statement and understand where they can access information on the organisation's arrangements, procedures, etc.
- General safety requirements that apply to the whole organisation, such as rules concerning the use of drugs and alcohol at work.
- The safety management structure, identifying safety as an issue at the highest level in the organisation and involving every employee. The new employee should be given the identities of safety advisors and means of contacting them.

- Employees' responsibilities and the organisation's general responsibilities under the Health and Safety at Work, etc Act 1974, and subsequent regulations.

Local Health and Safety Information

8.639 Local health and safety information might include:
- local fire and emergency arrangements, including the required action in the event of a fire or emergency, escape routes, muster points, identities of fire wardens, etc
- first-aid and accident-reporting procedures, such as the identities and locations of first-aiders, location of the accident book, and action to be taken in the event of an accident
- welfare arrangements, eg the location of facilities for obtaining food and water
- local rules and instructions, including any particular requirements for that location, over and above general company rules
- details of safety representatives and safety committees.

Job-specific Information Necessary to Allow the Person to Begin Working Safely

8.640 Job-specific information necessary to allow employees to begin working safely includes:
- the nature of the hazardous substances employees are using, and the relevant controls
- any hazards related to the equipment employees are using, along with instructions on safe working procedures
- any other hazards to which employees might be exposed, and the appropriate safe systems of work.

Safety Induction Training Methods

8.641 The training method used will depend on the length of the induction programme, and how hazardous the working environment is. In low-risk environments, eg offices, the training might be only a few hours long and could be conducted in the office itself. In more hazardous environments, a combination of methods might be necessary.

8.642 The principal methods appropriate for safety induction training are:
- lectures and discussions
- videos/DVDs
- computer-based training
- tours
- documents and literature.

Lectures and Discussions

8.643 For induction purposes, these must be informal and of short duration. They must give new employees the opportunity to ask questions. It is preferable to have a senior supervisor or manager attend, as well as the trainer, as this adds weight and importance to the subject matter.

Videos/DVDs

8.644 Many suitable videos/DVDs are available off-the-shelf that cover general topics relevant to induction training, such as:

- display screen equipment use
- general attitudes to safety and an introduction to the Health and Safety at Work, etc Act 1974
- fire safety
- office safety, etc.

8.645 Many organisations will have their own induction videos/DVDs produced, which not only have the advantage of covering site and company-specific issues, but also reinforce the safety culture by emphasising the importance of safety to the organisation. However, it is important not to rely wholly on videos/DVDs for induction training purposes — some interaction is also required within the training programme.

Computer-based Training

8.646 This involves the new employee undergoing training via interactive computer software, covering a particular health and safety topic. Normally these programmes are tailored to the needs of an organisation. The programme usually includes some form of evaluation, such as a test at each stage. It is also possible to monitor the time that employees spend on each subject area, to ensure that they do not neglect any of the material.

Tours

8.647 An escorted tour of the work premises is essential for the general familiarisation and orientation of new employees. It is also an opportunity to meet key personnel, such as safety advisors, first-aiders and fire wardens. During the tour, any hazardous areas, eg noise hazard zones, should be pointed out, and particular restrictions explained.

Documents and Literature

8.648 During induction, it is useful to give new employees appropriate documentation. However, it is important not to overload them with material that is either irrelevant or too detailed. Appropriate documentation might include:

- third-party information, eg leaflets produced by the Health and Safety Executive (HSE)
- company information, such as:
  - leaflets, eg a summary of the safety policy or general rules and information
  - specific safety instructions for carrying out certain tasks or working in hazardous areas, etc
  - purpose-designed induction information, which might also include an evaluation test to ensure understanding and retention of the contents of the programme.

Planning Induction Training

8.649 Whichever method, or combination of methods, is used, it is important that the training is planned, and that new employees are given a timetable of induction activities. The induction might be spread over the first few months of employment. However, certain information must be given at the earliest opportunity. Therefore, a typical induction programme might look like the following.

Typical Induction Programme

Timescale	Content
First day	• Fire evacuation procedures. • "No smoking" policy and general safety rules. • Names and locations of first-aiders. • Any job-specific information needed by employees immediately so they can carry out their jobs safely.
Within the first week	• The health and safety policy. • The safe systems of work applicable to their employment. • Names and locations of key staff, such as the safety advisors, safety representatives, etc.
Within the first month	• Formal induction to the organisation's safety arrangements. • More details of the safety culture of the organisation, and the standards expected of employees. • Evaluation of new employees' understanding and retention of the induction information.

Evaluating Induction Training

8.650 In simple, low-risk environments, eg offices, it may not be critical to evaluate the degree to which new employees have absorbed and understood the safety induction material. The greater the hazards faced, however, the more important it is to be reassured that new employees have fully understood what is required of them in terms of safety. Induction training, as opposed to other health and safety training, can be evaluated using:

- simple written tests, usually following a multiple-choice format, or requiring single word/short sentence answers
- tests incorporated within computer-based training, or company induction booklets
- close observation of behaviour by the supervisor or manager, to ensure that the induction material is reflected in new employees' conduct.

8.651 It is important that new employees are informed of the evaluation. They should also be informed that failure to achieve the minimum standard may result in their having to repeat the training or retake tests — and might ultimately impact on whether or not they are retained after the probationary period has expired.

Safety Induction for Non-employees

8.652 Employers have a legal duty to provide information to non-employees. However, this duty varies, according to the relationship between the employer and non-employee. Non-employees fall into three main categories:

- contractors, eg for maintenance or installation work
- temporary agency staff, provided by an employment business
- people working on fixed-term contracts.

Contractors

8.653 Organisations should have separate arrangements for the engagement and control of contractors. Hence, separate induction training is usually necessary. This training will include the organisation's requirements for contractors. Contractors may bring their own hazards onto the site, and both the contractors themselves and the client's employees could potentially be exposed to these hazards. Contractors may also be exposed to the client's hazards.

8.654 Induction training for contractors should include, as a minimum:

- health and safety rules for contractors
- specific site restrictions
- hazards of the client's undertaking, as far as they might affect the contractors
- security arrangements
- incident and accident reporting arrangements
- welfare arrangements — clients may restrict the use of their own facilities to employees, and require contractors to make their own arrangements
- fire and emergency arrangements for the site.

8.655 It is common for large clients to have booklets containing rules and arrangements specifically for contractors. These cover the above points. Induction training for contractors can be delivered via many different methods. In hazardous environments contractors should be tested on their understanding of health and safety issues before they commence work.

Temporary Agency Staff

8.656 Temporary agency staff are employees of the employment business ("agency"). Therefore, both the agency and the host employer have a legal duty to provide the following information to them.

- The host employer must inform the agency of any skills required by workers in order for them to work safely.
- Agencies must supply staff with the appropriate skills.
- Host employers need to ensure agency staff have the appropriate skills.
- The host employer must provide induction training detailing arrangements and work procedures for the health, safety and welfare of agency workers.

People Working on Fixed-term Contracts

8.657 People working on fixed-term contracts should be treated as employees, although a simplified induction training programme may be used for very short-term contracts.

List of Relevant Legislation

- Construction (Design And Management) Regulations 2007
- Management of Health And Safety at Work Regulations 1999
- Health and Safety at Work, etc Act 1974
- Regulatory Reform (Fire Safety) Order 2005

Further Information

Publications

HSE Publications

The following are available from *www.hsebooks.co.uk*.

- HSE27E (rev1) *Your Health, Your Safety — A Guide for Workers* (also available in Bengali/Punjabi/Gujarati/Urdu/Hindi)
- INDG36(L) (rev 2003) *Working with VDUs*
- INDG173(L) *Officewise*
- INDG259 (rev 2003) *An Introduction to Health and Safety — Health and Safety in Small Businesses*

Health Surveillance

- Health surveillance is defined as "putting in place procedures to detect early signs of ill health among workers exposed to certain risks, and acting on the findings".
- Not all health surveillance requires the involvement of health professionals.
- Employers must identify all aspects of work which may expose employees to the risk of occupationally-acquired ill health.
- If a health risk is identified in an activity or process, a risk assessment must be undertaken.
- Through risk control measures, the employer must reduce the risk to employee health so far as is reasonably practicable.
- Employers must determine if health surveillance is required and identify the criteria for undertaking the surveillance.
- Employers must also identify the procedures to be followed.
- Employers must identify the elements of health surveillance needed, and identify appropriate personnel to complete the surveillance.
- Employers must allocate adequate resources to implement health surveillance and act on the findings.
- Employers must keep employees and their representatives informed at all stages of the process, encouraging participation and involvement in the development and delivery of the programme.

8.658 Health surveillance is required when:

- there is good reason to believe employees, in the course of their work, will be exposed to substances or other hazards likely to damage their health
- there are techniques to detect health effects that are both valid and likely to benefit the employee.

8.659 For example, if noise levels cannot be reduced below the lower exposure action value and hearing protection is provided, regular audiometric surveillance provides a check for the individual and the employer on hearing ability. It detects any reduction in hearing ability, and provides a means for the employer to monitor the effectiveness of the hearing conservation programme.

8.660 The purpose of health surveillance is to:

- protect the health of individual employees, eg to minimise the risk of employees developing occupational asthma
- detect any adverse health effects at an early stage, eg to detect hearing loss at the mildest stage and take action to minimise any further hearing loss
- assist in the evaluation of control measures, eg regular surveillance will confirm whether anyone using personal protective equipment (PPE) provided has any signs of dermatitis.

8.661 Health surveillance data may also be used in the detection of hazards and the assessment of risk.

8.662 Health surveillance is a useful way of ensuring that the control measures put in place, following risk assessment, are effective. It must not be used as an alternative to the proper control of exposure.

Employers' Duties

8.663

- Employers have a general duty to ensure, so far as is reasonably practicable, the health, safety and welfare at work of all employees under the Health and Safety at Work, etc Act 1974.
- Under the Management of Health and Safety at Work Regulations 1999, employers must assess the risks to the health of their employees presented by all aspects of work.
- Employers must, before they employ a young person (under the age of 18), undertake a specific risk assessment in relation to the health risks the young person will be exposed to in the course of his or her work.
- If more than five people are employed, the significant findings of the risk assessment must be recorded.
- Employers must take appropriate steps, following the risk assessment, to reduce the risks to the health of their employees as far as is reasonably practicable, ie implement appropriate risk control measures.
- Employers must inform their employees, and others affected by employees' work, of the findings of the risk assessment and the steps being taken to reduce the risks to their health at work.
- Employers must provide suitable training and information to their employees to enable them to comply with the risk control measures being implemented.
- Employers must monitor the implementation of risk control measures and manage non-compliance to reduce the risks to employee health to the lowest practicable level.
- Employers must provide appropriate health surveillance if the risk assessment identifies such a need.
- Employers must act on the significant findings of any health surveillance, and reduce the risks to employee health from exposure to health hazards.
- Employers must appoint one or more competent persons to assist them in carrying out health risk assessments and health surveillance.

Employees' Duties

8.664

- Under the Health and Safety at Work, etc Act 1974 employees must:
  - comply with the employer's arrangements to ensure their health at work, including using equipment or carrying out work activities according to the training given, and complying with health surveillance arrangements
  - inform their employer of any health concerns they have in relation to their work, eg reporting symptoms, or reporting any incident which may have resulted in unusual or abnormal levels of exposure to a known health hazard.

In Practice

8.665 Not all health hazards are visible. For example, radiation cannot be seen or smelled, but over-exposure can result in serious illness. Micro-organisms are invisible to the naked eye, but some airborne bacteria and viruses can cause severe infection. Many vapours and dusts can be produced in sufficient quantity to pose a risk of respiratory damage to those inhaling particles, but the particles are not visible.

8.666 In practice, employers need to:

- work through the risk assessment process step by step
- identify the health hazards in the workplace and assess the possible impact on employee health
- consult and establish which employees, or groups of employees, are exposed to health hazards, the extent and frequency of exposure and the specific health risks of the substances and activities involved
- use every means possible to eliminate exposure to the health hazard, eg through substitution or implementation of risk controls
- assess whether health surveillance is needed, following the risk control process
- recognise that where PPE is required as part of the risk control process, health surveillance is likely to be required — PPE protects the wearer only if it is worn properly, and regular health surveillance to detect early symptoms of ill health is likely to be needed to fulfil the employer's statutory duties.

8.667 Assuming that health surveillance is required, employers must:

- identify the criteria for undertaking health surveillance
- identify the procedures to be followed
- identify which elements of health surveillance will be needed, and identify appropriate personnel to do the work
- allocate adequate resources to implement health surveillance and to act on the findings
- use the available references, advice and support to develop an appropriate health surveillance programme
- keep employees and their representatives informed at all stages of the process, encouraging participation and involvement in the development and delivery of the programme
- consult with employees on the following:
  - the aims of the programme, how the programme fits with the company health and safety policy and other means used to protect employees, and whether it is required by law
  - how employees can raise health and safety issues
  - the benefits to individuals, ie it is not in their long-term health interests to conceal symptoms
  - what is involved, including any referral procedures
  - what information will be given to the employer, how it will be conveyed and how the results will be used, including employment consequences
  - confidentiality
  - how the programme will be monitored and evaluated.

Criteria for Undertaking Health Surveillance

8.668 If any of the following criteria apply, health surveillance should be carried out.

- There is an identifiable disease or other identifiable adverse health outcome.
- The disease or health effect may be related to exposure.
- There is a likelihood that the disease or health effect may occur.
- There are valid techniques for detecting signs of the disease or health effects.

8.669 There is no requirement for health surveillance of workers who carry out manual handling, despite the fact that many people develop health problems associated with manual handling activities over time. There is no valid technique at present for detecting early signs of damage to the musculoskeletal system. However, while there is no requirement to undertake health surveillance, there is a requirement to reduce the risk of injury to the lowest practicable level.

Health Surveillance Process

8.670 Health surveillance, which is used to detect early signs of occupational ill health among employees exposed to certain risks, includes the implementation of procedures which are:

- systematic
- regular
- appropriate.

Ensuring Surveillance is Systematic

8.671 Procedures should be systematic and follow an agreed format, ie everyone undertaking surveillance should:

- follow the same protocol
- undertake the same tests
- use the same or comparable equipment
- keep records in one agreed format.

8.672 Occupational health service providers will either use an existing, professionally agreed protocol or modify a standard protocol to match the needs of the business for which they are providing health surveillance.

Ensuring Surveillance is Regular

8.673 Surveillance should take place at regular intervals, and every effort should be made to avoid slippage in the programme.

8.674 Surveillance may take place daily or, as in the case of hearing surveillance, at intervals of up to three years. The length of the intervals between health tests will depend on how quickly health effects may be noticed, eg skin reactions may appear overnight, but hearing deterioration is a much slower process.

Ensuring Surveillance is Appropriate

8.675 The surveillance procedure must be:

- appropriate to the exposure
- undertaken correctly
- timely.

8.676 Some tests, such as x-rays, are potentially harmful in themselves so employees must understand the reasons for them and give their consent.

8.677 Inaccurate results may lead to a false sense of security for both the employer and employee if they appear normal, or a great deal of worry if they are abnormal and repeat testing is required.

Elements of Health Surveillance

8.678 Health surveillance may involve one or more of the following elements.

Responsible Person

8.679 A responsible person is usually a trained employee who systematically checks for signs or symptoms of a reaction to the identified health hazard. For example, a responsible person may check a group of workers for:

- skin damage
- early signs of dermatitis.

8.680 A responsible person can also check responses to simple questionnaires about symptoms of respiratory problems. Responsible persons need to have access to qualified people when symptoms or abnormalities are detected by the surveillance methods they use.

Qualified Person

8.681 A qualified person is a healthcare professional or specialist technician who asks employees detailed questions about symptoms of ill health and conducts specific examinations or tests to measure or assess health. Examples include:

- an occupational health nurse carrying out lung function tests on employees exposed to respiratory hazards
- an audiology technician carrying out hearing tests.

Medical Surveillance by a Doctor

8.682 Medical surveillance by a doctor will usually involve a clinical examination. Some exposures, eg asbestos and diving, require periodic medical examination by a doctor approved by the Health and Safety Executive (HSE) to undertake such examinations.

Biological and Biological Effect Monitoring

8.683 Biological and biological effect monitoring involves taking samples, eg of blood, urine, hair or breath, to measure the take-up of exposure to certain substances, eg lead or other chemicals. Only appropriately trained people may carry out this form of monitoring. Accredited laboratories must be used for the testing.

Keeping Individual Health Records

8.684 There is a requirement for health surveillance records to be kept (in a form that does not include confidential medical information) in addition to any occupational health records maintained on the individual employee.

Self-checks by Employees

8.685 Self-checks by employees are extremely important, as is the training of employees in what to look out for and report. However, self-checks are not enough to comply with regulations and must be supported by periodic checks by a responsible or qualified person.

Baseline Health Assessments

8.686 Baseline health assessments form part of a health surveillance programme when they are undertaken for the specific purpose of establishing baseline data, eg on commencement of exposure, against which future health surveillance results will be assessed. A qualified person undertakes baseline health assessments.

Health Surveillance Techniques

8.687 Health surveillance techniques should be:

- sensitive
- specific
- easy to perform and interpret
- safe
- non-invasive
- acceptable.

Ensuring Surveillance is Sensitive

8.688 Health surveillance techniques should be sensitive and sufficiently detailed or focused to be able to detect early changes in employees. Questionnaires must be sufficiently detailed for the reviewer to pick up a change immediately, and the equipment used for lung function testing must be able to record slight changes in volumes.

Ensuring Surveillance is Specific

8.689 Health surveillance techniques should be specific and relate to a particular aspect of health or health risk. If the risk assessment identifies a requirement for health surveillance, the techniques involved must relate to the risk.

Ensuring Surveillance is Easy to Perform and Interpret

8.690 The majority of employees should be able to do the tests or complete health surveillance questionnaires without difficulty.

8.691 Tests should not normally require prolonged absence from work. Appropriately trained people should be able to interpret the results quickly and accurately, and provide appropriate information to the employee on the findings.

Ensuring Surveillance is Safe

8.692 The techniques themselves should not put the employee's health at risk. For example, repeated x-rays could expose the employee to dangerous levels of ionising radiation.

Ensuring Surveillance is Non-invasive

8.693 Health surveillance techniques should be non-invasive, ie techniques should not involve:

- puncturing the skin
- inserting instruments into the body.

8.694 The vast majority of health surveillance techniques are non-invasive, but there are a few exceptions, eg blood tests for levels of chemicals. If invasive techniques must be used, there is a greater need for written employee consent to the procedure, and appropriate levels of confidentiality regarding the results.

Ensuring Surveillance is Acceptable

8.695 Techniques must not offend or upset employees. If surveillance involves undressing, the provision of a chaperone may be required to make the examination acceptable.

Common Health Surveillance Tests

Noise

8.696 Hearing tests use an audiometer and acoustic muffs and are normally conducted in a soundproof hearing booth. The audiometer tests hearing in each ear at different frequencies, from low (500Hz) to high (8000Hz), and produces a graph showing the results. Pre-test examination of the ear canals is essential to ensure the result is not invalidated by wax in the ear canal. Most audiometers can produce a graph comparing test results with previous results.

Vibration

8.697 The HSE has produced a detailed symptom questionnaire for completion by individuals exposed to vibration. Assessment is made of vascular, sensory and nerve impairment using the Stockholm Scale. A responsible person may administer the questionnaire, but anyone reporting symptoms must be referred to a qualified person for detailed assessment and recommendations on work practices.

Respiratory Hazards

8.698 In some cases, a respiratory symptoms questionnaire is adequate as the initial health surveillance for those exposed to:

- dusts
- fumes
- vapours
- other respiratory sensitisers or irritants.

8.699 The lung function test is the most common test for employees exposed to respiratory hazards. It is capable of measuring a range of capacities, scoring these against both the normal range for age and gender and against previous test results.

Skin Hazards

8.700 A symptom questionnaire and visual inspection by a responsible person allows employers to monitor employees' skin on a frequent basis.

8.701 As many skin irritants or sensitisers are also respiratory hazards, qualified people undertaking lung function testing will commonly assess skin at the same time.

The first signs of occupational dermatitis may appear virtually overnight (after days or years of exposure without symptoms) so if skin surveillance is required, regular and frequent assessment by a responsible person offers the best support to employers.

Other Health Monitoring

Health Monitoring that is Legally Required

8.702 Various pieces of legislation require the periodic monitoring of individuals' health, eg:

- pre-placement and annual health assessment of fitness to work is required under the Diving at Work Regulations 1997
- vision screening for display screen equipment users is required under the Health and Safety (Display Screen) Regulations 1992
- annual health checks must be offered to night workers under the Working Time Regulations 1998.

Health Monitoring that is Not Legally Required

8.703 Health monitoring procedures which are not legally required include:

- pre-employment health screening where there is no statutory duty but the employer considers it appropriate or beneficial to make a formal assessment of fitness for work
- the monitoring of reasons for absence and sickness data
- well-person screening, where employees are offered a general health check, advice on healthy choices and information on maintaining optimum health
- testing for drug or alcohol misuse, unless those being tested are in safety-critical jobs, where screening is a legal duty to ensure the safety of the employee and others.

Training

8.704 There are many types of training that may be required in relation to health surveillance. Examples include training for:

- managers/supervisors, in carrying out risk assessments in relation to health hazards
- responsible persons, in undertaking relevant health surveillance within the workplace
- employees, in self-checks and in understanding the potential health effects of inappropriate exposure to health hazards
- qualified people, in undertaking the various surveillance tests and in interpreting and communicating the results
- managers, in understanding the implications of health surveillance results and in identifying appropriate management strategies.

8.705 Health and safety training that deals with risk assessments and the principles of health surveillance may be available locally through trade organisations, local colleges or through professional bodies such as the Institution of Occupational Safety and Health.

8.706 Suppliers, such as those supplying skincare products, are often able to provide short, relevant, in-house training for employees on the skin hazards associated with chemicals and sensitisers, thereby increasing employees' ability to carry out self-checks and recognise symptoms.

8.707 Companies specialising in occupational hygiene or occupational health can often provide suitable training for responsible persons.

8.708 Giving employees verbal and written information to help them to understand and to maintain safe working practices to guard their health is standard in any occupational hygiene surveillance.

8.709 Employers should encourage supervisors, etc to review manufacturers' safety data sheets, and present the information to employees through team meetings, toolbox talks or as part of other in-house methods of raising and maintaining awareness of safe working practices.

8.710 Running periodic poster campaigns, using materials available from the HSE, on health surveillance topics is another way of keeping health surveillance in employees' minds. In some organisations, posters are designed and laminated in-house, with small competitions between teams or shifts to design the best poster.

8.711 Any occupational hygiene service should be able to demonstrate the competence of its workforce to undertake relevant health surveillance tests, and should have various levels of expertise to interpret results and recommend action plans.

List of Relevant Legislation

- Control of Noise at Work Regulations 2005
- Control of Vibration at Work Regulations 2005
- Control of Substances Hazardous to Health Regulations 2002
- Management of Health and Safety at Work Regulations 1999
- Working Time Regulations 1998
- Diving at Work Regulations 1997
- Health and Safety (Display Screen) Regulations 1992
- Health at Safety at Work, etc Act 1974

Further Information

Publications

HSE Publications

The following are available from *www.hsebooks.co.uk*.

- EH44 (rev 1997) *Dust: General Principles of Protection*
- EH66 (rev 1998) *Grain Dust*
- HSG61 (rev 1999) *Health Surveillance at Work*
- HSG122 (rev 2002) *New and Expectant Mothers at Work: A Guide for Employers*
- HSG137 *Health Risk Management: A Practical Guide for Managers in Small and Medium-sized Enterprises*
- HSG165 (rev 2000) *Young People at Work: A Guide for Employers*

- HSG167 *Biological Monitoring in the Workplace — A Guide to Its Practical Application to Chemical Exposure*
- INDG126 *Hand–arm Vibration — Employees*
- INDG175 (rev 2008) *Control the Risks from Hand-arm Vibration: Advice for Employers on the Control of Vibration at Work Regulations 2005*
- INDG232(L) *Consulting Employees on Health and Safety: A Guide to the Law*
- INDG296 (rev 1) *Hand–arm Vibration: Advice for Employees*
- INDG304 *Understanding Health Surveillance at Work: An Introduction for Employers*
- INDG362 *Noise at Work: Guidance for Employers on the Control of Noise at Work Regulations 2005*
- L5 (rev 2005) *The Control of Substances Hazardous to Health Regulations 2002 (as amended). Approved Code of Practice and Guidance*
- L55 *Preventing Asthma at Work: How to Control Respiratory Sensitisers*
- L140 *Hand-arm Vibration. The Control of Vibration at Work Regulations 2005. Guidance on Regulations*
- MS24 (reprinted, with amendments 2004) *Medical Aspects of Occupational Skin Disease*
- MS25 (rev 1998) *Medical Aspects of Occupational Asthma*
- MS26 *A Guide to Audiometric Testing Programmes*
- *Health Surveillance Under COSHH: Guidance for Employers*

Organisations

- British Occupational Hygiene Society (BOHS)
  Web: *www.bohs.org*
  The BOHS seeks to promote and protect occupational and environmental health and hygiene and offers training in occupational hygiene and related subjects.
- Institution of Occupational Safety and Health (IOSH)
  Web: *www.iosh.co.uk*
  IOSH is Europe's leading body for health and safety professionals and provides guidance on health and safety issues.
- National Examination Board in Occupational Safety and Health (NEBOSH)
  Web: *www.nebosh.org.uk*
  Founded in 1979, NEBOSH is an independent awarding body.
- Trades Union Congress (TUC)
  Web: *www.tuc.org.uk*
  The TUC describes itself as "the voice of Britain at work". With 71 affiliated unions representing nearly 7 million working people from all walks of life, it campaigns for a fair deal at work and for social justice at home and abroad.

Insurance

- Duty holders must consider three different types of insurance policies.
- It is important to distinguish between claims that result from a criminal prosecution and those which result from a civil action.
- Almost all companies are required to take out employers' liability insurance to cover liability for the death of, or personal injury to, their employees.
- Public liability insurance provides cover in respect of injury to persons (other than employees) and damage to property.
- Professional indemnity insurance covers liability arising from breach, or alleged breach, of professional duty by professionals in the performance of their work.
- Under the Construction (Design and Management) Regulations 2007, designers are responsible for identifying and minimising risks to the health and safety of construction and maintenance workers.

8.712 Regarding the Construction (Design and Management) Regulations 2007 (CDM Regulations) and insurance, the various duty holders will have to consider three different insurance policies, depending upon the role being undertaken.

1. Professional indemnity insurance.
2. Employers' liability insurance.
3. Public liability insurance.

8.713 The insurance policy of most concern to designers and CDM co-ordinators will be professional indemnity insurance.

8.714 For the principal contractor or contractor, the insurances that will be most affected by the CDM Regulations are employers' liability insurance and public liability insurance.

8.715 However, should principal contractors assume either the role of designer or of CDM co-ordinator, then they may also have to consider professional indemnity insurance.

Employers' Duties

8.716

- Under the Employers' Liability (Compulsory Insurance) Act 1969, it is a duty of almost all companies to take out employers' liability insurance to cover liability for the death of, or personal injury to, their employees.
- Employers must display a copy of the certificate of insurance at the place of business, in such a position that it can easily be seen and read by employees.
- Under the Construction (Design and Management) Regulations 2007, designers are responsible for identifying and minimising risks to the health and safety of construction and maintenance workers, as well as those who may be affected by the work.
- Designers must co-operate with the CDM co-ordinator and any other designers involved in the project.

In Practice

Insurance of Liability

8.717 While certain insurance policies will provide a measure of protection from claims which may arise as a consequence of the introduction of the Construction (Design and Management) Regulations 2007, it is important to distinguish between claims that result from a criminal prosecution and those which result from a civil action.

Criminal Liability

8.718 It is a general principle of English law that, as a matter of public policy, a punishment imposed by a court for committing an offence should not be shifted onto another party. Any insurance which covers the amount of criminal fines imposed by courts is not illegal but it is unenforceable. It is therefore impossible for an insured party to attempt to enforce legally an insurance policy which purports to cover such liabilities.

8.719 It is possible, however, to arrange insurance to cover the costs and expenses incurred in connection with the defence of a criminal prosecution up to the point when a judgment is delivered.

Civil Liability

8.720 Insurance for claims which are the result of civil actions is widely available. It should not be overlooked that insurance is a form of contract and, as with most contracts, there are terms, conditions and exclusions.

8.721 The decision whether to insure against civil action is largely still a question of choice, although employers are required to maintain insurance against liability for the death of, or personal injury to, their employees, in accordance with the Employers' Liability (Compulsory Insurance) Act 1969.

8.722 Particularly in the construction industry, it is becoming increasingly common to insist that, as part of the contractual terms, the party performing the work has to maintain public liability and, in certain circumstances, professional indemnity insurance.

Insurance Policies

Employers' Liability

8.723

- By virtue of the Employers' Liability (Compulsory Insurance) Act 1969, almost all companies are required to take out employers' liability insurance to cover liability for the death of, or personal injury to, their employees.
- Historically, the indemnity limit was unlimited but a minimum level of cover has been introduced, ie £5 million for any one event including costs and expenses. (This amount was prescribed by the Employers' Liability (Compulsory Insurance) Regulations 1998.)
- Cover is available above £5 million by purchasing additional tranches of cover.

- An indemnity limit of £5 million may not be sufficient in all situations and contractors should consider higher limits for unusual or large projects.
- The employer must display a copy of the certificate of insurance at the place of business of the firm, in such a position that it can easily be seen and read by employees.
- Health and Safety Executive inspectors are empowered to enter the employer's premises and to demand to see the policy document or a copy.

8.724 The aim of such compulsory insurance is to cover civil claims made against employers by any employees injured or made ill through the course of their work.

Employers' Liability — Loopholes

8.725 It has recently been reported that employers are increasingly taking advantage of a number of loopholes in the 1969 Act.

- The Act excludes:
  - students working on an unpaid basis
  - persons taking part in a youth or adult training programme
  - those on work experience.
- The law is not clear in relation to employees with no regular workplace and does not cover the self-employed. Case law on the Employers' Liability (Compulsory Insurance) Act 1969 has indicated that the insured employer cannot recover from the insurance company unless the employer's liability has been established by "action, arbitration or agreement".

Employers' Liability — Failure to Insure

8.726 The application of the 1969 Act was considered by the Court of Appeal in the case of *Richardson v Pitt-Stanley and Others* [1994]. The facts were that R, an employee of B Ltd, was injured during the course of his employment. He claimed damages from the directors of the company on the basis of their breach of the 1969 Act, in that they had failed to take out insurance against liability for injuries suffered by employees. The key issue in the case was whether the 1969 Act created civil liability for compensation as well as criminal liability.

8.727 The Court of Appeal ruled as follows.

- The appeal should be dismissed. The claim failed.
- The 1969 Act did not expressly provide that employers or directors should be civilly liable.
- The Act was intended to be a criminal statute.
- Employees had a remedy for negligence or breach of statutory duty. They did not have a cause of action in the civil courts under the 1969 Act.

Employers' Liability — Failure to Display

8.728 In August 1996, at Alnwick Magistrates' Court, a mining company was fined £2000 for failing to take out employers' liability insurance. Both the company and its chairman were charged with offences under the 1969 Act.

8.729 The facts were that during a visit by the Mines Inspectorate to a colliery in Northumberland, it was noticed that no certificate of liability insurance was on display. Seven notices were served on the firm operating the mine, requiring it to produce its liability policy. It failed to do so.

8.730 The result of the lack of such a policy was that if any of the firm's employees had been injured, there would have been no insurance cover for payouts resulting from a possible compensation claim.

8.731 In mitigation for the company and its chairman, it was stated that the insurance policy had not been kept up-to-date because of an oversight caused by pressure of business. As well as the fine imposed on the company, the chairman was fined £250.

Public Liability

8.732

- Public liability insurance is also known as third party liability insurance.
- Public liability insurance provides cover in respect of injury to persons (other than employees) and damage to property.
- It is usual practice for clients to insist that contractors take out and maintain public liability insurance (for sizeable limits of £10,000,000 or £20,000,000) before they commence work on-site.
- Public liability insurance policies typically contain an exclusion relating to professional work performed for a fee, ie that covered under a professional indemnity policy.

8.733 Designers and CDM co-ordinators usually consider that they have little need for public liability insurance, other than as an extension to the office-combined policy for the public liability risks associated with owning and/or occupying their own property. Care should be taken to ensure that this does not result in gaps in insurance. For instance, a claim for property damage or bodily injury can be the result of a designer's or CDM co-ordinator's non-professional activity on-site, eg the dropping of a cigarette causing a fire.

8.734 Designers and CDM co-ordinators should therefore consider purchasing public liability insurance as well as professional indemnity insurance.

Professional Indemnity

8.735 Professional indemnity insurance is taken out by most professionals to cover liability arising from breach, or alleged breach, of professional duty by them in the performance of their work.

8.736 Designers and CDM co-ordinators usually take out and maintain such insurance, as do contractors if they assume the role of designer or CDM co-ordinator. Even if that role is then subcontracted, the contractor may still have the prime responsibility to the client for any breaches of professional duty.

8.737 Unlike employers' liability and public liability policies, cover for professional indemnity insurance is only available in the UK on what is known as a "claims made" basis. This means that cover has to be in force at the time the claim is actually made against the designer or CDM co-ordinator. Thus, not only must cover be maintained during the construction period, it must also continue in force for the period in which the designer or CDM co-ordinator could face a potential claim of liability (usually 6 years or 12–15 years if the terms of the Latent Damage Act 1986 apply). It is not usual, therefore, to arrange professional indemnity insurance on a contract-by-contract basis.

8.738 Professional indemnity is a retrospective insurance. Key factors considered by underwriters when deciding whether to underwrite a risk include:

- the previous claims' record of the prospective insured
- the calibre of the individuals within the company.

8.739 The CDM 2007 may affect professional indemnity insurance in three ways.

1. Increased liability exposure of the designer.
2. The role of the CDM co-ordinator.
3. Criminal prosecution.

Increased Liability Exposure of the Designer

8.740

- The Construction (Design and Management) Regulations 2007 (CDM Regulations) clearly impose a responsibility on designers to identify and minimise risks to the health and safety of construction and maintenance workers, as well as those who may be affected by the work.
- Designers must ensure that the design includes adequate information about any aspect which might affect health and safety.
- Designers must co-operate and co-ordinate with the CDM co-ordinator and any other designers involved in the project. The requirement to design with safety in mind, including the process of construction, is tempered by a state of the art defence (ie that designers cannot be liable for alleged negligence when risks are unforeseeable due to scientific or technical information being unavailable or unsubstantiated) and by the extent that it is reasonably practicable to address these considerations.

8.741 Underwriters are not currently proposing to increase the premium rates for designers, but this may be a logical progression if the CDM Regulations result in more claims against designers.

Role of the CDM Co-ordinator

8.742 Underwriters who have researched the role of CDM co-ordinator mostly agree to insure the new discipline, subject to certain underwriting criteria. In general, the approach is to consider the relevant experience of the company or individual wishing to be insured. For example, an architect with experience of light industrial building projects would not be the ideal CDM co-ordinator for a heavy civil engineering project.

8.743 The regulations now require all parties with duties under the regulations to be "competent". Now that the HSE have set criteria for measuring this competence (reference Appendices 4 and 5 of the ACOP (L144)) it may be that unless the CDM co-ordinator has the requisite qualifications, professional memberships and requisite experience they may become uninsurable in terms of professional indemnity.

8.744 The fees earned from the role of CDM co-ordinator will probably have to be itemised separately on the insurance proposal to renew the existing professional indemnity policy. It is considered that the activities of the CDM co-ordinator are similar to those of a project manager. The rate applied to the fees declared for CDM co-ordinators may therefore be approximately the same as the rate charged for project management.

8.745 The requirement for the fees to be declared separately suggests that it would be advisable for the CDM co-ordinator's appointment to be separate from that of the designer. This may also be good risk management, given the different roles attaching to the designer and the CDM co-ordinator.

Criminal Prosecutions

8.746 While breach of the CDM 2007 may not automatically give rise to a civil action, such a breach may be admissible as evidence in a civil case. The probability of claims for breach of contract or negligence following criminal prosecution may therefore be quite high.

8.747 Given the likelihood of this situation occurring, some underwriters have indicated that they are prepared to extend the usual cover provided by professional indemnity policies to include costs and expenses incurred in the defence of any criminal proceedings which arise out of the activities and duties which are covered under the policy. This means that it would be prudent to have the activity of CDM co-ordinator included within the policy. This extension of cover will usually be operative if, in the reasonable belief of underwriters, the criminal proceedings may otherwise give rise to a claim under the policy, ie a civil action.

List of Relevant Legislation

- Construction (Design and Management) Regulations 2007
- Management of Health and Safety at Work Regulations 1999
- Employers' Liability (Compulsory Insurance) Regulations 1998
- Latent Damage Act 1986
- Health and Safety at Work, etc Act 1974
- Employers' Liability (Compulsory Insurance) Act 1969

Further Information

Publications

HSE Publications

The following are available from *www.hsebooks.co.uk*.

- HSE4 *Short Guide to the Employers' Liability (Compulsory Insurance) Act*
- HSG150 (3rd ed, 2006) *Health and Safety in Construction*
- L144 *Managing Health and Safety in Construction. Construction (Design and Management) Regulations 2007. Approved Code of Practice and Guidance*

Other Publications

- *The Law of Health and Safety at Work*, Selwyn, N. Croner Publications

Organisations

- Construction Industry Training Board (CITB) and Construction Skills
  Web: *www.citb.org.uk*
  The CITB and Construction Skills provide assistance in all aspects of recruiting, training and qualifying the construction workforce.
- Health and Safety Executive (HSE)
  Web: *www.hse.gov.uk*
  The HSE is responsible for the regulation of almost all the risks to health and safety arising from work activity in the UK.

Permits to Work

- A permit-to-work system is a formal written system which specifies the precautions which need to be taken before certain high-risk activities are undertaken.
- Employers are required by the Health and Safety at Work, etc Act 1974 to provide and maintain systems of work that are, as far as reasonably practicable, safe and without risks to health. This includes the implementation of permit-to-work systems where appropriate.
- Activities which require a permit-to-work system will be determined by a risk assessment. A permit-to-work should be considered for all activities where the potential risk is high and cannot be minimised sufficiently by the implementation of control measures.
- Permit-to-work systems are particularly important where failure to implement safety measures could allow serious and immediate risks to health and safety. Examples include working in confined spaces, maintenance of high voltage electrical apparatus and hot work.
- A typical permit to work should include:
  - permit title
  - permit number
  - the location of work
  - plant identification
  - a description of the work to be carried out
  - hazard identification
  - necessary precautions
  - protective equipment
  - authorisation
  - acceptance
  - extension arrangements
  - completion of work arrangements
  - cancellation section.
- It is vital to determine who will be responsible for issuing the permit to work, who will be responsible for putting in place the necessary precautions and who will be responsible for the work that is carried out.
- Contractors working on-site are the employer's responsibility. If the employer has permit-to-work systems in place, contractors should use them.
- Records must also be kept of all permits to work.
- Training should be given to all staff, including those who will issue permits and those working under them, before a permit-to-work system is implemented.

8.748 A permit-to-work system is a formal written system used to control high-risk activities. A type of safe system of work, permit-to-work systems specify the precautions which need to be taken to control the risks. Examples include working in confined spaces, maintenance of high-voltage electrical apparatus and hot work.

8.749 Permit-to-work systems allow work to begin only after safe procedures have been defined and carried out.

Employers' Duties

8.750

- Employers have a general duty to ensure, so far as is reasonably practicable, the health, safety and welfare at work of all employees under the Health and Safety at Work, etc Act 1974 (HSWA).
- Employers must provide and maintain systems of work that are, as far as reasonably practicable, safe and without risks to health, under the HSWA and under the Management of Health and Safety at Work Regulations 1999 (MHSWR).
- Employers must carry out their activities, as far as reasonably practicable, in such a way as not to harm the health and safety of those not in their employment, under the HSWA.
- Under the MHSWR, employers must make an assessment of the risks to the health and safety of their employees while at work, and a similar assessment of the risks to the health and safety of persons not in their employment that may arise as a result of their work activities.

Employees' Duties

8.751

- Employees have a duty to take reasonable care of their own health and safety and that of other people who may be affected by their work under the Health and Safety at Work, etc Act 1974.
- Employees must use machinery, equipment, dangerous substances, transport equipment, means of production and safety devices in accordance with any instruction and training given by the employer, under the Management of Health and Safety at Work Regulations 1999 (MHSWR).
- Employees must inform their employer of any danger to health and safety posed by a work activity, under MHSWR.
- Employees must also inform their employer of any shortcomings in the employer's protection arrangements under MHSWR.

In Practice

Risk Assessments and Safe Systems of Work

8.752 Risk assessment will identify activities for which formal safe systems of work are necessary.

Which Safe Systems of Work Require Permit-to-work Systems?

8.753 Permit-to-work systems need to be introduced for all activities where the potential risk is high and if risks cannot be eliminated. They are particularly important where failure to implement safety measures could allow serious and immediate risks to health and safety. Examples of activities where permit-to-work systems are necessary include:

- work in confined spaces, such as vessels or sewers
- hot work, such as cutting and welding in premises under refurbishment
- working on high-voltage electrical equipment, such as jointing 33,000v cables
- maintenance of machinery on production lines, such as power presses
- working at height while using fall-arrest systems
- lone working in isolated areas, such as lift motor rooms
- work with asbestos, eg removal of insulation from hot water systems
- work on lifts
- excavation work.

8.754 Many industries, such as chemical and offshore operations, regularly use permit-to-work systems for a host of activities.

8.755 Once the activities which require a permit-to-work system have been established, the details should be entered on a master sheet.

Developing a Suitable Permit-to-work System

8.756 The first consideration is to determine the essential contents of a typical permit-to-work system, which can be tailored to suit any particular activity. Employers must also ensure the system:

- is unambiguous and avoids misleading statements
- identifies all potential hazards
- specifies all necessary isolation of plant and electrical equipment
- clearly determines the personal protective equipment (PPE) necessary
- clearly identifies responsible persons
- adequately deals with the hand-back of plant and equipment after work.

Permit Title

8.757 The title should describe the type of work to be carried out, eg entry into confined space.

Permit Number

8.758 The permit number is a unique reference, for cross-referencing and record purposes.

Location of Work

8.759 There should be clear identification of the work location, eg the exact location of a piece of plant within a factory, rather than the name of the factory.

Plant Identification

8.760 A clear identification of the specific plant will also be necessary, eg when carrying out maintenance on an electrical circuit breaker, identify the particular circuit breaker rather than just the switchboard.

Description of Work to be Carried Out

8.761 It is important to not only describe the exact work to be carried out, but also to outline the limitations of the work which can be carried out under the permit.

Hazard Identification

8.762 Any residual hazards or additional hazards introduced by the work itself should be identified, eg when working in a confined space, the use of certain plant will introduce fresh hazards.

Necessary Precautions

8.763 The precautions necessary to control the risks should be outlined, such as any isolation required, as well as how the isolation will remain secure. For example, when working on machinery the precautions may specify temporary guarding from other machines, or other parts of that machine, and how guarding will be retained in place, such as locking off.

8.764 If the individual issuing the permit effects such isolation, the permit should specify who is responsible and that person should sign the permit once the procedure has been carried out.

Protective Equipment

8.765 PPE, such as respiratory protective equipment, should be specified, as should any barriers, notices or earthing equipment required.

Authorisation

8.766 The signature of the person issuing the permit is necessary to confirm that all safety precautions have been met.

8.767 In some cases, not all the measures can be taken before the issue of the permit. For example, some isolators may be assessable only after work has commenced.

Acceptance

8.768 The person in charge of carrying out the work generally signs the acceptance section. It should not be signed until after the person issuing the permit has described the hazards involved and the precautions required. The signature should indicate that the details have been explained to all those who will carry out the work.

8.769 On occasions, the person issuing the permit may be the person carrying out the work, in which case the same process needs to be completed.

Extension Arrangements

8.770 If the work needs to be carried out after the time specified on the permit, any changes of personnel or arrangements should be included. A new time of expiry will be needed.

Completion of Work Arrangements

8.771 The person responsible for the work should sign to acknowledge that the equipment can be returned to service and specify any special precautions which may need to be taken. The person responsible should also confirm that all tools and/or temporary connections have been removed and that the equipment is safe for recommissioning and/or for return to normal service.

Cancellation

8.772 The person issuing the permit signs the cancellation section, which confirms that the plant has been successfully returned to service and that it is no longer safe to carry out work on it. A general permit-to-work form could be used or a more specific permit-to-work form may be required, eg for:

- electrical work
- work in confined spaces
- hot work
- lone work.

Control of Permit-to-work Systems

8.773 An essential element of a permit-to-work system is the management of the system. It will be necessary to determine who will devise the system, who will be responsible for issuing the permit, who will be responsible for putting in place the relevant precautions (if not the person issuing the permit) and who will be the person responsible for the work that is to be carried out.

8.774 Each type of permit will have different control features, however the essentials should be the same.

Who Devises the Permit to Work?

8.775 It is rare for a single person to devise a permit-to-work system, unless the activity is relatively simple. In most cases, a safety advisor or officer will be involved to outline the principles.

8.776 In most cases, the means of making apparatus and systems safe are technical and an engineer with knowledge of the task should be involved. For example, for a permit to work on production machinery, a mechanical engineer who understands the principles behind the machinery will need to have input.

8.777 The employees who will, or do, operate the equipment can also make valuable contributions to the process.

8.778 Information from the equipment manufacturer or maintenance manual may be helpful. In practice, the development of a permit-to-work system is normally a team effort.

Who Issues the Permit to Work?

8.779 A competent person, with enough technical knowledge to understand the hazards involved and how to control the risks, should issue the permit to work. The person should have specific training in the operation of permit-to-work systems. Such persons are sometimes known as nominated persons.

8.780 For example, a person nominated to issue permits to work for high-voltage electrical equipment may be a qualified electrical engineer, and a person nominated to issue such permits for hot work on a construction site might be a site manager or works foreman.

Who Puts in Place the Precautions?

8.781 The person who puts in place the precautions needs to be a competent person, working under the direction of the person issuing the permit (if this is not the same person).

8.782 For tasks which have to be carried out from more than one location, or if there is a multitude of tasks that require different skills, a second person will put in place some of the necessary precautions.

8.783 The person issuing a permit to work for a confined space may not have the necessary skill or instrumentation to declare that the atmosphere is safe, hence a second person or organisation may be employed.

Who is Responsible for the Work Carried Out Under the Permit?

8.784 The person responsible for the work carried out under the permit is not always involved in making the work safe, but would normally be competent to carry out the work.

8.785 The person must understand the principles of the permit-to-work system, as he or she has to sign it and explain its contents to staff. The person should also know what to do in an emergency.

8.786 On occasions, the person carrying out the work can also be responsible for issuing the permit. This is more usual for less hazardous tasks, such as hot work.

What About Contractors?

8.787 Contractors working on-site are the employer's responsibility. If the employer has permit-to-work systems in place, contractors should use them. The employer should ensure the contractor's staff are adequately trained and then issue the permit.

8.788 If an employer hires an expert contractor, eg for the disposal of fuel tanks which have contained other hazardous materials, the employer should ensure the contractor is competent and operating a satisfactory permit-to-work procedure of its own. In this example, if the tanks were cut up prior to removal from the site, a hot work permit would be implemented by the employer's own staff prior to the contractor's work.

Implementing Permit-to-work Systems

8.789 Permit-to-work procedures are used to control high-risk activities and it is vital to get it right first time. When the systems are first used, there should be adequate supervision to ensure the process is stopped if it proves to be unsafe. It is sensible to test the procedures, if possible, before they are used for the first time.

8.790 Employees' views should be sought before permit-to-work systems are implemented. Employees' views on working the systems once they are in place should also be sought.

Monitoring Permit-to-work Systems

8.791 As with most health and safety arrangements, there will be a need for monitoring. The monitoring should consist of two elements.

Reactive Monitoring

8.792 Reactive monitoring should include a review of accidents and near misses while operating the new systems. Those using the permit-to-work systems should be encouraged to report any near misses.

Active Monitoring

8.793 Active monitoring should include workplace inspections while the activity is being carried out.

Training

8.794 Training should be carried out before a permit-to-work system is implemented. All staff, including those who will issue permits and those working under them, should have sufficient, relevant training.

8.795 It should be noted that some of those involved will need to be qualified engineers or have some supplementary training on the control of particular hazards. This will be necessary where technical knowledge is needed to avoid danger.

8.796 Training specifically for the use of permit-to-work procedures falls into two categories.

All Staff Required to Work Under Permit-to-work Systems

8.797 Training for this group should involve:

- a basic understanding of permit-to-work systems
- employees' responsibilities under the systems
- what tasks can only be carried out under a permit
- what to do in an emergency.

8.798 All persons working under a permit-to-work procedure should have a briefing on the safety precautions before work commences.

8.799 The training can be relatively simple, such as a briefing or as part of the induction process. It is recommended that all staff are given a copy of the employee factsheet.

Staff Responsible for Issuing Permits and Putting in Place Precautions and Staff in Charge of Work

8.800 Training for this group should include:

- principles of permit-to-work systems
- hazards involved in the particular work
- implementation of the necessary precautions, eg means of effecting isolation
- how to test to ensure it is safe for work to commence, eg testing atmospheres in confined spaces or using live line detectors before work commences on electrical apparatus
- what to do in an emergency.

List of Relevant Legislation

- Management of Health and Safety at Work Regulations 1999
- Health and Safety at Work, etc Act 1974

Further Information

Publications

HSE Publications

The following are available from *www.hsebooks.co.uk*.

- HSG65 (rev 1997) *Successful Health and Safety Management*
- HSG250 *Guidance on Permit-to-work Systems: A Guide for the Petroleum, Chemical and Allied Industries*
- HSG253 *The Safe Isolation of Plant and Equipment*
- INDG98 (rev 1997) *Permit-to-work Systems*

Safe Systems of Work

- A safe system of work is a written procedure detailing how a particular work activity should be carried out in order to minimise the risks (to both employees and non-employees) involved in performing that activity.
- Employers must provide and maintain systems of work that are, as far as is reasonably practicable, safe and without risks to health under the Health and Safety at Work, etc Act 1974.
- The risk assessment process can be used to identify activities that require a safe system of work.
- Safe systems of work are normally adopted for work activities still carrying a residual risk, despite the implementation of control measures. Examples of activities that might require a safe system of work include working at height, maintenance of electrical systems, control of vehicle movements on-site, etc.
- A safe system of work will need to take several factors into account. These include:
  - the work location, equipment and plant to be used
  - staff undertaking and supervising the work
  - the necessary safety precautions
  - the sequence in which the work activity should be carried out.
- Some activities carry such a high residual risk that a more stringent safe system of work might be necessary, such as a permit-to-work system.
- When implementing safe systems of work, employers should ensure that employees are consulted, training is provided prior to implementation and there is an appropriate level of supervision, particularly if systems are in use for the first time.
- Safe systems of work should be monitored and reviewed regularly to make sure they are being followed. This is particularly the case in the event of any accidents, incidents or near misses.
- Employers should keep records of all safe systems of work in a place that is fully accessible to employees.

8.801 A safe system of work is a formal procedure resulting from an examination of any workplace activity in order to identify its hazards and assess its risks. This exercise culminates in the identification of safe methods of work to ensure that the hazards are eliminated or residual risks minimised.

8.802 Safe systems of work were the forerunner of the requirement to carry out risk assessments in the workplace. Many of the elements of formulating a safe system of work duplicate the risk assessment procedure.

Employers' Duties

8.803 The Health and Safety at Work, etc Act 1974 requires employers to:

- ensure, as far as is reasonably practicable, the health, safety and welfare of their employees while at work
- ensure, as far as is reasonably practicable, the health, safety and welfare of non-employees
- provide and maintain systems of work that are, as far as is reasonably practicable, safe and without risks to health

- carry out their activities, as far as is reasonably practicable, in such a way as not to harm the health and safety of those not in their employment.

8.804 The Management of Health and Safety at Work Regulations 1999 require employers to:
- make an assessment of any risks to the health and safety of their employees while at work
- make a similar assessment of the risks to health and safety of persons not in their employment that may arise as a result of their (ie the employer's) work activities.

Employees' Duties

8.805 The Health and Safety at Work, etc Act 1974 requires employees to:
- take reasonable care of their own health and safety and that of other people who may be affected by their activities while at work.

8.806 The Management of Health and Safety at Work Regulations 1999 require employees to:
- use machinery, equipment, dangerous substances, transport equipment, means of production or safety devices in accordance with any instruction and training given by the employer
- inform their employer of any danger to health and safety posed by a work activity
- inform their employer of any shortcomings in the employer's protection arrangements.

In Practice

Identifying the Activities That Require Safe Systems of Work

8.807 All work activities will require safe systems of work. However, for many activities, a formal system will not be required. For example, tasks such as operating a PC, sweeping the floor or changing a light bulb might not require a formal procedure, although they may require a risk assessment and control measures. However, there will be few residual risks involved following the implementation of control measures.

8.808 A formal safe system of work will be required for activities that have been assessed and control measures introduced, but where significant residual risks remain. Examples of such tasks include:
- maintenance of assembly-plant machinery
- window cleaning
- lone working
- vehicle loading-bay activities
- construction work
- working at height
- working in confined spaces
- maintenance of electrical systems and equipment.

Identifying the Risks

8.809 This will be a part of the normal risk assessment process. If possible, where hazards have been identified with a particular activity, any risks should be eliminated or minimised. Where this is not possible, a formal safe system of work should be developed for that activity.

8.810 The residual risks identified through this process will determine the control measures outlined in the safe system of work.

8.811 For example, a work activity is to be carried out in close proximity to an excavation-site. The hazard, ie the excavation, cannot be eliminated. There is a risk of persons falling into the excavation-site since access to the site is necessary for other work being carried out at the same time. A safe system of work will be necessary to control this risk.

Development of the Safe System of Work

8.812 Having identified the residual risks that need to be controlled with the safe system of work, it is now necessary to look at the elements of such a system.

Work Location

8.813 This element will address the physical layout of the workplace and take into account such things as:

- access and egress
- the space available for the activity, allowing for obstructions such as plant and equipment
- hazards such as excavations, low ceilings, overhead power lines
- access hazards, such as working at height or working in a confined space
- whether there are adequate means of fire protection
- whether the lighting is suitable or, alternatively, what additional lighting is required
- whether the temperature is suitable for the location and the task
- if there is suitable provision of welfare facilities.

Equipment and Plant

8.814 Factors to take into account when considering equipment and plant include the following.

- Is it suitable for the task?
- Can it be operated safely in the work environment?
- Are there any noise factors to be taken into account?
- Are the persons allocated to the task trained, and if necessary, certified to use the equipment?
- Has it been maintained to the required standard and is the necessary certification on-site?
- Where is the fuel, ie diesel, petrol or liquefied petroleum gas, to be stored?
- How is the power for the equipment to be safely used?
- Is there adequate security on-site to prevent unauthorised use of equipment?

Materials

8.815 This element includes substances as well as the more traditional materials used for the task. Factors to take into account include the following.

- Is the material suitable for the task?
- Can the material be safely handled manually, or is there a need for mechanical handling?
- Have there been any manual handling assessments carried out for the task?
- Are there facilities for materials to be stored safely?
- Have assessments under the Control of Substances Hazardous to Health Regulations 2002 (COSHH) and/or the Dangerous Substances and Explosive Atmospheres Regulations 2002 (DSEAR) been carried out on substances to be used or created (eg dusts)?
- Are there arrangements for the safe disposal of any waste materials?

Personnel

8.816 This is, arguably, the most important factor. Considerations include the following.

- Who is to manage the work?
- Does the work need to be supervised and, if so, by whom?
- Who will be responsible for the various tasks, ie who issues personal protective equipment (PPE) and who is responsible for safety briefings?
- Are all persons competent for the tasks they are to undertake, eg plant operators?

Safety Precautions

8.817 As a result of all the preliminary personnel considerations, what special safety precautions are needed? For example, consider the following.

- Is there a need to isolate plant/equipment/systems from power sources?
- Is PPE needed? If so, what type of equipment and to what standard?
- What safety signage is necessary, eg caution or prohibition notices?
- Are there any special instructions or safety briefings necessary before work commences?
- Are emergency procedures in place in the event of a failure of the safe system of work?

Sequence of Operations

8.818 A sequence of operations should be formulated to ensure the safe completion of the activity. For example, electrical equipment must be isolated before work commences.

Example of a Typical Formal Safe System of Work

8.819 This is only an example of what a typical safe system of work should include. Many of the elements are not applicable to more simple activities or for activities in a fixed location, such as a factory. The system will need to be adapted to the particular activity being undertaken.

Example of a Safe System of Work

Activity	Include a description of the work.
Location	Name the location, access and egress arrangements, any hazards posed by the location and any control measures.
Equipment and Plant	List plant and equipment to be used. List any hazards posed by the use of such plant in this environment and the subsequent control measures.
Materials	List materials to be used on the activity and any risks posed by their use. List any chemical substances used or produced, eg dusts and fumes, and refer to assessments conducted under the COSHH and/or the DSEAR. List any necessary control measures in relation to substances, as well as any manual handling considerations.
Personnel	List key personnel who will perform the activity. For example, who will manage the task? Is there a need for supervision? How many people are needed? What skills do they need?
Safety Precautions	Safety precautions should be in addition to those control measures listed under the individual headings above. Included here will be items such as additional PPE, safety briefings and the requirement for any permit-to-work systems.
Sequence of Operations	This section would list, in chronological order, the elements of the task, together with necessary control measures. It should be a logical sequence, starting with access arrangements and the necessary safety arrangements. It should then describe, in an orderly fashion, the sub-elements of the activity.

8.820 This table could be adapted for use as a form, in order to have a written record of a safe system of work. Alternatively, a slightly more simplified version could be used, covering the location of work, its description, hazards and the safe methods to be adopted.

8.821 A formal, written safe system such as this is sometimes known as a method statement. Many clients ask contractors for method statements before work is carried out on their behalf. Indeed, they are often part of a tendering process and the formulation of a sound safety method statement can be the key to obtaining a contract.

Permit-to-work Systems

8.822 Some work activities carry such high residual risks that they require a more stringent safe system of work, such as a permit-to-work system. Examples include entry into confined spaces and hot work.

Safe Systems of Work Register

8.823 All tasks that have a formal safe system of work should be entered onto a central register. This register must be available to all staff so that they are aware of the safe system of work they should use while carrying out a task.

Implement the Systems

8.824 Successful implementation of safe systems of work will depend on two things — their correct formation and the competence of people using them. Be wary of implementing safe systems of work that are unnecessarily complicated and include measures not strictly required. Such systems will not only lower productivity but the people using them will lose confidence in the system. Once this happens, it may put the whole process into disrepute — with a consequent risk to health and safety.

8.825 The exact procedure carried out to implement safe systems of work will vary according to the type of activity and the organisation. A construction company formulating a method statement for a one-off activity will have to closely monitor the system and make any subsequent changes as that activity is being carried out.

8.826 The implementation of a safe system of work for a new production activity in a factory, on the other hand, can be more measured. As the facility is being built, safe methods of work can be developed by both the manufacturer of the new plant and the employees who will be working at the new plant/on the new activity.

8.827 General points to consider when implementing safe systems of work are to:

- fully involve employees in their creation and implementation — staff need to "buy in" to the system for it to work
- ensure training is provided prior to implementation
- make employees aware of which activities are subject to formal safe systems of work
- ensure that when systems are in use for the first time, there is a level of supervision and expertise available to stop or change the process if it proves unsafe
- ensure written procedures are updated to reflect any necessary changes to the system.

Monitor the Arrangements

8.828 Once the safe systems of work are in place and working, there will be a need to monitor the arrangements. The monitoring should consist of two elements: reactive monitoring and active monitoring.

Reactive Monitoring

8.829 Reactive monitoring involves conducting a review of accidents and near misses while operating the new safe systems. The employer should already have accident reporting procedures in place and these may well need to be reinforced for this particular purpose. Careful analysis of accident data may result in changes needing to be made to the safe systems of work.

Active Monitoring

8.830 Workplace inspections should initially include a special study of new safe systems of work. Again, as a result of these, changes may be necessary.

Training

All Employees

8.831 An outline of what a safe system of work is should be given to all employees within induction training and general health and safety refresher training.

8.832 This should cover:

- the principles of safe systems of work

- which employee activities are subject to formal safe systems of work, including reference to the safe systems of work register
- employees' individual responsibilities in the working of such systems
- what to do when things go wrong
- how to report concerns with safe systems of work.

Those Who Supervise and Implement Safe Systems of Work

8.833 Such employees should receive the same training as all employees with the addition of:

- a review of the risk assessments that led to the formulation of the safe system of work
- residual risks and how they are to be controlled by the safe system of work
- how the system will work in practice and how the control measures will control the risks
- individual roles within the system
- emergency procedures.

8.834 Technical knowledge, such as that held by qualified engineers, might be necessary to implement some safe systems of work. Tasks such as isolating electrical supplies or some machinery may need specialist knowledge, which could be supplemented by training such as that outlined above.

List of Relevant Legislation

- Management of Health and Safety at Work Regulations 1999
- Confined Spaces Regulations 1997
- Health and Safety at Work, etc Act 1974

Further Information

Publications

HSE Publications

The following is available from *www.hsebooks.co.uk*.

- HSG65 *Successful Health and Safety Management*

Safety Signs and Signals

- Safety signs provide information or instruction about health and safety by means of signboards, safety colours, illuminated signs, acoustic signals, verbal communications and hand signals.
- The Health and Safety (Safety Signs and Signals) Regulations 1996 require employers to provide and maintain safety signs where risk assessment indicates that risks cannot be avoided or adequately reduced without them.
- The regulations contain specific requirements for the different types of safety signs in terms of colour, spacing, visibility and specific hand signals to be used.
- Employers should make sure that signage is used in the most appropriate and effective way, eg by not placing signboards with different messages close together.
- Typical uses of safety signs might be to indicate that a machine guard should be in place before starting work or as a warning about dangerous substances in pipes and containers.
- Fire safety signs must be provided to a specific standard and be maintained where necessary to comply with fire safety legislation.
- Suitable safety signs may also be required by other legislation, eg for the packaging, storage and transport of dangerous goods.
- Employers must provide employees with information, instruction and training on the meaning of safety signs.

8.835 The term "safety sign" is relatively broad and basically means a sign that covers a specific situation, object or activity and gives information or instructions on health and safety.

8.836 The Health and Safety (Safety Signs and Signals) Regulations 1996 only require employers to use safety signs if they have taken all other appropriate safety measures, but the risks cannot be adequately reduced without them. However, safety signs can also be useful as additional control measures and are widely used in this manner.

Employers' Duties

8.837 The Health and Safety (Safety Signs and Signals) Regulations 1996 require employers to:

- provide and maintain safety signs where risk assessment indicates that those risks cannot be avoided or adequately reduced without them
- provide safety signs to a specified minimum standard
- provide suitable and sufficient instruction and training for employees on the meaning of safety signs.

8.838 The Health and Safety at Work, etc Act 1974 requires employers to:

- provide such information, instruction, training and supervision as is necessary to ensure, as far as reasonably practicable, the health and safety at work of all employees.

8.839 The Management of Health and Safety at Work Regulations 1999 require employers to:

- make an assessment of risks to health and safety of their employees while at work
- make a similar assessment of the risks to health and safety of persons not in their employment that may arise as a result of their (the employers') work activities.

Employees' Duties

8.840 The Health and Safety at Work, etc Act 1974 requires employees to take reasonable care of their own health and safety and that of other people who may be affected by their activities while at work

8.841 The Management of Health and Safety at Work Regulations 1999 require employees to:

- use machinery, equipment, dangerous substances, transport equipment, means of production or safety devices in accordance with any instruction and training given by the employer
- inform their employer of any danger to health and safety posed by a work activity
- inform their employer of any shortcomings in the employer's protection arrangements.

In Practice

Identifying the Need for Safety Signs

8.842 Safety signs do not remove risk. They are a warning that the risk exists and, in some cases, advise what actions should be taken. Safety signs should be used as a "last resort", ie only where other health and safety measures have been taken, such as engineering controls, but it was found that they did not adequately control the risks.

8.843 If other health and safety measures reduce the risks so that no significant risk remains, safety signs are not necessary. For example, if there is a risk of falling from a height, then other health and safety regulations generally require measures other than safety signage such as edge protection. In these circumstances signage is not an alternative control measure.

8.844 On the other hand, it may be necessary on occasions to remove such fencing temporarily. In these circumstances signage will be necessary to indicate alternative protection measures that must be in place when the fencing is removed.

8.845 In reviewing the risk assessments for the need of safety signs, existing signage should be checked to ensure that it meets current standards and fulfils its purpose effectively.

Determining the Appropriate Signage

8.846 This will be determined by the requirements of the regulations and the relevant British Standards.

Safety Colours

8.847 There are four different colours specified, each of which have a particular meaning as follows.

Red

8.848 This colour indicates a prohibition showing actions that are not allowed, eg stop signs and no smoking signs.

8.849 Emergency stop devices are also coloured red to indicate the means of stopping a process or equipment.

8.850 Red is also used for fire-fighting equipment, such as extinguishers and to show the location of fire-fighting equipment.

Yellow

8.851 This colour is a warning colour indicating a risk of danger. It is used to identify hazards such as fire, explosion or chemical or to indicate where care should be taken to avoid a hazard such as obstacles or dangerous passages.

Blue

8.852 This colour indicates a mandatory action, such as wearing a hard hat or ear protectors in a certain area. It is therefore, in effect, an instruction to use hard hats and ear protection in such areas.

Green

8.853 This colour indicates a safe condition. It is used to show such things as emergency escapes and the provision of first-aid.

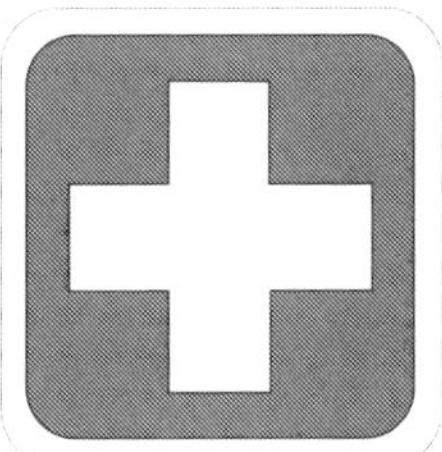

Signboards

8.854 Signboards contain pictograms to give information. They will be coloured according to the safety colour scheme.

8.855 Signboards should be large and clear enough to be easily noticed and understood. They should be well positioned so that it is obvious what area they relate to or where a piece of equipment is located.

8.856 There must be enough light for the sign to be seen easily, which may involve providing artificial lighting and/or making the sign reflective. Avoid having too many signboards too close to each other, as this may cause confusion or allow important information to go unnoticed.

8.857 The signboards should be durable and be fastened securely in the appropriate place. A signboard that has deteriorated or has become unfastened will not provide the information needed. They should be cleaned regularly to make sure they remain clear, and be replaced or resurfaced where necessary.

8.858 In general, signboards should be permanent, although in some cases portable signboards are appropriate, eg for use only on occasions where the floor is wet.

8.859 The pictograms used on signboards should be as simple as possible. The shapes and colours that must be used on the different types of signboard are shown in the following table.

Shapes and Colours for Different Types of Signboards

Safety Colour	Meaning	Contrasting Colour	Symbol Colour	Shape and Colour Details
Red	Prohibition	White	Black	Circular red band and cross bar. Red to cover at least 35% of sign area.
	Fire	White	White	Red rectangle or oblong. Red to cover at least 50% of sign area.
Yellow	Caution	Black	Black	Triangle with black band. Yellow to cover at least 50% of sign area.
Blue	Mandatory	White	White	Circular blue disc. Blue to cover at least 50% of sign area.
Green	Safe condition	White	White	Green rectangle. Green to cover at least 50% of sign area.

8.860 Sometimes it is useful to put a supplementary signboard containing text next to the pictogram signboard, particularly when installing a new sign. The background colour on the supplementary signboard should be the safety colour of the main signboard with the text in the contrasting colour.

8.861 It is best to use simple text, keeping it as short as possible to make it quick and easy to read and understand, eg the words ″no smoking″ should be used rather than ″smoking is prohibited in this area″.

Typical Situations Where Signs Are Used

Storage and Use of Hazardous Materials

8.862 Where hazardous substances of quantities that present a hazard are stored in rooms, enclosures, cupboards and other areas, safety signs are likely to be needed to warn employees of these dangers.

8.863 In the case of some containers, the warning labels required by the Chemicals (Hazard Information and Packaging for Supply) Regulations 2002 will be sufficient signage to control the risk.

8.864 Warning signs or labels should be put near the storage area or affixed to the door leading to a storage room. If there is only one class of substance kept in the storage area, the sign should indicate the hazard, eg flammable. However, where a number of different substances are kept, eg flammable, toxic and corrosive, a general danger (exclamation mark) warning sign may be used.

8.865 Most of the signage will consist of warning signs, ie they will be yellow. However, some may be:

- prohibitive, ie red, such as no smoking
- mandatory, ie blue, requiring the use of personal protective equipment
- safe, ie green, indicating the escape route.

8.866 Warning signs for sites where 25 tonnes or more of dangerous substances are kept, are required by the Dangerous Substances (Notification and Marking of Sites) Regulations 1990.

Machinery

8.867 Signage should rarely be used as the sole measure to prevent contact with the dangerous parts of machinery. Signage is not an alternative to guarding or other measures where these are practicable.

8.868 Signage is, however, extensively used to control subsidiary hazards or to supplement more rigorous control systems. For example, machinery is often accompanied with mandatory signs requiring the use of hearing protection.

8.869 Sometimes a combination of signs can be used — a yellow caution sign may be combined with a mandatory sign containing the text, "all guards must be in place before starting".

8.870 Other examples include prohibition signs on machine guards, which prohibit the removal of the guard. Although, in this case the sign should only supplement a secure guard.

Traffic/Pedestrian Routes

8.871 Caution yellow tape or paint should be used to delineate walkways from road traffic where barriers or a more permanent separation cannot be made. Other walkways through, eg machine shops or stores, can be signed similarly.

8.872 Obstacles on pedestrian routes, such as low headroom can be indicated with yellow caution tape or paint in the form of diagonal yellow and black strips.

8.873 Walkways can also be delineated with the use of signboards, such as prohibitive no entry signs or mandatory signs, indicating that the prescribed walkway must be used.

8.874 Traffic signs prescribed in Road Traffic Regulation Act 1984 should be used where there is a risk to employees from traffic movement or presence, even though strictly speaking the Act may not apply to a place of work.

Dangerous Substances In Pipes and Containers

8.875 Where containers or pipes contain dangerous substances, it is generally necessary to affix signs or labels to them. In some cases, this is not appropriate, such as where:

- a very short pipe connects to a container with a sign on
- the contents of a container change frequently (in which case there must be other measures in place to inform staff of the hazards).

8.876 The previous colour coding outlined is clearly not appropriate for the identification of the substance in a pipe since there are a multitude of such substances. BS 1710:1984 recommends a system of identification through colouring and examples include:

- purple for acids and alkalis
- dark brown for mineral oils
- light brown for gases
- green for water
- red for water in association with fire fighting
- blue for air.

8.877 The direction of the flow of the substance should be indicated with a directional arrow placed on the colour. The arrow should be black or white depending on the pipe colouring.

8.878 In the case of marking for pipes, the previous principles do apply to subsidiary warning signs indicating, eg that the contents are flammable, corrosive or toxic. These would be yellow warning signs attached to the pipe over the pipe's particular colour code.

8.879 Containers used for the storage of dangerous substances must be labelled using the appropriate signs. This must also include text giving the name of the dangerous substances and/or the nature of the hazard.

8.880 Signs or labels should be durable and affixed where they will be visible. Labels may either be self-adhesive or painted onto the surface.

8.881 Too many signs too close together may cause confusion. The signs or labels should therefore be positioned at points where employees could be exposed to the dangerous substance, eg filling points and drain valves.

Acoustic Signals

8.882 Acoustic signals must be loud enough for workers to hear them. The recommended level is 10dB above the level of ambient noise at that frequency, although it should not be painfully or excessively loud. When setting the volume of the signal, it is important to take into account anything that may impair the hearing of workers, eg wearing personal protective equipment.

8.883 It should be easy to recognise the signal, ie by length of the pulse and the length of time between pulses. Avoid using more than one acoustic signal at the same time as this is too confusing.

8.884 If a device that has an acoustic signal is capable of producing both variable frequencies and constant frequencies, the variable frequencies should be used when the danger is greater or the action is needed more urgently.

8.885 The code for evacuation, however, must always be continuous. The signal should sound for as long as the danger exists or until a specified action is taken to acknowledge the danger.

8.886 It is important to maintain acoustic signals, which will include checking them at regular intervals to make sure they are still functioning. The frequency of testing will depend on how hostile the environment is and the manufacturer's instructions.

Illuminated Signs

8.887 Illuminated signs must be bright enough to be seen, but not so bright that they cause a glare.

8.888 They should be located far enough away from other light sources and from other illuminated signs, so that confusion is not caused.

8.889 The lit up area of the sign may contain a pictogram or just consist of a single safety colour. Some signs are capable of being both constantly illuminated or flashing. In this case, the flashing should indicate a greater danger or the need for more urgent action.

8.890 The frequency and duration of flashes should be set up so that the message is clear and will not be confused with other illuminated signs. Also, some fast flash rates may trigger epilepsy in susceptible people.

8.891 Where acoustic signals and illuminated signs are used together, pulses and flashes should be synchronised.

8.892 It is important to maintain illuminated signs, which will include checking them at regular intervals. The frequency of testing will depend on how hostile the environment is. If a flashing sign is used to indicate imminent danger, it is vital that any malfunction, eg the need for a new bulb, is either detected rapidly or prevented, eg by using duplicate bulbs.

8.893 Arranging for a fail-safe system will depend on the degree of risk. In the case of a failure of the flashing sign, it may be necessary for the automatic closure of the activity the sign is protecting.

Hand Signals

8.894 Hand signals are used for giving directions during certain operations, such as directing vehicles or cranes.

8.895 Before the operation begins, it is important that both the operator and the signaller are clear what the hand signals mean. The signals should be straightforward, clear and easily made and understood. The signaller should be competent in signalling. This includes being suitably trained.

8.896 The basic minimum requirements for hand signals are given in Part IX of the first schedule to the Health and Safety (Safety Signs and Signals) Regulations 1996 and these should be followed for all activities involving hand signals. These hand signals are shown below.

General Signals

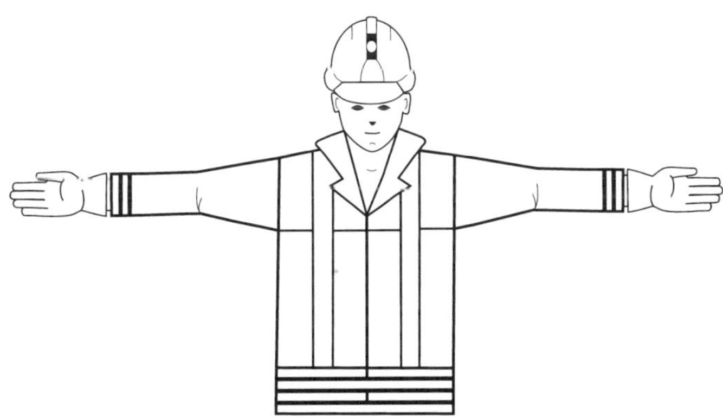

Hand Signal (General): START, Attention, Start of Command

Description: both arms are extended horizontally with the palms facing forwards.

Hand Signal (General):STOP, Interruption, End of Movement

Description: the right arm points upwards with the palm facing forwards.

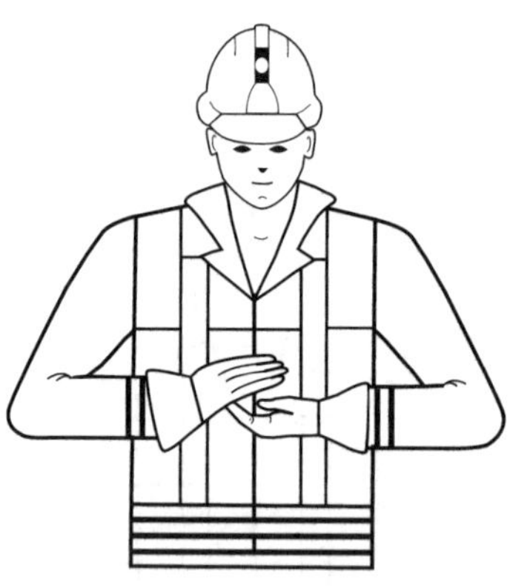

Hand Signal (General): END of Operation

Description: both hands are clasped at chest height.

Vertical Movements

Hand Signal (for vertical movements): RAISE

Description: the right arm points upwards with the palm facing forward and slowly makes a circle.

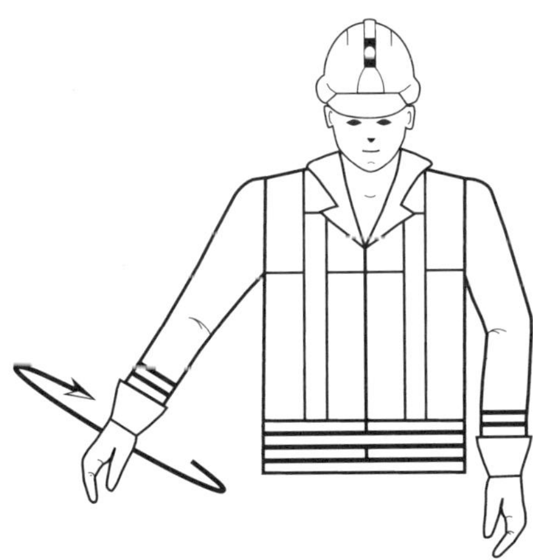

Hand Signal (for vertical movements): LOWER

Description: the right arm points downwards with the palm facing inwards and slowly makes a circle.

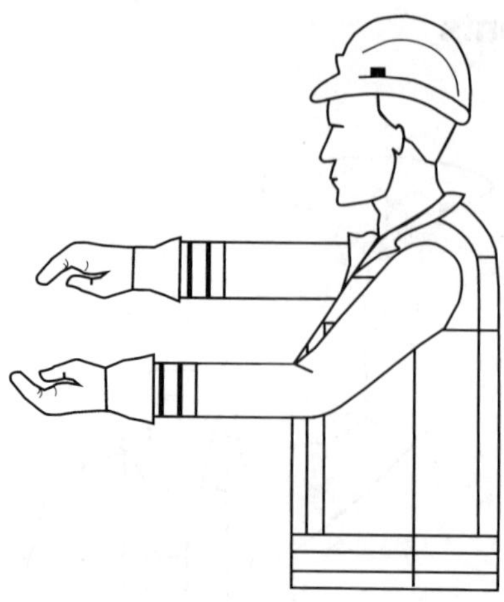

Hand Signal (for vertical movements): VERTICAL DISTANCE

Description: the hands indicate the relevant distance.

Horizontal Movements

Hand Signal (for horizontal movements): MOVE FORWARDS

Description: both arms are bent with the palms facing upwards and the forearms make slow movements towards the body.

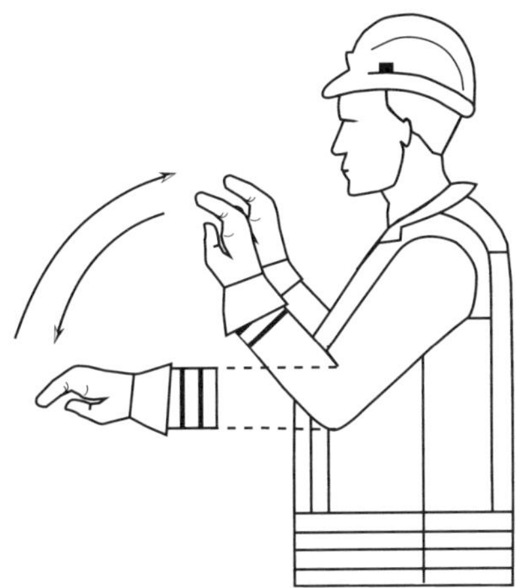

Hand Signal (for horizontal movements): MOVE BACKWARDS

Description: both arms are bent with the palms facing downwards and the forearms make slow movements away from the body.

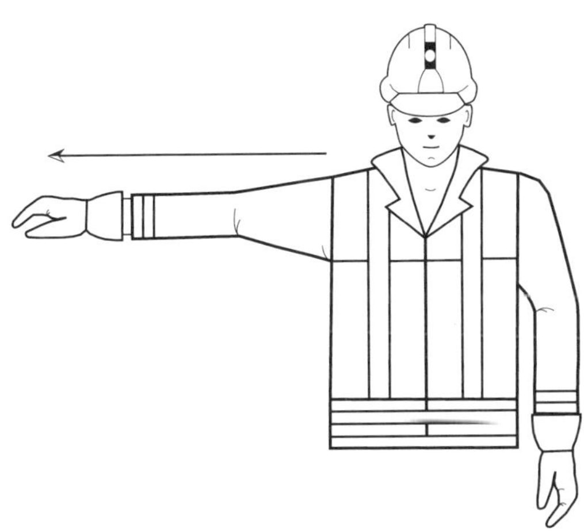

Hand Signal (for horizontal movements): RIGHT, to the signalman's

Description: the arm is extended more or less horizontally with palm facing downwards and slowly makes small movements to the right.

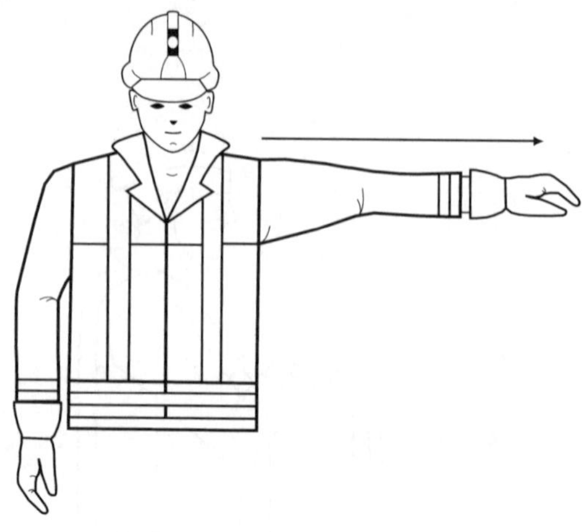

Hand Signal (for horizontal movements): LEFT, to the signalman's

Description: the arm is extended more or less horizontally with palm facing downwards and slowly makes small movements to the left.

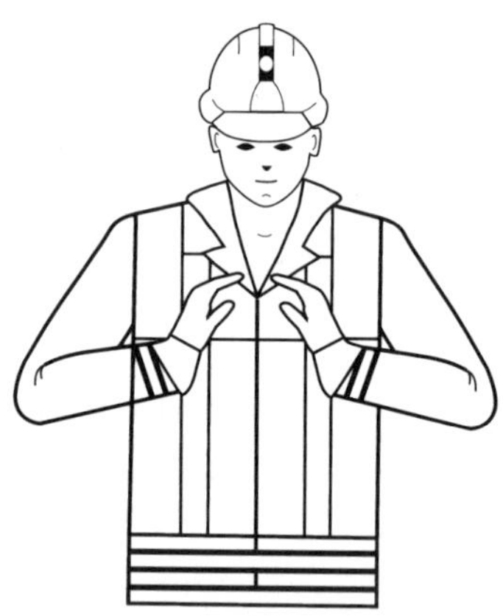

Hand Signal (for horizontal movements): HORIZONTAL DISTANCE

Description: the hands indicate the relevant distance.

Danger

Hand Signal: DANGER, Emergency Stop

Quicker and Slower

8.897 To indicate "quicker" all signals should be performed faster and to indicate "slower" all signals should be performed more slowly.

8.898 In some cases, the operation will require signals that are not given in these regulations or other codes, and so additional signals should be devised and used based on the principles outlined.

8.899 It is essential that the same system of hand signals is used throughout the workplace or organisation. New recruits may have used a different code and therefore may need training on the system used in their new workplace.

8.900 While the operation is taking place, the signaller should be positioned so that he or she has a clear view of all the manoeuvres without being at risk, eg of being struck by the vehicle.

8.901 The signaller should only be responsible for signalling and any measures necessary for protecting people during the operation. He or she should not be distracted by having to perform other tasks. He or she may sometimes need help, such as other signallers.

8.902 The operator must be able to see the signaller clearly. The signaller must wear one or more distinctive items, such as a high-visibility jacket. This must be distinctive enough to distinguish him or her from others in the vicinity who may also be seen wearing such jackets.

Verbal Signals

8.903 Verbal signals are spoken messages, either human or electronic, and may be either instantaneous or recorded. They may be used to give directions during an operation or to warn of a hazard, eg the familiar railway warning of "mind the gap". Verbal messages should be clear, simple and easy to hear and understand.

8.904 People who give or receive verbal signals should have a good grasp of the language. Before the operation begins, it is important for both the person giving verbal directions and receiving them to know what code will be used.

8.905 If most staff on the site speak a first language other than English, it is acceptable for the code to be in the other language (so long as the staff speak the same alternative).

8.906 Examples of verbal signals are given in Part VIII of the Appendix to the Health and Safety (Safety Signs and Signals) Regulations 1996 and these include:

- "start" — begin the operation
- "stop" — interrupt or end the operation
- "end" — stop the operation
- "raise" or "lower" — to have a load raised or lowered
- "forwards" or "backwards"
- "quickly" — to speed up a movement for safety reasons
- "danger" — perform an emergency stop
- "right" or "left" — move to the signaller's right or left.

Fire Safety Signs

8.907 The provision of fire safety signs is covered by fire safety legislation, but the Health and Safety (Safety Signs and Signals) Regulations 1996 specify the types of signs that must be used for fire safety, eg to show the location of fire-fighting equipment.

8.908 The requirement for fire safety signs will be determined as part of the fire safety risk assessment required under the Regulatory Reform (Fire Safety) Order 2005.

8.909 Fire safety signs:

- indicate escape routes if there is a fire, eg illuminated signs
- show the location of fire-fighting equipment, eg signboards
- warn that there is a fire, eg acoustic signals.

8.910 Signs that show the location and type of fire-fighting equipment must be red.

8.911 Those that show fire doors, emergency exits and escape routes must be green.

8.912 The Health and Safety (Safety Signs and Signals) Regulations 1996 (SSSR) transposed into UK legislation the requirements of EEC Directive 92/58 and in particular the requirements concerning signboard design. This resulted in the "Euro-sign" design, which may be supplemented with directional arrows as detailed in the regulations. However the regulations also state that signs that conform to the British Standard BS5499 will also meet the requirements as they follow the same basic pattern. The basic principle is one of uniformity and that the same design is used throughout the premises. However, text-only fire safety signs must not be used, they must be displayed in conjunction with pictograms.

Escape Routes and Emergency Exits

8.913 Emergency exits should have the sign fixed immediately above the door, where possible, or in a location where it is easy to see and least likely to be obscured by smoke.

8.914 For areas where there could be some uncertainty about what route to take to escape during an emergency, eg where the exit cannot be seen from the place of work, fire safety signs should be put up at suitable places along the escape route. These signs would incorporate directional arrows and a pictogram.

8.915 Although the selection of the most appropriate signage is important, the correct application and positioning of the signs is critical. To this end the BSI has produced a Code of Practice for escape route signing in the form of BS5499 Part 4.

8.916 The code provides "guidance on the application and siting of escape route signs together with advice on the use of arrows to provide directional information". Interestingly, the code recommends the use of the "internationally agreed" symbol as contained in BS5499 and not the Euro-sign.

8.917 The code states that "an escape route system should provide simple identification of the means of escape to allow people to escape without assistance, possibly under conditions of stress". To do this the code recommends that seven elements have to be addressed as follows.

- *Sign design*: Signs should provide clear, unambiguous instruction that will lead people directly to a final exit out of a building.
- *Sign location*: The escape route sign should indicate the shortest travel distance when a choice of routes is available. Any changes in direction should be marked with signs and escape route signs must take priority over other signs. Signs must be clear to see and unobstructed. The code gives further information and guidance on these matters.
- *Mounting height*: Escape routes signs should not be mounted at unusual heights, rather they must be in the normal field of view. The code details various aspects of sign mounting.
- *Supplementary text*: As already mentioned the BSI recommends the use of supplementary text but it should not overpower any symbols.
- *Arrows*: Arrows are commonly used particularly when confusion may arise.
- *Size*: Sign size is calculated according to the furthest distance from which it is required to be read. The code offers tables that enable correct calculation for sign size for internally and externally lit signs.
- *Position of signs*: Using illustrations, the code details the type of sign required and how it should be positioned. The code also gives advice in respect of sign construction in terms of its durability and suitability. It also recommends regular servicing and maintenance of escape route signs.

Lighting

8.918 Emergency lighting, illuminated signs and other appropriate measures should be used where the level of natural light may be low during an emergency. These methods of illumination should allow for failure of the main power source.

Fire-fighting Equipment

8.919 A red signboard or safety colour should be used to show the location and type of fire-fighting equipment. In some cases, eg where fire-fighting equipment is located in a place that is difficult to see, an appropriate sign incorporating a direction arrow should be used.

Fire Alarms

8.920 Fire alarms are a type of acoustic signal and should meet the standards for such signals. The sound level of the signal should be significantly higher than the ambient noise in the workplace, to ensure that it is easy for everyone to hear it from everywhere in the building. It should be obvious that it is a fire alarm and a continuous signal should be used to indicate evacuation is necessary.

8.921 Noise from a fire alarm and verbal signals should not get in the way of communication with the emergency services, eg where a sounder is next to the telephone used to alert the fire brigade.

Signage Required by Other Legislation

Packaging of Hazardous Substances

8.922 Suppliers of hazardous substances have a duty to label such substances together with some safety information.

8.923 The Chemicals (Hazard Information and Packaging for Supply) Regulations 2002 require safety symbols with an orange background indicating the hazard that the product poses. Each sign is accompanied by some text giving risk and safety phrases.

Storage of Large Amounts of Dangerous Substances

8.924 Where 25 tonnes or more of dangerous substances are present on a site, the Dangerous Substances (Notification and Marking of Sites) Regulations 1990 apply.

8.925 The aim of these regulations is to provide information on the dangerous substances present to the emergency services if they attend an incident at the site, ie by displaying warning signs. Although there is some overlap between these regulations and the Health and Safety (Safety Signs and Signals) Regulations 1996, the main differences are that:

- there is no specified quantity of dangerous substances that must be present before the Health and Safety (Safety Signs and Signals) Regulations 1996 apply
- the Dangerous Substances (Notification and Marking of Sites) Regulations 1990 aim to inform the emergency services, while the Health and Safety (Safety Signs and Signals) Regulations 1996 aim to inform employees.

8.926 Generally, if warning signs are displayed on a site in order to comply with the Dangerous Substances (Notification and Marking of Sites) Regulations 1990, they will also serve to comply with the Health and Safety (Safety Signs and Signals) Regulations 1996 in relation to storage of dangerous substances in rooms, enclosures and other areas.

Explosive Atmospheres Signage

8.927 There is a specific requirement under the Dangerous Substances and Explosive Atmospheres Regulations (DSEAR) to display a warning sign at the entrance into areas where an explosive atmosphere will, or may, exist. The regulations state that the sign should be triangular and that at least 50% of the background of the triangle should be yellow. The Regulations do not mandate the additional inclusion of the words “Danger Explosive atmosphere” that are shown below.

Transport of Dangerous Goods

8.928 Employers should be aware of safety signs required for the transport of dangerous substances.

8.929 The requirements for such signage on packaging and vehicles are contained in the Carriage of Dangerous Goods and Use of Transportable Pressure Equipment Regulations 2007.

8.930 If employees are involved with such goods and rely on the information on the packaging and vehicles to safely handle the goods, they need to be trained in the meaning of the relevant signage.

Using Safety Signs Effectively

Proximity of Signboards

8.931 Do not place signboards with different messages close together. This will dilute the impact of the different individual signs. Poor examples include those often seen outside construction sites where a mass of individual signage is included on a single signboard.

8.932 In this manner a mass of signs including mandatory signs, such as "hard hats must be worn" are alongside warning signs such as "danger, vehicles reversing". Such signage is necessary but posting them on one board with many others will diminish their impact.

8.933 The placement of signboards at locations such as these does pose a challenge since the entrance to most construction sites is the point at which the signage is needed. Careful planning of the access and staging of the necessary signage will be necessary.

8.934 It may be acceptable for signage referring to the same hazard to be placed together on occasions, eg a warning notice of a confined space may be combined with a notice prohibiting unauthorised entry.

Unambiguous Signage

8.935 Signage must be unambiguous. Conflicting signage, such as a prohibition notice forbidding entry to a compound placed together with a mandatory notice requiring the use of eye protection beyond that point, is clearly confusing. The real message is possibly that only authorised personnel should enter and then only when wearing eye protection!

Enforcement of Signage

8.936 Signage, particularly prohibition signage, must be enforceable. For example, placing "no smoking within 10m" signage on a flammable material store within 10m of a boundary fence adjoining a public highway is unenforceable.

8.937 Prohibition notices must be enforced if they are to remain believable. No entry signs on doors that are habitually left open will not only render that sign useless, but also tend to diminish the meaning of all signage on the site generally.

Relevance of Signage

8.938 Signage must be relevant. Warning signs warning of hazards that have long since been removed bring signage into disrepute at that particular location.

Training

8.939 In many cases, the meaning of safety signs is easy to understand. However, some basic training and information should be provided, particularly with regard to safety signs used on the site.

8.940 Explanations of safety signs should be given during induction and refresher health and safety training. It should include an explanation of:

- colour coding of signage and its implication
- basic hand signals
- acoustic signals and what they indicate
- fire safety signage
- basic verbal signals
- what to do if signage is missing or damaged.

8.941 Emphasis of the mandatory significance of red and blue safety signage and the implications of ignoring signage or misunderstanding it must be made.

8.942 Further training will be necessary where employees are involved with dangerous goods, using symbols and/or instructions required for the supply or transport of such substances.

List of Relevant Legislation

- Carriage of Dangerous Goods and Use of Transportable Pressure Equipment Regulations 2007
- Chemicals (Hazard Information and Packaging for Supply) Regulations 2002
- Dangerous Substances and Explosive Atmospheres Regulations 2002
- Management of Health and Safety at Work Regulations 1999
- Health and Safety (Safety Signs and Signals) Regulations 1996
- Dangerous Substances (Notification and Marking of Sites) Regulations 1990

Further Information

Publications

HSE Publications

The following are available from *www.hsebooks.co.uk*.

- CAIS16 *Safety Signs in the Catering Industry*
- INDG184(L) *Signposts to Health and Safety (Safety Signs and Signals) Regulations 1996*
- L64 *Safety Signs and Signals: Guidance on Regulations 1996*
- HSE Video *Best Signs Story: Safety Signs at Work*

British Standards

The following are available from *www.bsi-global.com*.

- BS 5378–2: 1980 *Safety Signs and Colours: Specifications for Colorimetric and Photometric Properties of Materials*
- BS 5499–4: 2000 *Safety Signs, Including Fire Safety Signs. Code of Practice for Escape Route Signing*
- BS 5499–5: 2002 *Graphical Symbols and Signs. Safety Signs, Including Fire Safety Signs. Signs with Specific Safety Meanings*
- BS 6736: 1986 *Code of Practice for Hand Signalling for Use in Agricultural Operations*
- BS 7121–2: 2003 *Code of Practice for Safe Use of Cranes: Inspection, Testing and Examination*
- BS 7121–3: 2000 *Code of Practice for Safe Use of Cranes: Mobile Cranes*
- BS 1710: 1984 *Specification for Identification of Pipelines and Services*
- BS EN 60849: 1998 *Sound Systems for Emergency Purposes*

Organisations

- British Standards Institution (BSI)
  Web: *www.bsi-global.com*
  Founded in 1901, the BSI develops and provides supporting information on national and international standards in a wide range of areas.
- Health and Safety Executive (HSE)
  Web: *www.hse.gov.uk*
  The HSE is responsible for the regulation of almost all the risks to health and safety arising from work activity in the UK.

Site Security

- The security requirements of a construction site should be based on a suitable and sufficient assessment of the risks presented by the site.
- Consideration should be given to the location of the site access and of the site office.
- It is good practice to segregate a construction from the general public. On many construction sites, the location of the site perimeter will be obvious.
- Suitable signs should be displayed, warning of any hazards and informing visitors to report to the site office.
- The site should be secured using suitable fencing, or other suitable materials or structures.
- Security should be extended to include suitable gates that can be locked securely when the site is not occupied.
- All fencing, gateways and notices must be checked regularly and maintained as required to ensure the security of the site.
- Suitable secure areas must be provided for the safe storage of plant, equipment and materials.
- Barbed wire and razor wire should only be deployed in extreme circumstances and after ample consideration of the risks and its location.
- CCTV systems may be used to improve site security.
- In certain cases, it may be necessary to employ the services of security guards or of a specialist security firm.
- Ladders and other access equipment should be secured and protected from use by appropriate means, when the site is unoccupied.
- When work takes place inside occupied premises, consideration needs to be given to (accidental) intrusions by other people using the premises. This includes work in schools, hospitals, care homes and many other environments.

8.943 Construction site security is an important aspect of controlling the risks to members of the public and others who may be affected by the operation of the site. It is also well established that construction sites provide a special lure to children and so additional steps must be taken to deny them access. In most cases, this is achieved by fencing the site off. Site security is also an important financial issue for construction sites as over £1 billion worth of plant is stolen from them every year.

Employers' Duties

8.944 The Health and Safety at Work, etc Act 1974 places a duty on all employers (and the self-employed) to take reasonably practicable steps to ensure the health and safety of people who are not in their employment, such as members of the public.

8.945 The Construction (Design and Management) Regulations 2007 requires that construction sites, so far as is reasonably practicable and in accordance with the level of risk posed:

- have their perimeter identified by suitable signs, so arranged that their risk extent is readily identifiable
- are fenced off.

8.946 Under the Management of Health and Safety at Work Regulations 1999, employers have to assess the risks to the health and safety of their employees and of others who may be affected by the work activity. This is for the purpose of identifying the necessary preventive and protective measures. The risk assessment will help to determine the appropriate levels of site security needed. In addition, employers are required to co-operate and share information with any other employers who share the same workplace (construction site).

8.947 Under the Occupiers' Liability Act 1956, the occupiers of the construction site have a common law duty of care to those who may access the site. This duty is extended to trespassers by the Occupiers' Liability Act 1984. In many cases, these duties will be best met by having appropriate levels of site security in place.

Employees' Duties

8.948 Under the Health and Safety at Work, etc Act 1974, employees have a duty to ensure that they, or their employer, reinforce safe working procedures and do not put others at risk. They are required to:

- take reasonable care of their own health and safety and that of other people who may be affected by their activities at work
- co-operate with their employer to enable the employer to comply with health and safety duties.

8.949 Under the Management of Health and Safety at Work Regulations 1999, employees are required to:

- use any machinery, equipment, dangerous substances, transport, safety devices and means of production in accordance with any training and instructions provided by the employer
- inform the employer of any serious and imminent dangers to health and safety
- inform the employer of any shortcomings in the employer's health and safety arrangements.

In Practice

Risk Assessment

8.950 It is recommended that the findings of a suitable risk assessment are considered when determining the security provisions for a construction site. Consideration needs to be given to the risks associated with the site and the work, and consideration should be given to the level of enticement that the site may hold for children.

8.951 The risk assessment should consider the location and number of access gates, the location of the site office, etc. It should also consider the location of scaffolds and excavations and their potential attraction to children.

8.952 In the case of work in occupied premises, the risk assessment should consider the potential for accidental intrusion from common areas (such as lifts and stairs) and the nature of others using the premises, especially in the cases of work in schools, hospitals and care homes, etc.

Site Access and Site Office

8.953 The task of controlling access to the construction site begins at the site entrance. The entrance should be clearly marked and should have separate access routes provided for pedestrians and vehicles.

8.954 Visitors and workers are required to report to the site office as soon as they arrive. Appropriate signs should be displayed to direct them to the site office or other fixed control or reporting-in point. A safe means of access the to site office must be provided for all visitors. The site office should be located close to the (main) site entrance and manned by either site security staff or the (principal) contractor.

8.955 Visitors and visiting workers must not be allowed to walk around unaccompanied unless they have undergone the site induction process and are familiar both with the site and the risks they may be exposed to. They should be made aware of typical and specific site-related hazards, restriction and emergency procedures.

8.956 It is recommended that the number of gates and other access routes into the site are minimised, so as to reduce the chances of unauthorised access to the site.

Site Perimeter and Site Boundary

8.957 On many construction sites, the location of the site perimeter will be obvious. The precautions to be taken to secure the site perimeter should reflect the level of risk.

Displaying Warning Signs

8.958 Suitable notices should be displayed warning that the site is a construction site and that access is prohibited. The signs should be as clear and unambiguous as possible and should be sited so as to be readily visible. Warning signs should make it explicit that access to the site is restricted to authorised persons and that all visitors must report to the site office. It is recommended that visitors are not allowed access to the site until they have completed recorded site induction training.

8.959 Warning notices cannot be used to avoid liability for death or personal injury caused by negligence. Under the Unfair Contract Terms Act 1977, any exclusion of liability (whether in a contract term or in a notice) is void if used for the purpose of evading liability for death or personal injury caused by negligence.

Use of Existing Structures

8.960 Where construction work is being carried out inside an existing building, then the walls and doors of the existing structure may be used to provide security. It may be necessary to supplement this security by boarding over windows and upgrading the door for the duration of the construction work.

8.961 In other cases, the layout of the site and the site characteristics will influence the position of the site perimeter fencing, etc. In some cases it may be possible to use existing permanent features, such as walls and fences.

Phased Construction Work

8.962 For some projects, it may be possible to phase the construction of new structures so they will form an effective barrier as the work progresses. As walls or new buildings are constructed, they may be used to form part of the perimeter fencing.

8.963 In the case of phased construction work, the site may need to be expanded and/or moved as the work develops. As sections of the site are finished, they may be released to the client or to the sales department (perhaps to allow viewings of the finished parts of the development). In such cases, consideration needs to be given to safe means of access to, and egress from, the released parts of the site.

Site Perimeter Fencing

8.964 A two-metre high fence usually provides a reasonable site barrier. Site perimeter fences may be made from a wide range of materials, including metal mesh, and should be difficult to climb. It is recommended that a close mesh is used that prevents children getting their hands and feet through it and gaining hand or foot holds.

8.965 Sectional fencing should be locked together so that it cannot be separated easily. Separating the fencing should:

- require the use of a tool
- be done from the inside of the site.

8.966 Gaps underneath the fence (and the gate) should be kept as small as possible in order to stop anyone gaining access from under the fence. Efforts should be made to ensure that children cannot gain access through gaps in or under (temporary) fencing. In the case of uneven ground, it may be necessary to level the surface in order to prevent gaps under the fence.

8.967 In many cases, sectional fencing is supported in moveable feet. Where the feet of this type of fencing points into pedestrian areas, these should be highlighted to help to avoid tripping hazards.

8.968 Where the fencing is solid (such as plywood, and other similar sheeting materials) the effects of the wind need to be considered. The perimeter fencing needs to be designed to be secure and safe in all foreseeable weather conditions, including high winds. Solid fencing has the added advantage of keeping the site obscured, reducing the potential level of enticement that it may hold for children.

Gates and Other Access Points

8.969 Site access points should be securable. It is recommended that gates should form part of the fence and, as such, should be of the same size. Gates should be locked whenever the site is unoccupied. Consideration needs to be given to use of the gate in windy conditions. The gate should not slam shut or be blown open.

Maintenance of Fencing

8.970 It is recommended that there are regular checks on the condition and continued suitability of the perimeter fencing. General weathering, vandalism, attempted unauthorised site access, land movement and changes to the nature of the site may lead to deterioration in the effectiveness of the perimeter fencing. Records should be kept of all checks and of corrective work (such as replacing fence sections, tightening fence ties, refooting, etc.).

Storage Compounds

8.971 It may be necessary to have a locked and secured compound. Whether it is on or off-site, the storage compound(s) should be big enough to accommodate all of the plant, equipment and materials out of working hours securely and safely.

8.972 Special consideration needs to be given to the storage of plant, equipment and materials in the site compound. Suitable arrangements must be made for the safe and secure storage of flammable materials and explosives, etc.

8.973 Storage containers may also be provided as a means of providing additional security, especially for small or valuable items. Storage containers may also be a suitable means of providing safe storage for higher risk materials, such as (highly) flammable liquids, toxic materials, etc.

Additional Security for Fencing

8.974 Additional security measures — such as the use of barbed wire and razor wire — should only be made under carefully controlled conditions and after careful consideration. Such measures should only be sited at heights in excess of two metres and in locations where accidental contact can be prevented. Suitable warning notices must be displayed.

CCTV

8.975 CCTV may be used to improve site security. To be effective, and to operate within the law, suitable signs and notices must be posted warning of the use of CCTV on the site. When using a CCTV system, it is essential to the validity of the material gained that the Data Protection Act 1998 and the Human Rights Act 1998 requirements are complied with. Failure to act within these guidelines could result in action being taken against the CCTV user or hamper the police's ability to use the footage gained to investigate a crime and to prosecute offenders.

Lighting

8.976 Suitable site and perimeter lighting can be a useful deterrent. It is also possible to incorporate infra-red devices — such as motion detectors — that will switch the lighting on if the site security is compromised.

Security Guards

8.977 Under certain circumstances, it may be necessary to improve the out of hours' site security by the use of security guards. These may be based at a particular site or may travel between sites in a locality. In many cases, it is possible to contract the services of a security company, rather than to employ the guards directly.

8.978 If security guards are expected to work alone, consideration needs to be given to their safety. A lone worker risk assessment should be completed and appropriate control measures introduced.

8.979 If the guards are expected to patrol the site, then suitable defined walkways need to be provided and suitable lighting is required. Security guards must be made aware of any site-specific hazards, such as excavation, etc.

Plant and Equipment

8.980 Plant and equipment should be stored in a safe and secure compound when not in use. The keys to mobile plant and equipment should be removed and stored in a separate, secure location.

8.981 The risks of theft from the site and of damage to materials, equipment and plant while on-site can be reduced by controlling the deliveries. It is recommended that all deliveries are signed for and that staff are aware that they are signing to state that the goods were received in good condition. Where possible, deliveries should also be timed to coincide with them being needed on-site.

8.982 Expensive plant and equipment should be on-site for the minimum amount of time — not only to reduce the costs, but to reduce the potential for theft and damage, including damage to the site due to attempted theft.

Ladders and Scaffolds

8.983 The risks from unauthorised access to ladders, scaffolds and types of access equipment should be assessed and suitable precautions put in place. This may necessitate the removal of ladders from position at the end of the working day. Ladders may be secured away and chained and padlocked for added security. In other circumstances, access to the ladders may be prohibited by the use of ladder boards secured over the first 1.5 to 2 metres of the rungs to prevent access.

Work Within Occupied Premises

8.984 When work is being carried out inside occupied premises, it may be necessary to consider the security of the construction from accidental intrusion, such as from someone alighting from the lift at the wrong floor and accidentally coming into the construction area. Alternatively, access from common staircases may need to be considered.

8.985 Consideration also needs to be given to construction site security in certain special cases, such as when construction is within schools or hospitals. In the case of schools, it is preferable to arrange to carry out the work when the schools are not occupied, ie during school holidays. In the case of hospitals and (residential) care homes, consideration needs to be given to the safety and also to the dignity of (potentially) confused people.

Staff Awareness

8.986 All employees should be made aware of the organisation's policies with respect to theft from site. It should be made clear that it is company policy to prosecute whenever possible in cases of theft or criminal damage.

8.987 It is recommended that security measures are discussed at the top level within the organisation and that all senior staff understand fully the implications of poor site security.

8.988 Employees are expected to report suspicious incidents and should be reassured that anything they report will be treated in confidence.

8.989 Good control of staff and vehicles on-site is also essential. Security staff should regularly check and search all employees' lockers and contractors' vehicles. Employees' private vehicles should be kept off-site.

Police

8.990 It is recommended that close liaison is maintained with the local police in the event of anticipated criminal activities, etc. If any plant is stolen, the theft should be reported to the local police immediately, giving as much information as possible about the missing item, including all serial numbers.

List of Relevant Legislation

- Construction (Design and Management) Regulations 2007
- Management of Health and Safety Regulations 1999
- Management of Health and Safety at Work Regulations 1999
- Data Protection Act of 1998
- Human Rights Act of 1998
- Occupiers' Liability Act 1984
- Unfair Contract Terms Act 1977
- Health and Safety at Work, etc Act 1974
- Occupiers' Liability Act 1956

Further Information

Publications

HSE Publications

The following are available from *www.hsebooks.co.uk*.

- HSG151 *Protecting the Public — Your Next Move*
- L144 *Managing Health and Safety in Construction. Construction (Design and Management) Regulations 2007. Approved Code of Practice*

Other Publications

- BS 7958: 2005 *Closed Circuit Television (CCTV). Management and Operation Code of Practice*
- *Traffic Signs Manual,* Department for Transport
- *The Safety of Loads on Vehicles: Code of Practice*

Organisations

- Health and Safety Executive (HSE)
  Web: *www.hse.gov.uk*
  The HSE is responsible for the regulation of almost all the risks to health and safety arising from work activity in the UK.

Chapter 9

Machinery and Work Equipment

Abrasive Wheels

- An abrasive wheel is defined as a wheel made of abrasive particles, which have been bonded together, using organic or inorganic substances, for example resin.
- Abrasive wheels and the tools which drive them are classified as work equipment by the Provision and Use of Work Equipment Regulations 1998.
- Accident statistics indicate that nearly half of all accidents involving abrasive wheels are due to an unsafe system of work or operator error.
- Abrasive wheels technically only include wheels consisting of abrasive particles bonded together. For practical purposes the precautions for safe use of abrasive wheels should be applied to all other types of wheel including diamond wheels.
- The choice of tool on which to mount an abrasive wheel should only be made after consideration of the environmental conditions of operation, eg flammability — any solvents in the vicinity, conductivity — is there a likelihood of cutting through a cable, ventilation — will the petrol exhaust fumes be inhaled.
- Cutting or grinding will always require the wearing of eye protection and hearing protection.
- Only authorised and trained persons should operate abrasive wheels.
- Risk assessments may be generic but should consider variables related to length of use such as exposure to vibration and noise.
- When working at height with abrasive wheels, never allow operation from a ladder or stepladder. A stable work platform complete with full edge protection is necessary wherever the height of work would bring the tool otherwise above shoulder height.

9.1 The misuse of abrasive wheels has led to some serious accidents. Wheels can easily maim or otherwise seriously injure a person. Ejected material is usually hot and may ignite flammable liquids and vapours. High vibration levels can be encountered, particularly with older equipment, and noise levels are also very high.

Employers' Duties

9.2

- The Dangerous Substances and Explosive Atmospheres Regulations 2002 (DSEAR) are concerned with removing the risk of fire and explosion when working with or close to materials which are classified as dangerous. A suitable risk assessment linked to the proposed method of work must address all aspects of the operation to decide how the work needs to be controlled to prevent the risk of a fire or explosion.
- The Provision and Use of Work Equipment Regulations 1998 (PUWER) require tools and associated consumables to be suitable for the intended purpose. Tools must be maintained sufficiently to remain safe in use and users must be trained.
- The Electricity at Work Regulations 1989 stipulate that all electrical equipment must include protective devices to prevent overload. Maintenance is required to anticipate breakdown of insulation and security of all connections. The Institute of Technology (formerly the Institute of Electrical Engineers) is taken as the authority for a suggested frequency of tests. These are known as PAT tests and for most

construction work the recommended interval between tests is three months. However, an electrical engineer can decide to shorten or lengthen the interval according to risk. Risk is determined by carrying out tests on a fleet and interpreting the occurrence and causes of faults.
- The Management of Health and Safety at Work Regulations 1999 require employers to assess risk and communicate relevant information arising from the assessment to employees.
- The Control of Noise at Work Regulations 2005 regulate exposure to high levels of noise in order to prevent hearing loss and tinnitus, an incurable ringing in the ears. Employees exposed to a daily average of 85 decibels or more must participate in a scheme of audiometric testing, and must wear hearing protection. Employers must seek to provide quieter tools, give information and reduce exposure.
- The Control of Vibration at Work Regulations 2005 set limits on daily exposure to prevent nerve and blood circulation problems which can be associated with long-term disability. Employees might need health surveillance. Employers must provide information and select methods and tools which reduce exposure. Measurement is rarely needed. However, knowledgeable interpretation of published data is essential in order to provide written exposure assessments. Although grinder use for two hours would be considered high risk, it is normally found that actual trigger time (the time of actual operation) is much shorter than estimated by workers.

Employees' Duties

9.3 The Management of Health and Safety at Work Regulations 1999 require employees to:
- use machinery, equipment, dangerous substances, transport equipment, means of production or safety devices in accordance with any instruction and training given by the employer
- Inform their employer of any shortcomings in the employer's protection arrangements.

9.4 In general, employees must ensure that they have the opportunity to check the equipment they will be using.

In Practice

9.5 All of the legislation applicable to working with abrasive wheels refers to a strict duty for employers to examine the competency of staff who operate abrasive wheels. For construction use, where the environment will vary considerably, the duty to assess the employee is partially covered by the employee having a suitably endorsed training certificate. The certificate might refer to the class of machine in which the individual was trained. More recently issued certificates will record the type of equipment, including the wheel diameter. As these tools are known for the danger they are associated with, the employer should also follow up training by investing in a system of monitoring actual working practices.

9.6 When examining older certificates the classes referred to are:

Class 1	Straight-sided wheels not exceeding 250 diameter
Class 2	Straight-sided wheels exceeding 250 diameter (including relieved and or recessed wheels)
Class 3	Tapered wheels
Class 4	Bonded abrasive discs
Class 5	Cylinder wheels
Class 6	Cup wheels including dish and saucer
Class 7	Cone wheels
Class 8	Depressed centre wheels
Class 9	Cutting-off wheels.

9.7 All tools which drive abrasive wheels need to be regularly examined. Electrical tools must have PAT tests which identify electrical safety faults. Air powered tools will need hose checks and governor speed checks. Petrol driven tools must have engine maintenance, filter changes etc. Engineers should use these services to double check that guards are secure, the tool is operating safely and on/off switches are reliable.

9.8 Anti-vibration mountings must be replaced at suitable intervals not exceeding the manufacturer's recommended intervals. This is an important part of a general strategy to prevent ill health from exposure to vibration.

9.9 The checklist overleaf can be used to help ensure the safe use of equipment.

Risk Assessment of Abrasive Wheels

Noise

9.10 This is a difficult problem to accurately assess because of the wide variation in circumstances. The material being cut, the structure around the work and the length of time will all affect the level of exposure. In practice it is acceptable, except for specialist contractors (eg chasing, diamond cutting), to use a single approximate value and relate this to the SNR rating in dB. It should be attempted to specify hearing protection that will bring the exposure to below 80dB. Normally site signage will indicate the hearing protection zones.

Vibration

9.11 While it is acknowledged that there is no safe level of vibration, the manufacturer's level should be doubled to bring it closer to field conditions. The HSE website at *www.hse.gov.uk/vibration/readyreckoner.htm* can be used to calculate how long it is permissible to run each type of abrasive wheel. Wearing CE-marked anti-vibration gloves are likely to increase the risk as they may restrict circulation. They are capable only of reducing exposure to high-frequency vibration such as that encountered by dentists. Gloves might help in keeping hands warm, but this is more safely achieved by having a nearby heat source and massaging the hands. There should be evidence that low-vibration tools have been selected and alternative methods of work with lower exposure considered.

Training	Is the user trained in the specific class of equipment?
Fuel	Is there any means by which the fuel or power source could cause fire?
Workpiece	Is the best means of preventing uncovenanted (unplanned) movement of the workpiece being adopted?
Other persons	Are suitable measures in place to protect other persons from the effects of the operation, eg noise, sparks, dust?
Dust	Are suitable means such as water suppression or local exhaust ventilation employed to reduce the amount of dust inhaled by the user?
Fumes	If fumes are generated (eg from cutting galvanised material) is there an effective means for removing the concentration of fumes? Alternatively, are suitable face masks in use?
Fire	Where relevant, have all materials at risk of ignition been removed or suitably covered?
Injury	Are the operator and any assistants wearing personal protective equipment such as safety eyewear, safety footwear and hearing protection? Is the guard properly fitted and adjusted? Does the operator have long hair or loose clothing or jewellery?
Ventilation	If petrol is in use, is there adequate ventilation to remove the exhaust fumes?
Electrics	Are electrical services protected or identified against damage by the blade?
The tool	If electrical, has the tool been PAT tested? Are the guard and handles securely in place?
Air	Are the hoses secured,therefore preventing whip injury?
Vibration	Is the operator aware of the precautions in place to prevent over exposure? Are the anti-vibration mountings maintained adequately?
Noise	Is the operator wearing suitable and properly fitting hearing protection?
Free running	Is the operator allowing approximately one minute of free, no-load operation after changing a new wheel?
Wheel speed and condition	Can the machine drive the wheel faster than the maximum safe speed marked on the wheel? Is the wheel in good condition?
Stopping	Is the wheel only stopped by allowing it to come to rest when in a no-load situation?
Environment	Are other activities taking place nearby likely to cause distraction or injury?
Task	Is the wheel being used for a purpose intended by the manufacturer? (Cut-off wheels should not be used for grinding.)
Clothing	Is the operator or assistant wearing clothes that will not get caught in the rotating blade or become burnt?
Wet or dry	Is the wheel designed for the type of cutting being carried out, ie dry or wet?
User inspection	Does the user carry out a pre-use inspection?
Slipping and tripping	Is the work area kept dry and free from trailing leads? Could a cordless tool be used?

Fire

9.12 Fire risk assessments consider the potential ignition sources which would include sparks from ejected particles of steel, etc and hot engine exhausts. This is related to the number of people at risk and the fuels that might be available for a fire, eg packing material. The risk is reduced by fire drills, fire-fighting equipment, the provision of protected escape routes and management procedures such as a hot work permit.

Electrical

9.13 Electrical risks arise from the danger of cutting into cables or experiencing an electric shock from the tool itself.

Manual Handling

9.14 Manual handling risks might include the difficulty in carrying equipment to the work site, but mainly the handling of products before and after cutting or grinding.

Clamping the Work

9.15 Work which is secured in place is safest. Therefore anything which requires a person to hold it manually, such as holding bricks with the foot, should be considered risky.

Use of Water

9.16 Water can make the work area slippery so hosing down wastes or closing the area until the dust has dried might be effective.

Wheel Burst

9.17 Modern wheels correctly mounted normally fail only when they get pinched by material under compression. It is important to match the wheel to the product — good manufacturers who have been in the field for many years have a wide variety. If the work piece is under compression then consider alternative cutting methods such as oxy propane.

Ejected Material

9.18 Screens (eg welding screens) can be used to good effect when it is not possible to divert other people away from the work area.

Dusts

9.19 Often natural ventilation is all that will be to hand. However, if wet cutting is not possible, eg inside a building, then on-tool extraction will be essential.

Control Measures

9.20 In addition to the controls suggested above, user inspection of both tool and task is essential. Paper risk assessment will not pick up all risks. Damage can occur if the manufacturer's spanners are not used for mounting the wheel (eg securing it to the tool). It is important to have readily accessible first-aid staff and facilities.

Further Information

Publications

The following are available from *www.hsebooks.co.uk.*

- HSG17 (rev 2000) *Safety in the Use of Abrasive Wheels*
- INDG296 (rev 1) *Hand–arm Vibration: Advice for Employees*

Cranes and Hoists

- Cranes and Hoists are used on most construction sites at some point.
- Generally, every unique lift requires a safe system of work established that should result in a lift plan derived from risk assessment.
- Where there are hire terms and conditions of usage/hire these should be explicit and agreed (standard conditions are usually applicable). Crane transport and operating conditions should be established in advance.
- Those involved in lifting operations must be competent, trained and assessed, able to present a record of training and assessment, and be physically able to carry out the work.
- Duties of personnel can overlap, or more than one role may be taken on by an individual person, so it is critical that the duties are defined and clarified on-site with those involved.
- The safe positioning of a crane is important during operation as many serious crane accidents have resulted from the crane overturning.
- All cranes must be thoroughly examined, and tested if required, at regular intervals.
- A permit-to-work system should be used to control maintenance work on the machine, and lock-off systems put in place to control/stop crane movements while maintenance/repairs are done.
- There must be an effective procedure for reporting crane defects and incidents along with a record of action taken, and subsequent authorisation for the crane to return to service.
- Only a competent crane driver familiar with the model should operate the crane. However, maintenance personnel may also operate it when it is in their care provided they are competent and trained in the crane motions they will carry out.
- Hoists on construction sites are used primarily to lift building materials to higher levels on high-rise construction.

9.21 Most construction sites utilise a crane or hoist at some point, and in many cases they are in use for the majority of the construction process. Cranes and hoists are primarily devices for lifting materials. Cranes for construction take many forms, including truck mounted, crawler, tower, lorry-mounted, pedestal, floating, etc.

Employers' Duties

9.22 In the UK, safety in the application and operation of cranes is covered both through regulations derived through the Health and Safety at Work Act, etc 1974, and through British Standards.

9.23 The Management of Health and Safety at Work Regulations 1999 require employers to:

- make an assessment of the risks to the health and safety of employees while at work
- make a similar assessment to the health and safety of persons not in their employment, but who may be affected by the work activities.

9.24 The Provision and Use of Work Equipment Regulations 1998 (PUWER) require the risks to people's health and safety from equipment they use at work to be prevented or controlled. The general duties placed on an employer are to ensure that:

- work equipment is suitable for the purpose and conditions in which it is to be used
- work equipment is maintained in a safe condition for use
- work equipment is inspected by a competent person to ensure it continues to be safe for use, and the inspections recorded
- risks posed by the equipment are controlled by technical measures (guards, alarms, markings, system devices and similar) and by safe systems of work for both operation and maintenance.

9.25 The Lifting Operations and Lifting Equipment Regulations 1998 (LOLER) aim to reduce health and safety risks from lifting equipment provided for use at work. The general requirements are for employers to ensure that such equipment is:

- stable and strong enough for intended use with safe working loads displayed
- set up correctly to minimise risks
- used correctly, and therefore safely
- subject to thorough inspection, testing and examination by competent persons before being put to use for the first time, or after significant alteration or erection (certified on Form F97), and inspected and examined at least annually thereafter or every six months if it carries personnel.

9.26 The Reporting of Injuries, Diseases and Dangerous Occurrences Regulations 1995 require employers to report the following incidents to the Health and Safety Executive (HSE):

- collapse, overturning or failure of load-bearing parts of lifting equipment
- plant or equipment coming into contact with overhead power lines.

9.27 The main standard for cranes is BS 7121, which is a 14-part series covering all crane types and recommendations for safe working systems, management, planning, selection, erection, dismantling, inspection, testing, operation and maintenance of cranes, as well as the planning and management of lifting operations.

Employees' Duties

9.28 The Management of Health and Safety at Work Regulations 1999 require employees to:

- use machinery, equipment and safety devices, and to transport equipment, as per instruction and training received from the employer
- inform their employer of any danger to health and safety posed by a work activity and of any flaws in protection arrangements.

In Practice

Crane and Hoist Selection

9.29 Crane selection considers:

- plans on how the crane will be used
- operational characteristics of different cranes.

9.30 Planning how the crane will be used ensures the crane is sufficiently strong, stable and suitable for the proposed use and that the operational characteristics consider the safety requirements for operating and maintaining the crane.

9.31 Without adequate regulation, control, knowledge, skills and experience, the use of cranes and hoists has the potential to be extremely hazardous and accidents may result from:

- instability, causing overturning
- the failure of structural elements
- the failure of mechanical components
- falling loads
- falling parts of loads striking objects or personnel
- the failure of lifting components, eg ropes, hooks, shackles, slings and strops
- connecting with overhead power lines
- misuse in operation.

Cranes

Management of Crane Operations

9.32 A risk assessment should be undertaken on every unique lift and, from this information, a safe system of work established, including a lift plan. This should include:

- method statements
- information on the suitability of equipment
- site preparation prior to the lift taking place
- personnel requirements and competency
- documentation requirements
- the system of reporting defects and incidents
- relevant toolbox talks
- required security of the crane.

9.33 However, where a series of lifts are repetitive and of a generally similar nature, a single generic assessment may be applied. Tandem lifts involving more than one crane may require particularly detailed consideration.

Arrangements

9.34 Where there are hire terms and conditions of usage/hire these should be explicit and agreed (standard conditions are usually applicable). Crane transport and operating conditions should be established in advance.

9.35 Insurance should be in place and where applicable cover agreed between the parties (standard conditions are usually applicable).

9.36 Certificates of competence and health for crane erectors and drivers, and certification for the crane and auxiliary items of lifting equipment should be available for verification.

9.37 Where cranes are being utilised in public areas third-party approvals may be required from the highways authority, local council, service authorities who may be affected, etc.

Personnel Requirements

9.38 The categories of personnel likely to be involved in any lifting operation are a crane supervisor, crane driver, slinger, signaller, crane erector and maintenance personnel, who must be:

- competent
- trained and assessed
- able to present a record of training and assessment
- physically able to carry out the work.

Duties of Personnel

9.39 Duties of personnel can overlap, or more than one role may be taken on by an individual person, so it is critical that the duties are defined and clarified on-site with those involved.

Crane Supervisor

9.40 Crane supervisors direct and supervise the lifting operation as per the lift plan.

Crane Driver

9.41 Crane drivers operate the crane to the manufacturer's instructions and within the lift plan.

Slinger

9.42 Slingers attach/detach the load and use the correct lifting accessories and equipment as per the lift plan. They also direct initial movements of the crane and ensure the load is stable and secure.

Signaller

9.43 Signallers relay the signal from the slinger to the crane driver. They may also direct crane movement in conjunction with, but not instead of, the slinger.

Crane Erector

9.44 Crane erectors erect and test the crane as per the manufacturer's instructions. Tower and crawler cranes need to be erected/dismantled for each particular site/location, and mobile cranes can arrive at site with separate components that need to be put together prior to use, eg jib extensions, counter-weights.

Maintenance Personnel

9.45 Maintenance personnel keep the crane in a safe condition and carry out maintenance in accordance with the manufacturer's instructions and schedules.

Selection of Cranes

9.46 Cranes for construction are available in many forms. In addition, most crane types come with differing jib configurations, including:

- main jib only
- fly jib (a portable extension to the main jib)
- luffing fly jib
- extended main jib.

9.47 The crane provider can usually advise on the most suitable crane type based on the following operating conditions.

- Mode of transport to site.
- Time on-site.
- Load characteristics, particularly weight, shape and centre of gravity.
- Path of the lift.
- Number and frequency of lifts.
- Site conditions for access, setting up and lifting.
- Prevailing weather conditions.
- Any particular requirements or limitations imposed.

Positioning

9.48 The safe positioning of a crane is important during operation as many serious crane accidents have resulted from the crane overturning due to failure of the inadequate ground or structure on which the crane is propped. The following should therefore be considered.

- The ground should be as level as possible and firm, without voids or other weakening conditions below.
- Structures such as bridges or decks should be capable of carrying the loads from the crane outriggers, tracks, rails or wheels in the worst loading condition envisaged.
- Outrigger pads should not be placed over voids, manholes, service covers or close to free edges. Where necessary, incompressible spreader mats may be placed below the outrigger feet to distribute the resulting ground forces evenly.
- Mobile cranes should be of a suitable chassis and drive type for the terrain they are to traverse.
- The presence, proximity and potential impact of surrounding hazards, such as electric lines, buildings, other cranes, railways, ships, bridges (and similar) must be allowed for.
- The tail of the crane may protrude during slewing and care should be taken to ensure the tail path is clear of objects and personnel.
- The effect of high winds during working, and also when out of use, as the wind effect on the load can increase the effect on the crane, or even destabilise the crane directly because of the sail effect on the crane and jib surfaces.
- The suitability of layout for positioning or erecting the crane, and subsequent demobilisation, eg the crane may need to be erected in the space in which it is to work, otherwise it may not gain access in the completed state.

General Safety Considerations

9.49 Fundamental safety arrangements that need to be covered when operating cranes in construction are as follows.

- The organisation or individual with overall control of the place of work, and the employers of personnel involved in the lifting operation, are those that are responsible for safety.
- Personal protective equipment (PPE) must be worn by the driver and by all personnel in the vicinity of lifting operations. This should include safety helmets, high-visibility clothing, safety footwear, safety harnesses when at height and sunglasses where necessary to reduce glare for the driver.
- There should be safe access and suitable emergency escape arrangements for the crane, site and personnel.
- Suitable fire extinguishers should be located onboard and at the site.

- Integrated crane safety equipment such as load indicator/limiter, radius indicator, motion-limit devices and wind speed anemometers should be functioning correctly and paid suitable attention.
- All cranes must be thoroughly examined, and tested if required, at regular intervals.
- Records and documentation should be kept for each crane so it may be used safely within its specified limits. These should include operating instructions, test certificates, examination reports, significant repairs and hours worked.
- All ancillary lifting devices, such as slings, strops, shackles and lifting beams, must be registered, subject to regular testing and inspection, certified and have their safe working loads (SWL) marked on.
- Utilising a crane to be used for personnel should be done only if there is no other practicable means necessary for those personnel to gain access. If personnel are to be carried it must be in a man-rider basket which should be load tested and certified every six months and have its SWL and the limited number of personnel it can carry clearly marked. Cranes lifting personnel must be inspected and certified every six months.

Maintenance and Repair

9.50 Maintenance usually takes place with the crane out of service. A permit-to-work system should be used to control working on the machine, and lock-off systems put in place to control/stop crane movements while maintenance/repairs are done. All corrective actions and works carried out should be logged. All cranes should be maintained and serviced in accordance with the manufacturer's instructions and schedules.

Reporting of Defects and Incidents

9.51 There must be an effective procedure for reporting crane defects and incidents along with a record of action taken, and subsequent authorisation for the crane to return to service.

Driver Operating Precautions

9.52 The following precautions are aimed at the crane driver, however, some may also apply to the other team members appointed to a lifting operation.

- Only a competent crane driver familiar with the model should operate the crane. However, maintenance personnel may also operate it when it is in their care, provided they are competent and trained in the crane motions they will carry out.
- Scheduled periodic checks should be carried out by the driver/competent person/maintenance personnel, including a full visual inspection over the crane and components as well as basic operating checks on motions and devices. Results should be recorded.
- The driver should make a general visual inspection of the crane and components before using it for the first time every day.
- Before making the first lift the driver should run through all crane motions, including braking, to ensure correct functioning. Additionally, all safety devices should be checked for correct functioning.
- The driver and banksman should be aware of the site layout, configuration and any limitations to the operation of the crane.
- The appointed banksman and slingers should be clearly identifiable to the operator as there may be several personnel on-site dressed similarly.
- All crane motions must be executed in a controlled and smooth manner so as to avoid load destabilisation, shock loading, swinging of the load and similar.

- The crane should never be used to pull loads by luffing, slewing, winching or any other means.
- The load path should avoid personnel or public activities below.
- Irregular shaped loads should be configured on the hook with slings/strops/chains of different lengths to achieve the centre of gravity of the load below the hook.
- Awkward, ungainly loads or loads that may be affected by light winds should be manually restrained from the ground by at least two stay ropes.
- Wheeled or crawler cranes should not be used to carry loads when travelling, unless their duty specifically permits.
- When not in use, even for short periods, the crane must be shut down and locked off to avoid unauthorised use.
- A crane must not be unattended for any period with a load suspended.
- When a crane is left out of service, even overnight, it should be left as specified by the manufacturer. Mobile telescopic cranes are usually fully retracted and parked safely. If a telescopic crane must be left with the jib extended this should be left only after consideration of weather conditions, and in the fully propped condition. Some manufacturers recommend tower cranes are left with the slew unlocked and free to allow movement of the jib with the wind.

Hoists

9.53 Hoists on construction sites are used primarily to lift building materials to higher levels on high-rise construction. They may be either vertical or inclined, depending on the geometry of the available access, and a hoist should therefore be selected which is capable of lifting the maximum load and is suitable for the type of construction.

General Safety of Hoists

9.54 Safety is important when operating hoists.

- Hoists should be arranged for operation from a single control position.
- Operators should be able to see all landing positions or communicate with a person at each position.
- Contact with a moving hoist should be prevented by:
  - enclosing the hoist-way at platforms or openings to prevent contact with personnel or materials (open mesh is sufficient)
  - the provision of gates at all access points, usually on landings where loading/unloading takes place.

9.55 The pathway of a hoist is effectively like a lift shaft and should be protected by:

- enclosing the hoist-way
- ensuring that gates are self-closing and do not impede the pathway of the hoist.

9.56 The gap between the hoist platform and working platform must be minimal enough to prevent a person falling through.

9.57 The risk of falling materials should be prevented by:

- stopping parts of loads falling from the hoist (ensure loose materials are stable and/or contained and that there are toe boards on the platform if not already enclosed)
- ensuring the platform has been examined, load tested, certified and has its SWL clearly marked and that the SWL is not exceeded.

9.58 Additional operational precautions should include:

- ensuring hoist erection is carried out by trained, competent personnel and in line with the manufacturers instructions or training programme
- sufficiently anchoring the hoist to its supporting structure
- ensuring the hoist operator is trained and is competent
- locking off the hoist controls when not in use to avoid unauthorised use
- placing loads in the centre of the hoist or evenly distributing them across it as far as possible
- subjecting the hoist to further examinations and ensuring that it is tested after substantial alteration or repair — at least at six-month intervals
- carrying out weekly checks and recording results in the register or on Form 91.

List of Relevant Legislation

- Management of Health and Safety and Work Regulations 1999
- Lifting Operations and Lifting Equipment Regulations 1998
- Provision and Use of Work Equipment Regulations 1998
- Reporting of Injuries, Diseases and Dangerous Occurrences Regulations 1995
- Health and Safety at Work, etc Act 1974

Further Information

Publications

HSE Publications

The following are available from *www.hsebooks.co.uk*.

- HSE31 (rev1 1999) *RIDDOR Explained: Reporting of Injuries, Diseases and Dangerous Occurrences Regulations*
- HSG136 *Workplace Transport Safety: An Employers Guide*
- INDG163(L) *5 Steps to Risk Assessment*
- INDG199 (rev 1) *Workplace Transport Safety: An Overview*
- INDG271 *Buying New Machinery*
- INDG275 *Managing Health and Safety: Five Steps to Success*
- INDG291 *Simple Guide to the Provision and Use of Work Equipment Regulations*
- L73 (rev 1999) *A Guide to the Reporting of Injuries, Diseases and Dangerous Occurrences Regulations 1995*
- MISC156 *Hiring and Leasing Out of Plant: Application of PUWER 98, Regulations 26 and 27*

British Standards Publications

The following are available from *www.bsi-global.com.*

- BS 7121–1: 2006 *Code of Practice for Safe Use of Cranes: General*
- BS 7121–2: 2003 *Code of Practice for Safe Use of Cranes: Inspection, Testing and Examination*
- BS 7121–3: 2000 *Code of Practice for Safe Use of Cranes: Mobile Cranes*
- BS 7121–4: *Code of Practice for Safe Use of Cranes: Lorry Loaders*
- BS 7121–5: *Code of Practice for Safe Use of Cranes: Tower Cranes*
- BS 7121–11: 1998 *Code of Practice for Safe Use of Cranes: Offshore Cranes*
- BS 7121–12: 1999 *Code of Practice for Safe Use of Cranes: Recovery Vehicles and Equipment*
- BS 7121–14: 2005*Code of Practice for Safe Use of Cranes: Side Boom Pipelayers*
- BS 7212: 2006 *Code of Practice for Safe Use of Construction Hoists*
- BS ISO 15513: 2000 *Cranes — Competency Requirements For Crane Drivers (Operators), Slingers, Signallers and Assessors*

Organisations

- Health and Safety Executive (HSE)
  Web: *www.hse.gov.uk*
  The HSE is responsible for the regulation of almost all the risks to health and safety arising from work activity in the UK.

Lifts and Lifting Equipment

- Lifting equipment is the general term for work equipment used for lifting or lowering loads. It includes attachments for anchoring the equipment, fixing or supporting it.
- When selecting lifting equipment, employers should take note of the working conditions, the risks that already exist to the health and safety of persons in the premises and any additional risks caused by the use of the lifting equipment.
- Employers should also consider whether the equipment has adequate strength for the proposed use.
- All lifting operations must be properly planned and supervised. The person doing the planning/supervising should be competent to do so.
- A risk assessment must be carried out of all lifting operations to determine the level of risk.
- A thorough inspection of the lifting equipment should be undertaken by a competent person.

9.59 Lifting equipment is the general term for work equipment used for lifting or lowering loads and includes attachments for anchoring the equipment, fixing or supporting it.

9.60 There are general guidelines to follow for the selection, use and maintenance of lifting equipment. It should be noted that there is an important link between the Lifting Operations and Lifting Equipment Regulations 1998 (LOLER) and the Provision and Use of Work Equipment Regulations 1998 (PUWER), which apply to all work equipment including lifting equipment. Therefore, it must be remembered that, for the provision and use of lifting equipment, compliance with all relevant aspects of PUWER is required as well as complying with LOLER.

Employers' Duties

9.61 Employers have a general duty to ensure, so far as is reasonably practicable, the health, safety and welfare at work of all employees under the Health and Safety at Work, etc Act 1974.

9.62 The Management of Health and Safety at Work Regulations 1999 require a risk assessment to be carried out to identify the nature and level of risks associated with lifting operations, including risks from the lifting equipment and operation and from other factors such as overhead power cables. The regulations also require that appropriate precautions are taken to control these risks.

9.63 The Lifting Operations and Lifting Equipment Regulations 1998 (LOLER) require that the risk assessment considers:

- the type of load being lifted, its weight, shape and what it consists of
- the risk of the load falling and striking a person or object and the consequences
- the risk of the lifting equipment striking a person or some other object and the consequences
- the risk of the lifting equipment failing or falling over while in use and the consequences.

9.64 In addition, the Management of Health and Safety at Work Regulations 1999 require that young persons should be protected from using high-risk lifting equipment unless they possess the necessary maturity and competence, which includes having completed appropriate training.

Employees' Duties

9.65 Employees have a general duty to take reasonable care of their own health and safety and that of other people who may be affected by their work under the Health and Safety at Work, etc Act 1974.

9.66 Also under the Health and Safety at Work, etc Act 1974, employees have a duty to co-operate with their employers on health and safety matters.

9.67 The Management of Health and Safety at Work Regulations 1999 place a duty on all employees to use machinery or equipment in accordance with any instructions and training received, and to report any work situation that represents a serious and immediate danger to health and safety.

In Practice

Selection of Equipment

The Workplace Environment

9.68 When selecting lifting equipment, employers should take note of:

- the working conditions
- the risks that already exist to the health and safety of persons in the premises
- any additional risks caused by the use of lifting equipment.

Strength and Stability

9.69 Employers should also consider whether the equipment has adequate strength for the proposed use. Account should be taken of the combination of forces to which the lifting equipment will be subjected, as well as the weight of any associated accessories used in the lifting operation. Employers should also consider the stability of the equipment for the proposed use.

9.70 Lifting equipment which is mobile, eg hydraulic lifts for automobiles, fork-lift trucks and mechanical excavators, or which is dismantled and re-assembled at different locations, should be stable during use under all foreseeable conditions. Particular account should be taken of the nature of the ground and other surfaces on which the equipment might be used.

Marking of Lifting Equipment

9.71 If lifting equipment can be set to different configurations, eg the extension of a lifting arm, the safe working load for each configuration should be clearly marked. Additional marking considerations are that:

- accessories for lifting should be marked to identify characteristics necessary for their safe use
- lifting equipment designed for lifting persons should be appropriately marked, including the maximum number of persons to be carried
- equipment not designed to carry persons should be clearly marked to that effect.

Planning of Lift Operations

9.72 The person planning and/or supervising the operation should be competent to do so, ie they should have adequate practical and theoretical knowledge and experience of lifting operations.

9.73 Where two or more items of lifting equipment are used simultaneously to lift a load, eg two lorry loaders, a written plan should be drawn up and applied to ensure safety, eg good co-ordination on the part of the operators.

9.74 Work should not be carried out under suspended loads and loads should not be carried over work areas where practicable. Consideration must be given to the path of falling loads before work starts.

Risk Assessment

9.75 The Management of Health and Safety at Work Regulations 1999 require a risk assessment to be carried out to identify the nature and level of risk associated with a lifting operation. The risk assessment will need to take note of:

- how often the lifting equipment is to be used
- where the lifting equipment is to be used
- the nature and characteristics of the load to be lifted
- any limitations on use specified by the manufacturer/supplier.

9.76 The following factors should be considered as part of a risk assessment:

- the type of load being lifted, its weight, shape and what it consists of
- the risk of the load falling or striking a person or object, and the consequences
- the risk of the lifting equipment failing or falling over while in use and the consequences.

9.77 A risk assessment checklist may be useful.

Thorough Examination and Inspection

9.78 A thorough examination should be carried out by a competent person. They must have appropriate practical and theoretical knowledge of the lifting equipment so that they are able to detect defects or weaknesses. The person should also be able to assess the importance of any defects or weaknesses in relation to the continued safety of the lifting equipment.

9.79 Lifting equipment that is exposed to conditions which are likely to cause deterioration should be thoroughly examined for any defect at the following intervals:

- after installation and before being put into service for the first time
- after assembly and before being put into service at a new site or in a new location, to ensure that it has been installed correctly and is safe to operate
- at least every six months for lifting equipment that lifts persons or lifting accessories
- at least every 12 months for other lifting equipment
- in accordance with an examination scheme.

9.80 Inspections must be carried out at suitable intervals between thorough examinations. These inspections are to ensure that health and safety conditions are maintained, and that any deterioration can be detected and remedied in good time.

9.81 No lifting equipment that has been hired or supplied by a contractor is to be used without physical evidence that the last thorough examination required under LOLER has been carried out.

9.82 When first installed, new lifts do not require any initial thorough examination as long as they have been manufactured and installed in accordance with the Lifts Regulations 1997 and have a current declaration of conformity, ie made not more than 12 months before. A new lift is one where:

- no lift previously existed
- an existing lift has been completely replaced
- only the existing guide rails and their fixings or the fixings alone have been retained.

Lifting Equipment Requiring Examination and Suggested Frequency

Main Item	Considered to Include	Maximum Intervals of Examination (Months)
Cranes	Jib cranes, static and rail mounted	12
	Container cranes	12
	Crawler cranes	12
	Derrick cranes	12
	Lorry loaders	12
	Overhead cranes	12
	Pillar jib cranes	12
	Portable jib cranes	12
	Tower cranes	12
	Excavator	12
Elevator	Bucket elevator	12
	Elevator	12
Hoists and lifts	Ash/coal, skip hoist	12
	Builders' hoist, goods only	12
	Builders' hoist, passenger	6
	Passenger hoist/lift	6
	Goods only hoist/lift	12
	Passenger/goods hoist/lift (ie multi-use)	6
	Inclined material hoist	12
	Service lift	12
	Home lift	6
	Paternoster goods only	6
	Scissor lift	12
	Stair lift	6
	Teagle hoist	12

Main Item	Considered to Include	Maximum Intervals of Examination (Months)
Patient hoist	Patient hoist	6
Winches	Winches, if used for lifting loads	6/12
	Vehicle winches	12
Shear legs blocks, ropes and tackle	Shear legs with winch	6/12
	Rope	12
	Hoist	12
	Manual	12
	Powered	12
	Pulley	12
	Snatch	12
	Chain	12
	Ratchet	12
	Gin wheels	12
	Hook hoist	12
Safety and rescue equipment for supporting, raising and lowering persons	Arborists equipment	6
	Bosuns chair	6
	Safety belts	6
	Fall arrestor/retractable liftline	6
	Safety harness with lanyard	6
Miscellaneous items (provided for the support of lifting gantry equipment)	Anchorage, suspension point	12
	Tracks	12
	"A"– frame	12
	Overhead gantry	12
	Davits	12
	Gantry	12
	Jib arms	12
	Overhead crane bridges	12
	Runway tracks & beams	12
	Trolleys	12

Main Item	Considered to Include	Maximum Intervals of Examination (Months)
Lifting accessories	Eyebolts	6
	Cradle	6
	Fork attachments	6
	Girder clips	6
	Grabs	6
	Hooks	6
	Lifting beams/frames	6
	Magnets	6
	Plate clips	6
	Rigging screws	6
	Running out block/pole carrier	6
	Shackles	6
	Slings	6
	Vacuum lifting devices	6
Access equipment	Suspended access equipment	6
Suspended work platforms	Window cleaning rig	6
	Work platforms	6
	Mast climbers	6
	Mast hoists	6
	Bridge maintenance access equipment — where lifting is involved	6
Platform stacker	Platform stacker	12
Car parking systems	Car parking systems	6/12
Vehicle recovery equipment	Vehicle recovery equipment	12
	"Spectacle" frame	6
Tailboard hoist/lift	Tailboard hoist/lift	6/12
Jacks	Jacks, multistage	12
	Jacks, trolley	12
Vehicle lifts	Motorcycle lifts	12
	(lifting attachments to/lifting accessories within motorcycle lifts)	[6]
	Vehicle lifts/hoists	12
	(lifting attachment to/lifting accessories within vehicle lifts/hoists)	[6]
Vehicle skip hoist	Vehicle skip hoists	12
Road vehicle wheel lifter	Road vehicle wheel lifter	12
Drum, coil, roll lifting device	Drum, coil, roll, lifting devices, including where fitted to process plant	12

Main Item	Considered to Include	Maximum Intervals of Examination (Months)
Fork-lift truck	Fork-lift truck	6/12
	(lifting attachments to/lifting accessories within fork-lift truck)	[6]
Pallet truck	Pallet (lift) trucks, manual & powered	12
Loading shovel	Loading shovel	12
Straddle carriers	Straddle carriers	12
Telescopic load handler	Telescopic load handler	6/12
Lighting mast	Aerial mast	12
	Lighting rigs where lifting is involved	12
	Articulated lamp post	12
Balancers	Tool balancers	12
Cable drum raising system	Cable drum lifters	12
Stage equipment hoist	Stage equipment hoist, utilities, scenery, stages	6/12
	Camera boom	6/12
	Flying ropes	6/12
Gymnasium equipment	Gymnasium equipment, if used at work	12
Piledriver	Piledriver	12
Drilling rigs	Drilling rig, where lifting is involved	12
Miscellaneous items	Drop sections	12
	Gangways and bridges	12
	Lock gates	12
	Roller shutter doors	12
	Tidal booms	12
	Service gates	12
	Suspended furnace/process plant doors	12

Alternative Examination Schemes

9.83 As an alternative to thorough examinations at the prescribed statutory intervals, the competent person may devise an examination scheme. The scheme may specify periods which are different from the statutory intervals, but this examination scheme must be based on a rigorous assessment of the risks. An examination scheme may be particularly appropriate for equipment that is used infrequently for light loads.

Defects in Lifting Equipment

9.84 The competent person who carries out a thorough examination under LOLER is required, on finding defects, to:

- notify the employer immediately
- make a report to the employer and to the person from whom the equipment has been hired as soon as practicable
- send a copy of the report to the relevant enforcing authority where the defect in the equipment presents an imminent risk of serious injury.

9.85 The relevant enforcing authority means, in cases where the defective equipment has been hired or leased, the HSE. Otherwise the relevant enforcing authority is the enforcing authority for the premises where the defective equipment was examined, eg the local authority.

9.86 An employer who receives a report on defects in equipment is required under LOLER to ensure that the equipment is not used before the defect is rectified.

Training

9.87 Employers should ensure that any employees or contractors who work with lifting equipment are competent and have received appropriate information and training. Employees using lifting equipment should be trained to the level of skill necessary to operate the lifting equipment safely and efficiently. The training of operators should include three stages:

- basic
- job specific
- familiarisation.

9.88 Basic training should cover the basic skills and knowledge required to operate equipment safely and efficiently. Job-specific training should be tailored to the employees' or contractors' needs. Familiarisation training needs to be done on the job, under close supervision.

List of Relevant Legislation

- Management of Health and Safety at Work Regulations 1999
- Provision and Use of Work Equipment Regulations 1998
- Lifting Operations and Lifting Equipment Regulations 1998
- Reporting of Injuries, Diseases and Dangerous Occurrences Regulations 1995
- Workplace (Health, Safety and Welfare) Regulations 1992
- Health and Safety at Work, etc Act 1974

Further Information

HSE Publications

9.89 The following are available from *www.hsebooks.co.uk*.

- HSG6 (rev 2000) *Safety in Working with Lift Trucks*
- INDG290 *A Simple Guide to the Lifting Operations and Lifting Equipment Regulations 1998*
- INDG339 *Thorough Examination and Testing of Lifts: Simple Guidance for Lift Owners*
- INDG398 *Are You Making the Best Use of Lifting and Handling Aids?*
- L22 (rev 1998) *Safe Use of Work Equipment. Provision and Use of Work Equipment Regulations 1998. Approved Code of Practice and Guidance*
- L24 *Workplace (Health, Safety and Welfare) Regulations 1992. Approved Code of Practice*
- L113 *Safe Use of Lifting Equipment. Lifting Operations and Lifting Equipment Regulations 1998. Approved Code of Practice and Guidance*

British Standards

The following are available from *www.bsi-global.com*.

- BS 5655 Part 10: 1986 *Specification for the Testing and Inspection of Electric and Hydraulic Lifts*
- BS 7121–1: 2006 *Code of Practice for Safe Use of Cranes: General*

Machine Guarding

- Types of machine guarding include fixed guards, moveable or adjustable guards and protective devices like pressure mates, trip bars, light guards, jigs and clamps and "hold-to-run controls".
- The Provision and Use of Work Equipment Regulations 1998 detail the hierarchy of machine guarding that needs to be considered when guarding the dangerous moving parts of machines.
- Employers have a general duty to provide machinery, equipment and other plant that is safe and to maintain it in a safe manner.
- Under the Provision and Use of Work Equipment Regulations 1998, employers must ensure that equipment provided for work is suitable for its intended purpose.
- Employers should take the necessary steps to prevent access to dangerous parts of machinery and that suitable controls have been installed.
- Risk assessments should be carried out on all machinery used within the workplace to identify hazards, including unsuitable or missing guards.
- Adequate information, instruction and training must be provided on the safe use of machinery.
- Anybody who designs, manufactures, imports or supplies any machinery for use at work must ensure that it is so designed and constructed as to be safe and without risks to health when properly used, set, cleaned and maintained.
- Adequate information must be provided by suppliers about the use of the machinery for which it is designed.
- Machinery should be CE marked to demonstrate the supplier's intention to supply a safe piece of machinery.

9.90 Several persons have responsibilities in relation to machinery guarding. During machine production, both the machine designer and the machine inspector will need to consider the need, style and adequacy of guards. A machine cannot be awarded a CE mark by the supplier unless it has been fitted with suitable guards where this is necessary to prevent danger.

9.91 When a machine is purchased, the user needs to ensure that the guarding system is adequate, that machine operators, maintainers and others are able clearly to identify the guarding principles used and types of guards fitted. The purchaser then needs to determine the training needs that will support such an understanding and to provide such training.

9.92 During the machine's lifetime, the user will also need to monitor guarding safety in order that corrective measures may be taken where necessary to ensure that all processes relating to that machine continue to be undertaken safely.

Employers' Duties

9.93

- The Health and Safety at Work, etc Act 1974 requires the supplier of articles intended to be used at work to ensure that they are safe whether or not CE marking procedures apply to a particular machine. Employers must provide only safe articles to site visitors who will use that equipment while on-site. Work equipment must be suitably guarded to be considered safe.

- The Management of Health and Safety at Work Regulations 1999 require an employer to conduct risk assessments to identify work-related hazards including unsuitable or missing guarding provisions.
- The Provision and Use of Work Equipment Regulations 1998 require the user to ensure that all machines in use are fitted with suitable guards to prevent danger. This is the main legislation that applies to machinery guarding.
- The Supply of Machinery (Safety) Regulations 1992 require suppliers of machinery to fit suitable guards to prevent danger.
- The CE Marking Regulations require machinery to be CE marked by the supplier to show intention to supply a safe machine. To be safe, the machine must be suitably guarded.

Employees' Duties

9.94 Employees have a duty to take reasonable care of their own health and safety and that of other people who may be affected by any acts or omissions in relation to their work under the Health and Safety at Work, etc Act 1974.

9.95 Employees also have a duty not to interfere with anything that has been provided for reasons of health and safety (such as machine guards) and to co-operate with their employer on health and safety matters.

9.96 The Management of Health and Safety at Work Regulations 1999 require employees to:

- use machinery, equipment, transport equipment, means of production or safety devices in accordance with any instruction and training given by the employer
- inform their employer of any danger to health and safety posed by a work activity
- inform their employer of any shortcomings in the employer's protection arrangements.

In Practice

Machine Guarding Design

9.97 During machine design, the designer should conduct structured and systematic risk assessments to determine the need to fit guards. Guards must be fitted to all dangerous parts, such parts being defined as those that, if left unguarded, will present a reasonably foreseeable risk to a person. Guards must present a physical barrier to access to the danger zone or be designed to stop the machine before a person can enter the danger zone.

9.98 The type of guards to be fitted will depend upon how the machine has been designed to carry out its function, the nature of the hazard from the unguarded machine and whether whole or broken parts of the machine or workpiece are likely to be ejected.

9.99 A hierarchy of four levels of guarding are required by legislation and, in order of choice, they are:

- fixed enclosing guards
- other guards or protection devices
- protection appliances (jigs, holders, push sticks, etc)
- provision of information, instruction and training.

9.100 The designer must be able to show that the method selected was chosen because a more stringent method was not possible.

9.101 It is important to remember that, in addition to normal use, the designer must also consider cleaning, setting, machine adjustment and repair. This does not mean that all operations must be performed with all guards in place. However, the removal of any guard for any of the operations above must be fully justified.

Different Types of Guard Designs

Fixed Guards

9.102 Fixed guards have no moving parts and are fixed in a constant position relative to the danger zone. The primary aim must be to fit permanent, fixed and fully enclosing guards wherever possible. To be considered permanent, such guards must only be removable with the use of a tool.

9.103 Fixed guards should ideally be fitted so that if the fixings are removed, the guard will fall away in order that the guard's absence will become obvious. It is good practice to label fixed guards with an appropriate hazard warning sign so that all are aware that if such a cover is removed, a dangerous part or parts will be exposed.

9.104 It is permitted to have openings in fixed guards provided that such openings satisfy the safe reach distances described in the standard BS EN 294 such that any part of the human body that can penetrate the opening still cannot touch dangerous parts. In practice, this means that as the openings become further away from the dangerous part, they can be designed larger.

9.105 Where the danger is of an electrical nature, any conductive (ie metallic) protective cover should be connected to earth. It is allowable to have holes in protective covers, typically to facilitate ventilation or to provide a means of inserting test probes to take necessary measurements, provided that the covers still exclude parts of persons.

9.106 The definitive guidance on this subject is contained within EN 60529: Protection Afforded by Enclosures. This document contains the IP Index (the Index of Protection), which contains the appropriate design data in tabular form.

Other Guards

9.107 Other guards can include movable guards, adjustable guards, automatic guards and fixed guards that often do not fully enclose the dangerous part. Where this type of guard is used the designer should also consider whether safety would be enhanced by the use of supplementary devices such as jigs, holders, push sticks, etc and fit them or specify them where required.

9.108 It is often necessary to fit movable guards with interlocks to improve safety. A control guard is a special type of movable guard that will allow materials to be fed into the machine or allow machine adjustment or cleaning. The use of such a guard should be limited to circumstances where it is absolutely necessary and can be fully justified.

Protection Devices

9.109 Protection devices do not enclose machinery but ensure that the body cannot make contact with the dangerous part, or stop the dangerous part before the danger zone is entered. Such devices as pressure mats, trip bars, light guards and hold-to-run controls (either one-handed or two-handed) fall into this category.

Pressure Mats

9.110 Pressure mats ensure that a machine will only run when the work person is standing on the mat (ie in a predetermined safe position). Such a device may be used to define the position of the maintainer if they need to diagnose a fault from inside the operator's safety barrier. If this device is used, then instructions for its safe use must be written into the maintenance instructions supplied with the machine.

9.111 A typical example of the use of a pressure mat would be where it was used in conjunction with a barrier around a machine. When it was necessary for the maintainer to move to a position inside the safety barrier to observe a machine function for diagnostic purposes, a correctly placed pressure mat in conjunction with the provision of an inching button would place such a task under safety control. The maintainer should also have received adequate training on the use of such a safety system.

Trip Bars

9.112 Trip bars or pressure bars are often used as a "sensitive edge" to the work area and are positioned so that if the operator leans too far over the machine and enters the danger zone, bodily contact with the bar will stop the machine. Trip bars are often used in addition to other guards.

Jigs and Clamps

9.113 Jigs, clamps and work holders together with push sticks fall within the description of other protective devices and should be provided or their use advised where appropriate.

Light Guards

9.114 Photoelectric devices, often referred to as "light guards", may be a viable safety guarding strategy. Such devices are often used where the operator needs to approach the machine to perform an operation such as removing a completed workpiece and inserting new material. Where fitted, they are intended to stop the machine before the operator can approach the danger. Such a function means that the time taken for the machine to stop becomes a design factor, together with the speed at which the operator will move.

9.115 The physical position of the light guard(s) will also be determined by the requirement to ensure that they are effective so that they cannot be easily circumnavigated. It also should not be possible for any person to step inside the light guard system (ie to enter the danger zone) and then to restart the machine from such a position.

Hold-to-run Controls

9.116 Hold-to-run controls ensure that the operator's hand(s) cannot be in the danger zone since that operator will need to hold the run button constantly for the machine to operate. It should be remembered that this type of control may be easily defeated and will not protect a "third party" in any event.

Interlocks

9.117 Where movable guards are used to allow some defined degree of operator access, there will almost always be a requirement to fit a safety interlock to stop the machinery when the guard is opened. Similarly, where fixed barriers are used around a machine and those barriers are fitted with gates to allow controlled access, the gates in question will be fitted with interlocks.

9.118 An interlock is a safety critical component because it represents a single point of failure in the guarding system; if the interlock fails to operate, then the person operating the guard or gate will be exposed to danger. It follows that the integrity of interlock systems must be high. It is therefore logical that the interlock switch design should be such that the device's reliability is high and the interlock should automatically fail to safety if it develops a fault.

9.119 Any software programme that is used within an interlock system must also be of high integrity and must fail to safety. All interlocks must be installed in such a way that they are not easily defeated.

9.120 Safety interlocks are addressed by EN 1088: *Construction and Selection of Safety Interlocks* and any device chosen should conform to this standard. Every interlock should carry a CE mark that must be supported by a declaration of conformity.

Machine Guarding During Use

9.121 The user (in law, the employer) has the prime responsibility to provide only safe machinery for use in the workplace; such a responsibility extends to the provision of adequately guarded machinery.

9.122 It is not unusual to find imported machinery that carries a CE mark and yet is inadequately guarded. Irrespective of the origin of a machine and of the markings and certifications that it may carry, the employer should conduct systematic risk assessments on all procured machinery before it is first put into use.

Risk Assessments of Machinery

9.123 Risks to be considered should include:

- the adequacy of the guarding system
- the position and environment into which it will be installed and how this may affect the safety provisions designed into the machine
- the suitability and training adequacy of all employees who will be involved with the machine during its entire life cycle within the organisation.

9.124 Risk judgment criteria should be established and recorded. A prompt list suggesting the type of issues to be considered is provided in the table below. There can be no universal checklist that will cover all machines and risk assessments should be specific to the machine and task in question.

Risk Assessment of Machinery: Prompt List

No	Risk to be Considered	Comments
1	Have all necessary guards been fitted?	Consider different types of guard design.
2	Do the chosen guard types appear to comply with the hierarchical structure of the Provision and Use of Work Equipment Regulations 1998?	Check whether a more stringent guarding system could have been selected.
3	Do all guards appear to be of adequate design and of robust construction and properly labelled?	Consider mainly upper limbs, lower limbs and crush injuries.
4	Can the operator still observe the machine cycle to the extent necessary to prevent danger?	This requirement may conflict with the need to ensure that very small objects or fluids should be contained by the guarding system.
5	Do I have a maintenance plan for the guarding systems?	Regular inspections and/or testing are required.
6	Can the maintenance be performed with the machine shut down and isolated?	Always a first choice since all injuries associated with machinery movement are eliminated.
7	Do I have a written safe system of work for the control of maintenance activities?	A permit system may be necessary.
8	If I need to isolate the machine, what type(s) of isolation must I use?	It may be necessary to employ electrical, hydraulic and mechanical isolation or isolation from thermal energy, eg steam.
9	Have I provided a competent person to undertake the isolation?	If the system that requires isolation is not within the maintainer's primary discipline, you may need to appoint another competent person to undertake the isolation task.
10	Are the energy sources adequately identified?	To prevent the wrong energy system feed from being isolated, accurate, clear and unambiguous labelling of energy systems and the machine is essential.
11	If machinery movement is necessary for diagnostic purposes, can I limit the speed or distance of travel?	Inching buttons may prove useful.
12	Will I need to fit additional temporary protection devices?	Pressure mats will ensure the maintainer is in a set position during machine movement.
13	Will the maintainer still need to take precautions against limited machine movement?	Loose clothing may still be a hazard.
14	If inching buttons have been provided, are they of the recommended design?	They must be recessed to prevent inadvertent operation.

No	Risk to be Considered	Comments
15	Will the maintainer be in proximity to dangerous fluids, mist, fume, noise, heat/cold levels, etc?	Surfaces wet with cutting fluid overspill or product overspill may cause slips and consequent injury. Personal protective equipment may be required for hands/arms/eyes/ head or whole body
16	Does the maintainer need to override an interlock?	This process must be controlled so that only permitted persons are allowed to do this.
17	When an interlock is overridden, will others be prevented from entering the danger zone?	It may be necessary to train others and to monitor compliance with the rules.
18	Does the maintainer need to be close to moving parts?	A system such as a pressure mat would ensure that the machine only moved when the maintainer was in a specific area.
19	Even if the machine is shut down, are there still hazards to consider?	Rough surfaces, sharp edges, noise and other hazards may still be present
20	Is the maintainer suitably trained?	If the danger zone is being entered, will special hazards be present?
21	Has the maintainer training addressed safety procedures?	Special managerial controls (eg interlock key withdrawal) must be understood.
22	Is a permit-to-work system required?	Often a useful way of formalising safety procedures.
23	Are the safe systems of work in a written format?	Where special maintenance safety procedures are required, they must be in a written form.
24	Are the written instructions up-to-date?	Do they consider machinery modifications or process/substance changes since the machine was first installed?
25	Is the maintainer familiar with the written instructions?	A check should be undertaken.
26	If diagnosis or maintenance is carried out when the machine is not shut down, will fixed guards need to be removed?	A procedure to ensure only necessary guards are removed should be in place and the fitting of temporary "part guards" should be considered to create minimum danger consistent with the task.
27	Will the removal of covers expose live electrical parts?	Check whether temporary insulating covers can be fitted to live electrical parts if the area is to be exposed.

9.125 Where several separate machines or sub-assemblies are integrated to function as a whole (eg in a production line), the design philosophy may be that parts of one machine or sub-assembly will form the guard of an attached assembly. This philosophy is acceptable; however, the employer must ensure that such separate sub-assemblies or machines are not operated on their own in a stand-alone mode since their guarding system may not be complete in such circumstances.

9.126 Where a machine such as a lathe may have part of the workpiece protruding beyond the headstock, there is a specific requirement for workpiece guards to be fitted in addition to any other machine guards already in place.

If Machine Guards Supplied are Inadequate

9.127 Where guards are found to be absent or inadequate and the machine carries CE certification, the recommended course of action is to inform the Health and Safety Executive (HSE) rather than to enter into a potentially heated debate with the supplier. Such supplier contact may result in the manufacturer agreeing to modify your machine. However, there is no way to be sure that similar machines supplied to others will also be modified.

9.128 If the HSE inspector agrees that the machine is dangerous, he or she will approach the supplier with a view to ensuring that all similar machines in all European Union Member States are satisfactorily modified.

Protection Appliances

9.129 Where movable guards have been provided, the machine supplier may also have supplied protection appliances or advised that such devices should be used. Push sticks are often used in conjunction with hand-fed woodworking machines and jigs or work clamps used with machines that cut metal. The employer should ensure that such devices are:

- suitable for the purpose for which they have been provided
- of good construction, sound material and adequate strength for their purpose
- adequately maintained to support their safety and efficient working order and to ensure that they remain in good repair.

9.130 Protection appliances should:

- not be easily bypassed or disabled
- not give rise to any increased risk to health and safety
- be situated a sufficient distance from the danger zone
- not unduly restrict the view of the machine's operating cycle (where such a view is necessary for safety)
- be constructed or adapted so that any operations required to replace parts or conduct other maintenance are conducted under conditions where access is restricted to areas where the work will be carried out and, wherever possible, such work can be undertaken without having to dismantle the guard or protection device.

Training

9.131 All work persons who use a machine must be fully conversant with all protective systems including all guarding arrangements. It would not be acceptable to leave a set of operating instructions with the user in the understanding that they will be read. The employer must deliberately train staff in all relevant safety matters and must monitor that such training has been effective.

9.132 After training is complete, a written set of safety instructions should be issued. These written instructions should include warnings not to remove guards, not to defeat interlocks or circumnavigate either hard barriers or light curtains.

9.133 Most accidents at machines occur during maintenance activities since the maintainer will often need to disable or remove guarding systems and individual safety covers or work inside the safety barriers designed normally to exclude people. In such cases, written instructions in the form of method statements and permits to work (where applicable) will prove helpful and are most strongly recommended. Such a management system will supplement the special training required.

9.134 Where permits are used, they should include checks which ensure that all guards that had been removed for maintenance have been replaced, all defeated interlocks made fully operational again and all staff involved in the maintenance have left the danger zone before motive power is restored.

List of Relevant Legislation

- Management of Health and Safety at Work Regulations 1999
- Provision and Use of Work Equipment Regulations 1998
- Supply of Machinery (Safety) (Amendment) Regulations 1994
- Supply of Machinery (Safety) Regulations 1992
- Health and Safety at Work, etc Act 1974

Further Information

Publications

HSE Publications

The following are available from *www.hsebooks.co.uk*.

- INDG229 *Using Work Equipment Safely*
- L22 (rev 1998) *Safe Use of Work Equipment. Provision and Use of Work Equipment Regulations 1998. Approved Code of Practice and Guidance*
- L114 *Safe Use of Woodworking Machinery: Approved Code of Practice and Guidance*

British Standards Publications

The following are available from *www.bsi-global.com*.

- BS EN 294: 1992 *Safety of Machinery — Safety Distances to Prevent Danger Zones Being Reached by Upper Limbs*
- BS EN 349: 1993 *Safety of Machinery — Minimum Gaps to Avoid Crushing of Parts of the Human Body*
- BS EN 811: 1997 *Safety of Machinery — Safety Distances to Prevent Danger Zones Being Reached by the Lower Limbs*
- BS EN 953: 1998 *Safety of Machinery — Guards — General Requirements for the Design and Construction of Fixed and Moveable Guards*
- BS EN 999: 1999 *Safety of Machinery — The Positioning of Protective Equipment in Respect of Approach Speeds of Parts of the Human Body*

- BS EN 1088: 1995 + A1: 2007 *Safety of Machinery — Interlocking Devices Associated with Guards — Principles for Design and Selection*
- BS EN 1760–1: 1998 *Safety of Machinery — Pressure Sensitive Protective Devices — General Principles for the Design and Testing of Pressure Sensitive Mats and Pressure Sensitive Floors*
- BS EN 1760–2: 2001 *Safety of Machinery — Pressure Sensitive Protective Devices — General Principles for the Design and Testing of Pressure Sensitive Edges and Pressure Sensitive Bars*
- BS EN 11161: 2007 *Safety of Machinery — Integrated Manufacturing Systems. Basic Requirements*
- BS EN ISO 12100–1: 2003 *Safety of Machinery — Basic Concepts, General Principles for Design — Basic Terminology, Methodology*
- BS EN ISO 12100–2: 2003 *Safety of Machinery — Basic Concepts, General Principles for Design — Technical Principles and Specification*
- BS EN 13849–1: 2006 *Safety of Machinery — Safety Related Parts of Control Systems. General Principles of Design*
- BS EN 13850: 2006 *Safety of Machinery — Emergency Stop. Principles of Design*
- BS EN 14121–1: 2007 *Safety of Machinery — Risk Assessment. Part 1: Principles*
- BS EN 60825–4: 2006 *Safety of Laser Products. Laser Guards*
- PD 5304: 2005 *Guidance on Safe Use of Machinery*

Portable and Mobile Access Equipment

- Portable and mobile access equipment includes a broad range of equipment used for working at height.
- The HSE reports that there has been a significant increase in the number of accidents involving alloy towers. These accidents are the result of inappropriate selection, installation and use.
- Only trained persons should erect and use portable access equipment.
- All portable access equipment must be given a user inspection before each use.
- Only industrial class stepladders, ladders, hop-ups and towers are suitable for construction work.
- Safety helmets must be worn in almost all circumstances when portable access equipment is used.
- Rescue arrangements must be decided before work at height commences.
- Every work platform must have at least 600mm of clear width free of materials.
- Short duration work is regarded by the HSE as less than 30 minutes.

9.135 Portable and mobile access equipment includes a broad range of equipment used for working at height. This can include anything from ladders to window cleaning gantries and mobile-elevated working platforms, etc. Only industrial class stepladders, ladders, hop-ups and towers are suitable for construction work.

9.136 All portable access equipment must be given a user inspection before each use and can be erected only by trained persons.

Employers' Duties

9.137

- Under the Work at Height Regulations 2005, dutyholders must:
  - avoid work at height where possible
  - where work at height cannot be avoided, use work equipment to prevent falls
  - where the risk of falls cannot be eliminated, take measures to minimise the distance and consequence of any falls.

9.138 In addition to adherence to these basic principles, dutyholders must ensure that:

- work at height is properly planned and takes into account weather conditions
- all those who work, plan or supervise such work are competent and properly trained
- the place of work at height is safe
- equipment such as scaffolding and ladders are inspected in accordance with the specific requirements of the regulations or guidance provided by the HSE or the manufacturer
- the risks associated with fragile surfaces are controlled
- the risks from falling objects are properly controlled.

- Under the Health and Safety at Work, etc Act 1974 employers have a general duty to ensure, so far as is reasonably practicable, the health and safety of all employees at work.
- Under the Management of Health and Safety at Work Regulations 1999 employers have a duty to undertake suitable and sufficient assessments of risks, including those associated with working at height.
- The Construction (Design and Management) Regulations 2007 (CDM) require designers to critically assess their design proposals to identify health and safety issues and to take steps to reduce risks.

Employees' Duties

9.139

- Under the Health and Safety at Work, etc Act 1974 employees have a duty to:
  - take reasonable care of their own health and safety and that of other people who may be affected by their activities at work
  - co-operate with their employer to enable the employer to comply with health and safety duties.
- Under the Management of Health and Safety at Work Regulations 1999 employees are required to:
  - use any machinery, equipment, dangerous substances, transport, safety devices and means of production in accordance with any training and instructions provided by the employer
  - inform the employer of any serious and imminent dangers to health and safety
  - inform the employer of any shortcomings in the employer's health and safety arrangements.
- Under the Work at Height Regulations 2005, employees must:
  - report to their employer or person in control of the work, any activity or defect relating to work at height that may endanger the safety of themselves or others
  - use any equipment provided for their use in accordance with any training and/or instructions relating to its use to enable their employer or person in control of the work to comply with their own health and safety duties.

In Practice

Working at Height

9.140 Falls are the most common cause of death in the construction industry. As a result, the Work at Height Regulations 2005 require all work at height to be risk assessed. The first step in such risk assessment is the selection of suitable staff.

9.141 Some conditions will need careful evaluation when considering the suitability of staff for working at heights, including:

- severe lung conditions
- significant impaired joint function
- alcohol and drug abuse
- heart condition
- psychiatric conditions (including fear of heights)

- epilepsy
- recurrent dizziness
- medication (any medication that recommends that the recipient should not operate machinery could be a problem).

9.142 The second most important issue is selecting, installing, using and dismantling safely the most suitable equipment. Factors to consider include:

- cost and convenience
- the work activity
- external factors which might cause overturn, fall from height or fall of tools and materials from the access equipment.

9.143 As a general guide, more complex equipment is necessary where the work requires movement. This includes movement of the sort experienced during shaking/forcing activities and movement where tools and materials are used which require firm control, such as drilling masonry where a reaction force is experienced. It is also necessary to have very stable equipment for longer duration work. Of course the equipment must be used strictly in accordance with the designer's intentions.

9.144 The following table gives some examples of activities and the equipment required to perform them safely.

Activity	Equipment Required
Painting a fairly small barge board on a bungalow	A footed ladder
Work on gutters or downpipes, eg replacing a simple rubber seal	A tied ladder
Replacing a length of gutter to a third storey building	A tower with jumbo stabilisers or a tower fixed to special anchors drilled into the building
Drilling a 125mm diameter hole for a vent in brickwork at first floor level	A mobile elevated work platform or steel scaffold — not a tower
Replacing a plastic window, eg 2m wide by 1350mm high (not glazed at time of installation) with access for raising and lowering the window within the property	A tower
An electrician installing a light bulb in an office with ceiling at 2.7m	A pair of stepladders
A painter steaming off wallpaper in a normal bedroom	A hop up without a handrail
A carpenter installing a 4m length of picture rail in a room with a ceiling height of 4.2m	Builders trestles with two 600mm wide stagings (reinforced planks) linked side by side by a patent bracket and handrail system utilising alloy handrails

User Inspection

Ladders and Steps

9.145 Ladders and steps should be manufactured to a suitable standard such as Class 1 industrial or BS EN131. Equipment made to Class 3 domestic, or no standard at all, should not be used. Normally a ladder or pair of steps will have a label affixed indicating the class.

9.146 Pre-use checks for ladders and steps must be carried out by the user each working day. There is no need to record this inspection.

9.147 A detailed visual inspection should be undertaken at intervals of not more than six months if in constant use, and this inspection should be recorded. For less frequently used equipment this interval could be increased to not more than annually. In addition to these inspections, regular maintenance is required in accordance with the manufacturer's recommendations.

9.148 Additional checks are required following a change in circumstances, eg after the ladder has been dropped, or moved from a dirty area to a clean one.

9.149 The following is a checklist for the safe use of ladders and steps.

- Stays and ties must be tight, free of movement and straight.
- All rungs and treads must be:
  - of the original cross section
  - not worn smooth
  - not repaired
  - free of cracks
  - secure at both ends.
- Stiles (the sides of the ladder) should be straight and negligible wear only is permitted on ladder feet.
- Defects can be detected by attempting to rotate the rungs or by trying to shake, widen or twist the stiles (without excessive force) while looking for any significant displacement.
- Sagging or swaying when in use can indicate an unsafe ladder.
- For double or triple extenders, all brackets and hooks should be securely fitted and free from unintended movement.
- Ropes and pulleys must be completely free of defect, secured and free running.
- Serial numbers or company identification should be legible.
- The manufacturer's tolerances or play between different sections must not be permitted to increase through wear or looseness.

Alloy Tower Inspection

9.150 The Work at Height Regulations 2005 require towers to be inspected and a report written before first use after installation, and also following alterations (except moving). A new report is also required for each seven-day period, and if there has been any outside influence which might have affected the safety of the tower, eg a collision.

9.151 The following is a checklist for the safe use of alloy towers.

- Wheels must be free running, and brakes must all be functional.
- All structural joints need to be sound and examined for looseness or fracturing.
- Safety feet on outriggers need to be in good condition.
- All welds must be free of cracks.
- Any texturing on rungs to provide grip must be sound.
- All members should be straight and free from significant dents or fractures of any kind.
- Safety devices should be compatible and securely fitted.
- Working platforms' end hooks must be free from defect and decking must be securely fixed.
- The installation should follow the manufacturer's diagram exactly.
- Serial numbers or company identification should be legible.

Staffing

9.152 Staff should be available in sufficient numbers for the equipment being used.

- Most steel ladders require two persons.
- Alloy towers require two persons.
- Extending ladders above 6m opened generally require two or more persons.

In Use

Selection

9.153

- For very small jobs, roof ladders should be used only by strong individuals with a high degree of experience.
- Handling materials weighing between 10kg–25kg must be risk assessed with a site-specific manual-handling assessment.
- Materials of over 25kg are not to be carried up a ladder or stepladder.

Usage

9.154 The following points should be followed when using ladders.

- Users should not stand higher than the third rung from the top of a ladder unless, in the case of stepladders, there is a handrail.
- Sufficient time should be allowed to keep the work area tidy.
- For podium steps, there is no limit on the duration of use.
- Materials must never be thrown down from access equipment.
- Both feet should be kept on the same rung or tread and the belt buckle or navel should be kept within the stiles.
- Safety footwear and helmets must be worn for almost all work on portable access equipment.

Installation

9.155

- Ladders should be at approximately a 1 in 4 angle of inclination.
- Extension ladders must not be split down and used as separate ladders.
- Access equipment must not be used on slippery surfaces or balanced on top of boxes or van roofs, etc.
- For dangerous work, stepladders should be tied between the second and third step from the top, eg near traffic or where the work will require some significant sideways forces.
- Where necessary items may be joined together with special joining brackets, eg for connecting roof ladders to access ladders. Connections should not be made with the rungs.
- Components should connect together as indicated on the instruction manual.
- The ground should not be too soft, and should be level enough, for the equipment to be accommodated.
- The method of preventing overturning must be in accordance with the instruction book.
- A risk assessment must have been prepared.
- Procedures need to be in place to prevent unauthorised alteration, use or theft.

- The weather conditions should be suitable and not above the design maximum for the equipment.

Storage

9.156 Access equipment should be stored out of the weather and away from strong chemicals like lime or wet cement.

Inspection

9.157

- Ladders and steps should be inspected by the user after any alteration to location, height or task. Hired or loaned ladders or steps must have proof of inspection with them.

Loading

9.158

- Not more than one person should be on a ladder or stepladder at any one time (with the exception of someone footing the ladder).
- Most towers are designed for two persons and their hand tools. They are not designed for the storage of materials other than in small quantities and lightweight.
- Towers must not be moved unless they have been unloaded and lowered to the height dictated in the instruction manual.

Legislative Management Issues

9.159 The purpose of a work at height risk assessment is to identify suitable equipment and have controls in place to make sure it is used properly. The choice should be directed by reference to the hierarchy laid down in the Work at Height Regulations 2005 as follows.

1. Avoid work at height where possible.
2. Carry out the work using an existing workplace.
3. If there is a risk of falling, put in place effective means to prevent falls which protect everyone present ("collective methods") and which require no particular action to activate the protection.
4. Where work allows a risk of falling then use precautions which reduce the consequences of a possible fall and protect everyone present, and which require no particular action to activate the protection.
5. Where "collective" methods are not possible, use methods which allow protection to a person who activates the protection (eg by wearing a harness and securing the lanyard).
6. As a last resort use a fall-arrest harness which will stop a person who has fallen before they land.

9.160 In practice this process of choice can be inefficient and tedious to document. Possession of a clearly stated company policy would assist an experienced and suitably informed supervisor.

9.161 The risk assessment required by the legislation must show how all of the influences which might make the arrangement unsafe have been addressed. This assessment should be relevant to all aspects of the task including the installation and use of the access equipment and its dismantling after the work is finished. For example, the risk of falling tools and materials could be reduced by a toe board and safety helmet and by prohibiting storage of materials on the working platform.

9.162 Precautions should reflect the likely outcome from a fall in terms of distance of travel, and the dangers of the area where the person would land. In all cases, the work should be properly planned and appropriately supervised. Those working at height should be competent to do so or, if being trained, they should be supervised by a competent person.

9.163 A competent person would have the right level of formal training and experience for them to be able to do their job. Competence includes involvement in the organisation, planning, supervision and inspection, and in the supply of access equipment.

9.164 Work at height also includes avoiding the danger of falling through fragile surfaces by:

- not working near them
- preventing approach to them
- reducing the degree of injury which would result from falling through.

Working at Height — Control Measures

9.165 There are several controls which apply to all work at height, whatever the type of equipment used.

9.166 The following steps must be considered to avoid danger from overhead cables.

- Move the work to a safe distance away from the cables.
- Require the owner of the cables to remove them.
- Require the owner of the cables to isolate them from power for a specified period, in accordance with a suitable permit to work.
- Require the owner to make the cables visible and carry out only work which does not enter the zone of risk from arcing, as specified by the owner of the cable.
- Utilise in addition to one or more of the above controls, appropriate signage such as warning of the presence of overhead cables, installing site barriers such as goalposts at distances specified by the owner of the cables and providing verbal instructions to workers.

9.167 The following steps must be taken to remove the danger of vehicles colliding with the access equipment.

- Re-route traffic away from the work area.
- Provide warning signs and speed limits, ie signs which comply with the specification for signs and barriers to be used on works carried out on the highway and ensure workers wear high-visibility clothing.
- Park vehicles between the work area and the oncoming traffic.
- Carry out the work at times of low traffic flow.
- Provide robust barriers such as heavy, brightly painted baulks of timber.
- Provide appropriate instruction to workers.

9.168 The following steps must be taken to prevent the danger of materials falling from or onto the work equipment.

- Re-route pedestrians away from the work area using signs and barriers.
- Arrange for specialists to design and install the mobile access equipment to incorporate screens or fans.
- Substitute fixed scaffolding for portions of the work where there is a risk of falling materials. This would include the provision of fans and debris netting and possibly tunnels with hardboard and polythene below the working platform to catch fluids and dusts.

- Programme the work around periods of low pedestrian risk and favourable weather conditions.
- Ensure helmets are worn in connection with either of the other controls.
- Arrange for other work taking place in the vicinity to be re-scheduled.

9.169 The following steps must be taken to prevent the overturn of the equipment.

- Arrange for the access equipment to be physically secured to the structure.
- Utilise outriggers or other stability devices.
- Provide ballast (eg bags of gravel, heavy concrete blocks, etc) secured to the equipment (in most cases an engineer must design this).

9.170 In order to prevent falls, it is essential to:

- provide training to staff
- ensure that supervisory staff are aware of any reasons why workers would be temporarily unfit for work at height, eg medication, alcohol intake, etc
- allow sufficient time for the project to permit temporary suspension in inclement weather
- have control over the quality of equipment used, eg non-slip ladder rungs
- provide ladder gates at the top of ladders to close the access gap in scaffolding
- secure ladders with special fittings that require tools to remove
- substitute stairs in place of ladders.

9.171 It is important to identify (and rectify) situations where a worker may:

- make jerking movements
- over-reach
- make alterations to the equipment, such as removing handrails
- lean the ladder at the wrong angle
- use a stepladder as a ladder
- erect equipment on an unstable surface
- rush the job
- try to do heavy or difficult tasks from the ladder
- be distracted.

9.172 Employers should ensure that:

- workers wear safety footwear
- any spillage, mud and materials are cleared from and around the equipment.

List of Relevant Legislation

- Work at Height Regulations 2005
- Provision and Use of Work Equipment Regulations 1998

Further Information

Publications

HSE Publications

The following are available from *www.hsebooks.co.uk*.

- HSG150 *Health and Safety in Construction*
- HSG210 *Asbestos Essentials: Task Manual: Task Guidance Sheets for the Building Maintenance and Allied Trades*
- INDG403 *A Toolbox Talk on Leaning Ladder and Stepladder Safety*
- INDG405 *Top Tips for Ladder and Stepladder Safety*
- CIS10 *Tower Scaffolds*

British Standards

The following are available from *www.bsi-global.com*.

- BS 1129: 1990 *Specification for Portable Timber Ladders, Steps, Trestles and Lightweight Stagings*
- BS 2037: 1994 *Specification for Portable Aluminium Ladders, Steps, Trestles and Lightweight Stagings*
- BS EN 131–1: 1993 *Ladders. Specification for Terms, Types, Functional Sizes*

Other Publications

- *PASMA Code of Practice*, Prefabricated Aluminium Scaffolding Manufacturers' Association

Scaffolding and Falsework

- Scaffolding is defined (BS EN 12811) as a "temporary construction, which is required to provide a safe place of work for the erection, maintenance, repair or demolition of buildings and other structures and for the necessary access".
- Falsework is defined (BS EN 12812) as: "temporary support for a part of a permanent structure while it is not self-supporting, and for associated service loads".
- Scaffolds and falsework should always be designed, erected, altered and dismantled by competent people.
- In considering the design of the scaffold or falsework, due consideration must be given to the possible risks and hazards which may be caused.
- Risk assessment of scaffolding structures can be divided into two parts — risks to those who work on the structure and risks and hazards posed by the structure itself.
- When people are working on scaffolds or falsework either during their erection or use, a safe system of work should always be adopted. This may include fall-arrest systems, personal protective equipment and inspection and maintenance. Records should be kept wherever possible.
- There are also steps which need to be taken to prevent injuries to adjacent occupiers, passers by and other members of the public.
- There are a number of other legal considerations which need to be addressed depending on what sort of scaffold is used, and where it is located.
- All of the above information should be communicated to persons who are likely to come into contact with scaffolding and falsework. This is done by using training programmes involving communication and consultation.

9.173 Scaffolding is used in construction work to provide a safe place of work for the erection, maintenance, repair or demolition of buildings and other structures. It also provides the necessary access for workers undertaking this work.

9.174 When considering the use of scaffolding in a construction project, due consideration must be given to the possible risks and hazards which may be faced.

Employers' Duties

9.175 In addition to the legal requirements of the Health and Safety at work, etc Act 1974, the following legislation has an effect on the use of scaffolding. The list is not exhaustive, but represents some of the main areas of influence.

- The Construction (Design and Management) Regulations 2007 impose duties on employers, designers and contractors to consider the health and safety aspects of construction work at the design stage.
- The Work at Height Regulations 2005 require all work at height to be properly planned and organised, and all those connected with it to be competent.
- The Provision and Use of Work Equipment Regulations 1998 require all work equipment, including scaffolds, to be suitable for the purpose for which it is intended.
- The Lifting Operations and Lifting Equipment Regulations 1998 require testing and inspection of all equipment used in lifting operations.
- The Access to Neighbouring Land Act 1992 prescribes the use of access orders and conditions which may be applied to work.

Employees' Duties

9.176

- Employees have a duty to take reasonable care of their own health and safety and that of other people who may be affected by their work under the Health and Safety at Work, etc Act 1974.
- Employees must inform their employer of any danger to health and safety posed by a work activity under the Management of Health and Safety at Work Regulations 1999.
- Employees must also inform their employer of any shortcomings in the employer's protection arrangements under the Management of Health and Safety at Work Regulations 1999.
- Employees have a duty to co-operate with the employer's health and safety arrangements under the Management of Health and Safety at Work Regulations 1999.

In Practice

Relevant Activities

9.177 According to figures supplied by the Health and Safety Executive (HSE), the most common kinds of accident resulting in major injury to workers are falls from height and being struck by a moving or falling object.

9.178 The Work at Height Regulations 2005 define work at height as "work in any place at, above or below ground level where a person could be injured if they fell from that place. Access and egress to that place could also be work at height". They place duties on employers, the self-employed and any person who controls the work of others (eg facilities managers or building owners who may contract others to work at height) to the extent they control the work.

9.179 An employer has a general duty under the Health and Safety at Work, etc Act 1974, to ensure the health, safety and welfare of its employees, so far as is reasonably practicable. If the scaffold or falsework may be classed as a "workplace" for the purpose of the Act, then this duty extends to those places as well.

9.180 Before making a decision on appropriate equipment, it is necessary to consider the following issues.

- What does the job involve?
- Where does it need to be done — eg will access to adjacent areas be required?
- Who might be affected by it — could the public be in danger?
- Under what conditions will it be undertaken? Factors such as weather may have a bearing — extremes of wind, rain, heat or cold will all have different effects.
- How long will it take?

9.181 It is always tempting to pick something simple, like a ladder or a step, to access work at height, but as a general rule it is preferable to use ladders only as a means of getting to a safe, level and secure platform with appropriate edge protection. Obviously, this may not always be possible, for a variety of practical reasons, but it should be the first consideration from a health and safety perspective.

9.182 There are some activities which it is simply not appropriate to carry out from ladders. Masonry work such as brickwork, and demolition, are two obvious examples. However, this also applies to most activities which will involve the storage of materials near to the workplace, or those which will involve work duration of any significant length.

9.183 Similarly, where access would otherwise be difficult, or protection could be required from environmental conditions, a secure platform should always be the first option that is looked at.

9.184 The comfort and safety of the operatives carrying out the work should be of paramount importance. If there could be problems with having to reach from ladders, or the possibility that falls may result, a system that includes proper foundations, guard rails, edge protection, fall-arrest mechanisms, harnesses and nets should be included to physically prevent falls and falling objects.

Design and Materials

9.185 BS EN 12811 contains a requirement that scaffold should be designed, constructed and maintained to ensure that it does not collapse or move unintentionally and so that it can be used safely. Scaffold components shall be designed so that they can safely be transported, erected, used, maintained, dismantled and stored.

9.186 In considering the design of a temporary structure such as scaffolding or falsework the designer should address:

- the possibility of movement, and the lateral stability of the structure as a whole
- access both to the structure itself for maintenance and to the platform or place of work it provides
- the requirements for loads should include "dead" loads (those attributable to the structure) and "live" loads (those as a result of people, their building materials and equipment or the structure being supported). "Lateral" loads, which may result from such things as wind but can be exacerbated by other elements like rain or snow, should also be considered. (BS EN 12811 specifies six load classes and seven width classes of working areas.) A further consideration should be "dynamic" load, which may result from the operation of powered plant and machinery on the platform.

9.187 Where falsework is required to provide structural support, BS EN 12812 stipulates two separate design classes.

- Class A: where the structural integrity is derived from knowledge of the structural performance of components of the structure, such as adjustable props or formwork equipment — for simple constructions such as *in situ* slabs and beams.
- Class B: Falsework where a complete design is undertaken. This is subdivided into two further sub-classes, depending on the relevant European structural standards cited in the British Standard.

9.188 Both standards contain requirements for materials; only those which have established properties and which are known to be suitable for their intended use are acceptable — generally specified by reference to other British or European standards, or where properties have been established by appropriate testing.

Risk Assessment

9.189 The Management of Health and Safety at Work Regulations 1999 state that every employer must assess:

- the health and safety risks their employees are exposed to while they are at work
- the health and safety risks to any persons not in their employment but who may potentially be affected by the work being done.

9.190 This assessment must be carried out in order to identify what measures must be taken in order to comply with the requirements and prohibitions imposed by the regulations. The findings must be recorded if five or more people are employed.

9.191 In addition, the Work at Height Regulations 2005 require the risks from work at height to be assessed and properly controlled.

9.192 The HSE leaflet INDG163, identifies five hierarchical steps to risk assessment.

1. Look for the hazards.
2. Decide who might be harmed and how.
3. Evaluate the risks and decide whether the existing precautions are adequate or whether more should be done.
4. Record your findings.
5. Review your assessment and revise it if necessary.

9.193 There are several considerations which need to be taken into account, both in regard to the design of the scaffold and its subsequent use. Generally speaking, they are those items which are considered elsewhere, as follows.

- Falls from height. Guardrails, toe boards and similar barriers should be provided whenever someone could fall. These should be strong and rigid enough to prevent people falling to the ground. Additional guardrails may be needed where there is a risk that persons might fall between the main guardrail and the toe board or where persons might fall over the main guardrail.
- Falling materials. Scaffolds should be designed and constructed to ensure that materials used during construction cannot fall.
- Working platforms. Working platforms should be wide enough to allow people to pass safely and use any equipment or material necessary for their work.
- Stability. All scaffolds must be based on a firm level foundation capable of supporting the weight of the scaffold and the loads likely to be placed on it.
- Loading platforms. Scaffolds are not usually designed to support heavy loads on their working platforms. If there is any intention to load out platforms, the scaffolder must be informed in advance. Where concentrated loads are expected, a loading tower structure or heavy-duty scaffold may be required.
- Separating distance. Working platforms should be no further than 300mm away from the building or structure. Where a building involves cantilevered or curved elements, this is not always easy.
- Protecting the public. Great care is needed when scaffold is erected above a public thoroughfare. In particular, effective steps need to be taken to ensure that nothing can fall onto people below. As well as brick guards and high-visibility netting, fans or covered walkways may give extra protection.

Safe Use

9.194 A scaffold should be designed, erected, altered and dismantled by competent people, with all scaffolding work under the supervision of a “competent person”.

9.195 A safe system of work should always be adopted during the erection, altering and dismantling of scaffolds. This will usually include the use of fall-arrest equipment.

9.196 Scaffolds must be inspected by a competent person:

- before first use
- after substantial alteration
- after any event likely to have affected their stability, eg following strong winds or adverse weather
- at regular intervals not exceeding seven days.

9.197 Any faults found must be put right. In addition, before contractors allow their workers to use someone else's scaffold they must make sure that it is safe. The HSE information sheet no. 49 offers further guidance to people using scaffolds as follows.

- Do not take up boards, move handrails or remove ties to gain access for work.
- Changes should be made only by a competent scaffolder.
- Never work from platforms that are not fully boarded.
- Do not use the top platform of a stepladder unless it is designed with special handholds.
- Ensure stepladders are positioned on level ground and used in accordance with the manufacturer's instructions.

9.198 Records should be kept of all inspections and changes to the original design. This may be included as part of a quality assurance scheme, or in some cases as part of the health and safety file for the project.

Protecting the Public

9.199 There are many persons other than building workers who may be affected by scaffolding work. Because scaffolds are often erected on the boundaries of buildings, they are much more likely to come into contact with members of the public. There are several areas where statutory requirements may need to be observed. One example is the Access to Neighbouring Land Act 1992 where an agreement, the "access order", is required to be drawn up.

9.200 The access order may include conditions regarding:

- the way in which the specified work must be carried out
- when the work in question may be carried out, including specific hours on specific days
- who is permitted to carry out the specified works and who is permitted to enter the site where the work is taking place
- the taking of all necessary precautions as may be stipulated in the order.

9.201 Similarly, the Highways Act 1980 has a range of provisions which apply should there be a need to erect scaffolding on a public highway. Typically this will involve securing a licence from the Highways Authority, which will include a series of conditions, such as the provision of appropriate signage, lighting and public protection measures (and, invariably, a financial charge).

9.202 Other points which need to be addressed include the following.

- The scaffold should be designed to carry the load from stored materials and equipment.
- Scaffolds should be designed to prevent materials falling, including brick guards, netting or sheeting.
- Where the risk is high, eg during demolition or facade cleaning, extra protection in the form of scaffold fans or covered walkways may be required.

- In populated areas such as town centres, erecting and dismantling scaffolds should be undertaken during quiet times.
- People should be prevented, with suitable barriers and signs, from walking under the scaffold during erection or dismantling.
- Unauthorised access should be prevented, eg by removing all ladders at ground level, whenever the scaffold is left unattended.
- Never throw materials from a scaffold. Use mechanical hoists or rubbish chutes to move materials and waste.
- The platform of a general-purpose scaffold should be at least four boards wide. All scaffolds, including "independent" scaffolds, should be securely tied, or otherwise supported. More ties will be required if:
  - the scaffold is sheeted or netted due to the increased wind loading
  - it is used as a loading platform for materials or equipment
  - hoists, lifting appliances or rubbish chutes are attached to it.
- System scaffolds should be erected following the manufacturer's instructions and may require more tying than independent scaffolds.

Training

9.203 Employers must ensure that everyone involved in work at height is competent (or, if being trained, is supervised by a competent person). This includes involvement in organisation, planning, supervision, and the supply and maintenance of equipment.

9.204 Where other precautions do not entirely eliminate the risk of a fall occurring, employers must (as far as it is reasonably practicable to do so) train those who will be working at height on how to avoid falling, and how to avoid or minimise injury to themselves should they fall.

9.205 Those employees, or those working under someone else's control, must:

- report any safety hazard to them
- use the equipment supplied (including safety devices) properly, following any training and instructions (unless it would be unsafe to do so, in which case further instructions should be sought before continuing).

9.206 There are three classes of people who need to make use of these principles:

- those who are tasked with designing or erecting the scaffold or falsework
- those who need to work on it
- those, often members of the public, who will come into contact with it, and need to be aware of the dangers that it poses.

9.207 Competence is described by the HSE as, "the experience, knowledge and appropriate qualifications that enable a worker to identify the risks arising from a situation and the measures needed to be taken".

9.208 Those undertaking scaffolding activities should be trained in the relevant system of work and any particular type of work equipment which is to be used. Those responsible for their management should ensure that this is the case.

9.209 There are some specific requirements which should be observed in particular situations. For example, if scaffolds are to be used near power lines, special training may be needed to address the particular circumstances and hazards that they might present.

9.210 In Great Britain, the Construction Industry Scaffolders' Record Scheme aims to ensure that operatives erecting, altering or dismantling scaffolds are properly trained and have sufficient experience to carry out work safely and correctly. Similar schemes operate in Northern Ireland.

9.211 The Construction Industry Training Board and Construction Skills issue record cards to operatives who have completed the required training, gained appropriate experience and achieved the relevant NVQ/SVQ.

9.212 The scheme is affiliated to the Construction Skills Certification Scheme and currently has over 15,000 registered members. Four types of card are available, based on training, experience and position.

9.213 In addition to these schemes, the importance of on-site training, "toolbox" talks and ongoing supervision and advice should never be overlooked.

Further Information

Publications

HSE Publications

The following are available from *www.hsebooks.co.uk*.

- CIS10 (rev 4) *Tower Scaffolds*
- CIS49 (rev 2003) *General Access Scaffolds and Ladders*
- CIS56 *Safe Erection, Use and Dismantling of Falsework*
- HSG33 (rev 1998) *Health and Safety in Roof Work*
- HSG150 (3rd ed, 2006) *Health and Safety in Construction*
- INDG284 *Working on Roofs*
- INDG401 *The Work at Height Regulations 2005 (as amended): A Brief Guide*
- RR116 *Falls From Height: Prevention and Risk Control Effectiveness*
- HSE video *A Head for Heights: Guidance for Working at Height in Construction*

British Standards Publications

The following are available from *www.bsi-global.com*.

- BS EN 74–1: 2005 *Couplers, Spigot Pins and Baseplates for Use in Falsework and Scaffolds. Couplers for Tubes. Requirements and Test Procedures*
- BS EN 1004: 2004 *Mobile Access and Working Towers Made of Prefabricated Elements. Materials, Dimensions, Design Loads, Safety and Performance Requirements*
- BS EN 12810–1: 2003 *Facade Scaffolds Made of Prefabricated Components. Product Specifications*
- BS EN 12810–2: 2003 *Facade Scaffolds Made of Prefabricated Components. Particular Methods of Structural Design*
- BS EN 12811–1: 2003 *Temporary Works Equipment. Scaffolds. Performance Requirements and General Design*
- BS EN 12811–2: 2004 *Temporary Works Equipment. Information on Materials*
- BS EN 12813: 2004 *Temporary Works Equipment. Load Bearing Towers*

- BS EN 13374: 2004 *Temporary Edge Protection Systems. Product Specification, Test Methods of Prefabricated Components, Particular Methods of Structural Design*

Organisations

- Construction Industry Training Board (CITB) and Construction Skills
  Web: *www.citb.org.uk*
  The CITB and Construction Skills provide assistance in all aspects of recruiting, training and qualifying the construction workforce.
- International Powered Access Federation (IPAF)
  Web: *www.ipaf.org*
  The IPAF works to promote the combination of the capabilities of modern platforms and training for safe and effective work at height.
- National Access and Scaffolding Confederation Ltd (NASC)
  Web: *www.nasc.org.uk*
  The NASC is the national representative employers' organisation for the access and scaffolding industry.
- Prefabricated Access Suppliers' and Manufacturers' Association
  Web: *www.pasma.co.uk*
  The association provides training and strives to advance safety standards in the UK access industry.

Temporary Works

- Temporary works are structures or formations designed to provide support, access and protection during the construction phase.
- Drawings and sketches which illustrate and explain the design of temporary works must be produced by qualified and experienced people.
- A sequence should be determined for checking for adherence to specified tolerances and assembly.
- An area which is as clear as possible of other activities and pedestrians needs to be designated for the safe unloading of components.
- Suspect items should be marked with a distinctive spray paint and set aside in a suitable "quarantine area".
- Unless the temporary works installer has expertise and resources to draw up suitable lifting plans, the movement of materials should be considered a candidate for contract lifts.
- High-visibility clothing, of a form not likely to get snagged, and footwear which gives good ankle support, must be worn, as well as a suitable safety helmet.
- Designers should have access to the relevant British Standards, and be holders of structural engineering qualifications.
- Wherever possible, working platforms should be used to avoid hazardous unslinging operations.
- A lifting plan must be prepared when mobile cranes are involved and it should be an integral part of the method statement.
- The method statement should also indicate what the temporary works designers have allowed for as a loading sequence, and the rate of loading.
- Lack of proper attention to drainage can result in ponding and scour which can undermine ground support, or cause slip planes which are unrestrained.

9.214 Few construction, demolition or renovation projects take place without some form of temporary work taking place. The term "temporary works" can be applied to:

- piling
- access scaffolding for construction purposes
- excavation supports
- formwork and falsework for structures
- façade retention
- temporary underpinning and guying
- formwork to walls
- formwork to soffites
- shoring and temporary bracing for masonry walls and structural frames
- permanent shuttering
- precast flooring support
- cranes and crane foundations
- structural shoring.

9.215 The temporary nature of these installations can lead to designers and contractors not giving the full attention that is required. As a result, temporary works can fail leading to injuries and, at worst, loss of life.

Employers' Duties

9.216

- Under the Construction (Design and Management) Regulations 2007, all practicable steps must be taken to ensure that all structures are prevented from being in any temporary state of weakness or instability which could result in their collapse. Any temporary support or structure must be designed, installed and maintained to withstand any foreseeable loads — they must also be used only as they were designed, installed and maintained to be used. It is also a requirement that no part of the structure is loaded in such a way as to make it unsafe.
- The Work at Height Regulations 2005 require reasonably practicable methods to be adopted to prevent falls from height. A hierarchy is given in the regulations.
  - Avoid work at height where possible.
  - When work at height cannot be avoided, use work equipment to prevent falls.
  - When the risk of falls cannot be eliminated, use work equipment to minimise the distance and consequences of any fall.
  - The risk assessment required by the regulations must show how all of the influences which might make the arrangement unsafe are addressed including installation, use and dismantling after the work is finished. For example, the risk of falling tools and materials could be reduced by a toe board and safety helmet, and by prohibiting storage of materials on the working platform. The precautions should reflect the likely outcome from a fall in terms of distance of travel, and the dangers of the area where the person would land. In all cases the work should be properly planned and appropriately supervised. Those working at height should be competent to do so or if being trained they should be supervised by a competent person.
  - Work at height also includes avoiding the danger of fragile surfaces by not working near them, preventing approach or reducing the likely injury from falling through.
- The Lifting Operations and Lifting Equipment Regulations 1998 require a testing and inspection regime for all lifting equipment. There must also be a documented system showing that sufficient planning and precautions are in place to prevent danger. In practice a lift using a crane would either need to be arranged as a contract lift or the person directing and planning the operation should pass a four-day appointed persons course. All staff involved in handling, slinging or directing crane operations should be qualified.

Employees' Duties

9.217

- Under the Health and Safety at Work, etc Act 1974 employees have a duty to:
  - take reasonable care of their own health and safety and that of other people who may be affected by their activities at work
  - co-operate with their employer to enable the employer to comply with health and safety duties.
- Under the Management of Health and Safety at Work Regulations 1999 employees are required to:

- use any machinery, equipment, dangerous substances, transport, safety devices and means of production in accordance with any training and instructions provided by the employer
- inform the employer of any serious and imminent dangers to health and safety
- inform the employer of any shortcomings in the employer's health and safety arrangements.

- Under the Work at Height Regulations 2005, employees must:
  - report to their employer or person in control of the work, any activity or defect relating to work at height that may endanger the safety of themselves or others
  - use any equipment provided for their use in accordance with any training and/or instructions relating to its use so as to enable their employer or person in control of the work to comply with their own health and safety duties.

In Practice

Starting Point

9.218 All temporary works should be designed by qualified and experienced people. Drawings and sketches, which illustrate and explain the design of the temporary works must be produced. Such drawings should include the relevant calculations.

9.219 Within a temporary works project, supervision of a well thought-out planning, design, installation, working and dismantling regime is essential. That supervision must have specialist experience to provide the insight necessary to foresee the consequences of work that is out of plumb, is missing any bracing or is constructed from inferior materials.

9.220 System scaffolding and support materials are supported by standard design packages provided by the manufacturers design team or by site-specific design produced at the offices of the supplier.

Method Statements

9.221 The installer should follow a method statement to indicate the sequence of build. Consideration should be shown in the method statement to the following issues.

Pre-start Meeting with Designers

9.222 The sequence of work, the likely conflicts with other civil work and the proposed detailing, heavy-handling provision and the consequences of external variables such as water table and access availability should be discussed.

Reference to Soil/Structural Survey or Investigation

9.223 Sufficient boreholes or other structural information should be available to minimise the risk of unexpectedly encountering boulders or lenses of hazardous sand or slip planes within the bearing surfaces for the temporary works.

Drawing Management, Method of Communicating Design Changes

9.224 Arrangements should be sorted out before a contract to carry out the work is signed to ensure that there is a lead-in time and a procedure for issuing, and acknowledging, receipt of drawings. This should include discussion between the client and the contractor and between the contractor and the supervisor.

Instructions and Drawings for All Support Systems

9.225 Where the design has been provided by a specialist supplier then standard details should be available for those installing the components to see, including:

- the sequence of assembly
- the orientation
- any checks required to establish correct assembly.

9.226 A sequence should be determined for checking for adherence to specified tolerances and assembly.

9.227 A person should be appointed to check (prior to loading) that the construction:

- is true to line and level
- has suitable weight distribution
- has been correctly assembled in accordance with the manufacturer's and engineer's design, including the correct position and number of all struts, binders, etc.

9.228 The management and identification of all drawings and details is essential. Drawings should be uniquely numbered, referenced to location, and cross-referenced to any other information with which they should be used. It is helpful to set up a system of grid references to the project.

Unloading Area

9.229 An area which is as clear as possible of other activities and pedestrians needs to be designated for the safe unloading of components. This area should be level, hard and well lit.

Inspection of Materials Prior to Use

9.230 This is important not only when materials are unloaded and signed for, but also as an ongoing procedure during assembly, use and removal. The materials, particularly the welds, should be subject to spot-checking. If defects are apparent on spot-checking then full inspection is essential. Suspect items should be marked with a distinctive spray paint and set aside in a suitable "quarantine area". Contractors need to be shown this store and individually told not to touch any of it.

Traffic Management

9.231 All sites should have an explicit traffic management plan. If useful, the details can be sent to the supplier of materials to ensure that, where possible, one-way systems, unloading areas, waiting areas, etc can be easily found. A clear map is invaluable for this purpose.

Make-up Area

9.232 Pre-assembly of components at ground level can reduce the amount of craneage and working at height. This should be done in a safe area, where craneage can be controlled by the foreman or supervisor, and traffic movements are restricted to those actually involved with the temporary works. The use of clean 75mm stone for this area is not satisfactory. A blinding layer is necessary.

Storage Area

9.233 Materials which are not immediately needed should be stored away from the assembly area, sorted by type and preferably within the arc of the crane. Where materials are stacked this can cause slinging off ladders and involve climbing on or over stacks. It is important to have more than the minimum as large panels can swing somewhat erratically.

Movement of Materials

9.234 Unless the temporary works installer has expertise and resources to draw up suitable lifting plans, the movement of materials should be considered a candidate for contract lifts. There is in either case a need for suitable ground preparation for the crane and vehicles associated with it.

Zoning from Other Operations

9.235 It is preferable to programme the installation and use of temporary works as a distinct phase in the project, because removal is likely to have a significant influence on the programme. Where possible other activities should be either deferred or restricted during this phase.

Access Provision

9.236 Sufficient resources, space and surface preparation should be invested so that the worker can use the safest possible access provision. The use of ladders should be avoided wherever possible. Fixed scaffolding or mechanical access such as mobile elevating work platforms are the preferred option.

Power Supply

9.237 If battery nut runners and compressed air tools can be used, this obviates the need for a vulnerable power supply using a transformer or a generator. Achieving and maintaining a reliable earth path can be difficult. If power must be supplied it should be laid inside robust steel ducts or at the very least along well-marked routes with mechanical protection in addition to the use of armoured cable. Earth proving should be repeated as often as the competent person determines is necessary.

PPE

9.238 Noise, dust and occasional sparks are to be expected and suitable individual issue of comfortable personal protective equipment is essential. Some sites will insist on gloves, but due to the rough surface of some components a variety should be available. In all cases replacements should be available as there is usually a high loss rate from carelessness and weather conditions.

9.239 High-visibility clothing, of a form not likely to get snagged, and footwear which gives good ankle support, must be worn, as well as a suitable safety helmet. Where water is likely to be encountered, waterproof boots should be made available.

Qualifications and Training

9.240 Designers should have access to the relevant British Standards, and be holders of structural engineering qualifications. One and two-day courses on temporary works are available through the CITB. It is not advisable to rely exclusively on experience.

Inspection

9.241 Particular care should be demonstrated in adhering to tolerances for verticality (see BS 5975). Checklists should be prepared by the designer for each stage of installation and they should be customised to the individual project.

Lighting and Emergency Lighting

9.242 In anticipation of tight programme schedules, the need for glare-free illumination at an early stage of preparation is essential. It is usually necessary, particularly where there will be working at height, to have adequate lighting to kick in if there is a power cut or fault.

Slinging Points

9.243 To minimise the need for choking, slinging points should be incorporated either in the individual components or in the stillages used to move them.

9.244 If choking is unavoidable, restraint bars should be inserted to prevent slippage when the lifting operation starts.

Method of Unslinging

9.245 Pile-handling equipment is usually equipped with remote unslinging. Wherever possible, working platforms should be used to avoid hazardous unslinging operations. The method statement should indicate that measures are in place to provide temporary stability.

Positions for Banksmen

9.246 Banksmen have to be seen by the crane operator. They also need to be clear of possible over-swing.

Lifting Plan

9.247 A lifting plan must be prepared when mobile cranes are involved. An appointed person should prepare the plan and it should be an integral part of the method statement. This plan will determine the crane position.

Avoiding Contamination and Disruption

9.248 Ground loading, the movement of heavy plant and the disturbance of ground strata by all types of temporary works requires that all those involved are made aware of how the spread of contamination, and disruption of services, should be minimised or avoided. There is a range of possible solutions, including temporary diversion to full remediation. The use of geotextiles and blinding can help.

Loading Sequence

9.249 The method statement should indicate what the temporary works designers have allowed for as a loading sequence, and the rate of loading. Actual loading should, in most projects, not proceed until a permit to pour or load has been issued. The permit should relate to what is to be loaded, eg it may be ready for reinforcement only.

Dismantling or Exclusion Zone

9.250 As work proceeds more and more material is squeezed into the construction area including waste and access equipment. The zone for dismantling should be made clear of unnecessary obstacles prior to issuing a permit to dismantle or strip out.

Protection of Open Edges

9.251 Open edges may have temporary instability so crash decks, bean bags, netting or air bags may need to be installed, particularly if it is unsafe to access to install edge protection. Wherever possible, components should be assembled with pre-installed edge protection at ground level. If this is not possible then harness and line may be necessary.

Waste Handling and Housekeeping

9.252 Offcuts and damaged material should be skipped and removed regularly to avoid tripping and falling accidents.

Protection of the Works

9.253 Once a structure is in place the possibility of collision should be prevented by installing sleepers or concrete lane dividers, or by simply re-routing traffic and the movement of materials.

Contingency Plans

9.254 Breakdowns and alterations to plans are part of the construction process. They need to be accommodated by having access to alternative programmes, working areas, and handling equipment. Without this “disaster planning”, short cuts will be taken with an increase of risk.

Weather as a Limiting Factor

9.255 Inclement weather can influence stability and handling characteristics, even causing slipping over on smooth forms. Each day should be assessed by an authorised person who is able to suspend operations if necessary

Method Statement Briefing and Toolbox Talks

9.256 The sequence changes planned by the designer, the detailing, the edge protection and the access arrangements should all be explained so that all parts of the workforce understand the reasons for any precautions in place.

Installation of Any Anchor Points

9.257 Where push-pull type props are required, fixings should be made into structural concrete.

Drainage Requirements

9.258 Lack of proper attention to drainage can result in ponding and scour which can undermine ground support, or cause slip planes which are unrestrained.

Dismantling Sequence

9.259 The order of dismantling should not be left to on-site staff to determine. Almost all temporary works have self-loads which are significant and at some point in the proceedings the structure will become unstable.

List of Relevant Legislation

- Construction (Design and Management) Regulations 2007
- Work at Height Regulations 2005
- Lifting Operations and Lifting Equipment Regulations 1998

Further Information

Publications

HSE Publications

The following publication is available from *www.hsebooks.com.*

- CRR394 *Investigation into Aspects of Falsework*

Other Publications

- *Final Report of the Advisory Committee on Falsework (The Bragg Report)*, OPSI, 1975, ISBN 0 11 880347 6
- CS030 *Formwork: A Guide to Good Practice*

British Standards

- BS 5975: 1996 *Code of Practice for Falsework*
- BS EN 12811–2: 2004 *Temporary Works Equipment. Information on Materials*

Chapter 10

Occupational Hazards

Asbestos

- The characteristics that make asbestos such a versatile building and insulation product also make it very deadly to health.
- When the material is broken or damaged, some of the fibres that can be released can be small enough to penetrate deep into the lungs.
- Employers should prevent, or where this is not reasonably practicable reduce to the lowest level reasonably practicable, the exposure of employees to asbestos.
- The main diseases that are caused by asbestos are asbestosis, lung cancer, mesothelioma and pleural plaques.
- Employers must protect employees and other persons from exposure to asbestos.
- Duty holders (eg owners, occupiers and managing agents) have to assess whether asbestos is, or is suspected to be, on the premises.
- Where asbestos is identified or suspected, the duty holder must determine the risk posed by the asbestos.
- Where exposure to asbestos cannot be avoided, employers must introduce control measures to reduce the risk as far as reasonably practicable. A written plan must be prepared that identifies the areas of concern and the necessary control measures.
- Work with asbestos, subject to certain exemptions where the exposure is sporadic and of low intensity, cannot be undertaken unless the employer holds a licence granted by the Health and Safety Executive, if it considers it appropriate.
- Exposure to asbestos is subject to a control limit. Certain legal requirements are triggered if anticipated exposure is likely to exceed these levels.
- The size of the fibre, its diameter, the concentration of fibres and the length of time of exposure to asbestos will all affect the risk of developing disease.
- Employers must prepare procedures for protecting employees in the event of an asbestos incident, accident or emergency.
- Asbestos waste must be transported in sealed, labelled containers. The Hazardous Waste Regulations 2005 require all waste containing asbestos to be disposed of at licensed sites.

10.1 Asbestos is a generic term for a group of fibrous minerals formerly used for making fireproof items. Its incombustibility and strength led to wide-scale commercial use in the past.

10.2 The characteristics that make asbestos such a versatile building and insulation product, however, also make it very deadly to health. (Deaths attributed to asbestos exposure are currently in excess of 3000 per annum in the UK and are predicted to peak between the years 2010 and 2020, with possibly as many as 10,000 deaths per annum.) It is resistant to acid and while the fibrous structure of asbestos provides good insulation and binding strength, when the material is broken or damaged, fibres can be released in great quantities. Some of the fibres can be very small and it is the indivisible fibres that are small enough to penetrate deep into the lungs, which have the most significant impact.

Employers' Duties

10.3

- Employers have a general duty to ensure, so far as is reasonably practicable, the health, safety and welfare at work of all employees under the Health and Safety at Work, etc Act 1974.
- Under the Management of Health and Safety at Work Regulations 1999, employers must make a suitable and sufficient assessment of the risks to the health and safety of their employees to which they are exposed while they are at work and of the risks to the health and safety of persons not in their employment who may be affected by their work.
- Under the Control of Asbestos at Work Regulations 2006, employers must:
  - find out whether their building contains asbestos, what condition it is in and assess the risk, eg if it is likely to release fibres, employers must make a plan to manage the risk and protect employees and other persons from exposure to asbestos — if the employer is not responsible for maintaining and repairing all or part of the property, or does not have control of the building, the person who has these responsibilities must carry out the above actions
  - be licensed by the Health and Safety Executive (HSE) if involved in any work with asbestos, with the exception of work in which the exposure is sporadic or of low intensity
  - ensure that asbestos waste is transported in sealed, labelled containers with consignment notes, and disposed of at licensed sites
  - make no use of any product that contains asbestos
  - protect employees and anyone else who may be affected by their work from exposure to asbestos.
- Employers are obliged to prevent, or where this is not reasonably practicable reduce to the lowest level reasonably practicable, the exposure of employees to asbestos (by measures other than the use of respiratory protective equipment in accordance with the Personal Protective Equipment Regulations 2002). Similarly, employers should also ensure that the number of employees exposed is as low as possible.
- If it is not reasonably practicable to reduce the exposure of employees to below the applicable control limits, in addition to these obligations employers must provide suitable respiratory protective equipment.
- Where employees could be exposed to asbestos in connection with the installation of any product or manufacturing process, employers are required where practicable to prevent such exposure by substituting a safer alternative. If it is not reasonably practicable, employers must prevent exposure to asbestos through measures that avoid or minimise the release of asbestos (eg control exposure at source).
- Employers must ensure that adequate information, instruction and training is given to employees who are liable to be exposed to asbestos. To achieve this, employers should make employees aware of:
  - the significant findings of a risk assessment
  - the risk to health from asbestos
  - the precautions to be observed
  - the relevant control limits and action levels.

- Employers are required to monitor the exposure of their employees to asbestos unless the exposure is not liable to exceed the action level. Asbestos fibres present in the air must be monitored at regular intervals and when a change occurs that may affect exposure.
- Employees who are exposed to asbestos above the action level must have adequate medical surveillance.

Employees' Duties

10.4

- Employees have a duty to take reasonable care of their own health and safety and that of other people who may be affected by their work under the Health and Safety at Work, etc Act 1974.
- Employees also have a duty to co-operate with their employers on health and safety matters.
- The Management of Health and Safety at Work Regulations 1999 place a duty on all employees to follow health and safety instructions and to report danger.
- The Control of Asbestos Regulations 2006 place a duty on employees to:
  - make full and proper use of any control measures provided in accordance with the regulations
  - take reasonable steps to return control measure equipment after use
  - immediately report any defects to control measure equipment to the employer
  - report to the employer the disturbance of any asbestos, or the presence of any asbestos which is not already known.

In Practice

What is Asbestos?

10.5 Asbestos is a generic term for a group of fibrous minerals, the most commonly used types of which are chrysotile (white asbestos), amosite (brown asbestos) and crocidolite (blue asbestos). Up to 500,000 commercial, industrial and public buildings in the UK are still thought to contain asbestos materials.

Main Diseases Caused by Asbestos

10.6 The main diseases that are caused by asbestos are as follows.

Asbestosis

10.7 Asbestos fibres lodge in the lung tissue in such quantity that their presence causes scarring. As the scarring of the lung increases, the lung becomes less efficient at transferring oxygen. As the disease progresses, shortness of breath becomes more pronounced. Any effort to move objects, lift them, etc, causes shortness of breath and severely restricts the activity of the sufferer. This then makes breathing more difficult and increases stress on the heart. Eventually the lungs fail and heart failure results.

10.8 Asbestosis has a latent period from first exposure of up to 30 years and for which currently there is no effective treatment.

Lung Cancer

10.9 Most lung cancers start in the bronchi, eg the tubes into which the trachea or windpipe divide. With asbestos, the lung cancer may also start in the trachea, bronchioles or alveoli. While lung cancer usually develops slowly, the cancer cells may break off and spread to other parts of the body.

10.10 The most significant cause of lung cancer is smoking. However, if somebody smokes and is exposed to asbestos, the synergetic effect increases the risk of developing lung cancer significantly. Exposure to both can increase the risk of developing lung cancer by 50 times. Encouragement should be given to all employees who smoke to stop smoking.

10.11 Lung cancer from asbestos exposure has a similar latent period to asbestosis and also currently has no effective treatment.

Mesothelioma

10.12 Mesothelioma is a very aggressive form of cancer. There is no known threshold of asbestos exposure that can trigger the cancer. While it is unlikely that a single fibre will kill, some people with very low exposures to asbestos have still died from this vicious disease.

10.13 In the UK there is a specific mesothelioma register and around 1400 people die each year from this disease. Mesothelioma takes many years to manifest itself. The time from initial exposure to the onset of the disease can range from between 15–60 years. This causes serious problems in trying to deal with the disease. The people who are suffering now could have initially been exposed in the 1960s or 1970s. It is expected that mesothelioma deaths will continue to rise until 2010–2015 and then start to decline. However this decline is not likely to be sharp and it is estimated that a significant number of people will still be dying of mesothelioma in 2030.

10.14 The mechanism which leads to the cancer is complex and not clear. It is thought to involve several steps where the fibres in the lungs disrupt cells over a period of time. Chemical changes around the areas of lung tissue which have fibres in them then act on cells around the fibres. The changes appear to be complex. Over a long period of time, they lead to the growth of cancerous cells. Currently there is no cure.

10.15 Mesothelioma is a rare cancer and the tumour develops in the membranous lining of the chest. The most common type occurs in the lung lining. To a lesser degree the cancer occurs in the stomach lining and occasionally in the heart. As tumours develop in the lung lining, often the first sign is a build-up of fluid within the membrane — this is known as "effusion".

10.16 With pleural mesothelioma, the cancer develops in the fine lining around the lung. The pleural membrane surrounds the lungs and stomach. The gap between the lungs and the stomach is the pleural space. The pleural membranes help to protect the stomach and the lungs. They also produce fluids that fill the gap and thus aid breathing by acting as a lubricating fluid. As such, the lung is able to move smoothly as people breath in and out. It is clear that anything that interferes with the smooth operation of the lungs in breathing will be painful, bringing great discomfort.

10.17 Those suffering with the disease start to have trouble breathing and often feel pain. As the cancer grows, the tumour gets larger, reducing the efficiency of the lungs and making it more difficult to breathe. Sufferers then have difficulty moving around. In some cases, the tumour can come through the ribcage causing severe and debilitating pain.

10.18 Treatment of this cancer is mainly aimed at trying to reduce pain and improve breathing. For some people, pain may be limited, in others it can be severe. There remains no cure. Diagnosis to death is often within 18 months.

10.19 The latent period from first exposure can be between 15 and 55 years and is almost certainly fatal.

Pleural Plaques

10.20 Pleural plaques are a form of non-malignant scarring of the pleura, caused by asbestos fibres. If the scarring is more focused and well defined, it is referred to as pleural plaques. The scarring is similar to that found in asbestosis but in this case it is not in the lung tissue but in the lining of the lung.

10.21 Pleural plaques and pleural thickening can both cause shortness of breath and impair lung function. While the symptoms can be treated, the conditions cannot be cured. They are often associated with asbestosis and can take around 10 years to develop. As such, they may be an indicator of more serious diseases to come.

Importance of Extent of Contact with Asbestos

10.22 The size of the fibre, its diameter, concentration, and length of time people are exposed will all affect the risk of developing disease. There is strong evidence to show that blue asbestos is the most difficult fibre type for the lungs to deal with. This is usually identified by sampling. However, in sufficient quantity, all forms of asbestos can cause asbestosis, lung cancer and mesothelioma. Asbestosis is associated with continuous exposure to large amounts of asbestos. The exposure level for lung cancer is still under debate, but the figure is expected to be around one fibre per millilitre. No threshold level of exposure has been identified for mesothelioma.

The Duty to Manage Asbestos

10.23 Due to the very significant health risks, there is a specific duty to manage asbestos carefully.

10.24 Duty holders include those responsible for the maintenance and/or repair of non-domestic premises. This includes the owners of such premises, whether they are occupied or vacant.

10.25 The aim is to protect workers who may come across asbestos in the course of their day-to-day activities, since the major problem facing these workers is that they often do not know where and when the material may be encountered.

10.26 The duty holder must:

- conduct a survey of the premises
- take reasonable steps to locate materials containing or presumed to contain asbestos
- complete a risk assessment to discover the likelihood of the release of dangerous fibres
- make a plan to manage the above risk, including if necessary sealing or encapsulating the asbestos-containing material or, as a last resort, removing the asbestos
- keep in a well-maintained asbestos register a written record of the location and condition of the asbestos in the building

- put procedures in place to ensure that the employer's own employees or third-party contractors have full knowledge of where asbestos may be present to prevent accidental exposure.

10.27 The Health and Safety Executive (HSE) Approved Code of Practice (ACOP) L127: *The Management of Asbestos in Non-domestic Premises* explains the duties of building owners, tenants and any other parties who have any legal responsibility for the work premises. It also sets out what is required of people who have a duty to co-operate with the main duty holder.

Where is Asbestos Found?

10.28 Due to its unique fire-resistance properties, asbestos has been identified in over 3,000 different products, most extensively in buildings. Most buildings built between 1950 and 1980 are likely to contain some asbestos. Asbestos in the form of asbestos cement can also be found in buildings constructed or refurbished between 1980 and 1999.

10.29 The main uses of asbestos in buildings were as:

- sprayed insulating coating on steelwork and concrete
- lagging on pipes and boilers
- insulation boards on walls, doors and ceilings
- asbestos cement as structural sheets, pipes and tanks
- some ceiling tiles
- some decorative plasters.

The Practical Stages of the Duty to Manage

10.30 In order to assess and manage the risk of asbestos, the duty holder must:

- survey the work environment
- assess the risk
- manage the risk.

10.31 There is detailed guidance from the HSE on surveying, sampling and assessment of asbestos. It is important to select a competent surveyor. A full list of companies accredited by the United Kingdom Accreditation Service (UKAS) can be obtained via their website.

10.32 Once the asbestos-containing materials (or materials presumed to contain asbestos) have been located, the ability of these materials to release asbestos fibres must be assessed, and a written plan prepared to manage the risk. This should ensure that details of the location and condition of any asbestos is provided to every person likely to disturb it. It should be made available, if required, to the emergency services.

10.33 The plan should include a procedure for monitoring and recording any condition changes. The frequency of the checks to determine condition changes will depend on the type of material present.

The Survey

10.34 There are three types of survey that can be carried out.

1. *Location and Assessment Survey.* Type 1 is a location and assessment survey, often referred to as a presumptive survey. The purpose is to locate and assess materials that could contain asbestos. If materials could contain asbestos, it should be assumed that they do. No sampling of the material is required but the

records kept must identify the material as presumed asbestos and all necessary precautions should be taken. The advantage of this type of survey is that it is unlikely to involve specialist third-party costs. Its disadvantage is that all materials are to be treated as asbestos, even if they do not contain asbestos. An experienced, competent surveyor should be used to carry out a presumptive survey.

2. *Standard Sampling and Assessment Survey.* This type of survey requires inspection and sampling of the materials to determine if they contain asbestos. The samples must be analysed by a company accredited to ISO 17025 by UKAS. Depending on the likelihood of future disturbance of asbestos-containing material, in certain buildings it may be practicable to carry out a mixture of both type 1 and type 2 surveys. Type 1 and 2 surveys do not need to be fully intrusive.
3. *Full Access Sampling and Identification Survey.* This survey builds upon the type 2 inspection but requires access to all areas where asbestos may be present. This survey is a requirement prior to demolition or major refurbishment of buildings. As materials in a type 3 survey will be removed, there is no need to assess the condition of the asbestos.

10.35 It is important to select a competent surveyor to carry out the survey of premises. There is a UKAS accreditation for surveying companies to ISO 17020 (EN 45004) standard. Individual surveyors will be required to hold a certification from UKAS to EN 45013.

The Risk Assessments

10.36 The assessment necessary for survey types 1 and 2 needs to identify the hazard, which is the material assessment, and the risk, which is the priority assessment. The material (hazard) assessment has four components:

- type of material
- extent of damage and/or deterioration
- what surface treatment is present
- the asbestos type present in the material.

10.37 When combined, these variables will assist in determining the ability of the material to release asbestos fibres. The priority (risk) assessment looks at:

- the normal occupation level
- the likelihood of disturbance
- the human exposure potential
- the likely maintenance activities.

10.38 When combined these should be used to develop the action plan required to manage the risk.

Managing the Risk

10.39 Measures must be specified in a plan for managing the risk. The plan should include adequate measures for ensuring that the information about the location and condition of any asbestos is provided to every person likely to disturb it. It should be made available, if required, to the emergency services. The management of the risk therefore requires a written plan to be prepared, incorporating the register of the asbestos materials present. The plan should also include a procedure for monitoring and recording any condition changes. The frequency of the checks to determine condition changes will depend on the type of material present, ie fibrous products will require more regular inspection than securely bonded products. It may be necessary to introduce a permit-to-work system to prevent disturbance of the material by persons unaware of the existence of the register.

10.40 Asbestos condition surveys should be carried out on the premises, as part of asset management planning.

Licensing

10.41 Work with asbestos must not be undertaken unless the employer holds a licence granted by the HSE. This does not apply to work where:

- the exposure of employees to asbestos is sporadic and of low intensity
- it is clear from the risk assessment that the exposure of any employee to asbestos will not exceed the control limit
- the work involves:
  - short, non-continuous maintenance activities
  - removal of materials in which the asbestos fibres are firmly linked in a matrix, such as asbestos cement and asbestos-containing decorative coatings (eg Artex)
  - encapsulation or sealing of asbestos-containing materials which are in good condition
  - air monitoring and control, and the collection and analysis of samples to ascertain whether a specific material contains asbestos.

Asbestos Removal

10.42 Removal of asbestos-containing materials must be carried out by licensed asbestos removal contractors, unless the removal is of materials in which the asbestos fibres are firmly linked in a matrix, or the exposure to asbestos during the removal is likely to be sporadic and of low intensity.

10.43 Licensed work must be notified to the appropriate enforcing authority (HSE or local authority) 14 days prior to the work commencing. This notification must be accompanied by a plan of work specific to the contract involved.

Air Clearance Monitoring and Certificate of Reoccupation

10.44 Once removal of the asbestos has been completed, the premises must be assessed to determine whether they are thoroughly clean and fit to be returned to normal occupation.

10.45 Site clearance certification should be carried out in four stages, with the next stage begun only when the previous one has been satisfactorily completed. The four stages of site clearance certification include:

- a preliminary check on site conditions and job completeness
- a full visual inspection inside the enclosure/work area
- clearance air monitoring
- final assessment after enclosure/work area dismantling.

10.46 On satisfactory completion of all four stages, a certificate of reoccupation is issued to enable normal occupation of the area to take place. Air-monitoring work must be carried out by a specially trained UKAS analyst.

10.47 Any person who issues a site clearance certificate for reoccupation must be accredited by UKAS as competent to perform work in compliance with ISO 17020 and ISO 17025.

Asbestos Waste

10.48 All removed asbestos, both fibrous and securely bonded, must be disposed of at a licensed waste disposal site. Local authorities have information on licensed sites in their area.

10.49 The waste should be doubled-bagged in heavy duty polythene bags and clearly labelled, before being transported by a registered carrier to a licensed disposal site. The sale or gift of any waste resulting from removal work is forbidden.

Selecting a Licensed Contractor

10.50 Unless the exposure is likely to be sporadic and of low intensity, all work with asbestos and asbestos-containing materials requires the contractor to be issued with an HSE licence.

10.51 UK legislation places a duty of care on the employers of asbestos-removal contractors and it is therefore important that a careful selection procedure is in place. In addition to the possession of a licence, there are other factors to consider when selecting a contractor.

- Due to the wide range of applications for which asbestos was used, it is important to ensure the contractor selected has experience of the type of work to be undertaken and is therefore able to offer references from projects of a similar nature.
- The contractor should be able to provide appropriate insurance cover for asbestos-related work, together with proof that there is a sufficient number of suitably trained workers and supervisors. These people will be required to have current medical and training certificates. Refresher training on an annual basis is a legal requirement.
- Competence may be indicated by membership of a trade association that audits members' procedures.

Work with Asbestos

10.52 Sometimes asbestos becomes accidentally damaged or disturbed through maintenance or construction work.

10.53 For any work with asbestos, the employer is required to:

- identify whether asbestos is present
- evaluate risks to health
- implement control measures
- implement air monitoring and health surveillance

Identifying Whether Asbestos is Present

10.54 There are two basic types of sampling and analysis carried out in connection with asbestos. These are testing:

- bulk sampling (testing of materials which are suspected of containing asbestos)
- air sampling (testing of airborne asbestos fibres).

10.55 The sampling of bulk materials is carried out in order to determine if they contain asbestos:

- before work is done which may disturb the material
- as part of an ongoing programme of establishing the nature of a building's fabric.

10.56 Monitoring of airborne fibres is performed:

- to monitor levels of asbestos before buildings are reoccupied after asbestos has been removed
- to monitor the occupational exposure of employees working with asbestos.

Evaluating Risks to Health

10.57 The assessment by a competent person will enable decisions to be made on measures necessary to prevent or adequately control the exposure to asbestos in the workplace. During the risk assessment process, the competent person should consider:

- visual assessment of the material to determine its condition and whether it presents an immediate risk to health
- asbestos type, ie blue (most hazardous), brown, or white (usually determined by sampling)
- where the asbestos is located.

Implementing Control Measures

10.58 If it is necessary to work on or with asbestos, control measures to limit the exposure of employees and others must be implemented. Such control measures include the following.

- Enclosing the work area. Any work with asbestos that is likely to lead to the control limits being exceeded should be undertaken within a suitable enclosure. This should be fitted with filtered extraction to ensure that the enclosure is maintained at a pressure lower than atmospheric, to avoid the possibility of asbestos leaking into the surrounding environment.
- Minimising fibre release by control at source. Possibly the most effective technique for reducing fibre emission is by control wet stripping through wetting of the material which contains asbestos.
- Implementing local exhaust ventilation to capture any fibres released.
- Controlling debris. Wherever possible, debris should be collected directly into bags or other containers.
- Providing respiratory protective equipment and protective clothing.
- Implementing personal decontamination procedures.

Implement Air Monitoring and Health Surveillance

Air Monitoring

10.59 Monitoring of airborne fibres is performed:

- to monitor the occupational exposure of employees working with asbestos
- to monitor levels of asbestos before buildings are reoccupied after asbestos has been removed.

Health Surveillance

10.60 Health surveillance is an assessment of some aspects of the health of an employee by a trained and competent person, eg a medical doctor.

10.61 Health surveillance programmes should seek to:

- provide employees with information regarding the risks of exposure and the means of reducing the risks of developing asbestos-related diseases
- provide employees with objective information about their current state of health
- develop, over time, a comprehensive health record for each employee, including details of work performed, location, start and end dates and average duration of exposure in hours per week

- provide information to alert employees to any early indication of disease and to enable them to decide whether or not to continue working with asbestos
- enable health professionals to warn employees who are smokers and who work with asbestos of the greatly increased risk of lung cancer
- enable health professionals to reinforce education on the correct use, testing, maintenance and storage of personal protective equipment
- support the employer in maintaining safe systems of work
- enable doctors to advise workers with an asbestos-related disease to make a claim.

Minor Works

10.62 The HSE has recognised that certain works of a minor nature can be carried out by people other than licensed contractors. Such is the case for work carried out during maintenance operations, plumbing and electrical installations. However, for all such work, a risk assessment must be carried out prior to the work taking place and steps must be taken to protect against exposure to asbestos.

10.63 The HSE has published HSG210 *Asbestos Essentials: Task Manual*, which offers practical guidance on the way minor works should be carried out. For licensable material, minor works are defined as work that can be carried out by an individual working for no more than one hour in seven consecutive days, or work where the time spent by all those involved does not exceed two hours.

10.64 Tasks covered by the manual include drilling holes in asbestos insulating board, removal of a single asbestos insulating board ceiling tile, and drilling holes in asbestos cement and other bonded materials.

Training

10.65 Employers must ensure that adequate information, instruction and training is given to employees who are liable to be exposed to asbestos, so that employees are aware of the:

- significant findings of a risk assessment
- risk to health from asbestos
- precautions to be observed
- relevant control limits and action levels.

10.66 There are two main objectives to be achieved when organising the asbestos-related diseases training programme.

- To enable employees to be aware of the health hazards arising from the mishandling of asbestos materials as well as the relative hazards arising from different types of asbestos materials.
- To enable employees to work with asbestos in a safe manner to minimise the risks to themselves and others.

List of Relevant Legislation

- Control of Asbestos Regulations 2006
- Hazardous Waste (England and Wales) Regulations 2005
- Management of Health and Safety at Work Regulations 1999
- Health and Safety at Work, etc Act 1974

Further Information

Publications

HSE Publications

The following are available from *www.hsebooks.co.uk*.

- EH10 (rev 2001) *Asbestos: Exposure Limits and Measurement of Airborne Dust Concentrations*
- HSG210 *Asbestos Essentials: Task Manual: Task Guidance Sheets for the Building Maintenance and Allied Trades*
- HSG213 *Introduction to Asbestos Essentials: Comprehensive Guidance on Working with Asbestos in the Building Maintenance and Allied Trades*
- HSG247 *Asbestos: The Licensed Contractor's Guide*
- HSG248 *Asbestos: The Analyst's Guide for Sampling, Analysis and Clearance Procedures*
- INDG187(L) *Asbestos Dust — The Hidden Killer*
- INDG188(C) *Asbestos Alert for Building Maintenance, Repair and Refurbishment Workers*
- INDG223(L) (rev 3) *A Short Guide to Managing Asbestos in Premises*
- INDG289 *Working with Asbestos in Buildings: Asbestos: The Hidden Killer! Are You at Risk?*
- L127 *The Management of Asbestos in Non-domestic Premises. Regulation 4 of the Control of Asbestos Regulations 2006. Approved Code of Practice and Guidance*
- L143 *Work with Materials Containing Asbestos. Control of Asbestos Regulations 2006: Approved Code of Practice and Guidance*
- MDHS100 *Surveying, Sampling and Assessment of Asbestos Containing Materials*
- MS13 (rev 2005) *Asbestos: Medical Guidance Notes*

Organisations

- Asbestos Control and Abatement Division (ACAD)
  Web: *www.tica-acad.co.uk*
  ACAD, a division of the Thermal Insulation Contractors Association (TICA), is a trade association that represents organisations specialising in asbestos and its removal.
- Asbestos Removal Contractors Association (ARCA)
  Web: *www.arca.org.uk*

The ARCA aims to promote and maintain the safe working standards required for the handling and removal of asbestos and other hazardous materials.

- United Kingdom Accreditation Service (UKAS)
  Web: *www.ukas.com*
  UKAS is the national accreditation body which assesses organisations that provide certification, testing, inspection and calibration services against internationally agreed standards.

Hand-arm Vibration

- Exposure to vibration through the use of hand-held tools which vibrate may cause a range of debilitating chronic medical conditions, which collectively are known as Hand-Arm Vibration Syndrome (HAVS). Among these conditions is Vibration White Finger (VWF).
- The HSE estimate that around five million workers are exposed to hand-transmitted vibration in the workplace in the UK. Two million of these workers are exposed to levels of vibration where there are significant risks of developing disease.
- The risk is dose-dependent. It is a function of the level of vibration produced and the time for which the equipment is used (known as the "trigger time").
- Many tools used in construction, such as poker vibrators, breakers, plate compactors and scabblers, generate significant levels of vibration, which may present a risk to workers if used for substantial periods.
- The Control of Vibration at Work Regulations 2005 establish an action level and limit value. Employers are required to assess the risk from hand-transmitted vibration, put controls in place where necessary, and provide health surveillance for those exposed to significant doses of vibration.
- Buying vibration-reduced tools and limiting usage time are two of the main controls on exposure.

10.67 Hand-Arm Vibration Syndrome (HAVS) describes a range of medical conditions caused by:

- the regular and frequent use of hand-held power tools
- work processes which expose workers to vibration transmitted through the hands.

10.68 Conditions which fall within this term include vibration white finger (VWF) and carpal tunnel syndrome, although there is a great range of possible vascular, neurological and musculoskeletal effects in the hand, arm and shoulder.

10.69 The Control of Vibration at Work Regulations 2005 impose duties on the employer to assess the risk of damage to employees, and put in place adequate controls where necessary. The risk is dependent on the dose received by an individual. For this reason, a risk assessment should use data on the vibration emission of the tool, combined with the time it is used by an individual each day. A control programme may include:

- avoiding the use of hand hold tools
- selecting tools for low vibration output
- managing exposure time
- training.

10.70 There are many things workers can do themselves to reduce the risk.

Employers' Duties

10.71 The Control of Vibration at Work Regulations 2005 (CVWR) place the following duties on employers.

- Any employer who carries out work which is liable to expose their employees to vibration has to make a suitable and sufficient assessment of the risk created by that work. The purpose of the assessment is to identify appropriate control measures to reduce the vibration exposure of individuals below the relevant exposure limits.
- Risk can be evaluated using two new criteria introduced in CVWR — the Exposure Action Value (EAV) and the Exposure Limit Value (ELV).
- The regulations require that, where the risk assessment shows a risk to health of employees from vibration or that an EAV is likely to be exceeded, the relevant employees are placed under a programme of health surveillance.
- Where employees are exposed to vibration which presents a risk to them, or the EAV is likely to be exceeded, their employer must provide suitable and sufficient information, instruction and training.

Employees' Duties

10.72 The Control of Vibration at Work Regulations 2005 place no additional duties on employees. However, under the Health and Safety at Work, etc Act 1974, employees have a duty to take reasonable care of their own health and safety and that of other people who may be affected by their work.

10.73 Under the Management of Health and Safety at Work Regulations 1999, employees have a duty to:

- use any machinery, equipment, transport, safety devices and means of production in accordance with any training and instructions provided by the employer
- inform the employer of any serious and imminent dangers to health and safety
- inform the employer of any shortcomings in the employer's health and safety arrangements.

In Practice

Effects of Hand-transmitted Vibration

10.74 Hand-arm vibration can cause a range of symptoms collectively known as hand-arm vibration syndrome (HAVS), as well as specific conditions, such as carpal tunnel syndrome. The effects of HAVS can broadly be sub-divided in to the following categories.

- Vascular (affecting the blood vessels) — The principal vascular symptom of excessive exposure to hand-transmitted vibration is finger blanching. Blanching typically occurs when the fingers are cold, and victims suffer significant pain on re-warming. After many years exposure, permanent discolouration of the fingers may occur. This collection of symptoms is known as vibration white finger (VWF), or secondary Raynaud's disease.
- Neurological (affecting nerves and sensory nerve endings) — Damage to nerve tissue in the hand caused by vibration may lead to symptoms such as:
  - tingling
  - numbness
  - loss of sensation

  - loss of manual dexterity
  - painful throbbing.
- Musculoskeletal (affecting muscle, bone and soft tissues, such as tendon, ligament and cartilage). Effects include:
  - muscle fatigue and loss of grip strength
  - disorders of the bones, such as cysts and vacuoles
  - joint disorders of the upper arm, such as tennis elbow and rotator cuff degeneration.

10.75 In addition, carpal tunnel syndrome (CTS) is a common result of excessive exposure to hand-transmitted vibration. Symptoms arise due to entrapment or compression of the median nerve in the palm. CTS caused by hand-held vibrating tools was made a prescribed disease from April 1993, and is therefore reportable under the Reporting of Injuries, Disease and Dangerous Occurrences Regulations 1995 (RIDDOR).

10.76 These symptoms, as well as being painful or uncomfortable, can cause a loss of amenity affecting work and home life. Inability to do fine work (eg assembling small components) or everyday tasks (eg fastening buttons) is typical.

Sources of Hand-transmitted Vibration in Construction

10.77 There is a wide range of equipment used in construction which generates significant levels of vibration at the hand. Such equipment includes:

- chainsaws
- concrete breakers/road breakers
- hammer drills
- scabblers
- power sanders
- poker vibrators
- hand-held grinders
- impact wrenches
- power hammers and chisels.

Risk Assessment

10.78 Under the Control of Vibration at Work Regulations 2005, any employer who carries out work which is liable to expose their employees to vibration has to make a suitable and sufficient assessment of the risk created by that work. The purpose of the assessment is to identify appropriate control measures to reduce the vibration exposure of individuals below the relevant exposure limits.

10.79 A fundamental element of the assessment is an evaluation of daily exposure of workers to vibration. This evaluation will include:

- observation of work practices to determine exposure times
- reference to information provided by equipment and tool manufacturers regarding the likely magnitude of vibration levels
- measurement of vibration levels from tools in their actual conditions of use.

10.80 The results of the risk assessment are to be recorded, and the assessment itself reviewed or repeated whenever circumstances change which could affect the outcome of the original assessment. Such circumstances could be a change of equipment, way of working or time of exposure.

10.81 The Control of Vibration at Work Regulations 2005 introduced two new criteria for the evaluation of hand-arm vibration and whole-body vibration.

10.82 The new criteria are as follows.

- Exposure Action Value (EAV) — where this value is likely to be reached or exceeded, the employer must reduce exposure to as low a level as is reasonably practicable by establishing and implementing a programme of organisational and technical control measures. The daily EAV for hand-arm vibration is 2.5ms-2 A8.
- Exposure Limit Value (ELV) — employers must ensure that the ELV is not exceeded. If it is exceeded, they must:
  - reduce exposure below the limit value
  - identify the reason for the limit being exceeded
  - modify the control measures to make sure that the limit is not exceeded again.

10.83 The daily ELV for hand-arm vibration is 5ms-2 A8.

10.84 Establishing vibrational levels produced by a tool is far from straightforward. Under the Supply of Machinery (Safety) Regulations 1992 manufacturers must supply information regarding levels of vibration for all tools and equipment producing more than 2.5ms-2. While this information may be useful for selection purposes, as it allows direct comparisons of equipment to be made, it should not be taken as an accurate indication of the levels which will arise during use. It has been suggested that actual operating levels may be as much as 35% higher than manufacturer's published levels. It is therefore often necessary to obtain measurements directly. This task must be carried out by a competent person, as there are many opportunities for errors in the gathering and interpretation of data. Useful data on levels of vibrational acceleration from tools has been published and can also be found on the Internet. This data may be used to determine presumptive doses, although in some cases, where appropriate existing data cannot be found, direct measurement may be necessary.

Evaluation

10.85 Where vibration levels from a particular tool or item of equipment are known, or have been directly measured, it is necessary to evaluate the risk to individuals using the equipment against the levels given in the Control of Vibration at Work Regulations 2005. The following descriptors are often used in this evaluation.

- *Frequency weighted vibration level (m/s^2):* The measured rms (root mean squared amplitude) equivalent continuous level in each of three orthogonal axes, for each handle.
- *Trigger time:* The operator's estimate of the actual time in an average day when the tool is in use. Only includes time when the tool is actually operating, and excludes setting up, etc.
- *Vector sum (m/s^2):* The sum of the individual measurements in each axis, calculated by the root-sum-of-squares method (see L140 *Hand Arm Vibration: Control of Vibration at Work Regulations 2005. Guidance* for formulae and guidance on this value).
- *A(8):* The A(8) is the vibration dose, eg a combination of the measured vector sum value and the estimated trigger time, according to the formula, $A(8) = a_{hv}\sqrt{T/T_o}$, where:

a_{hv}	= vector sum vibration value
T	= daily duration of exposure to the vibration value
T_o	= reference duration of 8 hours

Single-tool time to EAV (mins): This is the number of minutes for which the tool could be used each day by a single operator before the EAV of A(8) 2.5m/s^2 is reached. It assumes that the operator does not carry out any other tasks using vibrating hand tools during the day. It is useful for the management of exposure times.

1-hour exposure points: Since the exposure time is not directly proportional to A(8), it is difficult to evaluate operator exposures for different periods of use or where operators use more than one tool in a day. The "exposure points" method is useful for assessing the effect of combined exposures, eg where operators use a number of vibrating tools throughout the day. An exposure points system simplifies the risk assessment procedure. The exposure points values equivalent to the action and limit values in the Control of Vibration at Work Regulations 2005 are as follows:
- Exposure Action Value of 2.5m/s^2 A(8) = 100 points
- Exposure Limit Value of 5.0m/s^2 A(8) = 400 points

10.86 One method of managing hand-arm vibration is to calculate an exposure point value per hour of use, which is proportional to the measured vibration magnitude. For example, if two tools with 1-hour exposure points of 20 and 80 are in use for 2 and 2.5 hours respectively, then it is simple to calculate the combined exposure for multiple tool use using the 1-hour points system. The points score for each tool is 2 x 20 = 40 points and 2.5 x 80 = 200 points, so the combined exposure gives a total points score of 40 + 200 = 240 points. Hence the operator's daily exposure is 240 points, which is above the EAV of 100 points but below the ELV of 400 points.

10.87 Various ready-reckoners available on the HSE website may be used to calculate the descriptors described above. If data is to be obtained by direct measurement, then it will be necessary to either engage a competent vibration consultant, or purchase/hire suitable equipment and ensure the user attends and passes an HSE-approved hand-arm vibration assessment course (these courses are typically of 2 days duration).

Control Programme

10.88 The Control of Vibration at Work Regulations 2005 require the application of organisational and technical control measures where the EAV is likely to be reached or exceeded. They also require that a hierarchical approach to control is adopted whereby the first option when deciding on control measures is to utilise alternative work methods which eliminate or reduce exposure to vibration. For example, an excavator-mounted pecker breaker avoids the use of hand-held road breakers. If this approach is not reasonably practicable, then consideration should be given to the following.

Action by the Manufacturer at the Design Stage

10.89 Tool manufacturers are obliged to provide vibration values for their tools and, ideally, equipment and machinery should be designed so as to be low vibration. Chain-saws, chipping hammers and pneumatic screwdrivers are examples of tools that have had vibration designed out during their manufacture. Anti-vibration handles have been incorporated in chain-saws, stone-cutters and other tools for many years.

Selection of Tools and Machinery with Low Vibration Levels

10.90 To meet the general requirements of the Control of Vibration at Work Regulations 2005, employers should have in place a suitable purchasing strategy so as to eliminate or reduce at source the risk of vibration exposure. The Supply of Machinery (Safety) Regulations 1992 make explicit the general duties of designers, manufacturers, importers and suppliers under the Health and Safety at Work, etc Act 1974 (HSWA).

10.91 To comply with the regulations, machinery must meet the specified essential health and safety requirements, as defined in the regulations. These are wide-ranging and include an assessment of the risk to operators from the effects of vibration. The manufacturer must also supply clear instruction and maintenance handbooks. These must draw attention to any foreseeable hazards that may be encountered, whether by normal or abnormal use of the equipment.

Correct Installation of Machinery

10.92 The Supply of Machinery (Safety) Regulations 1992 also state that the supplier must provide instructions to ensure that vibration is controlled by correct installation. These instructions must be passed on to the installing engineer and adhered to.

Regular Maintenance of the Tool or Machinery

10.93 Even a correctly installed and smooth-running machine will eventually begin to vibrate as parts wear out, especially if they are not replaced until a breakdown occurs. Worn bearings, shaft misalignment, unbalanced rotating parts, loose bolts, damaged gear teeth, blunt cutting tools and neglected lubrication all combine to increase vibration levels.

10.94 The condition of machines should be monitored so that faults can be detected and then rectified at a convenient time, thus avoiding the disruption caused by breakdown. This can be achieved by selecting suitable measuring points on the machine for regular monitoring. Vibration measurements at these points should be recorded on installation, repeated at regular intervals and compared with the baseline values. If a frequency analyser is available, a vibration fingerprint for the machine can be established and any change in the pattern will indicate where the fault lies.

Reducing Vibration at Source by Modification After Installation

10.95 Where old machinery is still in use, or second-hand items are purchased, sources of vibration should be identified and suitable modifications made by employing current knowledge of vibration-damping techniques. The original manufacturer may be able to advise or supply the necessary improvements. It may be possible to retrofit anti-vibration handles to old tools, but this should be carried out by a specialist.

Reducing Vibration Transmission to the Hand

10.96 Training in the correct use of tools is essential. The tighter the grip on the handle, the greater the transmission of vibrations to the hand and arm. It is important to grip the handle as lightly as possible and let the tool do the work. If the tool can be supported to relieve the weight on the hand and arm, the grip force can be reduced. Users must be shown how to handle and use the tool with a light but safe grip. Most hand-held power tools are now fitted with anti-vibration handles. Tools bought before these devices were generally introduced (before 1980), which cannot be modified to incorporate them retrospectively, should be replaced with modern tools which possess anti-vibration handles.

Personal Protective Equipment

10.97 Anti-vibration gloves do not generally reduce vibration transmission to the hand at the most damaging frequencies, eg below 100Hz. Although the gloves are very efficient at frequencies above 500Hz, little attenuation is provided at lower frequencies and it is thought unlikely that any glove could be developed to provide attenuation below 200Hz. In fact, it has been shown that at certain low frequencies, vibrations are amplified by the gloves. Furthermore, the wearer usually finds it necessary to increase grip pressure, thus increasing the coupling and thereby the efficiency of energy transmission to the hand. However, by keeping the hands warm, some benefit is derived from wearing gloves. In the more advanced stages of hand-arm vibration syndrome (HAVS), the whole body should be kept warm.

Note: So-called "anti-vibration" gloves often provide limited protection. Their performance is frequency-dependent, and is often very poor at the frequencies to which humans are most susceptible.

Communication

10.98 Assessment information should be available and accessible in a form that is easily understood.

10.99 The simplest way to present HAV data is to convert the vibration values into "safe" working times, ie the usage time before the action level is reached. Plant and processes then become a "34-minute grinder" or a "98-minute process" which provides an intuitive indication of the relative risks. Alternatively, use the "daily usage limit" ("safe" working time) or the exposure points system as a risk indicator in the field as these are much more easily understood than actual vibration values.

Health Surveillance

10.100 Health surveillance refers to the employment of methods to detect the health effects of a particular agent as early as possible, and thus prevent the progression of health conditions. The Control of Vibration at Work Regulations 2005 require that, where the risk assessment shows a risk to health of employees from vibration or that an EAV is likely to be exceeded, the relevant employees are placed under a programme of health surveillance. There are numerous varied methods for the detection of health effects from hand-arm vibration.

10.101 The health surveillance is to be undertaken by a suitably qualified occupational health professional, such as an occupational health nurse or doctor. If the results of the surveillance identify any employees with health effects or disease which are considered to result from exposure to workplace vibration, then the employer must:

- consider assigning the relevant employees to alternative work which does not expose them to risks from vibration
- review the risk assessment
- ensure a suitably qualified person informs the affected employees of the outcome of the surveillance
- review the control measures.

10.102 Employees are required to present themselves for health surveillance when required by their employer, provided it is during working hours and at the employer's expense.

Training

10.103 Where employees are exposed to vibration which presents a risk to them, or the Exposure Action Value (EAV) or Exposure Limit Value (ELV) is likely to be exceeded, their employer must provide suitable and sufficient information, instruction and training. As a minimum, this must include:

- details of the EAVs and ELVs relevant to the situation (eg whole-body values or hand-arm values)

- the significant findings of the risk assessment
- the organisational and technical control measures
- why and how to detect and report signs of injury
- arrangements for health surveillance
- safe working practices employed to minimise exposure to vibration
- the collective results of health surveillance, in a form that prevents any individual from being identified.

10.104 The information, instruction and training must be updated to take account of any significant changes to working methods or types of work carried out.

10.105 In addition, the employer has to ensure that any person, whether or not an employee, who carries out work in connection with the duties in the regulations, has suitable and sufficient information, instruction and training. For example, if an employer engages an external occupational health contractor to provide health surveillance for his employees, he must ensure that the contractor's staff are adequately trained in the services they are providing.

10.106 While the Control of Vibration at Work Regulations 2005 only talk broadly about training duties towards employees and "others who carry out work in connection with the employer's duties", it is useful to further sub-divide and consider those who come within these categories. Therefore, the distinct groups that should be considered in terms of training requirements are:

- operators of equipment which poses a risk to health due to vibration, or is likely to reach or exceed the EAV
- supervisors and first-line managers of operators
- senior managers
- those who carry out vibration assessments
- those who carry out vibration measurements.

10.107 The agenda for a hand-arm vibration awareness training event for operatives and supervisors may include:

- introductions and clarification of course objectives
- the health effects of hand-arm vibration, explaining clearly how quickly permanent effects may occur, and how they can cause loss of amenity and interference with daily life
- whether employees are at risk, and if so whether the risk is high (likely to reach the ELV), medium (above the EAV) or low
- the risk factors (eg the levels of vibration, daily exposure duration, regularity of exposure)
- how to recognise and report symptoms
- the need for health surveillance, including how it can help workers remain fit for work, what it involves, the arrangements for conducting the surveillance, and how confidentiality of results will be maintained
- means by which risk is controlled, including:
  - correct techniques for equipment use, such as minimisation of grip force
  - correct selection and use of equipment
  - the need to maintain equipment properly
  - the importance of regular breaks from use of vibrating hand-held equipment
  - maintaining good blood circulation at work by keeping warm and massaging fingers and, if possible, stopping or cutting down smoking, which has an adverse effect on circulation

- a question and answer session and end-of-course assessment — a simple multi-choice test is useful for monitoring the assimilation of information, both at an individual level, and also as a monitor for the overall effectiveness of the training event.

10.108 Training should be a continuous process, and the initial awareness sessions should be reinforced by any of the following methods.

- Toolbox talks: where short sessions with small groups of workers are conducted at the workplace. A narrow range of subjects is normally covered, such as looking at techniques of using equipment that minimise the transmission of vibration from the equipment to the hand.
- Posters: strategically placing posters around the workplace with messages that support the initial awareness training. It is important to note, however, that posters lose their impact after a time, and should not be left permanently.
- Written method statements or safe operating procedures.
- Verbal guidance provided by occupational health staff during health surveillance. For example, occupational health nurses may explain the effects of vibration and the importance of maintaining good circulation while they carry out cold provocation tests to determine whether workers have vibration white finger.

List of Relevant Legislation

- Control of Vibration at Work Regulations 2005
- Management of Health and Safety at Work Regulations 1999
- Provision and Use of Work Equipment Regulations 1998
- Reporting of Injuries, Disease and Dangerous Occurrences Regulations 1995 (RIDDOR)
- Supply of Machinery (Safety) Regulations 1992
- Health and Safety at Work, etc Act 1974

Further Information

Publications

HSE Publications

The following are available from *www.hsebooks.co.uk*.

- HSG88 *Hand–arm Vibration*
- HSG170 *Vibration Solutions — Practical Ways to Reduce Hand–arm Vibration Injury*
- INDG175 (rev 2008) *Control the Risks from Hand-arm Vibration: Advice for Employers on the Control of Vibration at Work Regulations 2005*
- INDG242 (rev 1) *Control Back-pain Risks from Whole-body Vibration — Advice for Employers on the Control of Vibration at Work Regulations 2005*
- INDG296 (rev 1) *Hand–arm Vibration: Advice for Employees*
- INDG404 *Drive Away Bad Backs — Advice for Mobile Machine Operators and Drivers*

- L140 *Hand-arm Vibration. The Control of Vibration at Work Regulations 2005. Guidance on Regulations*

British Standards Publications

The following are available from *www.bsi-global.com.*

- BS EN ISO 5349–1: 2001 *Mechanical Vibration — Measurement and Evaluation of Human Exposure to Hand Transmitted Vibration. General Requirements*
- BS EN ISO 5349–2: 2002 *Mechanical Vibration — Measurement and Evaluation of Human Exposure to Hand-Transmitted Vibration. Practical Guidance for Measurement at the Workplace*

Legionnaires' Disease

- Legionnaires' disease is caused by bacteria that exist naturally in water and can be inhaled in the form of water droplets from outlets such as taps, showers and cooling towers.
- Employers have a legal duty to assess the risks of legionnaires' disease and to prevent or control the risk.
- Water systems and outlets must be maintained in good physical condition and kept clean.
- Water temperatures between 20°C and 45°C should be avoided. Where possible, water should be stored at the right temperature — <20°C or >60°C.
- Stagnation and the presence of foreign matter and biological nutrients in the water must be avoided.
- Water must be regularly tested and treated by professional, competent people.
- Prompt, appropriate action must be taken in the event of a legionnaires' disease outbreak.
- Legionnaires' disease is covered by the Health and Safety at Work, etc Act 1974 and the Control of Substances Hazardous to Health Regulations 2002.

10.109 Legionellosis is a group of diseases, which includes legionnaires' disease. The potentially fatal infection has symptoms similar to flu and pneumonia. *Legionella Pneumophila*, the bacteria responsible for legionnaires' disease, exist naturally in external watercourses, and can easily transfer to water used in buildings, via air-conditioning and recirculated hot and cold water systems. In certain conditions, bacteria can multiply to critical levels in stored water.

10.110 Legionnaires' disease is caused when water droplets containing the bacteria are inhaled. Typical sources of such water droplets include shower sprays and the exhausts from wet cooling systems and evaporative condensers. Industrial cooling towers and evaporative condensers may create the risk of off-site cases of legionnaires' disease.

Employers' Duties

10.111

- Employers and those responsible for building maintenance must carry out an assessment of the risk from legionella, and take steps to prevent or minimise such risks under the Health and Safety at Work, etc Act 1974 and Control of Substances Hazardous to Health Regulations 2002.
- The Notification of Cooling Towers and Evaporative Condensers Regulations 1992 require employers who have a cooling tower or evaporative condenser on-site to notify the local authority in writing of where it is located. Notification forms are available from local environmental health departments.
- If an employee contracts legionellosis, and if they have worked on cooling towers or hot water systems likely to be contaminated with legionella, it must be reported under the Reporting of Injuries, Diseases and Dangerous Occurrences Regulations 1995. In Scotland, legionellosis is a notifiable disease.

Employees' Duties

10.112

- Employees have a duty to take reasonable care of their own health and safety and that of other people who may be affected by their work under the Health and Safety at Work, etc Act 1974
- The Management of Health and Safety at Work Regulations 1999 require employees to inform their employer of any danger to health and safety posed by a work activity.

In Practice

Hazard Identification and Risk Assessment

10.113 The hazard identification and risk assessment process will help to establish the following.

- If any physical aspect of the water system could support the harbouring of legionella bacteria.
- Whether any water storage conditions exist under which legionella bacteria might multiply.
- Which water outlets might release a spray.
- The HSE's Approved Code of Practice *Legionnaires' Disease: The Control of Bacteria in Water Systems* offers practical guidance for the control of legionella bacteria in the workplace where water is used or stored in a way which may create a reasonably foreseeable risk of legionnaires' disease, including:
  - cooling towers
  - evaporative condensers
  - hot water systems in non-domestic premises, except where the volume of water does not exceed 300 litres
  - hot and cold water services, irrespective of the premises' size if the occupants are deemed particularly susceptible
  - humidifiers and air washers.

Physical Aspects

10.114 In hot and cold water systems, the initial design and any additions and adaptations will have a bearing on the likelihood of legionella bacteria being harboured. Hazard identification and risk assessment should ask whether:

- storage cisterns are unnecessarily large
- pipework is unnecessarily long or indirect or contains "deadlegs"
- anything prevents the system being completely drained or pumped out
- anything prevents easy access to cooling towers and evaporative condensers for visual inspection and maintenance
- cooling towers have efficient drift eliminators fitted.

Water Storage Conditions

10.115 Legionella bacteria need nutrients to support growth, so it should be determined if:

- sludge, scale or rust has accumulated in cisterns or pipework
- foreign matter can be or has been allowed into tanks through poor housekeeping or maintenance
- plumbing materials have been used which do not comply with water authority by-laws
- materials have been allowed to deteriorate
- algae, organic matter, insects or vermin have been allowed to enter and remain in tanks
- a bio film is coating hard surfaces or lying on the water surface.

10.116 Water temperatures are crucial to the existence of legionella bacteria.

- In water below 20°C, the bacteria remain dormant in low numbers.
- In water which is between 20°C and 45°C, the bacteria multiply, so water stored within this range is a hazard.
- In water above 45°C, bacteria growth slows.
- At 60°C, 90% of legionella will die within two minutes.

10.117 Temperatures should be monitored and insulation checked to prevent the growth of bacteria. The following should be considered.

- Insulation, or lack of insulation, which enables water to be stored at incorrect temperatures, is a hazard.
- Cold water tanks situated in warm parts of buildings are a hazard.
- Tanks where water is not uniformly heated are a hazard.

10.118 Water flow is another important consideration. Low or no flow results in stagnation, enabling bacteria to multiply undisturbed, so hazard identification and risk assessments should look for:

- cisterns or pipework which allow water to stand undisturbed for long periods
- deadlegs — pipework or tanks which are no longer used, but are still connected to the system.

Water Outlets That Might Release a Spray

10.119 Water outlets which might release a spray should be identified. These may include:

- taps
- shower heads
- spas or whirlpool baths
- pools, including hydrotherapy pools
- humidifiers
- fountains
- evaporative condensers
- wet cooling towers.

People at Risk

10.120 People at risk from legionella which originates from a hot and cold water system could be any building occupant or visitor to the building.

10.121 Cooling towers can create risks for additional people. In closed cooling tower systems, the vapour is recirculated within the building, so those at risk will be building occupants or visitors to the building. However, in open cooling tower systems, the vapour is released into the atmosphere and can infect neighbouring towers. Therefore, people in the surrounding area and those in nearby buildings served by their own cooling towers will be at risk.

10.122 Particularly vulnerable people include those over 40, especially if they are smokers, alcoholics, diabetics, have chronic respiratory or kidney disease, cancer, or if they are on renal dialysis or immunosuppressant drugs.

Control Measures

10.123 After the hazard identification and risk assessment process, including the recognition of at-risk people, attention should be given to control measures, including an assessment of the adequacy of existing controls and the identification of further actions needed to control the risks. Control measures fall into four categories:

- improvements to physical aspects of the system
- water temperature
- maintenance and cleanliness
- water testing and treatment.

Improvements to Physical Aspects of the System

10.124

- A complete and up-to-date schematic plan should exist of the hot and cold water system and wet cooling system.
- Main components and fittings, such as washers, gaskets, sealants, jointing coatings, linings, and hosing should be constructed from materials which comply with water authority by-laws.
- Water storage containers should be the right size to ensure uniform heating and to prevent stagnation.
- Where possible, tanks should be horizontal rather than vertical.
- Cisterns and storage tanks should have properly fitting covers.
- Cold water tanks should not be sited in warm areas of buildings.
- Insulation should be adequate. It should be possible to drain or pump out the system completely.
- All pipe runs should be as short and direct as possible.
- There should be no pipework deadlegs to cause stagnation.
- Redundant pipework or tanks should be isolated from the system.
- Vermin traps should be fitted where appropriate.
- Insulation should be adequate.
- Efficient drift eliminators should be fitted in cooling towers to reduce escape of spray.
- Where reasonably practicable, wet cooling towers should be replaced by dry cooling systems.
- It should be possible to drain or pump out the system completely.

Water Temperature

10.125

- Cold water should be stored below 20°C. Adequate insulation can assist.

- Hot water should be stored at 60°C. Thermostats need to be set and maintained at the appropriate level and adequate insulation installed.

Note: Water coming out of taps above 43°C poses the risk of scalding, and hot water pipes and radiators carrying water above 43°C could pose a hazard where vulnerable people may lean or fall against them, eg in healthcare environments.
Ideally, water should be circulated at 50°C, with scalding hazards controlled by fitting thermostatic control valves to radiators and to the water supply for baths, showers and hand basins.

Maintenance and Cleanliness

10.126 Maintenance and cleanliness are perhaps the most crucial aspects in the prevention of legionnaires' disease.

10.127 A detailed preventive maintenance schedule should incorporate regular visual inspection, cleaning, disinfection and physical maintenance. It should also include water storage tanks, pipework, water outlets, cooling towers, and all fittings.

10.128 A competent person (trained in the risks and nature of legionellosis and relevant prevention and maintenance techniques) should make regular inspections and keep a record. The following procedures should be included in the maintenance plan.

- Check cistern covers are in place.
- Check that dirt, debris and vermin have not entered cisterns.
- Clean and disinfect cisterns.
- Clean and disinfect heat exchangers.
- Clean and disinfect water filters.
- Check insulation is in good condition.
- Check water temperatures.
- Check that showers, shower heads and taps are clean and free from scale.
- Check that cooling towers, drift eliminators and fans are clean and in good condition.
- Clean cooling tower basin if slime, algae, or dirt are visible.
- Blow down direct chilled water risers.
- Flush, clean and disinfect the entire cooling tower system.
- Carry out any necessary repairs and improvements promptly.

Water Testing and Treatment

10.129 Since legionella bacteria are widespread in the environment, they cannot be prevented from entering water systems which means water needs to be regularly tested.

10.130 In chlorinated water systems, checking should be a continuous, automated process. Dip-slides can detect general bacterial growth, but not legionella specifically.

10.131 If automatic biocide systems are installed, an alarm back-up for malfunction should be installed and water regularly checked. Sampling of cooling towers for legionella should be done on a quarterly basis. Good record keeping is essential and records of inspections and tests should be kept for at least two years. If water services are outsourced, regular checks should be made of the provider's records.

10.132 Water treatment should be carried out on a regular basis, immediately after any work which requires shutting down or repressurising the system, and after any suspected outbreak of legionellosis. Water can be treated either manually or by automatic dosing.

10.133 Equipment must be properly installed, maintained and monitored and it is essential to ensure professional, competent people do all work. A Recommended Code of Conduct for Service Providers exists, supported by the British Association for Chemical Specialities, the Water Management Society and the HSE. It is a good idea to check if the provider has signed up to the code.

10.134 There are various types of treatment, each with their own advantages and disadvantages and some may be combined. Impartial, professional advice should always be taken to ensure that the most appropriate treatment method is used, including the following.

- Ozone, a high-energy form of oxygen, destroys legionella bacteria. It is produced electronically in a generator under vacuum and continuously drawn into the water. A build-up of sludge or sediment may need to be dealt with.
- Chlorination can be achieved by continuous injection of a chlorine compound or gas. It is hard to maintain constant levels, penetration of bio films may not be successful, and high concentrations of chlorine will corrode pipes.
- Ionisation treatment releases electrolytically generated copper and silver ions into the water through an intelligent control system, killing legionella bacteria. Care must be taken to remove scale build up from the electrodes and high pH levels will reduce effectiveness.
- Continuous UV light treatments involve mercury lamps within hydraulic chambers installed close to water outlets. It is easy to install, forms no by-products and does not harm water or plumbing. However, UV light cannot penetrate bio films, tubes must be regularly cleaned and water filtered to avoid build-up of scale.
- Thermal treatment by superheating and flushing water through the system is an emergency procedure adopted during an outbreak of legionnaires' disease. There is no residual protection.

Response to an Outbreak

10.135 If a case or cases of legionnaires' disease are reported or suspected among building occupants, or those who have been in the vicinity, immediate action must be taken. The source of the infection may be the employer's building or another building nearby, however all precautions must be taken as follows.

- Inform senior management, providing all relevant information.
- Consult the HSE immediately and follow its advice. In Scotland, legionellosis is a notifiable disease.
- Test water storage tanks and outlets to identify the possible source.
- Notify neighbours.
- Start water treatment.
- Consider closing the building. If this is done, staff should be given non-alarmist literature explaining the symptoms and what to do if they are concerned.
- Do not open the building again until test results show bacteria readings have dropped to a safe level.

10.136 As soon as an outbreak is suspected, a team of specialists including staff from the local environmental health department and the Public Health Laboratory Service will begin an investigation. This will be aimed at identifying the source and those likely to be affected, and ensuring the contaminated water system is treated as quickly as possible. The police and the HSE will also be involved.

Training

10.137 Employees involved in facilities management and building services maintenance with the responsibility for carrying out preventive maintenance, water checks and water treatments will need to be trained in the risks and nature of the disease as well as specific technical training.

10.138 Employees charged with responding appropriately in the case of a suspected outbreak will need to understand the risks and nature of the disease, who is at risk, and the appropriate response.

List of Relevant Legislation

- Control of Substances Hazardous to Health Regulations 2002
- Reporting of Injuries, Diseases and Dangerous Occurrences Regulations 1995
- Notification of Cooling Towers and Evaporative Condensers Regulations 1992
- Health and Safety at Work, etc Act 1974

Further Information

Publications

HSE Publications

The following are available from *www.hsebooks.co.uk*.

- IAC27 *Legionnaires' Disease — A Guide for Employers*
- INDG136 (rev 2005) *COSHH: A Brief Guide to the Regulations: What You Need To Know About the Control of Substances Hazardous to Health Regulations 2002 (COSHH)*
- L8 (rev 2000) *Legionnaires' Disease – Control of Legionella Bacteria in Water Systems – Approved Code of Practice and Guidance*

Organisations

- Chartered Institution of Building Services Engineers
  Web: *www.cibse.org*
  The institution is an international body which represents and provides services to the building services profession.
- Chartered Institution of Water and Environmental Management
  Web: *www.ciwem.org.uk*
  The institute is an independent, multi-disciplinary professional and examining body for scientists, engineers, other environmental professionals, students and those committed to the sustainable management and development of water and the environment.

Musculoskeletal Disorders

- Construction work, by its very nature, involves many manual handling activities, which means that construction workers are at particular risk of sustaining a musculoskeletal disorder (MSD).
- Symptoms of pain, aching and discomfort affecting the back, knees, neck and shoulders can often be closely related to the type of manual handling activity being undertaken. In each case long-term disability can result.
- If employers cannot avoid manual handling where there is a risk of an MSD-related injury, they must assess their manual handling operations and take steps to reduce the risk of injury to the lowest level reasonably practicable.
- Types of MSD include frozen shoulder, tennis elbow, carpal tunnel syndrome, repetitive strain injury and back pain.
- There are a range of trigger factors, from the repetitiveness of a task to psychosocial issues, that can all contribute to whether an individual will develop an MSD.
- Prevention is better than cure. MSD risks can often be avoided, or minimised, by carrying out the activity in a different way or by using different materials.
  - Employers and employees should have an understanding of MSDs and be committed to action on prevention.
  - Involving employees in the planning and organisational processes can be an important way of increasing the likelihood of success of any risk control strategy.
  - Key to any management programme is assessing the risk of MSDs.
  - Once risks have been assessed and prioritised a coherent process of risk reduction should be undertaken using an ergonomics approach.
  - Early medical management can stop established cases from deteriorating and also help the process of return to work.
  - The management programme needs to be reviewed on a regular basis to ensure that it continues to be effective.

10.139 Musculoskeletal disorders (MSDs) from manual handling tasks include repetitive strain injury (RSI), upper limb disorders (ULDs), and back pain, although MSDs can also include injuries sustained from hand-arm vibration and whole-body vibration.

10.140 Construction work, by its very nature, involves many manual handling activities, which means that construction workers are at particular risk of sustaining an MSD. Every year, one third of all construction industry accidents reported to the Health and Safety Executive (HSE) involve manual handling. These represent only a part of the actual problem, as many back injuries go unreported.

10.141 Many construction workers often experience symptoms of pain, aching and discomfort affecting their back, knees, neck and shoulders. These symptoms can be closely related to the type of manual handling activity. For example, work involving stooping and kneeling can lead to pain in the lower back and knees, while working with the arms raised above shoulder height can result in neck and shoulder pain. In each case long-term disability can result.

Employers' Duties

10.142 Under the Health and Safety at Work, etc Act 1974, employers must:

- ensure (so far as is reasonably practicable) the health, safety and welfare at work of all their employees
- prepare and revise a written statement of their general policy regarding the health, safety and welfare of their employees. This should, where appropriate, cover arrangements for the prevention of musculoskeletal disorders (MSDs).

10.143 Under the Manual Handling Operations Regulations 1992, if employers cannot avoid manual handling where there is a risk of injury, they must assess their manual handling operations and take steps to reduce the risk of injury to the lowest level reasonably practicable.

10.144 Under the Reporting of Injuries, Diseases and Dangerous Occurrences Regulations 1995, employers must report any injuries sustained in the workplace.

10.145 Under the Provision and Use of Work Equipment Regulations 1998, employers must:

- take account of ergonomic risk factors when selecting work equipment
- ensure that work equipment is suitable and safe for the work and does not present any health and safety risk to the user.

10.146 Under the Management of Health and Safety at Work Regulations 1999, employers must:

- assess the risk to their employees and others who may be affected by the company's activities
- carry out a risk assessment to take account of the task, load, working environment and the individual worker's ability to carry out the task
- make arrangements to put into practice measures to manage the risks identified in the risk assessment
- appoint competent people to help prepare and implement the measures needed to comply with the employer's duty of care
- ensure that employees have adequate health and safety information and training to avoid risks related to their work
- ensure that any temporary workers are given adequate health and safety information.

Employees' Duties

10.147 Under the Health and Safety at Work, etc Act 1974, employees have a duty to:

- take reasonable care for the health and safety of themselves and of other persons who may be affected by what they do or fail to do at work
- co-operate with their employers on health and safety matters.

10.148 Under the Manual Handling Operations Regulations 1992, employees must make full and proper use of any systems of work intended to reduce the risk of injury from manual handling operations.

10.149 Under the Management of Health and Safety at Work Regulations 1999, employees must follow all health and safety instructions and report any danger.

In Practice

Types of Musculoskeletal Disorders

Repetitive Strain Injury

10.150 Repetitive strain injury (RSI) is a term usually used to refer to a pain in the arms caused by computer use. It is a type of upper-limb disorder (ULD).

10.151 However, construction workers may also be at risk of RSI if engaged in an activity that uses the same muscles repeatedly. The more the task is repeated, the greater the risk. The speed at which the job is done also increases the risk. For example, a task that involves the movement of the whole arm at low speed may be just as risky as a task involving repeated small but quick movements.

10.152 Symptoms of repetitive strain injury include:

- tenderness
- aches and pains
- stiffness
- weakness
- tingling
- numbness
- cramp
- swelling.

Upper Limb Disorders

10.153 "Upper limb disorders" (ULDs) is an umbrella term for a range of disorders of the hand, wrist, arm, shoulder and neck. It covers those conditions with specific medical diagnoses (eg frozen shoulder, tennis elbow, carpal tunnel syndrome) and RSIs, where there is pain without specific symptoms.

10.154 Symptoms may include pain, swelling and difficulty moving. While the worse cases can result in permanent disablement if no action is taken, full recovery from ULDs usually occurs after appropriate rest.

10.155 It is estimated that 10.7 million working days (full-day equivalent) were lost in 2006/07 through musculoskeletal disorders mainly affecting the upper limbs or neck that were caused or made worse by work.

Back Strain

10.156 Back strain can arise in many work situations, but construction workers are at particular risk.

10.157 Back pain can involve damage to muscles, to the ligaments which bind the bones in the back, or to the discs which separate them. The most common type of injury is damage to the muscles and ligaments. Although temporary, the pain can be intense and prolonged.

10.158 However, the longer backs are mistreated, the more likely a "slipped" or prolapsed disc is. The result is sciatica — severe, long-term pain extending right down into the leg.

10.159 The exact cause of back strain is often unclear, but the reason back pain is common in construction workers is that it often results from day-to-day tasks that construction workers are involved in, such as:

- supporting loads, often in awkward positions
- moving heavy materials
- carrying loads over rough, uneven ground or within buildings
- carrying out highly repetitive tasks
- using handling equipment in difficult conditions, such as confined spaces.

Task-related MSD Trigger Factors

Repetition

10.160 Work is repetitive if it requires the same muscle group to be used repeatedly for a prolonged period (eg two hours) during the working day. Repetitive work does not allow the muscles time to recover. This can lead to muscle fatigue, inflammation and possible degeneration.

Working Posture

10.161 Poor working posture, which is awkward or fixed for long periods, increases the risk of injury. An ideal working position is one that is "neutral" (ie the trunk and head are upright, the arms are by the side of the body). A neutral position requires little muscle effort to maintain, so it carries a reduced risk of injury. An awkward posture is one where a part of the body is used outside its neutral position. When an awkward posture is adopted, muscle effort is required to keep the body in position. This increases the risk of injury. Static posture, where the body is held in position for extended periods, can restrict blood flow to muscles leading to rapid fatigue.

Force

10.162 The need to grip raw materials, products or tools is a potential risk factor as excessive force can lead to fatigue and injury. The force needed to grip can be influenced by a number of factors. These include:

- the type of grip used
- the posture of the wrist
- exposure to cold and vibration
- the effects of wearing gloves.

10.163 The force required to grip objects is also dependent on the material or item being gripped. For example, a screwdriver handle with a flexible grip requires less force to use than one with a harder handle. The size of the object being gripped can also affect the force required. For example, pliers with too wide or too narrow a span will be more difficult to grip. Muscle force is greatest when a power grip (eg gripping a handle in the palm with fingers and thumb) is used. This allows a large surface area of the hand to be used. The strongest grip strength occurs when the wrist is close to the "handshake" position and is slightly bent upwards.

Duration of Exposure

10.164 Duration refers to the length of time taken to perform a task. It includes:

- the length of time spent on the task in each shift
- the number of working days over which the task is performed.

10.165 Duration of the task is an important factor in assessing the risk of MSD. Many types of MSD, such as back pain, are cumulative in nature. Therefore, when duration time is increased, the risk of injury is increased. Short exposures are unlikely to create a significant risk, except where:

- the task is exceptionally demanding
- the worker has not been allowed to build up to the demands of the task over a period of time.

Environment-related MSD Trigger Factors

10.166 Some physical aspects of the work environment such as vibration and cold temperatures can increase the risk of MSDs.

10.167 Working in cold temperatures, handling cold products or having cold air blowing on parts of the body can place additional demands on the body. If the use of personal protective equipment is required for cold work, risk can be increased by the need for additional force to grip. Exposure to cold can result in:

- decreased blood flow to the hands and upper limbs
- decreased sensation and dexterity
- decreased maximum grip strength
- increased muscle activity.

Psychosocial-related MSD Trigger Factors

10.168 Psychosocial factors refer to:

- the way in which physical risk factors affect the physiological and biomechanical loading of the upper limb
- a worker's psychological response to work and workplace conditions.

10.169 These are important factors to consider when assessing risk as they have an influence on health and on how work is experienced by workers. Psychosocial risk factors include:

- the design, organisation and management of work
- the overall social environment.

10.170 Both physical and psychosocial risk factors can be experienced at the same time. The greatest benefit will be achieved when both are identified and controlled. Many of the effects of the psychosocial factors occur via stress-related processes. Psychosocial risk factors have been found to be common in job sectors where MSDs occur. Important aspects of work design include the:

- amount of control people have in their jobs
- level of work demands
- variety of tasks that they have to carry out
- support they receive from supervisors and co-workers.

Work-related MSD Trigger Factors

10.171 Some people may be more likely than others to develop MSDs. Individual differences may also have implications for employees reporting MSD-related conditions. Following a policy on MSDs should ensure that tasks are within the capabilities of the entire workforce. It is important that an individual is competent to perform the task required of them. In considering whether an individual is competent, issues such as training, skills, experience and motivation should be taken into account.

Managing MSDs

10.172 Prevention is better than cure. MSD risks can often be avoided, or minimised, by carrying out the activity in a different way or by using different materials. Ways in which this can be done include:

- reprogramming the order of work, eg external guttering can be fitted before the scaffold is dropped, removing the need to carry materials up, and then work from, a ladder
- using existing equipment on-site to mechanically handle large loads, eg attaching a lifting jib to a telehandler gives enough height to lift a full pack of roof trusses from ground to roof level, almost eliminating manual handling
- changing the design of the materials used, eg specifying lower weight building blocks; while not getting rid of the risks from manual handling altogether, this certainly reduces them.

10.173 It is advisable to take a systematic approach to the management of MSDs.

Understanding the Issues and Commit to Action

10.174 Employers and employees should have an understanding of MSDs and be committed to action on prevention.

10.175 This commitment may be expressed through positive leadership on the topic, by generating an effective health and safety policy on MSDs and by having appropriate systems in place. These actions will help to promote a positive health and safety culture in the workplace.

Creating the Right Organisational Environment

10.176 The organisational environment should foster active employee participation and involvement, have clear and open lines of communication and encourage partnership working in the following five steps. This will involve developing the competencies of employees, supervisors and employers for their differing roles.

10.177 Involving employees in the planning and organisational processes can be an important way of increasing the likelihood of success of any risk control strategy. Employees have first-hand knowledge and a unique understanding about particular aspects of the tasks they perform and will be able to provide key insights into activities and ideas of how tasks can be made easier.

Assessing the Risks of MSDs in the Workplace

10.178 A core feature of the management programme is to assess the risk of MSDs. This needs to be done in a systematic way so that the main risks in the workplace can be identified and prioritised for action. As risks are potentially widespread, simple checks, including a filter questionnaire, can be used to identify jobs which require a more detailed assessment.

Risk Assessment Filter

10.179 The HSE includes a guideline filter in its guidance on manual handling that provides a starting point for assessing manual handling tasks.

10.180 The filter is intended to assist in the screening out of straightforward, low-risk manual handling operations. If the filter shows that the load is within the numerical guidelines, and it is easy to grasp, the handler is in a stable position and the working environment is good, it is not normally necessary to perform a full assessment unless:

- an individual is at significant risk, eg due to pregnancy, previous injury or health condition, etc
- the activity is complex and requires greater preliminary assessment
- there are other considerations to take into account, such as psychosocial factors, eg high workloads, tight deadlines and a lack of control over working practices.

10.181 Application of the guidelines should provide a reasonable level of protection to around 95% of working men and women.

10.182 The figures used in the filter are based upon scientific literature and practical experience. However, the intention is to set an approximate boundary within which the load is unlikely to create a risk. It is important to note the guidelines should not be considered as safe weight limits for lifting. There are no limits below which manual handling activities can be regarded as safe. If in doubt, a more detailed risk assessment should be conducted.

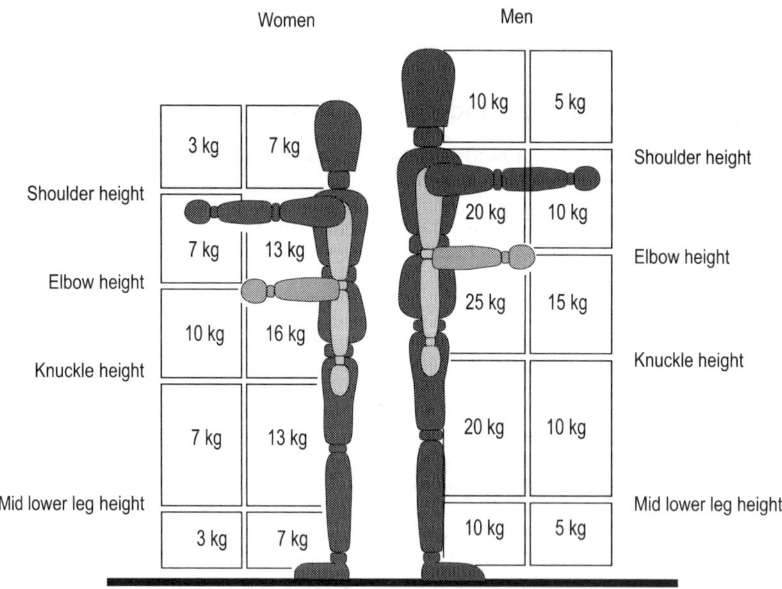

Reducing the Risks of MSDs

10.183 Once risks have been assessed and prioritised, a coherent process of risk reduction should be undertaken using an ergonomics approach. Possible risks should be reduced or eliminated at source. Implementation should include workforce participation as this is known to lead to better solutions and more effective, sustained changes.

An Ergonomics Approach

10.184 Ergonomics is concerned with ensuring work is designed to take account of people, their capabilities and limitations. Its objective is to optimise health, safety and productivity. An ergonomics approach is the most effective way of dealing with problems that lead to MSDs because it encourages all relevant parts of the work system to be taken into account and requires worker participation.

10.185 Having an adjustable chair in place of a "one-size-fits-all" chair is an example of an ergonomic approach.

Educating and Informing Employees

10.186 To enable participation and involvement of employees and for individuals to assume their proper responsibilities, provision of education and information is vital. Training will support all aspects of the management programme, and should be considered as an ongoing activity and not as a "one-off" task.

Managing Any Episodes of MSDs

10.187 Employees should be encouraged to identify any symptoms and to report them before they become persistent. Employers need to respond quickly by reviewing risks and introducing more effective controls, if necessary. They also need to reassure employees that reporting of symptoms will not prejudice their job or position. Early medical management can stop established cases from deteriorating and also help the process of return to work.

Carrying Out Regular Checks on Programme Effectiveness

10.188 The management programme needs to be reviewed on a regular basis to ensure that it continues to be effective.

Training

10.189 Adequate information and training should be provided to all employees who may be at risk from musculoskeletal disorders (MSDs). This will include:

- identifying all employees who require training
- designing an appropriate training programme
- implementing the training programme
- monitoring the effectiveness of the training
- keeping records of any training provided to employees.

10.190 All supervisors and employees should receive education and training on MSDs to help them identify the early warning signs. Training should raise general awareness of MSD issues and address specific needs of particular job tasks. The education and training of employees is an important part of ensuring their competency to perform their work.

10.191 Training should provide information on:

- the recognition of causes and symptoms of MSDs
- the risk factors that are present
- safe working methods to reduce or eliminate risk effects.

10.192 Training is important in the reduction of MSD risk. However, it should not be used as the primary control measure. It should be used to complement other risk reduction measures.

10.193 It is important that training is not seen as a one-off exercise. The training programme should take account of the output from the monitoring and review of the risk management system, addressing any specific risks identified.

List of Relevant Legislation

- Management of Health and Safety at Work Regulations 1999
- Reporting of Injuries, Diseases and Dangerous Occurrences Regulations 1995
- Health and Safety (Display Screen Equipment) Regulations 1992
- Personal Protective Equipment at Work Regulations 1992
- Manual Handling Operations Regulations 1992
- Health and Safety at Work, etc Act 1974

Further Information

Publications

HSE Publications

The following are available from *www.hsebooks.com*.

- HSG60 (rev 2002) *Upper Limb Disorders in the Workplace*
- HSG115 *Manual Handling. Solutions You Can Handle*
- HSG121 *A Pain in Your Workplace — Ergonomic Problems and Solutions*
- HSG144 *Safe Use of Vehicles on Construction Sites*
- INDG143(L) (rev 2, 2006) *Getting to Grips with Manual Handling: A Short Guide for Employers*
- INDG171(L) *Aching Arms (or RSI) in Small Business: Is Ill Health Due to Upper Limb Disorders a Problem in Your Workplace?*
- INDG398 *Are You Making the Best Use of Lifting and Handling Aids?*
- L23 (rev 2004) *Manual Handling. Manual Handling Operations Regulations 1992 (as amended). Guidance on Regulations*

Organisations

- BackCare
  Web: *www.backcare.org.uk*
  BackCare is a source of information about the causes, treatments and management of back pain and promotes best practice in the diagnosis, treatment and management of disorders of the spine and aims to prevent incapacity from back pain.
- Ergonomics Society
  Web: *www.ergonomics.org.uk*
  The Ergonomics Society aims to increase the general understanding of ergonomics and its importance.
- Health and Safety Executive (HSE)
  Web: *www.hse.gov.uk*
  The HSE is responsible for the regulation of almost all the risks to health and safety arising from work activity in the UK.

Occupational Asthma

- Asthma is a state of inflammation of the airways in the lungs which leads to narrowing of the airways and exaggerated airway reactivity. It is characterised by attacks of wheezing, chest tightness or breathlessness resulting from the constriction of airways.
- Asthma is almost always worse at night or in the early morning, provoked by exercise or cold air, and may be mild or infrequent, with potentially life-threatening attacks.
- Occupational asthma is caused by (rather than made worse by) exposure to hazardous substances in the workplace.
- Sensitisers that may be used on construction sites include certain glues and resins. In addition, wood dust can act as a sensitiser.
- Once a person has developed asthma, the threshold for reacting to a wide range of trigger factors drops considerably.
- Employers should prevent exposure to substances which cause allergic asthma and, if prevention is impossible, should aim to control substances well enough to prevent workers developing asthma.
- The hazard acts through inhalation, so agents which cause occupational asthma are usually present in dust or aerosol form.
- Employers should monitor workplace exposure limits to limit employees' exposure to substances known to be the cause of occupational asthma, also known as sensitisers.
- The earlier asthma is detected and the earlier exposure ceases, the better the long-term outcome, so employers should run health surveillance schemes on a regular basis.

10.194 Occupational asthma occurs if, as a result of exposure in the workplace, a substance causes a hypersensitive state in a person's airways and, hence, triggers a subsequent reaction in the airways. This reaction includes periodic wheezing, chest tightness or breathlessness. Substances causing occupational asthma are listed as hazardous substances under the Control of Substances Hazardous to Health Regulations 2002 (COSHH).

Employers' Duties

10.195

- Employers have a general duty to ensure, so far as is reasonably practicable, the health, safety and welfare at work of all employees under the Health and Safety at Work, etc Act 1974.
- The Control of Substances Hazardous to Health Regulations 2002 (COSHH) set out the legal requirements for protecting people in the workplace against health risks from hazardous substances, which include respiratory sensitisers.
- Employers must not expose employees to any potential sensitiser unless they have made a suitable and sufficient assessment of the risks created by work.
- Employers must ensure the exposure of employees to sensitisers is either prevented or, if this is not reasonably practicable, adequately controlled.
- Employers who provide control measures must take reasonable steps to ensure they are properly used or applied.

- Employers who provide control measures must ensure that exposure to the sensitiser is monitored in accordance with a suitable procedure.
- Employers must ensure employees exposed to sensitisers are under suitable health surveillance.
- Employees exposed to sensitisers must be provided with suitable and sufficient information, instruction and training.
- Employers must be prepared to react in the event of an accident, incident or emergency relating to the presence of a sensitiser.
- Employers must ensure those who carry out the assessment process and provide advice on the prevention and control of exposure are competent to do so, ie they should have:
  - adequate knowledge
  - training
  - experience in understanding hazard and risk
  - knowledge about how the work uses or produces sensitisers
  - the ability and authority to collate all the necessary, relevant information
  - the knowledge, skills and experience to make the right decisions about risks and precautions that need to be taken.

Employees' Duties

10.196

- Employees have a duty to take reasonable care of their own health and safety and that of other people who may be affected by their work under the Health and Safety at Work, etc Act 1974.
- Employees have a duty to co-operate with their employers on health and safety matters.
- The Management of Health and Safety at Work Regulations 1999 place a duty on all employees to follow health and safety instructions and to report danger.
- The Control of Substances Hazardous to Health Regulations 2002 (COSHH), place a duty on employees to:
  - make full and proper use of any control measure provided in accordance with the regulations
  - take reasonable steps to return control measure equipment after use
  - immediately report any defects to control measure equipment to the employer.

In Practice

Introduction

10.197 Asthma, as defined by the HSE is characterised by periodic attacks of wheezing, chest tightness or breathlessness resulting from the constriction of airways. A substance is considered to cause occupational asthma if, as a result of exposure in the workplace, it both:

- produces the biological change known as the hypersensitive state in the airways, and
- triggers a subsequent reaction in the airways.

10.198 All substances which cause occupational asthma are within the definition of substances hazardous to health under the Control of Substances Hazardous to Health Regulations 2002 (COSHH).

Sensitisers

10.199 Substances known to be a cause of occupational asthma, known as sensitisers, fall under several broad headings:

- wood dusts
- glues/resins
- soldering flux
- isocyanates
- flour/grain/hay
- laboratory animals
- gluteraldehyde.

10.200 The sensitisers that workers in the construction industry may be exposed to include wood dusts, isocyanates and epoxies in paints, glues and resins and possibly bacteria and biocides in cutting fluids.

10.201 Under the Chemicals (Hazard Information and Packaging for Supply) Regulations 2002 (CHIP) substances are assigned risk phrases. Substances assigned the risk phrase "may cause sensitisation by inhalation" or "may cause sensitisation by inhalation and skin contact" are considered to provide a risk in respect to occupational asthma.

Gathering Information About the Substances

10.202 The first stage is to compile a comprehensive list of all hazardous substances that might be present in the workplace. This should include all substances brought into the business from outside suppliers and consideration of how they are:

- worked on
- handled
- stored.

10.203 It is important to consider what additional substances might be generated by the process, such as:

- intermediates
- by-products
- finished products
- wastes.

10.204 Information on hazardous substances is available from a number of sources. Depending on the substance, it may be necessary to consult more than one source, including the following.

- The supplier, under the Chemicals (Hazard Information and Packaging for Supply) Regulations 2002, provides adequate information. This is normally provided in the form of a safety data sheet or package labels.
- The workplace exposure limits published by the HSE list substances which have been assigned occupation exposure limits.
- Trade associations and professional bodies also provide guidance publications setting out the best practice related to the Control of Substances Hazardous to Health Regulations 2002.

Evaluating the Risk to Health

10.205 The assessment will enable decisions about measures necessary to prevent or adequately control the exposure to sensitisers in the workplace to be made. The following should be considered during the risk assessment process:

- hazardous properties of the substance
- information on health effects provided by the supplier
- level and type of exposure
- circumstances of work, including the amount of substance involved
- activities such as maintenance, where there is the potential for a high level of exposure
- any relevant workplace exposure limits
- the effect of preventive and control measures
- the results of relevant health surveillance
- the results of monitoring of exposure
- if the work will involve the exposure to more than one substance hazardous to health, consider the risk presented by exposure to such substances in combination
- the approved classification of any biological agent
- additional information which may be needed.

Implementing Control Measures

10.206 If exposure cannot be avoided or prevented, the following hierarchy applies.

- Have a high level of inherent safety, by careful:
  - design
  - selection
  - use of appropriate work processes
  - use of systems and engineering controls
  - use of suitable work equipment and materials.
- Control exposure at source, by adequate ventilation systems and appropriate organisational measures such as reducing the:
  - number of employees exposed
  - level and duration of exposure.
- Use of personal protective equipment (PPE) should only be used to complement other methods. Adequate control cannot be achieved by PPE alone.

10.207 HSE guidance for wood dust gives the following key controls for wood dust control.

- Provide dust extraction at woodworking machines to remove dust before it can be inhaled.
- Maintain the extraction and collection system to ensure it continues to work efficiently.
- Have dust extraction equipment examined by a competent person at least every 14 months (as required by COSHH).
- Use a vacuum system with a HEPA filter to clear up wood dust. For particularly dusty tasks, such as sanding, use respiratory protective equipment as well as local exhaust ventilation.
- Do not use airlines or dry sweeping to clear dust away. This can cause high peaks of dust exposure and simply spread the dust around.

If the Sensitiser has a Workplace Exposure Limit

10.208 If the sensitiser has a workplace exposure limit, exposure by inhalation should be reduced to that limit or below.

10.209 If exposure exceeds the defined limit, control could still be considered adequate provided it can be demonstrated that appropriate steps are being taken to comply with the standard as soon as is reasonably practicable.

Duties under COSHH

10.210 Once the control measures have been provided, the employer is required under COSHH to take all reasonable steps to ensure it is properly used or applied. Employers must ensure employees use the control measures correctly.

10.211 If respiratory protective equipment and other protective equipment are used, employers must ensure they are used correctly and properly stored. If control measures are found to be defective, prompt remedial action should be taken.

10.212 The COSHH Regulations also require employers to maintain control measures in:

- an efficient state
- working order
- good repair
- a clean condition.

10.213 If engineering controls are provided, the employer must ensure they are maintained at appropriate intervals.

10.214 If working procedures are used as a control measure, they should be:

- observed regularly to ensure they are being followed
- subject to regular review to confirm if they are appropriate or could be improved.

Monitoring

10.215 Suitable procedures should be used if the risk assessment identifies the need to monitor exposure to respiratory sensitisers.

10.216 Valid and suitable occupational hygiene techniques should be used to measure the amount of employee exposure to hazardous substances. This would typically involve:

- collecting airborne samples form the employees' breathing zone
- other forms of periodic or continuous forms of monitoring as appropriate.

Health Surveillance

10.217 Although not strictly a control measure, health surveillance is appropriate for all employees who may be exposed to respiratory sensitisers. Under COSHH, health surveillance is appropriate if the exposure of the employee to a substance hazardous to health is such that an identifiable disease or adverse health effect may occur. The level of health surveillance must be related to the degree of risk of developing sensitisation.

10.218 If employee health surveillance is used, health records should contain only particulars approved by the HSE. A copy must be kept for at least 40 years from the date of the last entry. Other points to consider include the following.

- If there is only limited evidence of a hazard. Employees should be told about the symptoms to watch for and should report symptoms to a responsible person.

- If there is positive evidence of a hazard. The earlier asthma is detected and the earlier exposure ceases, the better the long-term outcome, so a health surveillance system should be introduced to monitor any symptoms in those employees who are exposed to the sensitiser. If symptoms develop, the employee should be referred to a doctor or nurse for further investigation.
- If there is a high risk of exposure. A high level of health surveillance of employees is required. This may also include regular examinations by a doctor or nurse.

Recording the Findings

10.219 Under COSHH the assessment needs to be suitable and sufficient and records need to be kept to demonstrate it has:

- considered all factors pertinent to the work
- reached an informed and valid judgment about the risks
- considered the steps needed to achieve and maintain adequate control of exposure if prevention is not reasonably practicable
- considered the need for monitoring exposure at the workplace and the need for health surveillance.

Reviewing the Assessment

10.220 The risk assessment should be reviewed at regular intervals, depending on the level of risk, or if:

- there is reason to suspect the risk assessment is no longer valid
- there has been a significant change in the work
- the results of monitoring show it is necessary.

Training

10.221 Under COSHH employees should be provided with suitable and sufficient information covering the:

- typical symptoms of asthma
- nature of any substances likely to cause occupational asthma they may be exposed to
- likelihood that once developed, occupational asthma could be permanent and what happens after further exposures
- procedures for reporting symptoms
- need to report immediately any symptoms that have occurred.

10.222 Employers should also give employees proper training, including induction training. Appropriate training should cover:

- correct use and maintenance of control measures provided
- work practices which prevent or reduce the emission of the substance into the atmosphere of both the workplace and general environment
- the use of respiratory protective equipment, where it is used as a control measure, and other control measures to further reduce exposure to the substance
- emergency procedures.

List of Relevant Legislation

- Chemicals (Hazard Information and Packaging for Supply) Regulations 2002
- Control of Substances Hazardous to Health Regulations 2002
- Management of Health and Safety at Work Regulations 1999
- Reporting of Injuries, Diseases and Dangerous Occurrences Regulations 1995
- Personal Protective Equipment at Work Regulations 1992
- Workplace (Health, Safety and Welfare) Regulations 1992
- Health and Safety at Work, etc Act 1974

Further Information

Publications

HSE Publications

The following are available from *www.hsebooks.co.uk*.

- EH40 *Workplace Exposure Limits* (revised annually)
- HSG53 *The Selection, Use and Maintenance of Respiratory Protective Equipment — A Practical Guide*
- HSG61 (rev 1999) *Health Surveillance at Work*
- HSG97 (rev 2004) *A Step by Step Guide to COSHH Assessment*
- HSG188 *Health Risks Management: A Guide to Working with Solvents*
- HSG193 (rev 2003) *COSHH Essentials: Easy Steps to Control Chemicals*
- INDG172(C) *Breathe Freely — A Workers' Information Card on Respiratory Sensitisers*
- L55 *Preventing Asthma at Work: How to Control Respiratory Sensitisers*
- WIS23 *LEV: General Principles of System Design*
- WIS24 *LEV: Dust Capture at Sawing Machines*
- WIS25 *LEV: Dust Capture at Fixed Belt Sanding Machines*
- WIS26 *LEV: Dust Capture at Fixed Drum and Disc sanding Machines*
- *Asthmagen?: Critical Assessments of the Evidence for Agents Implicated in Occupational Asthma*

Other Publications

- *Occupational Asthma: A guide for Employers, Workers and their Representatives*, Amicus, British Occupational Health Research Foundation, HSE and TUC

Occupational Dermatitis

- Occupational dermatitis is a skin condition that is usually caused by contact with certain chemicals or agents.
- The two main types are contact dermatitis (which causes dryness and cracking) and allergic dermatitis (sensitisation of the skin to a particular substance).
- Employers must assess the risks of exposure to substances in their workplace, including those in relation to dermatitis, under the Control of Substances Hazardous to Health Regulations 2002.
- High-risk substances should be eliminated from the workplace where possible, eg using water-based paints rather than solvent-based ones.
- Controls should be used that will prevent skin contact with the chemical, or reduce the number of workers involved.
- Protective gloves should only be used as a last resort if other methods of control do not provide effective protection.
- Some types of gloves such as powdered latex gloves may also act as a cause of dermatitis.
- Employers must provide adequate washing facilities and workers should wash thoroughly after working with sensitising chemicals.
- Employees who develop occupational dermatitis must be treated by a doctor. Some cases will also have to be reported to the enforcing authority.
- Health surveillance may be required in some circumstances.
- Anyone who may be exposed to an agent that causes dermatitis should receive suitable information, instruction and training.

10.223 Dermatitis is a skin condition that is usually caused by contact with certain chemicals or agents. The skin becomes dry, cracked and itchy. The likelihood of contracting dermatitis varies widely from person to person. Sensitisation to a particular chemical may cause dermatitis on exposure to relatively small amounts.

10.224 Employers must assess the risks of employees contracting dermatitis and implement suitable control measures to eliminate or minimise these risks.

Employers' Duties

10.225 There are no legal duties that specifically refer to dermatitis and skin management. However, the following general duties apply.

- Employers have a general duty to ensure, so far as is reasonably practicable, the health, safety and welfare at work of all employees under the Health and Safety at Work, etc Act 1974. This includes protecting employees from exposure to health hazards such as dermatitis.
- Employers must comply with the Control of Substances Hazardous to Health Regulations 2002, which apply to all processes and activities where employees are liable to be exposed to chemical substances. This includes:
  - assessing the risk of contracting dermatitis
  - implementing suitable control measures to eliminate or reduce the risk
  - conducting health surveillance if required.
  - Employers are required by the Management of Health and Safety at Work Regulations 1999 to provide employees with information on the risks to their

health and safety from handling substances that can cause dermatitis, and the preventive and protective measures that should be taken.

- Employers are required to provide employees with adequate and suitable washing facilities under the Workplace (Health, Safety and Welfare) Regulations 1992.
- Employers must report certain cases of dermatitis to the relevant enforcing authority under the Reporting of Injuries, Diseases and Dangerous Occurrences Regulations 1995.

Employees' Duties

10.226 Employees must:

- co-operate with their employers to enable their employers to comply with the Control of Substances Hazardous to Health Regulations 2002
- make proper use of control measures, including personal protective equipment
- return equipment after use to any storage place and report any defects found in the equipment
- attend medical examinations at the appointed time and give any information about their health as may be reasonable
- report any suspected cases of dermatitis to their employer.

In Practice

Types of Dermatitis

10.227 There are two main types of dermatitis.

Contact Dermatitis

10.228 Contact dermatitis is the most common form of the condition. It is caused by repeated contact with certain agents. This contact removes the skin's natural protective oils, causing drying and cracking. The skin then becomes more susceptible to other agents, making the condition worse.

Allergic Dermatitis

10.229 Allergic dermatitis is less common but can be more serious than contact dermatitis. It is sensitisation of the skin to particular chemicals, such as cement. Once the skin becomes sensitised, future exposure to the chemical will cause a reaction. Sometimes the reaction is so severe that the sufferer can no longer work with that chemical.

Risk Assessment and Dermatitis

10.230 Employers must assess the risk of exposure to substances in their workplace that may cause dermatitis. This is an integral part of a implementing a wider skin management programme as shown in the following flow chart.

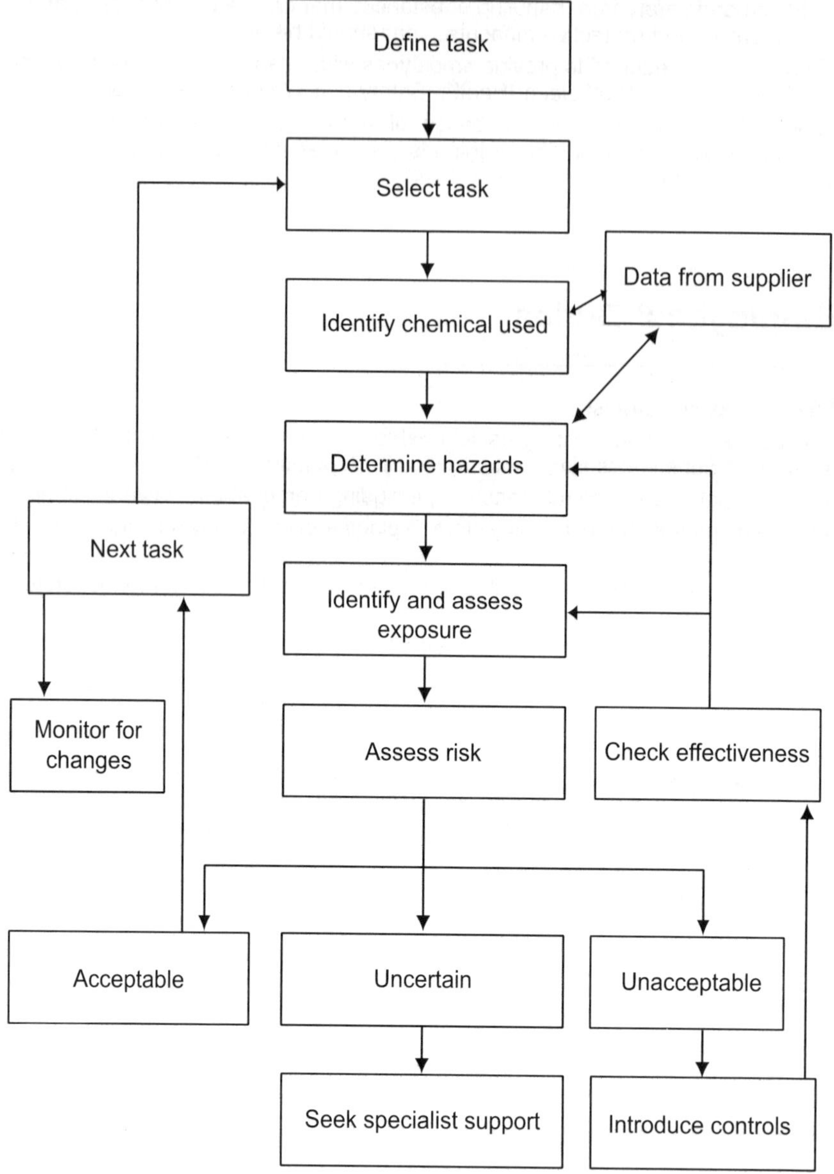

Identifying Potential Causes of Dermatitis

10.231 The first step is to identify any work activities involving chemicals that have the potential to cause dermatitis. As an example, consider the task of a worker laying an epoxy floor for which the task includes two main tasks:

- mixing the components
- spreading and levelling the resin.

10.232 The likelihood of skin contact should be assessed and steps taken to prevent any contact. In this case, these could be using pieced duo kits that allow mixing within the packet itself and using long-stemmed rollers and spreaders with splash guards.

10.233 The chemicals, agents and activities known to increase the risk of susceptible people contracting dermatitis. These include the following (the list is not exhaustive):

- cement
- chromates
- epoxies
- mineral wool glass fibre insulation
- solvents
- detergents
- irritant dusts
- nickel
- working in warm, dry environments
- wearing certain gloves, which trap sweat and make the hands hot
- x-rays
- prolonged or frequent wetting and drying of hands
- biological agents (eg plants, bacteria and fungi)
- mechanical abrasion (eg abrasive substances such as sand, and rough-edged surfaces and tools).

Cement

10.234 When wet cement containing chromium VI comes into contact with the skin, it can cause allergic contact dermatitis. The Control of Substances Hazardous to Health (Amendment) Regulations 2004 prohibited the supply or use of cement which has a chromium VI concentration of more than two parts per million. This prohibition should make allergic contact dermatitis — caused by skin contact with wet cement — a thing of the past, although wet cement can still cause serious burns if it comes into contact with the skin.

Chemical Labelling

10.235 Suppliers of chemicals are required to label the containers of any hazardous substances that they supply. The label on a chemical should be checked to see if it is classified as:

- sensitising on skin contact
- irritant
- corrosive.

10.236 A word of caution in consulting safety data sheets: many of them only list those chemicals that have risk or safety phrases. It may be necessary to obtain more information from the supplier.

Routes of Exposure

10.237 In addition to considering the potential risk of any chemicals used, the routes of exposure to such chemicals will also need to be considered. Skin exposure should be looked at alongside other routes of exposure, eg respiratory or ingestion.

10.238 There is growing evidence that dermal exposure can result in body uptake (where the chemical penetrates through the skin and circulates within the body) and may contribute to conditions such as asthma.

10.239 Routes for dermal exposure are not always easy to identify. The following table shows some possibilities.

Routes for Dermal Exposure

Gas	Airborne	• direct • indirect, eg absorption in substances such as clothing, filters in masks, etc
	Condensation to liquid	
Liquid	Direct	• through immersion • as spray • as condensation on cold surfaces
	Indirect	• contamination of clothing • absorption of liquid into substances
	Permeation	• through gloves or other types of personal protective equipment
Solid	Direct	• as solid, if sufficient free ions available • as dust from mechanical action • as fume or smoke particles from combustion, eg welding
	Indirect	• as contamination in clothing • as contamination in liquids, eg metals in metal-working fluids

Assessing the Risk of Exposure

10.240 If an agent is found which may cause dermatitis, the likelihood of someone contracting the condition should be assessed. Consider:

- how exposure occurs
- how much of the agent the workers are exposed to
- medical histories of the people who may be exposed to see if they have become sensitised to chemicals in the past.

10.241 If there is a risk that workers may develop dermatitis, suitable control measures, eg protective equipment, must be put in place, used and maintained.

Control Measures

10.242 The measure that should always be considered first is to eliminate any problem substances or work processes from the workplace. This may involve substituting a substance with a safer alternative or replacing any process that causes dermatitis with one that avoids the risks, eg the risk of exposure to wet cement from batch mixing can be reduced by using concrete pumps and cement silos; replace the painting or finishing of components on-site with pre-painted or assembled components.

10.243 Other ways to prevent skin contact with the chemical may be to use mechanical methods for applying paint or purchasing solvents in smaller containers to avoid the need for decanting.

10.244 The number of workers who are exposed to the agent should be minimised. It may be appropriate to allow only named individuals to carry out processes or work in areas that may lead to exposure.

10.245 Providing gloves or other clothing to protect workers against chemicals that cause dermatitis should only be done as a last resort.

Selection and Use of Gloves

10.246 Gloves are often a suitable control measure although it should be remembered that wearing gloves could make the work tasks more difficult. They can increase the risk of other hazards, such as dropping tools from a height or losing grip on machinery controls.

10.247 Many factors combine to affect how gloves perform as protection against chemical hazards. The data published by manufacturers on glove performance can only be taken as a indication of which glove will provide most satisfactory protection, not the length of time for which it can be used.

10.248 Due to permeation, gloves that appear to be in perfect condition may be allowing the chemical to come into contact with the hands, even though the wearer is unaware of this. Employers may need to consult a glove manufacturer or specialist on glove usage to ensure that the glove system is providing adequate protection.

10.249 Studies have shown that the length of a glove's useful life, used with the same chemical, may vary from several hours to a few minutes. It largely depends upon the nature of the task being carried out.

10.250 For mixtures of chemicals there may be no suitable glove, or the only glove that does provide protection will be expensive and offer only short-term protection. An example is the mixture of methylene chloride and methanol, often found as paint remover. The instructions may state that the appropriate glove be worn.

10.251 However, the only glove that provides any significant protection against this mixture costs in excess of £50 per pair. Even so, this is only a class-four glove, ie protection will be limited to a maximum of two hours, after which the gloves will have to be discarded.

10.252 Using the wrong glove, or one that is worn longer than its effective performance limit, may actually increase the damage being done to the skin. Gloves may also become a cause of dermatitis themselves due to lining powders and build up of sweat inside the glove. In particular, powdered latex gloves are now a recognised cause of dermatitis.

Barrier Creams

10.253 Barrier cream is sometimes provided to control exposure to agents that cause dermatitis. However, its effectiveness is very limited and the HSE no longer recommends its use.

Washing Facilities

10.254 Employers are required to provide adequate washing facilities. These facilities must have:

- hot and cold (or warm) running water
- soap
- nail-brushes
- a means of drying, ie towels or an electrical hand dryer.

10.255 It is vital that anyone who is exposed to an agent that may cause dermatitis washes thoroughly after finishing work. Soap itself may cause dermatitis, so it should be rinsed off with clean water after washing. Frequent handwashing with detergent is also a known cause of dermatitis, so a balance should be reached.

Emollients

10.256 It is important that skin is correctly cared for. Regular use of emollients (moisturisers, conditioners, etc) can play a major role in helping the skin to recover from the damage that has occurred during work, wearing gloves and hand cleansing. Such emollients should be applied at frequent intervals, preferably every time the hands are washed.

Health Surveillance

10.257 Health surveillance is required by the Control of Substances Hazardous to Health Regulations 2002 in some circumstances. In the case of dermatitis, there are methods of checking for development of the condition, and action that should be taken if it is found.

10.258 Skin health surveillance may be as simple as getting workers to check their skin for symptoms on a regular (eg daily) basis. The symptoms they should look out for include:

- red skin
- small blisters
- itching
- cracked skin, particularly between the fingers.

10.259 Anyone who thinks they may have symptoms should report them to a responsible person as soon as possible, and be referred to a doctor for confirmation and treatment.

10.260 Supervisors or other suitably trained persons should inspect their staff's hands for symptoms regularly, although this is likely to be done less frequently than self-examination.

10.261 In some situations it may be appropriate to consider biological monitoring or the use of specialised equipment. These can detect certain forms of skin damage before it becomes visibly apparent. Implementation of such a system should generally be done in consultation with either an occupational health nurse or an occupational physician.

Action if Dermatitis Develops

10.262 If employees develop dermatitis, a qualified doctor should treat them. The employer should take the following action:

- change the employee's work so that he or she is no longer exposed to the sensitising agent
- record the case of occupational dermatitis in the company's accident book
- report the case to the local enforcing authority within 10 days of finding out
- keep case details in employee's personal health records.

10.263 For contact dermatitis, a reduction in exposure is often effective in controlling the condition. However, if allergic dermatitis is contracted, exposure to even small amounts of the agent may cause a reaction. The severity of the reaction and the level of exposure that causes it will vary depending on the individual.

Reporting of Dermatitis

10.264 As soon as there is a written diagnosis of occupational dermatitis by a registered medical practitioner, it must be reported to the enforcing authority, if it occurs as a result of exposure to the following agents:

- acrylates and methacrylates
- antibiotics and other pharmaceuticals and therapeutic agents
- biocides, antibacterials, preservatives or disinfectants
- cement, plaster or concrete
- chromate (hexavalent and derived from trivalent chromium)
- colophony (rosin) and its modified products
- epoxy resin systems
- fish, shellfish or meat
- formaldehyde and its resins
- glutaraldehyde
- hairdressing products, particularly dyes, shampoos, bleaches and permanent waving solutions
- mercaptobenzothiazole, thiurams, substituted paraphenylene-diamines and related rubber-processing chemicals
- metal-working fluids
- organic solvents
- plants and plant-derived materials, particularly daffodil, tulip and chrysanthemum families, the parsley family (carrots, parsnips, parsley and celery), garlic, onion, hardwoods and the pine family
- soaps and detergents
- strong acids, strong alkalis, strong solutions, oxidising agents (eg bleach) and reducing agents
- sugar or flour
- any other known irritant or sensitising agent, particularly any chemical labelled "may cause sensitisation by skin contact".

Medical Screening

10.265 Before employing a new worker, many employers carry out medical screening to find out if they have any pre-existing conditions that may affect their ability to do the job. This screening should include enquiries about any history of skin sensitisation to the chemicals the worker is likely to encounter, should they be given the job.

Training

10.266 Training is an important part of an overall skin management programme. For maximum benefit, training must be relevant, ie it should be based on the chemicals and work activities that the organisation faces rather than being too generic. Training should also be matched to the needs of participants — managers and the workforce will require different training.

Management Training in Skin Management

10.267 Managers need awareness training to make sure they understand why skin management is important. Such training should include:

- the different legislative requirements relating to dermatitis
- the hazards and risks

- protective measures
- their role in skin management.

Workforce Training in Skin Management

10.268 Workforce training will tend to concentrate on good working practice and skin care. Such training should include:

- the consequences of skin damage
- symptoms of dermatitis and other skin conditions
- instruction on how to use protective equipment such as gloves correctly
- how to wash hands effectively
- use of emollients or reconditioning lotion
- the need to check the condition of their skin regularly.

10.269 Employees should be encouraged to raise any concerns they have following training and report any changes they notice in their skin's condition.

List of Relevant Legislation

- Chemicals (Hazard Information and Packaging for Supply) Regulations 2002
- Control of Substances Hazardous to Health Regulations 2002
- Management of Health and Safety at Work Regulations 1999
- Reporting of Injuries, Diseases and Dangerous Occurrences Regulations 1995
- Personal Protective Equipment at Work Regulations 1992
- Workplace (Health, Safety and Welfare) Regulations 1992

Further Information

Publications

The following are available from *www.hsebooks.co.uk.*

- INDG233(L) *Preventing Dermatitis at Work*
- L5 (rev 2005) *The Control of Substances Hazardous to Health Regulations 2002 (as amended). Approved Code of Practice and Guidance*
- L73 (rev 1999) *A Guide to the Reporting of Injuries, Diseases and Dangerous Occurrences Regulations 1995*
- MS24 (reprinted, with amendments 2004) *Medical Aspects of Occupational Skin Disease*

Noise

- Noise is any unwanted sound. Exposure to excessive levels of noise for a long period of time can be damaging to the hearing of employees.
- Noise at a level that is not loud enough to be physiologically damaging can cause stress and fatigue.
- If someone has to raise their voice to be heard over a distance of around 1.5 metres, a noise assessment is probably necessary.
- As a result of a noise assessment, measures may need to be taken by the employer to eliminate or reduce the risk. Risks from noise should be reduced or eliminated by technical and organisational means before considering control by the use of hearing protection.
- At or above the lower exposure action value, an employer must provide ear protection if requested by an employee.
- If the noise level reaches or exceeds the upper exposure action value, an employer must provide suitable ear protection, regardless of whether or not it has been requested by an employee.
- Clearly marked hearing protection zones should be established where upper exposure action values are reached. Suitable ear protection must be worn at all times in these zones.
- Employees must be provided with information, training and instruction on the risks of hearing damage, the control measures in place and the use of protective equipment to enable them to continue with their tasks with the least possible damaging effect to their health.
- The employee has a duty to wear any appropriate ear protection provided, and use any other controls provided to reduce risk.
- Employees must also co-operate with the employer and attend hearing checks when required.

10.270 Hearing damage caused by exposure to noise at work is permanent and incurable. Research estimates that over two million people in the UK are exposed to noise levels at work that may be harmful. Hearing loss is usually gradual due to prolonged exposure to noise, although hearing damage can also be caused immediately by sudden, extremely loud noises. As well as causing permanent hearing loss, exposure to noise can also cause tinnitus, which is a sensation of noise in the ears, such as ringing or buzzing. Tinnitus may occur in combination with hearing loss.

Employers' Duties

- Under the Health and Safety at Work, etc Act 1974, employers have a general duty to ensure, so far as is reasonably practicable, the health, safety and welfare at work of all employees.
- Under the Control of Noise at Work Regulations 2005, employers must:
  - identify noise hazards in the workplace
  - estimate likely exposures to noise of employees
  - identify measures required to eliminate or reduce risks, control exposures and protect employees
  - make a record of what measures are to be taken in the form of an action plan
  - protect employees with hearing protection, making its use mandatory in high-risk cases

- inform, instruct and train employees on the risks from noise, control measures, hearing protection and safe working practices
- provide health surveillance (including hearing checks) for those at risk
- maintain any noise control equipment and hearing protection in order to control exposure.

10.271 The Control of Noise at Work Regulations 2005 also impose limits on exposure which the employer must ensure are not exceeded.

Employees' Duties

- Under the Control of Noise at Work Regulations 2005, employees have a duty to:
  - use any controls supplied to control exposure to noise and to report any defects in control measures
  - attend hearing checks when required, providing they are during working hours.
- Employees have a duty to take reasonable care of their own health and safety and that of other people who may be affected by their work under the Health and Safety at Work, etc Act 1974.

In Practice

Noise Action Levels

10.272 As a general rule, if someone has to shout in order to be heard over a distance of about 1.5m, a noise assessment is probably necessary (as required by the Control of Noise at Work Regulations 2005). The regulations impose the following exposure limit values and action values.

Action Levels of Exposure

Lower Exposure Action Value	• a daily or weekly personal noise exposure of 80dB (A-weighted) • a peak sound pressure of 135dB (C-weighted)
Upper Exposure Action Value	• a daily or weekly personal noise exposure of 85dB (A-weighted) • a peak sound pressure of 137dB (C-weighted)
Exposure Limit Value	• a daily or weekly personal noise exposure of 87dB (A-weighted) • a peak sound pressure of 140dB (C-weighted) Exposure limit values are to take into account any hearing protection worn by the employee.

10.273 The daily and weekly exposure levels are given as doses, which are time-weighted levels, expressed as an eight-hour equivalent level. These are A-weighted levels, meaning that they are weighted to compensate for the frequency-dependent sensitivity of the human ear.

10.274 The peak levels, however, are C-weighted levels to indicate the highest level to which someone is exposed. C-weighting is a near-linear weighting, since with peak levels the potential for damage is less associated with frequency of the noise than with noise doses. It is important to note that when measuring noise, the peak level is not the same as the maximum level. A noise meter which incorporates the appropriate circuitry to deliver a peak reading must be used.

Assessment of Noise Levels

10.275 Exposure to noise at work must be assessed where the noise level is likely to be at the first action level or above. If the noise levels reach the second action level, exposure should be reduced to the lowest level reasonably practicable.

10.276 An assessment of noise levels should satisfy the following objectives.

- Identify which employees are exposed to levels at or above the defined action levels.
- Provide information on:
  - appropriate steps to reduce noise exposure, other than by the use of hearing protection
  - suitable hearing protection
  - any areas that should be designated as ear protection zones and the information to be provided to those working in these areas.

10.277 The time involved in making an assessment and the level of skill and competence required will depend on the complexity of the exposure patterns. A competent individual must carry out the assessment. This does not mean that a particular formal qualification is required, although this may be a perfectly satisfactory way of obtaining and demonstrating the knowledge required.

Records of Assessment

10.278 Employers must maintain adequate records of noise assessments. It is recommended that the records of each assessment are kept for a reasonable period, both to assist in identifying trends in an area and as evidence of levels of exposure of individuals, should any subsequent claims be made for noise induced hearing loss.

10.279 In recognition of the variable nature of the workplace, no single standard format is specified in the regulations for the record of the assessment. Descriptive comment, tabulated data, calculations and plans of the workplace are all acceptable formats for the assessment.

Noise Control and Management

10.280 Every noise problem can be broken down into three distinct parts:

- the source, which radiates noise
- the path, along which the sound energy travels
- the receiver, ie the human ear.

10.281 Noise produced by a source will travel outwards in all directions. The sound level measured at any point in a room will be the sum of the direct sound from the source plus the sounds reflected (as reverberation) from various surfaces such as:

- walls
- floors
- ceilings
- other hard, reflecting surfaces.

10.282 Therefore reverberation has the effect of increasing the noise at a distance from a machine, for example, to a higher level than it would be at the same distance in the open air.

Control of Noise at Source

10.283 The main source of noise hazards in the workplace is machinery. Noise is mainly produced by vibration of parts of machines and by the materials being worked, but is also caused by motors and pneumatic systems. The first step to control the source should be to identify the sources of the noise. On identifying the noise sources, the noisiest parts must be dealt with first, followed by the others in order of decreasing magnitude. The treatment could include any of the following:

- machine design and replacement
- planned maintenance
- substitution of machine or process
- vibration isolation and damping (surface treatments)
- reducing impact noise
- silencers.

Machine Design and Replacement

10.284 The best way to prevent noise is to consider the issue at the design stage. Often small changes in machine design or in the material used can result in a significant reduction in noise levels.

10.285 Noise control undertaken on existing equipment is usually more difficult. On purchasing new machines, the opportunity should be taken to include a low noise level in the purchase specification. The Supply of Machinery (Safety) Regulations 1992, as amended, provide for certain essential health and safety requirements, including information on noise. All machinery supplied or put into service should be designed and constructed so as to minimise noise emissions to the lowest level possible, taking into account technical progress and the availability of means of reducing noise.

Low Noise Purchasing Policy

10.286 The aim is that when new equipment is bought or hired, the employer will ensure that the quietest equipment or machinery is obtained. It should be noted that the cost of introducing noise reduction measures is often reduced if quiet equipment is introduced into the workplace. An effective low noise purchasing policy will give consideration to:

- how new or replacement machinery could reduce noise within the workplace
- ensuring that realistic noise output levels are specified for all new machinery, and checking that those involved in the supply chain are aware of their requirements and of their legal duties
- checking with suppliers about the likely noise levels generated under the particular conditions under which the machinery will be operated, as well as under standard test conditions (noise output data is only a guide: many factors can affect the noise levels experienced by employees)
- purchase or hire only from suppliers who demonstrate a low-noise design, with noise control as a standard part of the machine
- recording important elements of the decision-making process that is followed when buying or hiring machinery.

Planned Maintenance

10.287 It is important to check that regular maintenance is carried out. Replacement or adjustment of worn or unbalanced parts of machines, lubrication of machine parts, tightening of loose bolts and the use of properly sharpened cutting tools can result in a significant reduction in the sound level.

Substitution of Machine or Process

10.288 Substituting a noisy machine or process for a quieter one is not always feasible, but different ways of working can sometimes avoid the need for such a noisy process. Examples of substitution include the use of:

- hydraulic presses instead of mechanical presses
- welding rather than riveting
- belt drives rather than gears
- low-noise air-lines.

Vibration Isolation and Damping

10.289 A metal panel forced into vibration, by a machine to which it is attached for example, can radiate high noise levels, particularly when forced to vibrate at its natural frequency. The resonance can be avoided by altering its frequency, which can be done by:

- stiffening the panel
- adding damping material
- increasing the mass of the vibrating panel.

10.290 For maximum effect, the damping material must be heavier than the panel to which it is applied and it should ideally be glued on, rather than screwed or fastened. Machine anti-vibration mountings can also be used to isolate the machine and to reduce the transmission of vibrations to the floors and walls. These are often used with engines, compressors and ventilating fans. Anti-vibrating mountings are not usually very effective for reducing noise close to the machine. However, they can be effective in reducing structure-borne noises causing nuisance in nearby rooms.

Reducing Impact Noise

10.291 Where noise is caused by impact, it can sometimes be reduced by use of impact-absorbing materials. Materials such as heavy-duty rubber or neoprene should be used between the impacting surfaces, for example where metal objects or scrap metals fall into a metal bin. The bin should be lined with rubber or replaced with a perforated metal container and the drop height of articles should be reduced. Alternatively, their fall should be broken. Conveyor systems should be designed so the components being transported do not rattle against each other.

Silencers

10.292 The silencer is a device designed to attenuate noise at different frequencies. Silencers are inserted into ductwork or pipework associated with fans, pumps or compressors, which cause noise due to turbulent air or liquid flows, to reduce the transfer of noise without restricting movement of the gas or liquid. The choice of silencer will depend on the frequencies to be attenuated.

Control of the Noise Path

10.293 Having made every effort to control the noise exposure at the source, the next step is to reduce the transmission of the acoustic energy from the source to the receiver. Examples include:

- total or partial enclosure of the noise source
- increasing the distance from the source
- reducing the exposure time.

Total or Partial Enclosure of the Noise Sound

10.294 The most obvious way of modifying the route of exposure from the source to the operator is to enclose or partly enclose the machine with suitable sound-blocking material. A typical enclosure could be made of panels fitted with airtight seals and have an inner lining of sound-absorbing material. The level of enclosure or noise attenuation achieved in practice is largely dependent on safety and process operation considerations such as:

- temperature, light and ventilation requirements
- movement of materials
- access for operators and space for movement
- observation windows
- ducts and pipes penetrating the enclosure.

10.295 Partial enclosures in the form of screens and barriers around the machine can also be useful. However, their value is limited because noise does not travel in precisely straight lines but tends to curve around obstacles. The effectiveness of a screen will depend upon its height and length in relation to its distance from the source and its distance from the receiver. In general a screen will be effective only if:

- it is higher than the source and higher than the receiver
- the source is close to the screen or the receiver is close to the screen
- its length is greater than its distance from the source or the receiver.

10.296 The screen must be solid and continuous with no gaps or openings. A reflecting surface such as a high wall behind the source will reduce the effectiveness of the screen.

10.297 Enclosures often appear to be a practical solution to noise problems, but in practice are often not as cost-effective or as practicable as methods of control at source.

Increasing the Distance of the Worker from the Source

10.298 For every doubling of distance from the source of a noise, the noise level will be reduced by 6dB. Therefore, a noise level of 90dB at a distance of 0.5m from the noise source will be reduced to 84dB at 1.0m from the noise source. Thus, increasing the distance from the source can provide considerable reduction in the noise level. However, in practice, the use of this control option is obviously limited by the task being performed, as well as by the contribution from other noise sources in the vicinity and the reverberant noise component.

10.299 Another option to consider is moving or re-siting the machine. This may achieve the segregation of noisy processes to restrict the number of people exposed to high noise levels. As well as the position and distribution of machines, the location and routes of other noise sources, such as pipework and exhausts, need to be considered. Ideally these should be positioned away from workstations. For example, pneumatically powered equipment can often be fitted with a flexible exhaust hose discharging several metres away from the operator.

Reducing the Exposure Time

10.300 The table illustrates the equal energy principle, which allows a trade-off between noise level and time of exposure. The table shows a series of noise exposures that are all equivalent to a noise dose of 85dB(A) for eight hours.

Noise Exposures Equal to an $L_{(A)EP,d}$ of 85dB

Exposure Level in dB(A)	Maximum Duration of Exposure
85	8 hours
88	4 hours
91	2 hours
94	1 hour
97	30 minutes
100	15 minutes
103	7.5 minutes
106	3.8 minutes
109	2 minutes
112	1 minute
115	30 seconds
Reproduced with permission of Blackwell Scientific Publications Ltd	

10.301 If the sound level is above 85dB(A) the daily personal noise exposure ($L_{EP,d}$) can be kept below 85dB(A) by reducing the exposure time. This can be achieved by job rotation. Exposure to 88dB(A) for four hours or even 91dB(A) for one hour is permissible, provided the remainder of the shift is spent in a quiet environment of less than 80dB(A).

Control at the Receiver — Hearing Protection

10.302 Ideally, hearing protection should be used only under the following conditions.

- Where the lower exposure action value of 80dB(A) is exceeded, hearing protection must be made available.
- As a last resort, if attempts to control the noise by other methods have failed to reduce the noise below the second action level of 85dB(A).
- In an area of high noise level, eg a plant room which may be entered infrequently or only for short periods.
- For immediate protection where a noise problem has been identified, until longer-term measures have been put into place.

Selection of Hearing Protection

10.303 Where hearing protection is required it is important that the hearing protection supplied meets certain criteria, namely that:

- the noise attenuation is sufficient for the noise exposure
- it does not increase the overall risk, such as by masking auditory warnings
- it is supplied in consultation with the employees and their representatives
- employees have a choice of equipment, where practicable, so that they can choose a form that they find most comfortable and easy to use
- employees are trained in the correct use of the hearing protection, and its maintenance, and procedures for replacement.

10.304 Where use of hearing protection is compulsory, ie where exposure is above the upper exposure action value, employers should ensure that records are kept of the issue of equipment.

10.305 There is a vast array of hearing protection available, and this can be broadly divided into plugs and muffs. There are advantages and disadvantages to both types, and sub-divisions within the types.

10.306 Plugs may be corded or uncorded, or have plastic headbands, which are particularly useful in circumstances where it is necessary to regularly remove and reinsert the plugs. Plugs mounted on plastic headbands are known as semi-aural plugs. They generally only give a low level of attenuation (around 12dB overall), but are well-suited to environments with noise levels up to 85dB(A), where equipment giving higher attenuation may "over protect" leading to the masking of auditory warnings and difficulty with communication.

10.307 In general, plugs are more difficult to insert correctly, and have hygiene issues related to the ear canal, particularly on construction sites where there is significant potential for dirt to be introduced. It is also more difficult to tell at a distance if people are wearing ear plugs, making supervision more difficult. Custom-moulded plugs are gaining in popularity in industry. They give good performance, are comfortable and easy to fit. They are more expensive than other forms of protection, but by giving workers a custom-made product, they are generally more likely to wear them.

10.308 Ear muffs are easier to fit and give good performance, but can be uncomfortable to wear for extended periods. They are also unhygienic if not regularly cleaned.

Practical Implementation of Ear Protection

10.309 Whichever type of ear protector is used, it is important to remember that it will only provide the assumed protection if it:

- is in good condition
- fits
- is suitable for the individual
- is worn properly.

10.310 The theoretical and practical factors which need to be considered in selecting ear protectors include:

- the level and the frequency characteristics of noise
- the assumed protection of a given ear protector
- how the ear protector fits the individual worker
- compatibility of ear protectors with other protective equipment
- communication and safety requirements
- maintenance and inspection requirements
- the working environment
- the task performed
- the frequency of exposure
- costs in the short and long-term.

10.311 In practice, the main reasons for removing hearing protectors are physical discomfort and a feeling of isolation. Therefore, comfort is of equal importance to attenuation, and over-protection may be as detrimental as under-protection. It is also important to remember that if the ear protector is removed in noisy areas, even for short periods, the amount of protection provided will be reduced significantly.

Hearing Protection Zones

10.312 Every employer must designate any part of the premises where employees are likely to be exposed to the upper exposure action value or above as a "hearing protection zone".

10.313 Every hearing protection zone should be identified by means of a sign including text indicating that this is an ear protection zone and that employees must wear personal ear protectors while in the zone.

Training

Training of Competent Persons

10.314 The Control of Noise at Work Regulations 2005 require noise assessments to be carried out by competent persons.

10.315 In order to carry out the assessments, the competent person will not have to know how to choose and design the noise control measures, but should have a basic knowledge of:

- information needed and where to obtain it
- carrying out measurements
- calculating personal noise exposures
- maintenance and use of noise measurement equipment.

10.316 It is clear that the depth of knowledge required will vary widely according to circumstances, as will the training, and so there is no prescriptive definition of what constitutes a competent person. If they only need to deal with a limited area or types of machinery for example, less training will be necessary than for those who have to make decisions about a variety of control measures.

10.317 A variety of training courses are available to train individuals for the practical skills needed. If engineering control of noise is required, more in-depth training will be necessary.

10.318 An elementary training course should include subjects such as:

- the requirements of the Control of Noise at Work Regulations 2005
- official guidance, eg from the Health and Safety Executive
- the nature of sound
- the decibel scale
- noise measurement equipment
- carrying out a noise survey
- general principles regarding sources of noise
- general principles regarding ear protection types and capabilities.

Training of Operators

10.319 It is a requirement of the Control of Noise at Work Regulations 2005 that any employee exposed to noise at or above the lower exposure action value (80dB) is to be provided with sufficient information, instruction and training, which should include:

- a description of the nature of risks from exposure to excessive noise
- the organisational and technical measures taken by the organisation in order to reduce the risk from noise
- a description of the exposure action and exposure limit values contained in the regulations
- the significant findings of the noise assessment, including any measurements taken, with an explanation of the findings

- the arrangements for the provision, storage and maintenance of hearing protection, along with a description of the various types of equipment available
- how to use and maintain the various types of hearing protection
- why and how to detect and report signs of hearing damage
- the entitlement to health surveillance, and a description of what is involved
- safe working practices to minimise exposure to noise
- the collective results of any health surveillance, in a form which does not identify any particular person.

List of Relevant Legislation

- Control of Noise at Work Regulations 2005
- Management of Health and Safety at Work Regulations 1999
- Supply of Machinery (Safety) Regulations 1992
- Health and Safety at Work, etc Act 1974

Further Information

Publications

HSE Publications

The following are available from *www.hsebooks.co.uk.*

- HSG138 *Sound Solutions — Techniques for Reducing Noise at Work*
- INDG362 *Noise at Work: Guidance for Employers on the Control of Noise at Work Regulations 2005*
- INDG363 *Protect Your Hearing or Lose It!*
- L108 *Controlling Noise at Work – The Control of Noise at Work Regulations 2005. Guidance on Regulations*

Organisations

- British Occupational Hygiene Society (BOHS)
  Web: *www.bohs.org*
  The BOHS seeks to promote and protect occupational and environmental health and hygiene and offers training in occupational hygiene and related subjects.
- Institute of Acoustics
  Web: *www.ioa.org.uk*
  The institute administers and sets the standards for its Diploma in Acoustics and Noise Control and for a one-week course leading to a certificate of competence in workplace noise assessment. Various classes of membership are available.
- Institution of Occupational Safety and Health (IOSH)
  Web: *www.iosh.co.uk*
  IOSH is Europe's leading body for health and safety professionals and provides guidance on health and safety issues.

Radiation

- Ionising radiation is the transfer of energy in the form of particles or electromagnetic waves of a wavelength of 100nm or a frequency of $3x10^{15}$Hz or more and capable of producing ions either directly or indirectly.
- Non-ionising radiation is where the individual energy is too small to ionise matter.
- Ionising radiation is the emission of energy either by certain substances as a result of radioactive decay or by the use of certain electrical equipment. The emissions have the energy to ionise atoms (ie to produce positive and negative ions in materials that they irradiate).
- The degree of harm caused by ionising radiation depends on the type of radiation, the dose and dose rate, and the part of the body that is irradiated. However, there is no safe level of radiation exposure, so exposures should be kept to as low a level as possible.
- Every employer working with radiation must take all reasonable steps necessary to restrict, as far as is reasonably practicable, the extent to which employees and other persons are exposed.
- Non-ionising radiation may be categorised as either optical radiation or electromagnetic fields.
- As with ionising radiation, non-ionising radiation also has the potential to cause harm to people who may be exposed to it either in the course of their work or accidentally. The risk, ie the likelihood that the harm from this type of hazard will occur, has to be assessed.
- Every employer must ensure that those employees who are engaged in work with radiation are given appropriate training in the field of radiation protection.

10.320 Radiation is energy which is emitted, transmitted or absorbed either as particles or as electromagnetic waves. All humans are constantly being exposed to radiation either from artificial sources or from sources occurring naturally on earth or in space.

10.321 There are two distinct types of radiation: ionising and non-ionising. Risks to health from exposure to both ionising and non-ionising radiation can be very serious.

Employers' Duties

10.322

- Employers have a general duty to ensure, so far as is reasonably practicable, the health, safety and welfare at work of all employees under the Health and Safety at Work, etc Act 1974.
- The Management of Health and Safety at Work Regulations 1999 require employers to:
  - make an assessment of the risks to health and safety of their employees while at work
  - make a similar assessment of the risks to health and safety of persons not in their employment that may arise as a result of their (ie the employer's) work activities.

- The Ionising Radiations Regulations 1999 define ionising radiation as the means of the transfer of energy in the form of particles or electromagnetic waves of a wavelength of 100nm or less, or of a frequency of $3x10^{15}$Hz or more, and capable of producing ions directly or indirectly. The regulations require employers to:
  - obtain prior authorisation in writing from the Health and Safety Executive (HSE) to use electrical equipment intended to produce x-rays (exceptions apply)
  - notify the HSE of any work with ionising radiation 28 days prior to work commencing (exceptions apply)
  - carry out a suitable and sufficient assessment of the risk to employees and to any other persons before commencing any work with ionising radiation, in order to identify the measures that must be taken to restrict their exposure
  - ensure that doses of ionising radiation to employees do not exceed the limits specified
  - consult a radiation protection advisor for advice on the steps required to comply with the regulations
  - designate areas as controlled/supervised and introduce local rules
  - make an immediate investigation if they suspect a person is likely to have been overexposed and notify the HSE, the appointed doctor/employment medical advisor and the person affected if this turns out to be the case.
- The Radioactive Substances Act 1993 controls the acquisition, storage and disposal of radioactive materials. The Act requires employers to:
  - obtain a registration before obtaining radioactive materials, specifying both the radionuclides or class of radionuclides and the total activities allowed
  - store radioactive materials under conditions outlined, including use of a lockable store that is sufficiently shielded to reduce the radiation dose normally to below 7.5μSv/hour outside the store
  - dispose of waste subject to an authorisation.
- Additional registration requirements for high activity sources (as defined in the HASS Directive 2003/122/EURATOM) are described in the High-activity Sealed Radioactive Sources and Orphan Sources Regulations 2005.

Employees' Duties

10.323

- The Health and Safety at Work, etc Act 1974 (HSWA) places a duty on all employees to take reasonable care of the health and safety of themselves and of other persons who may be affected by what they do or fail to do at work.
- Under HSWA, employees have a duty to co-operate with their employers on health and safety matters.
- The Ionising Radiations Regulations 1999 require employees to co-operate with the employer and doctor over medical surveillance and to inform the employer about actual or suspected incidents.
- The Management of Health and Safety at Work Regulations 1999 place a duty on all employees to follow health and safety instructions and to report anything that represents a serious and immediate danger.

In Practice

10.324 Radiation can be divided into two distinct types.

- Ionising radiation — the transfer of energy in the form of particles or electromagnetic waves of a wavelength of 100nm or less or a frequency of more than $3x10^{15}$Hz and capable of producing ions either directly or indirectly.
- Non-ionising radiation — radiation where the individual energy is too small to ionise matter.

10.325 The human body reacts in different ways, often adversely, to the various types of radiation.

Ionising Radiation

10.326 Ionising radiation is the emission of energy, either naturally by certain substances as a result of radioactive decay or artificially by the use of certain electrical equipment.

10.327 The emissions have the energy to ionise atoms (ie to produce positive and negative ions in materials which they irradiate) and may be:

- beta particles
- alpha particles
- neutron sources
- gamma rays or x-rays.

10.328 The ions produced can cause molecular changes that may lead to biological damage in tissue.

Types of Ionising Radiation

Alpha Particles

10.329 Alpha-emitting substances are often encountered in the nuclear industry and are sometimes used in the elimination of static. Certain smoke detectors also emit alpha particles.

Radon

10.330 A naturally-occurring alpha emitter is radon gas, which comes from the minute amounts of radium present in all earth materials such as rocks, soil, brick and concrete. In some areas of the UK radon is present in relatively high levels in every workplace. Confined spaces, particularly underground — such as basements, caves, mines and ground-floor buildings — can contain significant amounts of radon gas which have seeped in from the ground.

10.331 Most radon gas breathed in is immediately exhaled and presents little radiological hazard. However, the decay products of radon are solid and are themselves radioactive. They can become attached to atmospheric dust and water droplets which, if inhaled, can become lodged in the lungs and airways and irradiate the respiratory tract. Some decay products are alpha emitters and can cause significant damage to the cells of the respiratory tract.

10.332 Radon is now recognised to be the second largest cause of lung cancer in the UK after smoking. Radon surveys should be conducted in any workplace where its location and characteristics suggest that radon may be found. Inexpensive surveys can be carried out by leaving small plastic passive detectors in rooms of interest. The Health Protection Agency (HPA) website provides details of validated laboratories capable of supplying detectors for undertaking radon measurements. A map showing radon-affected areas in England, Wales and Northern Ireland is also available from the HPA.

Beta Particles

10.333 Radioactive isotopes emitting beta particles are most frequently used in medical research. Sealed sources of beta emitters are also used for thickness gauges.

Neutron Sources

10.334 Neutrons are uncharged particles that are produced usually by fission or from nuclear reactions. Neutron sources are unlikely to be used outside the nuclear industry other than by geophysical exploration companies to measure the water content of geological strata.

X-rays and Gamma Sources

10.335 Many important radionuclides have gamma and x-rays associated with their radioactive decay. X-rays can also be generated by artificial sources.

10.336 The two main medical uses of x-rays are to diagnose disease or injury and to kill cancerous cells. Types of equipment emitting x-rays include familiar medical diagnostic equipment and x-ray optical equipment used by scientists. Other very large sources of this type of radiation used in industry are as follows.

- Industrial radiography is one of the most common methods to test the integrity of metal welds and other sources of metal failure.
- The pharmaceutical industry uses very large sources of radiation to sterilise dressings and other medical equipment.
- The treatment of cancer by radiotherapy requires large and powerful sources of x-rays or gamma rays.
- Certain radioactive sources are used in the construction industry to determine the water content in the ground.
- Radioactive sources of all types are used in educational establishments for demonstration purposes in science, and as tracers in research and practical exercises.

Radiation Dose Units

10.337 There are many different ways of measuring the radiation dose that a person or object receives, including the following.

- *Absorbed dose.* Radiation interacts with biological tissue by ionising the atoms of the tissue and therefore depositing energy. The amount of energy absorbed in unit mass of tissue is called the absorbed dose and is measured in joules per kilogram.
- *Dose equivalent (measured in sieverts, Sv).* The quantity of absorbed dose does not distinguish between the interactions with body cells of, eg an alpha particle and a gamma ray. Alpha particles are more readily absorbed than gamma rays. This effect can be characterised by the dose equivalent. The dose equivalent is the absorbed dose multiplied by a weighting factor, as shown in the table.
- *Effective dose equivalent.* This is the dose equivalent (in an organ) multiplied by a weighting factor related to a particular body organ. Organ weighting factors are based on the relative risk of harm to an organ. The relative risk of harm is based on animal and human studies.

Weighting Factor for Calculation of Dose Equivalent for the Different Types of Ionising Radiation

Radiation Type	Weighting Factor
β–particles γ-rays X-ray	1
α-particles	20
Neutrons	5-20 (energy dependent)

General Principles and Procedures

Radiation Protection Advisors

10.338 Every employer using ionising radiation must consult a radiation protection advisor for advice on the steps required to comply with the Ionising Radiations Regulations 1999.

Written Authorisation

10.339 Written authorisation from the Health and Safety Executive (HSE) must be obtained, prior to use, for any electrical equipment intended to produce x-rays.

10.340 However, the use of x-ray sets for routine analytical, medical and veterinary diagnostic or investigational purposes, or for electron microscopes, does not require prior authorisation.

Notify the HSE

10.341 Notification must be provided to the HSE at least 28 days prior to work commencing with ionising radiations. However, work with certain concentrations of radionuclides and specific types of apparatus are exempt from this constraint (see Schedule 8 of the Ionising Radiations Regulations 1999).

Risk Assessment and Control Measures

10.342 Work must not be started without first carrying out a suitable and sufficient assessment of the risk to employees and to any other persons in order to identify the measures to be taken to restrict their exposure. Although the risk assessment applies to those carrying out medical examinations, it does not apply to their patients.

Hazard Assessment

10.343 Where the hazard assessment shows that a radiation hazard to employees or other persons is likely to arise from the work being done, the employer must take all reasonably practicable steps to:

- prevent the occurrence
- limit its consequences
- provide employees with the information, instruction and training (including on/with equipment) necessary to restrict their exposure to ionising radiation
- make a contingency plan.

10.344 The risk assessment should consider the:

- identification of radiation hazards
- nature of the source
- estimated radiation dose rates to which anyone could be exposed
- likelihood of contamination
- any relevant results of previous dosimetry or area monitoring
- advice from manufacturers or suppliers about safe use and maintenance of equipment
- engineering controls in use or planned
- proposed systems of work
- estimated levels of contamination
- effectiveness and suitability of any personal protective equipment (PPE) to be used
- extent of unrestricted access to the working area
- possible accident situations
- steps to prevent or limit identified accident situations.

Control Measures

10.345 The risk assessment should enable the employer to decide what action is needed to keep exposure to as low a level as is reasonably practicable. These may include decisions on:

- engineering controls, design features, safety devices and warning devices
- systems of work
- provision of PPE
- whether dose constraints are appropriate
- altering the working conditions for pregnant or breast-feeding employees
- specified investigation levels to ensure that radiation exposure is restricted as far as is reasonably practicable
- maintenance and testing schedules
- training needs
- designation of controlled and/or supervised areas
- designation of employees as classified persons
- a suitable programme of dose assessment
- allocation of managerial responsibilities
- a programme of monitoring and auditing.

Restriction of Exposure

10.346 Every employer working with radiation must take all necessary steps to restrict, as far as is reasonably practicable, the extent to which employees and other persons are exposed to ionising radiation. Dose limits are specified in Schedule 4 of the Ionising Radiations Regulations 1999. A simplified list of these dose limits is reproduced in the table below. These apply to both internal and external doses and must not be exceeded in a single calendar year.

Radiation Dose Limits Measured in MSv

	Employees Aged 18 Years or Over	Trainees Aged Under 18 Years	Any Other Person	Notes
Dose limits for the whole body	20	6	1	The effective dose equivalent resulting from exposure to the whole or part of the body
Dose limits for the skin	500	150	50	Skin exposure is averaged over area of $1cm^2$
Dose limits for the lens of the eye	150	50	15	Dose equivalent from external and internal radiation delivered 3mm behind surface of eye
Dose limit for the abdomen of a woman of reproductive capacity			13	Must not be exceeded in any consecutive 3-month interval
Dose limit for a pregnant woman			1	Dose to foetus must not be exceeded during time from employer notification of pregnancy to end of pregnancy

10.347 Where an employer suspects that any person is likely to have been overexposed, the employer must investigate immediately and notify the HSE, the appointed doctor/employment medical advisor and the person affected.

10.348 The radiation protection advisor must be consulted about the prior examination of plans, acceptance testing of new or modified sources of ionising radiation and the adequacy of control measures to restrict exposure.

10.349 As far as reasonably practicable, the restriction of exposure should be achieved by the following hierarchy of control

- Engineering controls, eg:
  - shielding
  - containment and ventilation
  - safety features and warning devices.
- Procedural controls, eg:
  - restricting exposure by restricting access
  - systems of work, such as limiting exposure time to a few minutes, or introduction of a permit-to-work scheme
  - use of PPE.

Designated Areas and Classified Persons

10.350 The designation of controlled and supervised areas is required to restrict exposure to ionising radiation.

Controlled Area

10.351 An area is required to be designated as controlled where there is a likelihood of the dose exceeding an effective dose of 6mSv over the course of one year, or of the external dose rate in the area exceeding 7.5mSv/hour when averaged over the working day.

10.352 It is also necessary to designate an area as controlled if it is necessary for any person who enters the area to follow special procedures in order to restrict significant exposure to ionising radiation, or in order to prevent radiation accidents.

10.353 Access to any controlled area must be restricted by physical means and must be adequately described in the local rules.

Supervised Area

10.354 A supervised area must be so designated if between one-tenth and three-tenths of the dose limit (ie 2.5μSv/hour and 7.5μSv/hour) may be exceeded in one year. Such an area must be adequately described in the local rules.

Classified Persons

10.355 An employee likely to receive a dose of ionising radiation that exceeds three-tenths of any relevant dose limit is to be designated as a classified person. Only employees aged 18 years or over and medically certified to undertake such work can be designated as classified persons.

Local Rules

10.356 Local rules serve as statements of general principles and as the description of the means of complying with the Ionising Radiations Regulations 1999. They therefore contain details of the risk assessment, safe systems of work and the organisation's policy on radiation protection. Local rules must contain:

- a description and details of the controlled and/or supervised areas
- the procedures for restricting access to controlled areas
- contingency plans.

10.357 Local rules could also contain, if applicable:

- general procedures for keeping exposures to a level that is as low as is reasonably practicable
- name and location of the radiation protection advisor/supervisor
- expected radiological protection training standards for employees
- dosimetry arrangements for employees
- accounting procedures for radioactive materials
- storage procedures for radioactive materials
- monitoring procedures for dose rates and/or contamination
- reporting procedures for certain incidents involving staff and/or patients
- details of co-operation with other employers (eg service engineers)
- procedures for equipment handovers
- duties concerning critical examinations to be carried out for articles containing radioactive substances and/or radiation generators
- reference to operating manuals, protocols and other relevant instructions, etc.

Personal Monitoring

10.358 Arrangements must be made with one or more approved dosimetry services for:

- the measurement and assessment of personal doses

- the making and keeping of dose records
- giving summaries of dose records to the employer.

10.359 Information about this is available from the HSE.

Medical Surveillance

10.360 It may be necessary for employees working with ionising radiation to undergo medical surveillance carried out by an employment medical advisor or by an appointed doctor. Employees subject to this requirement are classified persons, persons who are to be classified, employees who have been overexposed and certain other employees.

10.361 Adequate medical surveillance involves:

- pre-employment medical examinations
- special medical examinations
- periodic health reviews (at least every 12 months)
- determination of whether further dose limit conditions are appropriate.

Non-ionising Radiation

10.362 Produced artificially, non-ionising radiation includes microwaves, laser light, static and power frequency fields, radiowaves, infrared, visible and ultraviolet radiation. These can be beneficial to a variety of industries, such as communications, catering and manufacture. It can also occur naturally as, for example, visible light.

10.363 Unlike ionising radiation, there is no legislation specific to non-ionising radiation. Non-ionising radiation may be categorised as:

- optical radiation:
  - ultraviolet
  - visible light
  - infrared
- electromagnetic fields:
  - static electric and magnetic fields
  - extremely low frequency (ELF) radiation
  - radiofrequency radiation (including microwaves and radiowaves).

10.364 In the optical region, the radiation emitted by bodies is of two types — coherent and incoherent radiation.

10.365 Incoherent optical radiation is so called because the waves propagating the radiation are "out-of-phase". Two types of incoherent optical radiation are:

- radiators such as a filament lamp or the sun
- sources such as the orange sodium lighting used in street lights.

10.366 The laser is another source of optical radiation, but lasers emit radiation of a coherent nature (ie in-phase waves).

Prevention and Control of Risk

10.367 The general principles of prevention and control of risk from the various hazards presented by the very wide range of non-ionising radiation emitters can be summarised as follows.

- Seek technologies that do not produce a radiation hazard.
- Reduce as much as possible the level of radiation required to achieve the desired result.

- Make sure that the level and properties of the radiation are understood and can be measured.
- Use engineering controls such as interlocks, shielding and non-reflective walls, etc to prevent accidental exposure.
- Remember that creating distance from the source, eg by erecting barriers, is an effective means of reducing risk in most cases; this may not apply to laser beams which are hardly attenuated by distance from the source.
- Introduce a safe system of work.
- Use PPE only where absolutely necessary and not as a substitute for other measures.
- Restrict access to the danger area so that it contains only personnel who are required to be there.
- Train and provide information to staff in the understanding of non-ionising radiation and its health effects.
- Monitor the health of staff who work with radiation and keep records.
- Institute emergency and first-aid procedures.
- Clearly identify means of turning off the radiation source and of preventing it from being reactivated accidentally.
- Appoint competent supervisors.
- Carry out full and sufficient risk assessments.

Risk Assessment

10.368 As with ionising radiation, non-ionising radiation has the potential to cause harm to people who may be exposed to it either in the course of their work or accidentally. The risk, ie the likelihood that the harm from this type of hazard will occur, has to be assessed.

Electromagnetic Fields

10.369 Electromagnetic fields can be further subdivided into three regions as follows.

- Static electric and magnetic fields.
- ELF and radiofrequency fields (RF) up to about 100kHz where the effects are due to electrically excitable tissue.
- RF fields above 100kHz where the effects are mainly due to the heating of tissue.

Specific Control Measures

10.370 For static magnetic fields:

- it is essential not to use (or have nearby) tools and instruments made of magnetic materials
- great care must be taken when using overhead cranes in the vicinity of static magnetic fields
- use tools made of non-magnetic materials such as beryllium copper alloy
- introduce a "high magnetic field" warning system that operates independently of the mains electricity supply and that is activated by the presence of the high field
- cleaners should be provided with non-metallic buckets.

10.371 For radio frequency dielectric heating equipment, a properly designed shield must be installed by a competent person.

10.372 For microwave and radiofrequency sources:

- ensure that microwave oven doors are regularly tested for radiation leakage

- the radiation should be attenuated to within about 30cm of the oven openings for commercial ovens that work on a production line system (without doors).

10.373 For mobile phones:

- avoid use near hospital intensive care departments and other sensitive areas
- avoid use on aircraft
- do not use at petrol stations
- use with consideration for others to avoid undue stress
- avoid use while driving but, if use is absolutely necessary, a hands-free set should be employed.

Ultraviolet Radiation

10.374 Ultraviolet radiation can be further divided into three regions: UVA, UVB and UVC.

10.375 The main source of ultraviolet radiation contributing to personal exposure is the sun. The major risk is associated with human behaviour related to solar radiation.

10.376 Prolonged exposure to ultraviolet radiation is well known to be a significant cause of cancers of the skin. It also causes premature ageing and can cause eye damage such as inflammation of the cornea and conjunctiva.

10.377 Generally, the risks to outside workers such as park wardens, gardeners, building site workers, etc must be considered and assessed.

10.378 Certain types of welding can present a hazard to the eyes and skin. Arc welding is especially hazardous.

Control Measures

10.379 For commercial ultraviolet tanning equipment the following applies.

- These should have adequate mechanical strength so that no part can fall on a customer or fail if a customer falls on it.
- The mechanism must be fail-safe so that no unwanted increase of ultraviolet radiation can occur.
- The timing device that controls the exposure to ultraviolet radiation should be accurate to within 10%.

10.380 For solar radiation the following applies.

- Exposure should be kept to a minimum, preferably by wearing clothing or being in the shade.
- Avoid exposure around noon.
- Sun screens and sun blocks can be used.

10.381 For welding the following applies.

- Wear appropriate PPE including the approved types of eye protection.
- Erect screens around the work area.
- Display appropriate warning signs.

Lasers

10.382 Lasers exist in all three regions of optical radiation, ie ultraviolet, visible and infrared radiation.

10.383 Radiation from laser sources is capable of being very highly focused to a very fine spot and has a potentially high power density. The focusing system of the eye can concentrate the beam onto the retinas and cause it to burn.

10.384 All sources of laser radiation should be properly marked.

Principles of Laser Safety

10.385 When working with lasers, it is vital to consider the following.

- Always use the lowest power output and the shortest beam distance necessary to achieve the desired result.
- Ensure that there are controls to prevent intra-beam viewing.
- Install lasers in restricted areas and prevent unauthorised access.
- Enclose the laser in a way that will prevent unnecessary exposure.
- Ensure that laser systems are always under the control of a competent person and are not left running unattended.

Laser Risk Assessment

10.386 Factors that should be taken into account when assessing the risks from laser radiation include the following.

- Is the beam intended to be viewed (eg a display laser), or would exposure only occur unintentionally (eg during maintenance work)?
- Is the design of the equipment such that there is no direct exposure to laser radiation during use?
- How accessible is the laser beam, what is the power density and what type of projection is employed (eg pulsed, scanned or static)?
- Is it necessary for workers to come into contact with the equipment at all, ie can the equipment be enclosed?
- Is there a reliable system for preventing exposure, such as the use of an interlock that will isolate the power supply once the cover is removed from the part of the equipment that presents the hazard?
- Is it possible to reduce the numbers of people potentially exposed to the risk, eg by installing the equipment in restricted areas?
- What systems are used to prevent unauthorised operation of the equipment, eg key controls or coded operation?
- Is there a device to show when the laser is energised, particularly during maintenance operations?
- Have those who will be working with or maintaining the equipment been issued with appropriate protective equipment?
- Have all workers been trained in all necessary aspects of laser safety?
- Are warning signs clearly displayed?
- Is there a system for regularly checking the condition of laser equipment?

Control Measures

10.387 Work areas should be marked so as to prevent unauthorised personnel from straying into them.

10.388 BS EN 60825: *Safety of Laser Products* specifies a range of control measures to be adopted when working with lasing equipment:

- use of a remote interlock connected to door circuits (for laser classes 3B and 4)
- key control, including removal of the key when the system is not being used (classes 3B and 4)
- prevention of unintentional exposure by use of a beam attenuator (classes 3B and 4)
- a device to indicate when the laser is energised (classes 3B and 4)
- warning signs (classes 3B and 4)
- termination of the beam at the end of the usual length (classes 2, 3A, 3B and 4)

- means to prevent unintentional reflections (classes 3B and 4)
- use of eye protective equipment where the maximum permissible exposure (MPE) is exceeded and other controls do not adequately reduce the risk (classes 3A, 3B and 4)
- use of protective clothing (sometimes necessary for class 3B, and specific requirements for class 4)
- training of all operators and maintenance personnel who may be affected by the use of class 3A, 3B and 4 laser products.

10.389 All laser protective eyewear must be clearly marked with information adequate to ensure proper choice of eyewear with particular optical power sources. Exceptions to this are:

- where engineering and administration controls substantially eliminate potential exposure in excess of the relevant MPE (the level of radiation exposure without suffering adverse effects)
- when, owing to unusual operating requirements, the use of eye protection is not practical.

Effects on Health

10.390 The two different types of radiation can penetrate the human body and affect it in a number of ways.

Ionising radiation

10.391 Despite having many practical uses in applications such as medicine, research and industry, ionising radiation presents a serious health hazard if used improperly. It can interact with living cells and may have sufficient energy to damage these and cause irreversible changes in biological organisms.

10.392 The degree of harm caused by ionising radiation depends on the type of radiation, the dose and dose rate, and the part of the body that is irradiated. However, there is no safe level of radiation exposure, so exposures should be kept to as low a level as practicably possible. These exposures might be:

- acute
- a short exposure to a high dose
- chronic where exposure is prolonged.

10.393 Symptoms can develop quickly depending on the initial exposure which can also determine their severity. Radiation exposure is cumulative and can cause skin reddening, radiation sickness, cancer and death. Its effects are determined by:

- the amount of radiation absorbed
- the type of radiation
- the route of exposure
- exposure duration.

10.394 Inadvertent ingestion or inhalation of a radioactive substance or contamination can be particularly serious as it could be placed close to an internal organ or tissue where its destructive effect could be concentrated.

10.395 Some effects do not present themselves until many years after initial exposure, therefore risking a mutating effect on future generations. Young children and foetuses are particularly susceptible to ionising radiation so special precautions should be taken by expectant or breast-feeding mothers, who may be exposed to ionising radiation at their workplace.

Non-ionising Radiation

10.396 Non-ionising radiation generally does not have sufficient energy to dissociate biological tissue but can still present a serious health hazard. The many types of non-ionising radiation can have various effects on the human body, including the following.

- Ultraviolet radiation can burn the skin causing skin thickening and eventually skin cancer. It can also affect the eyes, causing "arc-eye", blindness and cataracts.
- Infrared radiation can burn the skin and the thermal effects can cause dehydration and heat stroke.
- Microwave and radiowave radiation can induce heating in body tissue which causes serious burning.
- Static and low-frequency electric and magnetic fields can induce currents in the body which, if sufficiently large, can cause muscle spasm and nausea.
- Visible light is not usually a problem, although laser light or high intensity lights can cause serious eye damage.

Risks Specific to Construction Workers

10.397 The danger of construction workers being exposed to radiation is considerably high. Activities in the construction industry which increase the risk of exposure include the following.

- Radiographic inspection or the non-destructive testing of products, operating plant, engineered structures or pipelines, etc. This requires the use of intense radiation sources which can expose workers to large amounts of ionising radiation.
- Work in confined spaces, particularly underground — such as basements, caves, mines and ground-floor buildings — can contain significant amounts of radon gas (now the second biggest cause of lung cancer next to smoking in the UK) which have seeped in from the ground. Survey and exploration work may involve the use of radioactive sources.
- Work at nuclear installations involving construction or decommissioning at sites where there may be risk of exposure to radiation and radioactive contamination.
- Work with fire detectors. Some types of sensor contain a radioactive source which could, if the device was damaged, expose workers to ionising radiation.
- Outdoor work through which employees could be exposed to the risk of skin damage by ultra violet radiation from the sun. Even mild reddening of the skin is a sign of damage, which could lead to thickening of the skin and skin cancer.
- Welding work in which the welder could be exposed to significant magnetic fields generated by the welding current. Arc welding in particular produces ultra violet radiation which can burn the eyes and skin; chronic exposure can result in cataracts and skin cancer. Exposure is also a risk to assistants and bystanders if adequate precautions are not taken.
- Roofing and work on other tall structures may expose employees to radiation from communications antennae. For example, mobile phone antennae are sometimes sited in unlikely places and can be disguised as flagpoles so proximity to them may not be immediately obvious.
- Survey and installation alignment work could expose workers to laser light.
- Demolition work employing explosives near to radio transmitters, radar installations or power lines may expose workers to the risk of uncontrolled detonation. This is caused by currents induced in the firing circuits by electro magnetic radiation from the installations.

Legislative requirements

10.398 General duty sections of the Health and Safety at Work, etc Act 1974 apply to all forms of radiation with the potential for harm.

10.399 Specifically, the Ionising Radiation Regulations 1999 are the main control measure together with specific duties in the Management of Health and Safety at Work Regulations 1999. Two European Directives — Physical Agents (EMF) EC/40/2004 and Physical Agents (Artificial Optical Radiation) EC/25/2006 — also exist but have yet to be implemented in UK legislation.

10.400 There is no specific legislation governing work with non-ionising radiation, although various organisations do produce exposure guidelines. Two of these are the International Commission for Non-Ionising Radiation Protection and the Health Protection Agency.

Training

10.401 Every employer must ensure that those employees who are engaged in work with radiation are given appropriate training in the field of radiation protection. There should be provision of suitable and sufficient information and instruction about:

- the risks to health and safety created by exposure to radiation
- the precautions to be taken
- the importance of complying with legislation.

10.402 Persons other than employees who may be exposed to radiation should be provided with adequate information to ensure their health and safety so far as is reasonably practicable.

List of Relevant Legislation

- High Activity Sealed Radioactive Sources and Orphan Source Regulations 2005
- Ionising Radiation (Medical Exposure) Regulations 2000
- Ionising Radiations Regulations 1999
- Management of Health and Safety at Work Regulations 1999
- Provision and Use of Work Equipment Regulations 1998
- Health and Safety (Display Screen Equipment) Regulations 1992
- Personal Protective Equipment at Work Regulations 1992
- Electricity at Work Regulations 1989
- Radioactive Substances Act 1993
- Health and Safety at Work, etc Act 1974

Further Information

Publications

HSE Publications

The following are available from *www.hsebooks.co.uk*.

- HSG94 (rev 1998) *Safety in the Design and Use of Gamma and Electron Irradiation Facilities*
- HSG95 *The Radiation Safety of Lasers Used for Display Purposes*
- HSG122 (rev 2002) *New and Expectant Mothers at Work: A Guide for Employers*
- INDG209(L) *Controlling Health Risks from the Use of UV Tanning Equipment*
- INDG224(L) *Controlling Radiation Safety of Display Laser Installations*
- INDG337 *Sun Protection Advice for Employers of Outdoor Workers*
- IRIS1 *Industrial Radiography*
- IRIS2 *Radiation Doses — Assessment and Recording*
- L121 *Work with Ionising Radiation: Ionising Radiation Regulations 1999. Approved Code of Practice*
- PM77 *Fitness of Equipment Used for Medical Exposure to Ionising Radiation*

Other Publications

- *1990 Recommendations of the International Commission on Radiological Protection*, International Commission on Radiological Protection, publication no. 60

Organisations

- Health Protection Agency
  Web: *www.hpa.org.uk*
  The Health Protection Agency is a national organisation for England and Wales dedicated to protecting people's health and reducing the impact of infectious diseases, chemical hazards, radiation hazards and poisons.
- International Commission for Non-Ionising Radiation Protection (ICNIRP)
  Web: *www.icnirp.de*
  The ICNIRP is a body of independent scientific and consulting experts in the field of epidemiology, biology, dosimetry and optical radiation. The commission addresses the issues of possible adverse effects on human health of exposure to non-ionising radiation, and offers advice and information on the hazards of radiation.
- National Radiological Protection Board (NRPB)
  Web: *www.nrpb.org*
  The NRPB, members of which are appointed by health ministers, exists to advance the acquisition of knowledge about the protection of mankind from hazards from radiation and to provide that knowledge to government departments and other bodies and persons in the UK with relevant responsibilities.

Whole-body Vibration

- Regular, long-term exposure to vibration can lead to ill health, eg back pain caused by driving vehicles along unmade roads.
- The Control of Vibration at Work Regulations 2005 aim to protect persons against risks to their health and safety arising from exposure to vibration at work.
- These regulations also set exposure limit values and action values and require the employer to reduce exposure to as low as reasonably practicable.
- Employers are required to carry out a risk assessment to identify persons who may be at risk of exposure to vibration at work.
- Employers must take action to eliminate or control exposure to vibration at work.
- Health surveillance must be provided to employees liable to be exposed to vibration at work.
- Suitable information, instruction and training must be provided to employees and their representatives.

10.403 Exposure to vibration, eg contact with a "shaking" object, is a common occupational hazard which, if left uncontrolled, can cause:

- physical discomfort
- a reduction in productivity
- adverse health effects.

10.404 The two transmission routes for vibration exposure are hand-arm vibration and whole-body vibration.

Employers' Duties

10.405

- Under the Control of Vibration at Work Regulations 2005, employers are required to:
  - protect employers and other persons who may be affected by work carried out that may expose them to vibration
  - assess the risks to health created by vibration at work
  - eliminate or control exposure to vibration at work
  - provide health surveillance where necessary
  - provide information, instruction and training. It should be noted that the self-employed must also meet the above requirements except in relation to the provision of health surveillance.
- Under the Management of Health and Safety at Work Regulations 1999, employers are:
  - required to protect young persons and should not employ such a person where there is a risk to their health from vibration
  - ensure that when assessing the risks to new or expectant mothers, the risks from vibration are taken into account.

Employees' Duties

10.406 Under the Health and Safety at Work, etc Act 1974, employees have a duty to take reasonable care of their own health and safety and that of other people who may be affected by their work.

10.407 Under the Management of Health and Safety at Work Regulations 1999, employees have a duty to:
- use any machinery, equipment, transport, safety devices and means of production in accordance with any training and instructions provided by the employer
- inform the employer of any serious and imminent dangers to health and safety
- inform the employer of any shortcomings in the employer's health and safety arrangements.

In Practice

10.408 Occupational exposure to vibration arises in a number of ways, and if left uncontrolled can cause discomfort, a reduction in productivity and adverse health effects. Exposure to whole-body vibration via shaking or jolting of the human body through a supporting surface (usually a seat or the floor) can result in long-term back pain. One of the aims of the Control of Vibration at Work Regulations 2005 is to protect employees and other persons who may be exposed to risk from exposure to whole-body vibration at work.

10.409 The regulations came into force on 6 July 2005 with some exceptions. For work equipment provided prior to 6 July 2007 (that does not permit compliance with the exposure limit values contained within the regulations), employers do not have to ensure that an employee's exposure is below the limit value until 6 July 2010. However, employers will still be expected to take other measures to reduce exposure to as low as reasonably practicable. For the agricultural and forestry sections this date is 6 July 2014.

Exposure Limit Values and Action Values

10.410 The Control of Vibration at Work Regulations 2005 are based around the risk assessment process along with limit and action values. Exposure action values (EAVs) are daily exposure levels, which if reached or exceeded require action to be taken to reduce the risk. Exposure limit values (ELVs) are daily exposure levels that (with some exceptions) must not be exceeded. "Daily exposure" is defined as "the quantity of mechanical vibration to which a worker is exposed during a working day", which is normalised to an eight-hour reference period.

10.411 For whole-body vibration:
- the daily ELV is 1.15 m/s^2 A(8)
- the daily EAV is 0.5 m/s^2 A(8).

10.412 Where it is not reasonably practicable to eliminate risks at source and an EAV is likely to be reached or exceeded, the employer is required to reduce exposure to as low a level as is reasonably practicable by establishing and implementing a programme of organisational and technical measures which is appropriate to the activity.

10.413 The employer must also ensure that his or her employees are not exposed to vibration above an exposure limit value. Where an ELV is exceeded, the employer must:

- reduce exposure to vibration to below the limit value
- identify the reason for that limit being exceeded
- modify the measures taken in accordance with paragraphs (1) and (2) to prevent it being exceeded again.

10.414 The daily ELV for whole-body vibration is $1.15m/s^2$ *A*(8).

Identifying and Assessing Hazardous Operations

10.415 Any employer who carries out work, which is liable to expose any of his or her employees to risk from whole-body vibration is required to make a suitable and sufficient assessment of the risk created by that work to the health and safety of those employees.

10.416 In conducting the risk assessment, the employer must assess daily exposure to vibration by means of:

- observation of specific working practices
- reference to relevant information on the probable magnitude of the vibration corresponding to the equipment used in the particular working conditions
- measurement of the magnitude of vibration to which his or her employees are liable to be exposed.

Identification

10.417 Any activities and tools that may be hazardous should be identified and listed. These could include:

- poor design of controls, making it difficult for a driver to operate the machine or vehicle easily or to see properly without twisting or stretching
- incorrect adjustment by the driver of the seat position and hand and foot controls, so that it is necessary to continually twist, bend, lean and stretch to operate the machine
- sitting for long periods without being able to change position
- poor driver posture
- repeated manual handling and lifting of loads
- excessive exposure to whole-body vibration, particularly to shocks and jolts
- repeatedly jumping down from a height.

10.418 Once completed, procedures should be put in place to ensure that this list is kept up-to-date.

Risk Assessment

10.419 An employer who carries out work that is liable to expose employees to risk from vibration is required to make a suitable and sufficient assessment of the risk created by that work. All assessments must be recorded and reviewed on a regular basis. The risk assessment should take into consideration:

- the magnitude, type and duration of exposure, including any exposure to intermittent vibration or repeated shocks
- the effects of exposure to vibration
- any effects of vibration on the workplace and work equipment
- any information provided by the manufacturers of work equipment
- the availability of replacement equipment designed to reduce exposure to vibration
- any extension of exposure at the workplace to whole-body vibration beyond normal working hours
- specific working conditions such as low temperatures
- appropriate information obtained from health surveillance including, where possible, published information.

Control Measures for Whole-body Vibration

10.420 Employers are required to ensure that risk from whole-body exposure to vibration is either eliminated at source or, where this is not reasonably practicable, reduced to as low a level as is reasonably practicable.

10.421 Where it is not reasonably practicable to eliminate risk at source and an EAV is likely to be reached or exceeded, the employer must reduce exposure to as low a level as is reasonably practicable by establishing and implementing a programme of organisational and technical measures which is appropriate to the activity. Measures to take include:

- adjusting the driver weight setting on suspension seats, where it is available, to minimise vibration and to avoid the seat suspension "bottoming out" when travelling over rough ground
- adjusting the seat position and controls correctly, where adjustable, to provide good lines of sight, adequate support and ease of reach for foot and hand controls
- adjusting the vehicle speed to suit the ground conditions to avoid excessive bumping and jolting
- steering, braking, accelerating, shifting gears and operating attached equipment, such as excavator buckets, smoothly
- following worksite routes to avoid travelling over rough, uneven or poor surfaces
- choosing machinery suitable for the job, ie selecting vehicles and machines with the appropriate size, power and capacity for the work and the ground conditions
- ensuring paved surfaces or site roadways are well maintained, eg potholes are filled in, ridges are levelled, and rubble is removed
- maintaining vehicle suspension systems correctly
- replacing solid tyres on machines such as fork-lift trucks before they reach their wear limits
- obtaining advice from manufacturers when replacing vehicle seats
- consulting a trade association for advice.

10.422 The employer must also ensure that his or her employees are not exposed to vibration above an ELV. Where an ELV is exceeded, the employer must:

- reduce exposure to vibration to below the limit value
- identify the reason for that limit being exceeded
- modify the measures taken to prevent it being exceeded again.

10.423 An additional benefit of controlling vibration is the reduction of the noise it produces. It follows that many noise control techniques involve the control of vibration.

Action by the Manufacturer at the Design Stage

10.424 Manufacturers of vehicles and machines, other than agricultural tractors and road vehicles, are required to design and construct them so that whole-body vibration is reduced to the minimum level that can be achieved.

10.425 Manufacturers are also required to provide technical handbooks giving information on:

- the safe use of the machine in its intended application
- vibration emissions
- any maintenance procedures to maintain the performance of any vibration-reduction features
- whether there is any remaining risk from vibration
- how to use the equipment to avoid risk from whole-body vibration.

Health Surveillance

10.426 If the risk assessment indicates that there is a risk to the health of employees who are, or are liable to be, exposed to vibration (or exposed to vibration above an action value), employers must ensure that these employees are placed under suitable health surveillance.

10.427 Health surveillance intended to prevent or diagnose any health effect linked with exposure to vibration is deemed to be appropriate where the exposure of the employee to vibration is such that:

- a link can be established between that exposure and an identifiable disease or adverse health effect
- it is probable that the disease or effect may occur under the particular conditions of the employee's work
- there are valid techniques for detecting the disease or effect.

10.428 Suitable records must be maintained which the employee and enforcing authority can have access to.

10.429 If, as a result of health surveillance, an employee is found to have an identifiable disease or adverse health effect which is considered by a doctor or other occupational health professional to be the result of exposure to vibration, the employer must:

- ensure that a suitably qualified person informs the employee accordingly and provides them with information and advice regarding further health surveillance, including any which should be undergone following the end of the exposure
- review the risk assessment
- review any measure taken to eliminate or reduce exposure to vibration, taking into account any advice given by a doctor or occupational health professional or by the enforcing authority
- consider assigning the employee to alternative work where there is no risk from further exposure to vibration, taking into account any advice given by a doctor or occupational health professional
- provide for a review of the health of any other employee who has been similarly exposed, including a medical examination where such an examination is recommended by a doctor or occupational health professional or by the enforcing authority.

Training

10.430 Operators, supervisors and managers should be trained and educated to the appropriate level. Training relating to vibration is very important as personal behaviour can have a substantial impact on the risk to individuals from a given vibration dose. The three main areas that need to be covered are as follows.

- Management briefings and assessment competency training. Employers should establish and update the company policy and ensure that managers understand their responsibilities under the programme. In addition, in many cases it will also be necessary to train some staff up to full risk assessment competency in order to manage the process or to carry out risk assessments (live or "virtual" measurements).
- Supervisor briefings. As above but these should be cut down and slanted towards practical implementation and procedures.
- Operator toolbox talks. These should be short, memorable, educational and motivational for users of vibrating equipment.

10.431 The Control of Vibration at Work Regulations 2005 make specific requirements in relation to training. If the risk assessment indicates that there is a risk to the health of employees who are, or who are liable to be, exposed to vibration (including vibration at or above an action value), the employer is required to provide those employees and their representatives with suitable and sufficient information, instruction and training.

10.432 The information, instruction and training provided under that paragraph should include:

- the organisational and technical measures taken in order to eliminate or reduce exposure
- information in respect of the ELVs and EAVs
- the significant findings of the risk assessment, including any measurements taken, with an explanation of those findings
- why and how to detect and report signs of injury
- the entitlement to appropriate health surveillance and its purposes
- the safe working practices to minimise exposure to vibration
- the collective results of any health surveillance undertaken in a form calculated to prevent those results from being identified as relating to a particular person.

List of Relevant Legislation

- Control of Vibration at Work Regulations 2005
- Control of Vibration at Work Regulations (Northern Ireland) 2005
- Control of Noise at Work Regulations 2005
- Management of Health and Safety at Work Regulations 1999
- Provision and Use of Work Equipment Regulations 1998
- Reporting of Injuries, Diseases and Dangerous Occurrences Regulations 1995
- Personal Protective Equipment at Work Regulations 1992
- Supply of Machinery (Safety) Regulations 1992

- Social Security (Industrial Injuries) (Prescribed Diseases) Regulations 1985
- Health and Safety at Work, etc Act 1974

Further Information

Publications

HSE Publications

The following are available from *www.hsebooks.co.uk*.

- INDG242 (rev 1) *Control Back-pain Risks from Whole-body Vibration — Advice for Employers on the Control of Vibration at Work Regulations 2005*
- INDG296 (rev 1) *Hand–arm Vibration: Advice for Employees*
- L140 *Hand-arm Vibration. The Control of Vibration at Work Regulations 2005. Guidance on Regulations*
- L141 *Whole-body Vibration. The Control of Vibration at Work Regulations 2005. Guidance on Regulations*

British Standards

The following are available from *www.bsi-global.com*.

- BS EN ISO 5349–1: 2001 *Mechanical Vibration — Measurement and Evaluation of Human Exposure to Hand Transmitted Vibration. General Requirements*
- BS EN ISO 5349–2: 2002 *Mechanical Vibration — Measurement and Evaluation of Human Exposure to Hand-Transmitted Vibration. Practical Guidance for Measurement at the Workplace*

Organisations

- British Standards Institution (BSI)
  Web: *www.bsi-global.com*
  Founded in 1901, the BSI develops and provides supporting information on national and international standards in a wide range of areas.
- Health and Safety Executive (HSE)
  Web: *www.hse.gov.uk*
  The HSE is responsible for the regulation of almost all the risks to health and safety arising from work activity in the UK.
- Industrial Noise and Vibration Centre Limited (INVC)
  Web: *www.invc.co.uk*
  The INVC specialises in best practice and innovation in all aspects of noise and vibration, from assessment to control and training to specialist instrumentation.

Chapter 11

Protective and Safety Equipment

Fall-arrest and Fall-restraint Systems

- Falls from height are the greatest cause of death and serious injury in construction and there are legal requirements for controlling the risk of falls.
- Personal means of controlling the risk of falls are to prevent the fall or to mitigate the effects by using fall-arrest or fall-restraint systems.
- Selection, condition, poor fixing, swinging after a fall, inconvenient to use and lack of suitable rescue are the main hazards when using equipment.
- It is particularly important that there is suitable and sufficient use of the fall-protection equipment.
- Fall-protection equipment is more likely to be ready for use if it has been suitably checked.
- The Work at Height Regulations 2005 specify the need for easy and timely evacuation and rescue in an emergency.
- Proper use of fall-protection equipment is more likely if there is a suitable means of ensuring that supervisors and users have been suitably trained.

11.1 In 2005/6, 46 fatal injuries were caused by falls from a height, the highest cause of death and serious injury in the construction sector. In total, 2202 of major injuries were also caused by falls from height, with 2202 of these incidents due to a fall from below just 2 metres. Figures from 2004/05 also show that 39% of workplace deaths and 28% of major injuries were caused by such falls. These figures are for reported incidents and do not take into account the known high levels of underreporting.

11.2 The Construction (Design and Management) Regulations 2007 (CDM 2007) require that, so far as is reasonably practicable, all work sites must be safe for the people working there. In this context this relates to the need to protect people from the very foreseeable risk of falls from height.

Employers' Duties

11.3 Under the Health and Safety at Work, etc Act 1974, employers have a general duty to ensure, so far as is reasonably practicable, the health, safety and welfare at work of all employees. This includes ensuring the health and safety of those employees using any means of fall-arrest system as part of their work.

11.4 The Management of Health and Safety at Work Regulations 1999 require a risk assessment to be carried out to identify the nature and level of risks associated with working at height, including assessing the risk if fall-arrest systems are used and putting in place the appropriate precautions to control those risks.

11.5 The Work at Height Regulations 2005 specify the need for easy and timely evacuation and rescue in an emergency, including putting in place any appropriate measures to ensure a safe rescue in an emergency.

11.6 Further, under the Work at Height Regulations 2005, users of personal fall protection (from lanyards to rope access) are required to be trained in rescue techniques and to plan work with the consequences of a fall in mind. For example, employees need to know how and when orthostatic shock suspension trauma can happen, what to do to prevent it and how to treat it.

Employees' Duties

11.7 Under the Health and Safety at Work, etc Act 1974, employees have a general duty to take reasonable care of their own health and safety and that of other people who may be affected by their work. They also have a duty to co-operate with their employers on health and safety matters.

11.8 Under the Work at Height Regulations 2005, employees must report to their employer, or person in control of the work, any activity or defect relating to work at height that may endanger the safety of themselves or others. They must also use any equipment provided for their use in accordance with any training and/or instructions relating to its use, so as to enable their employer, or person in control of the work, to comply with their own health and safety duties.

In Practice

The Requirement to Control Falls

11.9 To expand on the requirements for controlling the risk of falls it is necessary to refer to the Work at Height Regulations 2005.

11.10 Following the general protocol in the Work at Height Regulations 2005 the aim is to avoid work at height were possible, or to control the risk by the use of safe means of access and work. Where this is not reasonably practicable, suitable work equipment must be provided to control the risks of falls when working at height. Where this is not reasonably practicable the distance of the fall must be restricted to reduce danger or means must be taken to minimise the consequences of the fall.

11.11 If personal fall-protection systems are considered as a means of controlling falls, it is a requirement of the Work at Height Regulations 2005 that:

- preferred ways have been considered and ruled out as not reasonably practicable
- the personal fall-protection system is suitable for the task, properly fits the user, is being worn and is properly and securely anchored
- the user is sufficiently trained in its correct use.

Fall-arrest or Fall-restraint

Personal Fall-protection System

11.12 In general, the personal fall-protection system consists of a harness worn by the user and a means to attach that harness to a suitable anchoring system. The harness should be suitable to hold the user and support them in an upright position if used. There are different types, but the type most likely to be suitable is a full body harness with thigh and shoulder straps, a central body belt and attachment rings at the front or rear or (preferably) both. The harness is connected to the anchorage point by some sort of textile webbing or rope system. These also come in different types, mainly a fixed length, known as a lanyard or "strop", which in some cases has a shock absorber system within its length, or a cable on a self-retracting "reel" that allows movement, but stops if there is sudden movement, ie a fall.

Fall-restraint System

11.13 A fall-restraint system, also known as a work-restraint system, aims to restrict the wearer to prevent a fall occurring. It will do this by ensuring that, at full extent, the wearer cannot go past the edge of the safe platform or area in which they are working. The advantage of this system is that the user is never in any significant danger if it is used properly and thus the risk of trauma as a result of a fall is completely prevented. However, this system does have two main disadvantages.

- In use, it is constantly under tension. People using it will quite often be at the limit of the extent of the lanyard, and so the lanyard and harness will be under more strain than fall-arrest systems, and so could fail earlier than expected.
- It does restrict the wearer — as it is intended to do. Thus it is quite possible that a person using the system may ill-advisedly disconnect the lanyard temporarily to reach something just beyond its maximum extent, thus completely negating its risk control ability.

Fall-arrest System

11.14 A fall-arrest system aims to minimise the consequences of the fall by restricting the distance the person can fall or by limiting the speed of the fall. In both cases the aim is that the person either does not hit something hard, or does not hit it hard enough to cause significant injury. The main advantage of a fall-arrest system is that, in normal use, it does not really restrict the wearer in any significant way and thus is more likely to be used correctly.

Hazards of Use of the Equipment

Causes of Danger from the Use of the Equipment

11.15 In general, the most likely causes of danger from the use of fall-protection equipment will include:

- incorrect selection of fall-protection equipment
- poor condition of fall-protection equipment
- inadequate fixing or anchorage points
- swinging at the end of the lanyard and striking something
- the inconvenience of the system possibly causing it not to be used
- inadequate means of rescue and suspension trauma.

Inadequate Fixing Points

11.16 Where a fixed anchorage point is not readily available or useable, quite often a railing or balustrade is used by clipping a lanyard back on itself after making a loop around a baluster or stanchion (the upright). In some cases this may be suitable; however, usually this can introduce a tight curve in the lanyard or possible sharp edges could damage it. A more serious problem is the strength of the railing or balustrade. It will have been designed to resist a reasonable load applied against it — to prevent a person falling over the edge.

11.17 The best solution to this is to install suitable numbers of fixed anchor points — usually eyebolts — in suitably strong parts of the fabric of the building in positions that are appropriate for the work at hand. Where the installation of fixed anchor points is not reasonably practicable, then some form of suitable temporary anchorage must be used.

11.18 If the work is on or near a roof, there are mobile anchor devices, that consist of heavy weights spread wide on "legs" to give a device that will not move when "jerk" loaded, but can be installed on a flat roof and dismantled and moved easily and without damage to the fabric of the building. Alternatively, a suitable structural member can be identified, such as a girder, ensuring that it can take the loading that it is generally intended to along with the loading caused by the use of the work-at-height risk control in operation. It can also be used without excessive lengths of lanyard or the lanyard pulling "round a corner" because a straight path is not available.

11.19 To this suitable structural member a properly protected lanyard can be affixed, possibly by winding around it more than once, leaving the ends available for connection to the working lanyard. As mentioned, care must be taken to ensure that no sharp edges or excessively tight turns compromise the strength of this temporary fixing point.

Inconvenience of the System

11.20 Many items of equipment used to protect a person when working at height are not entirely comfortable or convenient to use. As a result, people will commence work using them but will soon tire of fighting against them and dispense with them altogether with the consequential unacceptable increase in risk.

11.21 The safety harness can cause discomfort when it stops a fall and if the thigh straps are insufficiently tight. To function correctly, the harness must be suitably fitted and worn properly. If the harness is properly tight it can be uncomfortable when worn over a full shift period, which may result in it being slackened off to achieve greater general working comfort, but this will compromise its function if used in a fall.

Inadequate Means of Rescue

11.22 In general, the means of rescue must be easy to use correctly and readily available — in some cases there will be less than 20 minutes to rescue an unconscious fallen person if they are suffering from suspension trauma.

11.23 The main considerations when undertaking a rescue is that the fallen person may well be injured or unconscious, is almost certainly frightened and may be panicking. Also, depending on how the personal fall-protection system functions, the person to be rescued may be in a place that is difficult to get to. In addition, if the person being rescued is actually unconscious, it is foreseeable that they may be in a position that blocks the airway, and thus may need basic resuscitation before lowering, and possibly other first-aid if they are injured.

11.24 For these reasons suspending the rescuer(s) on a rope is unlikely to be a suitable means of rescue. Also the means of rescue must be capable of being deployed with return for the next person in a reasonable time so that the last person to be rescued will not be unduly put at risk.

Suspension Trauma

11.25 Orthostatic shock suspension trauma is the term given to the body's inability to cope with prolonged periods of immobility in a vertical position. For this reason, construction workers using suspension harnesses are at particular risk from this condition.

11.26 The main cause of the trauma is initially due to the vertical suspension of the person, causing a pooling of blood in the legs. In normal activities, even when sitting at a desk, the movement of the blood in the lower limbs is helped by the valves in the veins and the pumping action of the feet and calf muscles. Their movement pumps the blood back towards the heart, where the pressure in that part of the circulatory system enables it to return normally to the heart for another cycle.

11.27 However, if the person is not, or cannot, move their lower legs, the blood is restricted in its ability to return to the rest of the circulatory system and starts to collect or "pool" in the legs and feet. This collection of blood in the lower limbs restricts the amount of blood in the rest of the system and in particular restricts the amount of blood in the brain. When this happens, the person usually faints, and the consequent horizontal position then allows gravity to assist the blood back up from the legs and eventually provides more blood for the brain.

11.28 Unfortunately, when the body is suspended upright, the body's natural remedy — causing the body to faint — is no longer effective. In fact, fainting while suspended can worsen the situation by blocking the airway, making death likely in less than 5 minutes, rather than 20. Medical help should be called for immediately, to handle the situation.

Further Complications

11.29 Once the person is rescued, there is a further complication if blood has become trapped in a limb. The blood pooled in the legs loses oxygen and the nutrients it carries around the body. Instead, it becomes filled with waste products from metabolism and also toxic products from injuries and other cellular activity. As a result it is likely to be cold, have no oxygen and be highly toxic.

11.30 If the person is laid flat on rescue, the toxic blood suddenly hits the central circulatory system, where the cold blood exacerbates the shock to dangerous levels.

11.31 An additional side effect of the static nature of the blood pooling in the lower limbs is also likely to cause scattered clotting of the blood potentially giving rise to deep vein thrombosis (DVT). Competently trained personnel must handle the rescue situation.

Correct Use of a Harness

11.32 Where the use of a harness is the only reasonably practicable means to control the risks of working at height, the person responsible must carefully consider:

- the suitability and fit of the harness
- the condition of the person
- the position of the suspension point on the harness
- means for the person in the harness to raise their legs
- encouraging a person suspended in a harness to "pump" their legs and feet regularly.

11.33 It is vital that prior to any work at height where people use harnesses, there is a suitable and immediately available means to rescue them if they fall and are suspended in the harness. This must be capable of rescuing them within 10 to 15 minutes from the time of the fall and preferably within 5 minutes.

11.34 If the comfort and proper fit of the harness is not properly considered, especially in heavier wearers, the pain caused by straps that are too narrow for the size of the wearer or by incorrectly adjusted thigh straps cutting into the legs or the genitalia, can increase the likelihood of fainting or suspension trauma as a result of the increased shock.

11.35 The first control therefore is to ensure that the harness is suitable for the wearer, and that they are properly trained in its correct adjustment.

11.36 To ensure that the bodily functions of the person that may need to be rescued are fit to cope, no one should undertake work at height — where the control of risk is by a harness in which they could be suspended — if they are fatigued, dehydrated, hypothermic or low on energy.

11.37 It is also advisable to consider the position of the suspension point of the harness. Where the suspension point is at the rear of the harness, the head will hang forward possibly causing a blocked airway; if positioned at the front of the harness, the head will be allowed to fall back, opening the airway. Similarly the lower the suspension point, the more horizontal the person will hang and thus reduce the likelihood of pooling and suspension trauma.

11.38 If the person, while hanging and awaiting rescue, can raise their legs to achieve a more horizontal position, the pooling is again reduced, and thus the likelihood of suspension trauma. This is assisted by loops which, although unused normally, can be used to alter the hanging position of a conscious person, which will extend the period before they are likely to faint, and the situation becomes serious. Similarly if the person can use a foot loop to "stand on" and exercise their legs by "pumping" them, they can increase the blood movement and thus reduce the likelihood of suspension trauma.

Correct Use of the Fall Protection Equipment

11.39 The Work at Height Regulations 2005 require that a fall-arrest system must suitably absorb the energy of the fall and only be used where the lanyard cannot be cut. There must also be sufficient space available so that the person falling does not directly strike a solid object.

11.40 Before each use the equipment should be given a visual check for wear, paying particular attention to lanyard for cuts and wear, and that any shock absorber system or reel is not damaged and is likely to function correctly.

11.41 After use the equipment should be cleaned (if necessary) in accordance with the manufacturer's instructions and stored out of harm's way, preferably not in direct sunlight. Many oils and solvents will significantly reduce the working life of the webbing of the equipment and should be removed, using a method approved by the manufacturer.

Checking the Equipment

11.42 A personal fall-protection system is used where there is a significant risk of falling during work operations. As a result it is important where this equipment is used to ensure that the safety of the personal fall-protection system is subject to suitable and sufficient inspection and testing by an independent competent person.

11.43 This need is reinforced by the requirements of the:

- Personal Protective Equipment at Work Regulations 1992, which fall-arrest equipment falls under
- Provision and Use of Work Equipment Regulations 1998, as fall-arrest equipment is work equipment.

11.44 The requirements for the inspection and testing of fall-arrest equipment are given in BS EN 365: 2004 *Personal Protective Equipment Against Falls from a Height.* General requirements for instructions for use, maintenance, periodic examination, repair, marking and packaging, and the method of testing must be equal to, or better than this.

11.45 In BS EN 365, the testing frequency is given as at least every 12 months; however, some textile webbing may need to be tested more frequently than this. Thus it is advised that testing should take place every 6 months, or every 3 months if equipment is used in arduous environments.

11.46 The testing must be carried out by an independent, impartial competent person, who may be an employee of the owning organisation, but they must be permitted to be objective. To prove that these tests have been carried out, suitable records must be held and shown on request. To ensure the accurate relation of the records to the equipment, each item of fall-arrest equipment must be uniquely identified in some indelible way that will not become detached or damaged, ie labels on string are not suitable.

11.47 The test regime for fall-arrest equipment should be drawn up by a competent person and include:

- a description of the equipment — including its unique identification
- the frequency and type of inspection
- the person designated to carry out the test
- the action to be taken in the event of damage or test failure
- the means of recording the inspections
- the training of the users
- the means of monitoring to verify the inspections.

11.48 Full information on the recommendations for testing fall-arrest equipment are contained in the HSE publication *Inspecting Fall Arrest Equipment.*

11.49 To ensure that the means of suspension and the lanyard, strop or rope are in sound condition, the Lifting Operations and Lifting Equipment Regulations 1998 require a thorough examination and inspection before the equipment is put into use for the first time. In particular, the regulations require thorough six-monthly examinations, or more frequent tests if the competent person checking it deems appropriate.

11.50 To ensure that the safe working loads are not exceeded, the regulations give a requirement to mark all parts of the personal fall-protection system with the safe working load, and the fact that it is intended for lifting people. In a similar way it is necessary to be sure that the fixing points in the area where the work is being carried out are actually suitable. Thus it is equally necessary that the records of the tests of the securing points — eye bolts, etc — required under the regulations are available.

Rescue

11.51 The Work at Height Regulations 2005 specify the need for easy and timely evacuation and rescue in an emergency. All rescue planning and operations should address:

- the safety of the persons carrying out or assisting with the rescue
- the anchor points to be used for the rescue equipment
- the suitability of equipment (anchors, harnesses, attachments and connectors) that has already arrested the fall of the casualty for use during the rescue
- the method that will be used to attach the casualty to the rescue system
- the direction that the casualty needs to be moved to get them to the point of safety, ie raising, lowering or lateral
- the first-aid needs the casualty may have with respect to injury or suspension trauma
- the possible needs of the casualty following the rescue.

11.52 It should be noted that treating someone suffering from orthostatic shock suspension trauma is not the same as standard first-aid because of the complications that can arise from pooling of blood.

11.53 It is vital that prior to any work at height where people use harnesses, there is a suitable and immediately available means to rescue them if they fall and are suspended in the harness.

11.54 Rescue procedures should include the following.

- Ensure that the worker receives standard trauma resuscitation once rescued. Some authorities recommend that the patient be transported with the upper body raised.
- If the worker is unconscious, keep the worker's air passages open and obtain first-aid.
- Monitor the worker after rescue, and ensure that the worker is evaluated by a healthcare professional. The worker should be hospitalised when appropriate. Possible delayed effects, such as kidney failure, which is not unusual in suspension trauma cases, are difficult to assess on the scene.

11.55 The main considerations when undertaking a rescue is that the fallen person may well be injured or unconscious, frightened and may be panicking, and, depending on the means of fall-restraint or mitigation functions, the person to be rescued may well be in a place that is difficult to get to. However, there are ways to achieve a good rescue safely and effectively.

11.56 Firstly, the means of effecting a rescue must be readily available wherever the work at height is being carried out. The means of rescue then needs to be large enough for more than one person — even if only one is intended to be rescued. If the person being rescued is injured or unconscious, it is unlikely that the rescue will be feasible by one person. In addition, if the person being rescued is unconscious, it is foreseeable that they may be in a position that blocks the airway, and may need basic resuscitation before lowering, and possibly other first-aid if they are injured.

11.57 The equipment providing the means of rescue should — if the safe system of work is good enough — not actually be used that often. Therefore, it is important to subject the equipment to a regular and rigorous system of checking and inspection. These checks should include ensuring that all the appropriate parts of the equipment are there and available, and inspections to ensure that the equipment functions as it is intended to; it should have no observable defects or faults that may cause it to fail or not operate correctly during a rescue.

Training

11.58 All users and people supervising the users of fall-arrest and restraint equipment must receive suitable instruction, information and training to ensure that equipment is used and looked after properly. This training should be given before the people are first involved in the use of the personal fall-protection system, and at regular intervals afterwards to "refresh" the information.

11.59 It is particularly useful, when explaining the checks on the equipment, to have examples of the harness and lanyard available so that the attendees of the course can see and practise these checks.

11.60 In general, the information that users will need to be provided with is likely to include:

- how the fall-protection system functions and the different types
- the correct way to use it — particularly fitting the harness and anchorage points
- the dangers caused by its misuse
- the regular checks they should be undertaking
- the correct means to care for and store it
- the means of rescue of someone that has used the fall-arrest system
- the monitoring necessary to ensure its correct use.

List of Relevant Legislation

- Construction (Design and Management) Regulations 2007
- Work at Height Regulations 2005
- Management of Health and Safety at Work Regulations 1999
- Provision and Use of Work Equipment Regulations 1998
- Lifting Operations and Lifting Equipment Regulations 1998
- Personal Protective Equipment at Work Regulations 1992
- Health and safety at Work, etc Act 1974

Further Information

Publications

HSE Publications

The following publications are available from *www.hsebooks.co.uk.*

- CRR 451/2002 *Harness Suspension: Review and Evaluation of Existing Information*
- INDG367 *Inspecting Fall Arrest Equipment Made From Webbing or Rope*

British Standards Publications

The following are available from *www.bsi-global.com.*

- BS EN 354: 2002 *Personal Protective Equipment Against Falls From a Height: Lanyards*
- BS EN 355: 2002 *Personal Protective Equipment Against Falls From a Height: Energy Absorbers*
- BS EN 358: 2000 *Personal Protective Equipment for Work Positioning and Prevention of Falls From a Height: Belts for Work Positioning and Restraint and Work Positioning Lanyards*
- BS EN 361: 2002 *Personal Protective Equipment Against Falls From a Height: Full Body Harnesses*
- BS EN 365: 2004 *Personal Protective Equipment Against Falls from a Height. General Requirements for Instructions for Use, Maintenance, Periodic Examination, Repair, Marking and Packaging*

Organisations

- Construction Industry Training Board (CITB) and Construction Skills
  Web: *www.citb.org.uk*
  The CITB and Construction Skills provide assistance in all aspects of recruiting, training and qualifying the construction workforce.
- Fall Arrest Safety Equipment Training (FASET)
  Web: *www.faset.org.uk*
  FASET is a trade association and a training body in the fall-arrest industry.
- International Powered Access Federation (IPAF)
  Web: *www.ipaf.org*
  The IPAF works to promote the combination of the capabilities of modern platforms and training for safe and effective work at height.
- National Access and Scaffolding Confederation Ltd (NASC)
  Web: *www.nasc.org.uk*
  The National Access and Scaffolding Confederation is the national representative employers' organisation for the access and scaffolding industry.
- Prefabricated Access Suppliers' and Manufacturers' Association
  Web: *www.pasma.co.uk*
  The association provides training and strives to advance safety standards in the UK access industry.

Personal Protective Equipment: Principles of Use

- The Personal Protective Equipment at Work Regulations 1992 require employers to provide suitable personal protective equipment (PPE) to employees if it is necessary.
- PPE must be regarded as a last resort and only used when risks cannot be adequately controlled by more effective measures.
- Employers must choose PPE which is suitable by carrying out an assessment.
- To be suitable, PPE must:
  - be appropriate for the risks and working conditions
  - fit the wearers
  - provide adequate protection
  - meet the relevant standards.
- If more than one item of PPE is worn, the different types must be compatible.
- PPE must be maintained properly and appropriate storage must be provided.
- Employees must be provided with suitable information, instruction and training so they can use the PPE effectively.
- Employers must not charge employees for PPE.
- All types of PPE are covered by the Personal Protective Equipment at Work Regulations 1992 apart from those covered by other specific legislation, such as:
  - hearing protection (covered by the Control of Noise at Work Regulations 2005)
  - hazardous substances (covered by the Control of Substances Hazardous to Health Regulations 2002).
- Types of PPE include:
  - head protection
  - foot protection
  - hand and arm protection
  - protective clothing for the body
  - eye and face protection

11.61 Personal protective equipment (PPE) is all equipment worn or held by people at work to protect them against one or more risks to their health or safety. It includes:

- protective clothing such as:
  - aprons
  - weather protection
  - gloves
  - footwear
  - helmets
  - high-visibility wear
- protective equipment such as:
  - eye and face protectors
  - life jackets
  - respirators
  - breathing apparatus
  - safety harnesses.

11.62 It does not include:

- corporate uniforms
- ordinary working clothes
- clothing worn primarily for food hygiene.

11.63 The Personal Protective Equipment at Work Regulations 1992 place duties on employers and employees. They apply to all PPE, except hearing protection, most respiratory protective equipment and some other types of PPE. These are excluded from the Personal Protective Equipment at Work Regulations 1992 because they are covered by other regulations, for instance:

- hearing protection is covered by the Control of Noise at Work Regulations 2005
- respirators to protect people against hazardous substances are covered by the Control of Substances Hazardous to Health 2002.

11.64 There are many different British and European Standards that apply to specific types of PPE. Manufacturers and suppliers should provide information on which standards their products meet. They should also inform organisations of which standard applies to them and is suitable for their requirements.

Employers' Duties

11.65 Under the Personal Protective Equipment at Work Regulations 1992, duties include the following.

- Employers must provide suitable PPE to employees if there is a risk to their health or safety that cannot be adequately controlled by other means. To be suitable, PPE must:
  - be appropriate for the risks involved, the conditions where the risk occurs and the time for which it is worn
  - take account of the ergonomic requirements and the state of health of each person who may wear it and the characteristics of their workstations
  - be capable of fitting the wearers correctly
  - effectively prevent or adequately control the risk(s) involved without increasing overall risk, so far as is reasonably practicable
  - be CE marked to show it conforms with the relevant standards.
- Employers must consider PPE the last resort and must use other, more effective means to control risks, such as engineering controls or safe systems of work, as far as possible.
- If more than one item of PPE needs to be worn together, the employer must ensure they are compatible and adequately control risks.
- Before choosing PPE, employers must assess whether it will be suitable. The consideration must include:
  - an assessment of the risks to health and safety that have not been avoided by other means
  - a definition of the characteristics which the PPE must have in order to effectively control these risks, taking into account any risks that the PPE may create
  - a comparison of the characteristics of the PPE available with the characteristics it needs to have
  - an assessment as to whether it will be compatible to other PPE with which it will need to be used simultaneously.
- Employers must review these assessments if they may no longer be valid, or if there have been any significant changes.

- PPE must be maintained in an efficient state, in efficient working order and in good repair. This includes cleaning or replacement, as appropriate.
- Appropriate storage facilities for PPE must be provided for when it is not in use.
- Employees must be provided with adequate, relevant information, instruction and training that includes:
  - the risks the PPE will avoid or limit
  - the purpose it is to be used for and how it is to be used
  - any actions employees need to take to make sure it is properly maintained.
- If appropriate, employers must organise demonstrations in how PPE should be worn.
- Employers must take reasonable steps to make sure employees use PPE properly.

11.66 The Health and Safety at Work, etc Act 1974 prohibits employers from charging employees for PPE.

11.67 The Employment Act 1989 gives members of the Sikh religion who wear turbans exemption from the requirement to wear head protection on construction sites, as they choose to use a lower level of protection and a provision is made to limit an employer's liability should a claim occur after injury.

Employees' Duties

11.68 Duties under the Personal Protective Equipment at Work Regulations 1992 include the following.

- Employees must use PPE provided by the employer in accordance with the training and instructions they have received.
- The PPE should be returned after use to the storage places provided, if possible.
- Employees must report any loss or obvious defect in their PPE to the employer as soon as possible.

In Practice

Identifying the Need for PPE

11.69 Employers must assess all significant risks at work to decide what needs to be done to remove or adequately control them. The risk assessment may identify that PPE is necessary because there are no more effective ways of controlling risk.

11.70 For instance, personal protective equipment (PPE) may be required as a temporary measure to prevent people bumping their heads, until the obstructions that are creating the risk are removed.

11.71 As well as identifying when employees need PPE, consideration of who else might be at risk is necessary. For instance, are there visitors, maintenance contractors or cleaners who may need to wear PPE to protect them from risks on-site? If there are, decisions must be made on how best to protect them.

11.72 Contractors, cleaners and other people working on-site must be informed about the risks they may meet and what they need to do to protect themselves. This includes any PPE that they may need. For visitors, PPE may need to be provided, eg safety helmets for people walking around a construction site.

Using the Most Effective Means of Controlling Risks

11.73 PPE is the last resort when choosing how to control risks at work. It protects only the person who is wearing it, so it is better to use controls which protect everyone. No PPE provides 100% protection and the actual level of protection provided is difficult to assess.

11.74 PPE can restrict the people wearing it by limiting their mobility and ability to see or hear properly. It only provides effective protection if it is correctly fitted, maintained and used.

11.75 For all these reasons, other ways of controlling risks should be used in preference to PPE whenever possible. Removing the need for PPE by altering how something is done is one option. Alternatively, it may be possible to find a way of protecting everyone, for example by enclosing a dusty process. However, if risks cannot be fully controlled in other ways, suitable PPE must be provided.

11.76 PPE should usually be provided on a personal basis, unless sharing does not present any hygiene problems or other health risks. This is only likely to be true in a few circumstances, eg high-visibility waistcoats, life jackets, as most PPE should not be shared.

Assessing Whether PPE will be Suitable

11.77 The assessment is to ensure the PPE provided is correct for the particular risks involved and for the circumstances of its use. This should follow an assessment which identified the need for PPE but should not duplicate it. The assessment at this stage is not to decide whether PPE is required, but to ensure it will be suitable.

11.78 The questions listed can be used to help reach a decision on whether the PPE is suitable and on which type of PPE should be used.

- What hazards do people need protection against? Chemical splashes? Moving vehicles? Wet and cold weather? Sunshine? Drowning? Slips? Hot or cold surfaces? Welding fumes and light?
- What exactly is the nature of the hazard? For instance, if gloves are chosen for construction workers, will they provide protection from sharp materials while retaining dexterity? The right gloves can only be selected if enough is known about what they need to protect people against.
- What is the nature of the job and what demands does it place on the people doing it? The people doing the job need to be involved. The following may also need to be considered.
  - How much physical effort is required?
  - How is the job done, ie what methods are used?
  - How long does the PPE need to be worn for?
  - Are there any communication requirements, audio and visual? What about in an emergency?
  - Are there any features of the area and workstations where the PPE will be used that will affect its use, eg a confined space?
  - Are there any specific requirements that the job imposes on the PPE?
- What part of the body needs to be protected? Eyes? Head? Hands? Arms? Torso? Legs? Feet? The whole body?
- Who will be using the PPE? What is the range of sizes and styles required to make sure it will fit all of them?

- Do any of the PPE users have any health conditions which could affect their ability to use the equipment? For example, people who need to wear spectacles won't be able to use all types of eye protection.
- Is there any way the PPE might increase the overall risk? For example, gloves can reduce grip strength and make it harder for people to use equipment. Safety shoes with steel toe-caps can increase the risk of someone tripping on steps.
- What other PPE does it need to be compatible with?

11.79 Once answers to these questions are gathered, PPE can be chosen and its suitability can be checked.

11.80 It will be necessary to identify possible suppliers and find out what they can provide that will match the PPE requirements. Suppliers can be found via the Internet, telephone directories and trade associations, etc.

11.81 Those selecting PPE should select a range of styles and sizes that meet their requirements, from more than one supplier if possible. All PPE selected should be CE marked.

11.82 Once a range of types of PPE possibilities has been selected, the assessment is nearly completed. The final stage is a trial period for staff to try it out in practice.

11.83 As part of the trial, a check that the PPE provides the expected protection should be made. Has it reduced the risks to an acceptable level? For example, do safety shoes provide better grip in slippery areas? Do gloves provide an impervious barrier and keep the users' hands dry and clean?

11.84 This approach should help those responsible to choose PPE that is not only effective, but which will give minimum discomfort to the wearer.

Recording and Reviewing

11.85 Unless it is simple and obvious, and can be easily repeated at any time, the assessment should be recorded. There are no set requirements about how to do this, but all records need to be readily accessible.

11.86 Information on what has been done and how PPE has been chosen should be recorded. Employers may need to demonstrate, eg to an HSE inspector or an environmental health officer, that a proper assessment has been carried out.

11.87 Risk assessments need to be kept up-to-date. To do this, it will be necessary to review both the assessments that identified PPE was required and the assessment made to choose the PPE. In particular, a review is necessary:

- if there are any significant changes in the job
- if there is any reason to suspect that it is no longer valid, eg, if an accident, ill health or near miss occurs
- at regular intervals, depending on the nature of the risks and how likely things are to change in the job.

Involving PPE Users

11.88 Those who do a job are usually best placed to know what is involved and they should always be consulted when choosing PPE.

11.89 It is necessary to consult with employees and involve them as soon as possible when selecting suitable PPE. They should have played a part in the risk assessment in which PPE requirements were identified. They should continue to participate in the assessment to choose the PPE they will have to use.

11.90 Wearers should help select possible types of PPE, as long as they only choose from PPE that can provide adequate protection.

11.91 Once a range of possible types of PPE has been chosen, the next step is for it to be tried out in practice. Employers should provide samples to staff and ask them to try them out in their work, if possible comparing different styles, sizes and suppliers, etc. As wide a range of people as possible, with different sizes, shapes and attitudes should be included.

11.92 The final choice should be based on the feedback from employees who have tried it out, including how effective they found it in controlling the risks.

11.93 If the people who will wear or use the PPE are included in the selection process, they are more likely to find the PPE acceptable.

Making Sure Different Types of PPE are Compatible

11.94 If people have to wear more than one type of PPE at the same time, the equipment must be compatible, eg, some respirators may not fit properly and protect people if they also have to wear a safety helmet.

11.95 Compatibility should be checked with the PPE suppliers if possible. Staff should try out PPE in combination with any other PPE they need to wear. Many suppliers now provide PPE that combines different types of protection in one piece of equipment.

Availability, Storage, Maintenance and Cleaning

11.96 Once PPE is selected it will be necessary to establish a system to ensure it is readily available to staff. The system should be as simple as possible to make it easy for employees to obtain new PPE if they need replacements. If necessary, it may also need to include the provision of spare parts. Once the decision is made on where new PPE is going to be stored and how people will obtain it, staff will need to be told.

11.97 When PPE is not being used, it needs to be stored safely. Suitable storage that will protect the PPE from contamination, loss or damage, eg by hazardous substances, damp or sunlight, will need to be provided. This could simply be pegs for hanging up waterproof clothing or a case for safety glasses. If the PPE becomes contaminated during use, it should be stored separately from other clothing. If the PPE can contain hazardous materials, such as asbestos, it will need special storage arrangements.

11.98 It is necessary to establish a system for properly maintaining PPE, so it continues to protect people effectively. For example, mechanical fall-arrest equipment requires a regular, planned, preventive maintenance system, including thorough examinations, testing and overhauls. Something simple like gloves may only need regular inspections by the user, depending on what they are being used to protect against.

11.99 The manufacturer or supplier of the PPE should provide information on how it should be maintained, including recommended replacement periods, shelf lives, etc. The information should be followed — something should only be done differently if it has been discussed and agreed with the supplier.

11.100 Before PPE is used, it should be examined to ensure it is in good working order. This includes both before it is used for the first time, and before each time it is put on. Staff should be trained to do this. If PPE users are trained, they can carry out simple maintenance, but more intricate repairs will need to be done by specialists. For some equipment, contract maintenance services may be needed.

11.101 The decision may be made to provide disposable PPE, thereby removing the need for maintenance procedures. If so, those responsible must ensure the users know when and how it should be discarded and replaced.

11.102 There also needs to be a system in place for staff to report losses and defects. This should include steps to ensure defective PPE is repaired or replaced before it is needed again.

11.103 For all the systems put in place to support the use of PPE, it is necessary to specify:

- who is responsible
- what they should do
- when and how.

11.104 There also needs to be a monitoring system to make sure that arrangements work satisfactorily, eg that PPE is properly used and maintained. This could be part of other health and safety monitoring procedures.

Providing Information, Instruction and Training

11.105 For PPE to be effective, a systematic approach to training is needed. This will ensure that everyone involved in the use of PPE receives the appropriate information and instruction.

11.106 Employers have to take reasonable steps to make sure that PPE is properly used. Providing training and information is part of this, however simply telling staff to wear it is not likely to be enough.

11.107 If staff do not use PPE properly, it is vital to find out why it is not being used and take steps to resolve any problems. This may include the use of disciplinary procedures where appropriate, but positive actions such as discussing the issues with staff are more likely to be successful. If staff have been involved in the selection of PPE, they are much more likely to use it properly than if they have had it imposed on them.

11.108 All managers and supervisors must set a good example by always wearing their PPE properly, even if they are only going to be in the area where it is required for a few seconds. They should also demonstrate that they expect the workforce to use it, by not ignoring anyone who is not wearing PPE properly.

11.109 Where necessary, safety signs should be used to show what PPE must be worn. These are circular white signs within a blue circle.

Types of PPE

Head Protection

Does the Activity Require Head Protection?

11.110 The many processes and activities requiring head protection include the following.

- Building works, particularly working on, beneath or near to scaffolding, elevated workplaces, erection work on or stripping formwork or falsework, work on scaffolding and demolition work.
- Construction work on bridges, buildings, towers, masts, large structures and plant within power stations.
- Work in pits, trenches, shafts, tunnels, underground workings, mineral preparation.
- Work with blast furnaces, within steelworks, rolling mills, drop forging.

- Railway shunting work, other transport activities with a risk of falling objects, driving lift trucks or working in warehouse and storage areas.
- Building or repairing ships or offshore structures.
- Slaughterhouses or other activities with hazards created by hanging objects, sharp hooks or low obstructions.
- Tree felling or tree surgery.
- Work from bosun's chairs, suspended access systems, etc.

Selecting Suitable Head Protection

11.111 First, the basic type of head protection that is suitable for the work hazards needs to be chosen. Next, its fit should be considered. Head protection should be provided in a variety of sizes to meet the needs of everyone who requires it.

11.112 Safety helmets, bump caps and climbing helmets must fit the wearer well and should:

- be an appropriate shell size for the wearer
- have an easily adjustable headband, nape and chin straps, if relevant.

11.113 The range of size adjustment also needs to allow for the use of thermal liners in cold weather. Helmets should also be as comfortable as possible. Comfort can be improved by including:

- a flexible headband wide enough and contoured both vertically and horizontally to fit the forehead
- an absorbent sweatband, that can be easily cleaned or replaced
- textile cradle straps.

Is the Head Protection Compatible with the Work to be Done?

11.114 Wearing head protection can get in the way of the work to be done. It is important to involve the people who will be using the head protection when checking that it is compatible. Problems may only be apparent when the PPE is in use in the workplace.

11.115 For example, although safety helmets usually have a peak, this can restrict upward vision, which can present a problem for scaffolders and surveyors, etc. Bulky helmets can also be a problem for people working in small spaces, for example when maintaining plant and equipment.

11.116 Head protection often needs to be worn at the same time as hearing protection and eye or face protection. Compatibility problems can be overcome by selecting PPE that combines more than one type of protection, eg safety helmets that have earmuffs or a face shield attached.

Types of Head Protection

11.117 The six main types of head protection are:

- industrial safety helmets
- bump caps
- firefighters' helmets
- transport helmets
- leisure helmets
- lightweight head coverings.

Industrial Safety Helmets

11.118 Industrial safety helmets can protect against falling objects or impact with fixed objects and are the most commonly thought of type of head protection and the one normally found on building sites.

11.119 Helmets are also available which give protection against impact at high or low temperatures, against electrical shock from brief contact up to 440v ac and against molten metal splash.

11.120 On building sites, there will be designated hard hat areas and it is mandatory to wear safety helmets in them. The only exception to this rule is turban-wearing Sikhs, who are exempt.

11.121 Safety helmets must be worn correctly to give the correct protection. Wearing them backwards reduces their ability to protect. They should always be worn the right way round. A chin strap should be worn if the job involves bending down or forward or working in windy conditions. The helmet must be worn so the brim is horizontal, not on the back of the head, as this again will reduce the protection.

11.122 Safety helmets are primarily designed to protect against falling objects or impacts rather than injuries resulting from a fall of the wearer. However, if a chin strap is being worn then some protection may be offered in the event of a fall.

11.123 Helmets must never be modified by drilling holes or being cut and the internal harness or headband must never be modified.

Bump Caps

11.124 Industrial scalp protectors or bump caps give protection against striking fixed obstacles, scalping or entanglement and are intended to protect against lower energy impacts from walking into obstructions or when working beneath them.

11.125 Fitters working beneath vehicles on lifts are at risk from injuries to the scalp and bump caps prevent this and aid hygiene by keeping hair free from oil and dirt.

11.126 If worn with long hair held inside the cap, the risk of hair entanglement with moving parts of machines (or fan belts and pulleys of vehicles) is prevented.

Firefighters' Helmets

11.127 Firefighters' helmets are similar to industrial safety helmets, but cover more of the head and give greater protection against impact, heat and flame.

Transport Helmets

11.128 Transport helmets protect the head against injury when falling from a motorcycle or bicycle. The regulations only apply to such head protection when used "off road", eg when a risk assessment has determined the use of head protection is necessary for driving all-terrain vehicles.

Leisure Helmets

11.129 Leisure activity risks such as horse riding, canoeing or climbing require the wearing of helmets. When selecting this type of head protection, it is important to assess the risk of injury and then select the appropriate helmet based on its performance.

Lightweight Head Coverings

11.130 Caps, hairnets and lightweight head coverings can protect against scalping or entanglement and, as with bump caps, it is necessary for hair to be fully contained within the cap or hairnet to give protection.

11.131 Hairnets are frequently used for hygiene control and may be combined with beard snoods for similar reasons. These regulations do not cover PPE used for hygiene purposes, but often the two go hand-in-hand.

Storage, Maintenance and Cleaning

11.132 Storage for head protection can simply consist of a safe place, such as a peg or a cupboard.

11.133 Equipment should not be stored in direct sunlight or in hot, humid conditions or in any place where it might be contaminated. Head protection must not be stored on the rear window shelf of a vehicle — this exposes it to damaging UV light and may allow it to act as a missile in the event of heavy braking or collision.

11.134 Further reduction of protection can result from exposure to chemicals, exposure to heat or sunlight and ageing due to heat, sunlight, humidity or rain.

11.135 All head protection needs regular visual inspections to check for damage or deterioration. For caps and hairnets, a quick check before use is adequate.

11.136 For safety helmets and other head protection that protects against impact or falling objects, a more thorough examination is required to check the shell and straps for signs of damage and deterioration. Being dropped or thrown, striking it against objects and having objects falling on it can all damage the shell of a helmet.

11.137 All safety helmets and similar head protection should be replaced within the timeframe recommended by the manufacturer or supplier, even if no deterioration is apparent. This period is usually every two to three years.

11.138 Head protection should be replaced immediately if:

- there is sufficient damage to the shell or harness
- it has undergone a strong impact
- there are any visible cracks.

11.139 Cleaning should be carried out using soap and water (always following the manufacturer's instructions). Avoid solvents, harsh detergents and the use of abrasive material likely to cause surface scratches.

Foot Protection

Does the Activity Require Foot Protection?

11.140 Many different activities or jobs can involve risks to the feet that require the use of PPE. These include:

- work with mechanical handling equipment such as hoists, lift trucks, cranes, etc where there is a risk of objects falling on the feet or crushing the front of the foot
- manual handling where there may be a risk of objects falling on the feet, or where good grip is required
- all types of building and construction work, work in mines, quarries, pits, trenches, etc, where there may be a variety of hazards that feet need to be protected against, including falling objects, sharp objects on the ground, slipping hazards, etc
- work with hazardous chemicals where there is a risk of slipping or of chemicals coming into contact with footwear
- work outside where there is a risk of slipping or getting wet
- work in hot or cold conditions.

What Type of Foot Protection is Needed?

11.141 There is a wide range of foot protection available. Determining what type is suitable will depend on the risks involved. Suppliers of PPE should be able to provide further information for employers.

11.142 The main types of foot protection are outlined below.

- Safety boots and shoes can provide protection against falling objects, slipping, sharp objects on the floor, temperature, etc depending on the features they include. They usually have slip-resistant soles, steel toecaps to protect against falling objects or crushing, and can have steel mid-soles to protect against nails and other sharp objects. Boots provide ankle protection.
- Wellington boots provide protection against water and wet conditions. They are available in different materials, some of which provide insulation or chemical resistance. They can include steel toecaps and mid-soles, padding, cotton linings and are available from ankle boots to waders.
- Anti-static or conductive footwear prevents the build-up of static electricity and reduces the risk of igniting a flammable or explosive atmosphere.
- Foundry boots are heat resistant and are designed to keep out molten metal and to be taken off easily. In addition, foundry workers need leg protection, eg gaiters.

Selecting Suitable Foot Protection

11.143 The type of footwear used depends mainly on the hazards involved. The fit, comfort, style and durability of the PPE needs to be considered.

11.144 Safety footwear should be as comfortable as possible and should be available in different sizes, including different width fittings where necessary. It should be as light and as flexible as possible, to reduce the fatigue of wearing it for long periods.

11.145 The less expensive materials used for waterproof footwear are good at keeping water out, but are not permeable. As a result, wearing footwear made with these materials for long periods can mean that feet get hot and sweaty. Breathable, water-resistant materials are available and are more comfortable and hygienic. They are also more expensive.

Is the Safety Footwear Compatible with the Work to be Done?

11.146 If someone is on their feet all the time at work, then even greater attention should be paid to making sure that their safety footwear is comfortable.

11.147 If people are working in areas where there may be oils, solvents or other chemicals around, then the soles of their footwear need to be resistant to the relevant materials. In addition, the soles and uppers should be bonded or moulded together, not just stitched or glued.

11.148 There is unlikely to be a problem of compatibility with other PPE. However, safety footwear may need to be worn with leg protection (eg when using a chain-saw) so it may be necessary to check that they are compatible.

Safety Footwear: Storage and Maintenance

11.149 This is straightforward for most safety footwear. It should be stored in any suitable location where it will not be damaged or contaminated. If it needs to dry, then there should be enough ventilation.

11.150 Footwear should be cleaned regularly and checks should also be made on a regular basis to make sure that it (including the laces) is in good condition. If necessary, footwear used in wet conditions should be treated with suitable protective polishes, etc.

11.151 Where anti-static footwear has been provided the soles of this footwear must be kept clean so that good contact can be made with the floor. Dirty footwear (such as when the soles are covered with paint, resin or some other insulating material) will not discharge the static charge from the employee and so a spark may be generated in the workplace.

Hand and Arm Protection

Does the Activity Require Hand/Arm Protection?

11.152 Many activities involve risks to the hands that require PPE, and some require arm protection as well. These include:

- construction work outdoors where cold hands can result in a loss of dexterity
- other outside work, such as agriculture, gardening, forestry, etc
- manual handling, where there may be a risk of sharp, hot, cold, abrasive or contaminated loads
- work with hazardous substances, including handling, storage, maintenance, cleaning, etc
- work with vibrating tools such as chain-saws or pneumatic drills, as this is more likely to cause vibration white finger if hands are cold
- handling hot or cold materials or equipment, including working in freezers, catering, welding, etc.

11.153 Gloves should not be worn by anyone working with moving machinery (eg drills, lathes, etc) because of the risk that the gloves may get caught in the machine and cause the wearer to be drawn in.

Deciding What Type of Hand/Arm Protection is Needed

11.154 Safety gloves are available in a wide range of types. The protection they provide depends not only on the material that the gloves are made from, but also on how they have been made. Suppliers of PPE should be able to provide further information on what type is suitable.

11.155 The choice depends on the exact nature of the hazard(s) involved. The most common types of gloves include the following.

- Chemical protection gloves protect against contact with substances such as acids, alkalis, solvents, irritants, etc. Protection can be provided by many different materials.
- Gloves that protect against excessive high and low temperatures can be provided, made of materials such as Kevlar, terrycloth, glass fibre and leather, etc.
- Gloves providing protection against sharp edges, including splinters and abrasives, etc and made of leather, chain-mail, knitted Kevlar, etc.
- General use gloves are made from materials such as rubber, plastic and knitted fabric. They can resist cuts, repel some liquids and provide a good grip. Some thin gloves can protect the hands while still providing good dexterity and touch, such as those used by surgeons. In most cases, general-purpose gloves should only be used against low risks.

11.156 When providing protection against chemicals, it is important to know the exact nature of the substances in order to identify suitable glove materials. The safety data sheets on the substances may provide some information on which types of gloves are acceptable.

11.157 The degree of protection provided by gloves will depend on:

- the material used
- the thickness of the gloves
- how they have been made.

11.158 When selecting gloves it may be necessary to consider the "breakthrough" time. This is the length of time that the gloves will withstand exposure to the relevant chemical before it is able to penetrate the material.

11.159 In addition to hand protection, the provision of arm protection may be needed. This could be in the form of long gloves or as separate arm protectors, which can complement shorter gloves for some risks, such as heat protection.

11.160 Barrier cream can be useful in some circumstances where it is not possible to use gloves. For example, gloves are not suitable in a machine shop but there may be coolants and other substances that can affect the skin. However, barrier creams are not a substitute for suitable gloves.

Selecting Suitable Hand/Arm Protection

11.161 As well as providing protection, gloves must fit well and be comfortable. A wide range of sizes may be needed to cater for all employees.

11.162 Latex gloves can be very useful at work, however, some people are allergic to latex and are unable to use gloves made of this material. They may react to the powder present on the gloves. In this case, other employees should not use this type of glove near to anyone with a latex allergy.

11.163 Working for long periods wearing impervious gloves can adversely affect the skin, causing sweating and irritation. It might also aggravate existing skin conditions. Wearing a pair of thin cotton-lined gloves under the impervious ones can help reduce this problem.

Is the Hand and Arm Protection Compatible with the Work to be Done?

11.164 Although some gloves have little impact on dexterity and grip, those that are badly fitting or bulky can present problems. They can interfere with the wearer's sense of touch and the way they work. If gloves are too thick or stiff and this prevents people from doing their job easily, they may decide not to wear them.

11.165 There should not be a gap between the cuff of the gloves and the wearer's sleeves.

11.166 There are unlikely to be any compatibility issues for people using gloves in conjunction with other PPE.

Hand and Arm Protection: Storage and Maintenance

11.167 Gloves should be stored somewhere clean and free from contamination. There should be adequate facilities for the disposal of contaminated protective gloves.

11.168 Glove cleaning should be carried out following the manufacturer's recommendations. People using non-disposable gloves for chemical protection should clean the gloves before taking them off and then take them off in a manner that does not risk them contaminating themselves or the insides of the gloves. If they are grossly contaminated, they should be thrown away.

11.169 People using disposable gloves should take them off so that they do not touch the outside of the gloves. This can be done by pulling from the cuff and turning the glove inside out while removing it, providing the cuff is not contaminated (in which case, longer gloves are probably required).

11.170 Safety gloves should be kept in good condition, checked regularly and replaced when they are worn or damaged. They should be free of holes and cuts, etc and should not be distorted.

Protective Clothing for the Body

Does the Activity Require Body Protection?

11.171 Many different jobs and activities may require some kind of protective equipment or clothing for the body. These include:

- outdoor work, including on roads, railways, agriculture and conservation, etc
- work with moving vehicles, such as cranes, etc
- construction and building work
- hot processes, such as welding and foundry work, etc.

Deciding What Type of Body Protection is Needed

11.172 There are two main types of protective clothing — clothing to protect the body and clothing to protect the whole person. Suppliers of PPE should be able to provide further information on exactly which type will be suitable.

11.173 The main types of clothing used for body protection include:

- clothing providing protection against cold, such as quilted, insulted jackets and full-body suits, etc
- clothing providing protection against wet weather, such as jackets, trousers and leggings, etc
- clothing providing protection against heat, for example special flame-retardant clothing for welding and foundries
- clothing providing protection against chemicals
- clothing providing protection against chain-saws that covers the most vulnerable parts of the body (eg the fronts of the legs)
- chain-mail clothing for butchery (eg aprons).

11.174 The two main types of clothing worn to protect the whole person are:

- high-visibility clothing, which is fluorescent so that people can be seen easily, eg by drivers in road works or in a car park
- life jackets or buoyancy aids for people at risk of drowning because they are working on or near water.

Selecting Suitable Body Protection

11.175 The basic type of body-protecting PPE will depend on the hazards involved. Once the type has been chosen, it will be necessary to consider other issues such as fit, comfort, style and durability.

11.176 For waterproof clothing, breathable material is usually more comfortable for people who have to wear it for long periods, as it allows perspiration to escape. Layers of thin clothing may be more effective and comfortable for protection against the cold than bulky, thick coats.

Is the Body Protection Compatible with the Work to be Done?

11.177 Some of the different types of body protection will restrict movement and may make jobs harder to carry out. For example, full body suits are inevitably going to affect how the people wearing them move as well as their ability to work.

11.178 Some other kinds of PPE can be combined with body protection. For example, full body suits will have to include some form of breathing apparatus. Jackets to keep people warm and dry can incorporate high-visibility clothing.

Body Protection: Storage and Maintenance

11.179 Most clothing, unless it becomes contaminated during use, only requires simple storage facilities such as pegs and lockers, etc.

11.180 Any clothing will need to be washed regularly. If it is contaminated with hazardous substances, then washing may need to be done by specialist cleaners and should not be taken home to be done.

11.181 All protective clothing should be regularly checked and kept in good condition.

11.182 Other types of body protection will need more rigorous inspections and tests. For example, vapour suits need to be air tested with the manufacturer's test kit and stored in a protective case.

Eye and Face Protection

Does the Activity Require Eye or Face Protection?

11.183 Any activities that include risks to the eyes that cannot be eliminated are likely to require eye protection. Activities that may involve splashes of chemicals or molten metals, dusts, gases, mists, sprays, bright light, etc can all damage the eyes.

11.184 Activities that might require PPE to protect the eyes and face include:

- work with hazardous substances that cause burns and irritation, such as corrosives (acids and alkalis) or irritants (solvents, paints, bleach, etc)
- work with hazardous substances that can be absorbed through the skin, such as some solvents
- work with molten metals
- work with metal-cutting machinery, such as lathes, mills and drills, etc that emit swarf
- work with power-driven tools and equipment, where there may be chippings or abrasives ejected
- welding that gives off intense light or other optical radiation that could damage the eyes (eg causing arc eye)
- use of UV light sources
- work on any process with equipment that produces light amplification (eg lasers) or radiation
- using any gas or vapour under pressure, including air guns.

11.185 Eye protection may be required not only for people carrying out the work, but also for others in the area who may be affected by it. For example, visitors to a machine shop may need eye protection as well as the operators. This should be identified when assessing risks.

Deciding What Type of Eye/Face Protection is Needed

11.186 There is a wide range of eye and face protection. Which type is suitable will depend on the risks involved. Suppliers of PPE should be able to provide further information on exactly which type will be suitable.

11.187 The main types of eye and face protection are as follows.

- Safety glasses or spectacles provide protection against impact from small objects. Different levels of impact resistance are available. They are similar to prescription glasses, however they have side shields that provide lateral protection. They are suitable for general working conditions where there may be minor dust, chips or flying particles. They provide little or no protection against liquids or vapours.
- Eye shields are similar to safety glasses, however they have a single frameless one-piece lens. These provide a similar level of protection to safety glasses. Some eye shields can be worn over prescription glasses.
- Safety goggles provide protection for the eyes from all angles as they provide a seal around the entire area of the eyes. They are used when the eyes need to be

completely covered but the rest of the face does not need to be protected. Different types of goggles are available to provide protection from liquids, dusts, gases, vapours, molten metal and high impact levels. There are different designs to help prevent problems with fogging, however they need to be chosen carefully to ensure they are suitable for the work. Goggles can also be obtained with a range of filters to provide protection against lasers and welding.

- Face shields protect the face but do not fully enclose the eyes. Therefore, they do not provide protection against dusts, mists or gases, but can provide protection against impact, spraying, chipping, grinding or chemical splashes. They are frequently used in conjunction with eye protection, as they are not by themselves protective eyewear. They can include welding filters or reflective metal screens that deflect heat. These are useful in blast and open-hearth furnaces and other work involving radiant heat.

11.188 For protection against light and other non-ionising radiation (eg lasers, ultraviolet, welding flashes), it is important to make sure that you choose the correct type of filter. This will depend on the exact type of radiation, for example the class of laser or the type of welding.

11.189 For welding, helmets are available that include lenses which darken automatically when an arc is struck, providing clear vision when not actually welding.

Selecting Suitable Eye/Face Protection

11.190 The basic type will depend on the hazards involved. Once the basic type is chosen, comfort, style and durability will need to be considered.

11.191 Safety glasses are available in a variety of styles, weights and sizes, etc. They can include adjustable side arms and be fitted with prescription lenses for people who need them. They should be treated to reduce fogging problems.

11.192 Eye shields can be useful for visitors and other people who only need eye protection for short periods as some styles can be worn over prescription glasses. However, wearing both an eye shield and prescription glasses is unlikely to be comfortable for long periods and so it is usually better to provide prescription safety glasses to employees.

11.193 Goggles are heavier and less comfortable than glasses, however they provide much better protection. They are more prone to misting and should be treated with anti-mist coatings.

11.194 Face shields are the heaviest and bulkiest, however they should be comfortable if they are fitted with an adjustable head harness.

Is the Eye Protection Compatible With the Work to be Done?

11.195 Eye and face protection can interfere with how well people can see and this needs to be considered when selecting it.

11.196 It is often necessary to wear eye and face protection with other types of protection, especially respiratory protection. The best solution can be to use PPE that combines the different types of protection required. For example, welding helmets combine eye, face and respiratory protection and full-face masks combine goggles with respiratory protection. Safety helmets are available with visors or face shields.

Eye and Face Protection: Storage and Maintenance

11.197 Safety glasses should be kept in a suitable case when they are not in use.

11.198 The lenses of any eye protecting equipment must be kept clean as dirty lenses restrict vision. It will be necessary to find out from the suppliers the best way of cleaning them. (In most cases, this will simply mean wetting the lenses well and drying them properly.) Anti-misting cleaning fluids may be needed if misting is a problem.

11.199 If eye shields or other eye protection are provided for visitors, they should be thoroughly cleaned before they are reissued.

11.200 Lenses must also be kept free from damage. Scratched or pitted lenses should be replaced as they may not only impair vision but may also not provide enough protection against impact.

11.201 Face shields should be replaced when they are scratched, or if they warp or become brittle. Headbands should be replaced when they are damaged or worn out.

Training

11.202 The amount and level of training and information required varies with the complexity of the personal protective equipment (PPE). If it is simple to use and easily maintained, eg safety helmets, users may only need basic instructions. However, anyone using or maintaining fall-arrest equipment will need much more extensive training and information.

Who Needs Training?

11.203 Training must be provided to everyone involved, including:

- anyone using PPE, who needs to know:
  - how to use and fit it correctly
  - its limitations
- managers and supervisors who need to know:
  - why PPE is needed
  - how it should be used
- anyone maintaining, testing, repairing or cleaning PPE
- anyone selecting PPE, so they are competent to choose it properly.

What Type of Training is Best?

11.204 Practical training will be needed as well as the provision of theory and information. The training could be a combination of explanation, practical demonstrations by the trainer and practice by the participants.

11.205 It is important to ensure employees attend training *and* understand what they have been taught. Their capabilities must be taken into account when designing the training. Do they speak English well enough to understand properly? How will risks be explained if they are complex and difficult to understand?

11.206 One way of ensuring employees have understood their training is to watch them inspect, test and use the PPE.

What Should Training Include?

11.207 Training should be based on the recommendations and information provided by the manufacturer or supplier of the PPE. Many manufacturers of PPE can provide training for users, employers should check this will cover their particular requirements.

11.208 In general, employers should ensure training and information includes:

- an explanation of the risks involved
- why PPE is needed
- the operation, performance and limitations of the PPE
- instructions on the selection, use and storage of the PPE related to how it will be used
- any factors that could alter the effectiveness of the PPE, such as:
  - working conditions
  - fitting problems
  - damage and wear
  - other PPE
  - personal factors
- how to recognise defects in the PPE, and how to report these and losses
- practice in putting on, wearing and removing the equipment
- practice and instruction in inspection and, if necessary, testing of the PPE before use
- practice and instruction in the maintenance that will be done by the user, such as cleaning and replacing parts
- instruction in how to store it safely.

When Should Training be Provided?

11.209 Training should be provided *before* people need to use the PPE. This may be as part of an induction for new staff or before new PPE is introduced.

11.210 To fully involve staff, training for new PPE should be provided in stages. If employers start by ensuring staff understand the need for PPE, they will be better placed to help select suitable suppliers and types. Staff need to be trained in the practical aspects of using the PPE before they try it out in the workplace.

11.211 Refresher training may be needed from time to time. All training should be recorded.

Head Protection

11.212 Training and information should show users how to adjust safety helmets to fit them properly. It should also emphasise the importance of wearing them correctly (eg not over other headwear such as baseball caps and not back to front).

11.213 Users of safety helmets also need to understand how they can be damaged, especially by heat and sunshine, and how to inspect them for damage.

11.214 Any training must adequately cover:

- correct adjustment
- the need to wear protection correctly
- how to clean and inspect for damage
- how to obtain replacements
- correct storage, eg not in car rear windows.

Foot Protection

11.215 Training for people wearing foot protection should include instruction on how to clean and check them properly. For example, anything lodged in the tread should be completely removed. If necessary, safety footwear should be treated regularly with protective wax to keep it in good condition.

Hand and Arm Protection

11.216 Anyone wearing gloves to protect them against contact with hazardous chemicals needs to be trained in:

- how to put them on properly
- the importance of not touching any other parts of the body, equipment, etc while wearing them
- how to clean them
- how to remove them safely (especially disposable gloves that are contaminated)
- how to check them for damage.

11.217 Users also need to know how to take care of their hands, including the importance of washing them regularly, drying them properly and using a hand cream to keep the skin from drying out. They should also be instructed to keep cuts and abrasions covered with proper dressings.

Protective Clothing for the Body

11.218 The amount of training will vary greatly depending on the type of clothing being used. Little information is required for people wearing coats to keep them warm and dry. More information is required for people using life jackets who need to know how to wear them properly and maintain them. Much more information and training will be required for people wearing chemical or vapour suits.

Eye and Face Protection

11.219 Training should include providing information on how to care for safety glasses, including minimising the risk of scratching them, and how to clean and store them, etc.

11.220 Advice should also be provided on how to prevent the insides of the glasses and goggles from becoming contaminated by poor handling and other habits. For example, someone working with dusts that get into their hair might push their goggles up onto their forehead. This could lead to dust getting inside the goggles and into the wearer's eyes when they put them back on.

List of Relevant Legislation

- Control of Asbestos Regulations 2006
- Control of Noise at Work Regulations 2005
- Control of Lead at Work Regulations 2002
- Control of Substances Hazardous to Health Regulations 2002
- Personal Protective Equipment Regulations 2002
- Ionising Radiations Regulations 1999
- Management of Health and Safety at Work Regulations 1999
- Personal Protective Equipment at Work Regulations 1992
- Construction (Head Protection) Regulations 1989
- Health and Safety at Work, etc Act 1974

Further Information

Publications

HSE Publications

The following are available from *www.hsebooks.co.uk.*

- INDG174(L) *Personal Protective Equipment at Work*
- INDG330 *Selecting Protective Gloves for Work with Chemicals*
- L5 (rev 2005) *The Control of Substances Hazardous to Health Regulations 2002 (as amended). Approved Code of Practice and Guidance*
- L25 (rev 2005) *Personal Protective Equipment at Work. Guidance on Regulations. Personal Protective Equipment at Work Regulations 1992*
- L102 (rev 1998) *A Guide to the Construction (Head Protection) Regulations 1989*

Respiratory Protective Equipment

- There are two types of respiratory protective equipment (RPE), according to the Health and Safety Executive:
  - respirators that are designed to filter or clean contaminated air from the workplace before the RPE wearer inhales it
  - breathing apparatus that delivers breathable air from an independent source to the wearer.
- The need for RPE should be identified when risks in the workplace are assessed, such as through a COSHH risk assessment.
- The most effective means of controlling risks should be used. RPE should only be provided when it is not possible to control risks adequately in other ways.
- Where RPE is required, an assessment should be made to decide whether it is suitable for the risks, work and people.
- Suitable RPE should be selected after consideration of the types available.
- Work should be planned so that it is clear when RPE can be used.
- Those who will actually wear the RPE should be included in the process of the selection of suitable equipment.
- RPE must be available, properly stored, maintained, and replaced and cleaned when necessary.
- Employees who will be using and maintaining the RPE must receive suitable and sufficient information, instruction and training.
- Employers must ensure that RPE is used properly.

11.221 The need for respiratory protective equipment (RPE) should be identified in the workplace risk assessment. RPE should only be used when it is not possible to control risks adequately in other ways. Employers have a duty to provide suitable and sufficient protective equipment, including RPE.

Employers' Duties

11.222

- Employers have a general duty under the Health and Safety at Work, etc Act 1974 to ensure, so far as is reasonably practicable, the health and safety at work of all employees.
- The Management of Health and Safety at Work Regulations 1999 require employers to assess risks to employees and anyone who may be affected by their operations.
- The Control of Substances Hazardous to Health Regulations 2002 require employers to provide employees with suitable personal protective equipment, including RPE, if other control measures that are in place are not sufficient to control exposure to hazardous substances. Where RPE is provided, the employer must provide adequate training in its selection, use, maintenance and storage. Similar duties exist under the Control of Lead at Work Regulations 2002 and the Control of Asbestos Regulations 2006.

Employees' Duties

11.223

- The Health and Safety at Work, etc Act 1974 places a duty on all employees to take reasonable care of their health and safety and the health and safety of other persons who may be affected by what they do, or fail to do, at work.
- Employees also have a duty under the Health and Safety at Work, etc Act 1974 to co-operate with their employers on health and safety matters.
- The Management of Health and Safety at Work Regulations 1999 place a duty on all employees to follow health and safety instructions and to report danger.
- Employees have a duty under the Control of Substances Hazardous to Health Regulations 2002 to make full and proper use of the control measures identified in the risk assessment, including use of RPE.

In Practice

Types of RPE

11.224 Respiratory protective equipment (RPE) is designed to be worn in contaminated atmospheres and to provide its wearer with a supply of air that is safe to breathe. There are two classes of RPE.

1. Respirator (filtering device): this filters or cleans air from the workplace before the wearer inhales it. Respirators are not suitable for use in environments that are immediately dangerous to life or health, eg oxygen-deficient atmospheres.
2. Breathing apparatus (BA): this delivers breathable air or oxygen to the wearer from an independent source. Breathing apparatus may be suitable for environments that are immediately dangerous to health or life.

Types of Mask

11.225 There are numerous types of masks available. The main types are described here.

Filtering Half Masks to Protect Against Particles

11.226 These masks are often called "disposable respirators". They consist either entirely or substantially of filter material (or the main filter element is an inseparable part of the device). Filtering half masks should conform to BS EN 149: 2001 *Respiratory Protective Devices. Filtering Half Masks to Protect Against Particles. Requirements, Testing, Marking*. They are classified as FFP1, FFP2 or FFP3, according to filtration efficiency. The respirators should normally be used for a maximum of a single shift, but this depends on the manufacturer's instructions. These respirators may incorporate inhalation and exhalation valves, or exhalation valves only, or they may have no valves. Where the respirator has no valves, both the inhalate and exhalate pass through the filter material.

Half Masks Without Inhalation Valves

11.227 These devices (conforming to BS EN 1827: 1999 *Respiratory Protective Devices: Half Masks Without Inhalation Valves and With Separable Filters to Protect Against Gases or Gases and Particles or Particles Only. Requirements, Testing, Marking*) are comparatively new on the market. They consist of a lightweight half mask with separable and replaceable filter(s). The filter should be used for a maximum of a single shift, and the half mask itself may need replacement after a relatively few uses. The filter/mask combination is "dedicated", ie the mask is intended for use only with filters specified by the manufacturer. The mask has no inhalation valve and reduced strength requirements, but should not be confused with half masks meeting BS EN 140: 1999 *Respiratory Protective Devices: Half Masks and Quarter Masks. Requirements, Testing, Marking.* The filters can be used to protect against particles, gases/vapours or both. Complete devices are designated according to filter type used, and have the prefix FM. Examples of gas/vapour and combined filters for gases/vapours and particles are FMA1, FMB2P3, FMAX and FMSXP2.

Half Masks Reusable

11.228 This is a moulded facepiece of rubber or plastic. It covers the nose and mouth of the wearer and is held in place with adjustable straps. Air is drawn through the appropriate filter(s) by the wearer's lung power, a powered unit or suitable breathing apparatus attached to the mask. The exhaled air is discharged through an exhalation valve. Filters are available for particulates, gases/vapours or as a combination. A specification for half masks is given in BS EN 140: 1999 *Respiratory Protective Devices: Half Masks and Quarter Masks. Requirements, Testing, Marking.*

Full Face Masks

11.229 A full face mask covers the eyes, nose, mouth and chin and seals against the face of the wearer. It is held in place with adjustable straps. Air is drawn into the mask either through an appropriate filter by the wearer's lung power or from a powered unit or suitable breathing apparatus. Exhaled air is discharged through an exhalation valve. Most masks have an inner half mask, which in unpowered devices can reduce the re-inhalation of exhaled carbon dioxide and assist comfort. The visor provides protection against particulates and gases. Special-grade visors are required for protection against chemical splashes and impact.

11.230 There are three classes of full face masks.

- Class 1 is of light duty construction and is intended for respirators and light duty, compressed airline breathing apparatus.
- Class 2 is more robust and offers greater resistance to flammability.
- Class 3 offers greater protection against flame and radiant heat. This type of mask may be suitable for fire fighting.

11.231 The specification for full face masks is given in BS EN 136: 1998 *Respiratory Protective Devices. Full Face Masks. Requirements, Testing and Marking.*

Powered Respirators Incorporating Helmets and Hoods

11.232 These devices are commonly called "powered helmets" or "powered visors". Any form of loose-fitting facepiece can be employed, eg:

- simple transparent visors with a soft rubber or fabric seal against the face
- hoods which seal around the neck
- helmets
- full suits.

11.233 Filtered air is directed to the facepiece from a battery-powered fan unit. For correct performance, as well as having the appropriate filters fitted, these devices rely on a minimum airflow being exceeded in the facepiece area (the level being defined by the manufacturer). Equipment provided must meet current standards, and there must be a means for the user to check that the manufacturer's minimum design flow is exceeded prior to use. Batteries should have a minimum duration of four hours when charged and users need to ensure that these are properly maintained and checked.

Devices Meeting BS EN 12941

11.234 Devices meeting BS EN 12941: 1999 *Respiratory Protective Devices. Powered Filtering Devices Incorporating a Helmet or a Hood. Requirements, Testing, Marking* are for use against particles or gases/vapours or a combination of both, as specified by the manufacturer. A typical classification is TH2B1P.

Power-assisted Respirators Incorporating Full Face or Half Masks

11.235 These devices are often referred to incorrectly as "positive pressure-powered respirators". Filtered air is directed to the facepiece from a battery-powered fan unit. In most current devices the airflow is too low for the pressure inside the face mask to remain positive throughout the whole of the wearer's breathing cycle, especially at high work rates. These devices rely for their performance on the combination of air supply and tight-fitting face seal. Some devices interact with the wearer's breathing rate and supply increasing amounts of air in response to lower pressure inside the face mask as inhalation peaks.

11.236 For correct performance, as well as having appropriate filters fitted, these devices rely on a minimum design condition being present in the facepiece area. This level (either airflow or pressure) is defined by the manufacturer, and below it there is a possibility of higher leakage and higher levels of re-breathable carbon dioxide.

Devices Meeting BS EN 12942: 1999

11.237 Devices meeting BS EN 12942: 1999 *Respiratory Protective Devices: Powered-assisted Filtering Devices Incorporating Full Face Masks, Half Masks or Quarter Masks. Requirements, Testing, Marking* are for use against gases and vapours or particles, or a combination of both, as specified by the manufacturer. A typical classification is TM3A2P.

Types of Filters

11.238 Respirators should have the correct type of filter(s), determined by the substance(s) from which the wearer needs protection. All filters have a shelf-life, and once the filter has been unsealed the shelf-life can be seriously reduced. The manufacturer's instructions on how to use, and how often to replace, filters should be followed. The manufacturer's information should specify the filter's application, and the markings on the filter will include:

- the CE mark
- the type and class
- the colour code
- the identity of the manufacturer, eg name and trademark
- a European Standard number (if appropriate)
- the filter's shelf-life, if appropriate
- a "see instructions for use" note
- any additional marking relevant to the particular type of filter.

11.239 There are two basic types of filter: particle filters and gas/vapour filters. Gas/vapour filters are not designed to remove particles. However, combined filters are available which remove both.

Particle Filters

11.240 Particle filters are marked "P" and colour-coded white. They are available in three classes based on their efficiency. P1 refers to low efficiency, P2 to medium efficiency and P3 to high efficiency. The breathing resistance of a particle filter can increase substantially if it becomes clogged.

Gas/Vapour Filters

11.241 Gas/vapour filters are available for use against different types of gaseous contaminants. They are also available as multi-type filters, which may be used against more than one type of gas, as specified by the manufacturer. Where a multi-type filter is to be used against more than one type of gas at a time, assurance should be sought from the manufacturer that the filter will perform adequately.

11.242 Gas filters are further divided into three classes based on how much gas or vapour the filter can hold, rather than on the concentration that passes through. Class 1 filters have the lowest capacity and class 3 filters the highest.

Criteria for the Use of RPE

11.243 Whenever RPE is used at work, the following criteria must be met.

- The RPE must be "suitable" for the purpose for which it is to be used. This means that the equipment should provide effective protection to the wearer for the defined operation and in the specified environment. It should be capable of providing a sufficient quantity of clean air for the wearer to breathe and the device should fit the wearer properly.
- The RPE must either be CE-marked or HSE-approved. HSE approval of RPE ceased on 30 June 1995 but HSE-approved equipment made before 1 July 1995 can continue to be used as long as it is suitable and maintained to perform correctly.
- The RPE must be maintained in an effective state.
- Facilities must be provided for the safe storage of the equipment, in accordance with the manufacturer's instructions.
- Those responsible for the selection and maintenance of RPE must be properly trained for their work.
- Users of RPE must be given training and instruction on its safe use.

When the Equipment Can be Used

11.244 Legislation directs that personal protective equipment, including RPE, should be the last option for controlling (inhalation) exposure to substances hazardous to health. This means that other ways of controlling exposure must be implemented, where it is reasonably practicable, before a decision to use personal protective equipment is made. There are a number of reasons for this as follows.

- RPE can only protect the wearer, while other measures, such as enclosures and local exhaust ventilation, protect everyone by preventing the material entering the atmosphere in the first place.
- If RPE is used incorrectly or is badly maintained, the wearer may receive no protection.

Assessing the Requirements for the Equipment

11.245 It must be remembered that no form of RPE can provide complete protection against exposure to a hazardous substance. There is always the potential for leakage through and around RPE. In deciding which type of RPE to use, the following points should be considered.

- Will the equipment be used in an atmosphere of reduced oxygen?
- What hazardous substances are likely to be present and what are their properties?
- What effects would the identified substances have on the body?
- What form will the contaminants be in, ie dust, fibre, mist, fume, micro-organism, gas or vapour?
- What will be the likely concentration of the identified hazardous substances?
- What are the workplace exposure limits or safe exposure levels of the relevant hazardous materials?
- Are there other hazards associated with the job/task?
- For how long would RPE need to be worn?
- How easy would it be to decontaminate or dispose of the RPE?

11.246 The above points should be considered as part of the risk assessment process.

Selection of Suitable Equipment

11.247 Suitable RPE may be used to provide protection against a wide variety of substances, which can be split into two broad categories. These are:

- dusts, fibres, mists, fumes and micro-organisms
- gases and vapours.

11.248 Respirators are typically designed for only one of the above categories, although combination types that are suitable for both categories do exist. Breathing apparatus may be used for both categories and, if suitable, for environments which are immediately dangerous to life and health, including oxygen-deficient atmospheres. If there is no suitable RPE that would allow an individual to carry out the work safely, that individual's health and safety should not be put at risk.

11.249 Further information on the types of RPE available, and personal and work-related factors, is provided in HSG53 *The Selection, Use and Maintenance of Respiratory Protective Equipment.*

Use of the Equipment

11.250 When using RPE, other factors, which are not directly related to respiratory protection, should be taken into consideration, including the following.

- Materials used in some RPE can produce frictional or electrostatic sparks, which may ignite flammable or explosive atmospheres. There is suitable equipment available for this type of environment.
- If other types of personal protection are also needed (eg eye, skin or head protection), it should either be compatible with the RPE or incorporated into its design. Modifications of any form of protective equipment should not be carried out without the knowledge and consent of the manufacturer. Otherwise the wearer may be put at risk and the equipment certification invalidated.
- The need for free movement may rule out equipment with trailing tubes or hoses.

11.251 Breathing apparatus should only be used in accordance with the manufacturer's instructions and only by those who have received adequate training in its use. Airline breathing apparatus must not be used for rescue purposes, unless under specific site procedures. Where this type of breathing apparatus is used for rescue as part of pre-planned work, a time check must be maintained on any person entering a confined space, with the exception of the emergency services. The user must leave the contaminated area immediately if the low contents whistle blows.

Work Planning

11.252 Prior to work being carried out in a confined space, a risk assessment must be carried out to determine whether there is a need to enter the confined space, and what precautions are required. Entry into confined spaces will require a permit to work.

11.253 Work must be planned in good time and the use of breathing apparatus, eg for emergency rescue, must be the subject of a written procedure, including a checklist if required. The checklist can be used to confirm that equipment is operating and has been checked, and that equipment and persons are in the correct location. The provision of spare sets of breathing apparatus should be considered in the risk assessment.

Precautions When Using the Equipment

11.254 A person wearing breathing apparatus cannot move as freely or as easily as usual. Therefore it is important that there is an easy exit route from the confined space. RPE should always be examined, and the manufacturer's pre-use checks should always be carried out, before it is worn or used. It should not be used if it is found to be defective in any way.

11.255 The risk assessment carried out for the particular job/task to be carried out may highlight the need for resuscitation equipment. Rescue equipment provided should be appropriate for the likely emergencies identified in the risk assessment. Where this equipment is provided it must be properly maintained and used only by competent persons who have been adequately trained in its use.

11.256 Rescue equipment will often include:

- lifelines and lifting equipment (since even the strongest person is unlikely to be able to lift or handle an unconscious person on their own using only a rope)
- additional sets of breathing apparatus
- first-aid equipment.

Maintenance

11.257 Effective maintenance of respiratory protection and resuscitation equipment is required under health and safety legislation. This is to ensure that the equipment continues to provide the degree of protection for which it was designed. The maintenance programme will vary according to the type of equipment and how it is used, but the manufacturer's maintenance schedules and instructions must always be followed.

11.258 Reusable RPE requires cleaning and disinfecting after each use. Rubber facepieces can usually be cleaned with soap and lukewarm water. Cleaning can be carried out by hand. After washing, a thorough rinsing is essential to remove soap or detergent from the equipment. Only disinfectants recommended by the manufacturer of the equipment should be used. The equipment should be thoroughly dried, reassembled and placed in a protective container, such as a resealable plastic bag.

Rubber items should generally not be heated to more than 60°C, as damage may occur above this temperature. Chemical solvents should not be used for cleaning the equipment.

11.259 The wearer may carry out simple maintenance, such as replacement of filters, but only personnel trained for the purpose should perform more complex intricate repairs. Proprietary spare parts must be used in maintaining respiratory protective and resuscitation equipment. The use of non-original parts may invalidate certification and can compromise the health and safety of the wearer. An adequate stock of spare filters, head harness straps, batteries, etc should be maintained so that wearers can replace these parts as and when required.

Inspection, Examination and Testing

11.260 In general, respiratory protective and resuscitation equipment should be examined before use. Particular attention should be paid to rubber parts, such as facepieces, exhalation valves, breathing tubes and head harnesses. If the equipment is not in good working order it must not be used.

11.261 A thorough visual examination and test should be carried out once per month and as soon as possible after each use. Equipment must not be used unless it has had a recent thorough examination and test. Details of the requirements for thorough examinations are contained in L5 (rev 2002) *The Control of Substances Hazardous to Health Regulations 2002. Approved Code of Practice and Guidance.*

11.262 Records of examinations and tests should be kept.

Storage of the Equipment

11.263 Storage facilities should be provided for all respiratory and resuscitation equipment to protect it from:

- harmful contaminants
- excess moisture
- heat
- cold
- sunlight
- corrosive substances and damage.

11.264 The manufacturer's information will provide instructions for storage. There should be a clear segregation between equipment ready for use and that which is awaiting repair, maintenance or thorough examination and testing.

Medical Assessment Requirements

11.265 Employees are not required to have medical examinations before wearing and using compressed airline breathing apparatus or self-contained escape set breathing apparatus — normal fitness standards apply here. The design criteria for RPE take account of basic levels of performance, including limits on breathing resistance. However, employees with respiratory disorders such as asthma may have difficulty using RPE, and medical advice should be sought in such instances.

Cost Comparison: Purchase v Hiring

11.266 When deciding whether to purchase or hire RPE, employers should assess the costs, maintenance requirements, training and refresher training requirements. They should also assess how frequently breathing equipment will be used (the answer may be fairly infrequently). The following points should be considered.

- Where can equipment be purchased? Who will be responsible for maintaining it? Where will maintenance take place? Who will provide training (including refresher training) for those wearing and using it?
- In the event of needing to carry out a rescue, eg from a confined space environment, an employer might need to hire the services and equipment of a competent organisation. In this case, self-contained breathing apparatus might be used.
- If the hiring option is selected, designated employees will need to be adequately trained both in the use of the hired breathing apparatus and in confined space working. Companies specialising in hiring out breathing apparatus and associated safety equipment for use in confined spaces may also provide training in how to use the equipment and confined spaces awareness training. This relieves the site of the initial purchase, continued maintenance and refresher training of employees who may be required to carry out a rescue.
- Breathing apparatus, whether purchased or hired, must be fit for its purpose, ie correctly selected, matched to the job and matched to the individual. It must also comply with the Personal Protective Equipment at Work Regulations 1992, and display the CE mark (or be approved by the HSE). All new equipment purchased should carry the "CE mark".

Hiring Equipment

11.267 When hiring breathing apparatus, resuscitation and rescue equipment, it must have a valid test certificate to prove that it has been thoroughly examined and tested. In the case of self-contained breathing apparatus and resuscitation equipment, air supply cylinders must have a five-yearly pressure test certificate. Harnesses and lifting equipment must have their safe working loads marked on them. A competent person must visually inspect all equipment prior to use, and those wearing the equipment must be adequately trained in all aspects of its use.

11.268 Any person hired to carry out any aspect of confined spaces rescue work must be competent and adequately trained in the use of the equipment in the type of work environment envisaged.

Training

11.269 Employees who are required to use respiratory and resuscitation equipment must be adequately trained. Training should:

- be specific to the equipment
- cover both the theory and the practice of using the various types of equipment
- cover maintenance if the equipment is not hired.

11.270 A high standard of training is essential as the wearer may well rely on the equipment for life support. Initial training should be given to employees before they start to use the equipment, and there should be annual refresher training. This is particularly important where work requiring the use of breathing apparatus is not carried out very frequently.

11.271 Training in the use and maintenance of the equipment and training required for confined space entry and/or rescue can usually be combined by the training provider. If there have been changes in methods of work, safety procedures or equipment, it may be useful to arrange refresher training in the period immediately prior to the intended use.

11.272 It is important that training is not seen as a one-off exercise. The training programme should take account of the monitoring and review of the risk management system, and should address any specific risks identified.

11.273 HSG53 *The Selection, Use and Maintenance of Respiratory Protective Equipment* recommends that the training programme should cover:

- why RPE is required for the job and when to use it
- how RPE equipment works and what the wearer can and cannot do
- choosing the right RPE from the available types
- selecting the best type and model for individuals
- practice in putting on, wearing and removing the equipment
- instruction in obtaining a good face fit for RPE with a facemask
- factors that can affect the protection provided by RPE
- pre-use inspection and check of fit and performance
- the written procedures needed for safe working
- practical emergency procedures
- practice and instruction in simple user maintenance
- cleaning and inspection after use
- instruction in correct storage
- specific technical instruction for maintenance staff.

List of Relevant Legislation

- Control of Asbestos Regulations 2006
- Control of Lead at Work Regulations 2002
- Control of Substances Hazardous to Health Regulations 2002
- Ionising Radiations Regulations 1999
- Management of Health and Safety at Work Regulations 1999
- Provision and Use of Work Equipment Regulations 1998
- Confined Spaces Regulations 1997
- Reporting of Injuries, Diseases and Dangerous Occurrences Regulations 1995
- Manual Handling Operations Regulations 1992
- Health and Safety at Work, etc Act 1974

Further Information

Publications

HSE Publications

The following are available from *www.hsebooks.co.uk*.

- INDG255(L) *Asbestos Dust Kills — Keep Your Mask On*
- INDG289 *Working with Asbestos in Buildings: Asbestos: The Hidden Killer! Are You at Risk?*
- L5 (rev 2005) *The Control of Substances Hazardous to Health Regulations 2002 (as amended). Approved Code of Practice and Guidance*
- OC 282/28 *Fit Testing of Respiratory Protective Equipment Facepieces*

British Standards Publications

The following are available from *www.bsi-global.com*.

- BS EN 136: 1998 *Respiratory Protective Devices. Full Face Masks. Requirements, Testing and Marking*
- BS EN 140: 1999 *Respiratory Protective Devices: Half Masks and Quarter Masks. Requirements, Testing, Marking*
- BS EN 149: 2001 *Respiratory Protective Devices. Filtering Half Masks to Protect Against Particles. Requirements, Testing, Marking*
- BS EN 1827: 1999 *Respiratory Protective Devices: Half Masks Without Inhalation Valves and With Separable Filters to Protect Against Gases or Gases and Particles or Particles Only. Requirements, Testing, Marking*
- BS EN 12941: 1999 *Respiratory Protective Devices. Powered Filtering Devices Incorporating a Helmet or a Hood. Requirements, Testing, Marking*
- BS EN 12942: 1999 *Respiratory Protective Devices: Powered-assisted Filtering Devices Incorporating Full Face Masks, Half Masks or Quarter Masks. Requirements, Testing, Marking*

Index

Introduction

Please note the following points.

1. Index entries are to paragraph numbers.
2. Alphabetical arrangement is word-by-word, where a group of letters followed by a space is filed before the same group of letters immediately followed by a letter, eg "event tree analysis" will appear before "events". In determining alphabetical arrangement, initial articles and prepositions are ignored.

C

I

J

K

L

N

O

P

S

T

W